Sustainable Utilization of Natural Resources

Sustainable Utilization of Natural Resources

Edited by
Prasenjit Mondal and Ajay K. Dalai

CRC Press
Taylor & Francis Group
Boca Raton London New York

CRC Press is an imprint of the
Taylor & Francis Group, an **informa** business

MATLAB® is a trademark of The MathWorks, Inc. and is used with permission. The MathWorks does not warrant the accuracy of the text or exercises in this book. This book's use or discussion of MATLAB® software or related products does not constitute endorsement or sponsorship by The MathWorks of a particular pedagogical approach or particular use of the MATLAB® software.

CRC Press
Taylor & Francis Group
6000 Broken Sound Parkway NW, Suite 300
Boca Raton, FL 33487-2742

First issued in paperback 2020

© 2017 by Taylor & Francis Group, LLC
CRC Press is an imprint of Taylor & Francis Group, an Informa business

No claim to original U.S. Government works

ISBN 13: 978-0-367-57382-9 (pbk)
ISBN 13: 978-1-4987-6183-3 (hbk)

Library of Congress Cataloging-in-Publication Data

Names: Mondal, Prasenjit, editor. | Dalai, Ajay Kumar, 1959- editor.
Title: Sustainable utilization of natural resources / [edited by] Prasenjit
Mondal, Ajay K. Dalai.
Description: Boca Raton : Taylor & Francis, a CRC title, part of the Taylor &
Francis imprint, a member of the Taylor & Francis Group, the academic
division of T&F Informa, plc, [2017] | Includes bibliographical references.
Identifiers: LCCN 2016040254| ISBN 9781498761833 (hardback : acid-free paper)
| ISBN 9781498761840 (ebook)
Subjects: LCSH: Fuel. | Green chemistry. | Renewable energy sources. |
Conservation of natural resources. | Sustainable buildings.
Classification: LCC TP318 .S865 2017 | DDC 333.79/4--dc23
LC record available at https://lccn.loc.gov/2016040254

Visit the Taylor & Francis Web site at
http://www.taylorandfrancis.com

and the CRC Press Web site at
http://www.crcpress.com

Contents

Preface

This book aims to provide the scientific knowhow and orientation for the benefit of readers and society in the areas of the emerging technologies for the utilization of natural resources for sustainable development.

This book is organized into six sections and includes 22 chapters. The first section provides an overall introduction to the subject of the chapters to give the readers the stimuli and motivation, and it helps gain knowledge to advance their research and technologies. The second section deals with the utilization of conventional resources through cleaner sources. It includes chapters such as "Clean Coal Technologies," "Downstream Processing of Heavier Petroleum Fractions," "Hydrogen from Natural Gas," and "Liquid Fuels from Oil Sands." The third section is on the utilization of renewable biological resources. It includes chapters such as "Bioethanol Production from Lignocellulosic Biomass," "Butanol from Renewable Biomass," "Oil from Algae," "Bio-fertilizer and Bio-pesticides," Production of Life Saving Drugs from Marine Sources," "Production of Life Saving Drugs from Himalayan Herbs," and "Energy Production through Microbial Fuel Cells." The fourth section is related to the utilization of unconventional resources and includes chapters such as "Harvesting Solar Energy," "Utilization of CO_2 for Fuels and Chemicals," and "Hydrogen from Water." The fifth section deals with the optimization of resource utilization. Chapters in this section include "Integrated Water Resources Management," "Water Optimization in Process Industries," "Energy Integration in Process Plants," and "Green Building." The last section is on sustainability assessment. It includes the following chapters: "Life Cycle Analysis as the Sustainability Assessment Multicriteria Decision Tool for Road Transport Biofuels," "Optimization of Life Cycle Net Energy of Algal Biofuel Production," and "Sustainable Production and Utilization Technologies of Biojet Fuels."

Finally, this book deals with the sustainable utilization of natural resources. It is intended to help the young scientific professionals to upgrade their knowledge with the current thoughts and newer technology options along with their advances in the field of the utilization of natural resources for sustainable development.

The publication of this book was made possible by the researchers who contributed chapters to the six sections that comprise this book. We are indebted to them for all their efforts. We are also thankful to the all the reviewers who had put sufficient time and effort to review the respective chapters and for their valuable comments for revisions, which helped to improve the quality of this book. We also thank Arunima Soma Dalai for drawing the cover page of this book. We are thankful to the publishers for all the editorial work and for finally assembling the material in the most presentable format.

Dr. Prasenjit Mondal
Associate Professor, Department of Chemical Engineering,
IIT Roorkee, Roorkee,
Uttarakhand, India

and

Dr. Ajay Kumar Dalai, PEng, FRSC (Canada), FRSC (UK)
*Professor, Department of Chemical and Biological Engineering
and Canada Research Chair in Bio-energy and
Environmentally Friendly Chemical Processing,
University of Saskatchewan, Saskatoon, Canada*

Editors

Dr. Prasenjit Mondal is an associate professor in the Department of Chemical Engineering, Indian Institute of Technology Roorkee, India. He joined the institute in 2009 as an assistant professor. He has also worked as a process engineer in the industry for two years and as a scientist at the Centre for Scientific and Industrial Research, India, for three years before joining IIT Roorkee. His area of research is energy and environmental engineering (water/wastewater treatment through adsorption, electrocoagulation, and biological processes including phytoremediation, microbial fuel cells, oil from algae, energy from coal, biomass and wastes, and life cycle assessment). He has handled a number of R&D projects sponsored by the industry, Government of India, and international agencies. Presently, he is working on two international projects under Australia–India Strategic Research Fund and Indo-France Water Network Scheme. Dr. Mondal has published one book and more than 140 papers in international journals and conference proceedings. He is a recipient of NTSE scholarship, MHRD fellowship, Government of India, and S.J. Jindal Trust's scholarship, three-year membership award of the American Chemical Society. Dr. Mondal is the chairman of community dairy at IIT Roorkee. He is a reviewer of several international journals including *Environmental Progress & Sustainable Energy* and is a life member of the Indian Institute of Chemical Engineers and the treasurer of the Biological Engineering Society, India.

Dr. Ajay K. Dalai is a full Professor in the Department of Chemical Engineering at the University of Saskatchewan, Canada. In 2001, he was awarded a Tier 2 Canada Research Chair in Bioenergy and Environmentally Friendly Chemical Processing, and Tier 1 in 2009. From 2009 until 2014, Dr. Dalai was the associate dean of research and partnerships at the College of Engineering, in addition to taking his role as a professor and supervisor. His research focuses on the novel catalyst development for gas to liquid (GTL) technologies, biodiesel production and applications, hydrogen/syngas production from waste materials, hydroprocessing of heavy gas oil, and value-added products from biomass. He is currently working on the production and applications of activated carbon and carbon nanotubes (CNTs). The worldwide impact of this research is tremendous in terms of combating pollution and finding alternate energy resources; it has generated much interest and led to collaborative projects among research institutes and universities around the world. Dr. Dalai has published more than 300 research papers, mostly in heterogeneous catalysis and catalytic processes in international journals and conference proceedings. He has submitted several patent applications. He is an active board member, reviewer, and guest editor of several international journals, and is a life member of the Indian Institute of Engineers, the Indian Catalysis Society, the American Institute of Chemical Engineers, and an active member of the American Chemical Society and the Chemical Institute of Canada. Dr. Dalai is a fellow of CIC, CAE, EIC, AIChE and IIChE. He is the winner of 2014 CSChE Bantrel Award in Design and Industrial Practice. Dr. Dalai recently became a fellow of the Royal Society of Canada and Royal Society of Chemistry (UK).

Contributors

Syed Shaheer Uddin Ahmed
Department of Mechanical and Chemical
    Engineering
Islamic University of Technology
Gazipur, Bangladesh

Elvis Ahmetović
Faculty of Technology
University of Tuzla
Tuzla, Bosnia and Herzegovina

Kumar Anupam
Physical Chemistry, Pulping and Bleaching
    Division
Central Pulp and Paper Research Institute
Shahranpur, India

Sandeep Badoga
Catalysis and Chemical Reaction
    Engineering Laboratories, Department
    of Chemical Engineering
University of Saskatchewan
Saskatoon, Canada

P. Balasubramanian
Agricultural and Environmental
    Biotechnology Group, Department of
    Biotechnology and Medical Engineering
National Institute of Technology Rourkela
Rourkela, India

Ludger Beerhues
Institut für Pharmazeutische Biologie
Technische Universität Braunschweig
Braunschweig, Germany

Philip E. Boahene
Catalysis and Chemical Reaction
    Engineering Laboratories
University of Saskatchewan
Saskatoon, Canada

Ankur Bordoloi
CCPD
CSIR-Indian Institute of Petroleum
Dehradun, India

Pramod H. Borse
International Advanced Research Centre
    for Powder Metallurgy and New
    Materials
Hyderabad, India

Shri Chand
Department of Chemical Engineering
Indian Institute of Technology Roorkee
Roorkee, India

Amit Kumar Chaurasia
Department of Chemical Engineering
Indian Institute of Technology Roorkee
Roorkee, India

Raja Chowdhury
Department of Civil Engineering
Indian Institute of Technology Roorkee
Roorkee, India

Mariam Gaid
Institut für Pharmazeutische Biologie
Technische Universität Braunschweig
Braunschweig, Germany

Narayan C. Ghosh
National Institute of Hydrology
Roorkee, India

Ignacio E. Grossmann
Department of Chemical Engineering
Carnegie Mellon University
Pittsburgh, Pennsylvania

Suresh Gupta
Chemical Engineering Department
Birla Institute of Technology and Science
   (BITS)
Pilani, India

Fatin Hasan
Department of Chemical and Biomolecular
   Engineering, Melbourne School of
   Engineering
The University of Melbourne
Parkville, Australia

Nidret Ibrić
Faculty of Technology
University of Tuzla
Tuzla, Bosnia and Herzegovina

Kriti Juneja
Department of Biotechnology
Indian Institute of Technology Roorkee
Roorkee, India

P. Karthickumar
Department of Fish Process Engineering,
   College of Fisheries Engineering
Tamil Nadu Fisheries University
Nagapattinam, India

Janusz A. Kozinski
Lassonde School of Engineering
York University
Toronto, Canada

Zdravko Kravanja
Faculty of Chemistry and Chemical
   Engineering
University of Maribor
Maribor, Slovenia

Sudhir Kumar
Medicinal and Process Chemistry Division
Central Drug Research Institute
Lucknow, India

Sushil Kumar
Chemical Engineering Department
Motilal Nehru National Institute of
   Technology (MNNIT)
Allahabad, India

Preety Kumari
Department of Chemical Engineering
Indian Institute of Technology Roorkee
Roorkee, India

Rakesh Maurya
Medicinal and Process Chemistry Division
Central Drug Research Institute
Lucknow, India

Sonil Nanda
Lassonde School of Engineering
York University
Toronto, Canada

Ameeya Kumar Nayak
Department of Mathematics
Indian Institute of Technology Roorkee
Roorkee, India

Niren Pathak
School of Civil and Environmental
   Engineering
University of Technology
Sydney, Australia

M. Motiar Rahman
Civil Engineering Program Area, Faculty
   of Engineering
Universiti Teknologi Brunei
Gadong, Brunei Darussalam

Md. Mustafizur Rahman
Department of Mechanical and Chemical
   Engineering
Islamic University of Technology
Gazipur, Bangladesh

Vineet Kumar Rathore
Department of Chemical Engineering
Indian Institute of Technology Roorkee
Roorkee, India

Amiya Kumar Ray
Department of Polymer and Process
   Engineering
Indian Institute of Technology Roorkee
Saharanpur, India

Jhuma Sadhukhan
Department of Environmental Strategy
University of Surrey
Guildford, UK

A. K. M. Sadrul Islam
Department of Mechanical and Chemical
    Engineering
Islamic University of Technology
Gazipur, Bangladesh

J. N. Sahu
Petroleum and Chemical Engineering
    Program Area, Faculty of Engineering
Universiti Teknologi Brunei
Tungku Gadong, Brunei Darussalam

Shashank Sagar Saini
Department of Biotechnology
Indian Institute of Technology Roorkee
Roorkee, India

Sayedus Salehin
Department of Mechanical and Chemical
    Engineering
Islamic University of Technology
Gazipur, Bangladesh

Varun Sanwal
Department of Civil Engineering
Indian Institute of Technology Roorkee
Roorkee, India

Shubham Saroha
Department of Chemical Engineering
Indian Institute of Technology Roorkee
Roorkee, India

Mumtaj Shah
Department of Chemical Engineering
Indian Institute of Technology Roorkee
Roorkee, India

Shahriar Shams
Civil Engineering Program Area, Faculty
    of Engineering
Universiti Teknologi Brunei
Gadong, Brunei Darussalam

Ravi Shankar
Department of Chemical Engineering
Indian Institute of Technology Roorkee
Roorkee, India

Pratik N. Sheth
Chemical Engineering Department
Birla Institute of Technology and Science
    (BITS)
Pilani, India

Jyoti Singh
Department of Biotechnology
Indian Institute of Technology Roorkee
Roorkee, India

Rajesh P. Singh
Department of Biotechnology
Indian Institute of Technology Roorkee
Roorkee, India

Debabrata Sircar
Department of Biotechnology
Indian Institute of Technology Roorkee
Roorkee, India

Akter Suely
Institute of Biological Sciences
Faculty of Science
University of Malaya
Kuala Lumpur, Malaysia

Deepak Tandon
Residue Conversion Division
Indian Institute of Petroleum
Dehradun, India

Lokendra Singh Thakur
Department of Chemical Engineering
Indian Institute of Technology Roorkee
Roorkee, India

Prabu Vairakannu
Department of Chemical Engineering
Indian Institute of Technology
    Guwahati
Guwahati, India

Shobhit Verma
Department of Biotechnology
Indian Institute of Technology Roorkee
Roorkee, India

Paul A. Webley
Department of Chemical and Biomolecular
    Engineering, Melbourne School of
    Engineering
The University of Melbourne
Parkville, Australia

Hossain Zabed
Institute of Biological Sciences
Faculty of Science
University of Malaya
Kuala Lumpur, Malaysia

1

Sustainability Issues in the Twenty-First Century and Introduction to Sustainable Ways for Utilization of Natural Resources

Amiya Kumar Ray, Prasenjit Mondal, and Kumar Anupam

CONTENTS

1.1 Introduction

In the twenty-first century, the world is facing many social, environmental, and economic challenges of the time, but the greatest challenge remains to keep this planet in a better shape for future generations. To achieve this goal, the concept of sustainability is evolved, which emphasizes the overall development of a sustainable society. The following paragraphs outline the chronological evolution of the terms "sustainability" and "sustainable development" (Web 1).

The name sustainability is derived from the Latin *sustinere* (*tenere*, "to hold"; *sus*, "up"). "Sustain" can mean "maintain," "support," or "endure" (Stivers 1976; Meadows et al. 2004). The history of sustainability can describe human-dominated ecological systems from the earliest civilizations to the present time (Barbier 1987). It is reported that the concepts of sustainable development and sustainability were used in the past (~twelfth to sixteenth centuries) in forest management as "sustained yield," which is the translated form of German term *Nachhaltiger Ertrag* (Grober 2007; Ehnert 2009). However, over the past five decades, the concept has been significantly broadened (Ehnert 2009).

In the seventh decade of the twentieth century (the 1970s), the term sustainability was used to describe an economy in equilibrium with basic ecological support systems. In its classic report on the "Limits to Growth," the Club of Rome used the term "sustainable" for the first time in 1972, which has later been highlighted by scientists in many fields (Grober 2007; Finn 2009). To address the concerns over the impacts of expanding human development on the planet, economists have also presented an alternative term "steady-state economy."

Since the 1980s, sustainability has been used more in the sense of human sustainability on the planet earth. In 1980, sustainable development was first referred to as a global priority in "The World Conservation Strategy" published by the International Union for the Conservation of Nature (IUCN 1980). To guide and judge the human conduct affecting nature, five principles of conservation were raised by the United Nations World Charter for Nature in 1982 (UNWC 1982). The report "Our Common Future" also commonly known as "Brundtland Report" was released by the United Nations World Commission on Environment and Development in 1987. In this report, sustainability is defined as a part of the concept of sustainable development that ensures the needs of the present without compromising the ability of future generations to meet their own needs (Brundtland 1987; Smith and Rees 1998). *The Concept of Sustainable Economic Development* was also published in 1987 by the economist Edward Barbier, in which he had advocated that the goals of environmental conservation and economic development are not conflicting and can reinforce each other (Barbier 2006).

In 1992, the UN Conference on Environment and Development (UNCED) published the "Earth Charter," also known as the "Rio Summit," "Rio Conference," or "Earth Summit" (Portuguese: ECO92), which outlines the building of a sustainable peaceful global society in the twenty-first century. The action plan 21 Agenda identified information, integration, and participation as key building blocks to help countries achieve sustainable development (UNCSD 2012). To strengthen the implementation of sustainable growth, the UN Conference on Sustainable Development 2012 (UNCSD 2012), also known as Rio 2012, Rio+20, or Earth Summit 2012, was held in Brazil in 2012. After a debate on more than 90 environmental issues initially proposed, the United Nations Environment Programme (UNEP) also promulgated 21 emerging issues for the twenty-first century in 2012 (UNCSD 2012; UNEP 2012). In 2013, four interconnected domains, namely, ecology, economics, politics, and culture, were identified for reporting sustainability (Magee et al. 2013), which is related to our long-term cultural, economic, and environmental health as well as vitality. Sustainability can be achieved by linking these issues together rather than considering them as separate. An economically and environmentally viable process may not be sustainable until it is socially desirable as shown in Figure 1.1 (Smith and Rees 1998).

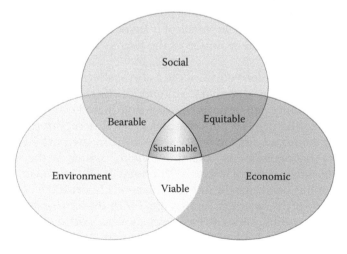

FIGURE 1.1
Venn diagram of sustainable development at the confluence of three constituent parts. (https://en.wikipedia.org/wiki/Sustainability.)

Sustainable development is a dynamic process or an action plan or a road map for a desirable future state for human societies in which living conditions and resource use continue to meet human needs without undermining the "integrity, stability, and beauty" of natural biotic systems (Web 1). It also ensures the carrying capacity of natural systems with the social, political, and economic stability (Stivers 1976). Hence, devising methods and mechanisms for utilization of natural resources for sustainable development of a human society is the primary goal of sustainability. A sustainable society focuses on the minimum use of nonrenewable resources such as minerals and fossil fuels; conservation of finite stocks of biodiversity; the maximum use of renewable resources such as freshwater, soils, and forests; and conservation of absorptive capacity of local and global sinks of wastes such as air and water resources. It also provides an opportunity for each human being to develop itself in freedom, within a well-balanced society, and in harmony with its surroundings. Thus, some important approaches for achieving sustainability can be lesser consumption of natural resources such as water and energy; reduction of energy wastage; more use of fuel efficient engines in vehicles and machines; more recycling and reusing of waste materials; and more efforts for protection of soil and forests.

This chapter mainly focuses on various sustainable ways of utilizing natural resources, which may be either exhaustible or renewable. The former includes minerals, coals, petroleum, nuclear, and natural gas, whereas the latter includes a variety of sources grown in or obtained from land, ocean, and air. Some examples of such renewable resources are biomass, water, wind, solar, microorganisms, herbs, and plants. Many of these resources are being utilized conventionally (without taking sufficient measures for the environmental, social, cultural, and health needs). These resources can be utilized in a more optimal and eco-friendly way to ensure sustainability. Further, some resources are not yet well exploited, although these have good potential for contribution toward sustainability. The following sections describe how sustainability can be achieved through various ways of utilization of natural resources. This chapter is organized as follows: Section 1.1.1 describes utilization of conventional resources via cleaner routes, Section 1.1.2 highlights exploitation of renewable biological resources, Section 1.1.3 explains utilization of unconventional resources, Section 1.1.4 illustrates methods of optimization of resource utilization, and Section 1.1.5 emphasizes sustainability assessment.

1.1.1 Utilization of Conventional Resources Adopting Cleaner Routes

Energy is one of the most significant inputs for economic growth and human development, and coal has been recognized as the most important source of energy. Today approximately 40% of the world's energy requirement can be achieved through coal. However, the major constraint in the utilization of coal is that ~65% of its global reserves contain low-rank coal, which on combustion produces serious ecological and environmental threats due to emission of obnoxious greenhouse gases. Extraction of coal from mines and its processing for application in power plant also add different types of pollutants in air and water. For the sustainable utilization of coal, emphasis is being laid on the development of suitable coal extraction technologies as well as energy recovery from the existing coal resources using clean coal technologies. Various clean coal technologies include coal beneficiation with ultrasound enhanced technology; coal gasification—both conventional and molten and plasma gasification; co-combustion with gasification; carbon capture and sequestration (CCS) using precombustion, postcombustion, and oxy-fuel combustion techniques; and clean combustion techniques including technologies for NO_x reduction, high-temperature air combustion, chemical looping combustion, and chemical looping reforming. Unconventional coal technologies such as underground coal gasification (UCG) and coal to liquid oil (CTL) production are also two important clean coal technologies. Integration of a mixture of these clean coal technologies in existing power generating systems is necessary to achieve a minimal energy penalty for CCS. According the National Energy Technology Laboratory, Morgantown, West Virginia, deployment of new advanced technologies can reduce emission as well as coal consumption for the production of a certain amount of electricity by increasing the overall plant efficiency (Web 2).

It has also been predicted that using clean power plants, the greenhouse gas emissions can be reduced by ~30% by 2030 with reference to the emission in the year 2005 (Web 3).

Petroleum crude oil, which produces useful products such as liquefied petroleum gas, naphtha, gasoline, kerosene, jet fuel, diesel, heating oil, and asphalt base, is another major source of energy after coal. The above products are produced from crude oil through atmospheric and vacuum distillation processes. Various conversion processes such as cracking and hydrotreating are also used to alter product distribution. A good amount of residue is also generated after vacuum distillation of crude oil called as vacuum residue (VR). The amount and quality of VR are dependent on the metal and sulfur content of crude oil as well as its viscosity. Utilization or disposal of this residue is a great concern of a refinery as it influences the economy and environmental requirements. Due to gradually degraded quality of crude oil, it is becoming heavier day by day, resulting in more residues related to more contaminants. Thus, upgrading of heavy residues is becoming a priority area for the sustainability of petroleum refinery. A number of approaches (both hydrogen addition and carbon rejection) are developed for upgrading the heavy residues. Some of these processes are cracking, hydrocracking, visbreaking, delayed coking, solvent deasphalting, and gasification (Ancheyta and Rana 2004; Speight 2013). Recently, nanotechnology and biological routes have also been investigated to explore the upgrading of heavy residue.

Natural gas is widely used as fossil fuel next to coal and petroleum crude. It mainly contains methane along with some higher hydrocarbons and CO_2. It is used for power and heat application in power plants as well as in the transportation sector as compressed natural gas. It can also be used to produce hydrogen through steam reforming. Hydrogen produced from natural gas can further be used to produce electricity in a fuel cell, or it can be used for the synthesis of many chemicals such as ammonia. Application of natural gas in a Honda Civic can reduce global warming pollution by ~15% than a conventional gasoline-powered Civic. However, ~30% emission reduction is possible if a gasoline–electric Civic hybrid is used in place of a conventional gasoline-powered Civic. Further, the conversion of natural gas to electricity or hydrogen and its subsequent application in plug-in vehicles or fuel cell vehicles can give ~40% saving of global warming emissions (Web 4 and Web 5).

Therefore, hydrogen production from natural gas is more sustainable than its other utilization routes. Conventional steam reforming units have very large capacity and hence are more economical. Recently, extensive research has been conducted to increase the efficiency of the small-scale reforming process using compact reactors such as microchannel reactors and monolith reactors.

Another nonconventional fossil fuel, which has attracted great attention in recent years, is oil sand, which is basically a mixture of clay, sand, water, and bitumen (a dense and extremely viscous form of petroleum). Oil sand deposits have been found in many countries around the world, including Canada (~169 billion barrels), the United States (~28 billion barrels), Venezuela (~100 billion barrels), Russia (~60 billion barrels), and some other countries (Web 6). However, the largest deposit of oil sands is found in Canada. More than 320 billion cubic meters (two trillion barrels) of global oil sand deposits has been estimated. Efforts are made around the world to extract and produce usable oil from oil sands, and the Athabasca deposit in Alberta is utilizing the most available technologically advanced production process. It is predicted that the oil sand production in Canada would increase 1.7 times by 2024 with respect to the production in the year 2014 (from 2.3 million barrels per day to 4 million barrels per day) (Web 5). Many international companies such as Shell, ExxonMobil, Sinopec, BP, Total, Chevron, and PetroChina are set to expand their oil extraction from oil sands in the upcoming years. However, developments of more cost-effective and eco-friendly methods are required for the utilization of this resource.

1.1.2 Utilization of Renewable Biological Resources

The major breakthrough in energy security can be established if renewable resources are utilized properly. Among the renewable resources, some are biological resources comprising mainly land biomass and biomass from marine and aquatic sources, whereas wind, ocean, hydro, and solar sources are some other important nonbiological sources of renewable energy. Microorganisms in wastes can also be used for energy production.

Biomass can be converted into liquid biofuels, which can be used in transportation. The major biofuel is bioethanol, which has high octane number and other desirable fuel properties. It is also environmentally friendly. It is mainly produced from abundant renewable biomass through a biochemical route in which low-cost and plentiful biomass from nonfood sources is first broken down into a number of sugars through pretreatment and chemical or enzymatic hydrolysis. This is then followed by fermentation in the presence of biocatalysts such as yeast, to produce bioethanol. Any sugary substances (derived from cane and beet), starchy agricultural crops, crop residues, lignocellulosic biomass, and algal biomass have the potential to be a feedstock for bioethanol production. The pretreatment steps become more important for handling lignocellulosic biomass. Many kinds of microbes such as *Saccharomyces cerevisiae*, *Zymomonas mobilis*, thermophilic *Thermobacter ethanolicus* or thermophilic ethanologen, thermophilic anaerobic bacteria, *Clostridium thermocellum*, *Themoanaerobactenum sachharolyticum*, aerobic mesophilic fungus, *Trichoderma reesei*, fungal glucoamylase, cellulase-producing fungus *Aspergillus niger*, and thermotolerant yeast *Kluyveromyces* sp. IIPE453 are used for bioethanol production (MTCC 5314) (Ray et al. 2013).

Production of bioethanol from edible biomass is under debate as the requirement of food crops is increasing due to increased population. Thus, application of lignocellulosic biomass for bioethanol production is a very important area for sustainable production of bioethanol. However, this process is not well established for commercial scale production due to low biodegradability of lignocellulosic biomass and complexity in product separation steps. Systematic efforts are required to develop a new technology to produce ethanol from lignocellulosic biomass. Currently, lignocellulosic ethanol can be competitive with fossil fuels at a crude price of $100 per barrel or more. However, in 2030 it is expected to be competitive at a crude price of $75 per barrel (Clixoo 2016). A combined technology based on the integrated biorefinery concept (BioGasol) for production of biogas, hydrogen, methane, ethanol, and solid fuel (lignin) from biomass must be employed to get sustainable low-cost production of lignocellulosic bioethanol (Ahring and Langvad 2008). The possibility of producing longer chain alcohol such as butanol, isopropanol, and 2,3-butanediol from lignocellulosic biomass should also be explored.

The energy density of bioethanol is ~40% less than that of regular gasoline, whereas for biobutanol, it is ~10% less than that of regular gasoline, which shows that biobutanol has more energy density than bioethanol. Physicochemical properties of these two alcohols, responsible for blending and antiknocking properties are also comparative (Mužíková et al. 2014). Thus, biobutanol can be more suitably used as gasoline blend than

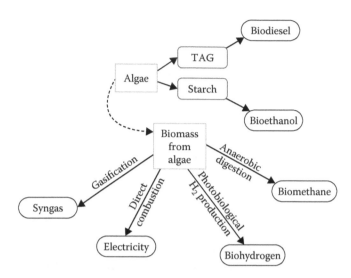

FIGURE 1.2
Path of bioenergy produced by algae. TAG, triacylglyceride or triacylglycerol. (From Biernat, K. et al., *The Possibility of Future Biofuels Production Using Waste Carbon Dioxide and Solar Energy*, 2013. With permission.)

bioethanol. It has been attracting strong attention in recent years (Jang et al. 2012). The yield and speed of production of biobutanol is dependent on the types of microorganisms and substrates used. Higher alcohol concentration (>3%) shows toxic effects to microorganisms (Qureshi and Maddox 1995). To overcome this limitation, new microbes are used and genetic modification of microbes is made (Durre 2007) to get high alcohol concentration withstanding strains. The mutant strains of *Clostridium acetobutylicum* and *C. beijerinckii* have been used to produce high concentrations of cellulosic *n*-butanol. Efforts are made to develop this process in commercial scale (Huang et al. 2010).

Among the nonconventional and renewable biosources, algal biomass seems to be the most promising one as its growth rate is very high. Through photosynthesis, algae produce carbohydrates, part of which is converted into lipid and protein through different metabolic pathways. The carbohydrate can form ethanol through fermentation, whereas lipid/oil can produce biodiesel through transesterification or other route. The proteins can be recovered and used for many applications. Algal biofuels such as bioethanol and biodiesel are highly biodegradable, are nontoxic, and contain no sulfur. Algae also help to reduce greenhouse gas by consuming CO_2. Figure 1.2 shows the necessary processing steps for production of various kinds of biofuels including bioethanol from algae.

Apart from bioethanol and biodiesel, biomethane, biohydrogen, electricity, and syngas can also be generated from algae. However, each fuel production requires a number of unit operations with different complexities. The major processing steps include pretreatment, conversion, final product separation, and purification. Pretreatment processes include oil extraction or thermochemical steps to produce algal oil or bio-oil along with the residues. The conversion steps consist of transesterification (biodiesel), hydroprocessing (green diesel, green jet fuel, and green gasoline), biochemical conversion (oxygenates-bioethanol, biobutanol, biomethane), and thermochemical gasification (syngas).

Biomass of plants and algae is proven as a competitive feedstock for energy production. These renewable resources can also be converted into biofertilizer, which has a significant potential for the improvement of soil properties for better crops. Although green revolution has improved the food security in the world by applying chemical fertilizer and pesticides, it has significantly affected the soil conditions and the crop yield is under threat in the twenty-first century. Further, it has also induced many contaminations in soil as well as groundwater. Use of biofertilizer and biopesticides can help to restore the soil properties and reduce pollution. Biofertilizers are generally produced from renewable resources, which utilize living microorganisms. These fertilizers when applied to seed, plant surfaces, or soil promote the plant growth by digesting large biopolymers such as proteins, carbohydrates, fibers, and fats, which increase the availability of primary nutrients to the host plant. Natural nitrogen fixation as well as phosphorus and potassium solubilization processes are promoted by living microorganisms to add nutrients (growth-promoting substances—amino acids, sugars, and fatty acids) to the soil. The main sources of biofertilizers are bacteria, fungi, and cyanobacteria (blue-green algae) and other natural plant resources such as neem. Some examples of living microorganisms for biofertilizers are *Rhizobium azospirillum, Azotobacter,* and *Acetobacter.* Biofertilizers can accelerate soil fertility, promote plant growth, reduce the use of chemical fertilizer, scavenge phosphates from soil layers, and increase the availability of nutrients. There are numerous uses of biofertilizers such as production of legumes, paddy crops including rice, wheat, corn, mustard, cotton, potato, and many other vegetables. Farm yard manure (a type of biofertilizer) can be produced by using raw materials such as cow dung, cow urine, and waste straw and dairy wastes after producing gobar gas through composting. Like biofertilizers, the biopesticides, which control pests, suppress the growth of other bacteria, fungi, and protozoa, are also derived from plants, bacteria, animals, and some minerals. Some examples of biopesticides are fermented curd water, cow urine extract, chilli–garlic extract, neem cow urine extract, baking soda, and canola oil. *Trichoderma viride* can act as a biofungicide. These biofertilizers and biopesticides are very effective in providing long-term benefits compared to chemical pesticides and other chemical products.

Like land biomass, renewable marine biomass has also high potential to support the sustainability of the society. The ocean represents a rich source for pharmaceutical products, nutritional supplements, cosmetics, agrichemicals, and enzymes (Vignesh et al. 2011). It can also provide many natural products having a unique structure, which can be used as a source of bioactive compounds suitable for combating deadly diseases such as cancer, osteoporosis, acquired immunodeficiency syndrome (AIDS), human immunodeficiency virus (HIV), Alzheimer's disease, and arthritis. Many compounds possessing analgesic, anti-infective, antimicrobial, antitumor, and anti-inflammatory properties have also been developed. Marine microorganisms, algae, and invertebrates are primarily found to be the source of such lifesaving compounds (Jha and Zi-rong 2004; Bhadury et al. 2006). The first marine-derived cancer drug Cytosar-U® (Medicines By Design 2011) is produced decades ago from a Caribbean Sea sponge, which is used to treat leukemia and lymphoma. Yondelis is another marine-derived cancer drug, which is isolated from *Ecteinascidia turbinata* and is under clinical testing (Bhadury et al. 2006). A growing number of marine fungi are the sources of novel and potentially lifesaving bioactive secondary metabolites. A nerve toxin Prialt™ Dublin, Ireland has been derived from cone snail and is being marketed by Elan Corporation, plc, in Dublin, Ireland. This drug jams up nerve transmission in the spinal cord and blocks certain pain signals from reaching the brain. The production of medicines from indigenous marine-pharmaceutical biomass with competitive price may stimulate new markets for the agriculture sector and can also create many job opportunities (Bruckner 2002).

Biomass from hills also possesses high potential for maintaining sustainability in the twenty-first century. The Himalaya regions in India have traditional knowledge of ayurvedic/herbal medicine. Some rare and endemic species of medicinal and aromatic plants are available in this region along with many other species, which are valuable to the pharmaceutical and cosmetic industries (Banerji and Basu 2011). Approximately 45% plants including 8000 species of angiosperms, 44 species of gymnosperm, 600 species of pterido-phytes, 1736 species of bryophytes, 1159 species of lichens, and 6900 species of fungi have medicinal properties (Samant et al. 1998). In 2008, the size of the global market of herbal drug was ~US$60 billion per annum (Sharma et al. 2008). The demand of herbal drugs is also increasing very fast and its global market size in 2017 may reach ~US$ 100 billion per annum (Sharma et al. 2008). Thus, the Himalayan medicine system, a vast treasure of herbal medicine, should be exhaustively explored and used for the economic regeneration of the local people as well as for the medical benefit of the whole world.

Based on chemical investigations on the herbs of traditional tribal folk available in the herb layer of the Himalayas for medicinal use, it has been established that a number of modern lifesaving drugs are the prominent constituents of these herbs. A number of lifesaving drugs such as reserpine, pilocarpine, ephedrine, theophylline, vincamine, atropine, aconite, and colchicines have been derived from traditional folk medicinal herbs. The following plants are the main constituents of the herb layer of the Great Himalayas: *Carex nubigena, C. muricata, Ainsliaea aptera, Viola canescens, Goldfusia dalhosiana, Stellaria monosperma, Bupleurum lanceo-latum, Valeriana jatamansi, Scutellaria angulosa, Justicia simplex, Oxalis corniculata, Rubia cordi-folia, Anaphalis contorta, Anemone rivularis, Swertia* spp., *Eupatorium* spp., and *Dipterocarpus* spp. Herbal drugs are used for the treatment of cancer, AIDS, malaria, liver diseases, kala-azar and other infectious diseases, hypertension, antiarthritic, bronchial asthma, and so on. These have also the antitumor, hepatoprotective, and ant fibrotic activities. These can also act as anti-inflammatory, sex hormone and oral contraceptive, an antidote to insect bites and snakebites, febrifuge, a stimulant to uterine contraction, and a sedative (Goswami et al. 2002).

Taxol obtained from yew tree, a rare Himalayan plant, is used for chemotherapy for cancer, breast, and ovarian cancer. The cost of this medicine is US$13 per milligram with a market potential of US$870 million. Ashwagandha (*Withania somnifera*) is also used for can-cer treatment. *Merremia peltata* and Malpighia emarginata, flavonoids of *Plantago asiatica* L., have been employed for AIDS/HIV inhibitors. There are noticeable advantages for herbal medicines compared to synthetic ones, and research is being conducted to further explore the herbal sources for medicinal use.

Not only the higher plants and herbs but also microorganisms have the potential to help sustainability in the twenty-first century. Microbial fuel cells (MFCs) provide new oppor-tunities for the sustainable production of energy from biodegradable, reduced compounds and effluents (Rabaey and Verstraete 2005). They convert biochemical metabolic energy into electrical energy. Thus, they can also be used simultaneously for wastewater treatment and electricity (Das and Mangwani 2010). Most recent developments in MFC technology include its use as microbial electrolysis cells, in which anoxic cathode is used with increased external potential at the cathode and hydrogen is produced. Phototropic MFCs and solar-powered MFC technology for electricity generation are also few latest developments. MFCs have the following operational and functional advantages over the technologies currently used for generating energy from organic matter (Rabaey and Verstraete 2005).

1. Substrate conversion efficiency is high.
2. They can operate efficiently at ambient and even at low temperatures.
3. They do not require gas treatment.
4. They do not need energy input for aeration when the cathode is passively aerated (Rabaey and Verstraete 2005).
5. They can be used in locations lacking electrical infrastructures.

The MFC technology is evaluated relative to current alternatives for energy generation. Though it is renewable, eco-friendly, sustainable, and useful for wastewater treatment, electricity generation, and bioremediation of toxic compounds at the same time, commercial exploitation is still awaited. Studies on economic feasibility for large-scale production are needed.

1.1.3 Utilization of Unconventional Resources

With the rapid depletion of fossil fuel reserves and the ever-increasing demand of energy with the associated cost of energy, it is an imperative necessity to explore unconventional renewable energy resources in the developing countries such as India. The major types of unconventional renewable energy sources include solar, wind, mini and micro hydroenergy, geothermal energy, wave energy, tidal power plants, and ocean thermal energy conversion; all of these have the potential to meet future energy needs. Among these resources, solar and wind energy are considered as the most important in India. Although solar energy production has some drawbacks such as low-energy density per unit area and uncertainty of availability with extreme seasonal variation, extensive efforts are made around the world to develop this technology as it is nonexhaustible and completely pollution free. Solar energy seems very promising for countries such as India, Pakistan, and China, as these fall in the solar zone (Enerco 2015). Presently, around 13.5% of the total energy production in India takes place through renewable routes and the solar energy contributes approximately 10% of the renewable energy production. It is also growing very fast with a highly ambitious target to produce 100 GW (100,000 MW) by 2022 (Chaurey and Kandpal 2010).

There are principally two methods of solar energy utilization: the thermal conversion of solar energy for heating applications and the photovoltaic (PV) conversion of solar energy into electric power generation. Solar heaters or thermosyphon solar collectors are used to raise the temperature of the fluid (water or air) flowing through the collector. However, poor heat transfer coefficient of air flowing through the collector results in low thermal efficiency (Saini and Saini 2005). Solar thermal systems also require a storage. In PV conversion, direct electricity is produced by means of silicon wafer PV cells called solar cells. It is easy to install and seems a better option for power generation if the life cycle energy use and greenhouse gas emission are considered, although large-scale exploitation of PV may lead to some undesirable environmental impacts in terms of material availability and waste disposal. Further, the cost and maintenance of solar cells are the greatest problems (Kellogg et al. 1998; Celik 2003; Arun et al. 2007; Hrayshat 2009; Singh et al. 2009; Bekele and Palm 2010; Chaurey and Kandpal 2010).

Combustion of fossil fuels emits high amount of CO_2 in the atmosphere, and globally around 30 Gt CO_2 has been emitted in a year in recent times. The CO_2 concentration in the atmosphere has increased from 300 ppm in preindustrial time to 400 ppm in the twenty-first century due to the emissions from fossil fuel combustion. Further, it is expected that in the next few decades also, fossil fuels will be the main source of energy; thus, it may be difficult to reach the target of CO_2 emission reduction even though the energy efficiency is improved and renewable energy is used. Therefore, efforts are made to capture CO_2 at its source of production as well as its utilization for the production of value-added products. Various technologies have been developed in recent years to capture CO_2 and its storage effectively and economically, although these are not matured yet for postcombustion power plants. The captured CO_2 can be used as a solvent or a working fluid, a storage medium for renewable energy as well as a feedstock for various chemicals. It can also be used for the growth of microalgae. It is a well-known fact that CO_2 helps the growth of autotrophic biomass through photosynthesis. Thus, properly designed method for the utilization of CO_2 can be applied for the growth of microalgae and other biomass for energy feedstock. Different energy storage chemicals, such as syngas, methane, ethylene, formic acid, methanol, and dimethyl ether, can be produced from CO_2. Various polymeric materials can also be produced by inserting CO_2 into epoxides. Approximately 0.3–0.7 Gt/year of CO_2 may be consumed through various chemical conversion pathways (Web 7). Conversion of CO_2 into inorganic minerals through electrochemical reactions followed by the necessary mineralization reaction has been a strong interest in recent years as these can be used in building materials. A preliminary estimate suggests that ~1.6 Gt/year CO_2 could be consumed if 10% of the world's building materials were replaced by such a source. Enhanced oil recovery, which can increase oil recovery by 10%–20%, is another important commercial technology for CO_2 and its storage. Similarly, methane recovery from unmined coal seams can also be done using CO_2.

Water, which is abundantly available in nature, can be a source of energy. In hydroelectric power plant, the kinetic energy of water is used to produce electricity; the water molecules are also used for the production of hydrogen through steam reforming, gasification, and so on. Hydrogen is also produced from water through other routes such as electrolysis and photocatalytic reactions. Hydrogen production from water through electrolysis is an old concept; however, it suffers with high capital and maintenance costs of the process (NREL 2009). A large number of investigations are being conducted to reduce the cost of this process so that water can be used more environmentally friendly for energy/hydrogen production.

Freshwater is a finite and vulnerable resource and essential to sustain life, development, and the environment. In a broader sense, water is linked up with all kinds of resources, namely, energy, agriculture, finance, industry, tourism, environment, and fisheries. A systematic process called integrated water resources management (IWRM) is evolved for the allocation and monitoring of water resources and their use in the context of economic, social, and environmental objectives to achieve sustainable development. IWRM considers interdependency among all the different uses of finite water resources. The final statement of the ministers at the International Conference on Water and the Environment in 1992 (so called the Dublin Principles) recommended the development of IWRM to promote essential changes in practices for improved water resources management (Web 8; Young et al. 2008). IWRM is hence a process, which promotes the coordinated development and management of water, land, and related resources in order to maximize economic and social welfare in an equitable manner without compromising the sustainability of vital ecosystems. The concept of IWRM is a way forward for the sustainable development and management of the world's limited water resources.

Different techniques, systems, or models such as geographic information system, database, and integrated water resources planning model are used for IWRM. Each model is complex in nature and needs a sophisticated software to solve. For example, time-series water balance model comprises water demand forecasting module, water balance module, water quality module, water allocation/costing module, resource management/development option module, and development scenario evaluation module. Similarly, watershed modeling is another interactive modeling employed for IWRM which consists of several components. Modeling through advanced techniques is being applied to achieve reliable water security, enhanced agricultural yield, improved living standard, and sustainable land-use and community consensus. However, because IWRM practices depend on the context, sometimes it becomes challenging to translate the agreed principles into concrete action.

1.1.4 Optimization of Resource Utilization

Optimum use of resources is another important approach for sustainable development. It improves the economy of a process as well as preserves resources for further application. In the process industry, the optimization of water and energy usage is an interesting area, which has a significant impact on the overall economy of the process. To maintain sustainability, reduction in water consumption as well as wastewater discharge is essential. Regeneration and recycling of wastewater in the process can reduce freshwater requirement while satisfying environmental regulations. Water minimization in the process industry (both freshwater and wastewater) can be made through improved operations and equipment as well as by eliminating or reducing water-intensive practices. Changes in heating and cooling methodologies, prevention of leakage from water-carrying systems and water spillover also help to reduce water consumption. Effective water monitoring and maintenance program is thus necessary. Optimization strategy is important in achieving not only freshwater minimization and minimum wastewater generation but also a subsequent reduction in the cost of operation. Integrated process design or process synthesis called pinch technology is one of such approaches based on recycle and reuse practices. The branch-and-bound method of optimization is also frequently employed to design the flow sheeting in chemical process engineering. To simultaneously target the minimum freshwater requirement, minimum wastewater generation, maximum water reuse, and minimum effluent treatment, a novel limiting composite curve also called as a source composite curve was proposed (Bandyopadhyay and Cormos 2008). To address the water management issues of the integrated process involving regeneration and recycle through a single treatment unit (single component), graphical representations as well as analytical algorithms can be used. In multicontaminant problems, the proposed methodology can be applied based on the limiting contaminant. Efforts are made to develop a general approach for dealing with multicomponent problems.

Chemical process industries such as sugar and paper, distillery are generally energy intensive. The energy costs in these industries are on the order of 25%–30% of the total cost of manufacture. Therefore, energy conservation in these industries is necessary, which can be achieved by reducing energy losses in the process. The issue of energy conservation is interrelated with steam generation, process steam demand, condensate extraction and blow-off, incorporation of other heat economy measures in the plant, proper choice of all equipment with conceptual optimal design features, design alternation of existing equipment and network, and so on. Efficient evaporator system design with vapor recompression and bleeding of vapors as well as splitting of multiple effect evaporator setups improve the energy economy. Improved heat exchanger network can also reduce energy consumption. Thus, energy optimization policies should be employed to evaluate the energy consumption in a specific industry, mainly steam usage, and then various modeling techniques should be utilized to optimize the energy consumption. Statistical multiple linear or nonlinear regression models have been used to study the effects of variables on the steam consumption (Raghavendra and Arivalagan 1993). Pinch technology has been used to optimize the heat exchanger network to reduce energy consumption. It also helps to achieve a targeted energy scenario with improved network.

Like process industries, water and energy optimization can be achieved in domestic and commercial complexes through implementing the concept of green building, which refers to a structure, and using a process that is environmentally responsible and resource efficient throughout a building's life cycle. The whole process consisting of design, construction, operation, maintenance, renovation, and demolition requires the cooperation of the design team, architects, engineers, and clients. The main goals of a green building are better sustainability index, sitting and structure design efficiency, energy efficiency, water efficiency, materials efficiency, indoor environmental quality, reduced wastes, and maintenance cost. The benefits of green buildings are evaluated based on environmental, economic, and human aspects, including thermal comfort, indoor environmental quality, health, and productivity (Hauge et al. 2011). Some important green building assessment tools are as follows: Leadership in Energy and Environmental Design, Building Research Establishment Environmental Assessment Method, Green Building Council of Australia, and Green Star (Zuo and Zhao 2014).

1.1.5 Sustainability Assessment

With increasing awareness on sustainability issue in the twenty-first century, extensive research is being conducted around the world to introduce new technologies/products, with less environmental impact, high economic benefit, and acceptable social impact as sustainability is related with these three major factors. Thus, to ascertain the sustainable development, it is imperative to quantify the sustainability aspect of a process/product before taking any decision on it by the policy maker. Many indicators have been developed for this quantification purpose, and some frameworks have been devised by many organizations. For example, sustainability metrics covering economic, environment, and social dimensions and involving a different set of indicators have been formulated by the Institution of Chemical Engineers (IChemE) (Labuschagne et al. 2005). The metrics as shown in Figure 1.3 was initiated to assess the sustainability performance of the process industry.

To assess the environmental impacts associated with all the stages of a product/process, the life cycle assessment (LCA) is used. The role of LCA is crucial in determining the values of various metrics and emissions along the entire chain of a product. These values are further used to overcome sustainability challenges through creative thinking, and the whole process is called sustainability life cycle assessment (SLCA). The SLCA is nothing

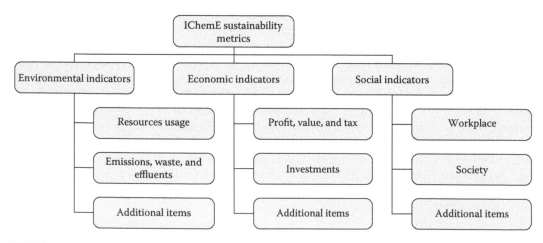

FIGURE 1.3
The IChemE metrics for sustainability. (From Singh, R.K. et al., *Ecol. Indic.*, 15, 281–299, 2012. With permission.)

but an assessment tool and an accompanying process to get a strategic overview of the full scope of social and ecological sustainability at the product level. An effective life cycle approach can identify where potential trade-offs may occur across different media and across the life cycle stages (Fava et al. 1993).

Considering the scarcity and degraded quality of fossil fuels, biofuels are attracting high attention as a renewable fuel. Extensive research is being carried out around the world to improve the quality and yield of biofuels from different biomass feedstocks. Although apparently it seems that biofuels are more attractive than fossil fuels, the strategies for sustainability have to be thoroughly assessed at several levels such as relevance of biofuels with respect to sustainable development, sustainable transport, main sustainable transport strategies, technology strategies, alternative fuel strategy, biofuel strategy from the first- to the fourth-generation biofuels, sustainability frameworks, standards, criteria and certification, theoretical perspectives, and methodology in respect of transfer effects, industrial ecology, and life cycle assessment to assess their suitability as transport fuels (Holden and Gilpin 2013). To be more specific, to assess the sustainability of biofuels for road transport, one should consider the key characteristics of biofuels as well as some essential criteria as stated below before adopting assessment policies (Curran 2013):

1. The four main dimensions for sustainable development must be satisfied if a biofuel is used.
2. Gains from biofuel strategies must be competitive to gains from other sustainable transport strategies, such as reducing transport volume and altering transport patterns.
3. Gains from using one generation of biofuels (e.g., first generation) must be compared favorably to gains from using other generations of biofuels (e.g., second through fourth generations).
4. Benefits from using biofueled vehicles must be competitive with those from using other alternative-fueled vehicles.

It is noteworthy to mention that no single strategy such as increasing the use of biofuels, reducing traffic volumes, improving public transport, increasing the use of plug-in hybrids, and long-range-battery electric vehicles can achieve sustainable transport; a full portfolio of strategies is required in this regard (Wang et al. 2009).

Further, because the potential feedstocks for biofuels are large in number and with different characteristics, it is challenging for the current LCA approach to apply a distributed decision-making methodology due to the vast scope of information needed to address so many alternatives (Halog and Bortsie-Aryee 2013). Thus, multicriteria decision analysis, such as the analytic hierarchy process, is used to determine the most critical criteria, variables, and indicators to stakeholders, which can represent their conflicting interests with respect to economic, environmental, technological, and social dimensions of systems sustainability (Holden and Gilpin 2013).

It is forecasted that biofuels will contribute 6% of the total fuel use by 2030 (Hannon et al. 2010), and algal biomass is emerging as a promising feedstock for biofuel production due to its high growth rate; however, a number of hurdles should be overcome by this technology to be competitive in the fuel market (Hannon et al. 2010; Sander and Murthy 2010). Some important challenges are identification of suitable strains and their improvement in terms of both oil productivity and crop protection, allocation and use of nutrient and resource as well as production of co-products to improve the economics of the entire system. Investigation is being conducted around the world on the LCA for algae biomass utilization with an aim to provide baseline information for the algae biodiesel process (Sander and Murthy 2010).

Like road transport biofuels such as bioethanol and biodiesel, the biomass-derived jet fuel (biojet fuel) is also becoming a key renewable fuel in the aviation industry's strategy. The biojet fuel has the potential to reduce operating costs, environmental impacts, and greenhouse gas emissions. Additionally, it must meet the American Society for Testing and Materials (ASTM) International specifications and potentially be a 100% drop-in replacement for current petroleum jet fuel. Such fuels can be produced through alcohol-to-jet, oil-to-jet, syngas-to-jet, and sugar-to-jet pathways. The main challenges for each technology pathway include conceptual process design, process economics, and LCA.

1.1.6 Conclusion

In this chapter, the sustainability issues in the twenty-first century along with various approaches for achieving sustainability and its assessment have been presented. The historical evolution of the term "sustainable development" in relation to sustainability has been briefly discussed. Attention has also been paid to the relation among environmental conservation, economic development, and population growth, which are in fact not conflicting and can reinforce each other with the interaction of four terms—economics, ecology, politics, and culture. The existing technology, research trend, improvement in processing, and future prospects regarding sustainability through utilization of conventional resources adopting cleaner routes have been addressed. The specific examples of clean coal technologies, downstream processing of heavier petroleum fractions, and hydrogen from natural gas and liquid fuel from oil sands are cited. Sustainability through utilization of renewable biological resources such as ethanol and butanol from lignocellulosic biomass, oil from algae, utilization of biofertilizer and biopesticides, exploring lifesaving drugs from marine sources as well as from Himalayan herbs, and bioenergy production through MFCs have been discussed. A critical analysis on sustainability through utilization of unconventional resources, specifically, utilization of solar energy production and PV cells, CO_2 for fuels and chemicals, hydrogen from water, and integrated water management, has been included. Sustainability through optimization of resource utilization such as water and energy optimization in process industries as well as green buildings has been assessed. Finally, life cycle analysis as the sustainability assessment, multicriteria decision tool for road transport biofuels, algal biofuel with special emphasis on residual biomass processing, and sustainable production and utilization technologies of biojet fuel have been highlighted.

References

Ahring, B.K., and N. Langvad. 2008. Sustainable low cost production of lignocellulosic bioethanol - "The carbon slaughterhouse." A process concept developed by BioGasol. *International Sugar Journal* 10(1311): 184–190.

Ancheyta, J., and M.S. Rana. 2004. Future technology in heavy oil processing. Encyclopedia of life support systems (EOLSS). http://www.eolss.net/sample-chapters/c08/e6-185-22.pdf, Accessed March 29, 2016.

Arun, P., R. Banerjee, and S. Bandyopadhyay. 2007. Sizing curve for design of isolated power systems. *Energy for Sustainable Development* 11(4): 21–28.

Bandyopadhyay, S., and C.C. Cormos. 2008. Water management in process industries incorporating regeneration and recycle through a single treatment unit. *Industrial & Engineering Chemistry Research* 47(4): 1111–1119.

Banerji, G., and S. Basu. 2011. Sustainable management of the herbal wealth of the Himalayas. https://www.pragya.org/doc/Pragya_Herbal-Wealth.pdf, Accessed March 29, 2016.

Barbier, E.B. 1987. The concept of sustainable economic development. *Environmental Conservation* 14(2): 101–110.

Bekele, G., and B. Palm. 2010. Feasibility study for a standalone solar–wind-based hybrid energy system for application in Ethiopia. *Applied Energy* 87(2): 487–495.

Bhadury, P., B.T. Mohammad, and P.C. Wright. 2006. The current status of natural products from marine fungi and their potential as anti-infective agents. *Journal of Industrial Microbiology and Biotechnology* 33(5): 325–337.

Biernat, K., A. Malinowski, and M. Gnat. 2013. *The Possibility of Future Biofuels Production Using Waste Carbon Dioxide and Solar Energy, Biofuels - Economy, Environment and Sustainability*, Prof. Zhen Fang (Ed.), InTech.

Bruckner, A.W. 2002. Life-saving products from coral reefs. *Issues in Science and Technology* 18(3): 39. https://coast.noaa.gov/funding/_pdf/NOAA-NOS-OCM-2016-2004570-ffo-posted-09.09.2015.pdf, Accessed March 28, 2016.

Brundtland, G.H. 1987. Report of the world commission on environment and development: "Our common future." United Nations. http://www.un-documents.net/our-common-future.pdf.

Celik, A.N. 2003. Techno-economic analysis of autonomous PV-wind hybrid energy systems using different sizing methods. *Energy Conversion and Management* 44(12): 1951–1968.

Chaurey, A., and T.C. Kandpal. 2010. A techno-economic comparison of rural electrification based on solar home systems and PV microgrids. *Energy Policy* 38(6): 3118–3129.

Clixoo. 2016. Future predictions of cellulosic ethanol production costs. http://www.clixoo.com/, Accessed March 28, 2016.

Curran, M. 2013. A review of life-cycle based tools used to assess the environmental sustainability of biofuels in the United States. U.S. Environmental Protection Agency, Washington, DC, EPA/600/R/12/709. https://cfpub.epa.gov/si/si_public_record_report.cfm?dirEntryId=262400, Accessed May 15, 2016.

Das, S., and N. Mangwani. 2010. Recent developments in microbial fuel cells: A review. *Journal of Scientific and Industrial Research* 69: 727–731.

Durre, P. 2007. Biobutanol: An attractive biofuel. *Biotechnology Journal* 2(12): 1525–1534.

Ehnert, I. 2009. *Sustainable Human Resource Management. A Conceptual and Exploratory Analysis from a Paradox Perspective*. Physica-Verlag, Berlin, Germany, 35–36.

Enerco. 2015. Renewable energy serve in India. http://www.econserve.in/solar_energy_latest_news, Accessed March 28, 2016.

Fava, J.A., F. Consoli, R. Denison, K. Dickson, T. Mohin, and B. Vigon. 1993. A conceptual framework for life-cycle impact assessment, workshop report, Society of Environmental Toxicology and Chemistry (SETAC), SETAC Foundation for Environmental Education, Pensacola, FL.

Finn, D. 2009. Our uncertain future: Can good planning create sustainable communities? Ph.D. Thesis, University of Illinois, Urbana-Champaign, IL. https://books.google.co.in/books?id=pS-kdAIX n8EC&printsec=frontcover&dq=Our+Uncertain+Future:+Can+Good+Planning+Create+Sust ainable+Communities&hl=en&sa=X&ved=0ahUKEwj1tfXs393MAhXMto8KHZ0QDNMQ6A EIKTAA#v=onepage&q=Our%20Uncertain%20Future%3A%20Can%20Good%20Planning%20 Create%20Sustainable%20Communities&f=false, Accessed May 15, 2016.

Goswami, A., P.K. Barooah, and J.S. Sandhu. 2002. Prospect of herbal drugs in the age of globalization - Indian scenario. *Journal of Scientific and Industrial Research* 61(6): 423–431.

Grober, U. 2007. Deep roots: A conceptual history of sustainable development (Nachhaltigkeit), WZB Discussion paper. http://econstor.eu/bitstream/10419/50254/1/535039824.pdf, Accessed May 15, 2016.

Halog, A., and N.A. Bortsie-Aryee. 2013. *The Need for Integrated Life Cycle Sustainability Analysis of Biofuel Supply Chains*. INTECH Open Access Publisher, Rijeka, Croatia.

Hannon, M., J. Gimpel, M. Tran, B. Rasala, and S. Mayfield. 2010. Biofuels from algae: Challenges and potential. *Biofuels* 1(5): 763–784.

Hauge, Å.L., J. Thomsen, and T. Berker. 2011. User evaluations of energy efficient buildings: Literature review and further research. *Advances in Building Energy Research* 5(1): 109–127.

Holden, E., and G. Gilpin. 2013. Biofuels and sustainable transport: A conceptual discussion. *Sustainability* 5(7): 3129–3149.

Hrayshat, E.S. 2009. Techno-economic analysis of autonomous hybrid photovoltaic-diesel-battery system. *Energy for Sustainable Development* 13(3): 143–150.

Huang, H., H. Liu, and Y.R. Gan. 2010. Genetic modification of critical enzymes and involved genes in butanol biosynthesis from biomass. *Biotechnology Advances* 28(5): 651–657.

IUCN. 1980. World conservation strategy. https://portals.iucn.org/library/efiles/edocs/WCS-004.pdf, Accessed March 28, 2016.

Jang, Y.S., A. Malaviya, C. Cho, J. Lee, and S.Y. Lee. 2012. Butanol production from renewable biomass by clostridia. *Bioresource Technology* 123: 653–663.

Jha, R.K., and X. Zi-rong. 2004. Biomedical compounds from marine organisms. *Marine Drugs* 2(3): 123–146.

Kellogg, W.D., M.H. Nehrir, G. Venkataramanan, and V. Gerez. 1998. Generation unit sizing and cost analysis for stand-alone wind, photovoltaic, and hybrid wind/PV systems. *IEEE Transactions on Energy Conversion* 13(1): 70–75.

Labuschagne, C., A.C. Brent, and R.P. Van Erck. 2005. Assessing the sustainability performances of industries. *Journal of Cleaner Production* 13(4): 373–385.

Magee, L., A. Scerri, P. James, J.A. Thom, L. Padgham, S. Hickmott, H. Deng, and F. Cahill. 2013. Reframing social sustainability reporting: Towards an engaged approach. *Environment, Development and Sustainability* 15(1): 225–243.

Meadows, D., J. Randers, and D. Meadows. 2004. *Limits to Growth: The 30-Year Update*. Chelsea Green Publishing Company, White River Junction, VT. http://www.unice.fr/sg/resources/docs/Meadows-limits_summary.pdf.

Medicines By Design. 2011. Drugs From Nature, Then and Now. https://publications.nigms.nih.gov/medbydesign/chapter3.html, Accessed March 12, 2016.

Meech, J.A. Oil sands mining and processing MINE 292 - Lecture 22. http://slideplayer.com/slide/3727964/, Accessed May 15, 2016.

Mužíková, Z., P. Šimáček, M. Pospíšil, and G. Šebor. 2014. Density, viscosity and water phase stability of 1-butanol-gasoline blends. *Journal of Fuels* 1–7. doi:10.1155/2014/459287.

NREL. 2009. Current (2009) state-of-the-art hydrogen production cost estimate using water electrolysis. https://www.hydrogen.energy.gov/pdfs/46676.pdf, Accessed May 15, 2016.

Qureshi, N., and I.S. Maddox. 1995. Continuous production of acetone-butanol-ethanol using immobilized cells of clostridium acetobutylicum and integration with product removal by liquid-liquid extraction. *Journal of Fermentation and Bioengineering* 80(2): 185–189.

Rabaey, K., and W. Verstraete. 2005. Microbial fuel cells: Novel biotechnology for energy generation. *TRENDS in Biotechnology* 23(6): 291–298.

Raghavendra, B.G., and A. Arivalagan. 1993. Applications of operations research techniques in paper industry. A review of the state of the art. *IPPTA* 5: 67. http://ipptaelibrary.org/DownloadPage.aspx?id=1126, Accessed March 29, 2016.

Ray, A.K., S.P. Yadav, and P. Sharma. 2013. Economic production of bio-ethanol from various indigenous sources in India - A review. *Proceedings of AIChE 2013 Annual Meeting*, San Francisco, CA, between November 3 and 8, 2013, Paper (181c).

Saini, S.K., and R.P. Saini. 2005. Solar air heaters with artificial roughned absorber plates. *Proceeding All India Seminar on Engineering Trends in Energy Conservation and Management*, The Institute of Engineers, University of Roorkee, Roorkee, India, pp. 267–282.

Samant, S.S., U. Dhar, and L.M.S. Palni. 1998. *Medicinal Plants of Indian Himalaya: Diversity, Distribution Potential Value*. Gyanodaya Prakashan, Nainital, India.

Sander, K., and G.S. Murthy. 2010. Life cycle analysis of algae biodiesel. *The International Journal of Life Cycle Assessment* 15(7): 704–714.

Sharma, A., C. Shanker, L.K. Tyagi, M. Singh, and C.V. Rao. 2008. Herbal medicine for market potential in India: An overview. *Academic Journal of Plant Sciences* 1(2): 26–36.

Singh, R.K., H.R. Murty, S.K. Gupta, and A.K. Dikshit. 2012. An overview of sustainability assessment methodologies. *Ecological Indicators* 15(1): 281–299.

Singh, S.N., B. Singh, and J. Ostergaard. 2009. Renewable energy generation in India: Present scenario and future prospects. In *Power and Energy Society General Meeting.* https://www.researchgate.net/publication/224598308_Renewable_energy_generation_in_India_Present_scenario_and_future_prospects, Accessed May 15, 2016.

Smith, C., and G. Rees. 1998. *Economic Development*, 2nd ed. Macmillan, Basingstoke.

Speight, J.G. 2013. *Heavy and Extra-heavy Oil Upgrading Technologies.* Gulf Professional Publishing, UK.

Stivers, R.L. 1976. The sustainable society: Ethics and economic growth. Westminster Press, Philadelphia, PA, 240.

UNCSD. 2012. Conference on sustainable development 2012. https://sustainabledevelopment.un.org/content/documents/Agenda21.pdf, Accessed March 28, 2016.

UNEP. 2012. Year Book 2012. http://www.unep.org/yearbook/2012/pdfs/UYB_2012_FULLREPORT.pdf, Accessed March 28, 2016.

UNWC. 1982. United Nations General Assembly. https://www.un.org/documents/ga/res/37/a37r007.htm, Accessed March 28, 2016.

Vignesh, S., A. Raja, and R.A. James. 2011. Marine drugs: Implication and future studies. *International Journal of Pharmacology* 7: 22–30. http://scialert.net/fulltext/?doi=ijp.2011.22.30&org=11, Accessed March 28, 2016.

Wang, J.J., Y.Y. Jing, C.F. Zhang, and J.H. Zhao. 2009. Review on multi-criteria decision analysis aid in sustainable energy decision-making. *Renewable and Sustainable Energy: Reviews* 13(9): 2263–2278.

Young, G., B. Shah, and F. Kimaite. 2008. Status report on integrated water resources management and water efficiency plans. UN Water Report. http://www.unwater.org/downloads/UNW_Status_Report_IWRM.pdf, Accessed March 28, 2016.

Zuo, J., and Z.Y. Zhao. 2014. Green building research–current status and future agenda: A review. *Renewable and Sustainable Energy Reviews* 30: 271–281.

https://en.wikipedia.org/wiki/Sustainability, Accessed March 28, 2016.

https://en.wikipedia.org/wiki/National_Energy_Technology_Laboratory, Accessed March 28, 2016.

http://insideclimatenews.org/carbon-copy/18052015/coals-future-facing-three-hurdles-and-steady-decline-projections-epa-clean-power, Accessed March 28, 2016.

http://www.ucsusa.org/clean_energy/our-energy-choices/coal-and-other-fossil-fuels/uses-of-natural-gas.html#Vvpoe2RsiSB, Accessed May 15, 2016.

http://www.energy.alberta.ca/ourbusiness/oilsands.asp, Accessed March 28, 2016.

https://issuu.com/dnv.com/docs/dnv-position_paper_co2_utilization, Accessed May 15, 2016.

http://www.sswm.info/es/category/concept/iwrm, Accessed May 15, 2016.

2

Clean Coal Technologies

Prabu Vairakannu

CONTENTS

ABSTRACT Coal is a primary fossil fuel and it is being used continuously as a main source for the production of electricity. It is a heterogeneous fuel, which contains carbon, hydrogen, sulfur, oxygen, nitrogen, moisture, and ash. Coal on combustion releases several pollutants such as CO_2, SO_x, NO_x, H_2S, and fly ash into the environment leading to global warming. Exposure of these gases to the atmosphere directly affects human health by causing lung and heart diseases. Therefore, it is crucial to develop efficient technologies for the production of clean energy from coal. Currently, research and developments are more focused toward clean coal technologies for commercialization. This chapter deals with the recent developments of coal technologies for a sustainable and clean energy recovery from the existing coal resources. It provides an overview of several clean coal technologies such as precombustion-, postcombustion-, and oxy-fuel combustion-based carbon capture and storage; coal beneficiation; NO_x and SO_x removal technologies; chemical looping combustion technologies; coalbed methane; underground coal gasification; supercritical and ultra-supercritical boiler-based IGCC technologies; and coal to liquid fuels and hydrogen. Integration of a mixture of these clean coal technologies in the existing power-generating systems is necessary to achieve a minimal energy penalty for carbon capture and storage. Therefore, the chapter also briefly discusses the feasibility of potential hybrid systems on clean coal combustion for clean energy recovery.

2.1 Introduction

Fossil fuels possess a vital role in the energy sector for the production of electricity. These fuels are present in three states such as coal (solid), petroleum crude (liquid), and natural gas (gas). The utilization of these fuels leads to the production of greenhouse gases such as CO_2, CH_4, and NO_x, which results in global warming. Therefore, research and development is essential toward the growth of clean fuel technologies. Of all the fossil fuels, coal satisfies 69% of the electricity demand of India (Ambedkar et al. 2011). Coal contains carbon, nitrogen, sulfur, and traces of mercury and other inorganic pollutants. On combustion of coal, these components are oxidized into CO_2, NO_x, SO_x, and inorganic oxides. Flue gas contains these global warming pollutants and contaminates the atmospheric air. Flue gas from coal combustion affects the life span of flora and fauna. Hence, the development of clean coal technologies is essential for the safe environment.

 Figure 2.1a and b shows the state-wise distribution of crude petroleum and natural gas in India, respectively (Energy statistics of India, 2014). It is estimated that 758.27 million tons of crude petroleum sources are found in India. The total quantity of natural gas resources of India is estimated as 1254 billion m^3 (Energy statistics of India, 2014). India also produces electricity using renewable resources. Figure 2.1c shows the percentage-wise distribution of electric power production from renewable energy. It is

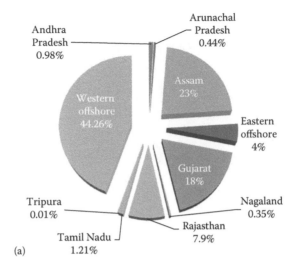

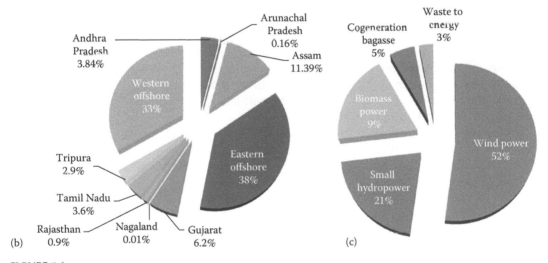

FIGURE 2.1
State-wise distribution of (a) crude petroleum and (b) natural gas resources of India; (c) percentage-wise distribution of renewable energy sources of India. (From Energy Statistics of India, Central statistics office, Ministry of statistics and programme implementation Government of India, New Delhi, 2014. With permission.)

evaluated that 94,125 MW (Energy statistics of India, 2014) of electricity are generated through wind, biomass, wastes, and so on.

Figure 2.2a shows the total quantity of electricity production using conventional energy sources of India (Energy statistics of India, 2014). Majorly, 8,17,225 GWh of electricity is generated from thermal power plants and 48,000 GWh of energy is produced from nonutility sources. Figure 2.2b shows the year-wise energy intensity of India (Energy statistics of India, 2014). It is estimated that an increasing trend of energy intensity of India is observed from 2005–2006 to 2012–2013. An estimate of 5.1% of energy deficit was calculated for the year 2014–2015. In order to reduce the energy scarcity, unconventional energy technologies such as underground coal gasification (UCG), coalbed methane (CBM), and enhanced oil recovery (EOR), should be commercialized in India.

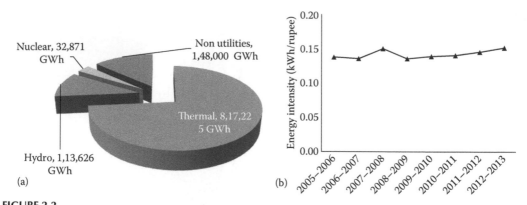

FIGURE 2.2
(a) Electricity production in India and (b) energy intensity of India. (From Energy Statistics of India, Central statistics office, Ministry of statistics and programme implementation Government of India, New Delhi, 2014. With permission.)

Coal is the primary fossil fuel in India. The major coal reserves of India are found in the states such as Jharkhand, Odisha, Chhattisgarh, West Bengal, Madhya Pradesh, Andhra Pradesh, and Maharashtra. Bituminous/sub bituminous coal and lignite reserves of India are estimated to be about 301.05 and 43.22 billion tons, respectively (Ministry of Coal, Annual Report 2013–2014). The state-wise bituminous/sub-bituminous coal resources of India are shown in Figure 2.3.

Coal production of India was estimated as 565.77 million tons during the year 2013–2014, which is 1.68% higher than the utilization during 2012–2013 (Ministry of Coal, Annual Report 2013–2014). However, the production of lignite in India is decreased by 4.7% in the year 2013–2014. The Energy Statistics report of India (Ministry of Coal, Annual Report 2013–2014) shows that coal and lignite produce a significant amount of energy of about 73.48% of the total energy production through primary fuel sources. Crude petroleum and natural gas produce 11.81% and 10.18% of the total energy, respectively. The quantity of raw coal and lignite utilization in various industries of India is listed in Table 2.1 (Ministry of Coal, Annual Report 2013–2014). Figure 2.4 shows the percentage-wise distribution of coal and lignite utilization in Indian Industries. It was found that 74.7% of raw coal and 83.09% of lignite were utilized for the production of electricity, and the rest amount of coal and lignite was used in other industries.

Further, it is estimated that 37% of coal resources of India are located at a higher depth of 300 m (Khadse et al. 2007). Indian coal has high ash content, and therefore, it is low calorific. Conventional mining is an uneconomical process for the extraction of deep mine coals. Thus, suitable coal extraction technologies need to be developed for the energy sustainability of India.

Consumption of huge reserves of coal for electricity production results in a massive production of global warming gases. In order to achieve a minimal pollutant environment, it is necessary to implement clean coal technologies in the existing power plants. Coal can be considered as a clean fuel, only if the final flue gas CO_2 is captured and stored underground without emitting it into the environment. CO_2 capture can be achieved using pre-, post-, and oxy-fuel combustion technologies. The conversion of carbon energy into hydrogen energy before combustion is referred to as precombustion CO_2 capture. This technology includes coal gasification (production of CO and H_2) and water–gas shift

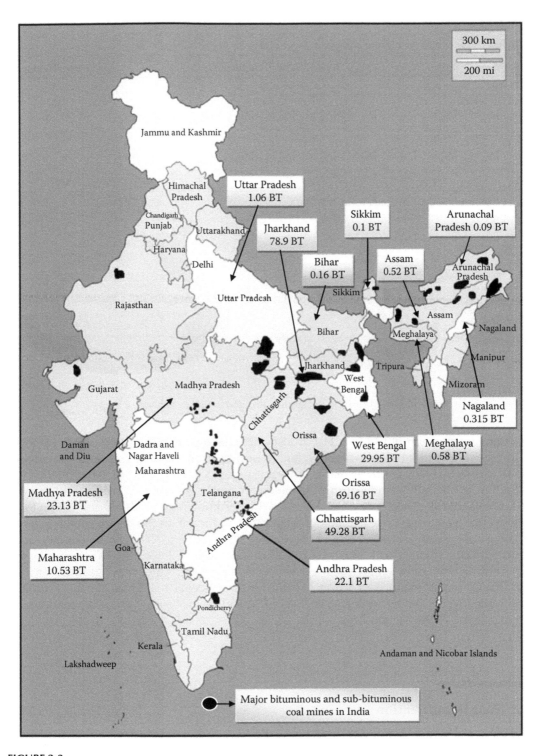

FIGURE 2.3
India's state-wise bituminous/sub-bituminous coal field. BT, billion tons. (From Energy Statistics of India, Central statistics office, Ministry of statistics and programme implementation Government of India, New Delhi, 2014. With permission.)

TABLE 2.1

Industry-Wise Coal and Lignite Utilization in India

Industries	Raw Coal (million tons)	Lignite (million tons)
Electricity	427.23	36.48
Steel and washery	23.13	0.03
Cement	11.96	1.4
Paper	1.67	0.66
Textile	0.36	2.83
Others	107.54	2.51
Total	571.89	43.91

Source: Ministry of Coal, Annual Report 2013–2014.

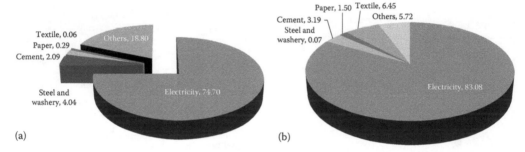

(a)   (b)

FIGURE 2.4

Percentage-wise distribution of (a) coal and (b) lignite utilization in Indian Industries.

reactions (conversion of CO into H_2). It also includes coal beneficiation, which removes sulfur and ash content prior to combustion. Precombustion cleaning technology increases the thermal efficiency of power plants. The capture of CO_2, NO_x, and SO_x from the flue gas is known postcombustion cleaning technology. Combustion of coal using pure oxygen produces a highly concentrated CO_2 stream, which is known as oxy-fuel combustion technology. Implementing these technologies leads to a reduction in the net thermal efficiency of the power plant. However, loss of thermal efficiency due to CCS can be compensated by integrating suitably with integrated gasification combined cycle (IGCC) technologies, supercritical and ultra-supercritical boiler technologies, and integrated gasification fuel cell (IGFC) technologies, and so on. Also, unmineable coal resources can be exploited using unconventional technologies such as CBM and UCG. These clean coal technologies are explained in the Section 2.6.

2.2 Coal Beneficiation

Precombustion fuel cleaning technology is the best option to avoid ash and sulfur handling problems during coal combustion. High ash and high sulfur coal require precombustion washing treatment for the production of clean fuel. As Indian coal contains high ash content, fly ash emissions and ash disposal are the major problems during power production. Indian coal contains 30%–40% of ash, which undergoes chemical and physical transformation on combustion, and a part of the fine ash is carried away with the flue gas. The fly ash needs to be captured using electrostatic precipitators or using other filtration techniques from the flue gas before it is exposed to the atmosphere. The residual bottom ash content of the boiler creates a disposal problem in thermal power plant stations.

In addition, Indian coal contains a significant amount of sulfur, especially from the northeastern region of the country. High sulfur coal leads to SO_x production on combustion. Release of SO_x into the atmosphere causes respiratory health issues and environmental impacts. Therefore, removal of ash and sulfur content of coal prior to combustion would result in clean energy and environment. Application of clean technologies in the precombustion stage reduces the complexity of operation and undesired side reactions. Conventional techniques for coal beneficiation are flotation, gravity settling, and acid treatment for the separation of ash and sulfur. However, these technologies consume a huge energy penalty for the separation of combustible and noncombustible matter. High carbon recovery from the noncombustible matter of low-rank coal cannot be achieved by the conventional method of flotation with normal oil collectors.

Ultrasound-enhanced beneficiation is a new technology for the separation of ash and sulfur from the mined coal (Ambedkar et al. 2011). Ultrasound enhances the separation of noncombustible matter and sulfur from coal through the acoustic streaming and cavitation process with a frequency above the human audible range. Ultrasonic treatment causes the formation and collapsing of bubbles, which produces high pressure and temperature differences resulting in turbulence and further removal of surface coatings of coal. Ultrasonic treatment can increase the hydrophobicity of coal and the hydrophilicity of sulfur during the coal flotation process. This technology is still at the research level and requires efforts for commercialization and to make economically feasible.

2.3 Coal Gasification

Coal gasification occurs in a gasifier, which converts solid coal into a gaseous mixture (syngas) having a good calorific value. The gas mixture is termed as syngas, which majorly consists of CO, H_2, CH_4, CO_2, and trace quantities of uncracked higher hydrocarbons.

$$C + H_2O \rightarrow CO + H_2 \tag{2.1}$$

$$C + CO_2 \rightarrow 2CO \tag{2.2}$$

$$C + 2H_2 \rightarrow CH_4 \tag{2.3}$$

The above reactions are gasification reactions, which occur between solid carbon and gasifying agents such as steam and CO_2. Conventional gasification is performed with the supply of a controlled amount of oxidant and a gasifying medium to the gasifier along with the feed. Advanced gasification technologies are molten gasification, plasma gasification, co-combustion, and so on, and these technologies are under development for commercialization.

2.3.1 Conventional Coal Gasification

In the conventional gasification process, coal is converted into syngas using moving bed, fluidized bed, and entrained bed gasifiers (Kivisaari et al. 2004). The choice of gasifiers is based on the nature of feed. As entrained bed gasifiers are operated at a high temperature with a higher feed velocity (less residence time), all types of feed can be processed. However, the size reduction of feed is an additional cost of the gasifier. In moving bed gasifiers, the size reduction of feed is not required, but the generated product gas is enriched with high tar. It is due to the noncracking of volatile matter, as it does not pass through the combustion zone in the gasifier. In fluidized bed gasifiers, the operating temperature of the reactor is maintained less than the ash melting point in order to sustain the fluidized condition of the bed. Therefore, high-ash Indian coals are suitably processed in the fluidized bed gasifiers.

Table 2.2 shows the comparison of the working conditions of the gasifiers (Smith and Klosek 2001). As fixed bed and entrained bed gasifiers are operating under the elevated temperature, low-ash coals are recommended due to ash slagging problems. High-ash coal can be gasified in fluidized bed gasifiers as they operate below the ash fusion temperature. Because of the low operating temperature of the fluidized bed gasifier, high reactivity coal is preferred for high conversion efficiency. Pure oxygen gas is preferred in fixed bed and entrained bed gasifiers in order to operate at high temperature.

2.3.2 Molten and Plasma Gasification

The waste heat of the slag from blast furnace is recovered by means of the molten gasification process. Li et al. (2015) performed thermogravimetric studies on the interaction of coal

TABLE 2.2

Comparison of Different Types of Gasifier

S. No.	Parameters	Moving Bed Gasifier	Fluidized Bed Gasifier	Entrained Bed Gasifier
1	Operating temperature (K)	1500–1800 (in the combustion zone)	1173–1323	1473–1873
2	Pressure (MPa)	2.8	2.5	2.5–4
3	Particle size	5–80 mm	0.5–5 mm	<500 μm
4	Tar content	High	medium	low
5	Cold gas efficiency (%)	88	70–85	74–81
6	Reaction zones	Distinct zones such as pyrolysis, combustion, and gasification	No distinct zones	No distinct zones
7	Slag formation	Yes	No	Yes (need refractory line coating)
8	Reaction time (s)	15–60 min (high-pressure operation), in the order of hours (atmospheric pressure operation)	10–100	On the order of seconds
9	Coal reactivity	All range of coal reactivity is suitable under slagging operation	High-reactivity coal is essential as it operates at a low temperature	All ranges of coal reactivity are suitable

Source: Kivisaari T. et al., *Chem. Eng. J.*, 100, 167–180, 2004.

with the blast furnace slag with CO_2 as a gasifying medium. It was found that the molten slag obtained as a waste from the blast furnace acts as a heat source as well as a catalyst for carbon gasification reaction. Plasma gasification is one of the recent technologies for the efficient conversion of coal into syngas. Coal is ionized into gases using an electric arc, which is energized by the electric current at an elevated temperature. Organic matter of coal is broken into elemental gases such as H_2 and CO. Inorganic substances are melted and removed as a slag at the bottom of gasifier (Janajreh et al. 2013).

2.3.3 Co-Combustion and Gasification

Co-combustion is a suitable clean technology for energy recovery from seasonal fuels (biomass) along with fossil fuels such as coal. Biomass is one of the renewable energy sources, and the energy production from such sources reduces the emission of CO_2. The biomass can be renewed in a short span of life with the absorption of atmospheric CO_2. Co-combustion of biomass with coal consumes less fossil fuels and thus reduces the level of global warming. There are several advantages of co-combustion. It is not only advantageous in terms of carbon reduction but also beneficial for the sustainability of flame front during combustion. Biomass contains high volatile matter and less carbon and ash content. The liberation of these gaseous components sustains the combustion front and enhances the gasification efficiency. However, the preliminary treatments such as feed storage, drying, and size reduction are the additional cost of the process.

2.4 Carbon Capture and Storage

Thermal power plants majorly produce CO_2 via electricity production. The emission of the carbon gases to the atmosphere leads to global warming. Capture and storage of CO_2 in a suitable reservoir under the earth and sea is the solution to maintain a clean environment. Flue gas contains the major portion of N_2, CO_2, and other trace gases in the conventional combustion process. The carbon capture method includes postcombustion, precombustion, and oxy-fuel combustion methods. In the postcombustion capture technology, the final flue gas contains a mixture of CO_2 and N_2, which can be separated by the amine absorption technique. Absorption of CO_2 from flue gas stream and regeneration of amine solution are the energy consumption steps in the postcombustion capture technology. The disadvantage of the technology is the handling of huge volume of gas for the gas treatment process. The precombustion technology includes gasification, high- and low-temperature water–gas shift technology, and pressure swing adsorption of CO_2 from raw gas. The sequential setup of reactors in the precombustion process consumes energy, which reduces the efficiency of the process. In the oxy-fuel combustion technology, the air separation unit (ASU) consumes the major portion of the energy for the separation of oxygen.

Pure stream of CO_2 from the carbon capture process can be stored in two ways such as geological and oceanic storage (Metz et al. 2005). Geological storage includes gas reservoir, oil reservoir, saline aquifers, and unexploited deep coal seams. EOR is an established technology for recovering the residual oil, which is left in the reservoirs after the primary oil recovery process using CO_2 stream. CO_2 can also be stored in saline aquifers in dissolved form. The dissolved CO_2 reacts with mineral contents of the surrounding rock strata and converted into solid carbonaceous material. Enhanced CBM (ECBM) technology uses CO_2 gas for replacing the leftover adsorbed methane gas from coal seam. As India contains 37% of coal reserves, which are found at a depth greater than 300 m, the storage of CO_2 in these unexploited seams is one of the potential options for the reduction of pollutants.

The formation of CO_2 lake and the dissolution of CO_2 on the sea are termed as oceanic storage. High-density liquid CO_2 settles and forms CO_2 lake on the seafloor. Absorption of CO_2 occurs naturally in seawater. As CO_2 dissolves in water, the concentration of CO_2 in air is found in equilibrium with seawater. An increase of CO_2 content in atmospheric air leads to absorption of the gas in seawater. Therefore, storing a large amount of CO_2 in the deep seawater can reduce the pollutants for several decades. Due to the acidity nature of CO_2, the pH of seawater may be low and this can be neutralized by the alkali mineral matter of the sea.

The level of CO_2 emission in India was estimated as 1610 million tons in 2007 (United Nations Statistics Division, 2007). CO_2 storage in unmineable coal beds of India is one of the potential sources of geological infrastructure. About 37% of Indian coal reserves is found too deep and it cannot be exploited economically by the conventional mining process. These coal seams are appropriate for long-term storage. Holloway et al. (2008) discussed the CO_2 storage potential of India and its subcontinent. It is theoretically assessed that 345 million tons of CO_2 can be stored in deep coal seams. However, it is pointed out that no single coal seam is available with CO_2 storage capacity, which is higher than 100 million tons. Also the CO_2 storage technology on coal seams is under demonstration phase. East Bokaro in Jharia, Barakar, and Raniganj coal fields contain 3.2, 6.7, and 0.97 billion tons of coal, respectively, at a depth below 1200 m. These deep coal seams are the potential options for CO_2 storage. The total estimated CO_2 storage capacity of coal between 300–600 m and 600–1200 m is 249 and 84 million tons of CO_2, respectively. However, the resource of coal for storage is insignificant compared to the level of CO_2 emission. Also the development of CO_2-ECBM technology is one of the feasible ways for the sequestration of CO_2. This technology is planned to initiate in Gondwana coal fields by the Directorate General of Hydrocarbons, India (Parikh 2010).

The geologic storage of CO_2 is also feasible in the Deccan Traps of India, which is extended from the region of Rajahmundry to the Gulf of Bengal (Parikh 2010). The basalt rock provides a cap for CO_2 storage and is stable for long-term storage. Sedimentary basins are capable of storing CO_2. Parikh (2010) estimated that 1,390,200 km² onshore and 394,500 km² offshore sedimentary basins are found in India. Further, in deep waters, 1,350,000 km² resource basins are evaluated and a grand total of 3,134,700 km² area of sedimentary basins is estimated. The sedimentary basins are porous in nature and can store CO_2.

Saline aquifer is one of the potential sources for CO_2 storage in shallow offshore areas of Gujarat, Rajasthan, Assam, Mizoram, and Cachar. It is recommended that the storage potential of CO_2 in saline aquifers and oil and gas fields should be evaluated and quantified field-wise (Holloway et al. 2008). The storage capacity of the oil fields in India is estimated as 1–1.1 Gt of CO_2 and that of the gas fields is evaluated as 2.7–3.5 Gt of CO_2. Particularly, the capacity of Mumbai oil basin reserves is estimated as 469 million tons of CO_2 (Holloway et al. 2008).

Many government organizations, public sector units, and private sector units of India are involving in the CCS research and development for commercialization (Parikh 2010). A rough estimate of CO_2 storage capacity of India is available in the literature, and a consistent storage capacity needs to be assessed for reliable data (Viebahn et al. 2014).

2.4.1 Precombustion Technique

This technique involves three stages. First, coal is processed to separate the ash and sulfur content using the coal beneficiation technique. Second, carbon-enriched coal is converted into hydrogen gas before the end use. This can be carried out by steam and CO_2 gasification of coal. After the conversion of solid fuel into syngas, it is further reformed into hydrogen through water–gas shift reaction (2.4):

$$CO + H_2O \rightarrow H_2 + CO_2 \tag{2.4}$$

Finally, CO_2 can be separated by the membrane separation technique or through the conventional amine absorption process. Hence, the final gas contains pure hydrogen and can be utilized as a clean fuel.

2.4.2 Postcombustion Technique

In this technique, coal is combusted directly without any pretreatment. If coal is combusted using air as oxidant, N_2- and CO_2-rich content flue gas is generated. Sulfur present in the coal is combusted into SO_x. The flue gas can be treated for the separation of CO_2 and SO_x. Amine absorption method is a conventional CO_2 separation technique. However, the postcombustion technique is expensive compared to other techniques.

2.5 Clean Combustion Techniques

Several modification technologies of combustion of fuel favor the efficient capture of CO_2 and reduce the formation of pollutants. These technologies are explained in Sections 2.5.1 through 2.5.5.

2.5.1 Oxy-Fuel Combustion Technique

In the oxy-fuel combustion technology, coal is combusted using pure oxygen as the oxidant. As a consequence, the generated flue gas contains a pure stream of CO_2 and H_2O. Further, CO_2 can be separated on condensation of steam and further removal of SO_x. In this technology, the adiabatic flame temperature of fuel in a boiler or a combustor is high and should be maintained within the operating limit. Steam-moderated oxy-fuel combustion (SMOC) and CO_2-moderated oxy-fuel combustion (CMOC) are the proposed techniques for moderating the flame temperature in the oxy-fuel combustion environment (Seepana and Jayanti 2010, 2012). The addition of steam or CO_2 or the recycling of CO_2-enriched flue gas to the combustor controls the adiabatic flame temperature. Hence, the flue gas contains CO_2-enriched gas and can be stored without any further treatment. As the oxy-fuel combustor operates at a high temperature, fuel NO_x may be formed as a pollutant in the flue gas. Oxy-fuel combustion is a matured technology at a pilot-scale level and reduces the cost of CCS compared to pre- and postcombustion CS technologies. The separation of oxygen from air using an ASU is major energy consumption factor in oxy-combustion technique. Several methods such as membrane technology, pressure swing adsorption technique, and cryogenic distillation method, are available for the separation of oxygen from air.

TABLE 2.3

Comparison of Different Air Separation Techniques

ASU Process	Purity Limit (vol.%)	Separation Material	Issues
Adsorption	93–95	Zeolite (nitrogen adsorption), carbon molecular sieve (oxygen adsorption)	Pretreatment is required to remove CO_2 and moisture in air.
Absorption	99	Molten salt absorbs O_2	Pretreatment of air is required. Corrosion issues related to use of salt.
Polymeric membrane	~40	Polymeric material	Continuous process at near-ambient conditions; membrane tolerates the presence of CO_2 and moisture in air.
Cryogenic separation	99	Distillation column	Refrigeration cycle is required. Liquid nitrogen and liquid argon are also produced as by-products.

Source: Smith A.R. and Klosek J., *Fuel Process. Technol.*, 70, 115–134, 2001.

Table 2.3 shows the comparison of different air separation techniques (Smith and Klosek 2001). Adsorption, absorption, and polymeric membrane technologies are under the developing stage. Adsorption process contains multiple packed beds, which can be interchanged during regeneration operation. Adsorbents can be regenerated by pressure swing and temperature swing operations. Absorption technique is a chemical process that uses an absorbent for the separation of oxygen. This absorbent can be regenerated by a stripper operation. Except the polymer membrane technology, other technologies require the pretreatment process for the removal of moisture and CO_2 from atmospheric air. The development of suitable polymeric membranes is under way for high purity air separation. Cryogenic separation technique is a matured technology that produces liquid oxygen, liquid nitrogen, and liquid argon. This technology works on the principle of refrigeration cycle.

2.5.2 Combustion Technologies on NO_x Reduction

NO_x emissions in pulverized-coal boilers have an unfavorable impact on the environment. NO_x reacts with water and other compounds and forms various acidic compounds, fine particles, and ozone. Ozone production at the ground level creates respiratory illness to the human beings. Excess oxidant in the pulverized boiler increases the carbon conversion as well as fuel NO_x generation. A suitable design of the burner and fuel–oxygen contacting pattern of feed may reduce the NO_x generation in the furnace. NO_x is formed under different reaction conditions of the fuel and oxidant. It is classified as follows:

1. *Thermal NO_x:* It occurs by the oxidation of N_2 and atmospheric oxygen at high temperature (>1500°C). The formation of thermal NO_x depends on the temperature and residence time of reactants.
2. *Fuel NO_x:* Nitrogen present in the coal is converted into NO_x during combustion. It depends on coal composition and the extent of pyrolysis reaction.
3. *Prompt NO_x:* This is due to the combination of CH portions (hydrocarbon fragments) of the fuel with nitrogen from air. HCN, H_2CN, and CN compounds are formed and they are easily oxidized further into NO.

The formation of NO_x can be reduced by combustion modification techniques. Introducing the fuel and oxidant in a combustor at several stages can reduce the NO formation. The details of combustion modification techniques are explained in Sections 2.5.2.1 and 2.5.2.2.

2.5.2.1 Design of Furnace and Combustion System

A tangentially fired utility boiler gas burners at the corners and coal particles are injected into each burner (Díez et al. 2008). The flame front is produced at all the corners and focused toward the center of the boiler. The flame fronts meet at the center, and a well-mixed zone is established. High carbon conversion is achieved due to intense mixing of fuel and oxidant. This ensures a complete combustion and uniform heat distribution around the boiler. As the peak flame temperature is minimized, the formation of thermal NO_x can be eliminated in the tangentially fired utility boiler.

The peak flame temperature can be reduced in a low-NO_x burner system using several modifications such as fuel/air staging, overfire air, reburning, flue gas circulation, and low excess air. The combination of these techniques greatly reduces the NO_x formation in the combustion system. The temperature-moderating agents such as steam, CO_2, or flue gas are injected along with the fuel for controlling the adiabatic flame temperature. Fuel/air staging is one of the modification techniques in which the fuel is burnt in an oxygen-deficient atmosphere at the early stage of combustion. However, the unburnt fuel is combusted by supplying the secondary oxidant at later stages through overfire air. Figure 2.5 shows the combustion modification technologies in a boiler working for the steam production.

In the primary combustion zone, low excess air is supplied for the combustion of coal (US DOE report 1999). Due to less flame temperature, the production of thermal NO_x is minimized in this zone. There is an adjacent reburn zone, where the additional fuel is

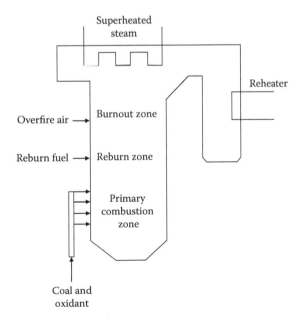

FIGURE 2.5
Schematic representation of combustion modification technologies in the boiler working for the steam production.

injected to create an oxygen-deficient condition (fuel rich). In the reburn zone, the generated thermal NO_x reacts with the CH portion of the injected fuel and reconverts NO_x into N_2 gas. Therefore, it is called as the reduction zone. A burnout zone, which is next to the reduction zone, is supplied with secondary air for the complete combustion of unconverted fuel, and it is known as overfire air. In the burnout zone, the overfire air is injected with 20% of the total oxidant requirement (US DOE report 1999). A low quantity of thermal NO_x is produced in the burnout zone. However, it is very small compared to the conventional combustion process.

$$NO_x \text{ formation: } N + 0.5\ O_2 \rightarrow NO \qquad (2.5)$$

Reduction in the reburn zone:

$$NO + CH \rightarrow NH \qquad (2.6)$$

$$NH + NO \rightarrow N_2 \qquad (2.7)$$

2.5.2.2 Inflame NO_x Reduction Burner

In the inflame NO_x reduction burner, the above-mentioned combustion modifications can be performed in the burner itself instead of carrying them to different stages of combustion. The contact pattern of oxidant and pulverized coal should be adjusted. Therefore, the design of coal nozzle tips and the tilting angle of the burner are crucial for designing the inflame low NO_x burner. Figure 2.6 shows the schematic representation of an inflame NO_x reduction burner.

Primary air is injected through the nozzle along with fuel in low excess quantity (Tsumura et al. 2003). The flame is classified into several zones. At the ignition point with low excess air, volatile matter is released and thermal NO_x is formed. In the next zone, hydrocarbon free radicals (CH*, NH*) from the volatile matter are generated (free radical formation zone). These free radicals reduce thermal NO_x into N_2 gas in the NO_x reduction zone. The secondary air is supplied in the side ways of the flame, and it meets the

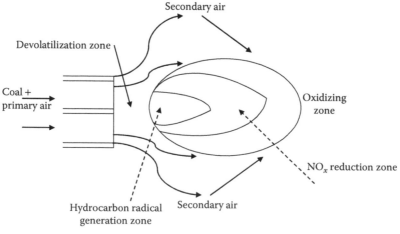

FIGURE 2.6
Schematic diagram of the inflame NO_x reduction burner.

unconverted fuel in the oxidizing zone, where the fuel is completely burnt out. Hence, the modification of the contacting pattern of oxidant and coal reduces the thermal NO_x in the flame itself instead of carrying it out to several stages of operation.

2.5.3 High-Temperature Air Combustion

High-temperature air combustion (HiTAC) is one of the promising techniques for NO_x reduction. The thermal stability of the combustion flame and the high efficiency of the boiler can be achieved using the HiTAC process. Air combustion of fuel leads to the formation of thermal NO_x due to the nonuniformity of thermal distribution in a boiler. Injection of cold air under the atmospheric condition into the boiler reduces the sustainability of the flame front. Further, it produces a high peak flame temperature, which leads to more NO_x formation. This can be reduced using the HiTAC technology. Figure 2.7 shows the schematic diagram of the HiTAC process.

Air is preheated in the regenerative heat exchanger using a portion of recovered thermal energy of hot flue gases from the boiler (Zhang et al. 2007). Oxygen concentration in air is reduced by mixing it with hot flue gas from the combustor. The temperature of preheated air is maintained nearly to the auto-ignition temperature of fuel, and therefore, the fuel is burnt uniformly in the combustor with high flame stability. Due to the reduction of oxygen concentration, thermal NO_x is minimized. Also due to uniform thermal distribution, high conversion of carbon is achieved, and therefore, the efficiency of the boiler is increased. Low-calorific fuels can be burnt easily. Hence, it is a feasible technology for utilizing high-ash Indian coals without the operation difficulty of flame instability.

2.5.4 Chemical Looping Combustion

In oxy-fuel combustion-based thermal power plants, pure oxygen is utilized for the combustion of fuel in order to produce CO_2-enriched flue gas for sequestration. This technology leads to a huge energy penalty in the ASU for the separation of oxygen from air. In the cryogenic technology, the compression cost of air leads to the consumption of huge energy in power plants. The ASU unit reduces 8%–10% (Prabu and Jayanti 2012a) of the net thermal efficiency of the power plant for oxygen separation. However, chemical looping combustion (CLC) is a clean fuel processing technology that processes fuel without an ASU unit (Sorgenfrei and Tsatsaronis 2014). It separates oxygen from air through metal oxide formation. The metal oxides such as NiO, Fe_2O_3, CuO etc. can be used for the combustion of fuel. As a result, 90%–98% of CO_2-enriched dry flue gas is obtained and can be sequestered without any further gas treatment (Prabu 2015). Therefore, it is an inherent clean coal technology for the implementation of CCS.

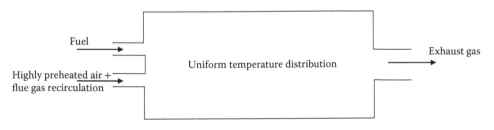

FIGURE 2.7
Schematic diagram of the HiTAC process.

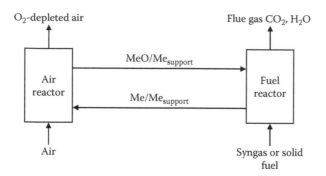

FIGURE 2.8
Schematic diagram of the CLC process.

The CLC technology can process any form of fuels such as solid, gas, and liquid fuels. Solid fuels such as coal, biomass, and wastes can be combusted in the CLC process. They can be processed in a direct or an indirect operation in the CLC process. However, it has a complex reactor arrangement for the fuel combustion. The CLC unit has two major reactors such as air and fuel reactors. Figure 2.8 shows the schematic diagram of the CLC process.

Atmospheric air is circulated in the air reactor and metal particles are oxidized under the fluidized bed condition. The outlet stream of the air reactor is depleted in oxygen and enriched with nitrogen. In the reactor,

$$\text{Metal (Me)} + \tfrac{1}{2}O_2(\text{air}) \rightarrow \text{MeO (metal oxides)} \tag{2.8}$$

The oxidized metal particles from air are circulated to the fuel reactor. In the fuel reactor, coal particles are burnt with metal oxides. For the combustion of solid fuel, a gasifying agent such as steam or CO_2 is supplied to the fuel reactor. First, coal is thermally decomposed into its constituents (Reaction [2.9]). Tar as a liquid product is further cracked into lower weight gas hydrocarbons. Carbon in coal is gasified by reacting with the supplied gasifying agents and converted into gaseous fuels such as CO and H_2. Second, the gaseous products formed as volatile matter (pyrolysis), tar gas (cracking), and carbon gas (gasification) are reacted with metal oxide particles and yield a CO_2- and H_2O-enriched flue gas. The reactions that are taken place in the fuel reactor are as follows:

Thermal decomposition of coal

$$\text{Coal} \rightarrow \text{volatile matter (CO, } H_2, CH_4, CO_2) + \text{char} + \text{moisture} + \text{ash} + \text{tar} \tag{2.9}$$

Thermal cracking

$$\text{Tar} \rightarrow \text{lower hydrocarbons} \tag{2.10}$$

In situ *coal gasification*

$$C + H_2O \rightarrow CO + H_2 \tag{2.1}$$

$$C + CO_2 \rightarrow 2CO \tag{2.2}$$

Metal oxide gas combustion

$$\text{MeO} + CO \rightarrow CO_2 + \text{Me} \tag{2.11}$$

$$MeO + H_2 \rightarrow H_2O + Me \tag{2.12}$$

$$MeO + CH_4 \rightarrow CO_2 + H_2O + Me \tag{2.13}$$

The metal oxides are reduced to metals and it is recycled to the air reactor for oxidation after the separation of ash particles. Thermodynamics of metal to metal oxides and reduction of fuel are the major issues that should be considered for the metal selection criteria.

2.5.4.1 Metal and Metal Oxides

Ni, Fe_2O_3, Cu, Mn, and Mg are the metal particles that are considered in the CLC process. The melting point of these metal particles and ash is the major constraint that decides the upper limit of the operating temperature of the reactors. These metal particles are supported in an inert medium such as Al_2O_3 and $FeAl_2O_4$.

2.5.4.2 Heat Balance in CLC

Heat balance is an important criterion in operating both the reactors. Mostly, the oxidation of metals in the air reactor is exothermic in nature. In the fuel reactor, the heat of reaction depends on the specific metal and gas that are involved in the reaction. If the reaction in the air reactor is highly exothermic, it will be reverse in the fuel reactor. Therefore, the heat should be transferred between the reactors. Heat exchange can be performed through the supply of excess metal particles. Inert supports are also useful in transferring the heat between the reactors.

2.5.5 Chemical Looping Reforming

Coal gas contains a higher proportion of CO and H_2S gas, which deteriorates the performance of fuel cell. Therefore, a highly purified syngas is essential for the fuel cell operation. However, the syngas cleaning process is an energy-intensive operation and lowers the efficiency of the power cycle. However, chemical looping reforming (CLR) technology is the suitable technology for the production of high pure hydrogen, which can be fed directly into the fuel cell system (Ersoz et al. 2006). It is a clean fuel production technology that indirectly converts carbon energy into clean fuel hydrogen energy. It is similar to the CLC process in which a steam reactor is added into the loop for hydrogen generation. In the steam reactor, the supplied steam oxidizes the metal particles to oxides with the production of hydrogen. In the steam reactor,

$$Me + H_2O \rightarrow MeO + H_2 \tag{2.14}$$

Hydrogen produced from the steam reactor is highly pure and free of coal gas contaminants, and it is suitable for the feed gas in the fuel cell system. The complete conversion in the fuel reactor may not be possible due to the limitation of thermodynamic and kinetic parameters. However, the unconverted metal particles from the steam reactor are further oxidized in the air reactor, and then, they can be recycled to the fuel reactor. This technology converts coal into pure hydrogen, and it can be supplied as a feed to the polymer electrolyte membrane fuel cell (PEMFC).

2.6 Unconventional Coal Technologies

Due to the increased energy demand, it is essential to explore new technologies for the efficient utilization of existing energy resources. Nowadays, unconventional clean coal technologies such as UCG, CBM, and EOR are emerging for the recovery of unmineable deep coal reserves.

2.6.1 Underground Coal Gasification

UCG is an inherent clean coal technology for exploiting deep coal resources. As 37% of Indian coal resources are found at a depth of more than 300 m (Khadse et al. 2007), it is the appropriate technology for the effective utilization of such resources. Coal mining, transportation, and storage are completely eliminated in the UCG process. Further, there is no requirement of surface reactors for the gasification process. As Indian coals are low calorific and contain 40% of ash (Iyengar and Haque 1991), emission of fly ash and handling of ash sludge from the boiler are the major problems in coal industries. These issues can be completely avoided in the UCG process. UCG is an *in situ* phenomenon that converts coal into calorific gas through combustion, gasification, and pyrolysis. The schematic diagram of various reaction zones of the UCG process is shown in Figure 2.9.

Injection and production wells are required to drill in the UCG site. These two vertical wells in the ground level are linked through a horizontal well in a coal seam. Therefore, the generated syngas can be escaped through the production well. A suitable oxidant and gasifying medium are sent through the injection well, and the syngas produced in the

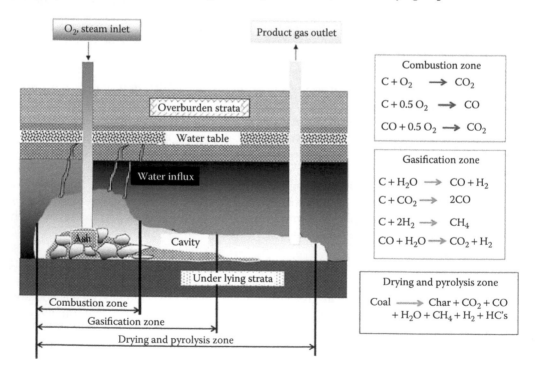

FIGURE 2.9
Schematic representation of the UCG process.

cavity are collected from the production well. The location of the injection spot of the oxidant is unchanged in the conventional UCG technique. However, this location is continuously changed in the controlled retractable injection point (CRIP) method. In the CRIP method (Hill and Shannon 1981), coal is burnt at various locations of the coal seam as soon as it gets consumed. Yang (2003) proposed a blinding-hole UCG technique, which has the potential to convert coal under buildings, railways, and so on. This technique requires an annular pipe, which is to be inserted into the coal seam. The oxidant and gasifying medium are injected into the inner tube, and the syngas is collected at the annular region of the outer pipe.

In the UCG process, the simultaneous solid–gas, gas–gas phase reaction occurs along the entire length of the coal seam. Coal is burnt at the location of oxidant injection, and the flame front is propagated toward the outlet well. The heat evolved in the combustion zone is utilized for the endothermic gasification reaction. As coal gets consumed, the empty space created inside the seam is called cavity. The solid–gas phase reactions (Equations 2.1 through 2.3, 2.15, and 2.16) occur at the surface of cavity. The gas phase reactions (Equations 2.4 and 2.17) occur in the UCG cavity:

$$C + O_2 \rightarrow CO_2 \tag{2.15}$$

$$C + \tfrac{1}{2} O_2 \rightarrow CO \tag{2.16}$$

$$CO + \tfrac{1}{2} O_2 \rightarrow CO_2 \tag{2.17}$$

The volatile matter of coal seam gets released over the entire seam, and significantly, it occupies a larger portion of the syngas volume. Although the UCG process is beneficial in extracting the deep coal resources, it is a more complex phenomenon. Visualization and control of the deep insight phenomenon of the coal seam are very difficult. Several physicochemical processes occur simultaneously during *in situ* gasification of coal seam. Water influx into the UCG reaction zone greatly affects the performance of reaction. The underground water body near the coal seam is the source of water influx into the cavity of coal seam. If water influx is uncontrolled, it consumes most of the chemical energy of coal in the form of steam. Chemical energy of coal has been used as the latent heat for the production of steam. Therefore, the outlet gas from the production well is enriched with steam. These problems could be avoided through the UCG operating pressure. However, if the UCG operating pressure is above the hydrostatic pressure of the water body, the gas from the coal seam may get escaped and contaminate the water body. Hence, an optimum pressure of the UCG cavity should be maintained for balancing both the phenomena. The hydrostatic pressure of the underground water body depends on the depth of coal seam and the density of overburden strata. It was reported that a coal seam at a depth of 860 m in Thulin, Belgium, had exerted a pressure range of 300–800 bar (Burton et al. 2006). In a French UCG pilot test, it was found that the coal seam at the depth of 860 m had a pressure of 450 bar (Burton et al. 2006). If a water body is found near the coal seam, the injection pressure of the feed gas is adjusted in such a way that the coal seam cavity pressure is slightly less than the hydrostatic pressure of water. This negative pressure allows a sufficient amount of water seepage into the cavity for the steam generation. Several other parameters such as coal inherent properties, permeability and porosity of coal seam, and geological properties of the surrounding strata affect the performance of the UCG process. In deep UCG process, gas loss is minimized because several impermeable stratas are embedded on the coal seam.

2.6.1.1 Effect of the Inherent Properties of Coal Seam

The conventional UCG technique is not a well-efficient technology for high-ash (~40%) Indian coals. Ash layer accumulation in the coal seam retards the diffusion of oxidant/gasifying medium to the coal cavity surface. Therefore, the CRIP technique is suitable for exploiting high-ash coals. High-moisture lignite coal (~55%) may not be suitable for the UCG process, as it produces a low-calorific gas (Prabu and Jayanti 2012b). High-volatile matter coal (~40%) is more suitable for *in situ* gasification. It would produce a well-sustained combustion flame front and enhances the efficiency of gasification. However, a high-carbon content coal (~50%) may not be more beneficial, as the ignition of such coal is difficult. It requires a high-temperature fuel source for ignition at the interrupted combustion stages.

2.6.1.2 Choice of the Oxidizing/Gasifying Medium

A suitable oxidizing/gasifying medium is essential for the effective exploitation of deep coal through the UCG process. Oxygen, air, O_2-enriched air, water/steam, and CO_2 are the various oxidants, and these should be chosen based on the geological and inherent properties of the coal seam. Prabu et al. (2012b) reported that oxygen gasification is suitable for the high-ash coal seams as it creates a high-temperature environment, which balances the ash diffusion barrier for the solid-phase reaction. However, a significant amount of unconverted oxygen is observed in the product gas. As a result, the energy penalty of separation of O_2 from air is huge, and it affects the economy of the UCG process. It is also proposed that if the UCG process is coupled with carbon capture and storage (CCS) based power generation system, the unutilized oxygen in the product gas would be beneficial for the oxy-combustion process. However, the installation and operation of the ASU increases the capital and operating cost of the UCG process. Air gasification also leads to uneconomic operation of UCG due to the energy penalty for the compression of air to high pressure. The compression cost of air gasification may be expensive as compared to oxygen separation cost of the ASU in O_2 gasification for a deep UCG process.

Air-, water-, and steam-based gasifying agents are favorable for a high-carbon coal seam. The use of high-temperature steam in the inlet feed of UCG requires a proper insulation of deep underground pipelines. It may be uneconomic, and particularly, the sustainability of flame front propagation in the high-ash UCG seam is difficult as the steam temperature drops. The deposition of ash heap on the cavity resists the inlet steam passage and the temperature of steam drops, which results in loss of latent heat. As a result of steam condensation, the flame front may extinguish. In such case, CO_2 gasification of UCG is beneficial (Prabu and Geeta 2015). As CO_2 gas is readily available and the occurrence of phase change during the diffusion of gas in the ash layer is avoided, it may be a suitable gasifying medium for high-ash coal seams. The flue gas enriched with CO_2 due to the oxy-fuel combustion can be recycled and utilized as a feed gas to the UCG seam. Further, the burnt UCG cavity is suitable for the sequestration of CO_2 in the un-burnt char of the coal seam.

2.6.1.3 Energy Losses in the UCG Seam

Energy losses of UCG include water influx, gas loss, and pressure drop in the cavity during the UCG operation. Pressure drop may occur in the UCG as coal gets consumed and the cavity enlarges with time (Prabu and Jayanti 2012a). Gas loss to the surrounding strata also results in pressure drop of the cavity. The Lawrence Livermore National laboratory, Livermore, California, performed several UCG real field run and reported their results

(Cena and Thorsness 1981). They performed several shallow UCG trials with air, steam, and oxygen as the gasifying medium. It was shown that 10%–12% of coal energy was lost in the form of latent heat (Cena and Thorsness 1981). Further, it was observed that a significant gas pressure drop was prevailed during the UCG operation.

2.6.1.4 Application of UCG Syngas

Electricity generation is one of the major applications of UCG syngas. The UCG gas can be used in a similar way of the conventional syngas generated from surface gasifiers. The operating pressure of UCG depends on several parameters such as the depth of coal seam, water influx conditions, and the nature of overburden strata. The inlet gas pressure to the injection hole decides the UCG cavity pressure, and there may be a pressure drop due to factors such as the growth of UCG cavity with time and gas loss to surrounding strata. If the outlet pressure of UCG gas is below 10 bar, it can be coupled with a steam turbine (ST) system for power generation. High-pressure UCG gas (>10 bar) can be coupled with the gas turbine system such as IGCC. In case of Chinchilla UCG plant (Queensland, Australia), the pressure of the UCG syngas has been increased by using compressors in order to integrate with the gas turbine system. Further, it is proposed to utilize the UCG syngas in a fuel cell system after the removal of gas impurities (Prabu and Jayanti 2012a). Gas purification may be a major issue in UCG syngas. As the released volatile matter does not undergo thermal decomposition in the coal seam, syngas gets enriched with higher hydrocarbons such as tar in vapor form. Liu et al. (2006) studied the volatilization of arsenic, mercury, and selenium in coal during UCG experiments. It is found that mercury and selenium show 90% volatility, and arsenic has 60% volatility to the syngas. Hence, it should be removed prior to the usage. After the removal of impurities, UCG gas can be utilized directly in a solid oxide fuel cell (SOFC), which operates under high temperature. The calorific gases, other than hydrogen gases, such as CO, CH_4, and C_2H_6, are internally reformed with steam and converted into hydrogen in a high-operating temperature fuel cell. However, this can also be achieved in a low-temperature fuel cell such as PEMFC using an external gas reformer. If the purification of UCG syngas is found to be more difficult or uneconomic, it can be converted into highly pure hydrogen using the CLR process. These hydrogen gases can be utilized as the feed source to fuel cells.

UCG gas at low pressure can be effectively transformed into electric power through supercritical CO_2 ($SCCO_2$) turbines. The UCG energy losses could be compensated if they would be integrated with suitable renewable energy sources. Integration of solar energy with UCG is one of the feasible ways of efficient power generation and could increase the efficiency of the thermal power plant system. The generation of steam through solar energy could be advantageous for steam-based UCG, and further, heating and reheating of turbine inlet fluid operating streams (steam/$SCCO_2$) using solar energy enhance the performance of power system.

CO_2-enhanced UCG process may generate CO-enriched syngas. Prabu (2015) proposed that CO-enriched UCG syngas is more suitable for producing power through CLC of fuel for achieving CCS. CO-enriched syngas eliminates the complexity of thermal balance between the fuel reactor and the air reactor in the CLC process. The result of power simulation studies (Prabu 2015) shows a higher efficiency of the CLC systems compared to the conventional power generation systems. UCG gas can be utilized as a feedstock for the manufacture of chemicals. Syngas can be converted into liquid hydrocarbon chemicals using the Fischer–Tropsch (FT) process. It is proposed that ammonia can be synthesized effectively using an integrated UCG process.

2.6.2 Coalbed Methane

CBM is a recent emerging clean coal technology in India. Fossil fuels such as coal, gas, and oil resources are formed due to the sediments of organic matter such as plants and animals. These sediments are converted into fossil fuels after a prolonged period of deposition. Inherently, coal contains a significant amount of volatile matter, which gets released on heating. Methane gas produced during coal formation process (coalification) is entrapped on the pores of coal due to high pressure in the underground deposits. On mining, methane gets desorbed and diluted with the atmospheric air. It may also lead to fire hazard. Methane is a greenhouse gas, and it can entrap more heat than CO_2. Thus, release of methane directly into the atmosphere causes global warming.

Capturing methane gas from coal seams prior to mining is a feasible solution and is known as coalbed methane. The CBM gas can be collected through dewatering of the coal seams by releasing the hydrostatic pressure of coal seam. Figure 2.10 shows the schematic diagram of the UCG integrated CBM process.

Water pumping reduces the pressure in coal seams, and the methane gets desorbed and collected at the surface. However, it does not desorb methane content completely. The residual methane content is desorbed through the ECBM. Injecting CO_2 into the coal seam desorbs the residual methane, and it gets adsorbed on the pores of the coal seam. This

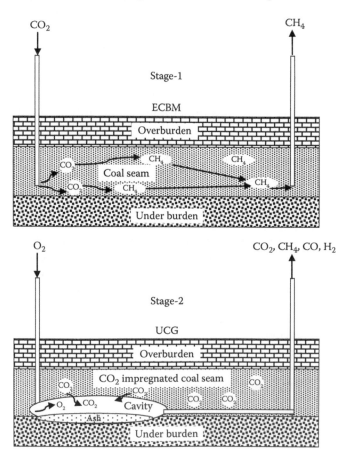

FIGURE 2.10
Schematic representation of the UCG-integrated CBM process.

is similar to the process called EOR, which is practicing in oil reservoirs. CO_2 has more adsorption affinity toward coal than methane. Hence, there is a mutual benefit that methane can be recovered upon CO_2 sequestration in the coal seam. It is estimated that two moles of CO_2 can be sequestered per mole methane desorption.

As India has 37% of its total coal reserves in the deep underground, ECBM is a feasible technology for methane recovery and CCS implementation. Prabu and Mallick (2015) estimated the CBM resources of India using Kim's correlations (Kim 1977). It is estimated that 1116.79 and 94.3 billion m^3 of methane gas are entrapped in bituminous coal and lignite coal, respectively (Prabu and Nirmal 2015). More than 40% of CBM is found at a depth of more than 300 m. Therefore, ECBM is a suitable technology that can exploit the methane gases effectively. Further, it is also estimated that 4400 million tons of CO_2 can be sequestered in the available coal resources. Prabu and Nirmal (2015) discussed the feasibility of integration of UCG with the ECBM technology. CO_2 sequestered through ECBM in a coal seam may act as a gasifying medium in a UCG operation. However, CO_2 gasification of UCG seam requires a high-temperature flame front (>900°C) to enhance the Boudouard reaction.

$$C + CO_2 \rightarrow 2CO \tag{2.2}$$

Air as an oxidizing medium does not sustain a high-temperature reaction front as it is diluted with 79% of nitrogen gas. Therefore, the supply of pure oxygen to the CO_2-impregnated coal seam maintains a high-temperature flame front, which enhances the Boudouard reaction. This may improve the reactivity of CO_2 and produce CO-enriched UCG syngas. Integration of these technologies would reduce the operating cost of both the processes. Methane can be utilized as a feedstock for chemical production. It can also be utilized as a source for power production through thermal and fuel cell power plants. Methane from the CBM source is known as sweet gas, as it is free from H_2S content. It contains more than 90% of methane and a less content of ethane, propane, and CO_2.

2.7 Coal-Based Power Generation Systems

The conversion of coal into a clean fuel consumes energy, and the implementation of clean coal technologies is energy intensive in terms of CCS. However, the energy penalty can be compensated by increasing the efficiency of power plants. The major use of coal is the production of electricity through thermal power plants. The power plant efficiency can be improved through the advanced power generating systems. The various coal-based power generation systems are given as follows:

1. Subcritical ST cycle
2. Supercritical and ultra-supercritical boiler STs
3. Supercritical carbon dioxide turbine system
4. IGCC
5. IGFC
6. Integrated CLC (ICLC) cycle
7. Integrated gasification solar combined cycle (IGSC)

The gross efficiency of these integrated advanced power generation systems is high and the energy penalty of CCS can be adjusted by integrating with the suitable technologies. Proper heat integration is necessary for the efficient usage of thermal energy. In case of CCS-integrated power system, the air separation system (ASU) and the CO_2 compression system (CCU) are the auxiliary units of the thermal power plants. The compression of CO_2 and air takes place in the ASU and the CCU, respectively. Multistage compression of the gases is carried out with intermediate cooling of these gases. Hence, a significant amount of heat energy generated during compression can be utilized in the power generation system. The preheating of feed water for the boiler input stream can be carried out using the leftover energy of the auxiliary units.

2.7.1 Subcritical ST Cycle

STs are operated based on the Rankine cycle (de Souza 2012). In this cycle, saturated liquid (water) is compressed using a water pump (isentropic process). High-pressure liquid is converted into a superheated steam by means of chemical energy of syngas in a boiler. Electrical energy is produced in the turbine system by isentropic expansion of the steam. Finally, the low-quality heat is rejected in the condenser. In conventional subcritical STs, the energy loss is high in terms of unutilized low-quality steam (through latent heat). The temperature of the ST is restricted to a maximum temperature of 540°C due to the operation limit. Therefore, less significant energy is produced in the ST-based system. In this system, solid coal or atmospheric pressure syngas is utilized as a fuel in the boiler for the generation of high-pressure steam. This high-temperature steam is supplied to the various STs operating at different pressures. High-pressure turbine (HPT), medium-pressure turbine (MPT), and low-pressure turbine (LPT) are integrated sequentially for the recovery of thermal energy in terms of electric power. Figure 2.11 shows the schematic diagram of the ST system.

Water is pressurized to 170 bar, and it is converted into steam in the boiler. The steam from the boiler drives the HPT, and the outlet steam is reheated in the reheater and sent to the MPT. The LPT is driven by the outlet steam of the MPT. There are several steam outlets in these turbines, and they are used for preheating the feed water to the boiler.

2.7.2 Supercritical and Ultra-Supercritical Boiler STs

Rankine cycle-based supercritical STs are the advanced power generation systems. Supercritical steam as the operating fluid enhances the net thermal efficiency of the power plant system. The supercritical pressure and temperature of steam are 22.06 MPa and 374°C, respectively (Breeze 2014). The fluids behave as an intermediate to the property of liquids and gases. In the subcritical steam boiler, water is pumped to a subcritical condition of 170 bar and heated to 540°C. STs, which are operating under subcritical, supercritical, and ultra-supercritical conditions, have maximum operating pressures of 170, 270, and 370 bar, respectively. In the ultra-supercritical boiler, the steam is heated to a temperature of 700°C, whereas it is 600°C in the supercritical boiler. Table 2.4 shows the net thermal efficiency of the UCG-based thermal power plants operating in the CLC mode of combustion (Prabu 2015).

The difference in the net thermal efficiency of the power plant system is increased by 3% on the order of subcritical, supercritical, and ultra-supercritical conditions of the boiler. The gain in thermal efficiency directly leads to a reduction in the unit cost of power and CO_2 emissions. However, the major concern of the supercritical technology

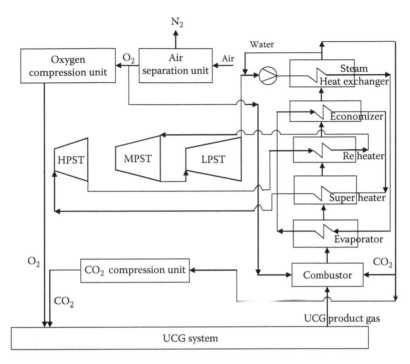

FIGURE 2.11
Schematic representation of the UCG-integrated ST.

TABLE 2.4

Overall Energy Analysis for the O_2/CO_2-Based UCG Integrated with CLC-Incorporated ST Power Plants

Power Plant System	Subcritical Condition	Supercritical Condition	Ultra-Supercritical Condition
AR temperature (°C)	916	1100	1200
FR temperature (°C)	885	1000	1100
Power consumption in the auxiliary units (MW)			
ASU load to UCG ($P_{O2,sep}$)	16.05	16.05	16.05
Oxygen compression to UCG ($P_{O2,com}$)	2.19	2.19	2.19
CO_2 compression to UCG ($P_{CO2,com}$)	7.18	7.18	7.18
CCU load for sequestration (P_{CCU})	30.05	30.05	30.05
Water pumping (P_{FWP})	3.55	5.06	6.19
Power production (MW)			
HPT	29.81	33.43	40.97
MPT	44.09	52.12	58.89
LPT	152.18	159.83	162.30
Gross power production	226.08	244.85	262.16
Efficiency (%)			
Gross efficiency	39.81	43.22	46.25
Net efficiency with CCS	29.42	32.55	35.40

Source: Reprinted from the publication *Appl. Energy*, 140, Prabu, V., Integration of *in-situ* CO_2-oxy coal gasification with advanced power generating systems performing in a chemical looping approach of clean combustion, 1–13, Copyright 2015, with permission from Elsevier.

is the availability of material for construction to withstand such a high temperature and pressure. High tube wall temperature increases the corrosion problem. Therefore, the development of materials to operate under supercritical conditions enables the technology for commercialization.

2.7.3 Integrated Gasification SCCO$_2$ Turbine Cycle (Brayton Cycle)

CO_2 acts as a supercritical fluid above the critical temperature of 303.98 K and the critical pressure of 7.38 MPa (Muñoz-Antón et al. 2015). In SCCO$_2$, the energy loss due to the change of phase from liquid to gas in a supercritical steam cycle is completely avoided. The CO_2 gas in the supercritical stage is compressed to a high pressure using compressors. And it is heated in the boiler using thermal energy of fuel. This SCCO$_2$ drives the gas turbine and the electric power is generated. CO_2 is removed from the turbine in the supercritical stage just above its critical point. Again it is recycled for the process. Atmospheric pressure syngas can be utilized efficiently for power production in SCCO$_2$ based Rankine cycle system.

2.7.4 Integrated Gasification Combined Cycle

The integration of both Rankine and Brayton cycles is termed as the combined cycle system. Figure 2.12 shows the UCG integrated combined cycle power generation system with CCS. The Brayton cycle (de Souza 2012) requires a high-pressure syngas and air, which is pressurized using the compressor. These gases are burnt in the combustor and expanded in the gas turbine, which generates electricity. The residual heat from the outlet gas from the gas turbine is used to operate the ST in the Rankine cycle.

High-pressure coal syngas generated from the conventional gasifier is utilized for the power generation system. Gas turbine requires the syngas purification system that removes the particulate matter, tar, and sulfur compounds. The particulate matter is removed using cyclone separators. Tar compounds are removed through the wet scrubbing process. Sulfur compounds can be removed through the amine absorption process. After the removal of impurities, the gas is sent to the combustor with the supply of high-pressure oxidant using the compressor, which is integrated with the gas turbine. Figure 2.12 shows the schematic representation of the IGCC-integrated UCG system (Prabu and Geeta 2015).

Table 2.5 represents the overall comparison of the net thermal efficiency of the UCG-integrated combined cycle system. It is reported that 4% efficiency gain is estimated for rise in the operating pressure from 9 to 27 bar in a UCG-integrated gas turbine.

High-pressure (>10 bar) and high-temperature gas (1000°C–1250°C) from the combustor drives the gas turbine and generates the electrical power. The residual heat energy from the exhaust gas of the turbine at atmospheric pressure is recovered in the heat recovery steam generator (HRSG). The final flue gas is sent to the sequestration unit. The gas turbine is operated based on the Brayton cycle. The steam generated in the HRSG is utilized to drive the ST through the Rankine cycle. The coupling of both cycles is termed as the combined cycle. IGCC produces a higher net thermal efficiency compared to the ST system. However, coal needs to be gasified, and further, it should be purified for the IGCC process. Hence, there is a significant loss of energy. Particularly, if syngas contains higher hydrocarbons in terms of tar, the sensible heat of the syngas is lost because of the wet scrubbing process for the removal of tar.

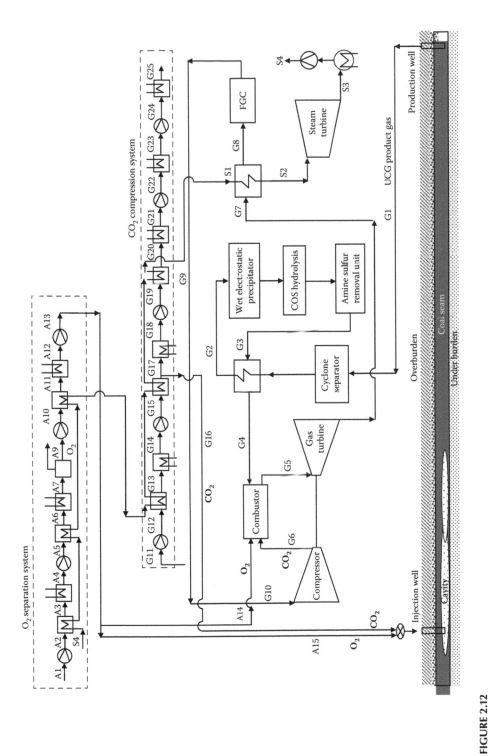

FIGURE 2.12

Schematic representation of the IGCC-integrated UCG system.

TABLE 2.5

Overall Energy Analysis for the CO_2-Blown UCG-Integrated Power Plant Systems

Power plant system				
Operating pressure, bar	9.5	9.5	27	27
CO_2/O_2 ratio	0.99	2.03	0.98	1.95
Power production (MW)				
Gas turbine	303.21	303.28	371.21	372.35
High-pressure ST	—	—	—	—
Medium-pressure ST	—	—	—	—
Low-pressure ST	53.32	53.57	35.73	35.88
Power supply from the gas turbine (MW)				
Oxygen compression to CMOC combustor	9.64	9.67	10.86	10.93
CO_2 compression to CMOC combustor	53.83	49.66	71.35	66.67
Gross electrical power production (MW)	293.06	297.52	324.73	330.63
Power consumption in the auxiliary units (MW)				
ASU load for UCG	18.67	16.67	21.42	17.52
ASU load for CMOC combustor	26.08	26.14	26.26	26.43
Throttle valve loss	3.24	3.25	3.25	3.25
Amine pumping cost in gas purification unit	0.018	0.018	0.051	0.051
Reboiler duty in H_2S stripper	1.05	1.06	1.05	1.06
Cooling load of turbo machinery	0.041	0.037	0.39	0.037
Oxygen compression to UCG	6.46	5.75	9.53	7.80
CO_2 compression to UCG	7.09	12.95	10.34	16.77
CCU load for sequestration	25.33	24.19	26.79	22.71
Water pumping	0.043	0.044	0.029	0.029
Total power consumption in the auxiliary units (MW)	88.02	90.11	99.11	95.66
Total useful output (MW)	205.04	207.41	225.62	234.97
Thermal energy of the coal (MW)	548.43	554.14	550.21	555.08
Gross efficiency	53.43	53.69	59.02	59.57
Net efficiency (with CCS)	37.39	37.43	41.00	42.33

Source: Reprinted from the publication *Energy*, 84, Prabu, V. and Geeta, K., CO_2 enhanced in-situ oxy-coal gasification based carbon-neutral conventional power generating systems, 672–683, Copyright 2015, with permission from Elsevier.

2.7.5 Integrated Gasification Fuel Cell System

In the IGFC system, coal is gasified into syngas using the conventional coal gasifier. Hydrogen is the feed for the fuel cell power generation. Carbon monoxide, methane, and ethane from syngas are converted into H_2 through water–gas shift reactors:

$$CH_4 + H_2O \rightarrow CO + 3H_2 \tag{2.18}$$

$$C_2H_6 + 2H_2O \rightarrow 2CO + 5H_2 \tag{2.19}$$

$$CO + H_2O \rightarrow CO_2 + H_2 \tag{2.4}$$

After the conversion of carbon fuel into hydrogen, the trace quantity of CO is converted into CO_2 through partial oxidation. The final gas contains pure H_2 and CO_2, which is sent as a feed to the fuel cell. A high-purity syngas is required for fuel cell operation. The produced syngas should be free of CO, sulfur compounds H_2S, SO_2, and NO_x. Also the trace component of Hg is also limited to parts per million level in fuel cell feed gas. The trace level of these gases severely affects the performance of fuel cell. The purified gases are sent to the fuel cell at the anode side. Air is supplied at the cathode side. Fuel cells are classified based on their electrolytes. In proton exchange membrane (PEM) fuel cell, hydrogen on the anode side is dissociated into electrons and protons. Electrons pass over the external load circuit and protons are permeated through the polymer electrolyte membrane. Hence, PEM fuel cell is otherwise termed as PEMFC. However, these fuel cells operate at low temperatures of 30°C–100°C (Larminie and Dicks 2003).

At the anode side:

$$H_2 \rightarrow 2H^+ + 2e^- \tag{2.20}$$

At the cathode side:

$$2H^+ + 2e^- + \tfrac{1}{2} O_2 \rightarrow H_2O \tag{2.21}$$

Alternatively, another fuel cell called solid oxide fuel cell (SOFC) operates at a high temperature in the range of 500°C–1000°C (Larminie and Dicks 2003). The electrolyte is a solid membrane, which has higher thermal stability. As it operates at high temperature, the syngas is reformed internally into hydrogen. Carbon monoxide, methane, ethane, and higher hydrocarbons are reformed into hydrogen fuel with excess supply of steam to the anode side along with the syngas. Therefore, the SOFC does not require any external reformer for water–gas shift reaction. Steam is sent in excess than stoichiometric to avoid the coking of membrane. In the SOFC, oxygen gets reduced at the cathode side, and the solid oxide electrolyte conducts the reduced oxygen ions to the anode side.

At the anode side:

$$H_2 + O^{2-} \rightarrow H_2O + 2e^- \tag{2.22}$$

At the cathode side:

$$\tfrac{1}{2} O_2 + 2e^- \rightarrow O^{2-} \tag{2.23}$$

At the anode side, hydrogen is converted into steam with the liberation of electrons. These electrons are passed through the external load for power production. As the chemical energy of fuel is directly converted into electrical power, energy loss is minimized in the IGFC cycle. Therefore, the IGFC cycle has higher efficiency than the IGCC and ST cycles. Figure 2.13 shows the schematic diagram of the IGFC-integrated UCG system (Prabu and Jayanti 2012c).

Table 2.6 shows the comparison of net thermal efficiency of the UCG-integrated fuel cell system with the conventional power generating system.

Integration of SOFC with UCG enhances the net thermal efficiency 6% higher than the ST system. Integration of the coal gasification unit with the fuel cell cycle leads to high electrical efficiency. It is beneficial in terms of the CCS. In the oxy-fuel combustion-integrated CCS system, the conventional boiler in the IGCC and ST systems requires an ASU system for the separation of oxygen from air. In the IGFC system, fuel cell electrolyte membrane acts a selective absorber of oxygen ions from atmospheric air. The membrane inherently separates oxygen from air, and the final flue gas is enriched with CO_2. Therefore, IGFC is inherently a carbon capture system without energy penalty.

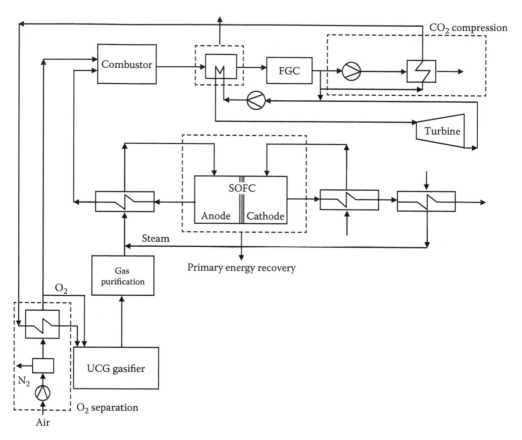

FIGURE 2.13
Schematic representation of the IGFC-integrated UCG system.

TABLE 2.6
Overall Energy Analysis for the UCG-Integrated SOFC and SMOC Power Plant

	UCG + SMOC (kW)	UCG + SOFC (kW)
Power production from SOFC	—	231,543
High pressure steam turbine (HPST)	47,979	—
Medium pressure steam turbine (MPST)	80,335	—
Low pressure steam turbine (LPST)	146,771	57,454
Gross electrical power production	275,085	288,997
Power consumption (ASU + O_2 compression)	59,939	37,010
CO_2 compression	27,638	26,125
Water pumping	6,395	62
Total power consumption	93,972	63,197
Total useful output	181,113	225,800
Thermal input to UCG (coal seam)	698,324	698,324
Gross efficiency	39.39	41.38
Net efficiency	25.94	32.33

Source: Reprinted from the publication *Int. J. Hydrogen Energy*, 37, Prabu, V. and Jayanti, S., 1677–1688, Copyright 2012, with permission from Elsevier.

2.7.6 Integrated Gasification Chemical Looping Combustion Cycle

In the ICLC system, coal is processed in two ways. First, it is gasified into syngas using the conventional gasifier. The generated syngas is sent as a feed to the fuel reactor. Alternatively, coal is fed into the fuel reactor directly, and it is gasified *in situ* (fuel reactor itself) with the supply of gasifying medium (CO_2/H_2O). Therefore, coal is converted into fuel gas in the reactor, and it reacts with metal oxides. Metal oxides such as NiO, CuO, Fe_2O_3, CoO, and Mn_3O_4 are used in the fuel reactor for the oxidation of syngas. The inert support material used for the entrapment of these metal oxides in the CLC process are Al_2O_3, SiO_2, TiO_2, ZrO_2, $NiAl_2O_4$, $MgAl_2O_4$, and so on (Fan 2010). Coal gasification and CLC occur simultaneously in the fuel reactor.

$$Coal \rightarrow volatile\ matter\ (CO, H_2, CO_2, CH_4) + char \tag{2.9}$$

$$C + CO_2 \rightarrow 2CO \tag{2.2}$$

$$C + H_2O \rightarrow CO + H_2 \tag{2.1}$$

These fuel gases are converted into CO_2 and H_2O in the fuel reactor, and the final flue gas is enriched with CO_2 and it is readily available for sequestration. In the air reactor, atmospheric air is used as an oxidant. After the oxidation of metal particles in the air reactor, oxygen-depleted air is sent to the atmosphere after the recovery of thermal energy. Figure 2.14 shows the schematic representation of CLC-integrated UCG system (Prabu 2015). Table 2.7 shows the net thermal efficiency of the CLC-based UCG integrated

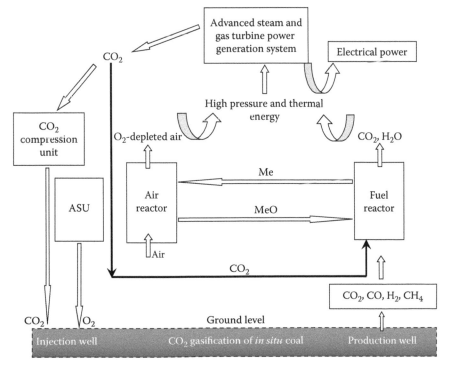

FIGURE 2.14
Schematic representation of the CLC-integrated UCG system.

TABLE 2.7

Overall Energy Analysis for the O_2/CO_2-Based UCG Integrated with CLC-Incorporated Combined Cycle Power Plant System

Power Plant System	IUGCC
Operating pressure (bar)	9.5
CO_2/O_2 ratio	0.99
Power production (MW)	
Gas turbine	89.72
Air turbine	254.62
Low pressure ST	65.88
Power supply from the gas turbine (MW)	
Air compression to CLC air reactor	115.98
CO_2 compression to CLC fuel reactor	4.24
Gross electrical power production (MW)	290.00
Power consumption in the auxiliary units (MW)	
ASU load for UCG ($P_{O2,sep}$)	18.67
Oxygen compression to UCG ($P_{O2,com}$)	6.45
CO_2 compression to UCG ($P_{CO2,com}$)	7.07
CCU load for sequestration (P_{CCU})	24.47
Water pumping (P_{FWP})	0.056
Total power consumption in the auxiliary units (MW)	56.76
Total useful output (MW)	233.24
Thermal energy of the coal (MW)	548.43
Gross efficiency	52.88
Net efficiency (with CCS)	42.53

Source: Reprinted from the publication *Appl. Energy*, 140, Prabu, V., Integration of *in-situ* CO_2-oxy coal gasification with advanced power generating systems performing in a chemical looping approach of clean combustion, 1–13, Copyright 2015, with permission from Elsevier.

with combined cycle power generation system. Integration of CLC with UCG enhances the net thermal efficiency of about 5% higher than the ST system.

2.7.7 Integrated Gasification Solar Cycle

Integration of renewable energy with coal greatly reduces CO_2 emission. Renewable energy sources such as sunlight and biomass are suitable for integrating with coal for power generation. Co-combustion of coal and biomass reduces the consumption of fossil fuel and promotes the utilization of seasonal biomass. Similarly, solar energy is another renewable energy, and the proper integration with the power generation cycle increases the net thermal efficiency of the system. In the ST system, solar energy can be utilized for preheating the feed water to the boiler (Suresh et al. 2010). Further, it is utilized for reheating the outlet steam of the HPT and MPT. In the IGSC system, the production of the gasifying medium such as steam and the preheating of feed water for the ST can be carried out using solar energy. In addition, syngas and oxidant can be preheated before sending them to the combustor. Hence, integration of solar energy with the coal power system is one of the feasible solutions for the reduction of CO_2 emission.

2.8 Coal to Liquid

Coal to liquid (CTL) is a clean coal technology for the conversion of solid fuel into liquid fuel. This can be carried out using direct CTL and indirect CTL methods. In the direct conversion method, coal is broken down into smaller molecules using suitable organic solvents (solvent dissolution methods) such as tetralin, quinoline, and naphthol (Hernández et al. 2012). These solvents dissolve the organic matter of coal and reject the mineral matter (ash content) and sulfur content of coal. Also hydrogen from these solvents is donated for the liquefaction process, and also the addition of pure hydrogen stabilizes the pyrolyzed lighter gas from coal into soluble products. Light, medium, and heavy oils are separated through vacuum distillation. However, further upgradation of the obtained liquid fuel is necessary for the removal of impurities. The obtained liquid product is enriched with a variety of hydrocarbons (naphtha, fuel oil, etc.), which is more aromatic in nature. Also it contains sulfur, oxygen, and nitrogen impurities, and requires further downstream processing for the purification. Approximately, the final product contains a yield of 28% of naphtha, 11% of C_3–C_4 liquified petroleum gas (LPG), 11% of light distillate, 19% of gas oil, and 31% of heavy distillate (Bellman 2007).

In the indirect CTL method, coal is converted into syngas using the conventional gasifier. After the removal of CO_2, sulfur, and acid gas impurities, syngas is converted into liquid by means of the FT synthesis process (Riazi and Gupta 2015). FT reactions are carried out at 150°C–300°C, and these reactions are exothermic. The removal of heat is essential to obtain the desired product. The liquid fuel obtained through the indirect method is free of contaminants and can be utilized directly. The final liquefied product contains 21% of naphtha, 14% of C_3–C_4 LPG, 18% of distillate, and 47% of wax. Hydroisomerization of these liquid products yields 20%–54% of gasoline, 17%–36% of jet fuel, and 16%–55% of diesel (Bellman 2007).

2.9 Conclusion

Coal plays a vital role in the energy sector for electricity production and will continue for several decades. Hence, it is essential to implement the developed clean coal technologies for efficient energy production as well as pollutant reduction. The integration of these technologies in appropriate way is the viable solution for the efficient utilization of the existing resources. It is essential to progress the basic and applied research on the conversion of coal into a clean fuel. Research level technological outcomes on clean coal technologies should be implemented to large scale for creating a clean environment.

Retrofitting of the existing technologies is another option for the economic recovery of fuel. The combustion modification technique should be implemented for the pollutant reduction. India contains huge coal resources, and nearly 37% of the resources are found at a great depth; the development of unconventional technologies is crucial for its recovery. Further, integration of coal technologies with renewable energy should be implemented for the reduction of pollutants.

Acknowledgment

The author expresses his gratitude to Elsevier for granting permission to reuse tables and figures from my publications with them.

References

Ambedkar B, Chintala T.N, Nagarajan R, Jayanti S. 2011. Feasibility of using ultrasound-assisted process for sulfur and ash removal from coal. *Chemical Engineering and Processing* 50: 236–246.

Bellman D.K. 2007. Coal to Liquids and Gas. Working Document of the NPC Global Oil and Gas Study, American Electric Power Co., Inc., National petroleum council, Washington, DC.

Breeze P. 2014. *Power Generation Technologies*, 2nd ed. Newnes, An imprint of Elsevier, Burlington.

Burton E, Friedmann J, Upadhye R. 2006. Best practices in underground coal gasification, U.S.DOE. Lawrence Livermore National Laboratory, Contract No. W-7405-Eng-48, Livermore, CA.

Cena J, Thorsness C.B. 1981. Lawrence Livermore national laboratory underground coal gasification data base. U.S.DOE. Lawrence Livermore National Laboratory, Livermore, CA, Report No. UCID-19169.

de Souza G.F.M. 2012. *Thermal Power Plant Performance Analysis*. ISBN 978-1-4471-2309-5. Springer-Verlag, London.

Díez L.I, Cortés C, Pallarés J. 2008. Numerical investigation of NOx emissions from a tangentially-fired utility boiler under conventional and overfire air operation. *Fuel* 87: 1259–1269.

Energy Statistics of India. 2014. Central statistics office. Ministry of statistics and programme implementation Government of India, New Delhi.

Ersoz A, Olgun H, Ozdogan S. 2006. Reforming options for hydrogen production from fossil fuels for PEM fuel cells. *Journal of Power Sources* 154: 67–73.

Fan L.S. 2010. *Chemical Looping Systems for Fossil Energy Conversions*. AICHE, Wiley, NJ.

Hernández M.R, Murcia C.F, Gupta R, Klerk A.D. 2012. Solvent-coal-mineral interaction during solvent extraction of coal. *Energy Fuels (American Chemical Society)* 26: 6834–6842.

Hill R.W, Shannon M.J. 1981. The controlled Retracting Injection Point System: A modified stream method for in situ coal gasification, U.S. DOE. Lawrence Livermore National Laboratory, Livermore, CA, Report No. UCRL-85852.

Holloway S, Garg A, Kapshe M, Khan S.R, Mahmood M.A, Singh T.N. et al. 2008. A regional assessment of the potential for CO_2 storage in the Indian subcontinent, IEA Greenhouse Gas R&D programme, Technical Study Report No. 2008/2.

Iyengar R.K, Haque R. 1991. Gasification of high-ash Indian coals for power generation. *Fuel Processing Technology* 27: 247–262.

Janajreh I, Raza S.S, Valmundsson A.S. 2013. Plasma gasification process: Modeling, simulation and comparison with conventional air gasification. *Energy Conversion and Management* 65: 801–809.

Khadse A, Qayyumi M, Mahajani S, Aghalayam P. 2007. Underground coal gasification: A new clean coal utilization technique for India. *Energy* 32: 2061–2071.

Kim A.G. 1977. Estimating methane content of bituminous coalbeds from adsorption data/by Ann G. Kim. Washington, DC: U.S. Department of the Interior, Bureau of Mines.

Kivisaari T, Bjornbom P, Sylwan C, Jacquinot B, Jansen D, Groot A. 2004. The feasibility of a coal gasifier combined with a high temperature fuel cell. *Chemical Engineering Journal* 100: 167–180.

Larminie J, Dicks A. 2003. *Fuel Cell Systems Explained*, 2nd ed. John Wiley & Sons Ltd, Chichester.

Li P, Lei W, Wu B, Yu Q. 2015. CO_2 gasification rate analysis of coal in molten blast furnace slagd—For heat recovery from molten slag by using a chemical reaction. *International Journal of Hydrogen Energy* 40: 1607–1615.

Liu S, Wang Y, Yu L, Oakey J. 2006. Volatilization of mercury, arsenic and selenium during underground coal gasification. *Fuel* 85: 1550–1558.

Metz B, Davidson O, Coninck H.D, Loos M, Meyer L. 2005. IPCC Special Report on Carbon Dioxide Capture and Storage. Working Group III of the Intergovernmental Panel on Climate Change, Cambridge University Press, New York.

Ministry of Coal, Annual Report 2013–2014. Government of India 2013, http://coal.nic.in.

Muñoz-Antón J, Rubbia C, Rovira A, Martínez-Val J.M. 2015. Performance study of solar power plants with CO_2 as working fluid. A promising design window. *Energy Conversion and Management* 92: 36–46.

Parikh J. 2010. Analysis of Carbon Capture and Storage (CCS) Technology in the Context of Indian Power Sector, TECHNOLOGY SYSTEMS DEVELOPMENT (TSD), PROGRAMME OF DST, DST Reference No: DST/IS-STAC/CO_2-SR-39/07.

Prabu V. 2015. Integration of in-situ CO_2-oxy coal gasification with advanced power generating systems performing in a chemical looping approach of clean combustion. *Applied Energy* 140: 1–13.

Prabu V, Geeta K. 2015. CO_2 enhanced in-situ oxy-coal gasification based carbon-neutral conventional power generating systems. *Energy* 84: 672–683.

Prabu V, Jayanti S. 2012a. Underground coal-air gasification based solid oxide fuel cell system. *Applied Energy* 94: 406–414.

Prabu V, Jayanti S. 2012b. Laboratory scale studies on simulated underground coal gasification of high ash coals for carbon neutral power generation. *Energy* 46: 351–358.

Prabu V, Jayanti S. 2012c. Integration of underground coal gasification with a solid oxide fuel cell system for clean coal utilization. *International Journal of hydrogen energy* 37: 1677–1688.

Prabu V, Mallick N. 2015. Coalbed methane with CO_2 sequestration: An emerging clean coal technology in India. *Renewable and Sustainable Energy Reviews* 50: 229–244.

Riazi M.R, Gupta R. 2015. *Coal Production and Processing Technology*. CRC press, Taylor & Francis Group, Boca Raton, FL.

Seepana S, Jayanti S. 2010. Steam-moderated oxy fuel combustion, *Energy Conversation and Management* 51: 1981–1988.

Seepana S, Jayanti S. 2012. Optimized enriched CO_2 recycle oxy-fuel combustion for high ash coals. *Fuel* 102: 32–40.

Smith A.R, Klosek J. 2001. A review of air separation technologies and their integration with energy conversion processes. *Fuel Processing Technology* 70: 115–134.

Sorgenfrei M, Tsatsaronis G. 2014. Design and evaluation of an IGCC power plant using iron-based syngas chemical-looping (SCL) combustion. *Applied Energy* 113: 1958–1964.

Suresh M,V,J.J, Reddy K.S, Kolar A.K. 2010. 4-E (Energy, Exergy, Environment, and Economic) analysis of solar thermal aided coal-fired power plants. *Energy for Sustainable Development* 14:267–279.

Tsumura T, Okazaki H, Dernjatin P, Savolainen K. 2003. Reducing the minimum load and NOx emissions for lignite-fired boiler by applying a stable-flame concept. *Applied Energy* 74: 415–424.

United Nations Statistics Division. 2007. http://unstats.un.org/unsd/environment/air_co2_emissions.htm.

U.S. Department of Energy. 1999. Reburning Technologies for the Control of Nitrogen Oxides Emissions from Coal-Fired Boilers. Topical report number 14.

Viebahn P, Vallentin D, Höller S. 2014. Prospects of carbon capture and storage (CCS) in India's power sector—An integrated assessment. *Applied Energy* 117: 62–75.

Yang L. 2003. Study of the model experiment of blinding-hole UCG. *Fuel processing Technology* 82: 11–25.

Zhang H, Yue G, Lu J, Jia Z, Mao J, Fujimori T, Suko T, Kiga T. 2007. Development of high temperature air combustion technology in pulverized fossil fuel fired boilers. *Proceedings of the Combustion Institute* 31: 2779–2785.

3

Downstream Processing of Heavier Petroleum Fractions

Shubham Saroha, Prasenjit Mondal, and Deepak Tandon

CONTENTS

ABSTRACT This chapter focuses on different processes for upgrading the heavy petroleum residues into valuable products. It highlights the growing demand for petroleum products every year and diminishing supplies of crude oil. It enlists the essential properties of these residues and provides the much needed selection criteria for adopting an upgradation technique. It discusses various upgradation techniques, that is, visbreaking, gasification, delayed coking, hydrocracking, and so on, and recent advancements that have been incorporated into them. It examines the merits and demerits of these techniques and compares them. It gives an overview of the biotechnological processes for the residue utilization.

3.1 Introduction

In a typical refinery, after desalting and atmospheric distillation, the petroleum crude is converted into petroleum gas, light and heavy naphtha, gasoline, kerosene, jet fuel, diesel oil, atmospheric gas oil, and atmospheric bottoms. The residue from the atmospheric distillation column (atmospheric bottoms) is further processed in a vacuum distillation column to produce light and heavy vacuum gas oil (VGO) as well as vacuum residue (VR). The heavier fractions such as VR, VGO, and atmospheric gas oil, further go through a

TABLE 3.1

Sulfur, Nitrogen, and Metal Content of Some Light and Heavy Crude Oils

Property	Light Crude Oil[a]	Heavy Crude Oil[a]
Nitrogen (wt.%)	0.1	Up to 1
Sulphur (wt.%)	0.05	Up to 5
Metals (ppm)	~3	Up to ~600

[a] Scherzer, J. and A.J. Gruia, *Hydrocracking Science and Technology*, Marcel Dekker, New York, 1996, p. 3.

number of catalytic and noncatalytic downstream processing steps to produce valuable products as well as to meet the market demand (Marafi et al. 2010). Gradual increase in sulfur and metal contents of petroleum crude as well as its viscosity make it heavier and difficult for processing through the conventional refinery unit. It produces higher amount of heavier fractions containing more metals and sulfur than light crude oil. The typical concentrations of sulfur, nitrogen, and metals (V and Ni) in light as well as heavy crude oils are provided in Table 3.1.

Therefore, the upgradation of heavier fractions through downstream processing is becoming more interesting day by day and attracting attention of refiners. It has also been reported that for utilizing heavy and extra heavy crude oils, refining and nonrefining processes need to be integrated (Marafi et al. 2010). Based on the work of Rana et al. (2008), a suitable flowsheet for heavy and extra heavy petroleum crudes is shown in Figure 3.1. From the figure, it seems that a part of heavy and extra heavy crude oils avoids the distillation step and is processed through conventional downstream processing steps.

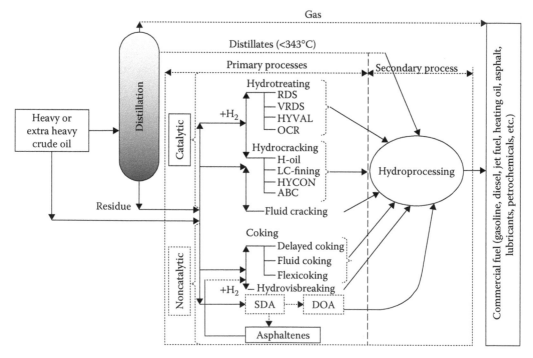

FIGURE 3.1

Advanced refinery for upgrading heavy and extra heavy feeds. (From Rana, M.S. et al., *Int. J. Oil Gas Coal Technol.*, 1, 250–282, 2008. With permission.)

Heavy crude oil is thicker with an American Petroleum Institute (API) gravity of less than 20°, more resistant to flow, and usually contains higher levels of sulfur and other contaminants than light crude oil. It gives more residues after the processing of crude oil from the vacuum distillation unit than light crude oil. According to the density, heavy crudes can be classified as heavy oils (API within 10°–20°) and extra heavy oils (API < 10°) (Manning and Thompson 1995). The *in situ* viscosity makes the distinction between extra heavy oils and bitumen. It is reported that during the past 15 years (2001–2015), the demand for petroleum products has grown with a rate of ~1.7% per annum, which is also expected to continue over the next 15 years (Hedrick and Seibert 2006)(Web 1). Further, the supplies of light crudes have diminished in recent years, whereas global heavy oil consumption is increasing gradually, and it is predicted to continue until 2030 (World Energy Outlook 2008). Therefore, both the heavy crude oil containing 40%–64% residues and the nondistillable heavy residues produced from it attract strong interest from the refiners to produce valuable products (Speight 2000; Marafi et al. 2010).

In addition, the market demand for different petroleum products varies with time, for example, in recent years middle distillates have been in more demand, whereas in the past (until 2007), gasoline was in higher demand (U.S.A. Energy Information Administration 2016). In the present scenario, conversion of heavy residues into light-value products and subsequent processing are more attractive than its conventional use as a heavy fuel component (Sarkar 1998). Many methods including high vacuum distillation, shortcut distillation (cutpoint extended up to 700°C), and sequential extraction fractionation have been developed for the fractionation of heavy crude oils (Chung et al. 1997). Some upgradation techniques such as thermal cracking, catalytic cracking, hydrocracking (HDC), and solvent extraction are also applied for processing heavy petroleum crudes as well as heavy residues generated by the fractionation. Figure 3.2 shows the contribution of different processes such as cracking/visbreaking, coking, HDC, hydrotreating (HDT), and deasphalting to the downstream processing of heavy residues.

For the upgradation of different types of gas oils (atmospheric, light vacuum, and heavy vacuum), generally hydrogen addition processes (HDT, HDC, etc.) are used, whereas for

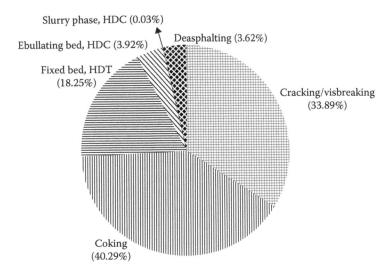

FIGURE 3.2
Worldwide distribution of commercial residue processing capacity. (From Castaneda, L.C. et al., *Catal. Today*, 220–222, 248–273, 2014. With permission.)

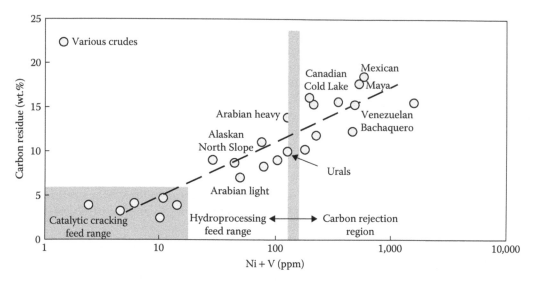

FIGURE 3.3
Feasibility region of the commercial residue conversion processes. (From Sawarkar, A.N. et al., *Can. J. Chem. Eng.*, 85, 1–24, 2007. With permission.)

processing vacuum residues, carbon rejection processes (thermal and catalytic cracking) are preferred. Thermal cracking may be visbreaking and delayed coking. Selection of a suitable process for the upgradation of residue depends upon the carbon residue and the metal content in the residue. As shown in Figure 3.3, residues having lower Conradson Carbon Residue (CCR) and metal content (<6% and 15 ppm, respectively) are suitable for catalytic cracking, whereas residues having higher carbon residue and metal content (>10% and 150 ppm, respectively) are suitable for carbon rejection.

A little heavier feedstock (CCR ≤ 10%) containing a metal content of ≤150 ppm can be processed through hydroprocessing. Some important technologies used in this process are Shell's hydroconversion (HYCON) and Chevron's on-stream catalyst replacement (OCR). Hydroprocessing has some advantages over catalytic cracking as it can handle relatively heavier feedstocks with more heavy metal content, and it can also produce good-quality products such as catalytic cracking. However, its processing/investment cost and the need for hydrogen limit its application for the feedstocks having a CCR value and a metal content of more than 10% and 150 ppm respectively, which can be upgraded through non catalytic carbon rejection process (Shen et al. 1998; Phillips and Liu 2002).

Heavy petroleum crude has higher impurities content, and it yields lower middle distillates, which have a high market demand. It also produces heavier residues due to the presence of more paraffins, naphthenes as well as aromatics, including asphaltenes and resins. The properties of heavy residues depend on the constituents of crude oil and are important to select/develop suitable upgradation processes because the chemical structure and complexity of the residues influence the upgradation process.

3.1.1 Properties of Heavy Petroleum Residues

A residual oil is composed of asphaltenes, resins, aromatics, and saturated hydrocarbons as shown in Figure 3.4 (Chrisman et al. 2012).

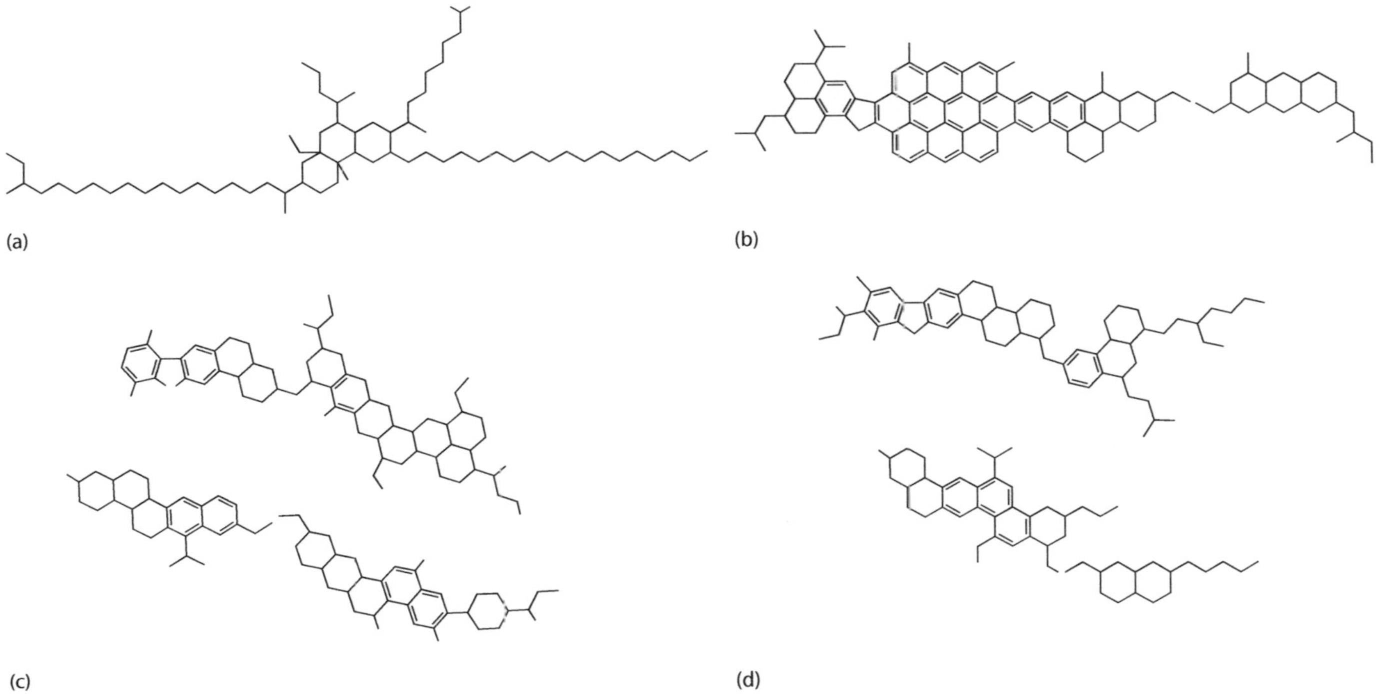

FIGURE 3.4

Typical composition of heavy residues: (a) saturated, (b) asphaltenes, (c) aromatics, and (d) resins. (From Chrisman, E. et al., Asphaltenes—Problems and Solutions in E&P of Brazilian Crude Oils, Chapter 1, in *Crude Oil Emulsions—Composition Stability and Characterization*, InTech, 2012, http://www.intechopen.com/books/crude-oil-emulsions-composition-stability-and-characterization/asphaltenes-problems-and-solutions-in-e-p-of-brazilian-crude-oils. With permission.)

TABLE 3.2

Some Important Properties of the VR

Properties	Value	Reference
Viscosity at 135°C (cSt)	2.E+02	
Gravity, API	10.0°	Jechura 2016 (Web 5)
Heating value (kcal/kg)	10310	
Boiling point (°C)	615	Altgelt and Boduszynski (1994)
H/C ratio	~1.38	Furimsky (1999)
Heavy metals (wt.%)	0.1	
Elemental carbon (wt.%)	79.75	
Hydrogen (wt.%)	9.37	Gupta and Gera (2015)
Nitrogen (wt.%)	0.47	
Sulphur (wt.%)	4.20	

It can be described as a colloidal system in which the dispersed phase comprises asphaltenes and the continuous phase represents maltene/oil (containing aromatics, resins, and saturated hydrocarbons). The resin molecules play the role of surfactants in stabilizing colloidal particles of asphaltenes in oil. During thermal cracking, the oil phase cracks, and after certain limit of cracking, the colloidal equilibrium is disturbed, which results in flocculation of asphaltenes. At this stage, cracked fuel oil becomes unstable in nature (Hur et al. 2014). Heavy residues contain higher viscosity, heavy metals, sulfur, nitrogen, and naphthenic acids. Some important physicochemical characteristics of VR are provided in Table 3.2.

Asphaltenes correspond to a crude oil fraction consisting of polar molecules having high molecular weight, which are capable of self-association at a critical concentration. The VR is also rich in aromatic content normally in the CCR range of 15–25 wt.%. In some cases, the CCR value is less (1%–6%), where naphthene quantity becomes more (Schulman and Dickenson 1991). The aromatic content influences the viscosity, and the coupling of viscosity with CCR helps to evaluate the average molecular weight of the feedstocks, which is related to the kinematic viscosity. The higher the aromaticity of a heavy oil, the higher is its viscosity (Speight 1991). Because the asphaltenes are the most viscous aromatic compounds in a heavy oil, their influence on the heavy oil viscosity is maximum among all other heavy oil constituents. The converted vacuum residual oils (from visbreaking and residue ebullated bed H-Oil HDC unit) demonstrate the lower dependence of viscosity on the asphaltene content. This could be a result of decreasing dimensions in the macrostructure of the converted asphaltene molecule. The mass ratio of asphaltenes and resins in a heavy residue influences its stability, whereas its surface properties are dependent on the association of nonpolar and polar constituents as well as their molecular composition. Asphaltene has a high molecular weight and is insoluble in *n*-pentane and other noncyclic hydrocarbons. It is nonvolatile and held in colloidal suspension in the hydrocarbon phase with the resins, which act as peptizing agents. Different tests such as merit number test (IFP-3024-82), Shell *P*-value test (1400-2), toluene/xylene equivalent test, and Shell hot filtration test (SMS 742) are used to determine the stability of a heavy residue.

3.2 Upgradation of Heavy Residues

Carbon rejection and hydrogen addition are the two types of processes used for the upgradation of heavy residues. Visbreaking, fluid catalytic cracking (FCC), steam cracking, delayed coking, and solvent extraction methods are the important conventional carbon rejection processes, whereas fixed bed and ebullated bed catalytic hydroconversion, HDC, hydrovisbreaking, and donor solvent processes are the important conventional hydrogen addition processes. Application of nanomaterials, gasification, and biological processes are some emerging technologies for the processing of heavy residues. The hydroconversion processes are mostly catalytic, whereas the carbon rejection processes may be catalytic or noncatalytic. In case of the solvent extraction method, some solvents are required to separate the asphaltenes and resins from the heavy residue. Hydrogen addition processes have higher product flexibility and produce good-quality products, although these are costlier than carbon rejection methods. In terms of the operating cost and simplicity, the noncatalytic carbon rejection processes are superior to others and have been widely used for the upgradation of heavy residues around the world. Delayed coking and visbreaking processes account for approximately 63% of the value addition processes of heavy residues (Sawarkar et al. 2007). Thermal cracking of heavy residue is an old process; however, due to it complex nature and nonavailability of analytical facilities, the kinetic modeling of this process was not explored initially. Some important residue upgradation processes are described in Sections 3.2.1 through 3.2.7.

3.2.1 Visbreaking

Visbreaking process has been developed to produce fuel oil from heavy residue to save the valuable lighter products called "cutter stocks." Today, visbreaking units are also used to prepare the feedstocks for secondary units such as FCC and hydrocracker. This process reduces the viscosity of petroleum residue due to heat application and also produces a small amount of light hydrocarbons such as liquefied petroleum gas (LPG), naphtha, and gasoline, heavy gas oil, and residual part with reduced viscosity.

A typical visbreaking unit consists of a furnace reactor, a flash chamber, and a fractionating column. The furnace is normally coil type, and the typical operating conditions for furnace are as follows:

Furnace inlet temperature 305°C–325°C and pressure 15–40 bar

Furnace outlet temperature 480°C–500°C and pressure 2–10 bar

Residence time 2–5 min

Steam injection 1 vol.%

To reduce the furnace temperature, a soaker unit is placed between the flash chamber/fractionator and the furnace as shown in Figure 3.5. Temperature and pressure maintained in the soaker unit are 440°C–460°C and 5–15 bar with a residence time of 20–30 min, respectively. Preheating of the feedstock to ~335°C is done by using visbreaking tar. The overhead product of the flash chamber gives naphtha, LPG, and gasoline, and the bottom liquid is further fractionated into gas oil and residual product (reduced viscosity). In this

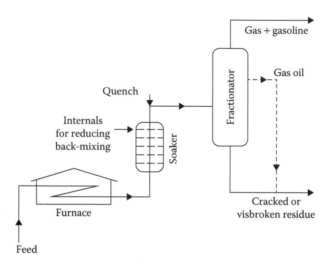

FIGURE 3.5
Visbreaking process in the soaker unit. (From Speight, J.G., Thermal cracking, Chapter 2, in *Heavy and Extra-heavy Oil Upgrading Technologies*, pp. 15–38, 2013. With permission.)

process, approximately fivefold viscosity reduction is possible when 5%–10% naphtha is produced. It is a complex process and the reaction mechanism for the production of light hydrocarbons can be expressed through a typical free radical mechanism (Equations 3.1 through 3.10):

Initiation

$$C_6H_{14} \longrightarrow C_2H_5\bullet + C_4H_9\bullet \tag{3.1}$$

Propagation

$$C_2H_5\bullet + C_6H_{14} \longrightarrow C_2H_6 + C_6H_{13}\bullet \tag{3.2}$$

$$C_4H_9\bullet + C_6H_{14} \longrightarrow C_4H_{10} + C_6H_{13}\bullet \tag{3.3}$$

Radical $C_4H_9\bullet$ can be cracked to different species as per the following equations, as big radicals are unstable and decomposed to form olefins and smaller radicals.

$$C_4H_9\bullet \longrightarrow C_4H_8 + H\bullet \tag{3.4}$$

$$C_4H_9\bullet \longrightarrow C_3H_6 + CH_3\bullet \tag{3.5}$$

$$C_4H_9\bullet \longrightarrow C_2H_4 + C_2H_5\bullet \tag{3.6}$$

Radical $C_6H_{13}\bullet$ may further be cracked as per the following equation:

$$C_6H_{13}\bullet \longrightarrow C_4H_8 + C_2H_5\bullet + \text{many other products} \tag{3.7}$$

Termination

The free radical chain reaction is terminated when the radicals combine:

$$C_2H_5\bullet + H\bullet \longrightarrow C_2H_6 \tag{3.8}$$

$$C_2H_5\bullet + CH_3\bullet \longrightarrow C_3H_8 \tag{3.9}$$

$$CH_3\bullet + H\bullet \longrightarrow CH_4 \tag{3.10}$$

Visbreaking can produce a visbroken product with lower viscosity and pour point. Depending upon the reactor design, this process can be classified into two categories: soaker visbreaking and coil visbreaking.

The soaker visbreaking process as shown in Figure 3.5 is a relatively low-temperature process with a relatively high-residence time (100–300 s for coil visbreaker and 1200–1800 s for soaker visbreaker). In this case, a two-stage conversion, a minor degree of conversion for a short period in the furnace and a major degree of conversion for a predetermined extended period in the soaker unit, takes place. The low-temperature operation saves fuel consumption; however, decoking in the soaker unit requires more equipment and handling cost. It also generates more wastewater as high-pressure water is used for decoking in the soaker unit. In the coil visbreaking unit, the furnace itself contains a dedicated soaker in which most cracking takes place. The cracked products from the furnace outlet are quenched by preheating the furnace feed or heating cold gas oil, which helps to cease cracking reactions. The quenched products yield LPG, gasoline, gas oil, and tar in the fractionator. The speed of the feedstock in the tubes of the furnace controls the extent of cracking reaction. Decoking is frequent but easy for coil-type visbreakers. In terms of the product yield, there is little difference between the two reactor options (soaker visbreaker and coil visbreaker). Fouling in the soaker unit is slower than in the tubes of the furnace, but it increases the downtime of the whole unit. Recently, improvement in heater design has been able to eliminate the downtime of coil visbreaking (Speight 2012). Similarly, the application of internals in the soaker visbreaker is an important development, which controls the back-mixing in the soaker unit. Back-mixing can enhance overcracking, which reduces the stability of fuel oil and can be avoided if radial gas holdup profile is flat. The application of soaker internals satisfies this condition and controls the stability of products. It also increases conversion, reduces heat duty, and enhances run length by preventing undesirable side reactions leading to coke formation. Table 3.3 shows the comparison of conversion and viscosity of the cracked products obtained from the soaker with and without internals using the VR from Arab Mix Short Residue–Hindustan Petroleum Corporation Limited, Mumbai, India (Kumar et al. 2004).

In the visbreaking process, feedstock composition (properties) as well as the operating parameters influence the product yields and specifications. Decomposition of

TABLE 3.3

Comparison of Performances of Internals in the Soaker Unit

	Temperature (°C)	Conversion at 150°C (wt.%)	Kinematic Viscosity (cSt) at 100°C (150°C + Residue)	Stability
Soaker without internals	420	2.90	248.43	Stable
Soaker with internals	420	4.12	230.67	Stable

Source: Kumar, M.M. et al., Residence time distribution studies for development of advanced soaker technology, *XIIth Refinery Technology Meet*, Goa, India, 2004.

different types of compounds present in the feedstock varies as follows: *n*-paraffins > *i*-paraffins > cycloparaffins > aromatics > aromatic–naphthanes > polynuclear aromatics. Heavy hydrocarbon oils and resins can easily crack thermally, but it is difficult to crack asphaltenes. First, they are partially cracked into lighter products, and then carboids and coke are formed through polymerization, condensation, dehydrogenation, and dealkyaltion reactions. The carboids and coke as well as asphaltenes remain in stable colloidal suspension unless visbreaking reactions proceed too far. After certain limit of visbreaking, they tend to precipitate from the oil and form deposits in the cracking furnace and produce unstable fuel oil (Kuo 1985).

An increase in temperature at constant pressure and conversion increases the amount of lighter material in the products and decreases the yield of heavy oil and coke (Safiri et al. 2015). The properties of cracked gasoline also change with temperature as unsaturates increase gradually with an increase in temperature.

With increase in cracking time, the gasoline fraction gradually increases, reaches a maximum value, and decreases thereafter because of secondary reactions converting some gasoline into smaller hydrocarbon as well as polymerization of gasoline to some higher molecules. Extended cracking time produces more saturates in products. An increase in pressure enhances the polymerization of light unsaturates in the vapor phase and yields more liquid products.

Although visbreaking reduces the viscosity, it reaches a limiting value with an increase in conversion, and hence, addition of diluent is required if the cracked product is required to be transported through pipelines. Further, the products of visbreaking are not stable due to the presence of unsaturated molecules. In spite of this disadvantage, this method is widely used for the upgradation of heavy residues around the world.

3.2.2 Delayed Coking

In the delayed coking process, the coke formation is obtained by providing a sufficient long residence time (delayed) in the coke drum. More lighters are obtained and coke is obtained as a by-product. Delayed coking consists of thermal cracking of heavy residue in the empty drum in which deposition of coke takes place. The product yield and quality depends on the type of feedstock processed. A typical delayed coking consists of a furnace to preheat the feed, a coking drum in which cracking and coke formation take place, and a downstream fractionator of the coke drum in which the fractionation of the vapor-phase products take place. A process flow diagram for delayed coking is shown in Figure 3.6. The feed is first preheated in the furnace in which the desired cooking temperature (around 490°C–510°C) is achieved and fed to the coking drums, which are maintained at a prefixed temperature and are normally installed in pairs where the cracking reaction takes place and the coke is deposited at the bottom of the drum. The pressure of the coke drum gradually builds up, and after attaining a required pressure (~3 kg/cm^2), the coke drum overhead vapor is allowed to flow into the fractionating column in which it is separated into overhead streams containing wet gas, LPG, and naphtha as well as two-side gas oil streams. Recycled stream from the fractionating column combines with the fresh feed in the bottom of the column and is further preheated in the coke heater and flows into the coke drum. When the coke drum is filled, the heated streams from the coke heater are sent to the other drum.

The reaction involved in the delayed coking is partial vaporization and partial cracking. It can process VR, visbreaker residue, and FCC residual, and produces gases, naphtha, fuel oil, gas oil, and coke. The product distribution depends mainly on the CCR value of the

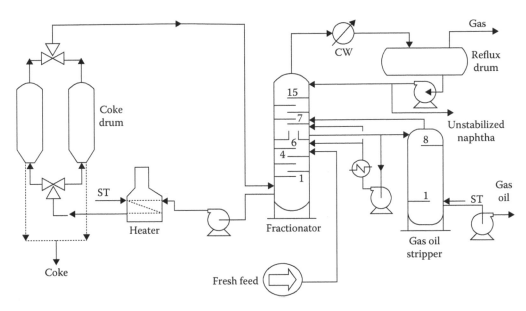

FIGURE 3.6
Process flowchart of delayed coking. ST, steam; CW, cooling water. (From Sawarkar, A.N. et al., *Can. J. Chem. Eng.*, 85, 1–24, 2007. With permission.)

feed. Some empirical relationships are available to predict the yield of various products as expressed in the following equations (Kumar 2009):

$$\text{Coke yield (wt.\%)} = 1.6 \times (\text{wt.\% feed CCR}) \tag{3.11}$$

$$\text{Gas } (C_4)(\text{wt.\%}) = 7.8 + 0.144 \times (\text{wt.\% feed CCR}) \tag{3.12}$$

$$\text{Gasoline (wt.\%)} = 11.29 + 0.343 \times (\text{wt.\% feed CCR}) \tag{3.13}$$

$$\text{Gas oil (wt.\%)} = 100 - \text{wt.\% coke} - \text{wt.\% gas} - \text{wt.\% gasoline} \tag{3.14}$$

Other feedstock properties that influence the product distribution are characterization factor, sulfur content, and metal content. The characterization factor is defined as ([Mean average boiling point in °R]$^{1/3}$/Specific gravity 60°F/60°F). Lower characterization factor gives higher coke yield (McKetta 1992). Operating parameters such as temperature, pressure, and recycle ratio also influence the product distribution as well as the quality of coke. Higher temperature increases the gas yield and reduces the coke yield. However, hardness of the coke increases with an increase in temperature. Higher pressure and recycle ratio increase gas as well as coke formation. Lower recycle ratio is desirable for more liquid products.

In the delayed coking process, three types of reactions such as dehydrogenation, rearrangement, and polymerization are involved simultaneously. Through the dehydrogenation step, an aromatic hydrocarbon produces one hydrogen radical and one aromatic free radical. The rearrangement step produces more stable aromatic ring system as a building block for graphite. The polymerization step of aromatic ring systems in the liquid phase produces coke.

In this process, coke yield increases with an increase in the coke drum pressure (keeping the reactor outlet temperature same). This is attributed to the fact that more condensation and polymerization reactions take place at higher pressure. Further, with an increase in the reactor outlet temperature as well as the coke drum temperature, coke yield decreases, which indicates that more cracking takes place at higher temperature. Different types of cokes such as sponge coke, needle coke, and shot coke are formed depending upon the feedstock quality as well as the operating parameters, which are discussed as follows:

Sponge coke: It is porous and has a sponge like appearance. It is formed when the asphaltene content in the VR is low to moderate. It is basically used as a fuel; few cokes with low metal and sulfur content (<2%) can be used as the anode material.

Needle coke: It is a high-value product and is obtained through the delayed coking process. Currently, in India, the delayed coking units are producing only sponge/fuel coke, and all the indigenous requirements of needle coke are being met through imports. Premium-grade needle coke production technology is a closely guarded secret and not easily available from licensors. Needle coke, which is an essential precursor for the ultrahigh-power graphite electrodes in electric arc furnaces, is produced by delayed coking. Characteristics such as low coefficient of thermal expansion (CTE), high density, high electrical conductivity, and low puffing are essentially required for a quality needle coke. Such qualifying characteristics have been recognized to be strongly influenced by the nature of feedstock and operating/carbonization conditions. Operating parameters also play an important role in establishing the crystalline structure of needle coke. It has a needlelike appearance and can be produced from the feedstock containing more aromatics. Decant oil from the FCC unit after hydrodesulfurization can be a suitable feedstock for needle coke. It can be used as a high-quality electrode due to its very low electrical resistance and CTE.

Shot coke: High asphaltene containing feedstocks at high coke drum temperature produces shot coke. It is not a desirable product and is used as a fuel with sponge coke.

3.2.3 Fluid Catalytic Cracking

It is an important conversion process when more gasoline production is required from heavier petroleum fraction such as VGO, deasphalted oil (DAO), and coker gas oil. It can produce higher quality liquid and gaseous products than thermal cracking. In this process, the feedstocks crack with the help of a fluidized catalyst powder in the reactor. The used (spent) catalyst is activated for reuse in the regenerator. The products are fractionated into different parts. Two types of design such as side by side and stack type are available in the FCC unit. In the first type, the reactor and the catalyst regenerator are two separate vessels, whereas in the second type, these are continued in a single vessel. The reactor and regenerator are considered to be the heart of the FCC unit.

Figure 3.7 shows the process flowsheet of a side-by-side configuration FCC unit. This unit operates in a closed circuit, and pressure balance plays an important role in the operation of this unit. Normally, the preheated feedstock injected into the catalyst riser (~540°C) is vaporized and cracked into smaller molecules by the catalytic action of hot catalyst powder in the riser within 2–4 s. The hydrocarbon vapors help to fluidize the catalyst powder, which is also mixed with the hydrocarbon vapors and enters the reactor at approximately 535°C and ~1.72 bar pressure. The reactor contains a two-stage cyclone that

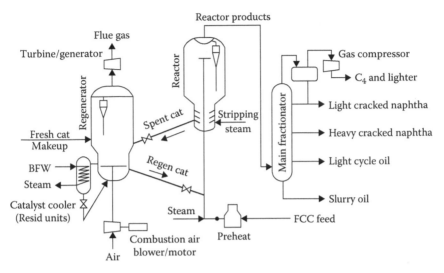

FIGURE 3.7
FCC unit. (From Talmadge, M.S. et al., *Green Chem.*, 16, 407–453, 2014. With permission.)

helps to separate the catalyst powder from the products and directs the deactivated (spent) catalysts to the regenerator through the steam stripper to remove hydrocarbon molecules attached with the catalysts. Deactivation of catalyst takes place due to coke deposition, and in the regenerator, the spent catalyst gets free from the deposited coke by oxidation at high temperature (715°C) and pressure (2.14 bar). The heat generated by exothermic combustion in the regenerator is partially carried over by the regenerated catalysts, which enter into the riser and mix with the feedstock as well as provide heat for endothermic cracking reaction. The hot vapor coming out from the reactor is further fractionated into different product streams. Use of riser internals and modification of the feed injection system have improved the performance of the FCC process. The riser internals promote ideal plug flow conditions in the riser by reducing the density and velocity variation inside the riser. Improvement in the riser termination technology, with the help of the two-stage cyclone system, results in controlled cracking and desired product yield. Development of new nozzles for injecting the FCC feedstock into the riser has been able to maintain proper mixing of it with catalyst and as a consequence the temperature profile in riser (Chen 2004).

The FCC process has been modified to produce gasoline- as well as olefin-rich products from heavy residues. The Universal Oil Products (UOP)'s Resid FCC (RFCC) process produces gasoline and lighter components from the VR using a two-stage stacked regenerator with their proprietary catalysts. The I-FCC^{SM} process, which combines Lummus Technology's FCC process with the proprietary INDMAX catalyst of Indian Oil Corporation Limited, Tamil Nadu, India, can selectively produce propylene and other light olefins. It also provides feedstock conversion (up to 45%). The operating conditions for this process are as follows: riser reactor 530°C–600°C and catalyst-to-oil ratio 12–20. The partial pressure of hydrocarbon is also lower than that of the conventional FCC process (Soni 2009).

3.2.4 Solvent Deasphalting Process

When asphaltene and metal content are high in the heavy residue, the solvent deasphalting (SDA) process becomes attractive for its upgradation, as these materials adversely

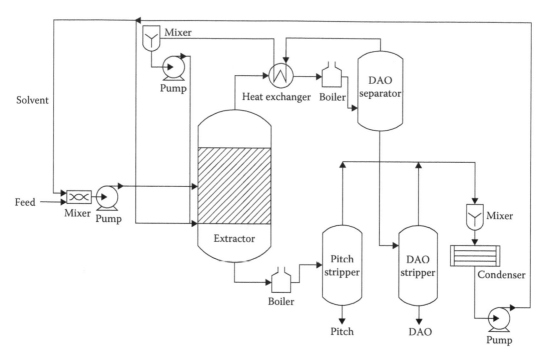

FIGURE 3.8
Solvent deasphalting process. (From Lee, J.M. et al., *Fuel Process. Technol.*, 19, 204–210, 2014. With permission.)

affect the performance of catalysts for hydrodesulfurization, HDC, or any catalytic conversion of the heavy residues. In the SDA process, asphaltenes from the heavy residue are
removed through liquid–liquid extraction procedure, in which a paraffinic solvent (C_3–C_7)
is used as an extracting solvent. The paraffinic components in the heavy residue are separated from the asphaltenes and come out as extracts with the solvent and form DAO. The
asphaltenes settle down in the extractor. In VR, about 80% asphaltene is present, which can
be converted into diesel after separation of asphaltenes and processing of DAO. Figure 3.8
shows a typical process flowsheet for an SDA plant. From this figure, it is evident that both
asphaltenes and DAO pass through the solvent recovery step; the recovered solvent is recycled and reused to mix the heavy residue. To recover the solvent from DAO, it is heated and
flashed, whereas the solvent entrained in asphaltenes is removed by heating it above the
minimum asphalt pumping temperature to ensure the flow ability of asphaltenes after solvent recovery. Solvent recovery above the critical conditions of the solvent can improve the
energy efficiency and reduce utility consumption and capital investment by introducing
a compact unit. The asphaltene is a feedstock for road making and gasification, whereas
the DAO after solvent recovery is hydrotreated to remove sulfur and acids as well as to
increase the yield in downstream cracking. DAO produced through propane deasphalting
can also be used to produce lube base stock.

The quantity and quality of DAO depends on the solvent used, solvent-to-feed ratio,
pressure, temperature, extractor type, and so on. Higher molecular weight solvent allows
heavier and resinous molecules to enter into DAO, which results in higher DAO yield
but reduces its quality as these heavier molecules contain impurities. Thus, to achieve a
proper DAO quality, solvent selection is important. Increasing the solvent-to-feed ratio,
the asphaltene precipitation can be increased up to 10 times; however, a further increase in

TABLE 3.4

Operating Conditions for SDA Using Different Solvents

Conditions Solvents Used in SDA Process	Temperature (°C)	Pressure (bar)	Solvent Ratio (v/v)
Propane	50–80	35–40	6–9
Butane	100–130	40	4–7
Pentane	170–210	40	3–5

Source: Siauw, H.N., *Energy Fuels*, 11, 1127–1136, 1997; Konde, M., Solvent Deasphalting of Short Residue: Indigenous Technology and further Development, *Symposium on Solvent Extraction Revisited*, New Delhi, India, 2010, http://slidegur.com/doc/1267180/solvent-deasphalting-of-short---vacuum.

this ratio does not affect asphaltene settling much (Gonzalez et al. 2004). An increase in the extractor temperature reduces the density of the solvent; as a result, DAO yield reduces. An increase in the pressure increases the density of the solvent; consequently, DAO yield increases. In refinery, a well-balanced approach is applied to select these parameters for optimum DAO production, and normally variation of pressure is not well accounted due to mechanical and process restrictions (Sattarin et al. 2006). Typical operating conditions for the SDA process using propane, butane, and pentane are provided in Table 3.4.

Kellogg Brown and Root (KBR), UOP, and Foster Wheeler are the major technology licensors for SDA. The French Institute of Petroleum, France; the Research Institute of Petroleum Processing, China; and Engineers India Limited with the Indian Institute of Petroleum, Dehradun, India have also transferred this technology to some industries. Deasphalting tower may contain baffle trays/louvre trays, rotating disc contractor, packing (random/structured), or parallel interceptor plate assemblies. The residuum oil supercritical extraction (ROSE) unit contains a proprietary tray-type Lubemax internal and produces a superior quality DAO suitable for the production of lube base stock.

To improve the economy of the SDA process, a gasification unit can be integrated with the SDA unit. In such case, value addition takes place through the production of syngas from asphaltenes in the gasifier unit. The produced syngas can be used as the source of sensible heat as well as many chemicals (Ogawa 1996).

3.2.5 HDC Technology

In this process, cracking of heavy petroleum fractions, including residues, takes place in the presence of hydrogen and a catalyst (Rogers and Stormont 1968), and produces lesser coke than thermal cracking. Different types of reactors such as fixed bed, moving bed, ebullated bed, and slurry-phase reactors are used (Cherzer and Gruia 1996). In the fixed bed reactor, normally multiple beds of catalysts are present, whereas in the moving bed reactor, expansion of the catalyst bed takes place, which gives lesser pressure drop (Ancheyta and Speight 2007). High API feedstocks, including the middle distillate, are hydrotreated in the fixed bed reactor, whereas for complex feedstocks, the moving bed or ebullated bed reactors are the preferred option. A combination of both fixed bed and moving bed/ebullated bed reactors in series may be used if the feedstock quality is very low (Ramirez et al. 2007). Although a mechanism is similar for all the reactor types, tolerance of impurities differs among them. Experimental conditions and catalyst properties also influence the product selectivity in these reactors. Nanostructured catalysts have been developed for the HDC of heavy residues (API gravity 2.3°) in recent years, which may improve the efficiency of

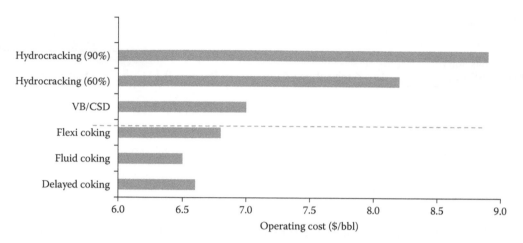

FIGURE 3.9
Operating cost for various processes. (From Gillis, D. et al., Upgrading Residues to Maximize Distillate Yields, UOP LLC, A Honeywell Company, Des Plaines, IL, 2009; Morel, F. and J.P. Peries, Use of solvent deasphalting process in residues conversion schemes, in *15th World Petroleum Congress*. World Petroleum Congress, 17, p. 7, 1997; Shen, H. et al., Thermal Conversion-An Efficient Way for Heavy Residue Processing, in *Proceedings of the 15th World Petroleum Congress*, Beijing, China, pp. 907–913, 1998. With permission.)

the process (Hur et al. 2014). Comparison of different technologies for the upgradation of heavy petroleum fractions is provided in Figure 3.9.

From this figure, it is evident that the delayed coking is a process with low capital cost and relatively high conversion. It produces petcoke that can further be used through different routes. However, the economic feasibility of this process may be tightened in future due to increased cost associated with strict environmental norms.

3.2.6 Gasification

The low-cost (sometimes negative cost due to environmental regulation) carbon-rich products such as petcoke and asphaltenes, generated through carbon rejection routes of residue upgradation, can be utilized through gasification to convert them into syngas, which can further be upgraded to various valuable products such as liquid fuel, hydrogen, and chemicals. Other heavy petroleum fractions can also be utilized through this route. In gasification, the carbonaceous feedstocks are heated in a controlled amount of oxygen/air to produce the syngas. A controlled amount of steam is also added to feedstocks such as petcoke where the moisture content of the feedstock is less. The major reactions involved in the gasifier are presented as follows:

$$C + \tfrac{1}{2} O_2 \longrightarrow CO \rightarrow \Delta H^0{}_{298} = -110.5 \text{ kJ/mol} \tag{3.15}$$

$$C + O_2 \longrightarrow CO_2 \rightarrow \Delta H^0{}_{298} = -393 \text{ kJ/mol} \tag{3.16}$$

$$C + CO_2 \longrightarrow 2CO \rightarrow \Delta H^0{}_{298} = 172 \text{ kJ/mol} \tag{3.17}$$

$$C + H_2O \longrightarrow CO + H_2 \rightarrow \Delta H^0{}_{298} = 131.4 \text{ kJ/mol} \tag{3.18}$$

The important minor reactions involved in the gasification are expressed as follows (Mondal et al. 2011):

$$CO + 3H_2 \longrightarrow CH_4 + H_2O \to \Delta H^0{}_{298} = -205 \text{ kJ/mol} \tag{3.19}$$

$$CO + H_2 \longrightarrow CO_2 + H_2 \to \Delta H^0{}_{298} = -40.9 \text{ kJ/mol} \tag{3.20}$$

$$C + 2H_2 \longrightarrow CH_4 \to \Delta H^0{}_{298} = -74.8 \text{ kJ/mol} \tag{3.21}$$

The gasification with CO_2 is also called as the Boudouard reaction, and Equation 3.17 is called as the water shift reaction. In the above equations, only carbon has been considered. However, in carbonaceous feedstocks, many other elements are also available in terms of many complex compounds. Thus, efforts are made to generalize the reactions involved in gasification. It is assumed that when the feedstock comes in contact with high heat in the gasifier, it produces volatilized hydrocarbons and char (Smoot and Smith 1985). The hydrocarbon vapor reacts with oxygen to give syngas as follows (Shoko et al. 2006):

$$C_nH_m + \frac{1}{2}nO_2 \longrightarrow \frac{1}{2}mH_2 + nCO \tag{3.22}$$

However, the gasification of char takes place as per the overall reaction (Smoot and Smith 1985):

$$CH_xO_y(\text{char}) + (1-y)H_2O \to (x/2 + 1 - y)H_2 + CO \tag{3.23}$$

A further generalization approach represents gasification reactions through Equations 3.24 through 3.27. It is assumed that char is composed of C, H, O, N, S, and metals.

$$CH_hO_{o+x}N_nS_sZ + aO_2 \to bCO_2 + cCO_2 + dH_2O + eH_2S + fN_2 + ZO_x \tag{3.24}$$

$$CH_hO_{o+x}N_nS_sZ + CO_2 \to 2CO + \frac{o}{2}H_2O + \left(\frac{h}{2} - s - o\right)H_2 + sH_2S + \frac{n}{2}N_2 + ZO_x \tag{3.25}$$

$$CH_hO_{o+x}N_nS_sZ + (1-o)H_2O \to CO + \left(1 - o + \frac{h}{2} - s\right)H_2 + sH_2S + \frac{n}{2}N_2 + ZO_x \tag{3.26}$$

$$CH_hO_{o+x}N_nS_sZ + \left(2 + o + s - \frac{h}{2}\right)H_2 \to CH_4 + oH_2O + sH_2S + \frac{n}{2}N_2 + ZO_x \tag{3.27}$$

The following equation is also considered to justify the presence of COS in syngas:

$$H_2S + CO_2 \to COS + H_2O \tag{3.28}$$

The properties of syngas produced through the gasification of a carbonaceous feedstock depend on the operating conditions and the type of gasifier used. A typical composition of syngas produced from petcoke and petroleum residues is as follows: 25%–30% H_2 (v/v), 30%–60% CO (v/v), 5%–15% CO_2 (v/v), and 2%–3% H_2O (v/v) (Gupta 2005; Gills 2006). A small amount of CH_4, H_2S, N_2, NH_3, HCN, Ar, COS, Ni, and Fe carbonyls may also be present (Gills 2006).

TABLE 3.5

Salient Features of Various Types of Gasifiers

Gasifier (1)	Technology (2)	Typical Process Conditions (3)	Remarks (4)
Fixed bed	BGL, Lurgi, Dry ash	Combustion temp.: 1300°C (slurry feed) and 1500°C–1800°C (dry feed) Gas outlet temp.: 400°C–500°C Pressure: 1.5–24.5 bar R.T.: 15–30 min Feed particle size: 2–50 mm O_2/feed: 0.64 Nm³/kg Gas heat value (MJ/Nm³): 10.04 Cold gas efficiency: High	• Can carbonize coal containing ~35% ash • Cannot use liquid fuels • Steam requirement is high • Syngas contains tar and phenolic compounds • More loss of fine particles of feed produced during feed preparation
Fluidized bed	HTW, IDGCC, KRW, Mitsui Babcock	Combustion temp.: 900°C–1200°C Gas outlet temp.: 700°C–900°C Pressure: 1–29.4 bar R.T.: 10–100 s Particle size: 0.5–5.0 mm O_2/feed: 0.37 Nm³/kg Gas heat value (MJ/Nm³): 10.71 Cold gas efficiency: Medium	• *In situ* sulfur capture when S content <2 wt.% • Preferred for high-reactive feedstocks such as waste fuels, biomass, and low-rank coal • Char particles need to be recycled for high conversion • Moderate steam requirement • Syngas does not contain tar or phenolic compounds • Reduced loss of fines
Entrained bed	BBP, Hitachi, MHI, PRENFLO, SCGP, E-Gas, Texaco	Combustion temp.: 1500°C Gas outlet temp.: 900°C–1400°C Pressure: 29.4–34.3 bar R.T.: 1–10 s Particle size: <200 mesh (90%) O_2/feed: 1.17 Nm³/kg Gas heat value (MJ/Nm³): 9.58 Cold gas efficiency: Medium	• Syngas does not contain any tar or phenolic compounds • Preferred for low-reactivity fuels such as petcoke • Suitable for cogasification of petcoke and high ash–coal mixture having ~22% ash • Steam requirement is moderate • *In situ* sulfur removal and no fine loss
Transport reactor	KBR	Combustion temp.: 900°C–1050°C Gas outlet temp.: 590°C–980°C Pressure: 2.9–14.7 bar R.T.: 1–10 s Particle size: <50 μm O_2/feed: 1.06 Nm³/kg Gas heat value (MJ/Nm³): – Cold gas efficiency: Medium	• Prevents the exposure of raw coal to the oxidant preventing combustion of the volatile matter released • Only char combustion and improved process efficiency • High throughput and conversion • Not well proven

Source: Mondal, P. et al., *Fuel Process. Technol.*, 92, 1395–1410, 2011. With permission from Elsevier.
R.T., residence time, temp., temperature.

Different types of gasifiers such as fixed bed, fluidized bed, entrained bed, and transport reactor are available, which are operated under different conditions with different particle sizes of feedstocks and show different efficiencies as well as product qualities. The salient features of these gasifiers are summarized in Table 3.5.

Among these gasifiers, the fixed bed gasifier is suitable for handling highly active feedstocks such as coal and biomass. In this gasifier, the feedstock enters from the top and the gasifying medium is sent from the bottom; hence, the feedstock comes in contact with hot gas with increasing temperature when the feedstock proceeds through different zones

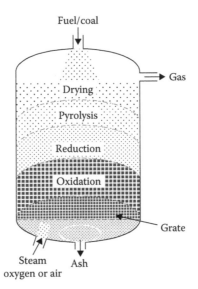

FIGURE 3.10
Schematic diagram of fixed bed gasifier and various reaction zones inside the gasifier. (From Mondal, P. et al., *Fuel Process. Technol.*, 92, 1395–1410, 2011. With permission.)

(from the drying zone to the combustion zone) of the gasifier as shown in Figure 3.10. Although a maximum temperature (~1800°C) is available in the combustion zone of this gasifier, the temperature along the height of the gasifier decreases with an increase in height, and at the exit of gas, the temperature reduces to 600°C. For this reason, the product gas contains more tar and phenolic compounds.

In case of the fluidized bed gasifier, both the feedstock and the gasifying medium are injected at the bottom of the gasifier (Warnecke 2000). A uniform temperature along the length of the gasifier is maintained due to high level of back-mixing. The temperature in the gasifier remains below the ash fusion temperature (900°C–1050°C); as a result, no ash melting and clinker formation take place. It is a more suitable gasifier type for coal gasification.

In case of the entrained bed reactor, the feedstock and gasifying medium enter into the gasifier from the top, and a high temperature (1500°C) is maintained in the gasifier throughout its length; thus, it produces good-quality syngas with negligible tar and phenolic. This type of gasifier is more suitable to gasify less reactive feedstocks such as petcoke and heavy petroleum fractions. Entrained bed gasifier produces slag due to very high temperature inside it, which shows an adverse impact on the burner and refractory life. To reduce the ash melting temperature of feedstocks, fluxes such as limestone can be added, which can also fix some amount of sulfur present in the feedstocks. Entrained bed gasifiers may handle feedstocks in the dry phase or slurry phase. Shell and Prenflo gasifiers are the best-known dry feeding gasifiers, whereas Texaco and Destec gasifiers are the best-known slurry feeding entrained bed gasifiers. The salient features of these gasifiers are summarized in Table 3.6.

Syngas produced from the gasification of the heavy petroleum fractions can be used in various downstream processing. In recent years, to make the process more economic, the concepts of integrated gasification combined cycle (IGCC) and polygeneration have been utilized. Integration of heat application with other syngas utilization routes such

TABLE 3.6

Comparison of Various Gasifiers

Parameters	Texaco	Shell	Destec
Feed system	Slurry	Dry	Slurry
Gasifier configuration	Single-stage downflow	Single-stage upflow	Two stage (horizontal and vertical)
Gasifier wall	Refractory	Membrane	Refractory
Pressure (bar)	41	30	27.6
Temperature (°C)	1250–1550	2000	1371–1537
Coal size	<0.1 mm	<0.1 mm	<100 µm

Sources: Zheng, L. and E. Furinsky, *Energy Convers. Manage.*, 46, 1767–1779, 2005; Breton, D.L., Improved Performance of the Destec Gasifier, *Gasification Technologies Conference*, Houston, TX, 1999. With permission from Elsevier.

as hydrogen production, Fischer–Tropsch synthesis, and chemical synthesis makes the process more economic. However, the requirements of purity of syngas for various downstream processing are different, and to meet these specie qualities, the conditioning and cleaning of syngas is essential. Conditioning is used to improve the H_2:CO ratio in syngas, which is carried out through water–gas shift reaction, whereas cleaning involves the separation of particulates and acid gases such as H_2S, COS, and CO_2. The shift reactions can be performed either before cleaning acid gases (sour shift) or after removing acid gases (sweet gas shift). A comparison of these two types of shift reactions is provided in Table 3.7.

For the removal of acid gases from the syngas, both wet and dry processes are used. The wet process is based on physical, chemical, and both physical and chemical absorption, whereas the dry process involves the adsorption of acid gas components in warm adsorbent bed. Some important wet processes for the removal of acid gases are Rectisol, Selexol, Sulfinol, and so on; metal oxides such as ZnO are used as an adsorbent in the dry process. The spent adsorbent is regenerated at high temperature. Among these processes, Rectisol is having the maximum removal capacity; however, it is the costliest process. Some important gasifier units running on the heavy petroleum fraction along with downstream application of syngas are summarized in Table 3.8.

TABLE 3.7

Comparison of Sweet and Sour Shift Reactions

Characteristics	Sweet Shift	Sour Shift
Typical flowsheet	Two High temperature shift (HTS) and one Low temperature shift (LTS) conversion stages with cooling between the reactors	Two to three conversion stages with heat exchangers and sometimes steam addition as required
CO conversion	Within two HTS steps, sweet gas shift can reduce CO conversion from 44.6% to 2.1%. The residual CO can be further converted into 0.5% in the LTS step	From 44.6% to 1.8% within two steps using a slightly higher amount of steam. To reach a CO level below 1%, large quantity of steam is added in the third reactor before shift reaction
Catalyst	Usually more expensive	Usually less expensive
Reactor	The size of catalyst bed is smaller	The size of catalyst bed is higher

Source: Mondal, P. et al., *Fuel Process. Technol.*, 92, 1395–1410, 2011.

TABLE 3.8
Major Electricity-Producing Gasification Plants around the World

Country (1)	Plant Name (2)	Technology (3)	Feedstock (4)	Products (5)	Syngas Cleaning Option	Year (6)
China	Dalan Chemical Industrial Corp.	Texaco	Visbreaker residue	Ammonia	Rectisol	1995
China	Inner Mongolia Fertilizer Co.	Shell	VR	Ammonia	Rectisol	1996
China	Juijang Petrochemical Co.	Shell	VR	Ammonia	Rectisol	1996
China	Lanzhou Chemical Industrial Co.	Shell	VR	Ammonia	Rectisol	1998
China	Fujian project	Shell	Deasphalted rock	Electricity and H_2		2006
Europe (unspecified)	Unspecified plant	Shell	Residue	Electricity		2005
France	Normandie IGCC plant	Texaco	Fuel oil	Electricity, steam, and H_2		2005
France	–	Texaco	Refinery residue	H_2	Selexol	2006
Germany	Hydro AgriBrunsbittel	Shell	Heavy VR	Ammonia	Rectisol	1978
Germany	Leuna methanol Anlage	Shell	Visbreaker residue	H_2, methanol, and electricity		1985
Germany	SARGmbH	Texaco	VR	H_2 and chemicals	Sulfinol	1988
Germany	Slurry/oil gasification	Lurgi MPG	Oil and slurry	Electricity and methanol		1968
Holland	Pernis refinery	Shell	Resid oil	Electricity		1997
India	Gujrat National Fertilizer Co.	Texaco	Refinery residue	Ammonia	Rectisol	1982
India	Bhatinda IGCC	Texaco	Petcoke	Electricity		2005
Italy	ISAB Energy Project	Texaco	ROSE asphalt/ heavy oil	Electricity, H_2, and steam	MDEA	2000
Italy	SARLUX GCC/H_2 plant	Texaco	Visbreaker residue	Electricity, H_2, and steam	Selexol	2001
Italy	ApiEnergis S.p.A	Texaco	Visbreaker residue	Electricity, H_2, and steam	Selexol	2001
Italy	Agip IGCC	Shell	Visbreaker residue	Electricity and H_2	Amine	2003
Italy	Sannazzaro GCC plant	Texaco	Visbreaker residue	Electricity		2005
Japan	Marifu IGCC plant	Texaco	Petcoke	Electricity		2004
Japan	Nippon Pet. Ref. CO.	Texaco	VR	Electricity	ADIP	2004
Japan	Yokohama Cogen/B	Texaco	VR	Electricity		2003
Netherlands	Pernis Shell gasifier. Hydrogen plant	Shell	Visbreaker residue	H_2 and Electricity	Rectisol	1997

(Continued)

TABLE 3.8 (Continued)

Major Electricity-Producing Gasification Plants around the World

Country (1)	Plant Name (2)	Technology (3)	Feedstock (4)	Products (5)	Syngas Cleaning Option	Year (6)
Poland	Gdansk IGCC plant	Texaco	Visbreaker residue	Electricity, H_2, and steam		2005
Portugal	QuimigalAducos	Shell	VR	Ammonia	Rectisol	2001
Singapore	Chawan IGCC plant	Texaco	Residual oil	Electricity H_2, and steam	Flexorb	2001
Singapore	Singapore Syngas	Texaco	Visbreaker tar	H_2 and CO (for acetic acid)		2000
Spain	Puertollano GCC plant	PRENFLO	Coal and petcoke	Electricity	MDEA	1997
Spain	Bilbao IGCC plant	Texaco	VR	Electricity and H_2		2005
Spain	PIEMSA	Texaco	Visbreaker tar	Electricity and H_2	MDEA	2006+
USA	Wabassh River Energy Ltd.	E-GAS (Destec/Dow)	Petcoke	Electricity	MDEA	1995
USA	Delaware clean energy Cogen. project	Texaco	Fluid petcoke	Electricity and steam	MDEA	2001
USA	Coffeyville Refinery	Texaco	Petcoke	H_2 and NH_5		2000
USA	Farmland Industries Lnc.	Texaco	Petcoke	Ammonia	Selexol	2000
USA	ExxonMobil Baytown Syngas project	Texaco	Deasphalter bottom	H_2 and CO	Rectisol	2001
USA	Valero Refinery	Texaco	Petcoke	Electricity, steam		2002
USA	Port Arthur GCC Project	E-GAS	Petcoke	Electricity		2005
USA	Lake Charles IGCC plant	Texaco	Petcoke	Electricity, H_2, and steam		2005
USA	Deer Park GCC plant	Texaco	Petcoke	Electricity, syngas, and steam		2006
USA	Polk Country gasification plant	Texaco	Petcoke	Electricity		2005

Source: Mondal, P. et al., *Fuel Process. Technol.*, 92, 1395–1410, 2011. With permission from Elsevier.

3.2.7 Biological Processing of Heavy Fractions

Biotechnological processes are being explored around the world to develop new routes for energy production due to their low cost and eco-friendly nature. Recently, it has been reported that some microbial strains can degrade various polynuclear aromatic hydrocarbons (PAHs) such as anthracite, mono-aromatic hydrocarbons such as toluene, or aliphatic hydrocarbons such as *n*-alkanes from petroleum-contaminated sites (Whyte et al. 1997; Kim et al. 2001; Östberg et al. 2007; Liu et al. 2010; Karimi et al. 2013).

Bacillus cereus has been used for the remediation of petroleum-polluted soils (Gupta and Gera 2015). Several integral parameters including the conditions for microbial degradation activity (e.g., presence of nutrients, oxygen, pH, and temperature); the quality, quantity, and bioavailability of the contaminants (e.g., particle size distribution); and the soil characteristics have been found to affect the rate of microbial degradation of hydrocarbons in soils (Gupta and Gera 2015). Therefore, the bacteria with high physicochemical endurance and degradation ability could be a proper choice not only in bioremediation but also in degradation of PAHs to smaller molecules. These PAHs are present in many heavy petroleum fractions. Thus, further development in this area may produce new biological routes for the upgradation of heavy residues in future. However, an extensive research and development is required in this field.

3.3 Concluding Remarks

The quality of petroleum crude is degrading day by day, which is producing heavier petroleum fractions and forcing the refiners to process the heavier fractions for value addition to the process. Different types of upgradation processes based on hydrogen addition and carbon rejection are used to utilize the heavier fraction. Various conventional methods such as visbreaking, delayed cracking, FCC, and HDC have their own advantages and disadvantages. The conventional FCC process uses VGO, DAO, and so on effectively; however, for handling the residue, the RFCC process can be used. Similarly, methods such as I-FCC[SM] have been developed to get more olefin-rich products from the heavy residue. Gasification followed by polygeneration with the IGCC concept can improve the economics of heavy residue utilization. Efforts are also made to develop new biotechnology and nanotechnology for the utilization of heavy petroleum fractions.

References

Altgelt, K.H., and M.M. Boduszynski. 1994. *Composition and Analysis of Heavy Petroleum Fractions*, Marcel Dekker, 23, 45.

Ancheyta, J., and J.G. Speight. 2007. *Reactors for Hydroprocessing of Heavy Oil and Residua*, Chapter 6, Eds., Taylor and Francis group, New York.

Breton, D.L. 1999. Improved Performance of the Destec Gasifier, *Gasification Technologies Conference*, Houston, TX.

Campbell, P.E., J.T. McMullan, and B.C. Williams. 2000. Concept for a competitive Coal Fired Integrated Gasification Combined Cycle Power Plant, *Fuel*, 79, 1031–1040.

Castaneda, L.C., J.A.D. Munoz, and J. Ancheyta. 2014. Current Situation of Emerging Technologies for Upgrading of Heavy Oils, *Catalysis Today*, 220–222, 248–273.

Chen, Y.M. 2004. Recent Advances in FCC Technology, Fluidization and Fluid Particle Systems, *Powder Technology*, 163(1–2), 2–8.

Cherzer, J., and A.J. Gruia. 1996. *Hydrocracking Science and Technology*, Marcel Dekker, New York.

Chrisman, E., V. Lima, and P. Menechini. 2012. Asphaltenes–Problems and Solutions in E&P of Brazilian Crude Oils, Chapter 1, in *Crude Oil Emulsions – Composition Stability And Characterization*, InTech, http://www.intechopen.com/books/crude-oil-emulsions-composition-stability-and-charac-terization/asphaltenes-problems-and-solutions-in-e-p-of-brazilian-crude-oils.

Chung, K.H., C. Xu, Y. Hu, and R. Wang. 1997. Supercritical Fluid Extraction Reveals Resid Properties, *Oil & Gas Journal*, 95, 66–69.

Furimsky, E. 1999. Gasification in Petroleum Refinery of 21st Century, *Oil & Gas Science and Technology*, 54(5), 597–618.

Gillis, D., M. VanWees, and P. Zimmerman. 2009. *Upgrading Residues to Maximize Distillate Yields*, UOP LLC, A Honeywell Company, Des Plaines, IL.

Gills, D.B. 2006. (UOP), Technology to Maximize Value of Synthetic Gas, *Proceedings: Conference on Clean fuels from Coal/Pet Coke*, CHT, New Delhi, India.

Gonzalez, E.B., C.L. Galeana, A.G. Villegas, and J. Wu. 2004. *AIChE Journal*, 50(10), 2552.

Gupta, R. 2005. Major Process Steps of Gasification Based Plants, *Proceedings: RTI-IIP Seminar on Technologies for Gasification of Carbonaceous Feedstocks & Syngas Utilization*, New Delhi, India.

Gupta, R.K., and P. Gera. 2015. Process for the Upgradation of Petroleum Residue: Review.

Hedrick, B.W., and K.D. Seibert. 2006. A New Approach to Heavy Oil and Bitumen Upgrading, UOP LLC. https://www3.kfupm.edu.sa/catsymp/symp16th/pdf%20papers/09%20opdorp.pdf.

Hur, Y.G., M.S. Kim, D.W. Lee, S. Kim, H.J. Eom, G. Jeong, M.H. No, N.S. Nho, and K.Y. Lee. 2014. Hydrocracking of Vacuum Residue into Lighter Fuel Oils using Nanosheet-Structured WS_2 catalyst, *Fuel*, 137, 237–244.

Jechura, J. 2016. Colarado School of Mines–Crude Oil Assay- Hibernia (from ExxonMobil Site). http://inside.mines.edu/~jjechura/Refining/02_Feedstocks_&_Products.pdf.

Karimi, A., F. Golbabaei, M. Neghab, P.M. Reza, A. Nikpey, K. Mohammad, and M.R. Mehrnia. 2013. Biodegradation of High Concentrations of Benzene Vapors in a Two Phase Partition Stirred Tank Bioreactor, *Iranian Journal of Environmental Health Science & Engineering*, 10, 10.

Kim, I.S., J. Park, and K.W. Kim. 2001. Enhanced Biodegradation of Polycyclic Aromatic Hydrocarbons using Nonionic Surfactants in Soil Slurry, *Applied Geochemistry*, 16, 1419–1428.

Konde, M. 2010. Solvent Deasphalting of Short Residue: Indigenous Technology and further Development, *Symposium on Solvent Extraction Revisited*, New Delhi, India. http://slidegur.com/doc/1267180/solvent-deasphalting-of-short---vacuum.

Kumar, B. 2009. Presentation on Delayed Coking Technology. Indian Oil Corporation Limited. petro-fed.winwinhosting.net/upload/25-28May10/BIRENDRA_KUMAR.pdf.

Kumar, M.M., D. Tandon, Yadram, R.P. Kulkarni, K.L. Kataria, A.B. Pandit, J.B. Joshi. 2004. Residence Time Distribution Studies for Development of Advanced Soaker Technology, *XII th Refinery Technology Meet*, Goa, India.

Kuo, C.J. 1985. Effect of Crude Types on Visbreaking Conversions, *Oil and Gas Journal*.

Lee, J.M., S. Shin, S. Ahn, J.H. Chun, K.B. Lee, S. Mun, S.G. Jeon, J.G. Na, and N.S. Nho. 2014. Separation of Solvent and Deasphalted Oil for Solvent Deasphalting Process, *Fuel Processing Technology*, 19, 204–210.

Liu, W., Y. Luo, Y. Teng, Z.Q. Li, and L. Ma. 2010. Bioremediation of Oily Sludge Contaminated Soil by stimulating Indigenous Microbes, *Environmental Geochemistry and Health*, 32(1), 23–29.

Manning, F.S., and R.E. Thompson. 1995. *Oilfield Processing of Petroleum: Crude Oil*, vol. 2, Pennwell Publishing Company, Tulsa, OK.

Marafi, M., A. Stanislaus, and E. Furimsky. 2010. *Handbook of Spent Hydroprocessing Catalysts*, 1st ed., Elsevier B.V., Amsterdam, the Netherlands, pp. 5–16.

McKetta Jr., J.J. 1992. *Petroleum Processing Handbook*, 1st ed., CRC Press, Boca Raton, FL.

Mondal, P., G.S. Dang, and M.O. Garg. 2011. Syngas Production through Gasification and Cleanup for Downstream Applications—Recent Developments, *Fuel Processing Technology*, 92(2011), 1395–1410.

Morel, F., and J.P. Peries. 1997. Use of Solvent Deasphalting Process in Residues Conversion Schemes, in *15th World Petroleum Congress*, vol. 17, World Petroleum Congress, p. 7.

Ogawa, A. 1996. Study on R.F.G. Application in Iranian Refinery, J.C.C.P. – N.I.O.C. Joint Seminar Program Reformulated Gasoline.

Östberg, T.L., A.P. Josson, and U.S. Lundström. 2007. Enhanced Degradation of N-Hexadecanein Diesel Fuel Contaminated Soil by the Addition of Fermented Whey, *Soil Sediment Contamination*, 16(2), 221–232.

Phillips, G., and F. Liu. 2002. Advances in Resid Upgradation Technologies Offer Refiners Cost-Effective Options for Zero Fuel Oil Production, *Proceedings of European Refining Technology Conference*, Paris, France, 2002.

Ramirez, J., M.S. Rana, J. Ancheyta, and J.G. Speight. 2007. Characteristics of Heavy Oil Hydroprocessing Catalysts, Chapter 6, Eds., Taylor and Francis, New York.

Rana, M.S., J. Ancheyta, S.K. Maity, and G. Marroquin. 2008. Comparison between Refinery Processes for Heavy Oil Upgrading: A Future Fuel Demand, *International Journal of Oil, Gas and Coal Technology*, 1(3), 250–282.

Rogers, L.C., and D.H. Stormont. 1968. Oil Scramblng to Unravel Sulphur-Curb Supply Knot, *Oil & Gas Journal*, 66(26), 41–44.

Safiri, A., J. Ivakpour, and F. Khorasheh. 2015. Effect of Operating Conditions and Additives on the Product Yield and Sulfur Content in Thermal Cracking of a Vacuum Residue from the Abadan Refinery, *Energy Fuels*, 29(8), 5452–5457.

Sarkar, G.N. 1998. *Advanced Petroleum Refining*, 1st ed., Khanna Publishers, New Delhi, India.

Sattarin, M., H. Modarresi, H. Talachi, and M. Teymori. 2006. Solvent Deasphalting of Vacuum Residue in a Bench-Scale Unit Integration of Solvent Deasphalting and Gasification, Texaco Inc.

Sawarkar, A.N., A.B. Pandit, S.D. Samant, and J.B. Joshi. 2007. Via Delayed Coking: A Review, *The Canadian Journal of Chemical Engineering*, 85, 1–24.

Scherzer, J., and A.J. Gruia. 1996. *Hydrocracking Science and Technology*, Marcel Dekker, New York, p. 3.

Schulman, B.L., and R.L. Dickenson. 1991. Upgrading Heavy Crudes: A Wide Range of Excellent Technologies Now Available. *5th UNITAR/UNDP International Conference on Heavy Crude and Tar Sands*, vol. 4, Caracas, Venezuela.

Shen, H., Z. Ding, and R. Li. 1998. Thermal Conversion-An Efficient Way for Heavy Residue Processing, in *Proceedings of the 15th World Petroleum Congress*, Beijing, China, pp. 907–913.

Shoko, E., B. McLellan, A.L. Dicks, and J.C.D. Costa. 2006. Hydrogen from Coal: Production and Utilization Technologies, *International Journal of Coal Geology*, 65, 213–222.

Siauw, H.N. 1997. Nonconventional Residuum Upgrading by Solvent Deasphalting and Fluid Catalytic Cracking, *Energy & Fuels*, 11, 1127–1136.

Smoot, L.D., and P.J. Smith. 1985. *Coal Combustion and Gasification*, Plenum Press, New York.

Soni, D., V.K. Satheesh, R. Rao, G. Saidulu, and D. Bhattacharyya. 2009. Catalytic Cracking Process enhances Production of Olefins, *Petroleum Technology Quarterly Magazine*, Q4, 95–100.

Speight, J.G. 1991. *Chemistry and Technology of Petroleum*, 2nd ed., Marcel Dekker, New York.

Speight, J.G. 2000. *The Desulfurization of Heavy Oils and Residua*, 2nd ed., Marcel Dekker, New York.

Speight, J.G. 2012. Visbreaking: A Technology of the Past and the Future, *Scientia Iranica*, 19(3), 569–573.

Speight, J.G. 2013. Chapter 2, Thermal cracking, in *Heavy and Extra-heavy Oil Upgrading Technologies*, pp. 15–38.

Tabatabaee, M., and M. Assadi. 2013. Vacuum Distillation Residue Upgrading by an Indigenous *Bacillus cereus*, *Journal of Environmental Health Sciences & Engineering*, 11, 18.

Talmadge, M.S., R.M. Baldwin, M.J. Biddy, R.L. McCormick, G.A. Ferguson, S. Czernik, K.A. Magrini-Bair, T.D. Foust, P.D. Metelski, C. Hetrick, and M.R. Nimlos. 2014. A perspective on oxygenated species in the refinery integration of pyrolysis oil, *Green Chemistry*, 16, 407–453.

U.S.A. Energy Information Administration. 2016. Modified from OSHA technical manual. https://
        www.eia.gov/tools/faqs/faq.cfm?id=23&t=10.
Warnecke, R. 2000. Gasification of Biomass: Comparison of Fixed Bed and Fluidized Bed Gasifier,
        Biomass Bioenergy, 18, 489–497.
Whyte, I.G., I. Bourbonniere, and C.W. Greer. 1997. Biodegradation of Petroleum Hydrocarbons by
        Psychrotrophicpseudomonas Strains possessing both Alkane (alk) and Naphthalene (nah)
        Catabolic Pathways, *Applied and Environmental Microbiology*, 63(9), 3719–3723.
World Energy Outlook. 2008. International Energy Agency. http://www.worldenergyoutlook.org/
        media/weowebsite/2008-1994/weo2008.pdf Accessed May 12.
Zheng, L., and E. Furinsky. 2005. Comparison of Shell, Texaco, BGL and KRW Gasifiers as part of
        IGCC Plant Computer Simulations, *Energy Conversion and Management*, 46, 1767–1779.

4

Hydrogen from Natural Gas

Mumtaj Shah, Prasenjit Mondal, Ameeya Kumar Nayak, and Ankur Bordoloi

CONTENTS

ABSTRACT Hydrogen is considered as a promising energy carrier, which can replace fossil fuels in most of their applications. Currently, a large number of technologies are available to produce hydrogen using different types of feedstocks, including water, biomass, and hydrocarbons. Steam reforming of natural gas is a well-established and mature technology to produce hydrogen at large scale. With immense development of natural gas resources, it has become the most economical option to generate industrial hydrogen. However, steam reforming of natural gas is an energy extensive process and produces a large amount of air pollutants. As a result, much research efforts are being made to develop the environmentally benign and energy-efficient hydrogen production technologies for large-scale operation. Some novel approaches to steam reforming such as membrane, plasma, and microchannel reforming have been proposed. Other reforming technologies such as partial oxidation, autothermal reforming, dry reforming, bi-reforming, and tri-reforming can also be used to generate hydrogen from natural gas, but their economic viability make them unfavorable for large-scale production. This chapter presents an overview of different natural gas reforming technologies for hydrogen production and their current state of art, R&D gaps, and scope of development.

4.1 Introduction

Global energy demand is growing day by day due to the rapid increase in population as well as the rising standards of living. The world annual energy consumption in the year 2014 was ~492 quadrillion Btu, and it is expected to reach a level of 700 quadrillion Btu by 2035 (British Petroleum Corporation [BP] statistics report 2015). Fossil fuels such as coal, petroleum, and natural gas meet more than 85% of global energy demand, in which the contribution of coal, petroleum, and natural gas is 18%, 34%, and 29%, respectively (AEO report 2015).

Natural gas is a clean, safe, and energy-efficient fossil fuel. It is considered as a vital fuel in the world energy supply. It is a mixture of hydrocarbon and non-hydrocarbon gases in varying amounts with methane contributing the lion's share. It is generally classified as dry, wet, and condensate depending on the proportion of hydrocarbons heavier than methane present in it and its phase change during production. It is considered as "dry" when it contains almost 99% of methane, "wet" when it forms the liquid phase during production, and "condensate" when it forms the liquid phase in the reservoir due to high proportion of higher hydrocarbons. It is also referred to as sweet and sour, depending on its sulfur and CO_2 content (Mokhatab et al. 2015). The composition of natural gas varies significantly from source to source. Table 4.1 shows the typical composition of natural gas. At the end of 2014, the total proven natural gas reserves around the globe are estimated to be 187.1 trillion cubic meters; only 30% is considered as a conventional resource (BP statistics report 2015).

For a long time, natural gas is known to be the most useful fuel for water heating and street lightning. With the advent of the industrial era, natural gas has a wide range of applications as both energy and non-energy usage. Figure 4.1 depicts the important applications of natural gas. Due to the environmental concern of other fossil fuels, the use of natural gas is growing rapidly. According to the BP statistics forecast, global natural gas demand is expected to reach a level of 129 billion Nm^3/day by 2035 with a rate increase of 1.9% per annum from 2014 onward. In petrochemical and ammonia industry, natural gas

TABLE 4.1

Composition of Natural Gas

Component	Typical Analysis (mole %)	Range (mole %)
Methane	87.76	70–95
Ethane	5.00	1.8–5.1
Propane	1.00	0.1–1.5
i-Butane	0.30	0.01–0.30
n-Butane	0.50	0.01–0.50
n-Pentane	0.20	Traces to 0.3
n-Hexane	0.30	Traces to 0.3
n-Heptane	0.26	Traces to 0.3
n-Octane	0.15	Traces to 0.2
n-Nonane	0.08	Traces to 0.1
Carbon dioxide	2.85	0.1–1
Nitrogen	1.4	1.3–5.6
Oxygen	0.02	0.01–0.1
Hydrogen	Trace	Traces to 0.02

Source: Demirbas, A., *Methane Gas Hydrate*, Berlin, Germany, Springer Science & Business Media, 2010.

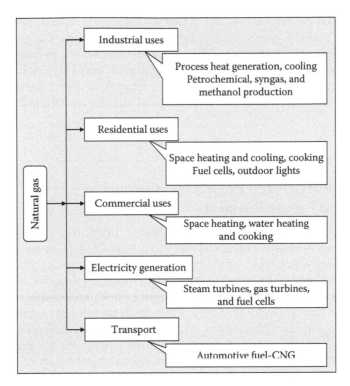

FIGURE 4.1
Applications of natural gas.

is basically used for generation of hydrogen, which is either used directly or converted to other chemicals. With the advent of the hydrogen economy, more attention is being paid toward the conversion of natural gas into more environmentally benign hydrogen fuel. The steam reforming of natural gas for syngas production and the gas-to-liquid (GTL) process for synthetic gasoline or diesel production have already been commercialized.

Global CO_2 emission from fossil fuel combustion and industrial processes is estimated to be around 35.7 billion tons (Bt) in 2014 (Olivier et al. 2015). In the late 2015, the United Nations Environment Programme (UNEP) has held a Climate Change Conference (COP 21) in Paris on the concern of the growing level of global warming. The member countries of the United Nations have agreed to set a goal of limiting the global temperature increase to less than 2°C compared to preindustrial levels and to achieve zero net anthropogenic greenhouse gas emissions during the second half of the twenty-first century (UNEP 2015). Therefore, the development of clean technology for fossil fuel utilization and the introduction of alternative greener fuels have become essential to reduce the adverse effects of greenhouse gas emissions and subsequent climatic changes. Among various alternative fuels, hydrogen is regarded as a sustainable energy carrier and offers near zero end-use emissions of pollutants and greenhouse gases (Ajayi-Oyakhire 2012).

Therefore, it seems that the utilization of natural gas through the hydrogen route may be more environment-friendly and suitable to meet the target. Its applicability has been technically proven and has been used commercially in fuel cells since 1965 (Pa 2008). Apart from natural gas, hydrogen can also be produced using a variety of feedstocks such as water, coal, biomass, wind, and solar energy through many different process routes. Currently, steam reforming of hydrocarbons is the widely used hydrogen production technology.

About 95% of global hydrogen demand is met by reforming of coal and different hydrocarbon feedstocks (natural gas, residual oils, high petroleum cuts, naphtha, etc.). However, the conventional technology of steam reforming of hydrocarbons generates a high amount of greenhouse gases (13.7 kg CO_2 equiv./1 kg H_2) (Spath and Dayton 2003). To reduce the carbon emission, steam reforming of natural gas with carbon capture and storage is seen as a viable technology, but economic and technological hurdles are still to be rectified.

4.2 Overview of Hydrogen Production: Current Scenario and Future Prospect

Hydrogen is widely distributed in a large number of substances in nature, such as freshwater and seawater, biomass, and fossil fuels. It is extracted from natural substances at the expense of energy. A variety of process technologies are available to extract hydrogen from these natural substances, including chemical, biological, electrolytic, photolytic, and thermochemical methods with an input of both renewable and nonrenewable energy resources (Dincer and Zamfirecu 2012). Each technology offers its own unique opportunities, benefits, and challenges, and each stands in a different stage of development. Technology selection, among various available options for economic hydrogen production, depends on numerous factors such as local availability of feedstock, market applications and demand, maturity of technology, costs, and policy issues (Muradov and Veziroğlu 2005). A comparison of different hydrogen production technologies is presented in Table 4.2. In the late 1920s, water electrolysis was introduced as the first commercial technology for pure hydrogen production. Later, with the introduction of low-cost fossil fuel-based reforming technology in the 1960s, the industrial production of hydrogen has started using natural gas reforming (IEA, Hydrogen report 2006).

In steam reforming of hydrocarbons, natural gas is the primary feedstock for industrial hydrogen generation, amounting to 49% among the other higher hydrocarbons. Water electrolysis is also used to produce pure hydrogen in food industry, fuel cell etc. However, due to high cost, hydrogen from the water electrolysis technology shares only 4% in total hydrogen production. The percentage share of feedstocks and technologies for commercial hydrogen production is depicted in Figure 4.2.

Fossil fuel-based hydrogen production routes are not considered as sustainable and eco-friendly because of the depleting nature of fossil fuel reserves and the high level of CO_2 emission associated with them (Midilli et al. 2005). To realize the hydrogen economy, green hydrogen production routes must be implemented in industrial practice.

Owing to the great potential of hydrogen as a fuel/energy carrier, the concept of "hydrogen economy" came to light in 1970. The main aim of the concept was to use hydrogen as an alternative fuel in a safe and inexpensive way, as well as to develop integrated hydrogen production, distribution, and storage facilities worldwide. Continuing research and development (R&D) effort around the globe was then initiated so that hydrogen can play an active role in the energy system. Hydrogen production is a growing industry. At the end of the year 2014, worldwide market of hydrogen was estimated to be $103.5 billion. It is expected to grow with a compound annual growth rate of 5.9% and will reach a level of $138.2 billion by 2019 (Hydrogen Generation Market Research Report 2015).

Total world production of hydrogen is estimated to be approximately 720 billion Nm^3 in 2013. This is equivalent to 3% of the total energy consumption, and it is anticipated

TABLE 4.2

Comparison of Important Hydrogen Production Technologies

Technology	Feedstock	Efficiency (%)	Maturity
Steam reforming	Hydrocarbons	70–85	Commercial
POX	Hydrocarbons	60–75	Commercial
ATR	Hydrocarbons	60–75	Near term
Plasma reforming	Hydrocarbons	9–85	Long term
Biomass gasification	Biomass	35–50	Commercial
MCD	Hydrocarbons	60–70	Long term
Coal gasification	Coal	60–70	Commercial
Biomass reforming	Biomass	35–50	Mid term
Plasma cracking	Hydrocarbons	70–80	Long term
Liquid phase reforming	Carbohydrates	35–55	Mid term
Thermolysis	Water+heat	50	Long term
Electrolysis	Water+electricity	50–70	Commercial
Photolysis	Water+sunlight	0.5	Long term
Thermochemical water splitting	Water+heat	40–50	Long term
Dark fermentation	Biomass	13	Long term

Source: Dincer, I. and C. Zamfirescu, *Int. J. Hydrogen Energ.*, 37, 16266–16286, 2012; Kalamaras, C.M., and A.M. Efstathiou, Hydrogen production technologies: Current state and future developments, in *Conference Papers in Science*, vol. 2013, Hindawi Publishing Corporation, 2013.

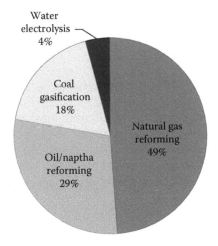

FIGURE 4.2
The percentage share of feedstocks and technologies for commercial hydrogen production. (Adapted from Ewan, B.C.R. and R.W.K. Allen, *Int. J. Hydrogen Energ.*, 30, 809–819, 2005. With permission.)

to grow at a rate of 3.5% per annum till 2018 (World Hydrogen-Freedonia Group 2015). Almost all the produced hydrogen is used as a feedstock in the chemical industry; only a fraction of it is currently used for energy purposes. Hydrogen-powered automotive vehicles are a reality now; Mercedes-Benz has developed hydrogen fuel cell-powered electric cars. This electric car technology is now a proven and mature technology. Refueling and distribution structure is being developed in Germany, the United States, and Japan. There are currently about 200 hydrogen fueling stations across the globe (Verhelst 2014).

Overall, hydrogen-fueled cars have much higher efficiency and less than 30% CO_2 emission than the standard gasoline cars (Midilli et al. 2005; White et al. 2006). With the advancement in fuel cells and hydrogen storage technologies, the global market share of hydrogen-fueled light-duty vehicles is growing, and it is projected to attain a share of 30%–70% by 2050, resulting in a hydrogen demand of approximately 16 EW (Dincer and Acar 2015).

4.3 Hydrogen Production from Natural Gas

Through reforming processes, natural gas is converted into syngas, which with further upgradation and purification gives hydrogen. Different types of reforming processes such as dry reforming, steam reforming, partial oxidation (POX), autothermal reforming (ATR), bi-reforming, and tri-reforming have been investigated for the conversion of natural gas into syngas. Thermal decomposition of methane can also generate hydrogen.

In dry reforming process, hydrocarbons of natural gas react with CO_2-producing syngas with an equimolar ratio of H_2 and CO. It is a highly endothermic process consuming 1.6 times more heat than the SRM. However, it overcomes the disadvantage of high CO_2 emissions in steam reforming by producing less CO_2. Thermodynamically, dry reforming is favored at high temperature (870°C–1040°C) and low pressure (close to atmospheric pressure). Ni-based catalysts get deactivated due to coke deposition under high-temperature conditions, which can be alleviated by the high-pressure operation (Zhang et al. 2014). The dry reforming of methane (DRM) can be represented by the following equation:

$$CH_4(g) + CO_2(g) \leftrightarrow 2CO(g) + 2H_2(g) \qquad \Delta H^0_{298K} = 247\ kJ/mol \qquad (4.1)$$

Steam reforming of natural gas involves a catalytic endothermic reaction between hydrocarbons such as methane and steam, producing syngas with high H_2/CO ratio (2.75–4.75). This reforming process is extensively used at industrial scale. Conventionally, SMR is accomplished in a vertically stacked multitubular reactor. The tubes in the reactor are filled with a highly active catalyst of suitable shape and size, and the reactor is operated at high pressure (15–30 bar) and temperature (700°C–850°C) (Wilhelm et al. 2001).

Steam reforming of methane involves the following reactions:

SRM1:

$$CH_4(g) + H_2O(g) \leftrightarrow CO(g) + 3H_2(g) \qquad \Delta H^0_{298K} = 206.2\ kJ/mol \qquad (4.2)$$

SRM2:

$$CH_4(g) + 2H_2O(g) \leftrightarrow CO_2(g) + 4H_2(g) \qquad \Delta H^0_{298K} = 164.7\ kJ/mol \qquad (4.3)$$

Water–gas shift (WGS) reaction:

$$CO(g) + H_2O(g) \leftrightarrow CO_2(g) + H_2(g) \qquad \Delta H^0_{298K} = -41.5\ kJ/mol \qquad (4.4)$$

Steam reforming is always associated with the WGS reaction (Equation 4.4), which generates a high amount of CO_2. At low steam/carbon (S/C) ratio, CO_2 can be converted into CO via Reverse Water-Gas Shift (RWGS) reaction. However, this is accompanied with a high concentration of unreacted methane (methane slip) in the outlet stream. This can be compensated by carrying out the reforming reactions at high temperature (Murphy et al. 2013). However, high temperature with low S/C ratio leads to coking via side reactions (decomposition of methane and Boudouard reaction). Therefore, an S/C ratio of 3 is usually preferred to avoid "coking" or coke deposition on the catalyst surface. A fraction of the incoming methane feedstock (up to 25%) or waste gases is combusted in a combustion chamber to provide the required process heat. Indirect preheating of reactants takes place through a heat exchanger.

POX is a noncatalytic exothermic reaction between hydrocarbons such as natural gas and oxygen to produce syngas (H_2/CO ratio < 2.0). The absence of a catalyst allows the deposition of coke; hence, the process can be carried out at a high temperature and without steam to lowering the CO_2 content of syngas. The exothermic nature of the reaction eliminates the indirect heat exchanger requirement, which makes POX reactors more compact than steam reformers. The size and cost of the reactor can be reduced by using pure O_2 instead of air. The POX unit has a relatively high efficiency of 70%–80% (Bharadwaj and Schmidt 1995). However, the POX-based reformer is less energy efficient than the steam reformer because rapid heat evolution leads to higher heat loss and makes heat recovery problematic. Compactness of POX system makes its fabrication and maintenance cost lower than steam reforming system. However, it requires an expensive air separation unit. POX of methane involves the following reaction:

$$CH_4(g) + \frac{1}{2}O_2(g) \rightarrow CO(g) + 2H_2(g) \qquad \Delta H^0_{298K} = -35.6 \text{ kJ/mol} \qquad (4.5)$$

ATR is a combination of SMR and POX. In ATR, hydrocarbons such as methane react with both air/oxygen and steam producing a hydrogen-rich syngas mixture. The heat required for steam reforming reactions is internally provided by POX of hydrocarbons in feed when fuel, air, and steam are fed in right proportion. The operating conditions are in between steam reforming and POX. Typical operating pressure and temperature ranges are 40–50 bar and 900°C–1100°C, respectively (Dahl et al. 2014). The use of air as a reactant in ATR results in high reactor volume and heat loss to the surroundings. Therefore, an economic ATR operation requires the use of oxygen in place of air. The air separation unit incurs high investment costs. The efficiency of ATR is higher than that of POX systems because the heat of combustion is utilized in carrying out the endothermic reactions of steam reforming. ATR of natural gas involves SMR1 reaction (Equation 4.2) and POX reaction (Equation 4.5).

Bi-reforming is a combination of steam reforming and dry reforming. Conventional reforming processes such as steam reforming and dry reforming require additional process units for H_2/CO ratio adjustment. POX and ATR systems require an expensive air separation unit and also suffer from hot spot formation due to excess heat generation. DRM is highly endothermic in nature and produces syngas with low H_2/CO ratio (~1). It also suffers from severe coke deposition at high reaction temperature and pressure (Armor 1998). Bi-reforming, however, requires less heat than dry reforming and generates less coke resulting in longer catalyst life (Jun et al. 2011). It also provides flexibility to manipulate the H_2/CO ratio in syngas by varying the H_2O/CH_4 and CO_2/CH_4 ratios. Low-grade natural gas containing a high CO_2 concentration and flue gas from industrial stacks can be used as feedstocks for dry reforming. Reactions involved in bi-reforming of methane are SMR1, SMR2, DRM, and WGS (Equations 4.1 through 4.4).

Tri-reforming is a synergic union of three reactions: (1) dry reforming, (2) steam reforming, and (3) POX (as expressed in Equations 4.1, 4.2, and 4.5) (Sun et al. 2010). Tri-reforming of methane (TRM) possesses a number of unique features. It is the thermoneutral process that does not require pure oxygen. It is expected to produce syngas having a H_2/CO ratio of ~1.5–2 by monitoring the molar ratios of feed components. The major benefits of TRM are its capability to eliminate coke deposition from the catalyst surface and its ability to directly convert the CO_2 from flue gas into valuable syngas (Pan 2002). The study of TRM reaction is not well documented; thus, efforts are being made to design the most suitable range of operating parameters and to develop a highly active and stable catalyst. Challenges that need to be resolved before industrial implementation of TRM include the minimization of air pollutants such as NO_x and SO_x in syngas. The air-blown TRM process contains a large amount of inert N_2, which leads to a larger reactor size and more heat loss. TRM reactions proceed in a competitive environment of reactants (CO_2, O_2, and methane); to achieve a high CO_2 conversion, a tailored TRM catalyst is required. However, this technology is not commercialized yet (Song and Pan 2004). The salient features of various types of reforming processes are outlined in Table 4.3.

From the table, it is evident that syngas produced through the SMR process has the highest H_2/CO ratio, whereas this ratio is the lowest in case of dry reforming (≤ 1). Other reforming processes produce syngas with a H_2/CO ratio of ~2. Further, among these reforming processes, steam reforming and POX are commercially used or near the commercial stage. A comparison of different reforming technologies is presented in Table 4.2.

4.3.1 Steam Methane Reforming

SMR is a well-established and commercial technology for hydrogen production from natural gas, which accounts for nearly 48% of the total world production (Holladay et al. 2009). In steam reforming of natural gas, steam and natural gas with a proper ratio are fed into a tubular reactor filled with a suitable catalyst. Syngas containing mainly hydrogen, carbon dioxide, and carbon monoxide is produced through a set of reactions. The produced syngas is upgraded to valuable chemicals such as methanol and Fischer–Tropsch synthesis products, or hydrogen is separated from the syngas to be used in ammonia production, fuel cell, and so on. Methane is a very stable compound; therefore, reforming reactions are accomplished at elevated temperature and pressure (800°C–1000°C and 10–30 bar). Methane and water reactions (Equations 4.2 and 4.3) primarily occur in SMR. SMR reaction (Equation 4.2) is a highly endothermic reaction. This reaction is associated with WGS reaction, represented by Equation 4.4, which is slightly exothermic. Globally, the process is endothermic. Although natural gas is mainly composed of methane (90%–95%), higher hydrocarbons may also be present (3%–5%). The overall reaction of natural gas reforming containing higher hydrocarbons can be written as follows:

$$C_nH_{2n+2} + nH_2O \leftrightarrows nCO + (2n+1)H_2 \quad \Delta H^0_{298K} > 0 \tag{4.6}$$

C_nH_{2n+2} in the above equation represents the higher hydrocarbons contained in natural gas. The natural gas steam reforming occurs normally in the alumina-supported Ni-based catalyst. SMR catalyst is vulnerable to deactivation due to the carbon deposition (coke) on the surface of it. Furthermore, certain natural gas contaminants such as sulfur, halogen compounds, and heavy metals such as lead, arsenic, and vanadium act as a poison to SMR catalyst (Häussinger et al. 1989). Coke formation in SMR takes place

TABLE 4.3

Comparison of Performances of Different Methane Reforming Technologies

S. No.	Process Technology	Reactions	Operating Conditions	Advantages	Disadvantages	Energy Efficiency (%)	H_2/CO ratio	CO_2 Emission	Technology Maturity
1	SMR	$CH_4 + H_2O \leftrightarrow CO + 3H_2$ $CO + H_2O \leftrightarrow CO_2 + H_2$	800°C–900°C, 10–30 bar	• Most extensive industrial experience • No oxygen required in feed • Lowest process operating temperature • The best H_2/CO ratio for production of liquid fuels	• Highest air emissions • More costly than POX and autothermal reformers • Recycling of CO and removal of the excess hydrogen by means of membranes or PSA	70–85	2.75–4.75	7 kg CO_2/ Kg H_2	Commercial
2	POX	$CH_4 + \frac{1}{2}O_2 \leftrightarrow CO + 2H_2$	1000°C–1500°C, 10–80 bar	• Feedstock desulfurization not required • Exothermic reaction	• Very high process operating temperature and hot spots • Usually requires oxygen plant	60–75	1.6–2	—	Commercial
3	DRM	$CH_4 + CO_2 \leftrightarrow 2CO + 2H_2$ $CO + H_2O \leftrightarrow CO_2 + H_2$	700°C–900°C, 1 bar	• Greenhouse gas CO_2 can be consumed instead of being released into the atmosphere • Almost 100% of CO_2 conversion	• Formation of coke on catalyst • Additional heat is required as the reaction takes place at 600°C		0.8–01	2.43 kg/ CO_2/Kg H_2	Demonstration

(Continued)

TABLE 4.3 (*Continued*)

Comparison of Performances of Different Methane Reforming Technologies

S. No.	Process Technology	Reactions	Operating Conditions	Advantages	Disadvantages	Energy Efficiency (%)	H$_2$/CO ratio	CO$_2$ Emission	Technology Maturity
4	ATR	$CH_4 + H_2O \leftrightarrow CO + 3H_2$ $CH_4 + \frac{1}{2}O_2 \leftrightarrow CO + 2H_2$	800°C–900°C, 20–70 bar	• Lower temperature required than POX • Syngas methane content can be tailored by the adjusting reformer outlet temperature	• Limited commercial experience • Usually requires oxygen plant	60–75	1.8–2.6	—	Commercial
5	Bi-reforming	$CH_4 + H_2O \leftrightarrow CO + 3H_2$ $CH_4 + CO_2 \leftrightarrow 2CO + 2H_2$	800°C–900°C, 10–30 bar	• The best H$_2$/CO ratio for production of liquid fuels • Coke deposition drastically reduced	• Separation of unreacted methane from SMR syngas • Project installation cost		1.5–2	—	Pilot plant
6	TRM	$CH_4 + H_2O \leftrightarrow CO + 3H_2$ $CH_4 + \frac{1}{2}O_2 \leftrightarrow CO + 2H_2$ $CH_4 + CO_2 \leftrightarrow 2CO + 2H_2$ $CO + H_2O \leftrightarrow CO_2 + H_2$	800°C–900°C, 10–30 bar	• Directly using flue gases, rather than preseparated and purified CO$_2$ from flue gases • Over 95% of methane and 80% CO$_2$ conversion can be achieved	• Usually requires oxygen plant • Low H$_2$/CO ratio limits its large-scale application for Fischer–Tropsch and MeOH synthesis		1.5–2	—	Lab. scale

Source: Gangadharan, P. et al., *Chem. Eng. Res. Des.*, 90, 1956–1968, 2012.

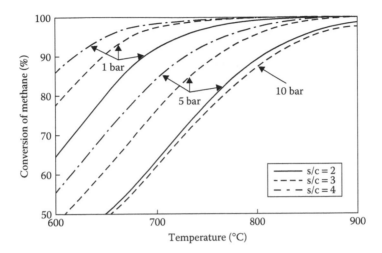

FIGURE 4.3
Equilibrium conversion of methane in SMR. (Reprinted with permission from Joensen, F. and J.R. Rostrup-Nielsen, *J. Power Sources*, 105, 195–201, 2002.)

due to CO disproportionation reaction (Equation 4.7) and methane decomposition reactions (Equation 4.8):

$$2CO \rightarrow C + 2H_2 \qquad \Delta H^0_{298\,K} = -175.5 \text{ kJ/mol} \qquad (4.7)$$

$$CH_4 \rightarrow C + 2H_2 \qquad \Delta H^0_{298\,K} = 74.9 \text{ kJ/mo} \qquad (4.8)$$

SMR requires steam and methane as feed in a stoichiometric S/C of 1:1. However, the higher S/C ratio of 3:1 is mostly used in industrial practice to achieve higher methane conversion. High S/C ratio also favors the reduction of coke formation but decreases the CO/CO₂ ratio in the product stream (Van Hook 1980). Because steam reforming is an endothermic process and conveyed by expansion in volume, the reaction is favored at high temperature and low pressure. The heat of endothermic reaction is provided through an external means. Product stream from SMR may contain 5%–10% of unconverted methane due to equilibrium limitation of natural gas conversion. Natural gas steam reforming is mainly influenced by three operating variables: temperature, pressure, and the molar S/C ratio. Joensen and Rostrup-Nielsen (2002) demonstrated the equilibrium conversion of steam reforming against temperature, pressure, and steam-to-methane ratio, as given in Figure 4.3.

It is clearly seen from the figure that high S/C ratio and low pressure favor high methane conversion even at the lower operating temperature. From another study, it has been found that high S/C ratio decreases CO selectivity and increases the amount of CO₂ in syngas. This is undesirable for downstream processing of syngas, that is, Fischer–Tropsch synthesis. Usually, SMR produces syngas with a H₂/CO ratio of 2.75–4. The steam reforming process has an efficiency of ~87.7% with steam production and ~84.3% without steam production (Korobitsyn et al. 2000).

4.3.1.1 Catalysts and Mechanism of Steam Methane Reforming

Owing to the great practical importance of SMR, a large number of studies are available in literature on the design and development of a highly active and stable catalyst system for SMR. SMR reaction is film resistance limited; hence, most of the reactions occur on the catalyst surface. Accordingly, catalysts having higher geometric surface area show higher catalytic activity.

SMR catalyzed using transition metals (Fe, Co, Ni, and Cu) and noble metals (Ru, Rh, Pd, Ir, and Pt) have been studied extensively in the past (Van Hook 1980; Rostrup-Nielsen and Bak Hansen 1993). Nickel and noble metals are found to be more catalytically active for methane reforming process. These metals are generally supported on materials having high surface area and high thermal stability. Common support materials for SMR catalyst are α- and γ-Al_2O_3, CaO-Al_2O_3, MgO, $MgAl_2O_4$ spinel, TiO_2, ZrO_2, and SiO_2. Among them, α-Al_2O_3, $MgAl_2O_4$ spinel, and CaO–Al_2O_3 are commercially available (van Beurden 2004). Ni is the most widely used metal for commercial reforming catalysts because it is cheap. However, it is less active than noble metals and usually more prone to deactivation by coke deposition and sulfur poisoning. In a study of relative catalytic activity of metallic catalysts (Ir, Ni, Ru, Rh, Pd, and Pt) in the SMR system operated at 550°C, 0.1 MPa, and an S/C ratio of 4, it was found that Ru is the most catalytically active metal, and the order of the other metals is Ru > Rh > Ir > Ni > Pt > Pd (Rostrup-Nielsen and Bak Hansen 1993). The activity of the catalyst is related to the degree of metal dispersion on the support surface. Catalyst having higher metal dispersion on the support has higher activity. Common dispersion of Ni-based catalysts with metal particles of 20–50 nm is 2%–5% (Rostrup-Nielsen 1984). Generally, in industrial practice, ~15–20 wt.% of Ni in the form of NiO is used for the highly active and stable catalyst system. Acidity and basicity of catalyst support also play an important role in the catalytic activity and stability. Acidic support accelerates CH_4 activation, but it also promotes cracking and polymerization. A metal having a strong interaction with support is found to be more resistant to deactivation by sintering and more resistant to coke formation. Promotion of Ni/Al_2O_3 with Pt and Ru produces more stable catalyst system than that of monometallic Ni catalyst. There is a kind of synergy between the type of support and metal, which promotes the activity and stability of the catalyst. Industrial catalysts are supposed to be stable for more than 5 years. Baden Aniline and Soda Factory (BASF), Haldor Topsøe (Denmark), and United Catalysts (Süd-Chemie) are some of the industrial SMR catalyst producers (Padban and Becher 2005).

The reaction mechanism of the SMR reaction is strongly affected by the nature of the catalytically active metal and support. In the early 1955, the first study on the kinetics of SMR was carried out by Akers and Camp (1995); they used Ni metal supported on Kieselguhr material as a catalyst. In literature, there are a number of studies on reaction mechanisms and kinetics of SMR using different catalytic systems. One of the important and realistic mechanistic studies of SMR system was carried out by Xu and Froment (1989). They investigated the intrinsic kinetics of the reforming reaction combined with the WGS reaction on the commercial Ni/$MgAl_2O_4$ spinel catalyst with some amount of α-Al_2O_3 as a diluent. They proposed a complex free radical-type mechanism (expressed in Equations 4.9 through 4.21) and derived the reaction model based on Langmuir–Hinshelwood–Hougen–Watson (LHHW) mechanism:

$$H_2O + {}^* \leftrightarrows O{-}^* + H_2 \tag{4.9}$$

$$CH_4 + {}^* \leftrightarrows CH_4{-}^* \tag{4.10}$$

$$CH_4{-}^* + {}^* \leftrightarrows CH_3{-}^* + H{-}^* \tag{4.11}$$

$$CH_3{-}^* + {}^* \leftrightarrows CH_2{-}^* + H{-}^* \tag{4.12}$$

$$CH_2{-}^* + O{-}^* \leftrightarrows CH_2O{-}^* + {}^* \tag{4.13}$$

$$CH_2O{-}^* + O{-}^* \leftrightarrows CHO{-}^* + H{-}^* \tag{4.14}$$

$$CHO{-}^* + {}^* \leftrightarrows CO{-}^* + H{-}^* \text{ (rate-determining step (r.d.s.): } R_1) \tag{4.15}$$

$$CO-* + O-* \leftrightarrows CO_2-* + * \text{ (r.d.s.: } R_2) \tag{4.16}$$

$$CHO-* + O-* \leftrightarrows CO_2-* + H-* \text{ (r.d.s.: } R_3) \tag{4.17}$$

$$CO-* \leftrightarrows CO + * \tag{4.18}$$

$$CO_2-* \leftrightarrows CO_2 + * \tag{4.19}$$

$$2H-* \leftrightarrows H_2-* + * \tag{4.20}$$

$$H_2-* \leftrightarrows H_2 + * \tag{4.21}$$

LHHW models of the rate of reaction based on the proposed mechanism are as follows:

$$R_1 = \frac{\dfrac{k_1}{P_{H_2}^{2.5}}\left(P_{CH_4}P_{H_2O} - \dfrac{P_{H_2}^3 P_{CO}}{K_1}\right)}{DEN^2} \tag{4.22}$$

$$R_2 = \frac{\dfrac{k_2}{P_{H_2}}\left(P_{CO}P_{H_2O} - \dfrac{P_{H_2}P_{CO_2}}{K_2}\right)}{DEN^2} \tag{4.23}$$

$$R_3 = \frac{\dfrac{k_3}{P_{H_2}^{3.5}}\left(P_{CH_4}P_{H_2O}^2 - \dfrac{P_{H_2}^4 P_{CO_2}}{K_3}\right)}{DEN^2} \tag{4.24}$$

where:

$$DEN = 1 + K_{CH_4}P_{CH_4} + K_{CO}P_{CO} + K_{H_2}P_{H_2} + \left(\frac{K_{H_2O}P_{H_2O}}{P_{H_2}}\right) \tag{4.25}$$

P_{H_2} = Partial pressure of Hydrogen
P_{CO} = Partial pressure of Carbon mono-oxide
P_{CO_2} = Partial pressure of CO_2
P_{CH_4} = Partial pressure of CH_4
k_1, k_2, k_3 are reaction constants and
KCH_4, KCO, KH_2, and KH_2O are the adsorption coefficients

Based on the available literature on the catalyst and reaction mechanism aspects, the important insights into SMR can be gained as follows:

1. Most of the kinetic studies are based on the Ni active metal catalyst. Pd, Rh, Ru, and Pt are also studied but to a smaller extent. Rh and Ru have shown a very good catalytic activity, thermal stability, and less coking compared to Ni-based catalysts.

2. Kinetic mechanism is studied when supports such as ZrO_2, Al_2O_3, CeO_2, and SiO_2 are used to anchor any of the aforementioned metals. It was found that reducible supports enhance the rate of reforming much higher than nonreducible substrates.

3. Most of the kinetic models indicate that steam reforming shows the first-order kinetics with respect to methane.

4. CH_4 adsorption or C–H activation is considered to be the rate-determining step.

5. The reaction between the adsorbed CH_x and O_x species is found to be responsible for the CO_2/CO selectivity.

4.3.1.2 Conventional Natural Gas Steam Reforming Process

SMR is a highly developed and mature industrial technology that forms an essential part in ammonia and methanol production plants. Figure 4.4 depicts the overall block diagram of hydrogen production plant through the SMR plant. The plant consists of five main blocks: natural gas input, desulfurization, reforming section, WGS section, and product gas separation/H_2 purification section.

4.3.1.2.1 Desulfurization Unit

Natural gas cleaning is the first stage in the SMR plant. Raw natural gas may contain small amount of sulfur, chlorides, other halides, and mercury. The SMR and WGS catalysts are very sensitive to sulfur poisoning. Chlorides and other halides, as well as mercury, are also considered as poisons for the SMR catalyst. A desulfurization unit (DSU) is required to eliminate these poisonous compounds from the SMR feed. In industrial practice, sulfur level for SMR is specified at 25 ppb or less. Desulfurization takes place in three steps: hydrogenation, chlorine removal, and sulfur removal. Sulfur organic compounds (e.g., thiophenes, carbonyl sulfide [COS]) first react with H_2 on the Co–Mo catalyst at 290°C–370°C generating H_2S (Scholz 1993). COS is converted into H_2S according to the following reaction:

$$COS + H_2O \rightarrow CO_2 + H_2S \tag{4.26}$$

This is followed by chlorine removal using amine solution. HCl is a common form of chloride presents in the SMR field, which accelerates sintering the catalyst metal crystallites. Halides (e.g., chlorides) are removed by passing the feed through alumina guards

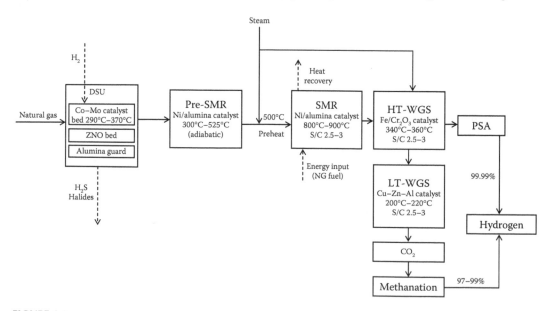

FIGURE 4.4
Block diagram of conventional hydrogen production plant through natural gas steam reforming route.

(Hawkins 2013). In the third stage, sulfur scrubbing (Equation 4.27) is accomplished at a temperature range of 340°C–390°C using two beds of ZnO operated in the alternate mode:

$$H_2S + ZnO \rightarrow ZnS + H_2S \tag{4.27}$$

ZnO bed is regenerated by blowing air through the bed at high temperature:

$$ZnS + 3/2O_2 \rightarrow ZnO + SO_2 \tag{4.28}$$

HCl reacts with ZnO to form $ZnCl_2$, which blocks H_2S absorption sites in the sulfur scrubbing unit. Therefore, it is better to remove HCl prior to sulfur scrubbing. Ni is a good absorber of sulfur; therefore, at a concentration higher than 5 ppm, sulfur can cause severe deactivation to the catalyst. Olefins in natural gas can promote the rapid formation of hot bands in the reformer tubes if present in more than 1%–2%. Most of the SMR catalysts can handle the olefins up to a certain limit. Olefins also promote the catalyst deactivation by coke formation.

4.3.1.2.2 *Reforming Section*

Prior to the reaction, the catalyst has to be activated by H_2 or natural gas feed, reducing nickel oxide to metallic nickel. In the reforming reaction Equation 4.6, pre-reforming may be employed, if the natural gas contains a high amount (>10%) of higher hydrocarbons (C_2+). The aim of pre-reforming is to eliminate these higher hydrocarbons by converting them into CH_4, CO_2, and H_2. The presence of higher hydrocarbons (C_2+) in the reformer feed can lead to coke deposition. The higher hydrocarbons compete with methane in the reformer, and at such an elevated temperature, they readily decompose into carbon and hydrogen. Pre-reforming is accomplished in an adiabatic reactor at a temperature of 300°C–525°C using alumina-supported high Ni with promoters. Pre-reforming of natural gas greatly increases the plant efficiency by lowering the required S/C ratio in the reforming section.

After pre-reforming, natural gas is mixed with steam (around 20 bar), and the mixture is preheated to 500°C and fed to the catalytic reforming reactor. Steam reformer is a heat exchanger, in which vertical tubes are arranged in a stack of 500–600 tubes with a length of 7–12 m and an inside diameter of 70–130 mm (Moulijn et al. 2013). The heat of the endothermic reaction is provided by burning a part of process feed (NG) at the shell side to heat up the process gas. Almost 50% heat is transferred to the process stream from the shell side, another 25% of heat can be recovered through steam production, and the rest is lost in the stack gas (Foulds and Gray 1995). The radiation is the dominant mode of heat transfer. The heat distribution is greatly affected by the shell side arrangements of burners and the pitch of tubes. Furnace off-gas heat recovery is accomplished by generating high pressure steam, or it can be used to preheat the process stream. The global process efficiency of SMR is between 85% and 95%. Reforming reaction products (mixture of CO, H_2, CO_2, CH_4, etc.) leave the reformer at 700°C–950°C; approximately 95% conversion of methane is achieved. The composition of syngas is greatly affected by the process conditions (Häussinger et al. 1989).

4.3.1.2.3 *Water–Gas Shift Reactor Section*

WGS reaction (Equation 4.4) is slightly exothermic, which is therefore thermodynamically preferable at low temperature. In the WGS reaction, carbon monoxide reacts with water to produce more hydrogen. The product stream from the reforming section is cooled down to about 350°C and introduced into the WGS reactor, where further enrichment of syngas takes place. Temperature affects CO conversion in the WGS reaction in a reverse manner; a decrease in temperature favors equilibrium in the right direction (Equation 4.2) but lowers the rate of reaction. Therefore, WGS is carried out in two separate reactors in series to get a high conversion of CO: high-temperature WGS (HT-WGS) reaction and low-temperature WGS (LT-WGS) reaction. CO conversion in the HT-WGS reactor is carried out in the presence of iron–chromium-based catalyst (FeO_3–Cr_2O_3) and at a temperature of 340°C–360°C. The product gas stream from HT-WGS is further cooled down to 220°C and entered into the LT-WGS reactor, where CO conversion occurs in the presence of Cu–Zn–Al or Cu–Zn catalyst (CuO: 15%–30%, ZnO: 30%–60%, and Al_2O_3). The CO concentration in the LT-WGS reactor leaving stream is about 0.05–0.5 vol.% (Hinrichsen et al. 2008).

4.3.1.2.4 H_2 Purification Section

Hydrogen gas separation and purification from a reformer product stream can be attained in two ways. One way is to use a pressure swing adsorption (PSA) unit, which is capable of producing extremely pure hydrogen of approximately 99.999% purity (Dybkjaer and Madsen 1998). The use of PSA also eliminates the use of LT-WGS reactor; the product stream of the HT-WGS reactor is first cooled down at ambient pressure to remove water, and then the cooled stream is introduced into the PSA unit. Pure hydrogen is collected and the off gas from the PSA unit can be used to fuel the reforming section. The second option is called the classical absorption process. The product stream of the LT-WGS reactor is cooled down to low temperature and fed to a carbon dioxide scrubber in which monoethanolamine or hot potash is used as a scrubbing medium. The catalytic methanation step is used next to the scrubbing unit to remove the remaining impurities in the hydrogen stream, traces of carbon dioxide, and carbon monoxide. Both HT-WGS and LT-WGS reactors are employed in this process. Hydrogen purity of 97%–99% can be achieved in this process (Häussinger et al. 1989).

4.3.1.3 Reactor Technologies for Steam Natural Gas (Methane) Reforming

4.3.1.3.1 Conventional Methane Steam Reformer

Conventional methane steam reformers are available in a wide range of sizes: from 25 to 100 million Nm^3/h of hydrogen for a large-scale operation and 0.1–1 million Nm^3/h of hydrogen for small-scale chemical processes such as oil refining. These are indirectly heated tube reactors consisting of long (12 m) catalyst-filled vertical tubes and set aside in a furnace (Figure 4.5). Process heat is generated through top, side, or bottom mounted burners inside the furnace. The high-temperature operation of SMR requires high-cost alloy steels for reactor tube manufacturing, which increases the capital cost of the SMR plant. The capital cost for the SMR plant having a hydrogen production capacity of 22 million Nm^3/h is approximately $200/kW hydrogen, whereas the capital cost for the SMR plant having a large production capacity of 220 million Nm^3/h of hydrogen is estimated to be approximately $80/kW hydrogen (DTI report 1997).

Large plant size and high capital cost investment are the disadvantages of the SMR process. Small-scale hydrogen production using conventional long tube SMR reactor is not economical for stranded natural gas reserves and for stand-alone hydrogen refueling stations. Many engineering companies design and build SMR plants; among them, M.W. Kellogg (London, UK), ICI (London, UK), Lurgi (Frankfurt, Germany), Howe-Baker (London, UK), KTI (Poland), Kværner (Lysaker, Norway), Foster Wheeler (London, UK) and Haldor Topsøe are the leading companies (Ogden 2001).

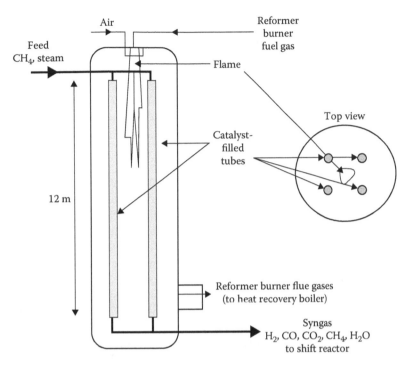

FIGURE 4.5
Conventional small-scale steam methane reformer design. (Adapted from Ogden, J.M., Review of small stationary reformers for hydrogen production, in *Report to the International Energy Agency*, 609, 2001.)

4.3.1.3.2 Compact Steam Methane Reformers "Fuel Cell Type"

Compact steam methane reformers are specially designed reactors, working at low temperature and pressure, and are built using low-cost materials. Small-scale steam methane reformers with a production capacity of 2000–140,000 Nm³ hydrogen per hour have been developed for a stand-alone hydrogen production station and to be used with fuel cells (Halvorson and Farris 1997). The capital cost per unit for compact reformers having a hydrogen production capacity of 0.1–1.0 million Nm³/h may fall in the range of $150–$180/kW hydrogen when 1000 such units are produced (DTI report 1997). However, the capital cost per unit of hydrogen production ($/kW H_2) would be similar for both compact reformers and conventional reformers. Moreover, these compact reformers have high energy conversion efficiencies possibly between 70% and 80%. In recent years, many companies have built and commercialized compact steam methane reformers for natural gas to hydrogen conversion for fuel cell applications. These include Haldor Topsøe, Osaka Gas Company (Osaka, Japan), Ballard Power Systems (Burnaby, Canada), UTC Fuel cell (Connecticut, UK), and Sanyo Electric (Moriguchi, Japan) (Barbir et al. 2000).

4.3.1.3.3 Plate-Type Steam Methane Reformers

"Plate-type" reformer is a compact steam methane reformer designed for fuel cell systems. In this reformer, steam reforming catalyst is coated on one side of the plate; on the other side of the plate, anode exhaust gas from the fuel cell undergoes catalytic combustion, which provides heat to drive the endothermic reforming process. A number of such plates are arranged in a stack, and the reformer is supplied with methane and steam. Plate-type reformers offer certain advantages over other compact reformers such as more compact, standardized low-cost design, better energy conversion efficiency, and faster start-up (Marrelli et al. 2011). Owing to high energy efficiency and compact design of plate reformers, many groups are working to build the commercial plate-type reformers (Shinke et al. 2000). No commercial plate-type steam methane reformers are available until now for fuel cell applications. However, a good number of patents have been awarded in recent years around the world. Osaka Gas Company, Air Products (Pennsylvania, US), UTC Fuel cells, Ztek Corporation (Massachusetts, US), and Ishikawajima-Harima Heavy Industries (Tokyo, Japan) are working to develop and design plate reformers.

4.3.1.3.4 *Membrane Reactors for Steam Reforming*

SMR using membrane reactor is a promising technology. In membrane reactors, a single reactor is used to accomplish all steps (SMR, WGS, and H_2 separation) from raw natural gas to hydrogen production. A hydrogen selective membrane typically made of Pd or Pd/Ag or other Pd alloy is allocated to the reactor, embedded in the catalyst bed (Figure 4.6). H_2 permeates through the membrane making the reaction zone hydrogen free and thus reducing the chemical equilibrium barrier. Membrane reactors are operated at rather lower temperature (400°C–600°C) in comparison with conventional steam reforming reactors. Highly pure hydrogen can be produced using membrane reactors. Membrane reactor catalysts must be very active at such low temperature. The membrane reactor offers many advantages such as higher CH_4 conversion at low temperature, high energy efficiency, low capital cost (fewer equipment required), and simple and compact design (Gallucci et al. 2013). Owing to potential benefits of the membrane reactor technology, this technology nowadays is considered as a good candidate for substituting the conventional SMR reactor, especially for the production of high purity hydrogen.

Since the last decade, many different types of membrane materials have been tested to make them more efficient and economic, these include polymeric, microporous ceramics, porous carbon, and proton-conducting dense ceramic (Iulianelli et al. 2016). A large number of industrial R&D activities of membrane technologies for syngas and hydrogen production and their commercialization are going on in many places of the world. Recently, a pilot plant study of a fluidized bed membrane reactor of 20 Nm³/h has been successfully conducted using a commercial membrane (Shirasaki et al. 2009). There is a large number of companies involved in the development and design of membrane technology such as CRI Catalyst Company of Shell (Netherlands), Energy Research Centre of the Netherlands, Eltron Research,

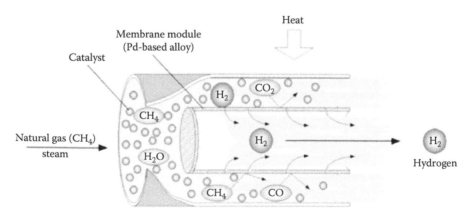

FIGURE 4.6
Principle of membrane reformer. (Adapted from Shirasaki, Y. et al., *Int. J. Hydrogen Energ.*, 34, 4482–4487, 2009. With permission.)

Inc., (Colorado, US) Green Hydrotec (Taoyuan, Taiwan), Hy9 (Massachusetts, US), Media and Process Technology, Inc., (Pennsylvania, US) Membrane Reactor Technologies (Vancouver, Canada), Pall Corporation (New York, US), and Tokyo Gas Co., Ltd (Tokyo, Japan).

4.3.1.3.5 Microchannel Reactor Technology

The microchannel process technology is known to be a promising alternative to many conventional chemical reactors. There is a growing interest in the development and design of microchannel reaction engineering and applications to compact chemical reactors in both academic and industrial sectors. Microreactors are fabricated as long parallel channels, having gaps on the order of microscale (typically <1000 μm) or mesoscale (1000 μm to a few centimeters). The most attractive benefit of using the microreactor technology in chemical processing is its efficiency to reduce mass and heat transfer resistance, offering a high surface-to-volume ratio, thereby increasing the overall conversion of reactants (Wegeng 1999). Velocys Company (Houston Texas, US) from Battelle Memorial Institute has designed and fabricated a microchannel process technology for a large-scale SMR process. Integrated microreactor heat exchanger, having a channel width of 0.25 mm, has been successfully tested for SMR process. Highly active 10% Rh on a gamma alumina support modified with 4.5% Mg is washcoated on the inner walls of channels, and methane fuel with low excess air (25%) is combusted to generate process heat. It is demonstrated from this study that the heat transfer rate to the endothermic reaction exceeds 18 W/cm² of interplanar heat transfer surface area and exceeds 65 W/cm³ of the total reaction volume of methane steam reforming with a contact time of 4 ms. The microchannel technology-based Velocys methane steam reformer has shown a high heat flux in comparison with conventional steam reformers, which have a typical volumetric heat flux of less than 1 W/cm³. The capital and operating costs of a commercial unit can significantly be reduced by the integration of multiple unit operations and using process intensification in the microchannel reforming technology (Tonkovich et al. 2004).

4.3.1.4 Advances in Steam Reforming of Natural Gas Technology

4.3.1.4.1 Sorption-Enhanced Reforming

Natural gas reforming to syngas and subsequent production of hydrogen is a chemical equilibrium-limited process. To achieve high conversion rates, the process must be operated at high temperature. Another disadvantage of SMR is the generation of large volume of greenhouse gases by the combustion of fuel to generate the heat required for the process. To enhance the conversion and product selectivity, *in situ* separation of one of the products from the reaction mixture is required. The sorption-enhanced reaction process (SERP) is a process in which CO_2 is selectively removed from the reaction by the addition of CO_2 acceptor (sorbent), and then CO_2 is converted into solid carbonate. Further, regeneration of the sorbent is done separately, which releases relatively pure CO_2 suitable for sequestration. As a result, the equilibrium is shifted toward the products according to Le Chatelier's principle. Han and Harrison (1994) studied the SERP using CaO to capture CO_2, achieving a complete CO conversion by overcoming the equilibrium limitation. The noncatalytic highly exothermic carbonation reaction (Equation 4.29) is included in the sorption-enhanced steam reforming in addition to reactions (Equations 4.2 through 4.4):

$$CaO(s) + CO_2(g) \leftrightarrow CaCO_3(s) \qquad \Delta H^0_{298\,K} = -178\ kJ/mol \qquad (4.29)$$

SERP offers many advantages over conventional SMR such as (1) lower reaction temperature, which may reduce catalyst coking and sintering; (2) increased energy efficiency;

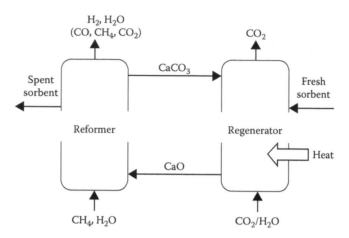

FIGURE 4.7
Schematic diagram of the sorption-enhanced SMR process. (From Johnsen, K. et al., *Chem. Eng. Sci.*, 61, 1195–1202, 2006. With permission.)

(3) fewer SMR processing steps; and (4) no need of shift reactor. However, regeneration of sorbent is an energy-intensive process, reverse reaction of carbonation reaction (Equation 4.29), which imposes a challenge in terms of heat transfer and reactor construction. Long-term performance of CaO sorbent in SERP is suspicious. The sintering of CaO sorbent may occur due to the high temperature of $CaCO_3$ regeneration. To maintain the activity of sorbents in a cyclic SERP process, a number of different kinds of sorbents have been tested. Sorbents can be natural such as limestone and dolomite; these calcium-based sorbents are available worldwide at low cost. However, natural sorbents are more prone to deactivate in multiple recarbonation cycles. Another category of sorbents are synthetic sorbents such as $CaO/Ca_{12}Al_{14}O_{33}$ and Li_2ZrO_3. Synthetic sorbents have better multicycle stability, but the reaction rate and their cost are constrained to compete with natural sorbents (Abanades et al. 2012). Figure 4.7 illustrates the simplified diagram of the SERP. In the figure, two fluidized bed reactors are used in a cyclic process. First, the fluidized bed reactor containing SMR catalyst and CaO sorbent is operated adiabatically in which three simultaneous reactions (i.e., reactions 4.1, 4.2, and 4.25) are taking place at 725°C. The heat required by the endothermic SMR reaction is balanced by the heat released by the exothermic shift and carbonation (Equation 4.25) reactions. Thermodynamically, about 88% conversion of methane is attainable, and the product gas contains 95% (v/v) hydrogen (Johnsen et al. 2006).

The spent CaO sorbent ($CaCO_3$) is regenerated in another adiabatic fluidized bed reactor at ~975°C. Regenerated and spent CO_2 acceptor is continuously recirculated between the reforming and regeneration reactors in a cyclic manner. SERP seems to be a promising technology for clean hydrogen production and for CO_2 sequestration and storage. However, it is still in the demonstration stage; no commercial implementation has yet been done. Commercialization of SERP needs answers of some key issues such as continuous separation of the CO_2 acceptor and the reforming catalyst, sorbent lifetime for multiple cycle operations, and reactor design. Air Products and Chemicals, Inc., demonstrated the SERP on a pilot plant, in which K-promoted hydrotalcite was used to separate CO_2 from the reaction zone. As a result, high conversion of methane (90% or more) at relatively low temperatures (300°C–500°C) was achieved (Hufton et al. 1999).

4.3.1.4.2 *Plasma Reforming of Natural Gas*

Plasma technology can be applied efficiently for conversion of methane into syngas and hydrogen. Plasma is a state of matter and characterized by a dense, highly ionic phase at temperatures of 3,000°C–10,000°C. Plasma can be applied to the reforming process to provide the required heat released by the endothermic reaction and to speed up the reaction kinetics of the reforming reactions even without a catalyst. The plasma is created by an electric arc using electricity alone or a combined electrical and chemical energy. Plasma arc technology is categorized into two groups: thermal versus nonthermal and oxidative versus nonoxidative plasma systems. In thermal plasma SMR, the reactant mixture (methane and steam) is introduced into the reactor (plasmatron), which consists of two electrodes (anode and cathode). A high temperature of ~5000°C is achieved by creating an electric arc in the plasmatron, which operates on high power densities. Gaseous reactants diffuse through this arc at high velocities and converted into ionized species and free radicals, and hydrogen-rich gas is produced (Bromberg et al. 1999).

Nonthermal plasma systems are also referred to as cold plasma systems, which operate under nonequilibrium thermal conditions. An electric discharge, similar to thermal plasmatron, generates the chemically active species, which can catalyze the methane and steam reaction. The use of nonthermal plasmatron offers several advantages over thermal plasmatron such as reduced electric consumption, lower temperature operation (near room temperature), insignificant electrode erosion, and compact reactor design. Application of different types of nonthermal plasma systems have been reported in literature for reforming of natural gas to hydrogen-rich gas such as (1) gliding arc technology, (2) corona discharge, (3) microwave plasma, and (4) dielectric barrier discharge.

Bromberg et al. (1998), at the Massachusetts Institute of Technology (MIT), Cambridge, Massachusetts, have developed and applied plasma reforming systems using a diesel fuel feedstock for two different sets of experiments using SMR and POX technology. At the best, the maximum attainable conversion of methane in an electric arc-enabled plasmatron is 95% with a specific energy consumption of 14 MJ/kg H_2. Currently, BOC Gases, the National Renewable Energy Laboratory, and researchers at the MIT are evaluating the potential of plasma reforming technology for small-scale hydrogen production (Ogden 2001).

4.3.1.4.3 *Steam Methane Reforming Using Alternative Energy Sources*

Owing to the endothermicity of SMR reactions, a large volume of fossil fuels is burnt to generate the process heat in a large-scale hydrogen production unit. However, fossil fuel-based hydrogen production routes are not considered as sustainable and eco-friendly because of high levels of CO_2 and air pollutant emission associated with them. From this perspective, the use of alternative energy sources with low carbon emission should be implemented in industrial practice of SMR. For a long time, it has been realized that high-temperature nuclear and solar heat sources could be the possible source of energy for SMR. Nuclear power is a low-carbon technology that has the potential to be utilized on a wide scale and in large capacities. Nuclear power-based steam reforming can significantly reduce the fossil fuel consumption and CO_2 emission level by about 30%. In a study of nuclear-assisted natural gas reforming, higher conversion efficiencies have been reported (85% in comparison with 75% in conventional reforming). Nuclear cogeneration of heat and power can be used for hydrogen production with electricity generation. High-temperature gas-cooled reactors (HTGRs) and very high-temperature reactors (VHTRs) can be the nuclear energy options for steam reforming and other hydrogen production processes with a typical capacity of 200–800 tons H_2 per day. Nuclear-based hydrogen production is not commercialized until today; development and demonstrations are under way through various projects such as the HTGR design project at Japan Atomic Energy Agency (Ogawa and Nishihara 2004).

Solar energy is a sustainable and abundant source of energy, which can be used to produce hydrogen in an eco-friendly and economical way. The thermochemical route is based on the concentrated solar energy in the range 500°C–1500°C, which can be used to provide heat to the SMR reaction. A good number of solar-based hydrogen projects are running for further development of technology to a commercial level, especially EU projects. Solar reforming of natural gas has been extensively studied in either solar furnace (noncatalytic pyrolysis) or solar simulators using different reactor configurations and catalysts. Ceramic pin-based "DIAPR-Kippod" volumetric-type reformer has successfully been tested in the SMR process in the presence of La and Mn oxides and promoted Ru catalysts supported on alumina at a temperature range of 500°C–1100°C. The catalysts showed a stable operation for 100 h at 1100°C (Berman et al. 2006). However, the global economics of alternative energy routes is not comparable with the conventional SMR.

4.3.2 Partial Oxidation of Natural Gas

POX of natural gas is another major route based on the dry oxidation of CH_4/O_2 to produce syngas and hydrogen. POX technology was developed and commercialized around 1964 (Arutyunov and Krylov 1998). In the POX process, methane and oxygen (or air) are mixed in a stoichiometric ratio, and the mixture is processed in a reactor at temperatures above 750°C and a pressure of 10–80 bar generating syngas with a H_2/CO ratio of 2:1, according to the following equation:

$$CH_4 + \frac{1}{2}O_2 \leftrightarrow CO + 2H_2 \qquad \Delta H_{298\,K}^0 = -44\,kJ/mol \qquad (4.30)$$

POX process can be accomplished either with a catalyst or without a catalyst. Many alternate versions of POX processes have been proposed depending on the type of feedstocks, catalyst, type of reactor, and use of pure oxygen or air.

4.3.2.1 Homogeneous Partial Oxidation of Natural Gas

Homogeneous POX is a direct oxidation of natural gas without using any catalyst. It is generally carried out at elevated temperatures of 1100°C–1500°C and at high pressure (10–80 bar). The high temperature of the reaction zone drives the POX reaction. The use of air as an oxidant increases the volume of the reactor and downstream processing. In POX (Equations 4.30, 4.31, and 4.32), the formation of H_2O and CO_2 occurs simultaneously along with CO and H_2, which affects the selectivity toward CO and H_2.

$$CH_4 + \frac{3}{2}O_2 \leftrightarrow CO + 2H_2O \qquad \Delta H_{298\,K}^0 = -519.2\,kJ/mol \qquad (4.31)$$

$$CH_4 + 2O_2 \leftrightarrow CO_2 + 2H_2O \qquad \Delta H_{298\,K}^0 = -799.1\,kJ/mol \qquad (4.32)$$

Owing to the exothermic nature of homogeneous POX reactions, the reactor based on POX reactions is much more energy efficient in comparison with the conventional steam reformer. POX reactions are also much faster than steam reforming reactions; as a result, smaller reactors and higher throughput would be required for syngas generation. Figure 4.8

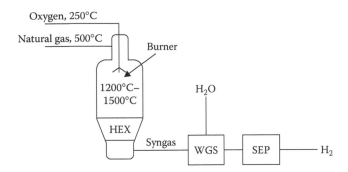

FIGURE 4.8

POX-based hydrogen production unit. (Modified from Korobitsyn, M.A. et al., Sofc as a gas separator, in *ECN-Fuels, Conversion & Environment,* 2000.)

depicts a hydrogen production plant based on homogeneous POX, which comprises a POX reactor, followed by a WGS reactor and a hydrogen separation unit. POX reactor is a refractory-lined pressure vessel in which preheated natural gas and oxygen are mixed in a burner and introduced into the reactor at a typical pressure of 40 bar. POX reaction occurs immediately in the combustion zone below the burner. The product stream exits the reactor at a temperature range of 1300°C–1500°C with a H_2/CO ratio between 1.6 and 2. Almost 100% conversion of methane is achieved in the POX process, and the efficiency of the POX unit is relatively high (70%–80%). A thoroughly mixed feed of natural gas and oxygen must be fed into the reactor, and the reaction temperature should not be lower than 1200°C to avoid carbon deposition and soot formation (Foulds and Gray 1995).

Large-scale POX of natural gas has not yet been used commercially. Because POX requires a premixing of preheated CH_4 and O_2, this can be flammable or even explosive. This makes it difficult to study the underlying phenomenon of POX fundamentally. Small-scale POX systems using natural gas and gasoline have recently been developed for fuel cell applications. On-board reformers based on POX from Epyx, Nuvera, and Hydrogen Burner Technology, Inc., are commercially available. However, these small-scale POX systems are still undergoing through R&D to make them more economic, safe, and efficient (Mitchell et al. 1995).

A large-scale industrial application of POX technology for generation of syngas from nonassociated natural gas is used in Shell Middle Distillate Synthesis plant of Bintulu, Malaysia. The plant is producing specialty chemicals, synthetic fuels, and waxes via Fischer–Tropsch synthesis on a scale of 12,000 barrels per day (Korobitsyn et al. 2000).

4.3.2.2 Catalytic Partial Oxidation of Natural Gas

Catalytic POX (CPO) process involves a heterogeneous catalyst to produce syngas from a reforming reaction of methane (NG) using oxygen/air as an oxidant (Equation 4.30). CPO of methane to synthesis gas was initially suggested by Liander (1929), for ammonia plant, but its detailed study was started in 1946 by Prettre and Perrin (1946). They studied the catalytic conversion of mixtures of methane and oxygen employing a reduced 10% Ni catalyst supported on a refractory material in the temperature range of 725°C–900°C and at 1 atm pressure. Experiments were performed using a superficial contact time of 6–42 s and 20–40 cm³ of catalyst at a bed depth of ~12 cm. Based on the temperature data along the catalyst bed under oxygen-deficient conditions (CH_4:O_2 = 2:1), they proposed a dual thermodynamic process. CPO in such conditions proceeds with an initial

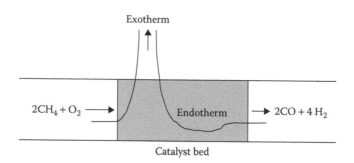

FIGURE 4.9
Schematic representation of the temperature profile in CPO catalyst beds. (From Zhu, Q. et al., *J. Nat. Gas Chem.,* 13, 191–203, 2004. With permission.)

exothermic reaction, which is followed by an endothermic stage. Later, Vermeiren et al. (1992) also confirmed a similar pattern of temperature profile along the catalyst bed (Figure 4.9). CPO reactions are governed by their thermodynamic chemical equilibrium. There are a number of simultaneous reactions involved in CPO to reach to a desired product. The product stream leaves the reactor in thermodynamic equilibrium. Heat and mass transport mechanism involved in CPO greatly affects the process. In the CPO reactor, formation of hot spots and larger temperature gradients within the catalyst bed are typically observed because of the difficulty in removal of exothermic heat from the catalyst bed. Using a catalyst with high thermal conductivity, such as monolith, may be helpful in reducing the hot spot formation.

Excessive heat due to hot spot can deactivate the catalyst by sintering the active metals. Another challenge is to avoid the potential explosion of CH_4 and O_2 mixture to safely run the CPO process.

4.3.2.2.1 Catalysts for Catalytic Partial Oxidation

The CPO of methane (NG) is generally carried out in a fixed bed reactor in the presence of suitable catalyst. In order to enhance the CPO reactions, the catalyst must possess some basic characteristics such as high activity, high selectivity, and operability at high gas hourly space velocity, and must allow a long operation time without appreciable deactivation. Transition metals such as Ni, Co, and Fe, and noble metals such as Rh, Pt, Pd, Ir, Ru, and Re are the most commonly used active metals in the catalyst systems for CPO reactions. These active metals are generally supported on a suitable support system made of alumina, or silica, and are usually used in the form of pellets, monoliths, or foams. Oxygen diffusion limitation due to the film surrounding the catalyst and deactivation of the catalyst due to the formation of $NiAl_2O_4$ were observed in a CPO experiment using Ni/Al_2O_3 catalyst conducted by Huszar et al. (1971). Noble metals are found to be highly active toward CH_4 conversion in CPO, but Ni is widely investigated as it is cheap and abundant. Ni catalyst is more prone to deactivation by coke deposition.

Carbon deposition on the supported metal catalyst was reported by Claridge et al. (1993) in the order: Ni > Pd > Rh > Ru > Pt. CO and H_2 selectivities and conversion are dependent on the active metals. Deposited carbon may have different morphologies; for example, whisker, encapsulated, and filamentous, depending on the operating conditions and the active metal. The complete conversion of methane into syngas at a lower temperature of approximately 650°C using a mixed metal perovskite catalyst with a general formula of $LaNi_{1-x}Fe_xO_3$ is reported in literature (Provendier et al. 1999). Ni-supported (1%–5%)Al_2O_3, SiO_2, SiO_2–ZrO_2, and SiO_2–Al_2O_3 catalysts were compared with Ni-supported H–Y zeolite catalyst by Vermeiren et al. (1992). It was observed that 5 wt.% nickel on H–Y zeolite exhibits slightly lower selectivity to CO compared to the other catalysts. The Ni-supported Al_2O_3 and SiO_2–Al_2O_3 show comparable results in terms of equilibrium composition.

CPO of methane over Co/Al_2O_3 typically yields results close to equilibrium over reduced catalyst. However, Co-based catalyst is less stable at higher temperatures in comparison with nickel-based catalyst due to oxidation of Co metal. Noble metal is reported to be a good promoter, which reduces the ignition temperature on Ni surface; the presence of boron with the active metal gives similar results (Chen et al. 2005). An addition of the second and third metals with Ni catalyst was found to be stable for more than 100 h due to promoter-supported gasification of carbon from the catalyst surface.

4.3.2.2.2 Reactors for Catalytic Partial Oxidation

CPO of natural gas on a laboratory scale is mostly studied in fixed bed microreactors as shown in Figure 4.10a. Monolith reactors (Figure 4.10b) and fluidized beds (Figure 4.10c) have also been used successfully on a laboratory scale.

Laboratory fixed bed microreactors were made of quartz, consisting of tubes with diameters of 1–4 mm filled with a powder catalyst (20–50 mg of weight) and placed in a furnace or a heater to thermostatically control the temperatures under adiabatic conditions. Monolith reactor consists of several extruded monoliths with a catalyst wash-coated on the inner surface of these monoliths. The fluidized bed reactor is a new entry in syngas production through the CPO process.

4.3.3 Autothermal Reforming of Natural Gas

ATR combines the SMR process and the non-CPO in one system. The combination of exothermic and endothermic processes resulted in a more energy-efficient process. The ATR process was developed in the late 1950s by Haldor Topsøe (Chauvel and Lefebvre 1989). Later in 1990, further developments were made in the ATR process, such as the steam-to-carbon ratio, designs of the burner, and safety issues (Christensen and Primdahl 1994). The produced syngas has a H_2/CO ratio of 2:1, which is a suitable ratio of syngas as a feedstock for methanol synthesis and GTL plant. In ATR, the pre-reformed or heated natural gas reacts with oxygen and steam at a temperature of ~1200°C, followed by equilibration of gas with respect to the steam reforming of methane and WGS reactions

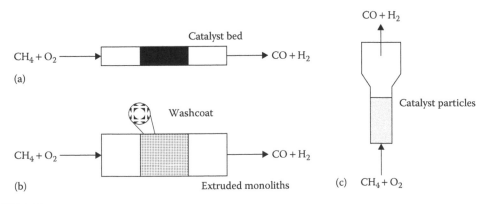

FIGURE 4.10
(a) Fixed bed microreactor, (b) monolith reactor, and (c) fluidized bed reactor. (From Bharadwaj, S.S. and L.D. Schmidt, *Fuel Process. Technol.*, 42, 109–127, 1995. With permission.)

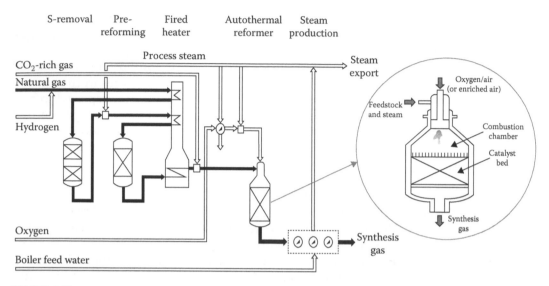

FIGURE 4.11

Schematic diagram of a process concept of syngas production with ATR reactor for GTL plant. (Adapted from Aasberg-Petersen, K. et al., *Fuel Process. Technol.*, 83, 253–261, 2003. With permission.)

TABLE 4.4

Process Economics of ATR at Various Steam-to-Carbon Ratios

Steam-to-carbon ratio	0.2	0.6	1
Relative energy consumption	0.97	1.0	1.03
Relative investment	0.9	1.0	1.1

Source: Dybkjaer, I.B., and T.S. Christensen, *Stud. Surf. Sci. Catal.*, 136, 435–440, 2001.

on the catalyst bed at temperatures of 850°C–1050°C (Pen et al. 1996). The syngas mixture leaves the reactor at temperatures of ~725°C. Chemical reactions are expressed in Equations 4.2, 4.4, and 4.5. The composition of the produced syngas depends on the thermodynamic equilibrium of the ATR reaction at the exit temperature and the pressure of the reactor. The overall ATR process can be operated as exothermic, endothermic, and thermoneutral, depending on the hydrocarbons:oxygen:steam ratio in the feed stream (Ahmed and Krumpelt 2001). The produced synthesis gas is completely free of soot and oxygen. Figure 4.11 depicts a schematic diagram of the ATR reactor within a conceptual process of syngas production for GTL. The ATR reactor basically consists of a ceramic-lined tube, similar to the POX reactor, and a fixed catalyst bed. ATR requires a low O_2/CH_4 ratio (0.55–0.60) in comparison with the POX process (Pen et al. 1996). The steam-to-carbon (H_2O/CO) ratio in ATR determines the economy of the plant; operation at low steam-to-carbon ratio improves the process economics (Table 4.4). A H_2O/C ratio of 0.6 has been proven industrially. Decreasing this ratio decreases the H_2/CO ratio in the product stream and increases the risk of soot formation in the ATR reactor and also the coke deposition in the pre-reforming reactor (Ernst et al. 2000).

According to the process concept of methanol plant using syngas from ATR, the flame core temperature may exceed 3000°C, which requires a catalyst with an excellent thermal stability. Ni catalyst supported on alumina or Mg–Al spinel is being used in the industrial setup.

Haldor Topsøe offers a two-layered catalyst system for the ATR process (Dahl et al. 2014). Ayabe et al. (2003) studied the activity of a series of 2% metal on alumina support and found that the activity of the catalyst system follows the order: Rh > Pd > Ni > Pt > Co. However, 10% Ni on an alumina support has shown higher activity than 2%Rh/Al_2O_3 with a little carbon deposition.

According to a report written by Haldor Topsøe, the capital investment of a syngas plant operated using the ATR route is only 75% of those operated using SMR (Dybkjaer and Madsen 1998). However, operational costs may be the same or even higher due to O_2 production unit required for ATR. Oxygen plant makes up to 40% of the total capital cost of a synthesis gas plant operated using the ATR route (Sogge et al. 1994). As a result, routes based on air to eliminate the air separation plant have been suggested. ATR technology is commercially available, but it still has a limited commercial experience. The main licensors are Haldor Topsøe, Lurgi, ICI, and Foster Wheeler.

4.3.4 Production of Hydrogen by Methane Decomposition

Hydrogen production technologies based on fossil fuel exploitation require the implementation of carbon capture and sequestration or decarburation due to environmental concerns regarding greenhouse gas emissions to the atmosphere. The technological development of steam reforming with carbon capture and sequestration at an industrial level is under way, and the estimated cost of hydrogen produced is nearly double to that without carbon capture. Therefore, the development of greenhouse gas emission-free technologies with fossil fuel utilization might boost the current scenario of sustainable hydrogen production. Innovative new solutions are being put into practice. Decarburation via dry natural gas (methane) cracking is a promising alternative and carbon-free technology (Abbas and Daud 2010). Methane can be thermally decomposed into its elemental species, hydrogen and carbon, according to the following reaction (Catón et al. 2001):

$$CH_4 \leftrightarrow C_{(S)} + 2H_2 \qquad \Delta H^0_{298K} = 75.6 \text{ kJ/mol} \qquad (4.33)$$

Methane decomposition reaction enables a direct production of pure hydrogen, free of carbon oxides, which is suitable for fuel cells or any other purposes. Methane decomposition requires only half of the energy (37.8 kJ/mol H_2) that is required for SMR (63.3 kJ/mol H_2) (Dunker et al. 2006). However, due to the high strength of C–H bond, this reaction generally takes place at a high temperature of ~1200°C (Abbas and Daud 2010). The temperature at which ΔG_0 is zero for the decomposition reaction (Equation 4.33) is 545°C, suggesting that cracking reaction occurs theoretically above such temperature. However, methane cracking has also been reported below 545°C. The reaction temperature appears to be the most critical issue in methane decomposition, which controls the conversion of methane into hydrogen as well as the formation of secondary hydrocarbons as acetylene or benzene (Gueret et al. 1997). CO_2 emission from the burning of process fuel (methane) is ~0.05 mol CO_2/mol H_2, which is much lower than that of 0.43 mol CO_2/mol H_2 in the SMR process. This CO_2 emission can further be decreased, if a part of produced hydrogen (theoretically 16%) is burnt to produce the process heat (Muradov et al. 2006).

The direct methane decomposition is not economically viable at moderate temperatures due to a low yield of hydrogen. However, reaction temperature can be reduced by the addition of supported active metal catalysts to decompose the hydrocarbons to produce hydrogen at more moderate temperatures. SiO_2 and Al_2O_3 supported the catalyst in active metal phase; Co, Ni, and Fe can be used for decomposition of methane (NG). Nickel is particularly active for natural gas cracking (Avdeeva et al. 1999). Methane catalytic decomposition (MCD) is reported to suffer from black carbon deposition on the active sites of the catalyst, which might inhibit the continuous operation as required by the industrial processes. Different reactor configurations for continuous methane cracking have been proposed, such as a set of parallel fixed bed reactors alternating between different conditions or a fluidized bed/regenerator combination. The fluidized bed reactor system offers more flexibility and ease in continuous operation (Lee et al. 2004). Catalyst regeneration is achieved by continuous catalyst circulation between the cracker and regenerator sections. The other advantages of fluidized bed CMD reactor are improved heat and mass transfer, better utilization of catalyst particles, and small pressure drop.

Methane decomposition is an energy-efficient process; net energy efficiency is evaluated as 58% in terms of H_2, and the remaining 42% is stored in the form of stored energy in carbon particles (Blok et al. 1997). Therefore, methane decomposition offers added advantages over SMR process such as similar energy conversion efficiency, carbon particles having a potential of nonenergy use, simplicity in process flow sheet, and absence of CO_2 in the product streams. Earlier literature on methane cracking reveals that methane decomposition is being primarily used for the production of carbon fibers and hydrogen is a by-product. The decomposition process was conducted at an elevated temperature of 1500°C in firebrick furnaces. Commercial hydrogen production using MCD was first came to light with the introduction of HYPRO process by Universal Oil Products in 1996 (Pohlenz and Scott 1996). The HYPRO process was based on a circulating fluidized bed using a 7% Ni/Al_2O_3 catalyst at an operating temperature of up to 980°C and at atmospheric pressure. A 90% conversion of CH_4 was achieved. The important feature of the HYPRO process is its catalyst regeneration scheme; the spent catalyst was burnt in a separate regenerator, and the energy produced from the burning is used for process energy requirements. Recent analysis of the economics of methane decomposition process indicates that the catalyst regeneration scheme and the selling price of produced carbon are the two major factors in the decision making of the economic viability of the process.

4.3.4.1 Thermodynamics of Methane Cracking

The mechanism of catalytic methane cracking is very complex in nature. The overall reaction mechanism is given by Equation 4.33. Each mole of methane decomposes into two moles of hydrogen and one mole of carbon. Decomposition reaction is moderately endothermic and favored at high temperature and low pressure. The coke formation is an integral part of methane cracking reaction. In methane cracking, the amount and type of coke formed depends on several factors such as methane flow rate, operating temperature and pressure, and the amount of H_2 present in the reactor. The effects of the operating temperature and pressure on the equilibrium conversion of CH_4 in the decomposition process are depicted in Figure 4.12. Quantitative thermodynamic analysis of the above cracking reaction is done by calculating its equilibrium constants as given by Equation 4.34. Assuming the activity of the carbon catalyst as unity, the equilibrium constant for methane catalytic cracking reaction is described in terms of partial pressures of CH_4 and H_2 (Rostrup-Nielsen 1972).

$$K_P = \frac{\left(P_{H_2}^2 \right)_{eq}}{\left(P_{CH_4} \right)_{eq}} \tag{4.34}$$

P_{H_2}, P_{CH_4} are the partial pressures of H_2 and CH_4 respectively.

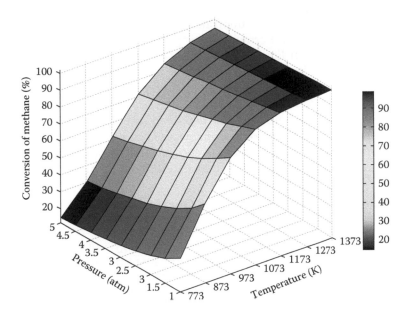

FIGURE 4.12
Equilibrium conversion of methane under different pressure and temperature. (Adapted from Li, Y. et al., *Catal. Today*, 162, 1–48, 2011. With permission.)

Mechanistic analysis of carbon formation on the surface and the interior of the solid catalyst show that carbon is involved in various activities of the active catalyst. It is well known that carbon deposition propagates in layers over Ni by making a solution with it. Carbon diffuses through Ni toward metallic support until supersaturation is reached. Therefore, carbon formation affects the equilibrium, and its activity seems to be inhomogeneous over active metals.

The equilibrium constant for the overall reaction, considering the above-described mechanism, is expressed as follows (Fukada et al. 2004):

$$K_{CH_4/H_2}^{sol} = \frac{\left(P_{H_2}^2\right)_{eq} c_{CH_4/H_2}^{sol}}{\left(P_{CH_4}\right)_{eq}} = \exp\left(\frac{-\Delta G_{CH_4/H_2}^0}{RT}\right) \tag{4.35}$$

where c_{CH_4/H_2}^{sol} is the solubility of carbon in the active metal (mol C/volume of active metal particles).

The Gibbs free energy for decomposition reactions is computed by Equation 4.36, which is proposed by Ginsburg et al. (2005), with the consideration of the graphite type of carbon deposition:

$$\Delta G_0 = 58886.79 + 270.55T + 0.0311T^2 - 3x10^{-6}T^3 + 291405.7/T - 54.598T\ln(T) \tag{4.36}$$

The temperature at which ΔG_0 is zero for decomposition reaction (Equation 4.36) is 547°C. This implies that methane decomposition reaction can only take place at a temperature of above 547°C. However, cracking of methane is also reported, operating at a temperature of lower than 500°C. The reason for this low-temperature methane cracking may be the type and surface structure of carbon deposited on the catalyst surface other than the graphite (Rahman et al. 2006). Many authors reported that the deviation of the calculated ΔG_0 and the experimental equilibrium data are considerable; a maximum of 35% deviation was reported. Deposited carbon may be amorphous, filamentous, and graphite platelets, all having different structures and thermodynamic properties (Ginsburg et al. 2005).

Here, K_P is defined through Equation 4.34, K_M is defined by the equation analogous to Equation 4.34 and K_M^* is called "coking threshold", a condition at which overall methane cracking is zero K_M^* can be estimated from Equation 4.34, if above said condition is satisfied.

Depending on the relative values of K_M^* and K_M, the following conclusions can be drawn: if $K_M > K_M^*$, no coking occurs at all, and if the deposited carbon is graphite, K_M^* will be equal to K_P. In case of encapsulated carbon or filamentous carbon, K_M is smaller than K_M^*. However, K_M^* does not determine whether filamentous or encapsulated carbon will be produced. Hence, K_M^* does not give a clue regarding the life of the catalyst under filamentous carbon formation.

4.3.4.2 Technological Alternatives of Methane Cracking

Methane cracking is a nonoxidative processing of natural gas for the production of hydrogen and carbon black. It is a scientifically proven technology, but its industrial implementation based on reliable, economic, continuous, and sustainable process is still a concern of R&D. Coke formation in the reactor is a main showstopper in the methane cracking process, because it either blocks the reactor in direct methane cracking or deactivates the catalyst in catalytic cracking. Innovative new ideas are being created for a successful development of a continuous methane cracking technology. Figure 4.13 shows different paths of methane cracking, starting with direct pyrolysis of methane in a tubular reactor and ending on a latest technology of methane cracking, methane cracking in molten salt media and shows a successive advancement in methane cracking technologies in time.

4.3.4.2.1 Thermal Cracking of Methane

When natural gas is heated alone in the reactor, it decomposes into its elemental species, hydrogen and carbon, according to Equation 4.33. The European Commission initiated the SOLHYCARB project in 2009. The SOLHYCARB project was aimed to coproduce hydrogen and carbon black from natural gas. The project was based on the noncatalytic route of methane cracking in which a 10 kW solar thermal industrial-scale reactor was tested for a temperature range of 1467°C–1997°C and a pressure of 1 bar. Methane conversion was found to

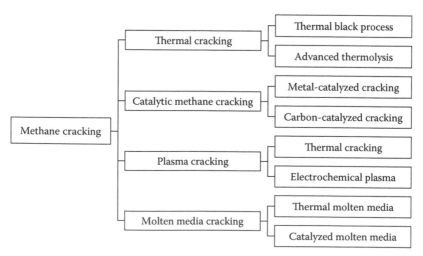

FIGURE 4.13
Different technological routes of methane decomposition.

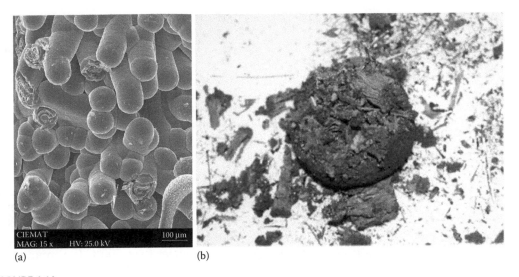

FIGURE 4.14
(a) Carbon particles formed during the test at 1700°C in the CEIMAT experiment and (b) carbon plug specimen obtained in one of the CEIMAT experiments. (Adapted from Abanades, A. et al., *Energy*, 46, 359–363, 2012. With permission.)

be approximately 97% (Rodat et al. 2011). The energy efficiency of this SOLHYCARB process is estimated to be approximately 55%. Abanades et al. (2012) studied the direct methane cracking in a tubular ceramic reactor at CIEMAT and showed that the deposition of carbon black is the major problem in continuous operation of methane cracking. Figure 4.14 shows the agglomeration of typical hard carbon particles in the CIEMAT experiment.

The mechanism of direct thermal cracking of methane has been extensively studied by several researchers (Dean 1990; Hidaka et al. 1990; Olsvik et al. 1995; Snoeck et al. 1997). Methane conversion and distribution of H_2, ethylene, benzene, and other polycyclic hydrocarbons in the product stream are governed by the operating temperature and pressure. The reaction mechanism is very complex due to a large number of intermediate steps involved in completion of the reaction which is not well understood. According to Holmen et al. (1995), the overall thermal decomposition of methane at high temperature can be expressed as a stepwise dehydrogenation reaction:

$$2CH_4 \rightarrow C_2H_6 \rightarrow C_2H_4 \rightarrow C_2H_2 \rightarrow 2C \tag{4.37}$$

Product formation follows the free radical mechanism. Production of hydrogen, ethane, ethene, and benzene is clearly understood, but the details of higher hydrocarbons and coke formation are not yet clearly understood. Free radical mechanism comprises primary and secondary reactions; initialization step involves the formation of H_2 and ethane.

$$CH_4 \rightarrow CH_3\bullet + H\bullet \tag{4.38}$$

$$CH_4 \rightarrow H\bullet + CH_3\bullet + H_2 \tag{4.39}$$

$$2CH_3\bullet \rightarrow C_2H_6 \tag{4.40}$$

The formation of acetylene might involve the reverse of reaction (Equation 4.40) and by the following secondary reactions of ethylene (Back and Back 1983).

$$C_2H_6 \rightarrow 2CH_3\bullet \tag{4.41}$$

$$C_2H_4 + H\bullet \rightarrow C_2H_3 + H_2 \tag{4.42}$$

$$C_2H_4 + CH_3\bullet \rightarrow C_2H_3\bullet + CH_4 \tag{4.43}$$

$$C_2H_3\bullet \rightarrow C_2H_2 + H\bullet \tag{4.44}$$

Most of the mechanistic studies of methane dehydrogenation confirm that the primary source of free radicals is the homogeneous dissociation of methane and it controls the rate of the overall process. After the formation of acetylene by a series of parallel reactions, it plays an important role in the formation of coke and those polynuclear aromatics.

4.3.4.2.2 Methane Catalytic Decomposition

Owing to the high strength of C–H bond and the low yield of hydrogen, direct methane decomposition is not economically viable at moderate temperatures (below 1200°C). However, the reaction temperature can be reduced by the addition of supported active catalysts to decompose natural gas at a temperature of lower than 500°C (Rahman et al. 2006). To make this process economically viable and to reduce the operating temperature, various metal and carbonaceous catalysts have been introduced into the methane decomposition process. Figure 4.15 shows the type of carbon deposited in MCD with and without catalyst.

Since 1960, numerous researches have been conducted in search of a suitable catalyst system and to reduce the impact of carbon formation in the MCD process. Activated carbon and carbon black have been in focus due to certain advantages over other carbon materials. Various supported or powdered transition metals such as Co, Fe, Mo, Ni, Cu, Ru, Rh, Pt, Re, Ir, Pd, and W have been studied for metal-catalyzed methane decomposition. However, Ni or Ni/alumina is found to have the highest catalytic activity (Ashik et al. 2015). They have been used as a single metal catalyst or a bi-metallic catalyst with and without promoters. Alumina, silica, and carbon (mainly AC) are mainly used as

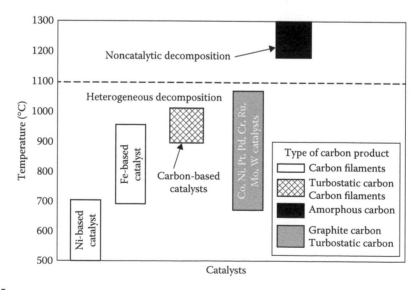

FIGURE 4.15
Summary of the types of deposited carbon and the corresponding temperature range related to catalytic methane decomposition reaction. (Modified from Muradov, N.Z. and T.N. Veziroğlu, *Int. J. Hydrogen Energ.,* 30, 225–237, 2005. With permission.)

the support material for catalysts in MCD. Along with transition metals, applications of various carbon materials such as activated carbon, carbon black, acetylene black, glassy carbon, carbon nanotube, graphite, diamond powder, and fullerenes are widely reported for methane thermal decomposition (Al-Hassani et al. 2014).

The mechanism of methane cracking of catalytic action mainly involves five steps (Suelves et al. 2009), which are as follows: (1) chemisorption of methane on the leading face of a catalyst particle; (2) progressive breaking of four C–H bonds of the chemisorbed methane molecules; (3) assembling of adsorbed atomic hydrogen into molecules followed by gas-phase emission; (4) atomic carbon aggregation into encapsulated carbon leading to progressive catalyst deactivation or atomic carbon diffusion through the bulk catalyst, from the leading face to the trailing face driven by the existing concentration gradients; and (5) carbon nucleation followed by formation and growth of carbon nanofibers in the trailing face of the catalyst particle. Numerous studies are being conducted around the globe in search of a suitable set of catalyst supports and reactor configurations to produce a high yield of hydrogen at moderated temperature. In every case, catalyst deactivation by coke deposition is the major problem.

4.3.4.2.3 Plasma Cracking of Methane

Plasma-assisted cracking of natural gas for the production of hydrogen with carbon is an energy-efficient and environment-friendly process. This method falls into two categories: thermal and nonthermal plasma cracking processes. The thermal plasma process uses an electrical power of ~1 kW and the reactor attains a temperature of approximately 5,000°C–10,000°C. Thus, this process is a highly energy-consuming process. One effective and clean way to carry out the thermal plasma process is to use the direct current and/or radio frequency excitation. Industrial implementation of thermal plasma-assisted methane decomposition was done by Kvaerner, Norway, in 1990 to produce hydrogen and carbon black at a high temperature through the development of carbon black and hydrogen process (Gaudernack and Lynum 1998). A plasma assisted methane cracking plant consumes 2100 kW of electricity with 1000 kW of high-temperature steam to produce 2000 Nm³/h of hydrogen and 500 kg/h of pure carbon. The benefits of using plasma arc in methane decomposition are fourfold: (1) high thermal efficiency (>80%), (2) high purity of hydrogen (98 vol.%), (3) simultaneous production of valuable by-product carbon, and (4) very low CO_2 emissions. However, the consumption of relatively high electrical energy in plasma processes can make their applicability restricted to large-scale processes. Muradov et al. (2009) practically applied the concept of plasma arc in methane decomposition and developed a decarburation reactor using a carbon-based catalyst. The experimental study shows that the carbon generated in the plasma arc has higher catalytic performance in comparison with other carbon-based catalysts. Nonthermal plasma can catalyze the decomposition reaction at rather low temperature (350°C–750°C) through the generation of active species of fast electrons. The use of plasma technology in decomposition of methane to produce hydrogen and carbon black has gained strong interest in recent decades. Most of the research is directed toward the development of more sophisticated and compact decarburation reactor for stand-alone applications such as hydrogen generation for fuel cell application.

4.3.4.2.4 Molten Salt Media Cracking of Methane

Methane cracking reaction is a scientifically proven technology for production of hydrogen and carbon black, but its industrial-scale implementation requires the design and development of continuous, sustainable, and economic process. Catalyst deactivation and blockage of decarburation reactor due to carbon deposition is a technological showstopper. Therefore, there is a need to develop a system that can remove the carbon particles from the reactor either by chemical cleaning mechanically or by controlling the process. In this way, the application of liquid media has been proposed to remove the carbon particles effectively from the reactor. Steinberg and Dong (1998) proposed the use of molten salts (NaCl, NaF) and molten metals (iron, tin) in a bubble column reactor. Molten media can catalyze the decomposition reaction as well as separate the carbon particles from the reaction zone by the differences in densities of black carbon and molten media. Separation of carbon from molten media is similar to the blast furnace operation, where the slag is skimmed off the surface of molten salt or metal. Wang et al. (2008), in their experiment of methane decomposition using molten media, demonstrated that molten magnesium was an efficient catalyst for homogeneous methane decomposition for the production of hydrogen and simultaneous separation of carbon particles from the reaction zone in a continuous operation mode. Methane cracking reaction using molten media is a promising one, although a continuous operation scheme for industrial implementation of methane cracking still needs to be justified through a lot of experimental work. The compatibility between hydrogen, carbon, and molten media, and the stability of molten media at various temperature ranges of the process might be the main source of difficulties. These questions have to be answered before thinking a large-scale methane cracking operation.

4.4 Conclusions

The long-term potential of hydrogen as an energy carrier in an environmentally sustainable energy system of the future is broadly accepted. Near to mid-term methodologies for the transition from the present fossil fuel-based energy system to the hydrogen-based energy system is extensively discussed in literature. This type of energy transition system should be based on potent energy conservation measures. Owing to the advancement of hydrogen production, distribution, and storage technologies, the demand for hydrogen is increasing continuously day by day. Reforming of natural gas is the most developed and the most economic method of hydrogen production in the present time. SMR technology is being employed in a large number of commercial hydrogen production plants. Natural gas will still play a major role in hydrogen production in the coming future due to its wider availability, cost competitiveness, and convenience of storage. Development of associated gas and methane hydrate resource would give a boost to the reforming technology. Although high greenhouse gas emissions and extensive energy utilization in hydrogen production using natural gas reforming technologies are raising the environmental concern, natural gas steam reforming with CO_2 sequestration seems to be a straight forward solution to environmental concerns. However, current knowledge of the state of the art does not predict the real green hydrogen production scenario. The challenges with CO_2 sequestration that are remained unanswered are the cost-effective operation over a long time, the understanding of reservoir systems, and the effect on the environment. Novel approaches to steam reforming such as membrane reforming, plasma reforming, and microchannel reforming are getting much attention in the last two decades. However, industrial applications of these technologies require high level of R&D on a pilot plant scale. Detailed modeling and simulation of these novel systems is needed to carry out the systematic parametric study. Total decarburation and thermal and catalytic cracking of natural gas might be potential alternatives to conventional SMR technology coupled with carbon capture. Methane cracking is an environmentally benign and scientifically viable technology, but it is not yet developed at an industrial level. There is a need to develop cheap catalysts and continuous reactor systems for methane cracking to make it economically viable. Other competitive reforming technologies such as POX and ATR of natural gas are also commercially implemented, but the limited industrial experience and low H_2/CO ratio in product stream make them unfavorable for hydrogen production. CPO is a cost-effective and reliable hydrogen production technology, but more R&D efforts are needed to implement this technology on an industrial scale. Dry reforming, bi-reforming, and tri-reforming of natural gas are also proposed in literature as an alternative to SMR technology. However, due to low H_2/CO ratio of the product, these technologies cannot be considered for industrial hydrogen production. However, reforming systems need to be renovated in terms of high energy efficiency with high economic performance. In this way, the use of alternative energy source such as nuclear energy and solar energy to generate process heat is likely to be implemented in the reforming technology of the future. Overall, the future hydrogen production from natural gas will take place in a more eco-friendly and energy efficient way.

Acknowledgment

The authors thank the Australia–India Strategic Research Fund for the financial support of this work, which forms a part of the Mini-DME project. The aim of the Mini-DME project is the conversion of natural gas to a clean and low carbon emission fuel, dimethyl ether.

References

Aasberg-Petersen, K., T. S. Christensen, C. S. Nielsen, and I. Dybkjær. 2003. Recent developments in autothermal reforming and pre-reforming for synthesis gas production in GTL applications. *Fuel Processing Technology* 83(1): 253–261.

Abanades, A., C. Rubbia, and D. Salmieri. 2012. Technological challenges for industrial development of hydrogen production based on methane cracking. *Energy* 46(1): 359–363.

Abbas, H. F., and W. W. Daud. 2010. Hydrogen production by methane decomposition: A review. *International Journal of Hydrogen Energy* 35(3): 1160–1190.

Ahmed, S., and M. Krumpelt. 2001. Hydrogen from hydrocarbon fuels for fuel cells. *International Journal of Hydrogen Energy* 26(4): 291–301.

Ajayi-Oyakhire, O. 2012. *Hydrogen - Untapped Energy?* The Institution of Gas Engineers and Managers (IGEM). Accessed August 5. http/www.igem.org.uk.

Akers, W. W., and D. P. Camp. 1995. Kinetics of the methane-steam reaction. *AIChE Journal* 1(4): 471–475.

Al-Hassani, A. A., H. F. Abbas, and W. W. Daud. 2014. Production of CO_x-free hydrogen by the thermal decomposition of methane over activated carbon: Catalyst deactivation. *International Journal of Hydrogen Energy* 39(27): 14783–14791.

Annual Energy Outlook (AEO). 2015 with projections to 2040. U.S. Energy Information Administration, U.S. Department of Energy Washington, DC. Accessed August 15. http/www.eia.gov/forecasts/aeo.

Armor, N. 1998. Important targets in environmental catalysis. *Research on Chemical Intermediates* 24(2): 105–113.

Arutyunov, V. S., and O. V. Krylov. 1998. Oxidative transformations of methane. Nauka Publishers, Moscow, Russia.

Ashik, U. P. M., W. W. Daud, and H. F. Abbas. 2015. Production of greenhouse gas free hydrogen by thermocatalytic decomposition of methane–A review. *Renewable and Sustainable Energy Reviews* 44: 221–256.

Avdeeva, L. B., D. I. Kochubey, and S. K. Shaikhutdinov. 1999. Cobalt catalysts of methane decomposition: Accumulation of the filamentous carbon. *Applied Catalysis A: General* 177(1): 43–51.

Ayabe, S., H. Omoto, T. Utaka, R. Kikuchi, K. Sasaki, Y. Teraoka, and K. Eguchi. 2003. Catalytic autothermal reforming of methane and propane over supported metal catalysts. *Applied Catalysis A: General* 241(1): 261–269.

Back, M. H., and R. A. Back. 1983. *Thermal Decomposition and Reactions of Methane*. Academic Press, New York.

Barbir, F., M. Fuchs, A. Husar, and J. Neutzler. 2000. *Design and operational characteristics of automotive PEM fuel cell stacks*. No. 2000-01-0011. SAE Technical Paper.

Berman, A., R. K. Karn, and M. Epstein. 2006. FUEL 140-Steam reforming of methane on Ru-catalysts for solar hydrogen production. In *Abstracts of Papers of The American Chemical Society*, vol. 232, American chemical society, Washington, DC.

Bharadwaj, S. S., and L. D. Schmidt. 1995. Catalytic partial oxidation of natural gas to syngas. *Fuel Processing Technology* 42(2): 109–127.

Blok, K., R. H. Williams, R. E. Katofsky, and C. A. Hendriks. 1997. Hydrogen production from natural gas, sequestration of recovered CO_2 in depleted gas wells and enhanced natural gas recovery. *Energy* 22(2): 161–168.

BP Statistical Review of World Energy June 2015. British Petroleum Corporation, James's Square, London. Accessed August 15. http://www.bp.com/en/global/corporate/energy-economics/statistical-review-of-world-energy.html.

Bromberg, L., D. R. Cohn, A. Rabinovich, and N. Alexeev. 1999. Plasma catalytic reforming of methane. *International Journal of Hydrogen Energy* 24(12): 1131–1137.

Bromberg, L., D. R. Cohn, A. Rabinovich, C. O'Brie, and S. Hochgreb. 1998. Plasma reforming of methane. *Energy & Fuels* 12(1): 11–18.

Catón, N., J. I. Villacampa, C. Royo, E. Romeo, and A. Monzón. 2001. Hydrogen production by catalytic cracking of methane using $Ni-Al_2O_3$ catalysts. Influence of the operating conditions. *Studies in Surface Science and Catalysis* 139: 391–398.

Chauvel, A., and Lefebvre, G. 1989. Petrochemical Processes, Part 1: Synthesis-Ga Derivatives and Major Hydrocarbons, Editions Technip, Paris, France.

Christensen, T. S., and I. I. Primdahl. 1994. Improve syngas production using autothermal reforming. *Hydrocarbon Processing; (United States)* 73(3): 39–46.

Chen, L., Y. Lu, Q. Hong, J. Lin, and F. M. Dautzenberg. 2005. Catalytic partial oxidation of methane to syngas over Ca-decorated-Al_2O_3-supported Ni and NiB catalysts. *Applied Catalysis A: General* 292: 295–304.

Claridge, J. B., M. L. H. Green, S. C. Tsang, A. P. E. York, A. T. Ashcroft, and P. D. Battle. 1993. A study of carbon deposition on catalysts during the partial oxidation of methane to synthesis gas. *Catalysis Letters* 22(4): 299–305.

Dahl, P. J., T. S. Christensen, S. Winter-Madsen, and S. M. King. 2014. Proven autothermal reforming technology for modern large-scale methanol plants. *Nitrogen and Syngas*: 12.

Dean, A. M. 1990. Detailed kinetic modeling of autocatalysis in methane pyrolysis. *Journal of Physical Chemistry* 94(4): 1432–1439.

Demirbas, A. 2010. *Methane Gas Hydrate*. Berlin, Germany, Springer Science & Business Media.

Dincer, I., and C. Acar. 2015. Review and evaluation of hydrogen production methods for better sustainability. *International Journal of Hydrogen Energy* 40(34): 11094–11111.

Dincer, I., and C. Zamfirescu. 2012. Sustainable hydrogen production options and the role of IAHE. *International Journal of Hydrogen Energy* 37(21): 16266–16286.

DTI report: Directed Technologies, Inc. (DTI), Air Products and Chemicals, BOC Gases, The Electrolyser Corp., and Praxair, Inc., July 1997, Hydrogen Infrastructure Report, prepared for Ford Motor Company under USDOE Contract No. DE-AC02-94CE50389, Purchase Order No. 47-2R31148.

Dunker, A. M., S. Kumar, and P. A. Mulawa. 2006. Production of hydrogen by thermal decomposition of methane in a fluidized-bed reactor—Effects of catalyst, temperature, and residence time. *International Journal of Hydrogen Energy* 31(4): 473–484.

Dybkjaer, I. B., and T. S. Christensen. 2001. Syngas for large-scale conversion of natural gas to liquid fuels. *Studies in Surface Science and Catalysis* 136: 435–440.

Dybkjaer, I. B., and S. W. Madsen. 1998. Advanced reforming technologies for hydrogen production. *International Journal of Hydrocarbon Engineering* 3: 56–65.

Ernst, W. S., S. C. Venables, P. S. Christensen, and A. C. Berthelsen. 2000. Push syngas production limits. *Hydrocarbon Processing* 79(3): 100C–100J.

Ewan, B. C. R., and R. W. K. Allen. 2005. A figure of merit assessment of the routes to hydrogen. *International Journal of Hydrogen Energy* 30(8): 809–819.

Foulds, G. A., and B. F. Gray. 1995. Homogeneous gas-phase partial oxidation of methane to methanol and formaldehyde. *Fuel Processing Technology* 42(2): 129–150.

Fukada, S., N. Nakamura, J. Monden, and M. Nishikawa. 2004. Experimental study of cracking methane by Ni/SiO$_2$ catalyst. *Journal of Nuclear Materials* 329: 1365–1369.

Gallucci, F., E. Fernandez, P. Corengia, and M. S. Annaland. 2013. Recent advances on membranes and membrane reactors for hydrogen production. *Chemical Engineering Science* 92: 40–66.

Gangadharan, P., K. C. Kanchi, and H. H. Lou. 2012. Evaluation of the economic and environmental impact of combining dry reforming with steam reforming of methane. *Chemical Engineering Research and Design* 90(11): 1956–1968.

Gaudernack, B., and S. Lynum. 1998. Hydrogen from natural gas without release of CO$_2$ to the atmosphere. *International journal of hydrogen energy* 23(12): 1087–1093.

Ginsburg, J. M., J. Piña, T. E. Solh, and H. I. De Lasa. 2005. Coke formation over a nickel catalyst under methane dry reforming conditions: Thermodynamic and kinetic models. *Industrial & Engineering Chemistry Research* 44(14): 4846–4854.

Gueret, C., M. Daroux, and F. Billaud. 1997. Methane pyrolysis: Thermodynamics. *Chemical Engineering Science* 52(5): 815–827.

Halvorson, T., and P. Farris. 1997. Onsite hydrogen generator for vehicle refueling application. In *Proceedings of the '97 World Car Conference*, pp. 19–22.

Han, C., and D. P. Harrison. 1994. Simultaneous shift reaction and carbon dioxide separation for the direct production of hydrogen. *Chemical Engineering Science* 49(24): 5875–5883.

Häussinger, P., R. Lohmüller, and A. M. Watson. 1989. Hydrogen. In *Ullmann's Encyclopedia of Industrial Chemistry*, vol. A13, 5th ed., edited by Elvers, B., Hawkins, S., Ravenscroft, M., and Schulz, G., pp. 297–442, Weinheim, Germany: VCH Verlagsgesellschaft.

Hawkins, G. B. 2013. *Steam Reforming-Poisons.* Technical presentation from GBH Enterprises Ltd. Accessed September 10. http://www.slideshare.net/GerardBHawkins/steam-reforming-poisons.

Hidaka, Y., T. Nakamura, H. Tanaka, K. Inami, and H. Kawano. 1990. High-temperature pyrolysis of methane in shock waves. Rates for dissociative recombination reactions of methyl radicals and for propyne formation reaction. *International Journal of Chemical Kinetics* 22(7): 701–709.

Hinrichsen, K. O., K. Kochloefl, and M. Muhler. 2008. Water gas shift and COS removal. In *Handbook of Heterogeneous Catalysis*, Vol. 6, edited by G. Ertl, H. Knozinger, F. Schuth, and J. Weitkamp, pp. 2905–2920. Weinheim, Germany: Wiley-VCH.

Holladay, J. D., J. Hu, D. L. King, and Y. Wang. 2009. An overview of hydrogen production technologies. *Catalysis Today* 139(4): 244–260.

Holmen, A., O. Olsvik, and O. A. Rokstad. 1995. Pyrolysis of natural gas: Chemistry and process concepts. *Fuel Processing Technology* 42(2): 249–267.

Hufton, J. R., S. Mayorga, and S. Sircar. 1999. Sorption-enhanced reaction process for hydrogen production. *AIChE Journal* 45(2): 248–256.

Huszar, K., G. Racz, and G. Szekely. 1971. Investigation of partial catalytic oxidation of methane. 1. Conversion rates in a single-grain reactor. *Acta Chimica Academiae Scientarium Hungaricae* 70(4): 287.

Hydrogen Generation Market by Generation & Delivery Mode (Captive, Merchant), Technology (Steam Methane Reforming, Partial Oxidation, Gasification, and Electrolysis), Application (Refinery, Ammonia Production, and Methanol Production), & Region - Global Forecast to 2021, Markets and Markets. Accessed November 15. http://www.marketsandmarkets.com/PressReleases/hydrogen.asp.

Iulianelli, A., S. Liguori, J. Wilcox, and A. Basile. 2016. Advances on methane steam reforming to produce hydrogen through membrane reactors technology: A review. *Catalysis Reviews*: 1–35.

Joensen, F., and J. R. Rostrup-Nielsen. 2002. Conversion of hydrocarbons and alcohols for fuel cells. *Journal of Power Sources* 105(2): 195–201.

Johnsen, K., H. J. Ryu, J. R. Grace, and C. J. Lim. 2006. Sorption-enhanced steam reforming of methane in a fluidized bed reactor with dolomite as CO_2-acceptor. *Chemical Engineering Science* 61(4): 1195–1202.

Jun, H., M. Park, S. C. Baek, J. WookBae, K. Ha, and K.-W. Jun. 2011. Kinetics modeling for the mixed reforming of methane over Ni-CeO_2/$MgAl_2O_4$ catalyst. *Journal of Natural Gas Chemistry* 20(1): 9–17.

Kalamaras, C. M., and A. M. Efstathiou. 2013. Hydrogen production technologies: Current state and future developments. In *Conference Papers in Science*, vol. 2013, Hindawi Publishing Corporation, Cairo, Egypt.

Korobitsyn, M. A., F. P. F. Van Berkel, G. M. Christie, J. P. P. Huijsmans, and C. A. M. van der Klein. 2000. Sofc as a gas separator. *ECN-Fuels, Conversion & Environment*.

Lee, K. K., G. Y. Han, K. J. Yoon, and B. K. Lee. 2004. Thermocatalytic hydrogen production from the methane in a fluidized bed with activated carbon catalyst. *Catalysis Today* 93: 81–86.

Li, Y., D. Li, and G. Wang. 2011. Methane decomposition to CO_x-free hydrogen and nano-carbon material on group 8–10 base metal catalysts: A review. *Catalysis Today* 162(1): 1–48.

Liander, H. 1929. The utilization of natural gases for the ammonia process. *Transactions of the Faraday Society* 25: 462–472.

Marrelli, L., M. D. Falco, and G. Iaquaniello. 2011. Integration of selective membranes in chemical processes: Benefits and examples. In *Membrane Reactors for Hydrogen Production Processes*, pp. 1–19. London, Springer.

Midilli, A., M. Ay, I. Dincer, and M. A. Rosen. 2005. On hydrogen and hydrogen energy strategies: I: Current status and needs. *Renewable and Sustainable Energy Reviews* 9(3): 255–271.

Mitchell, W. L., J. H. J. Thijssen, J. M. Bentley, and N. J. Marek. 1995. *Development of a catalytic partial oxidation ethanol reformer for fuel cell applications*. No. CONF-951210--. Society of Automotive Engineers, Inc., Warrendale, PA.

Mokhatab, S., W. A. Poe, and J. Y. Mak. 2015. *Handbook of Natural Gas Transmission and Processing: Principles and Practices*. Gulf Professional Publishing, Waltham, MA.

Moulijn, J. A., M. Makkee, and A. E. V. Diepen. 2013. *Chemical Process Technology*. John Wiley & Sons, Sussex, UK.

Muradov, N., F. Smith, G. Bockerman, and K. Scammon. 2009. Thermocatalytic decomposition of natural gas over plasma-generated carbon aerosols for sustainable production of hydrogen and carbon. *Applied Catalysis A: General* 365(2): 292–300.

Muradov, N., F. Smith, C. Huang, and T. Ali. 2006. Autothermal catalytic pyrolysis of methane as a new route to hydrogen production with reduced CO_2 emissions. *Catalysis Today* 116(3): 281–288.

Muradov, N. Z., and T. N. Veziroglu. 2005. From hydrocarbon to hydrogen–carbon to hydrogen economy. *International Journal of Hydrogen Energy* 30(3): 225–237.

Murphy, D. M., A. Manerbino, M. Parker, J. Blasi, R. J. Kee, and N. P. Sullivan. 2013. Methane steam reforming in a novel ceramic microchannel reactor. *International Journal of Hydrogen Energy* 38(21): 8741–8750.

Ogawa, M., and T. Nishihara. 2004. Present status of energy in Japan and HTTR project. *Nuclear Engineering and Design* 233(1): 5–10.

Ogden, J. M. 2001. Review of small stationary reformers for hydrogen production, in *Report to the International Energy Agency* 609.

Olivier, J. G. J., G. Janssens-Maenhout, M. Muntean, and J. A. H. W. Peters. 2012. Trends in global CO_2 emissions; 2012 Report. PBL Netherlands Environmental Assessment Agency. Institute for Environment and Sustainability of the European Commission's Joint Research Centre.

Olsvik, O., O. A. Rokstad, and A. Holmen. 1995. Pyrolysis of methane in the presence of hydrogen. *Chemical Engineering & Technology* 18(5): 349–358.

Pa, P. S. 2008. Performance assessment of proton exchange membrane fuel cell on toy design. *ECS Transactions* 16(2): 729–740.

Padban, N., and V. Becher. 2005. Clean Hydrogen-rich Synthesis Gas, Literature and State-of-the-Art review, (Re: Methane Steam Reforming), Report No. CHRISGAS October 2005_WP11_D89.

Pan, W. 2002. Tri-reforming and combined reforming of methane for producing syngas with desired H_2/CO ratios. PhD thesis, Pennsylvania State University.

Pen, M. A., J. P. Gomez, and J. L. G. Fierro. 1996. New catalytic routes for syngas and hydrogen production. *Applied Catalysis A: General* 144(1): 7–57.

Pohlenz, J. B., and N. H. Scott. 1996. Method for hydrogen production by catalytic decomposition of a gaseous hydrocarbon stream. U.S. Patent 3,284,161, November 8, 1966.

Prettre, M. C., and M. Perrin. 1946. The catalytic oxidation of methane to carbon monoxide and hydrogen. *Transactions of the Faraday Society* 42: 335b–339.

Provendier, H., C. Petit, C. Estournes, S. Libs, and A. Kiennemann. 1999. Stabilization of active nickel catalysts in partial oxidation of methane to synthesis gas by iron addition. *Applied Catalysis A: General* 180(1): 163–173.

Rahman, M. S., E. Croiset, and R. R. Hudgins. 2006. Catalytic decomposition of methane for hydrogen production. *Topics in Catalysis* 37(2): 137–145.

Rodat, S., S. Abanades, and G. Flamant. 2011. Co-production of hydrogen and carbon black from solar thermal methane splitting in a tubular reactor prototype. *Solar Energy* 85(4): 645–652.

Rostrup-Nielsen, J. R. 1972. Equilibria of decomposition reactions of carbon monoxide and methane over nickel catalysts. *Journal of Catalysis* 27(3): 343–356.

Rostrup-Nielsen, J. R. 1984. *Catalytic Steam Reforming*. Springer, Berlin, Germany.

Rostrup-Nielsen, J. R., and J. H. Bak Hansen. 1993. CO_2-reforming of methane over transition metals. *Journal of Catalysis* 144(1): 38–49.

Scholz, W. H. 1993. Processes for industrial production of hydrogen and associated environmental effects. *Gas Separation & Purification* 7(3): 131–139.

Shinke, N., S. Higashiguchi, and K. Hirai, October 30–November 2, 2000, *Fuel Cell Seminar Abstracts,* Portland, OR, pp. 292–295.

Shirasaki, Y., T. Tsuneki, Y. Ota, I. Yasuda, S. Tachibana, H. Nakajima, and K. Kobayashi. 2009. Development of membrane reformer system for highly efficient hydrogen production from natural gas. *International Journal of Hydrogen Energy* 34(10): 4482–4487.

Snoeck, J.-W., G. F. Froment, and M. Fowles. 1997. Filamentous carbon formation and gasification: Thermodynamics, driving force, nucleation, and steady-state growth. *Journal of Catalysis* 169(1): 240–249.

Sogge, J., T. Strøm, and T. Sundset. 1994. Technical and economic evaluation of natural gas based synthesis gas production technologies. *Report of The Foundation for Scientific and Industrial Research (SINTEF)*, STF21 A93106, Trondheim, Norway.

Song, C., and W. Pan. 2004. Tri-reforming of methane: A novel concept for catalytic production of industrially useful synthesis gas with desired H_2/CO ratios. *Catalysis Today* 98(4): 463–484.

Spath, P. L., and D. C. Dayton. 2003. *Preliminary screening-technical and economic assessment of synthesis gas to fuels and chemicals with emphasis on the potential for biomass-derived syngas*. No. NREL/TP-510-34929. National Renewable Energy Lab Golden Co.

Steinberg, M., and Y. Dong. 1998. Method for converting natural gas and carbon dioxide to methanol and reducing CO_2 emissions. U.S. Patent 5,767,165, June 16.

Suelves, I., J. L. Pinilla, M. J. Lázaro, R. Moliner, and J. M. Palacios. 2009. Effects of reaction conditions on hydrogen production and carbon nanofiber properties generated by methane decomposition in a fixed bed reactor using a Ni-Cu-Al catalyst. *Journal of Power Sources* 192(1): 35–42.

Sun, D., X. Li, and L. Cao. 2010. Effect of O_2 and H_2O on the tri-reforming of the simulated biogas to syngas over Ni-based SBA-15 catalysts. *Journal of Natural Gas Chemistry* 190(1): 369–374.

The 2015 United Nations Climate Change Conference (COP 21). 2015. United Nations Environment Programme (UNEP). Accessed January 12. http/www.cop21paris.org.

The International Energy Agency (IEA). 2006. *Hydrogen production and storage-R&D priorities and gaps.* https://www.iea.org/publications/freepublications/publication/hydrogen.pdf.

Tonkovich, A. Y., S. Perry, Y. Wang, D. Qiu, T. LaPlante, and W. A. Rogers. 2004. Microchannel process technology for compact methane steam reforming. *Chemical Engineering Science* 59(22): 4819–4824.

van Beurden, P. 2004. On the catalytic aspects of steam-methane reforming. *Energy Research Centre of the Netherlands (ECN), Technical Report I-04-003.*

Van Hook, J. P. 1980. Methane steam reforming. *Catalysis Reviews—Science and Engineering* 21(1): 1–51.

Verhelst, S. 2014. Recent progress in the use of hydrogen as a fuel for internal combustion engines. *International Journal of Hydrogen Energy* 39(2): 1071–1085.

Vermeiren, W. J. M., E. Blomsma, and P. A. Jacobs. 1992. Catalytic and thermodynamic approach of the oxyreforming reaction of methane. *Catalysis Today* 13(2–3): 427–436.

Wang, K., W. S. Li, and X. P. Zhou. 2008. Hydrogen generation by direct decomposition of hydrocarbons over molten magnesium. *Journal of Molecular Catalysis A: Chemical* 283(1): 153–157.

Wegeng, R. S. 1999. Reducing the size of automotive fuel reformers. In *IQPC Fuel Cells Infrastructure Conference*, Chicago, IL.

White, C. M., R. R. Steeper, and A. E. Lutz. 2006. The hydrogen-fueled internal combustion engine: A technical review. *International Journal of Hydrogen Energy* 31(10): 1292–1305.

Wilhelm, D. J., D. R. Simbeck, A. D. Karp, and R. L. Dickenson. 2001. Syngas production for gas-to-liquids applications: Technologies, issues, and outlook. *Fuel Processing Technology* 71(1): 139–148.

World Hydrogen—Demand and Sales Forecasts, Market Share, Market Size, Market Leaders. "Fredonia Group" Accessed November 2015. http://www.freedoniagroup.com/World-Hydrogen.html.

Xu, J., and G. F. Froment. 1989. Methane steam reforming, methanation and water-gas shift: I. Intrinsic kinetics. *AIChE Journal* 35(1): 88–96.

Zhang, Y., S. Zhang, H. H. Lou, J. L. Gossage, and T. J. Benson. 2014. Steam and Dry reforming processes coupled with partial oxidation of methane for CO_2 emission reduction. *Chemical Engineering & Technology* 37(9): 1493–1499.

Zhu, Q., X. Zhao, and Y. Deng. 2004. Advances in the partial oxidation of methane to synthesis gas. *Journal of Natural Gas Chemistry* 13(4): 191–203.

5

Liquid Fuels from Oil Sands

Sandeep Badoga and Ajay K. Dalai

CONTENTS

ABSTRACT The increasing energy demand is building an impetus for extraction of liquid fuels from oil sands. This chapter describes the process of extraction of liquid transportation fuels from oil sands. The hydrocarbon mixture called bitumen is extracted from oil sands mainly via surface mining. For deep bituminous reserves, the steam-assisted gravity drainage (SAGD) technique is used. The bitumen is separated from sand, clay, or other minerals and water via a series of processes described in this chapter. It is upgraded by fractionation in atmospheric and vacuum distillation units. The long-chain carbon molecules are broken into smaller value-added molecules in hydrocracker and coker units, and the excess carbon is removed in the form of coke in the coker unit. The main product streams from the primary upgrading include naphtha, light gas oil (LGO), and heavy gas oil (HGO). The sulfur and nitrogen content present in the bitumen is transmitted to HGO and LGO. HGO contains 3.5–4.5 wt.% sulfur and 0.35–0.45 wt.% nitrogen, whereas LGO contains 2.0–2.5 wt.% sulfur and 0.12–0.17 wt.% nitrogen. These impurities need to be removed by the hydrotreating process before transferring the HGO and LGO for downstream refining in order to obtain the product specifications of liquid transportation fuel. Hydrotreating is a catalytic process, and a thorough understanding of the catalyst is required to develop a highly efficient catalyst. The better the catalytic performance, the better is liquid fuel quality and process economics. This chapter discusses about various hydrotreating NiMo catalysts supported on a variety of mesoporous materials including SBA15, M-SBA15 (M = Al, Ti, Zr); mesoporous ZrO_2 and Al_2O_3; and mesoporous mixed metal oxide TiO_2–Al_2O_3, ZrO_2–Al_2O_3, and SnO_2–Al_2O_3. The reaction mechanisms and the role of the catalyst are explained by a thorough characterization of catalysts using X-ray diffraction (XRD), N_2 adsorption–desorption isotherms, Fourier transform infrared spectroscopy (FTIR), X-ray absorption near-edge structure (XANES), and high-resolution transmission electron microscopy (HRTEM). The hydrotreated LGO, HGO, and naphtha are blended to make synthetic crude, which is then processed in the existing refineries to obtain high-quality liquid fuels such as diesel, gasoline, kerosene, jet fuel, and petrochemicals. However, this industry faces many challenges, including bigger environmental footprint, tailing ponds, water usage, technological limitations, coke management, and land reclamation.

5.1 Introduction

The worldwide increasing development and flourishing economy needs more energy. Petroleum being the available major source of energy is being exploited to meet the high energy demand. The global oil consumption in 2015 was around 94 million barrels per day (mb/d), and its growth is expected to increase by 1.2 mb/d in 2016 (Oil Market Report 2015; Yardeni and Johnson 2015). To comply with this huge demand for oil, both conventional and nonconventional resources are exploited. The conventional resources are those that provide crude oil by drilling the well and the oil flows out of the reservoir. However, oil extraction from unconventional resources requires specialized techniques because the oil trapped in reservoirs have high viscosity and little or no ability to flow. This includes oil sands, oil shale, coal, and biomass-based liquid supplies, and Venezuelan extra-heavy oil and natural bitumen.

Oil sands are the major source of unconventional oil, and Canada is the world's largest commercial producer of crude oil from oil sands. Canada has 174 billion barrels of oil, of which 169 billion barrels are oil sands reserves. The total oil sands-derived oil production in 2014 reached about 2.3 mb/d (*Upgraders and Refineries Facts and Stats* 2015). Oil sands are a mixture of sand, clay or other minerals, water, and bitumen. The bitumen is a dense and viscous hydrocarbon mixture from which crude oil is extracted. It is as viscous as cold molasses at room temperature. The National Energy Board of Canada defines bitumen as "a highly viscous mixture of hydrocarbons heavier than pentanes which, in its natural state, is not usually recoverable at a commercial rate through a well because it is too thick to flow." The World Energy Council defines natural bitumen as "oil having a viscosity greater than 10,000 centipoise under reservoir conditions and an API gravity of less than 10° API."

Majorly, the bitumen is extracted via two techniques: surface mining and steam-assisted gravity drainage (SAGD). The latter is used in places where bituminous deposits are too deep to be mined. For surface mining, large trucks and shovels are used, and the oil sand is transported via big trucks to the extraction facility. It is then crushed and conveyed to the rotary drums (see Figure 5.1). In the rotary drum, hot water and chemical agent (caustic) are added. The hot water reduces the viscosity of bitumen and makes it flow. The water, sand, and bitumen mixture is then transferred to the gravity settler/separator. The bitumen froth being lighter stays on the top, and the water, sand, and chemical mixture in the bottom of the separator is sent to the tailing ponds for treatment. The bitumen froth is treated with hydrocarbon solvent (naphtha, hexanes) in the froth treatment unit, and the extracted bitumen is sent for upgrading (Sustainable Development of Oil Sands-Challenges in Recovery and Use 2006; *The Oil Sand Extraction Process* 2008; Diagram of Oil Sands Mining and Extraction 2016; Sensors for Mining and Bitumen Extraction 2016). The extraction of bitumen from oil sands using the hot water drum process is a big component of oil extraction, and bitumen recovery determines the economics of the entire oil sands to liquid fuels process. Masliyah et al. (2004) have discussed in detail the water-based extraction of bitumen from oil sands in their review paper.

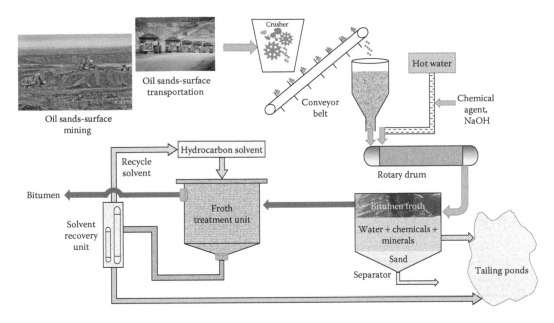

FIGURE 5.1
Bitumen extraction via surface mining technique. (From Sustainable Development of Oil Sands-Challenges in Recovery and Use 2006; *The Oil Sand Extraction Process* 2008; Diagram of Oil Sands Mining and Extraction 2016; Sensors for Mining and Bitumen Extraction 2016; Courtesy of Michael S. Williamson, *The Washington Post*/Getty and Dan Woynillowicz, *The Globe and Mail*.)

In the SAGD process, a high-pressure steam is injected into the underground formation as shown in Figure 5.2. The high temperature of the steam makes bitumen to flow, and it is pumped out of the surface and sent for upgrading (Talk about SAGD 2013; Chahal et al. 2015).

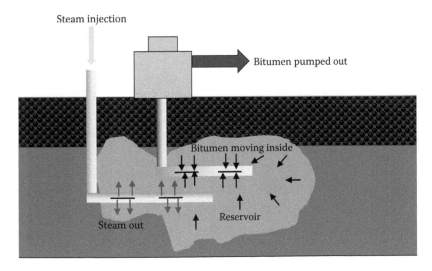

FIGURE 5.2
Bitumen extraction via SAGD technique. (From Talk about SAGD, *Alberta Energy*, http://www.energy.alberta.ca/OilSands/pdfs/FS_SAGD.pdf, 2013; Chahal, Danish et al., Steam Assisted Gravity Drainage, *Energy Education*, http://energyeducation.ca/encyclopedia/Steam_assisted_gravity_drainage, 2015.)

5.2 Bitumen Upgrading

Bitumen is a complex hydrocarbon mixture and highly viscous with an American Petroleum Institute (API) gravity of around 8.5. It contains 4–6 wt.% sulfur and about 0.3–0.6 wt.% nitrogen. In order to extract the usable crude oil from bitumen and process it in the existing refineries along with conventional crude oil, bitumen upgrading is required. The upgrading process is divided into two stages: primary upgrading and secondary upgrading. Primary upgrading includes the distillation of bitumen to obtain different boiling range cuts, whereas secondary upgrading includes the enrichment of product quality by removing the impurities such as sulfur, nitrogen, and metals.

5.2.1 Primary Upgrading

In this process, the distillates, including naphtha, kerosene, light gas oil (LGO), and heavy gas oil (HGO), were extracted from the bitumen in a series of operations such as atmospheric distillation, vacuum distillation, hydrocracking, and coker (see Figure 5.3).

5.2.1.1 Atmospheric Distillation Unit

The diluted bitumen (with naphtha) is sent to the atmospheric distillation unit (ADU) after preheating at 350°C (see Figure 5.3) to recover the naphtha and to extract the volatiles and LGO with nondestructive distillation. A temperature higher than 350°C results in thermal cracking of feedstock. This may cause the coke formation, and the coke deposits can cause the failure of distillation unit. Generally, The ADU contains 30–50 fractionation trays with high temperature at the bottom and low temperature at the top of the unit. The portion of the condensate from the top of the tower is sent back as reflux that helps in fractionation. The products are withdrawn from the side streams. Typically, naphtha has a boiling range of C_5 to 177°C so it is withdrawn from the upper section of the unit. The LGO, which

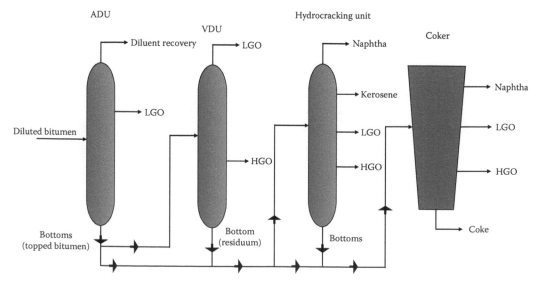

FIGURE 5.3
Primary upgrading of bitumen.

is in the boiling range of 177°C–343°C (Product Specification 2015), is collected from the middle lower section of the ADU. As the product side stream is in equilibrium with the vapor phase, some of the lighter ends are also withdrawn. Therefore, the side stream product is stripped for removing unwanted volatiles in a separate tower containing 4–10 trays (Gray et al. 2007). A fractionation tower (ADU) may be 13 feet in diameter and 85 feet high; however, the side stream stripper tower may be 3–4 feet in diameter and 10 feet high (Ancheyta and Speight 2007). The bottoms from the ADU called as topped bitumen are mainly sent to the vacuum distillation unit (VDU).

5.2.1.2 Vacuum Distillation Unit

Vacuum distillation is used to recover the high boiling fraction from the topped bitumen without using thermal decomposition. Typically, the VDU operates at 10–40 mmHg pressure, and the temperature is limited to 350°C–380°C (further higher temperatures result in cracking of the topped bitumen). The vacuum of 10–40 mmHg pressure is typically maintained by a three-stage steam ejector system. The boiling point decreases under vacuum and this causes a significant increase in the volume of the vapor. Therefore, the VDU is designed with large diameters to maintain the comparable vapor velocities. It may have a diameter of 40 feet or more (Gray et al. 2007). The internal of VDU is different from ADU. The vacuum distillation column internals consist of both trays and packing. To maintain the lower pressure drop, a packing material (such as Raschig rings or structural sheet metal) is used to facilitate the vapor liquid contact. However, the distillations trays are often used at the location of product withdrawals from the side of column. LGO is collected from the top section of the VDU and HGO is collected from the middle section of the VDU (Xu et al. 2005). HGO has a boiling range of 343+°C; however, it may be obtained at 250–350°C under vacuum. The bottom from the VDU is called as residuum and is sent to the coker and hydrocracker for further upgrading.

5.2.1.3 Hydrocracker Unit

The extraction of low-boiling, high valuable products (naphtha and gas oils) from high-boiling (>500°C) residues of the ADU and VDU can be achieved by hydrocracking. This helps in getting more valuable products from the barrel. In the hydrocracker unit, the long-chain heavy hydrocarbon molecules present in the topped bitumen and residuum are broken down into smaller valuable molecules but in the presence of catalyst and hydrogen. The catalyst helps in cracking the long-chain molecules, and the hydrogen saturates these molecules. Therefore, in comparison with cokers, hydrocrackers produce high-quality liquid fuels having an improved hydrogen/carbon (H/C) ratio. Hydrogenation is an exothermic process, whereas cracking is an endothermic process. The resultant heat released is quenched by passing hydrogen at multiple locations in the hydrocrackers. The hydrocracked feed is then sent to the fractionator to obtain HGO, LGO, kerosene, and naphtha. The hydrocrackers typically operate at high pressures (100–150 bar) and high temperatures (400°C–430°C) (Sahu et al. 2015). The conventional catalysts used for hydrocracking are bifunctional acid and metal sites. Acid sites help in cracking, whereas metal sites favor hydrogenation. Typically, the hydrocracking catalyst comprises a silica–alumina (zeolite)-based support material, which provides cracking sites, and metals such as platinum, palladium, tungsten, and nickel, which supports hydrogenation (Gray et al. 2007). There are mainly two types of hydrocrackers used for upgrading the residue from VDU (boiling point > 550°C) and ADU (boiling point > 350°C): fixed bed and ebullated bed. In fixed-bed reactors, the catalyst is packed in the bed, and after the catalytic deactivation due to coking and fouling, the reactor is shut down to remove the catalyst. The regenerated/fresh catalyst is loaded again in the reactor. However, in ebullated bed reactors such as LC-finer, the catalyst is in suspension or expanded with liquid and gas phases (just like fluidized-bed reactor). The feed and H_2 are pumped from the bottom of the reactor, and the effluent/products are withdrawn from the top of the reactor. The products are separated to remove undesirable gases such as H_2S and NH_3, and the liquid is sent for fractionation. The spent catalyst is continuously taken out and the regenerated catalyst is fed back to the reactor. The LC-finer hydrocracker unit operates for long periods of operations without shutdown (3+ years), provides the most efficient heat recovery, and achieves the maximum conversion to light clean products. Typical yields from the LC-finer unit are 17 wt.% naphtha, 10 wt.% kerosene, 19 wt.% LGO, and 37 wt.% HGO (Gray et al. 2007). The bottoms of the hydrocracker unit are sent to the cokers.

5.2.1.4 Coker Unit

The residuum (from the VDU) and some of the topped bitumen (from the ADU) are sent to the coker unit to remove excess carbon content. This results in increasing the H/C ratio of the gas oils derived from this unit and hence helps in generation of higher value products from high-molecular-weight, low-value residuum. The excess carbon is removed in the form of solid residue known as petroleum coke. The high temperature in the coker (480°C–550°C) helps in thermal cracking of long-chain bitumen molecules into more valuable short-chain molecules such as kerosene (boiling range 143°C–260°C), LGO, HGO, and naphtha. A typical coking unit consists of a reactor, where a high temperature of 480°C–550°C is maintained to facilitate the thermal cracking reaction (Gray 1994), and the cracked products (liquid and vapors) are transferred to the fractionation column to attain different fractions of products based on the boiling point (Gray et al. 2007). This primarily includes naphtha, kerosene, coker light, and HGO as shown in Figure 5.3.

There are mainly three types of coker units: delayed coker, fluid coker, and flexicoker. In the delayed coking process, the feed is preheated to a cracking temperature (~500°C) in a furnace/heater and sent to a coker drum, where the vapor products consisting of gases and distillates are removed from the top and quenched with colder feed in the fractionator. The solid residue called coke stays in the coker drum. The feed is switched to another coker drum, once the drum in use is filled with coke. The coker drum is emptied to remove the deposited coke, and the drum is used back in the cycle. Usually, the upgrading plants have two to four coker drums. In the fluid coking process, the feed is sprayed on the fluidized hot fine coke particles in the coker vessel. Coking occurs on the surface of these particles at a temperature of ~530°C. The cracked vapors are drawn off from the top of the vessel and sent to the fractionator, after passing through a cyclone to separate the coke particles. The coke formed in this process is continuously removed from the burner vessel where it is heated to generate the hot coke, which can then be sent back to the fluidized bed to keep coking reaction continuous. The excess coke produced is removed and 25–30% of the coke is burned to generate the heat. The shorter residence time in the fluid coking process helps to maintain high temperature, higher C_{5+} liquid yield, and lower coke formation. Flexicoking is the modified fluid coking process where in place of the burner vessel, a heater and a gasifier are integrated to produce hot coke and high-value flexigas (CO/H_2), which can be used in place of natural gas or fuel gas in refining operations. Delayed coking and fluid coking processes are mainly used for upgrading the bitumen-derived HGO. Typically, ~55 wt.% C_{5+} liquid yield and 35 wt.% coke are attained from the coking process.

TABLE 5.1

Physical Properties of Bitumen-Derived HGO and LGO

Characteristics	HGO	LGO
Sulfur (wt.%)	4.0	2.3
Nitrogen (wt.%)	0.37	0.14
Density (g/mL)	0.97	0.9
Aromatic content (%)	44.0	35
Boiling point distribution		
IBP (°C)	281	175
FBP (°C)	655	481
Boiling range (°C)	wt.%	wt.%
IBP–300	3.0	46.4
301–350	5.2	25.5
351–400	15.6	17.2
401–450	28.0	7.2
451–500	26.0	3.7
501–600	20.2	—
600–FBP	2.0	—

Sources: Narayanasarma, Prabhu, Mesoporous Carbon Supported NiMo Catalyst for the Hydrotreating of Coker Gas Oil, University of Saskatchewan, 2011; Badoga, Sandeep, Synthesis and Characterization of NiMo Supported Mesoporous Materials with EDTA and Phosphorus for Hydrotreating of Heavy Gas Oil, University of Sakatchewan, 2015. IBP, initial boiling point; FBP, final boiling point.

The details of the thermal cracking reactions and the mechanism of coke formation were discussed by M. R. Gray in 1994.

HGO, LGO, and naphtha obtained from bitumen upgrading are blended to create the crude oil called as synthetic crude. The synthetic crude is then sent downstream to the refineries for further processing to create final products. However, before blending, HGO and LGO are hydrotreated to remove or reduce sulfur and nitrogen, and other impurities such as Ni, V, and As, which are initially present in the bitumen. The physical properties of bitumen-derived HGO and LGO are shown in Table 5.1.

5.2.2 Secondary Upgrading—Hydrotreating

In this process, the impurities such as sulfur, nitrogen, and metals (As, V) were removed or reduced to make the gas oil suitable for further processing in existing refineries. The major process for secondary upgrading is hydrotreating. Hydrotreating is a catalytic process in the presence of hydrogen where typically the catalyst is loaded in a fixed-bed reactor operating at high temperatures (340°C–400°C) and moderate pressures (8–10 MPa). The feed gas oil containing impurities is allowed to react with hydrogen on the catalyst surface, and the multiple reactions that take place in the reactor include hydrodesulfurization (HDS), hydrodenitrogenation (HDN), hydrodeoxygenation, hydrodemetallization, and hydro-dearomatization (HDA). The conventional catalyst used for the hydrotreating reaction is molybdenum (Mo) or tungsten (W) supported on γ-alumina. The most common promoters used are nickel (Ni) or cobalt (Co) (Badoga et al. 2012). This catalyst works satisfactorily for

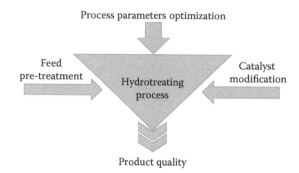

FIGURE 5.4
Factors governing the efficiency of the hydrotreating process.

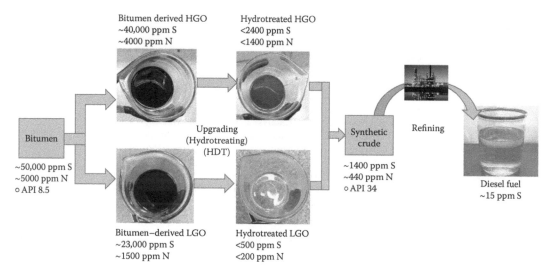

FIGURE 5.5
Process of reducing sulfur and nitrogen content present in crude oil derived from oil sands.

sweet crudes (sulfur < 0.5 wt.%). However, Table 5.1 shows that the bitumen-derived HGO contains 40,000 ppm sulfur, and as per current U.S. Environmental Protection Agency regulations, the sulfur content in diesel fuel should not exceed 15 ppm. Therefore, a highly active hydrotreating process is required to reduce the sulfur and nitrogen level of bitumen-derived gas oil to the level of conventional sweet crudes. The hydrotreating process activity can be improved by adjusting the three main parameters as shown in Figure 5.4: (1) feed pretreatment, (2) process parameter optimization, and (3) change in the catalyst. The treated oil is then further processed along with the conventional crude in existing refineries to get the final products such as diesel and gasoline. The systematic process for removal of sulfur and nitrogen from oil sands-derived gas oil is shown in Figure 5.5.

Feed pretreatment includes the removal of those impurities from oil which decreases the hydrotreating activity. For example, nitrogen, the nitrogen-containing compounds present in gas oil, severely inhibits the catalytically active sites and decreases the HDS rate. Also, as mentioned in Table 5.1, the bitumen-derived gas oil has higher nitrogen content so more inhibition is expected. Therefore, removal of nitrogen from gas oil before sending the oil for hydrotreating can increase the hydrotreating activity. Laredo et al. (2013) have discussed in theirs review paper about the various types of solid adsorbent materials such as silica gel, activated carbon, ion-exchange resin, molecular sieves, and metal organic frameworks for nitrogen removal. Misra et al. (2015) have synthesized a specialty polymer for selective removal of nitrogen compounds from gas oil to adsorb and remove the nitrogen from gas oil before sending for hydrotreating. However, high cost is involved in setting up a separate unit for feed pretreatment. The second criterion is to optimize the process parameters. Usually, high temperature and pressure will result in more activity, but it severely affects the catalyst life. It also adds to higher operating costs. Moreover, increasing the temperature (above 400°C) leads to cracking, which puts limits on tuning the process parameters. The third criterion to increase the hydrotreating activity is to modify/change the catalyst. This is the most acceptable solution because after catalyst life span it needs to be changed and changing the catalyst does not involve any costs related to fixed infrastructure. Therefore, worldwide, the research is being conducted to change/modify the catalyst so as to increase the hydrotreating activity under typical industrial conditions. The improved hydrotreating process will help in producing high-quality transportation fuel from oil sands.

The hydrotreating catalyst has four main components: support material, active metals, promoter, and additives. Conventionally, γ-Al$_2$O$_3$ is used as the support. The support material not only helps in dispersion of active metals but also participates in the modification of the electronic properties, the morphology of the active phase, and the bifunctional reaction with their acid sites, which could drastically affect the catalytic activity (Badoga 2015). The Bronsted acidity in the support helps in the isomerization of alkyl groups of alkyl dibenzothiophene (DBT), which favors the removal of sulfur from these otherwise difficult-to-hydrodesulfurize molecules. The basicity of support material prevents the coke formation. The active metals such as Mo or W were sulfided to MoS$_2$ or WS$_2$ before they participate in the hydrotreating reaction, and addition of promoters such as Ni or Co enhances the hydrotreating activity. Topsoe et al. (Topsøe 2007) suggested that the active site in the NiMo/CoMo catalysts is Ni (Co)–Mo–S phase. The building blocks of the Ni–Mo–S structure are small MoS$_2$ nano-crystals with Ni (or Co) promoter atoms located at the edges of the MoS$_2$ layers in the same plane of Mo atoms as shown in Figure 5.6. It was also concluded from their studies that the relative amount of Ni atoms present as Ni–Mo–S phase was found to correlate linearly with the HDS activity.

Further studies on NiMoS active sites have revealed two types of Ni–Mo–S structures: type I and type II. Type I NiMoS sites are single layered and may be partially sulfided. These types of structures were formed when the active metal and support interaction is strong.

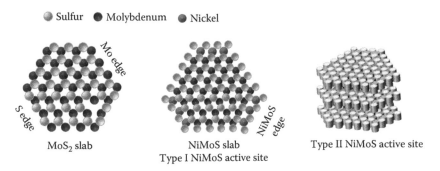

FIGURE 5.6
Different types of active sites on NiMo-supported catalysts. (From Badoga, Sandeep et al., *Appl. Catal. B: Environ.*, 125, 67–84, 2012.)

The strong interaction retards the reducibility/sulfidation of active metals and also results in fine dispersion of active metals. Type II NiMoS sites are two or more layered and fully sulfided (see Figure 5.7). These sites were formed when moderate-to-weak active metal and support interactions were expected. Type I and type II sites not only are different because of their structure but also favor different reactions. Type I sites tend to undergo hydrogenolysis reaction, whereas type II sites favor hydrogenation reaction (Besenbacher et al. 2008). The difference between the two reactions is shown in Figure 5.7 explaining the HDS reaction mechanism for 4,6-dimethyldibenzothiophene (4,6-DMDBT). The hydrogenolysis route involves the direct scission of C–S bond, and it is also called direct desulfurization. However, with alkyl-substituted bulky molecules such as 4,6-DMBDT, the accessibility of sulfur to the active sites is sterically hindered. Therefore, these molecules first undergo hydrogenation to saturate the bonds and weaken the C–S bond. It is then followed by C–S bond scission to remove sulfur (Jian and Prins 1998). This reaction route is called the hydrogenation route as shown in Figure 5.7 (Egorova and Prins 2004). The understanding of HDS, HDN, and HDA reaction mechanisms is important to design a catalyst to enhance the hydrotreating activity for obtaining a high-quality liquid fuel.

The catalytic activity can be increased by (1) changing the support material, (2) adding the chelating agents such as ethylenediaminetetraacetic acid (EDTA), and (3) using promoters such as phosphorus and boron. γ-Alumina has been widely used as the support for hydrotreating reaction; however, it shows a strong metal–support interaction resulting in the formation of tetrahedral MoO_2 species, which are difficult to sulfide. Also, this support has a limitation of the surface area and does not have uniform pore size distribution. However, neutral supports such as SBA-15 has weak metal–support interactions resulting in poor dispersion but easier sulfidation/reduction of active metals. Therefore, the extent of metal–support interaction is very important in determining the extent of metal sulfidation/reduction during catalyst activation. For instance, molybdenum oxide on γ-alumina support starts to sulfide at 150°C; however, niobium oxide on γ-alumina is hard to sulfide at even 350°C (Allali et al. 1995). Consequently, the need for better supports that can

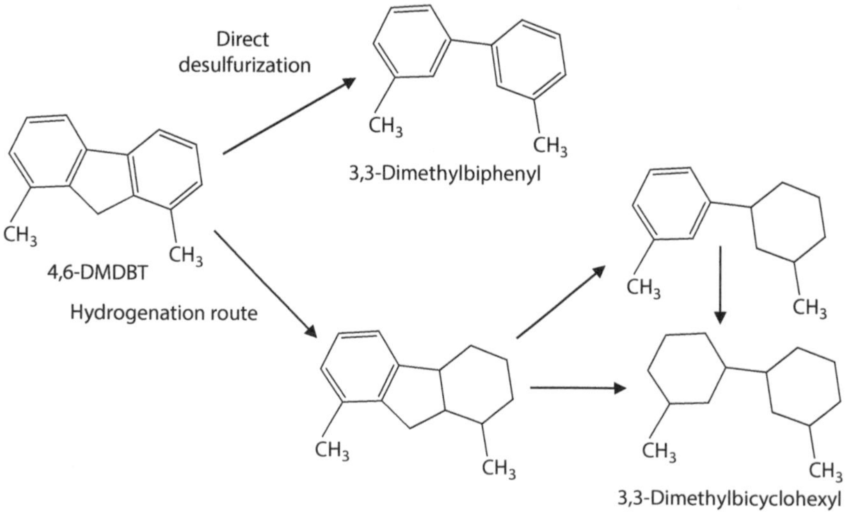

FIGURE 5.7
Hydrogenation reaction mechanism. (From Badoga, Sandeep, Synthesis and Characterization of NiMo Supported Mesoporous Materials with EDTA and Phosphorus for Hydrotreating of Heavy Gas Oil. University of Sakatchewan, 2015.)

provide optimum metal–support interactions, high surface area, optimum pore diameter (7–15 nm), and desired acidic properties results in synthesis and testing of a large number of new materials. These include TiO_2, ZrO_2, silica–alumina, mixed oxides, carbon, SiO_2, zeolites, and mesoporous materials (Duchet et al. 1991; Grzechowiak et al. 2001; Carati et al. 2003). Metal oxide supports such as TiO_2 and ZrO_2 have a low surface area that limits their use; however, the activity is 3–4 times more than that shown by alumina-based catalyst (Caero et al. 2003). To utilize the properties of metal oxides and to overcome the problem of small surface area, mixed oxides including TiO_2–Al_2O_3 (Breysse et al. 2003), ZrO_2–Al_2O_3, TiO_2–SiO_2 (Carati et al. 2003), and TiO_2–ZrO_2 are synthesized and tested by various research groups. The mixed oxides have a higher surface area and exhibit acid–base properties that are favorable for promoting desulfurization of alkyl DBTs.

Rana et al. (2003) have investigated the effects of Ni(Co)Mo catalysts supported on TiO_2–SiO_2 and ZrO_2–SiO_2 mixed oxides for hydrotreating of model compounds. They explained the role of titania and zirconia on metal dispersion and activity. In another work, Leyva et al. (2008) have synthesized and tested the $NiMo/SiO_2$–Al_2O_3 catalyst for hydrotreating of heavy oil and found it promising. Maity et al. (2006) have performed a significant work on the catalyst development for hydrotreating of Maya heavy crude. They have tested alumina–titania and titania–zirconia mixed oxides, and high surface TiO_2 as supports for hydrotreating catalysts. Ramirez et al. (2004) have developed the hydrotreating catalyst. They have also tested various support materials, including alumina, titania, silica–alumina for the Ni(Co)Mo(W) catalyst. In their work, they have described the electronic role played by titania in alumina framework and how it impacts the HDS activity of the supported Mo, CoMo, NiMO, and NiW catalysts. Dominguez-Crespo et al. (2008) have synthesized the NiMo catalyst using different alumina precursors and used spray and incipient impregnation methods. They have also studied the effects of pH on the catalytic activity. The results for the hydrotreating of Mexican HGO revealed an increase in HDS and HDN activities when the crystal size of the support was varied from 3 to 20 nm. They have also showed that the catalyst prepared in a basic medium performed better. Ji et al. (2004) have studied the effects of Co and Ni promotion on the hydrogenation of tetralin and the HDS of thiophene using various supported MoS_2 and WS_2 catalysts. They utilized ZrO_2, alumina-stabilized TiO_2, and pure alumina as support materials. The Ni promoted, ZrO_2 supported showed the highest activity. Therefore, it is evident from the literature that different supports have different metal–support interactions and hence different hydrotreating activities. Also, the catalyst performance is dependent on the type of feed gas oil.

In our work, we have synthesized various support materials, including SBA-15, M-SBA-15 (M = Ti, Al, Zr), mesoporous ZrO_2 and Al_2O_3, and mesoporous mixed metal oxides TiO_2–Al_2O_3, ZrO_2–Al_2O_3, and SnO_2–Al_2O_3 (Badoga et al. 2012; Badoga, Dalai, et al. 2014; Badoga, Sharma, et al. 2014a, 2014b; Badoga et al. 2015). The synthesis procedure of each material is mentioned in our previous work. The textural properties of each support are mentioned in Table 5.2. All the above-mentioned materials have been utilized to synthesize the NiMo-supported catalysts, and the catalysts have been tested for hydrotreating of HGO.

The hydrotreating reactions have been carried out in a laboratory-scale continuous fixed-bed reactor setup (see Figure 5.8). The setup consists of a feed tank for gas oil and a pump to transfer gas oil from the feed tank to the reactor. The hydrogen for the reaction is supplied from the tank via a mass flow controller. Post reactor there is an ammonia scrubber and then a two-phase (gas–liquid) separator. The liquid collected in the separator is removed and stripped with nitrogen before taken for analysis in Antek Nitrogen/Sulfur analyzer to measure the total nitrogen and sulfur content. The catalytic activity is

TABLE 5.2

Textural Properties of Support Materials

Material	BET Surface Area (m²/g)	Pore Volume (cm³/g)	Average Pore Diameter (nm)
SBA-15	550	1.10	6.8
TiSBA-15	560	1.20	7.0
ZrSBA-15	507	1.00	7.5
AlSBA-15	410	0.90	8.1
Meso ZrO_2	120	0.38	13.0
Meso Al_2O_3	470	1.00	6.5
Meso TiO_2–Al_2O_3	480	0.86	5.2
Meso ZrO_2–Al_2O_3	420	0.76	4.2
Meso SnO_2–Al_2O_3	450	0.60	4.1
γ-Al_2O_3	278	0.80	8.0

Sources: Badoga, Sandeep, Ajay K. Dalai, John Adjaye, and Yongfeng Hu, *Ind. Eng. Chem. Res.*, 53, 2137–2156, 2014;
Badoga, Sandeep, Rajesh V Sharma, Ajay K Dalai, and John Adjaye, *Ind. Eng. Chem. Res.*, 53, 18729–18739,
2014a; Badoga, Sandeep, Rajesh V. Sharma, Ajay K. Dalai, and John Adjaye, *Fuel*, 128, 30–38, 2014b.

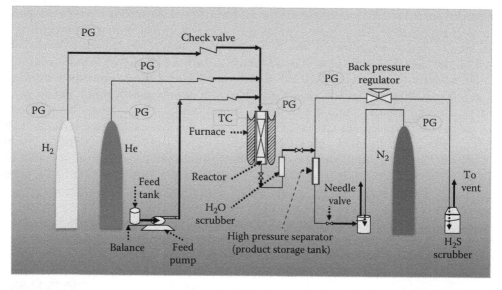

FIGURE 5.8

Hydrotreating reaction setup: Continuous fixed-bed reactor. (From Badoga, Sandeep, Synthesis and Characterization of NiMo Supported Mesoporous Materials with EDTA and Phosphorus for Hydrotreating of Heavy Gas Oil. University of Sakatchewan, 2015.)

reported in terms of HDN, HDS, and HDA. The aromatic content of gas oil is measured using carbon-13 nuclear magnetic resonance.

The activity of each catalyst for hydrotreating of HGO is reported in Table 5.3. It can be observed from the table that addition of heteroatoms such as Ti, Al, and Zr in otherwise neutral support SBA-15 has resulted in an increase in activity. This could be attributed to an increase in metal–support interactions as evident from an increase in metal dispersion (see Table 5.3). Considering the role played by heteroatoms in SBA-15 matrix, the mesoporous metal oxide support materials such as ZrO_2 and Al_2O_3 has been synthesized. The N_2 adsorption–desorption isotherm for ZrO_2 material has been found to be type IV with H1 hysteresis

TABLE 5.3

Textural Properties and Hydrotreating Activity of Supported NiMo Hydrotreating Catalysts with HGO at 395°C (Catalyst = 5 cm^3, Pressure = 8.8 MPa, Liquid Hourly Space Velocity = 1 Hour^{-1}, and H$_2$/Oil Ratio = 600 [v/v])

Material	Textural Properties			Metal Dispersion (%)	HDS (wt.%)	HDN (wt.%)	HDA (wt.%)
	BET Surface Area (m^2/g)	Pore Volume (cm^3/g)	Average Pore Diameter (nm)				
NiMo/SBA-15	441	0.85	6.6	6.5	68	25	22
NiMo/TiSBA-15	360	0.88	7.3	8.8	82	42	27
NiMo/ZrSBA-15	350	0.72	7.0	8.8	73	36	30
NiMo/AlSBA-15	320	0.70	7.2	8.3	75	38	26
NiMo/Meso ZrO$_2$	90	0.26	12	11.0	89	47	25
NiMo/Meso Al$_2$O$_3$	345	0.7	6.1	19.0	96	63	42
NiMo/Meso TiO$_2$–Al$_2$O$_3$	424	0.41	4.1	15.0	96	62	63
NiMo/Meso ZrO$_2$–Al$_2$O$_3$	398	0.42	4.2	10.0	92	58	43
NiMo/Meso SnO$_2$–Al$_2$O$_3$	225	0.29	4.5	6.5	75	40	38
NiMo/γ-Al$_2$O$_3$	240	0.60	7.2	12.0	90	52	45

Sources: Badoga, Sandeep, Rajesh V. Sharma, Ajay K Dalai, and John Adjaye, *Ind. Eng. Chem. Res.*, 53, 18729–18739, 2014a; Badoga, Sandeep, Rajesh V. Sharma, Ajay K. Dalai, and John Adjaye, *Fuel*, 128, 30–38, 2014b.

loop, which confirmed the mesoporous structure. Also, powder wide-angle X-ray diffraction (XRD) (see Figure 5.9) has confirmed the presence of the tetragonal phase of ZrO$_2$, which is mostly the catalytically active phase (Badoga, Sharma, et al. 2014b). The surface area of zirconia has been still less than that of γ-alumina, which limits the dispersion and activity. A variety of mesoporous aluminas with different textual properties have been synthesized using a triblock copolymer as a structure direction agent. These materials have been utilized as a support for NiMo hydrotreating catalyst. The textural properties of the best material obtained are mentioned in Tables 5.2 and 5.3. The synthesis procedure plays a critical role in determining the final structure of the material, which is related to the activity. During the synthesis of mesoporous alumina, the HNO$_3$/H$_2$O ratio has been varied from 0 to 2, and it was observed that by increasing the water content in the synthesis mixture, the structure changes from an ordered hexagonal to a wormlike/spongelike to a fibular and then to a corrugated platelet/rodlike structure (see Figure 5.10) (Badoga et al. 2015).

Another class of materials tested were mesoporous mixed metal oxides. Three different types of metal oxides—TiO$_2$, ZrO$_2$, and SnO$_2$—were selected based on their increasing Lewis acidic strength, and were mixed with alumina using the direct synthesis method to get materials such as TiO$_2$–Al$_2$O$_3$, ZrO$_2$–Al$_2$O$_3$, and SnO$_2$–Al$_2$O$_3$ (Badoga, Sharma, et al. 2014a). The temperature-programmed desorption (TPD) of ammonia was performed to determine the acidic strength of catalysts supported on these materials, and the results are shown in Table 5.4. The acidic strength follows the following order: TiO$_2$–Al$_2$O$_3$<ZrO$_2$–Al$_2$O$_3$<SnO$_2$–Al$_2$O$_3$. It can be seen from Table 5.3 that the activity also follows the same order. This indicates that the activity decreases with increasing the acidic strength. This could be attributed to the inhibition of the active site by nitrogen-containing compounds present in HGO. To confirm this, acridine Fourier transform infrared spectroscopy (FTIR) was performed, and the results are shown in Figure 5.11. Acridine represents the basic nitrogen molecule. It was adsorbed on all three mixed metal oxide-supported catalysts,

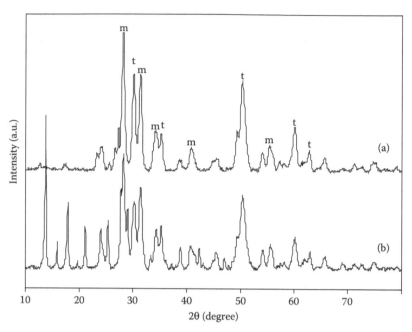

FIGURE 5.9

Powder wide-angle XRD pattern for (a) NiMo/Meso-Zr and (b) NiMo/Meso-Zr (EDTA). m and t represent monoclinic and tetragonal phases of zirconia, respectively. (Reprinted from *Fuel*, 128, Badoga, Sandeep, Rajesh V. Sharma, Ajay K. Dalai, and John Adjaye, Hydrotreating of heavy gas oil on mesoporous zirconia supported NiMo catalyst with EDTA, 30–38, Copyright 2014b, with permission from Elsevier.)

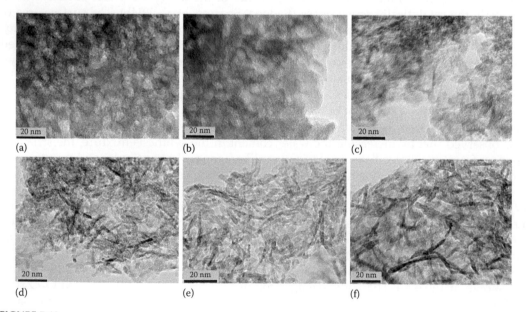

FIGURE 5.10

HRTEM micrographs for (a) Meso-Al-2, (b) Meso-Al-1.25, (c) Meso-Al-0.6, (d) Meso-Al-0.4, (e) Meso-Al-0.2, and (f) Meso-Al-0. The number at the end of each material name represents the HNO_3/H_2O ratio. (Reprinted from *Appl. Catal., A: Gen.*, Badoga, Sandeep, Rajesh V. Sharma, Ajay K. Dalai, and John Adjaye, 489, Synthesis and characterization of mesoporous aluminas with different pore sizes: application in NiMo supported catalyst for hydrotreating of heavy gas oil, 86–97, Copyright 2015, with permission from Elsevier.)

TABLE 5.4

NH₃-TPD Results of Mesoporous Mixed Metal Oxide Catalysts

Catalyst	Amount of NH₃ Desorbed (µmol/g)	
	at 140°C	at 350°C
NiMo/Al₂O₃	45	120
NiMo/Al–Ti	39	174
NiMo/Al–Zr	48	264
NiMo/Al–Sn	17	315

Source: Reprinted with permission from Badoga, Sandeep, Rajesh V. Sharma, Ajay K. Dalai, and John Adjaye, *Ind. Eng. Chem. Res.,* 53, 18729–18739. Copyright 2014a American Chemical Society.

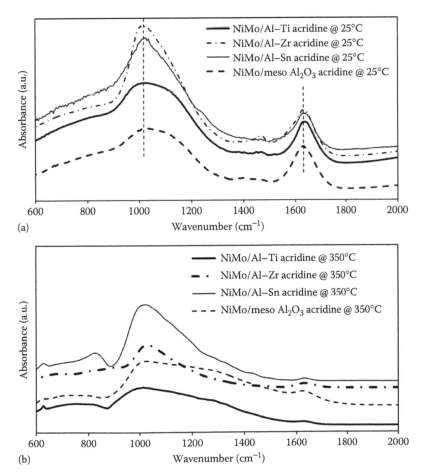

FIGURE 5.11
Acridine FTIR for mesoporous mixed metal oxide-supported catalysts at (a) room temperature and (b) 350°C. (Reprinted with permission from Badoga, Sandeep, Rajesh V. Sharma, Ajay K. Dalai, and John Adjaye, *Ind. Eng. Chem. Res.,* 53, 18729–18739. Copyright 2014a American Chemical Society.)

and the FTIR spectra were obtained. Then all the catalysts were heated up to 350°C to desorb the adsorbed acridine, and then again the FTIR spectra were obtained. It was observed that the ZrO_2–Al_2O_3- and SnO_2–Al_2O_3-supported catalysts still show the peaks related to acridine. This confirms that the nitrogen-containing compounds present in gas oil are inhibiting the active acidic sites resulting in an decrease in activity. Therefore, for nitrogen-rich feeds such as bitumen-derived HGO, very high acidic strength will create a negative impact on activity. However, a slight increase in the acidic strength as shown by TiO_2–Al_2O_3-supported catalyst in comparison with alumina-supported catalyst is helpful.

Further increase in the catalytic activity was achieved by using chelating ligands. As discussed previously that NiMoS is the active site with nickel atoms decorated in the corner and edges of the MoS_2 slab (see Figure 5.6). During sulfidation process, nickel starts to sulfide at 50°C and sulfidation completes at 150°C. However, molybdenum starts to sulfide at 150°C (Badoga et al. 2014). Therefore, early sulfidation of nickel prevents the movement of nickel atoms to the corner and edges of molybdenum sulfide slabs. Consequently, a chelating ligand is required which can delay the nickel sulfidation to the temperatures where molybdenum starts to sulfide. Chelating ligands are those molecules that have two or more donor atoms (dentates), which are available to bind a metal cation, for example, EDTA. Various other chelating ligands such as citric acid, glycol, nitriloacetic acid, 1,2-cyclohexanediamine-tetraaceticacid (CyDTA), and ethylenediamine were also tested for hydrotreating catalyst. In our previous work, we studied the mechanism of interaction between support–EDTA and EDTA–metallic species at different reaction conditions by using different characterization techniques such as X-ray absorption near-edge structure (XANES), high-resolution transmission electron microscopy (HRTEM), and XRD (Badoga et al. 2012). Ni K-edge XANES analysis was performed for a series of catalysts with and without EDTA at Canadian Light Source, Saskatoon, Saskatchewan (see Figure 5.12). Cat 0 is NiMo/SBA-15, Cat 1 is NiMo/SBA15/1EDTA with an EDTA/Ni molar ratio of 1, and Cat 2 is NiMo/SBA15/2EDTA with an EDTA/Ni molar ratio 2. All these catalysts were sulfided at 150°C, 250°C, and 350°C, respectively. The Ni K-edge for nickel oxide shows an intense peak at 8355 eV corresponding to 1s to 4p electronic transitions. It was shown in the figure that nickel oxide is converted into nickel sulfide in Cat 0 (catalyst without EDTA) at 150°C. However, nickel in Cat 1 and Cat 2 is still in the oxide phase. At 350°C, nickle in all catalysts is sulfided. This confirms that EDTA helps in delaying the nickel sulfidation temperatures to the point where molybdenum starts to sulfide, thereby resulting in the formation of more numbers of NiMoS active sites that increase the activity as shown in Table 5.5.

The Mo LIII-edge XANES analysis for NiMo/M-SBA-15 (M = Ti, Al, Zr) catalysts with and without EDTA (see Figure 5.13) has confirmed that EDTA not only helps in the formation of NiMoS sites but also increases the sulfidation and number of type II NiMoS sites. The Mo LIII-edge shows a peak at 2525 eV corresponding to 2p to 4d electronic transition (Badoga et al. 2014). The split in the peak is due to ligand field splitting of d-orbitals. The energy separation between the two split peaks gives the information about the structure of molybdenum oxide. If the peak energy separation (~2.0 eV) is smaller, the structure is tetrahedral; if the peak energy separation (~3.0 eV) is large, the structure of molybdenum oxide is octahedral. The octahedral structure is more desired because complete sulfidation can be easily achieved. Also, this type of structure favors the multilayered (type II) MoS_2 slab formation. It can be seen from Figure 5.13 that the catalyst NiMo/M-SBA-15 shows a smaller energy separation in the split peak confirming the presence of predominantly tetrahedral structure. However, while adding EDTA, the peak-to-peak energy separation in Mo LIII-edge increases, indicating the presence of predominantly octahedral molybdenum oxide.

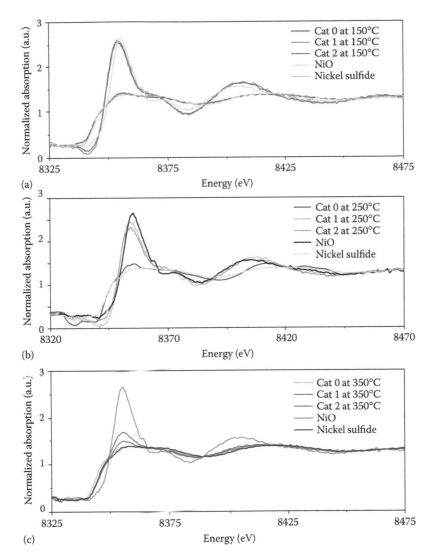

FIGURE 5.12
XANES spectra of sulfided Ni K-edge of (a) catalysts at 150°C, (b) catalysts at 250°C, and (c) catalysts at 350°C. (Reprinted from *Appl. Catal., B: Environ.*, Badoga, Sandeep, K. Chandra Mouli, Kapil K. Soni, a.K. Dalai, and J. Adjaye, 125, Beneficial influence of EDTA on the structure and catalytic properties of sulfided NiMo/SBA-15 catalysts for hydrotreating of light gas oil, 67–84, Copyright 2012, with permission from Elsevier.)

This concludes that EDTA helps in complete sulfidation and formation of type II NiMoS active sites, which increase the hydrotreating activity as shown in Table 5.5. Another role played by EDTA is the increase in dispersion as shown by HRTEM analysis. The HRTEM images were taken for sulfided catalysts, and the MoS_2 slab length and stacking degree distribution analysis were performed. The average slab length and stacking degree are shown in Table 5.6. It was observed that while adding EDTA, the average slab length and stacking degree decrease, which means the dispersion increases. Therefore, EDTA plays a variety of roles to enhance the hydrotreating activity. The additives such as phosphorus and boron were used to further boost the hydrotreating activity. Phosphorus is the most commonly used additive, and sometimes it is called the second promoter. Various other

TABLE 5.5

Hydrotreating Activities of Mesoporous Material-Supported NiMo Hydrotreating Catalysts with and without EDTA with HGO at 395°C (Catalyst = 5 cm^3, Pressure = 8.8 MPa, Liquid Hourly Space Velocity = 1 hour^{-1}, and H$_2$/Oil Ratio = 600 [v/v])

Material	HDS (wt.%)	HDN (wt.%)	HDA (wt.%)
NiMo/SBA-15	68	25	22
NiMo/SBA-15/EDTA	78	32	27
NiMo/TiSBA-15	82	42	27
NiMo/TiSBA-15/EDTA	95	54	32
NiMo/Meso ZrO$_2$	89	47	25
NiMo/Meso ZrO$_2$/EDTA	94	55	40
NiMo/γ-Al$_2$O$_3$	90	52	45
NiMo/γ-Al$_2$O$_3$/EDTA	94	58	50

Sources: Badoga, Sandeep, Ajay K. Dalai, John Adjaye, and Yongfeng Hu, *Ind. Eng. Chem. Res.,* 53, 2137–2156, 2014; Badoga, Sandeep, Rajesh V. Sharma, Ajay K. Dalai, and John Adjaye, *Fuel,* 128, 30–38, 2014b.

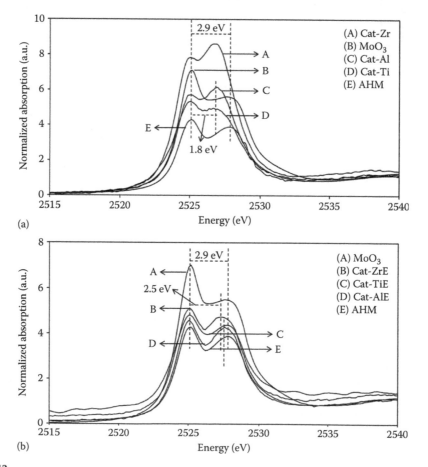

FIGURE 5.13

XANES spectra of Mo LIII-edge of (a) catalysts with EDTA and (b) catalysts without EDTA. (Reprinted with permission from Badoga, Sandeep, Ajay K Dalai, John Adjaye, and Yongfeng Hu, *Ind. Eng. Chem. Res.,* 53, 2137–2156. Copyright 2014 American Chemical Society.)

TABLE 5.6

Average Length and Number of the Stacking Layers of MoS_2 Slabs on Supported Sulfided Catalysts

Catalyst	Average Length (nm)	Average Stacking (Number of Layers)
NiMo/SBA-15	6.2	4.6
NiMo/SBA-15/EDTA	4.8	2.9
NiMo/TiSBA-15	4.8	3.2
NiMo/TiSBA-15/EDTA	3.6	2.0

Source: Badoga, Sandeep, Ajay K Dalai, John Adjaye, and Yongfeng Hu, *Ind. Eng. Chem. Res.*, 53, 2137–2156, 2014.

types of catalysts such as trimetallic (NiMoW, NoCoMo) supported sulfided catalysts and supported nickel tungsten phosphides were also tested for hydrotreating of gas oil, which have shown promising results.

Therefore, hydrotreating is a very important process in producing the clean liquid fuels from oil sands because it improves the properties of bitumen-derived crude to match with the properties of conventional crude oil. Hydrotreating process reduces the sulfur and nitrogen level of (1) LGO to 65 and 25 ppm from 23,000 and 1400 ppm, respectively, and (2) HGO to 2330 and 1330 ppm from 40,000 and 3700 ppm, respectively. The upgrading process also helps to increase the API gravity from 8.5° of bitumen to 36.2° for LGO and 22.5° for HGO (Product Specification 2015). The hydrotreated LGO and HGO and naptha were blended to make synthetic crude with an API gravity of 34°, and containing the sulfur and nitrogen level of 1400 and 440 ppm, respectively. The synthetic crude is sent to refineries where it is processed to obtain fuel fractions such as gasoline, diesel, kerosene, jet fuels, and petrochemicals. Hence, hydroprocessing makes the oil usable by the existing refineries to produce environmentally safe liquid fuels.

5.3 Challenges and Opportunities in Oil Sands Industry

The extraction of liquid fuels from oil sands, however, has many challenges, including bigger environment footprint, tailing ponds, water usage, technological limitations, land reclamation, and coke production. The extraction of oil from oil sands is energy intensive and adds the bigger environmental footprint compared to the conventional oil extraction process. Therefore, it is important to innovate and develop a technology which has minimal impact on environment. The research is being carried out to find effective ways to increase the efficiency of *in situ* extraction by modifying the steam system or by using an alternate heating method (*Impiantistica Italiana-Settembre-Ottobre* 2013). The oil extraction process generates a mixture of water, salts, chemicals (such as benzene, naphthalene, phenol, and hydrocarbons), clay, sand, and slit in large quantities, and it is stored or processed in tailing ponds. The process-affected water in the tailing ponds is being reused, which decreases the amount of freshwater required for oil extraction process. However, with an increase in oil production, the size of the tailing ponds is increasing. Currently, in Alberta, Canada, the tailing ponds covered an area of nearly 77 km² (Oil Sands Tailings 2013). The increasing size of the tailing ponds is the major and most difficult environmental challenge faced by an oil sands mining sector. The regulations are forcing to reduce the size of the tailing ponds and transform the nonused tailing ponds to the reclaimed land. Therefore, intensive research and development is being carried out in this area, which is mostly focused on separation of solids, biological, and chemical treatments.

The process of extracting liquid fuels from oil sands produces asphaltenes and coke as major by-products. The asphaltenes are used as paving materials on roads, roof shingles, and building foundation waterproofing. However, the challenges are associated with coke utilization, and its storage is a concern for upgraders (Andrews and Lattanzio 2013; Khosravi and Khadse 2013; Murthy et al. 2014). The coke produced during upgrading contains 6–7 wt.% sulfur and metals such as nickel and vanadium, which make it difficult to use by steel and aluminum industries, which require low sulfur and pure coke. However, majority of the petroleum coke is utilized as a fuel, and cement and power plants are the major consumers. Researchers are trying to find the alternate use of raw petroleum coke, which includes the gasification of petroleum coke to produce synthesis gas, which is then converted to liquid fuels via Fischer–Tropsch synthesis (Khosravi and Khadse 2013; Murthy et al. 2014), and the synthesis of activated carbon from coke (Rambabu et al. 2014).

5.4 Summary

Extracting liquid fuels from oil sands is an energy-intensive process compared to obtaining liquid fuels from conventional resources. The hydrocarbon content of oil sands called bitumen is extracted via two main techniques: (1) surface mining and (2) SAGD. For deep bitumen reservoirs, the SAGD process is preferred. The bitumen is very thick and highly viscous to flow, and has an API gravity of less than 10°. It also contains very high amount of sulfur (4–6 wt.%) and nitrogen (0.3–0.6 wt.%). To extract the crude oil from bitumen that can be processed in existing refineries, the bitumen is upgraded in two steps: primary upgrading and secondary upgrading. Primary upgrading includes the fractionation of bitumen mainly into naphtha, LGO, and HGO, and the removal of excess carbon in the form of coke and asphaltene. The obtained LGO and HGO still have very high sulfur and nitrogen content compared to conventional crude oil. Therefore, secondary upgrading of LGO and HGO was performed to reduce the impurities. This includes hydrotreating, which is a catalytic process. The higher hydrotreating efficiency determines the product quality and hence profitability. Therefore, intensive research is performed in catalyst modification and development, which includes modification in support materials, active metals, and use of additives. Various supports, including mesoporous materials (SBA-15, Al-SBA-15), mesoporous metal oxides (Al_2O_3, TiO_2, ZrO_2), and mixed metal oxides (TiO_2–Al_2O_3, ZrO_2–TiO_2, TiO_2–SiO_2, ZrO_2–Al_2O_3), have been extensively studied for hydrotreating of gas oils derived from oil sands. It was observed that alumina-based supports perform better in hydrotreating reactions. The addition of chelating ligands (EDTA) and additives (phosphorus and boron) have significantly improved the hydrotreating activity of bitumen-derived HGO. The difference in the hydrotreating activity of various catalysts is well explained using detailed characterization techniques such as, Brunauer-Emmett-Teller (BET), XRD, temperature programmed reduction (H_2-TPR), FTIR, RAMAN, NH_3-TPD, CO chemisorption, HRTEM, and XANES. After hydrotreating, LGO, HGO, and naphtha were mixed to form synthetic crude having less sulfur and nitrogen, and an API gravity of 34°. The synthetic crude is then processed in refineries to obtain final product liquid fuels such as diesel, gasoline, kerosene, and petrochemicals. The oil sands industry faces many challenges such as tailing ponds, water usage, technological limitations, land reclamation, and large amount of coke production. However, worldwide concerns for deteriorating environment and global warming bring the motivation (1) to explore non-fossil-based energies such as solar and wind, and (2) to innovate an improved technology for extracting and purifying liquid fuels derived from nonconventional resources such as shale oil and oil sands.

References

Allali, Nabil, Anne-Marie Marie, Michel Danot, Christophe Geanlet, and Michèle Breysse. 1995. Carbon-Supported and Alumina-Supported Niobium Sulfide Catalysts. *Journal of Catalysis* 156: 279–289.

Ancheyta, Jorge, and James G. Speight, eds. 2007. *Hydroprocessing of Heavy Oils and Residua*. CRC press Taylor & Francis Group, Boca Raton, FL.

Andrews, Anthony, and Richard K. Lattanzio. 2013. Petroleum Coke: Industry and Environmental Issues. Congressional Research Service.

Badoga, Sandeep. 2015. Synthesis and Characterization of NiMo Supported Mesoporous Materials with EDTA and Phosphorus for Hydrotreating of Heavy Gas Oil. University of Sakatchewan.

Badoga, Sandeep, K. Chandra Mouli, Kapil K. Soni, A.K. Dalai, and J. Adjaye. 2012. Beneficial Influence of EDTA on the Structure and Catalytic Properties of Sulfided NiMo/SBA-15 Catalysts for Hydrotreating of Light Gas Oil. *Applied Catalysis B: Environmental* 125 (August): 67–84. doi:10.1016/j.apcatb.2012.05.015.

Badoga, Sandeep, Ajay K. Dalai, John Adjaye, and Yongfeng Hu. 2014. Combined Effects of EDTA and Heteroatoms (Ti, Zr, and Al) on Catalytic Activity of SBA-15 Supported NiMo Catalyst for Hydrotreating of Heavy Gas Oil. *Industrial & Engineering Chemistry Research* 53: 2137–2156.

Badoga, Sandeep, Rajesh V. Sharma, Ajay K. Dalai, and John Adjaye. 2014a. Hydrotreating of Heavy Gas Oil on Mesoporous Mixed Metal Oxides (M–Al_2O_3, M = TiO_2, ZrO_2, SnO_2) Supported NiMo Catalysts: Influence of Surface Acidity. *Industrial & Engineering Chemistry Research* 53: 18729–18739. doi:10.1021/ie500840d.

Badoga, Sandeep, Rajesh V. Sharma, Ajay K. Dalai, and John Adjaye. 2014b. Hydrotreating of Heavy Gas Oil on Mesoporous Zirconia Supported NiMo Catalyst with EDTA. *Fuel* 128 (July): 30–38. doi:10.1016/j.fuel.2014.02.056.

Badoga, Sandeep, Rajesh V. Sharma, Ajay K. Dalai, and John Adjaye. 2015. Synthesis and Characterization of Mesoporous Aluminas with Different Pore Sizes: Application in NiMo Supported Catalyst for Hydrotreating of Heavy Gas Oil. *Applied Catalysis A: General* 489: 86–97.

Besenbacher, Flemming, Michael Brorson, Bjerne S. Clausen, Stig Helveg, Berit Hinnemann, Jakob Kibsgaard, Jeppe V. Lauritsen, Poul Georg Moses, Jens K. Nørskov, and Henrik Topsøe. 2008. Recent STM, DFT and HAADF-STEM Studies of Sulfide-Based Hydrotreating Catalysts: Insight into Mechanistic, Structural and Particle Size Effects. *Catalysis Today* 130 (1): 86–96. doi:10.1016/j.cattod.2007.08.009.

Breysse, Michèle, Pavel Afanasiev, Christophe Geantet, and Michel Vrinat. 2003. Overview of Support Effects in Hydrotreating Catalysts. *Catalysis Today* 86 (1–4): 5–16. doi:10.1016/S0920-5861(03)00400-0.

Caero, Luis Cedeño, Alma R. Romero, and Jorge Ramirez. 2003. Niobium Sulfide as a Dopant for Mo/TiO_2 Catalysts. *Catalysis Today* 78: 513–518.

Carati, Angela, Giovanni Ferraris, Matteo Guidotti, Giuliano Moretti, Rinaldo Psaro, and Cesara Rizzo. 2003. Preparation and Characterisation of Mesoporous Silica–alumina and Silica–titania with a Narrow Pore Size Distribution. *Catalysis Today* 77 (4): 315–23. doi:10.1016/S0920-5861(02)00376-0.

Chahal, Danish, Jordan Hanania, James Jenden, and Ellen Lloyd. 2015. Steam Assisted Gravity Drainage. *Energy Education*. http://energyeducation.ca/encyclopedia/Steam_assisted_gravity_drainage.

Diagram of Oil Sands Mining and Extraction. 2016. Accessed March 19. https://www.globalmethane.org/m2mtool/oilsands.html.

Domínguez-Crespo, Migual Antonio, Aide Minerva Torres-Huerta, L. Díaz-García, Elsa Miriam Arce-Estrada, and E. Ramírez-Meneses. 2008. HDS, HDN and HDA Activities of Nickel–molybdenum Catalysts Supported on Alumina. *Fuel Processing Technology* 89 (8): 788–796. doi:10.1016/j.fuproc.2008.01.004.

Duchet, Jean Claude, M.J. Tilliette, Daniel Cornet, Laurence Vivier, Guy Perot, L. Bekakra, Claude Moreau, and Georges Szabo. 1991. Catalytic Properties of Nickel Molybdenum Sulphide Supported on Zirconia. *Catalysis Today* 10 (4): 579–592. doi:10.1016/0920-5861(91)80040-G.

Egorova, Marina, and Roel Prins. 2004. Competitive Hydrodesulfurization of 4,6-Dimethyl-dibenzothiophene, Hydrodenitrogenation of 2-Methylpyridine, and Hydrogenation of Naphthalene over Sulfided NiMo/γ-Al$_2$O$_3$. *Journal of Catalysis* 224 (2): 278–287. doi:10.1016/j.jcat.2004.03.005.

Gray, James H., Glenn E. Handwerk, and Mark J. Kaiser. 2007. *Petroleum Refining Technology and Economics*. 5th ed. CRC press Taylor & Francis Group, Boca Raton, FL.

Gray, Murray R. 1994. *Upgrading Petroleum Residues and Heavy Oils*. Marcel Dekker Inc., New York.

Grzechowiak, Jolanta R., Jacek Rynkowski, and Iwona Wereszczako-Zielińska. 2001. Catalytic Hydrotreatment on Alumina–titania Supported NiMo Sulphides. *Catalysis Today* 65 (2–4): 225–231. doi:10.1016/S0920-5861(00)00562-9.

Impiantistica Italiana-Settembre-Ottobre. 2013. Alberta's Oil Sands: Challenges and Opportunities. albertainnovates.ca/media/19161/ab-oilsands-impiantistica-atricle.pdf.

Ji, Yaying, Pavel Afanasiev, Michel Vrinat, Wenzhao Li, and Can Li. 2004. Promoting Effects in Hydrogenation and Hydrodesulfurization Reactions on the Zirconia and Titania Supported Catalysts. *Applied Catalysis A: General* 257 (2): 157–164. doi:10.1016/j.apcata.2003.07.004.

Jian, Mou and Roel Prins. 1998. Mechanism of the Hydrodenitrogenation of Quinoline over NiMo(P)/Al$_2$O$_3$ Catalysts. *Journal of Catalysis* 179: 18–27.

Khosravi, Morteza, and Anil Khadse. 2013. Gasification of Petcoke and Coal/Biomass Blend: A Review. *International Journal of Emerging Technology and Advanced Engineering* 3 (12): 167–173. http://www.ijetae.com/files/Volume3Issue12/IJETAE_1213_29.pdf.

Laredo, Georgina C., Pedro M. Vega-Merino, Fernando Trejo-Zárraga, and Jesús Castillo. 2013. Denitrogenation of Middle Distillates Using Adsorbent Materials towards ULSD Production: A Review. *Fuel Processing Technology* 106: 21–32. doi:10.1016/j.fuproc.2012.09.057.

Leyva, Carolina, Mohan S. Rana, and Jorge Ancheyta. 2008. Surface Characterization of Al$_2$O$_3$–SiO$_2$ Supported NiMo Catalysts: An Effect of Support Composition. *Catalysis Today* 130 (2–4): 345–353. doi:10.1016/j.cattod.2007.10.113.

Maity, Samir Kumar, Jorge Ancheyta, Mohan Rana, and Patricia Rayo. 2006. Alumina–Titania Mixed Oxide Used as Support for Hydrotreating Catalysts of Maya Heavy Crude–Effect of Support Preparation Methods. *Energy and Fuels* 20 (2): 427–431. doi:10.1021/ef0502610.

Masliyah, Jacob, Zhiang Joe Zhou, Zhenghe Xu, Jan Czarnecki, and Hassan Hamza. 2004. Understanding Water-Based Bitumen Extraction from Athabasca Oil Sands. *The Canadian Journal of Chemical Engineering* 82 (4): 628–654. doi:10.1002/cjce.5450820403.

Misra, Prachee, Jackson M. Chitanda, Ajay K. Dalai, and John Adjaye. 2015. Immobilization of Fluorenone Derived Π-Acceptors on Poly (GMA-Co-EGDMA) for the Removal of Refractory Nitrogen Species from Bitumen Derived Gas Oil. *Fuel* 145: 100–108. http://linkinghub.elsevier.com/retrieve/pii/S0016236114012472.

Murthy, Bontu N., Ashish N. Sawarkar, Niteen A. Deshmukh, Thomas Mathew, and Jyeshtharaj B. Joshi. 2014. Petroleum Coke Gasification: A Review. *Canadian Journal of Chemical Engineering* 92 (3): 441–468. doi:10.1002/cjce.21908.

Narayanasarma, Prabhu. 2011. Mesoporous Carbon Supported NiMo Catalyst for the Hydrotreating of Coker Gas Oil. University of Saskatchewan.

Oil Market Report. 2015. https://www.iea.org/oilmarketreport/omrpublic/.

Oil Sands Tailings. 2013. *Oilsands.alberta.ca.* http://oilsands.alberta.ca/FactSheets/Tailings_FSht_Sep_2013_Online.pdf.

Product Specification. 2015. http://www.cdnoilsands.com/energy-marketing/ProductSpecifications/default.aspx.

Rambabu, Nedunuri, Sandeep Badoga, Kapil Kumar Soni, Ajay Kumar Dalai, and John Adjaye. 2014. Hydrotreating of Light Gas Oil Using a NiMo Catalyst Supported on Activated Carbon Produced from Fluid Petroleum Coke. *Frontiers of Chemical Science and Engineering* 8 (2): 161–170. doi:10.1007/s11705-014-1430-1.

Ramírez, Jorge, Gabriela Macías, Luis Cedeño, Aida Gutiérrez-Alejandre, Rogelio Cuevas, and Perla Castillo. 2004. The Role of Titania in Supported Mo, CoMo, NiMo, and NiW Hydrodesulfurization Catalysts: Analysis of Past and New Evidences. *Catalysis Today* 98 (1–2): 19–30. doi:10.1016/j.cattod.2004.07.050.

Rana, Mohan, Samir Kumar Maity, Jorge Ancheyta, Gudimella. Murali Dhar, and Turaga Prasada Rao. 2003. TiO_2-SiO_2 Supported Hydrotreating Catalysts: Physico-Chemical Characterization and Activities. *Applied Catalysis A: General* 253 (1): 165–176. doi:10.1016/S0926-860X(03)00502-7.

Sahu, Ramakanta, Byung Jin Song, Ji Sun Im, Young-Pyo Jeon, and Chul Wee Lee. 2015. A Review of Recent Advances in Catalytic Hydrocracking of Heavy Residues. *Journal of Industrial and Engineering Chemistry* 27: 12–24. doi:10.1016/j.jiec.2015.01.011.

Sensors for Mining and Bitumen Extraction. 2016. *Alberta Innovates Technology Futures.* Accessed March 19. http://www.albertatechfutures.ca/OurTeams/AppliedSpectroscopy/SensorsforMiningandBitumenExtraction.aspx.

Sustainable Development of Oil Sands-Challenges in Recovery and Use. 2006. *Western US Oil Sands Conference 21.* Alberta Research council. http://www.globaloilsands.com/Processing/index.shtml.

Talk about SAGD. 2013. *Alberta Energy.* http://www.energy.alberta.ca/OilSands/pdfs/FS_SAGD.pdf.

The Oil Sand Extraction Process. 2008. https://www.hatch.ca/News_Publications/Hatch_Report/HR_July2008/Fort_Hills_Phase_II.html.

Topsøe, Henrik. 2007. The Role of Co–Mo–S Type Structures in Hydrotreating Catalysts. *Applied Catalysis A: General* 322 (April): 3–8. doi:10.1016/j.apcata.2007.01.002.

Upgraders and Refineries Facts and Stats. 2015. www.energy.gov.ab.ca/Oil/pdfs/FSRefiningUpgrading.

Xu, Zhiming, Zhizhang Wang, Judy Kung, John Woods, Xinalex Wu, Luba Kotlyar, Bryan Sparks, and Keng Chung. 2005. Separation and Characterization of Foulant Material in Coker Gas Oils from Athabasca Bitumen." *Fuel* 84 (6): 661–668. doi:10.1016/j.fuel.2004.03.019.

Yardeni, Edward, and Debbie Johnson. 2015. *Energy Briefing: Global Crude Oil Demand & Supply.* www.yardeni.com.

6

Bioethanol Production from Lignocellulosic Biomass: An Overview of Pretreatment, Hydrolysis, and Fermentation

Hossain Zabed, J. N. Sahu, and Akter Suely

CONTENTS

ABSTRACT Bioethanol is an attractive and eco-friendly biofuel that can be used as an effective replacement for fossil fuels. It is produced from renewable sources through fermentation of sugars. Over the past several years, sugar and starch-based feedstocks have been widely used for generating ethanol on large scale. However, the ethical issues and rising debate on food versus fuel have led to finding out the promising ethanol feedstock from nonfood sources. As a result, current research efforts extensively focus on ethanol production from lignocellulosic biomasses (LCBs), which are mostly waste materials, inexpensive, abundant and noncompetitive with food, and grown on arable lands. LCBs originate from the nonfood sources such as agricultural wastes, forest biomass and residues, perennial grasses, dedicated energy crops, aquatic plants, municipal solid wastes, and industrial wastes. Ethanol production from LCBs basically includes (1) pretreatment of the biomass to release cellulose and hemicellulose from lignocellulosic complex, (2) hydrolysis of the pretreated biomass to generate fermentable sugars, (3) fermentation of the sugars by yeast to produce ethanol, and (4) recovery of ethanol through distillation. Much improvement has been made in recent years on the technological approaches for better conversion efficiency and enhanced ethanol yield. This chapter has been designed to present an overview of the ethanol production from LCBs that covers sources and composition of LCBs, pretreatment, hydrolysis, fermentation, microorganisms, and technological achievements.

6.1 Introduction

Currently, the world is mostly dependent on fossil fuels for meeting its energy demand, where more than 80% of the total global energy is obtained by burning fossil fuels, of which 58% alone is used in the transportation (Escobar et al. 2009). In particular, petroleum-based fuels are mainly used for energy that include gasoline, diesel, liquefied petroleum gas, and compressed natural gas (Demirbas 2009). However, there are three major challenges with these conventional fuels that are being faced globally:

- The rapid increase in the consumption rate of fossil fuels due to the growing industrialization and motorization of the world has resulted in the fast depletion of these nonrenewable energy sources (Agarwal 2007). Moreover, the limited reserves of fossil fuels have been anticipated to be exhausted by the next 40–50 years (Vohra et al. 2014).

- The use of fossil fuels has raised another concern, as it contributes to the greenhouse gas emissions and global warming that cause climate change, rise in sea level, loss of biodiversity, and urban pollution (Singh et al. 2010).

- The political crises, particularly in the Middle East, which resulted the incidence of oil supply disruption by the major oil-producing countries in the 1970s, have also led to rethinking our dependence on fossil fuels, because such crises are unsettling to the energy sector of both the developed and developing nations (Ogbonna et al. 2001).

Taking into consideration the above facts, it is necessary to find out an alternative source of energy for our industrial economies and consumer societies by using renewable, sustainable, efficient, and cost-effective sources with lower emissions of greenhouse gases. Four strategically important and sustainable options have been considered on either a small or a large scale, as attempts to find out alternative energy sources, which include biofuels, hydrogen, natural gas, and syngas (synthesis gas) (Nigam and Singh 2011). Among these four categories, biofuels have been proven to be the most eco-friendly, sustainable, and more promising to replace fossil fuels as they are biodegradable and can be produced from renewable sources (Bhatti et al. 2008). Biofuels include mainly bioalcohols (bioethanol and biomethanol), vegetable oils, biodiesel, biogas, biosynthetic gas, bio-oil, biochar, and biohydrogen (Demirbas 2008).

Ethanol (C_2H_5OH) is a liquid biofuel and is referred to as bioethanol when it is produced from biomass (Nigam and Singh 2011). Compared to other biofuels, bioethanol is known to be an attractive alternative to fossil fuel due to its ease of production and lack of toxicity. Many countries such as the United States, Brazil, China, and a number of EU member states have already proclaimed commitments to bioethanol programs (Balat and Balat 2009), where the former two countries have shown the largest commitments thus far (Larson 2008).

Bioethanol has been earmarked for a potential fuel due to its several advantageous properties over gasoline. Even though 1 L of ethanol affords 66% of energy provided by the same amount of gasoline, it has a higher octane number (108) than gasoline (90) that enhances the performance of the latter when blended with ethanol (Nigam and Singh 2011). Higher octane level of ethanol also allows it to be burnt at a higher compression ratio with shorter burning time, resulting in lower engine knock (Balat 2007; Kar and Deveci 2006). In addition, ethanol has higher evaporation enthalpy (1177 kJ/kg at 60°C) than gasoline (348 k J/kg at 60°C), higher flame speed, and wider range of flammability (Naik et al. 2010).

When environmental impact is considered, bioethanol is found to be an eco-friendly oxygenated fuel containing 34.7% oxygen, whereas gasoline contains no oxygen that results in about 15% higher combustion efficiency for the former than that of the latter (Kar and Deveci 2006), thereby keeping down the emissions of nitrogen oxides and other greenhouse gases (Malça and Freire 2006; Searchinger et al. 2008). Compared to gasoline, ethanol contains negligible amount of sulfur, and mixing of these two fuels helps to decrease the sulfur content in the fuel as well as the emission of sulfur oxide, which is a carcinogen and has a major contribution to acid rain (Pickett et al. 2008). Moreover, ethanol reduces the interference on the ozone layer due to having lower ambient photochemical reactivity (Lynd et al. 1991). The by-products of incomplete oxidation of ethanol are acetic acid and acetaldehyde, which are less toxic than those of other fuels (Vohra et al. 2014). Bioethanol is also a safer substitute to methyl tertiary butyl ether (MTBE), which is a commonly used octane enhancer for gasoline and is added to the latter for its clean combustion so that the production of carbon monoxide (CO) and carbon dioxide (CO_2) can be kept minimized (McCarthy and Tiemann 1998). MTBE has been reported to make its way into the groundwater that contaminates the drinking water causing severe detrimental effects on health (Green and Lowenbach 2001). The United States Energy Policy Act released an Advance Notice of Proposed Rulemaking in 2000 under the Toxic Substance Control Act to limit the use of MTBE as a gasoline extender (Yao et al. 2009).

Ethanol can be used as a fuel additive, an octane enhancer, and an oxygenate. It is used either as blended ethanol by mixing it with gasoline in the unmodified engines or directly in its pure form in the modified spark ignition engines (Nigam and Singh 2011). When ethanol is blended with gasoline, it is called "gasohol." The blended ethanol E-5 (5% ethanol and 95% gasoline by volume) is used as an oxygenate under the EU quality standard—EN/228 (Demirbas 2008). The most common use of ethanol in the United States is E-10 (10% ethanol and 90% gasoline by volume) and is known for fuel extender (Balat and Balat 2009; Demirbas 2008). Brazil, uses pure ethanol or blended ethanol in a combination of 24% ethanol and 76% gasoline by volume (De Oliveira et al. 2005). Ethanol can also be used as E-85 (85% ethanol and 15% gasoline by volume) in flexible fuel vehicles (Demirbas 2008). Another term for bioethanol being used as fuel is "diesohol" that comprises a blend of diesel, hydrated ethanol, and an emulsifier (84.5%, 15%, and 0.5% by volume, respectively) (Demirbas 2008).

Bioethanol is produced from a biomass containing free sugars or complex carbohydrates capable of converting them into soluble sugars (Aggarwal et al. 2001; Stevens et al. 2004). These biomasses can be divided into three major groups: (1) sugar crops (sugarcane, sugar beet, sweet sorghum) and sugar materials (cane molasses, beet molasses, cheese, whey), (2) starchy grains (corn, wheat, triticale), tubers (potatoes), and root crops (sweet potato, cassava), and (3) lignocellulosic biomass (LCB) (Zabed et al. 2016a). These three groups of the biomass differ considerably from each other with respect to the obtainment of sugar solutions (Mussatto et al. 2010; Sanchez and Cardona 2008; Vohra et al. 2014). Sugar crops require only the extraction process to get fermentable sugars, whereas starchy crops involve the hydrolysis step to convert starch into glucose using amylolytic enzymes. LCB has to be pretreated before hydrolysis in order to alter its cellulose structures for enzyme accessibility in the subsequent step (Mussatto et al. 2010).

6.2 Lignocellulosic Ethanol: An Overview

Almost all the current commercial production of bioethanol derives from sugar and starchy materials (Zabed et al. 2016b). However, these sources are not adequate to replace one trillion gallons of fossil fuels currently consumed each year in the world (Bell and Attfield 2009). Furthermore, the ethical concerns and debate on food versus fuel have been raised due to the utilization of food crops for generating fuel, which competes with the limited agricultural lands used for food and fiber production (Searchinger et al. 2008). The high costs of raw materials, as estimated roughly to be 40%–70% of the total ethanol production costs, also limits the sustainability of ethanol production from sugars and starchy crops (Claassen et al. 1999). In addition, it often requires government subsidies to compete with the conventional petroleum-based fuels (Doornbosch and Steenblik 2008; Sims et al. 2010). These concerns have led to encourage research efforts to be more focused on the ethanol production from LCBs as they are mostly waste materials, available with low and stable prices, rich in carbohydrates (cellulose and hemicellulose), versatile, and noncompetitive with food chain (Zhao and Xia 2010). Moreover, the conversion process of lignocellulose into ethanol generates low net greenhouse gases compared to the ethanol production from sugars and starch, thereby reducing the environmental pollution (Hahn-Hägerdal et al. 2006).

Production of fermentable sugars from LCBs before fermentation is, however, far more difficult than those obtained from sugar and starch-based feedstocks due to greater recalcitrance of cellulose and hemicellulose. Overall, lignocellulosic ethanol production limits its commercial implementation for several bottlenecks in the overall conversion process (Paulová et al. 2013):

- Requirement of costly and energy-consuming pretreatment process
- Necessity for a costly hydrolysis process
- Requirement of utilization of both pentose and hexose sugars
- Relatively lower fermentable sugars and ethanol yields than those of sugars and starch
- Inadequate separation of cellulose and lignin from the lignocellulosic matrix that reduces the conversion efficiency
- Formation of unwanted by-products, such as acetic acid, furfurals and phenolic compounds that are toxic to yeast cells
- High energy consumption
- Production of waste

However, recent technological achievements, particularly the development of enzyme technology by Genencor and Novozymes, have resulted in a dramatic drop of the costs for enzymes, which is accompanied by up to a 30-fold drop (2.6–5.3 cents per liter of ethanol) (Greer 2005). With the technological improvements, numerous commercial, pilot, and demonstration plants have been established and running worldwide to produce ethanol from LCBs (Table 6.1). The lignocellulosic ethanol plants are more extensive in the United States that started its commercial ethanol production from LCBs in 2014 (RFA 2015). Similar facilities are also under construction or even running in Canada, Brazil, Europe, Japan, and some other countries.

TABLE 6.1

Status of the Current Lignocellulosic Ethanol Production in the World

Company	Location	Feedstock	Type of Plant	Capacity (million gallons/year)	References
Abengoa Bioenergy	York, Nebraska	Wheat straw	Pilot	0.02	ABE 2015
Abengoa Bioenergy	Salamanca, Spain	Wheat and barley straw	Demonstration	1.3	ABE 2015
Abengoa Bioenergy	Hugoton, Kansas	Corn stover, wheat straw, switch grass	Commercial	16	ABE 2015
AE Biofuels	Butte, Montana	Multiple sources	Demonstration	—	Anonymous 2014
Alico, Inc.	La Belle, Florida	Multiple sources	Commercial	—	Anonymous 2014
ACE	Galva, Iowa	Corn kernel fiber	Commercial	3.75	Fuels-America 2014
Beta Renewables	Crescentino, Italy	Wheat and rice straw, and perennial cane	Commercial	19.8	EBTP-SABS 2014
Beta Renewables	Fuyang, China	Straw and corn stover	Commercial	30	EBTP-SABS 2014
Bluefire	Izumi, Japan	Wood waste chips	Commercial	79.2	BluFire 2015
BlueFire Ethanol, Inc.	Irvine, California	Multiple sources	Commercial	17	Anonymous 2014
Canergy	Imperial Valley, California	Perennial cane	Commercial	30	EBTP-SABS 2014
DuPont	Nevada, Iowa	Corn stover	Commercial	30	Anonymous 2014
Ethtec	Harwood, Australia	Wood and sugarcane residues	Pilot	—	Ethtec 2014
Gulf Coast Energy	Mossy Head, Florida	Wood waste	Commercial	70	Anonymous 2014
GranBio	Alagoas, Brazil	Straw and bagasse	Commercial	22	EBTP-SABS 2014
Iogen	Ottawa, Canada	Wheat straw, corn stover and bagasse	Demonstration	—	Iogen 2014
Inbicon	Fredericia, Denmark	Wheat straw	Demonstration	1.4	Inbicon 2015

(Continued)

TABLE 6.1 (*Continued*)

Status of the Current Lignocellulosic Ethanol Production in the World

Company	Location	Feedstock	Type of Plant	Capacity (million gallons/year)	References
Lignol Innovations, Inc.	Commerce City, Colorado	Wood	Pilot	2	Anonymous 2014
Mascoma Corp.	Lansing, Michigan	Wood	Commercial	40	Anonymous 2014
Mascoma Corp.	Vonore, Tennessee	Switch grass	Large-scale demonstration	5	Anonymous 2014
Pacific Ethanol	Boardman, Oregon	Mixed biomass	Demonstration	2.7	Anonymous 2014
POET–DSM	Emmetsburg, Iowa	Corn cobs, leaves, husks and stalks	Commercial	25	Fuels-America 2014
Pure Vision Technology	Ft. Lupton, Colorado	Corn stalks and grasses	Pilot	2	Anonymous 2014
Range Fuels	Treutlen County, Georgia	Wood waste	Commercial	20	Anonymous 2014
Raízen	Brazil	Bagasse	Commercial	40	EBTP-SABS 2014
Verenium Energy	Jennings, Louisiana	Wood waste	Demonstration	1.4	Anonymous 2014
Western Biomass Energy	Upton, Wyoming	Wood waste	Pilot	1.5	Anonymous 2014
Xethanol Corp.	Auburndale, Florida	Citrus peels	Commercial	8	Anonymous 2014

6.3 Sources of Lignocellulosic Biomasses

The major sources of LCBs can be divided into several groups: energy crops such as perennial grasses and other dedicated energy crops; aquatic plants such as water hyacinth; forest materials such as softwood, hardwood, sawdust, pruning, and bark thinning residues; agricultural residues such as cereal straws, stovers, and bagasse; and organic portion of municipal solid wastes (MSWs) (Figure 6.1) (Kumar et al. 2009; Limayem and Ricke 2012). These biomass resources seem to be the largest, promising, and abundant throughout the world (Table 6.2), which can be used to produce ethanol without necessitating any extra land or interference on food and feed crop production (Sims et al. 2010). The annual production of plant biomass in the world is accounted for approximately 200×10^9 tons, where nearly 8×10^9 to 20×10^9 tons can be used for biofuel production (Kuhad and Singh 1993; Saini et al. 2015). It has been anticipated that nearly 442 billion liters of ethanol can be produced each year from LCBs if both crop residues and wasted crops are taken into account (Kim and Dale 2004; Sarkar et al. 2012). The United States alone generates a total of 1.3×10^3 million tons (MT) of biomass available for bioethanol production, out of which 428 MT comes from agricultural wastes, 370 MT from forestry wastes, 377 MT from energy crops, 87 MT from grains, 58 MT from municipal and industrial solid wastes, and 48 MT from other wastes (Saini et al. 2015).

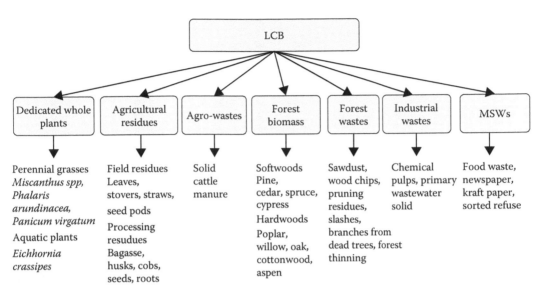

FIGURE 6.1
Different sources of LCB along with some examples.

The majority of the global LCBs as agricultural residues are obtained from four crops, which are corn, wheat, rice, and sugarcane, whereas the rest of agro-wastes make up only a little amount of the total world biomass (Saini et al. 2015). One of the most promising crop residues for lignocellulosic ethanol is corn stover. It is usually left unused after the harvest of corn grains and consists of stalks, leaves, cobs, and husks. The average estimated production rate of corn stover is 4.0 tons/acre (Kim and Dale 2004). Wheat straw is the residue of the harvested wheat, and it generates 1–3 tons/acre of biomass annually. Rice straw is one of the most abundant and promising LCBs in the world, which includes stems, leaf blades, and leaf sheaths left after harvesting the kernels, with a global production rate of 731 MT per year (Saini et al. 2015). Bagasse is a cheap residue left after the processing of sugarcane, which has been estimated to be 317–380 MT per year (Sánchez 2009). Most of the rice and wheat straws are produced in Asian countries, whereas the majority of the corn stovers and bagasse are generated in North and South America, respectively.

Biomass-based ethanol industry requires a continuous and reliable supply of raw materials to maintain a low-cost ethanol production. Energy crops are the promising ethanol feedstocks in this regard, which require the minimal use of water, fertilizer, and cultivable land. Energy crops used for bioethanol feedstocks may be either C_3 or C_4 plants, and include miscanthus (*Miscanthus* spp., C_4), switchgrass (*Panicum virgatum*, C_4), reed canary grass (*Phalaris arundinacea*, C_3), giant reed (*Arundo donax*, C_3), and alfalfa (*Medicago sativa*, C_3) as the most promising sources. In comparison with other sources (such as woods and C_3 crops), C_4 perennial grasses produce biomass more than two times higher each year in warm and temperate regions due to their more efficient photosynthetic pathway (Lewandowski et al. 2003).

Forest biomass includes mainly woody materials such as hardwoods and softwoods, whereas forest wastes are mainly sawdust, pruning, bark thinning residues, wood chips, and branches from dead trees (Limayem and Ricke 2012). Forest woody feedstock accounts for approximately 370 MT of biomass per year in the United States (Perlack et al. 2005). Other forest-rich countries are Canada, Russia, Brazil, and China that comprise together more than half of the total forest area of the world (Hadar 2013). Softwoods are derived from the evergreen species such as pine, cedar, spruce, cypress, fir, hemlock, and redwood,

TABLE 6.2

Potential of Different LCBs as Ethanol Sources

Sources	Brief Introduction	Potential	Potential Ethanol Yield (L/ton)	References
Perennial grasses	• Twenty perennial grasses have been reported as potential ethanol feedstocks	• High biomass yield; 0.9–37 ton/ha • High cellulose contents • Easy to grow and harvest • Capacity to provide 50%–70% of the total biomass for ethanol	160–460	Kallioinen et al. 2012; Lewandowski et al. 2003; Rengsirikul et al. 2013; van der Weijde et al. 2013
Aquatic plants	• Water hyacinth (*Eichhornia crassipes*)	• Widely prevalent aquatic weeds • Abundant in certain parts of the world making it a suitable feedstock for distributed ethanol production • They grow in waterbodies and do not compete with arable lands • Exceptionally fast-growing plants • Biomass productivity is very high	—	Aswathy et al. 2010; Kumar et al. 2009
Agricultural residues	• Agricultural field and processing residues after harvesting crops	• Easily available • Crop waste management to value-added products • Minimizing the reliance on forest woody biomass and thus reducing deforestation • Short harvest rotation period • Global crop residues are estimated (million tons/year): 2802 for cereal crops, 3107 for 17 cereals and legumes, and 3758 for 27 food crops	235–450	Kim and Dale 2004; Kreith and Krumdieck 2013; Lal 2005; Limayem and Ricke 2012
Forest biomass and wastes	• Biomass includes softwoods (conifers and gymnosperm trees) and hardwoods (angiosperm trees) • Wastes include sawdust, wood chips, pruning residues, slashes, branches from dead trees, and forest thinning	• One of the largest unexploited and underutilized LCBs, which is estimated, for example, nearly 60 million dry tons annually in the United States • High density makes the transportation more economical • Flexible harvesting times	220–275 for softwood, and 280–285 for hardwood	Gonzalez et al. 2011; Zhu and Pan 2010
MSW	• Mixed MSW consists of integrated cupboard, paper, food residues, garden waste, metal, glass, plastics, and textile	• The estimated MSW is 20.7 million ton/year for an urban area with 217 million population • Prospects for both energy development and waste management	154	Buah et al. 2007; Li et al. 2012; Prasad et al. 2007

which possess low densities and fast growth. Hardwoods are mainly found in the Northern Hemisphere and include trees such as poplar, willow, oak, cottonwood, and aspen.

MSWs are the recyclable cellulosic materials that originate from either residential or nonresidential sources such as food wastes and paper mill sludge (Limayem and Ricke 2012). The estimated global production of MSWs was 1.3×10^9 tons in 1990, which became almost double after 10 years, accounting for 2.3×10^9 tons in 2000 (Hadar 2013). Mostly, the organic fraction of MSWs is used as an ethanol feedstock (Schmitt et al. 2012). However, MSWs are not considered as ideal feedstocks for bioethanol production due to their wide diversification in composition and existence of microbial contaminations. Nevertheless, these feedstocks have the potential to be promising ethanol feedstocks in regional perspectives.

6.4 Composition of Lignocellulosic Biomasses

The LCBs consist of mainly cellulose, hemicellulose, and lignin, which are the structural ingredients of plant cell walls (Figure 6.2). However, the contents of these three components of LCBs significantly vary depending on the type of biomass and botanical origin (Table 6.3). Cellulose and hemicellulose constitute roughly two-third of the total dry weight and are the substrates for producing ethanol (Gírio et al. 2010). These two carbohydrate polymers are connected with the lignin by the covalent and hydrogen bonds, forming a lignocellulosic network that is highly robust and recalcitrant to depolymerization (Limayem and Ricke 2012). The recalcitrance of LCBs is due to several properties of the biomass, including the crystalline nature of cellulose, low available surface area, protection of carbohydrates by lignin, and heterogeneous nature of the biomass particles (Chang and Holtzapple 2000; Mosier et al. 2005). The composition and structure of the biomass determine the digestibility of LCBs to the chemical or biochemical treatments (Mosier et al. 2005).

Cellulose is a linear and crystalline structure containing linear chains of glucose units and joined together by β-1,4-glucosidic linkage (Limayem and Ricke 2012). The average molecular weight of cellulose is approximately 100kDa (Saxena et al. 2009). Extensive hydrogen bonds among the molecules create a crystalline network through cross-linkages of various hydroxyl groups and make microfibrils that give cellulose molecules more strength and rigid (Limayem and Ricke 2012). This complex structure of cellulose results in tightly packed backbone and high crystallinity, which in turn gives water insolubility to the cellulose polymer and resistance to depolymerization (Mosier et al. 2005).

Hemicellulose is a complex and heteropolymer, having a mean molecular weight of approximately 30kDa. It consists of short, linear, and highly branched chains of different monomers, including hexoses (β-ᴅ-glucose, α-ᴅ-galactose, and β-ᴅ-mannose) and pentoses (β-ᴅ-xylose and α-ʟ-arabinose), and may contain sugar acids (uronic acids), namely, α-ᴅ-glucuronic, α-ᴅ-galacturonic, and α-ᴅ-4-*O*-methylgalacturonic acid, which remain in association with cellulose in the plant cell wall (Limayem and Ricke 2012; Saxena et al. 2009). Sometimes, other sugars such as a small amount of α-ʟ-rhamnose and α-ʟ-fructose may also be present in hemicellulose. Its backbone chain is principally made up of xylan through β-1,4 linkages that include ᴅ-xylose (around 90%) and ʟ-arabinose (roughly 10%) (Gírio et al. 2010). The commonly known hemicelluloses are xylans and glucomannans. Xylans are the main hemicelluloses in hardwoods, forest wastes, agricultural residues,

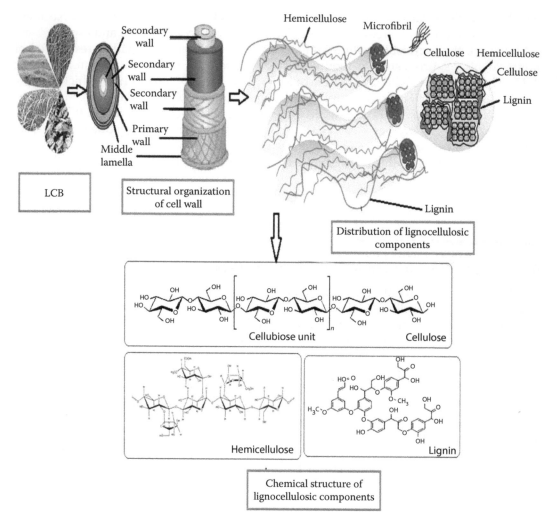

FIGURE 6.2
General composition of LCB and distribution of lignocellulosic components in the cell wall. (Adapted from Menon, V. and M. Rao, *Prog. Energy Combust. Sci.*, 38(4), 522–550, 2012.)

and municipal and industrial wastes, whereas softwoods are typically rich in glucomannans (Gírio et al. 2010; Limayem and Ricke 2012).

Lignin is a highly branched aromatic polymeric component present in the plant cell wall and is linked to cellulose polymers forming a lignocellulosic matrix (Drummond and Drummond 1996). The monomers of a lignin polymer are three phenolic compounds such as coumaryl, coniferyl, and sinapyl alcohol. Among the LCBs, woods contain the highest amounts of lignin with their cellulose contents. Softwood has been marked for the maximum lignin content, ranging between 30% and 60%, whereas the lignin content of hardwood varies between 30% and 55%. The grasses and agricultural wastes are the minimum lignin-containing LCBs that contain 10%–30% and 3%–15% lignin, respectively (Limayem and Ricke 2012).

TABLE 6.3

Chemical Composition of LCBs (%, dry basis)

Name	Cellulose	Hemicellulose	Lignin	References
Switch grass	5–20	30–50	10–40	McKendry 2002
Miscanthus	38.2–40.0	18.0–24.3	24.1–25.0	Brosse et al. 2010
Grass Esparto	33–38	27–32	17–19	Sánchez 2009
Elephant grass	22	24	23.9	Sánchez 2009
Bermuda grass	25	35.7	6.4	Prasad et al. 2007
Grasses (general)	25–40	25–50	10–30	Saini et al. 2015
Alfalfa stalks	48.5	6.5	16.6	Chandel et al. 2007
Sugarcane whole	25	17	12	Saxena et al. 2009
Napier grass	32	20	9	Saxena et al. 2009
S32 rye grass (early leaf)	21.3–26.7	15.8–25.7	2.7–7.3	Sánchez 2009
Orchard grass	32	40	4.7	Sánchez 2009
Water hyacinth	18.2–18.4	48.7–49.2	3.5–3.55	Kumar et al. 2009; Nigam 2002
MSW	33–49	9–16	10–14	Li et al. 2012; Saxena et al. 2009
Kraft paper	57.3	9.9	20.8	Schmitt et al. 2012
High-grade paper	87.4	8.4	2.3	Schmitt et al. 2012
Mixed/low-grade paper	42.3	9.4	15.0	Schmitt et al. 2012
Leaves and grass	15.3	10.5	43.8	Schmitt et al. 2012
Food waste	55.4	7.2	11.4	Schmitt et al. 2012
Office paper	68.6	12.4	11.3	Mosier et al. 2005
Newspaper	40–55	25–40	18–30	Howard et al. 2004
Waste papers from chemical pulps	60–70	10–20	5–10	Prasad et al. 2007
Sorted refuse	60	20	20	Sánchez 2009
Primary wastewater solids	8–15	0	24–29	Sánchez 2009
Solid cattle manure	16–4.7	1.4–1.33	2.7–5.7	Sánchez 2009
Coffee husk	43	7	9	Gouvea et al. 2009
Nut shells	25–30	25–30	30–40	Howard et al. 2004
Corn cob	42–45	35–39	14–15	Kuhad and Singh 1993; Prasad et al. 2007
Cotton seed hairs	80–90	5–20	0	Prasad et al. 2007
Corn stover	38–40	24–26	7–19	Saini et al. 2015; Zhu et al. 2005
Corn fiber	14.28	16.8	8.4	Mosier et al. 2005
Coir	36–43	0.15–0.25	41–45	Saini et al. 2015
Sugarcane bagasse	42–48	19–25	20–42	Kim and Day 2011; Saini et al. 2015
Rice straw	28–36	23–28	12–14	Saini et al. 2015
Wheat straw	33–38	26–32	17–19	Saini et al. 2015
Barley straw	31–45	27–38	14–19	Saini et al. 2015
Sweet sorghum bagasse	34–45	18–27	14–21	Saini et al. 2015
Banana waste	13.2	14.8	14	Medina de Perez and Ruiz Colorado 2006
Bast fiber Kenaf	31–39	22–23	15–19	Sánchez 2009
Leaf fiber Abaca (Manila)	60.8	17.3	8.8	Sánchez 2009
Sponge gourd fibers	66.6	17.4	15.5	Guimarães et al. 2009
Pineapple leaf fiber	70–82	18	5–1	Reddy and Yang 2005

(Continued)

TABLE 6.3 (*Continued*)

Chemical Composition of LCBs (%, dry basis)

Name	Cellulose	Hemicellulose	Lignin	References
Oat straw	31–37	27–38	16–19	Sánchez 2009
Rye straw	33–35	27–30	16–19	Sánchez 2009
Bamboo	26–43	15–26	21–31	Sánchez 2009
Bast fiber seed flax	47	25	23	Sánchez 2009
Bast fiber jute	45–53	18–21	21–26	Sánchez 2009
Leaf fiber Sisal (agave)	43–56	21–24	7–9	Sánchez 2009
Leaf fiber Henequen	77.6	4–8	13.1	Sánchez 2009
Coffee pulp	35	46.3	18.8	Sánchez 2009
Banana waste	13.2	14.8	14	Sánchez 2009
Rice husk	25–35	18–21	26–31	Ludueña et al. 2011
Softwood	27–30	35–40	25–30	McKendry 2002
Softwood barks	18–38	15–33	30–60	Saini et al. 2015
Softwood stems	45–50	25–35	25–35	Sánchez 2009
Hardwood	20–25	45–50	20–25	McKendry 2002
Hardwood barks	22–40	20–38	30–55	Saini et al. 2015
Hardwood stems	40–55	24–40	18–25	Sánchez 2009
Pine	44–46.4	8.8–26	29–29.4	Olsson and Hahn-Hägerdal 1996; Mosier et al. 2005
Poplar	47.6–49.9	27.4–28.7	18.1–19.2	Olsson and Hahn-Hägerdal 1996; Mosier et al. 2005

6.5 Conversion of Lignocellulosic Biomass into Ethanol

Bioethanol production from LCB requires degradation of the recalcitrant lignocellulosic matrix into its fragments, namely of cellulose, hemicellulose, and lignin (Limayem and Ricke 2012). The carbohydrates are then hydrolyzed into soluble sugars, which are converted into ethanol in the subsequent steps. Finally, ethanol is recovered from the fermentation media and purified to obtain a fuel grade ethanol. Therefore, the overall conversion process can be divided into four major steps that include pretreatment, hydrolysis, fermentation, and product recovery (Figure 6.3).

6.5.1 Pretreatment

The most complicated and costly step in the conversion of LCB into ethanol is the pretreatment. Cellulose can be enzymatically converted into glucose using cellulases before fermentation (Lynd et al. 1991). However, the availability of cellulose for enzymatic action in a raw LCB is difficult as cellulose fibers are embedded in a lignin–hemicellulose matrix (Esteghlalian et al. 1997). Furthermore, complete removal of holocellulose from the lignocellulosic matrix is required for avoiding the interference of lignin and hemicellulose on enzyme accessibility to cellulose during hydrolysis (Sun and Cheng 2002). Therefore, a suitable pretreatment method must be applied to LCBs for removing the protecting shield from the lignin–hemicellulose network. This kind of treatment alters the macroscopic, submicroscopic, and microscopic structures of the biomass. The major goals of pretreatment

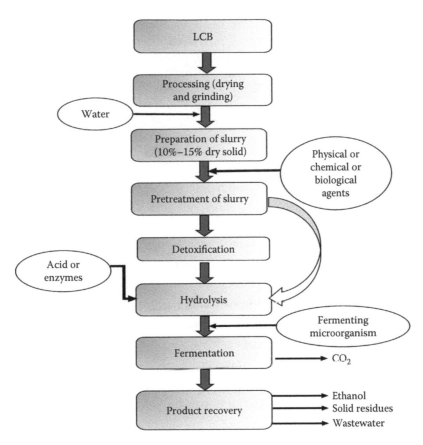

FIGURE 6.3
General steps in the conversion of LCB into ethanol. (Hamelinck, C.N., G. Van Hooijdonk and A.P. Faaij, *Biomass Bioenergy*, 28(4), 384–410, 2005.)

on the biomass are to remove lignin, decrease the crystallinity of cellulose, and increase the surface area and porosity (Wyman et al. 2005). Some pretreatment processes can also release fermentable sugars by hydrolyzing the hemicellulose portion of the bio-mass (Sørensen et al. 2008).

Even though pretreatment is one of the most expensive steps in lignocellulosic ethanol production particularly for requiring high-energy input. However, many efforts have been made in recent years to improve the overall efficiency and reduce the energy consumption as well as costs (Lynd et al. 1996; Mosier et al. 2005). Prior to considering a pretreatment method whether it is economically viable, its energy efficiency is evaluated. The energy efficiency can be determined by the ratio of the total soluble sugar yield to the total energy consumption (Zhu and Pan 2010):

$$\eta = \frac{TSY}{TEC} \qquad (6.1)$$

where:
   η is the pretreatment energy efficiency (kg/MJ)
   TSY is the total soluble (monomeric) sugar yield (kg)
   TEC is the total energy consumption during pretreatment

A good number of pretreatment methods have been investigated till to date, which are either simple or technologically more intensive. These methods can be broadly classified into four categories: physical, chemical, physicochemical, and biological (Galbe and Zacchi 2007; Sørensen et al. 2008). The processing conditions, the reaction time, and the mode of action of each method differ considerably (Mosier et al. 2005). Furthermore, each pretreatment method has some advantages and disadvantages as summarized in Table 6.4. To be effective and competitive from the economic perspective, a pretreatment method should meet some essential requirements (Alvira et al. 2010; Yang and Wyman 2008):

- Recovery of maximum cellulose and hemicellulose with improved digestibility.
- Minimum loss of hemicelluloses and cellulose while converting them into by-products.
- Minimum degradation of sugars into sugar-derived products.
- Production of minimum amounts of toxic compounds.
- Minimum requirements for size reduction of biomass before pretreatment.
- Feasibility to operate in reasonable size and moderate costs reactors.
- Minimum production of solid waste residues.
- Effective at the low moisture level.
- High fermentation compatibility of the products.
- High lignin recovery.
- Minimum requirements for heat and power.
- Low capital and operational costs.

Different types of physical processes can be used for pretreatment of LCBs that improve the enzymatic hydrolysis or biodegradability of the biomass (Taherzadeh and Karimi 2008). These treatments include milling such as ball milling, two-roll milling, hammer milling, colloid milling, and vibro-energy milling, and irradiation such as gamma rays, electron beams, or microwaves. Physical pretreatments exert their effects on the biomass by increasing the accessible surface area and pore volume, and decreasing the degree of polymerization of cellulose and its crystallinity, hydrolysis of hemicelluloses, and partial depolymerization of lignin (Szczodrak and Fiedurek 1996). The energy requirements for physical pretreatments depend on the final particle size and crystallinity of the biomass. In most cases, when physical pretreatment is the only option, the required energy is higher than the theoretical energy content available in the biomass. Physical pretreatments are expensive and usually cannot be used on a large-scale process (Brodeur et al. 2011).

Chemical pretreatment of biomass basically includes alkalis and acids that alter the biomass through delignification, decrease in the degree of polymerization and crystallinity of cellulose (Szczodrak and Fiedurek 1996). H_2SO_4, HCl, H_3PO_4, and HNO_3 are used during acid pretreatment of biomass. Among these chemicals, H_2SO_4 is the most commonly used acid, whereas NaOH is the major alkali. Acid pretreatment is applied to solubilize the hemicellulosic fraction of the biomass and make cellulose more accessible to enzymes. Organic acids such as fumaric or maleic acids are also used as alternative acids that enhance cellulose hydrolysis and reduce the production of inhibitors (Alvira et al. 2010; Mosier et al. 2005). Cellulose solvents are another type of chemical additives that include alkaline H_2O_2, ozone, organosolv, glycerol, dioxane, phenol, or ethylene glycol, which are

TABLE 6.4

Advantages and Disadvantages of Different Pretreatment Methods

Pretreatment Methods	Brief Description	Advantages	Disadvantages	References
Physical pretreatments				
Mechanical comminution	A combination of chipping, grinding, or milling to a final particle size of the material (10–30 mm after chipping and 0.2–2 mm after milling or grinding).	• Reduction of cellulose crystallinity • Increased surface area • Reduction in the degree of polymerization • Easy handling of feedstocks for subsequent processing steps	• High energy requirement for grinding • No lignin removal	Alvira et al. 2010; Mosier et al. 2005
Irradiation	Treatment of biomass with high-energy radiation, including gamma rays, ultrasound, electron beam, pulsed electrical field, ultraviolet, and microwave heating.	• Increased surface area • Reduction in the degree of polymerization • Reduction in cellulose crystallinity • Partial degradation of lignin	• Far too expensive • Slow rate of reaction • Energy intensive • Not eco-friendly	Zheng et al. 2009
Chemical pretreatments				
Dilute acid treatment	The effective concentration of acids is <4%. Works at high temperature (e.g., 180°C) during a short period of time or at low temperature (e.g., 120°C) for longer retention time (30–90 min). Both inorganic acids (H_2SO_4, HCl, H_3PO_4, HNO_3) and organic acids (fumaric and maleic acids) can be used.	• Increased accessible surface area • High reaction rate • Removal of hemicellulose • Alteration in lignin structure • Improved digestibility • More studied method	• Little lignin removal • Formation of inhibitors • Requirement of neutralization • Requirements for disposal of neutralization salts	Alvira et al. 2010; Mosier et al. 2005; Paulová et al. 2013
Concentrated acid pretreatment	The concentration of the acid ranges from 70% to 77% and the temperature from 40°C to 100°C. Mostly inorganic acids (H_2SO_4 and H_3PO_4) are used.	• Complete removal of the cellulose crystalline structure • Achievement of amorphous cellulose • Increased accessible surface area • High reaction rate	• Formation of inhibitors • Corrosion of equipment • Need for acid recovery • High operational and maintenance costs	Alvira et al. 2010; Paulová et al. 2013

(Continued)

TABLE 6.4 (*Continued*)

Advantages and Disadvantages of Different Pretreatment Methods

Pretreatment Methods	Brief Description	Advantages	Disadvantages	References
Alkali pretreatment	• It involves a base such as NaOH, KOH, NH₄OH, and Ca(OH)₂. • It can perform at room temperature • The reaction time ranges from seconds to days	• Low temperatures and pressures • Increased surface area • Removal of hemicellulose • Removal of lignin • Alteration in lignin structure • Lower degradation of sugars than acid pretreatment	• Conversion of alkali into irrecoverable salts or incorporation as salts into the biomass • Need for pH adjustment for subsequent processes	Alvira et al. 2010; Mosier et al. 2005
Ozonolysis	• Treatment with ozone gas usually at room temperature and normal pressure • The reaction time is several hours	• Efficient delignification • No formation of inhibitors • Ambient temperature and pressure	Need large amounts of ozone that make the process economically unviable	Alvira et al. 2010; Paulová et al. 2013
Ionic liquid (IL) pretreatment	• ILs are salts typically composed of large organic cations and small inorganic anions • They exist as liquids often at room temperature and have a tendency to remain liquid in a wide range of temperatures (<100°C)	• No toxic or explosive gases are formed • Carbohydrates and lignin can be simultaneously dissolved • Minimum degradation of desired products • Low temperature • High biomass loading • High lignin solubility	• Expensive • Requirement for washing before reuse • Lack of mature commercial IL recovery methods	Alvira et al. 2010; Paulová et al. 2013
Organosolv	• Treatment with a mixture of organic or aqueous solvents such as methanol, ethanol, acetone, and ethylene glycol • Solvent can be combined with acid catalysts such as HCl, H₂SO₄, oxalic, or salicylic to break hemicellulose bonds	• Recovery of relatively pure lignin as a by-product • Minimum cellulose loss (less than 2%) • High pretreated material yield • Low sugar degradation	• Requirement for removal of solvent from the system • High costs of chemicals • Formation of inhibitors	Alvira et al. 2010; Park et al. 2010

(*Continued*)

TABLE 6.4 (*Continued*)

Advantages and Disadvantages of Different Pretreatment Methods

Pretreatment Methods	Brief Description	Advantages	Disadvantages	References
Physicochemical pretreatments				
Uncatalyzed steam explosion	• Treatment with steam at high temperatures and pressures such as 160°C–260°C and 0.69–4.83 MPa, respectively. • The reaction time is several seconds to a few minutes. • No chemicals are added.	• Removal of hemicellulose • Improved enzyme accessibility and reduction of particle size • Increased pore volume	• Little lignin removal • Decomposition of sugars • High energy demand	Mosier et al. 2005; Paulová et al. 2013; Sun and Cheng 2002
Acid catalyzed steam explosion	• Steam explosion is catalyzed by addition of H_2SO_4 or SO_2. • The temperature range is 160°C–220°C.	• Removal of hemicellulose • Increased surface area • Increased enzyme accessibility • Low environmental impact • Less hazardous process	• Formation of inhibitors • Partial hydrolysis of hemicellulose	Alvira et al. 2010; Paulová et al. 2013
Liquid hot water	• Hydrothermal treatment of biomass with rapid decompression. • Pressure is applied to maintain water in the liquid state at elevated temperatures (160°C–240°C). • The pH range is 4–7 and the reaction time is up to 15 min.	• Structural and chemical alteration in lignin • Does not require any catalysts or chemicals • Increased accessible surface area • Removal of hemicellulose • No or little inhibitor formation • Low costs for reactors	• High water demand • High energy requirements • Hemicellulose degradation	Alvira et al. 2010; Mosier et al. 2005; Paulová et al. 2013
AFEX	• Treatment with anhydrous liquid ammonia at 60°C–120°C and above 3 MPa for 30–60 min, followed by rapid decompression.	• Increased surface area • Reduction in cellulose crystallinity • Removal of hemicellulose • Removal of lignin • Alteration of lignin structure • No inhibitor formation	• Less efficient for softwood-based biomass • Environmental concerns	Mosier et al. 2005; Paulová et al. 2013
ARP	• Aqueous ammonia at a concentration of 5%–15% passes through a reactor packed with biomass at 140°C–210°C, for up to 90 min with percolation rate about 5 mL/min.	• Removal of lignin • Reduction in cellulose crystallinity • Alteration of lignin structure • Low inhibitor formation	• High energy requirement • Environmental concerns	Alvira et al. 2010; Mosier et al. 2005; Paulová et al. 2013

(Continued)

TABLE 6.4 (*Continued*)

Advantages and Disadvantages of Different Pretreatment Methods

Pretreatment Methods	Brief Description	Advantages	Disadvantages	References
SAA	• An alternative to APR performs at a lower temperature (30°C–75°C).	• Formation of low inhibitors • Lower energy requirements	• Environmental concerns	Alvira et al. 2010
Wet oxidation	• An oxidative pretreatment of biomass using oxygen or air as catalyst. • The oxidation is performed for 10–15 min at 170°C–200°C and 10–12 bars.	• Solubilization of hemicellulose and lignin • Increased digestibility of cellulose • Formation of low inhibitors • Removal of lignin	• High costs for oxygen and catalysts • Requirement of high pressure and temperatures	Alvira et al. 2010
CO_2 explosion	• CO_2 is used as a supercritical fluid for pretreatment of biomass.	• Effective removal of lignin • Increased digestibility • Increased accessible surface area • Low sugar degradation • Low temperature	• Requirement of costly equipment	Alvira et al. 2010
Biological pretreatment				
Pretreatment with microorganism	• Treatment of biomass with microbes, mainly brown, white, and soft rot fungi, under normal temperatures and pressures.	• Lignin degradation • Reduction in the degree of polymerization of cellulose and hemicellulose • No chemical required • Mild environmental conditions • Low capital costs • Low energy demand • Low inhibitor formation	• Low efficiency • Loss of carbohydrates as consumed by fungi • Long residence time (10–14 days) • Requirement of carefully controlled growth conditions	Alvira et al. 2010; Paulová et al. 2013; Szczodrak and Fiedurek 1996

known to disrupt the cellulose structure and promote hydrolysis. Concentrated mineral acids (H_2SO_4, HCl), ammonia-based solvents (NH_3, hydrazine), aprotic solvents (DMSO, Dimethyl sulfoxide), and metal complexes (ferric sodium tartrate, cadoxen and cuoxan) have also been reported to reduce cellulose crystallinity and disrupt the association of lignin with cellulose, as well as dissolve hemicelluloses (Mosier et al. 2005).

Physicochemical pretreatment includes a good number of technologies, such as steam explosion, ammonia fiber explosion (AFEX), ammonia recycling percolation (ARP), soaking aqueous ammonia (SAA), wet oxidation, and CO_2 explosion. Like other pretreatment methods, physicochemical technologies also increase the accessible surface area of the biomass for enzyme accessibility, decrease the cellulose crystallinity, and remove hemicelluloses and lignin during pretreatment (Alvira et al. 2010; Mosier et al. 2005).

Biological pretreatment of LCBs can be carried out using microorganisms, particularly fungi such as white rot, brown rot, and soft rot fungi, where white rot fungi are the most efficient for this purpose (Sarkar et al. 2012). This kind of pretreatment alters the structure of lignin and cellulose. Brown rot fungi attack cellulose, whereas white and soft rot fungi attack both cellulose and lignin (Prasad et al. 2007). The biological conversion of different lignocellulosic feedstocks such as forest and agricultural residues, and dedicated energy crops offer numerous benefits in the conversion process (Sanchez and Cardona 2008). Lignin degradation occurs during biological pretreatment of biomass through the action of lignin-degrading enzymes secreted by the fungi. The major disadvantages are low rates of hydrolysis and prolonged pretreatment time compared to other technologies. Current efforts on biological pretreatments are made mainly to combine this technology with other processes and develop novel microorganisms for rapid hydrolysis (Brodeur et al. 2011).

As LCBs differ in their physicochemical characteristics and composition, appropriate pretreatment methods and conditions also vary among the biomass (Table 6.5). Furthermore, pretreatment methods have affect the hydrolysis and fermentation with respect to the changes in cellulose digestibility, production of inhibitors, and demands for wastewater treatment (Galbe and Zacchi 2007).

6.5.2 Detoxification

The severe conditions applied during thermochemical pretreatment often lead to a partial breakdown of lignin and hemicellulose, which results in the production of several unwanted by-products. These by-products include sugar acids, acetic acid, formic acid, levulinic acid, hydroxymethyl furfural (HMF), and furfural (Jönsson et al. 2013). Some of these products act as inhibitors for both fermenting microorganisms and cellulose-degrading enzymes when their accumulation is sufficiently high (Cavka and Jönsson 2013). The inhibitors are categorized into three major groups based on the origin: aliphatic acids, furan derivatives, and phenolic compounds (Larsson et al. 1999; Palmqvist and Hahn-Hägerdal 2000).

There are several ways to counteract the problems with inhibitors by removing or neutralizing them from the pretreated solution using biological, physical, and chemical methods (Palmqvist and Hahn-Hägerdal 2000). The usage of microorganisms for fermentation that are resistant to inhibitors would be one of the most promising biological detoxification methods. Chemical detoxification can be done by treating the pretreated biomass with certain chemicals, such as sodium dithionite, sodium sulfite, and sodium borohydride (Cavka and Jönsson 2013). However, the effective detoxification process depends on the type and quantity of the inhibitors present in the solutions. Detoxification may be essential when strongly inhibiting components are present in high quantities, or when a fermenting

TABLE 6.5

Pretreatment and Downstream Digestibility of Some Biomass

Biomass	Pretreatment	Catalysts	Pretreatment Conditions (Temperature in °C/time in min)	Downstream Digestibility (%)/ Sugar Yield (%)	References
Pine	Organosolv	50% ethanol + 1% H_2SO_4 50% ethanol + 1% $MgCl_2$ 50% ethanol + 2% NaOH	150–210/0–20	53–57[a] 57–61[a] 12–58[a]	Park et al. 2010
Corn cob	Dilute acid-alkali (combined)	2% H_2SO_4 2% NaOH	80–121/45	—	Zhang et al. 2010
Corn stover	Physicochemical	Electrolyzed water	195/10–30	83[b]	Wang et al. 2012
Corn stover	Two-step pretreatment process	1st stage: 1% (w/w) NaOH in 70% (w/w) ethanol; 2nd stage: hot water	135/30	83[b]	Saha et al. 2013
Corn stover	Hydrothermal	Water	200/5	72[b]	Saha et al. 2013
Aspen (wood chips); hardwood	Acid catalyzed steam explosion	0.7%–0.9% SO_2	205/3–10	75–78[b]	De Bari et al. 2007
Salix (wood chips); hardwood	Acid-catalyzed steam explosion	0.25%–0.5% H_2SO_4	200/8	80–81[b]	Sassner et al. 2008
Poplar (knife milled); hardwood	Acid-catalyzed steam explosion	3.0% SO_2	190/5	96[b]	Wyman et al. 2009
Lodgepole pine; softwood	Acid-catalyzed steam explosion	4.0% SO_2	200/5	60[b]	Ewanick et al. 2007
Spruce (wood chips); soft wood	Acid-catalyzed steam explosion	2.5%–4.0% SO_2	200–215/5	50–62[b]	Söderström et al. 2004; Monavari et al. 2009
Spruce (wood chips); soft wood	Two-stage acid-catalyzed steam explosion	First stage: 3.0% SO_2 Second stage: 0% SO_2	First stage: 190/2 Second stage: 210/5	62[b]	Söderström et al. 2004
Wheat straw	Organosolv without catalyst	50% ethanol	210/60	86[b]	Wildschut et al. 2013
Bamboo	Organosolv	75% ethanol (2% H_2SO_4)	180/30	77.1[b]	Zhang et al. 2016
Rice straw	Organosolv	75% ethanol (1% H_2SO_4)	150/60	46.2[b]	Amiri et al. 2014
Miscanthus	Organosolv	80% ethanol (1% H_2SO_4)	150–170/30–60	75[a]	Obama et al. 2012
Oil palm	Physicochemical	Hot-compressed water and wet disk milling	150–190/10–240	88.5–100[b]	Zakaria et al. 2015
Sunflower stalks	Physicochemical	Steam explosion	220/5	37.8[a]/72[b]	Ruiz et al. 2008
Elephant grass	Organosolv	Ethanol (3%–14%)	180/1–3	—	Carioca et al. 1985
Miscanthus	Alkali	NaOH (12%)	70/various	69[a]	De Vrije et al. 2002

[a] Digestibility

[b] Theoretical sugar yield

microorganism having low capacity to inhibitor tolerance is used during fermentation. An efficient detoxification method should selectively remove inhibitors, and be cheap and easy to integrate into the process (Palmqvist and Hahn-Hägerdal 2000).

In the 1940s, treatment with a reducing agent, such as sulfite, was suggested as a way to overcome the unfavorable reduction potential in lignocellulosic hydrolysates (Leonard and Hajny 1945). In the more recent study, hydrolysate obtained from dilute acid hydrolysis of spruce has been treated with sodium sulfite that showed a decrease in the concentration of furfural and HMF (Larsson et al. 1999). A combination of sulfite and over-liming has been shown to be more effective for detoxifying willow hydrolysate (Olsson et al. 1995). Furthermore, a threefold reduction in fermentation time could be occurred when hydrolysate is supplemented with 0.1% sodium sulfite and heated at 90°C for 30 min, along with over-liming (Olsson et al. 1995). However, a higher concentration of sodium sulfite has been reported for detoxifying the hydrolysate of sugarcane bagasse through adjusting the pH to 10 with KOH, followed by readjusting the pH to 6.5 with HCl and addition of 1% sodium sulfite at room temperature (Palmqvist and Hahn-Hägerdal 2000; Van Zyl et al. 1988).

Other promising chemicals for detoxification of biomass hydrolysates are sodium borohydride and sodium dithionite. According to a study conducted by Cavka and Jönsson (2013), treatment of the biomass hydrolysates with sodium borohydride resulted in the improved fermentability than what was observed with untreated hydrolysates. An increased fermentability of the hydrolysate was observed until the concentration of borohydride reached 47 mM, and a higher borohydride concentration at 55 mM gave slightly lower improvements than those observed between 31 and 47 mM. However, addition of as little as 5 mM sodium dithionite has resulted in improved fermentability compared to untreated control fermentation (Alriksson et al. 2011; Cavka and Jönsson 2013).

6.5.3 Hydrolysis

Cellulose and hemicellulose in LCB must be converted into soluble sugars through hydrolysis (Lamsal et al. 2010). In general pretreatment of LCB produces two fractions: a water-insoluble solid portion containing mainly cellulose and lignin, and a liquid fraction containing hemicellulose. In response to the pretreatment methods and conditions, hemicellulose of the biomass may be either completely hydrolyzed to its monomeric sugars or converted into oligosaccharides if it undergoes incomplete depolymerization that requires a further hydrolysis (Gírio et al. 2010; Lavarack et al. 2002). The dominant sugars in the hemicellulose are mannose in softwoods, and xylose in hardwoods and agriculture residues, in addition to the presence of small amounts of arabinose and galactose in all biomass (Taherzadeh and Karimi 2008). Even though a small portion of cellulose may also be converted into glucose under the pretreatment condition, most of this carbohydrate remains unreacted (Hamelinck et al. 2005), and can be converted into glucose through hydrolysis. The conversion of hemicellulose and cellulose into fermentable sugars can be done either by the acid or enzymatic hydrolysis (Chandel et al. 2007; Fockink et al. 2015; Maitan-Alfenas et al. 2015), and the outcomes usually depend on the type and compositions of the biomass as well as hydrolysis conditions (Table 6.6).

The major portions of cellulose and hemicellulose are glucan and xylan, respectively. Digestibility of both glucan and xylan can be determined from the yield of the respective monomers and disaccharides using Equations 6.2 and 6.3 (Wyman et al. 2005).

TABLE 6.6

Fermentable Sugar Yield during Hydrolysis of Some LCBs in Recent Studies

Feedstocks	Solid Load (%, w/w)	Substrate	Hydrolysis Mode	Hydrolysis Agents	Temperature (°C)	pH	Time (hr)	Sugar Yield	Reference
Sugarcane bagasse	18.3	Both H and C	Enzymatic	Cellulase from *Trichoderma reesei* ATCC 26921 β-Glucosidase (Novozyme 188)	45	5.3	72	Glu: 85.3 g/L Xyl: 18.8 g/L Man: 3.4 g/L Ara: 1.4 g/L Gal: 0.7 g/L	Cavka and Jönsson 2013
Spruce	12.1	Both H and C	Enzymatic	Cellulase from *T. reesei* ATCC 26921 β-Glucosidase (Novozyme 188)	45	5.3	72	Glu: 84.4 g/L Xyl: 8.0 g/L Man: 13.7g/L Ara: 1.9 g/L Gal: 2.0 g/L	Cavka and Jönsson 2013
Corn stover	10	Both H and C	Enzymatic	Cellulase (Celluclast 1.5 L) β-Glucosidase (Novozyme 188) Xylanase (Fiberzyme)	45	5	72	Glu: 24.9–36.3 g/L Xyl: 16.9–23.0 g/L Ara: 3.6–4.3 g/L	Avci et al. 2013
Water hyacinth	5.2–7.8	H	Acid	2% v/v sulfuric acid	RT	—	7	TRS: 18.8 %,w/w Pentose sugar 13.3%,w/w	Kumar et al. 2009
Water hyacinth	5.0–7.6	H	Acid	1% H_2SO_4	RT	—	7	Xyl: 12.4 %,w/w Glu: 1.7 %,w/w Ara: 2.2 %,w/w Gal: 1.2 %,w/w Man: 3.5 %,w/w	Nigam 2002
Corn stover	10	H	Alkali–acid	10% ammonia 1.5% H_2SO_4	RT–108	—	6–24	Xyl: 12.1 g/L Glu: 0.9 g/L Ara: 2.9 g/L	Zhao and Xia 2010
Wheat straw	5	H	Acid	1.85% H_2SO_4	90	—	18	Xyl: 12.8 g/L Glu: 1.7 g/L Ara: 2.6 g/L	Nigam 2001
Sugarcane bagasse	10	H	Acid	HNO_3 (2%–6%)	100–128	—	24	Xyl: 18.6 g/L Glu: 2.87 g/L Ara: 2.4 g/L	Rodírguez-Chong et al. 2004

H, hemicellulose; C, cellulose; RT, room temperature; Glu, glucose; Xyl, xylose; Man, mannose; Ara, arabinose; Gal, galactose; TRS, total reducing.

The sugar yield after hydrolysis can be used to determine the efficiency of the process and the potential of the biomass as ethanol feedstocks. The sugar yield during hydrolysis can be easily determined from the yields of xylose and glucose using Equation 6.4 or 6.5 (Zhu et al. 2008).

$$DGn\ (\%) = \frac{AG + 1.053 \times AC}{1.111 \times AGn} \times 100 \tag{6.2}$$

where:
  DGn is the digestibility of glucan (%)
  AG is the amount of glucose (g)
  AC is the amount of cellobiose
  AGn is the amount of glucan
  1.053 is the hydrolysis factor for cellobiose
  1.111 is the hydrolysis factor for glucan

$$DXn\ (\%) = \frac{AX + 1.064 \times AXb}{1.136 \times AXn} \times 100 \tag{6.3}$$

where:
  DXn is the digestibility of xylan (%)
  AX is the amount of xylose (g)
  AXb is the amount of xylobiose
  AXn is the amount of xylan
  1.064 is the hydrolysis factor for xylobiose
  1.136 is the hydrolysis factor for xylan

$$TSC = \frac{GY + XY}{TGY + TXY} \tag{6.4}$$

$$= \frac{([G] - [G_0]) \times V + ([X] - [X_0]) \times V}{\dfrac{W \times GC}{0.9} + \dfrac{W \times XC}{0.88}} \tag{6.5}$$

where:
  TSC is the total sugar conversion (%)
  GY is the glucose yield
  XY is the xylose yield
  TGY is the theoretical glucose yield
  TXY is the theoretical xylose yield
  GC is the glucan content
  XC is the xylan content
  $[G]$ is the glucose concentration in the hydrolysate (mg/mL)
  $[G_0]$ is the initial glucose concentration in the slurry (0 mg/mL)
  $[X]$ is the xylose concentration in the hydrolysate (mg/mL)
  $[X_0]$ is the initial xylose concentration (0 mg/mL)
  V is the initial volume of the slurry (mL)
  W is the initial dry weight of biomass (mg)
  0.9 is the conversion factor for glucose to equivalent glucan
  0.88 is the conversion factor for xylose to equivalent xylan

6.5.3.1 Acid Hydrolysis

Acid hydrolysis can be done by two approaches: dilute acid treatment at high temperatures and pressures with a short reaction time ranging between seconds and minutes, and concentrated acid treatment at low temperatures (Balat 2011; Chandel et al. 2007). The major problems with acid hydrolysis are the necessity for recovery or neutralization of the acids before fermentation and production of large amounts of wastes. Mostly, H_2SO_4 is used in either approaches of acid hydrolysis, although other inorganic acids as well as some weak organic acids such as hydrochloric acid (HCl), nitric acid (HNO_3), trifluoroacetic acid (TFA), phosphoric acid (H_3PO_4) have also been reported to be used for this purpose (Gírio et al. 2010).

Dilute acid hydrolysis is generally applied for hydrolysis of hemicellulose and pretreatment of cellulose. However, both carbohydrate polymers can be hydrolyzed into fermentable sugars using dilute acid, but this requires a two-stage hydrolysis process. In the first stage, acid hydrolysis is carried out at low temperature to maximize hemicellulose conversion. However, the second stage involves high temperatures that range between 230°C and 240°C to hydrolyze cellulose (Lee et al. 1999; Wyman 1999). The major drawbacks for cellulose hydrolysis using acids are high energy consumption and high risks for the formation of inhibitors through degradation of sugars. The most commonly used temperatures and concentrations of H_2SO_4 for dilute acid hydrolysis are 120°C–160°C and 0.5%–1.5%, respectively (Balat 2011; Gírio et al. 2010).

Concentrated acid hydrolysis can be applied for converting both hemicellulose and cellulose into their monomeric sugars. Like dilute acid hydrolysis, H_2SO_4, HCl, or TFA can be used for concentrated acid hydrolysis in the range between 41% for HCl and 100% for TFA (Fengel and Wegener 1983). The concentration of H_2SO_4 for hydrolysis of the LCBs is 70%–90% for concentrated hydrolysis (Hayes 2009). Reaction times required for this acid hydrolysis are much longer than the times needed for dilute acid hydrolysis. However, concentrated acid hydrolysis offers a complete and rapid conversion of cellulose to glucose, and hemicellulose to five carbon sugars (Balat 2007). Acid hydrolysis has been shown interest for to its moderate process temperature and nonrequirement for costly enzymes (Zhang et al. 2007). However, corrosion of the equipment with high acid concentration is a major drawback for this technique.

6.5.3.2 Enzymatic Hydrolysis

Enzymes can hydrolyze both cellulose and hemicellulose specifically and produce soluble sugars at low temperatures ranging from 45°C to 50°C. It does not have any corrosion problem for the equipment (Duff and Murray 1996). Compared to acid hydrolysis, sugar degradation and production of inhibitors are much lower in enzymatic hydrolysis. Enzymatic hydrolysis of cellulose and hemicellulose can be done by using cellulase and hemicellulase (xylanase) systems, respectively.

Cellulase is a mixture of different cellulolytic enzymes that act synergistically on the cellulose molecules (Reczey et al. 1996). A typical cellulase system includes endoglucanase, exoglucanase, and β-glucosidase. Endoglucanase exerts its effect on cellulose through the random breakdown of β-1,4 glycosidic linkages in the D-glucan chains of the amorphous regions of a cellulose molecule, which results in the production of free chain ends. Exoglucanase works on the cellobiose units in D-glucan chains from the nonreducing ends. The outcome of both endoglucanase and exoglucanase actions is the production of a disaccharide, namely, cellobiose, which is finally converted into glucose by the action of β-glucosidase (Lee 1997). An efficient cellulase system should be capable of degrading cellulose in its crystalline stage, as well as showing tolerance to acidic pH (4–5) and stress conditions (Bajaj et al. 2009).

Unlike cellulose, hemicellulose (xylan) is chemically quite complex and its degradation requires multiple enzyme system, which is known as the xylanase (hemicellulase) system. Typically, a xylanase system contains endo-xylanase, exo-xylanase, β-xylosidase, α-arabinofuranosidase, α-glucoronisidase, acetyl xylan esterase, and ferulic acid esterase (Saha 2004). Endo- and exo-xylanases act on the main chains of xylan and hydrolyze them into smaller chains. A β-xylosidase attacks xylo-oligosaccharides and produces xylose. α-Arabinofuranosidase and α-glucuronidase act on the xylan backbone and remove arabinose and 4-o-methyl glucuronic acid, respectively (Saha 2003). Acetyl esterases attack the acetyl substitutions on xylose moieties, whereas feruloyl esterases hydrolyze the ester bonds located between arabinose substitutions and ferulic acid. Feruloyl esterases also make it easy to release hemicellulose from lignin (Howard et al. 2004).

Sugar yield during enzymatic hydrolysis may be affected by a good number of factors, which are loosely divided into two groups: (1) enzyme- and process-related factors and (2) substrate-related factors, which may be interlinked with each other. The enzyme- and process-related factors are the type of enzymes, dosage, sources, and efficiency. The substrate-related factors are pointed out as follows (Alvira et al. 2010):

- Composition and structure of the feedstock
- Crystallinity of cellulose
- Degree of polymerization of cellulose
- Available surface area on the substrate
- Presence of lignin that may limit the rate of enzymatic hydrolysis by acting as a physical barrier
- Hemicellulose content in the pretreated biomass
- Feedstock particle size
- Pore size of the substrate in relation to the size of the enzymes
- Cell wall thickness (coarseness)

6.5.4 Fermentation

Certain bacteria and yeasts can ferment soluble sugars in the absence of oxygen and produce ethanol. Compared to starch-based ethanol production that uses hydrolysates of starch containing mostly glucose, the hydrolysates of LCBs contains a mixture of pentose and hexose sugars. Therefore, fermentation LCB needs to metabolize both pentose and hexose sugars simultaneously. Hexose sugars such as glucose, galactose, and mannose are readily fermented to ethanol by many naturally occurring microorganisms, traditionally, *Saccharomyces cerevisiae* and *Zymomonas mobilis*. However, most of these natural microorganisms cannot ferment pentose sugars, such as, xylose and arabinose (Hamelinck et al. 2005; Mosier et al. 2005). On the other hand several yeast (*Pichia stipitis*, *Candida shehatae*, and *C. parapsilosis*) can ferment xylose but not as efficient as hexose fermenting yeast.

The most common and traditional microorganism for fermenting sugars into ethanol is a yeast, *S. cerevisiae*, which has been proved to be very robust and well suited for fermentation of lignocellulosic hydrolysates (Galbe and Zacchi 2002). However, the natural strains of this yeast can ferment only six carbon sugars and hardly pentoses as it lacks enzymes that convert xylose to xylulose (Tian et al. 2008). Bacterial strains used for ethanol fermentation are *Z. mobilis*, *Escherichia coli*, and *Klebsiella oxytoca*, which have attracted a particular interest due to their capability to ferment hexose sugars rapidly (Balat 2011). Genetic modifications of bacteria and yeast have produced strains capable of co-fermenting both pentoses and hexoses (Mosier et al. 2005). In this context, research efforts have been carried out in two aspects: (1) genetically modification of yeast and other natural ethanol-producing microorganisms to include an additional pentose metabolic pathway and (2) enhanced ethanol yield using genetically engineered microorganisms (Chandel et al. 2007). The performance of any microorganism as ethanol fermenter depends on the process conditions, including temperatures, pH, and ethanol concentration. An ethanologenic microorganism should meet some criteria that include high ethanol tolerance, optimum growth rate of the cells, high ethanol yield and productivity, tolerance to osmotic stress, specificity, genetic stability, and tolerance to inhibitors (Balat 2011).

Microorganisms are frequently used as free cells during fermentation, which proliferate in the media and carry out their metabolic functions, producing ethanol from sugars. In recent years, there has been a growing interest in immobilization of cells on different inert supports that offer some technical and technological benefits over free cell system. Immobilized cells have a prolonged cellular stability and an increased tolerance to high substrate concentration. The major advantages of cell immobilization are increased ethanol yield; greater volumetric productivity; reduced end-product inhibition; reduced risk for contamination due to high cell densities; decreased energy demand and process expenses due to ease in product recovery, regeneration, and reuse of cells for extended periods; cell recycling in repeated batch fermentations; and protection of cells against toxic substances (Kourkoutas et al. 2004; Nikolić et al. 2010). Several carriers have been reported for cell immobilization, including apple pieces, k-carrageenan gel, polyacrylamide, g-alumina, chrysotile, calcium-alginate, sugarcane pieces, banana leaf sheath, and orange peel (Zabed et al. 2014).

Fermentation can be performed in three modes: batch, fed-batch, and continuous. Selection of an appropriate mode for fermentation depends on the kinetic properties of the microorganism to be used in fermentation and the type of the substrate to be converted into ethanol, in addition to the consideration of process economics.

The batch mode of fermentation is carried out in a closed culture system. The fermentation medium is supplemented with necessary ingredients at the initial stage and yeast or other bacterial inoculum added to this medium before starting fermentation. In the initial stage, the growth of yeast cells shows a lag phase that follows an exponential phase afterward. However, the cells compete for limited amounts of nutrients in the medium and finally enter into a stationary phase as nutrients are exhausted (Balat 2011). It is the simplest mode of fermentation and does not require anything to add in the medium, except acid or base solutions for pH adjustment during fermentation. In this situation, yeasts or bacterial cells work at a high substrate concentration in the initial stage, whereas they need to show the activities at a high ethanol concentration in the final stage (Olsson and Hahn-Hägerdal 1996).

The fed-batch system combines both batch and continuous modes of fermentation from a technological perspective and is widely used for ethanol production on a commercial scale (Saarela et al. 2003). During fed-batch fermentation, some of the ingredients are added to the medium while fermentation progresses. Microbial cells grow at a low substrate concentration in this system throughout the fermentation period and produce enhanced amounts of ethanol. The fed-batch system offers some advantages over the batch mode, such as high ethanol yield and volumetric ethanol productivity, maximum viable cells, prolonged lifetime of the cells, and tolerance to high ethanol concentration. Furthermore, fed-batch mode maintains the process conditions (temperature, pH, and dissolved oxygen level) at specific levels through a feedback control (Balat 2011). Nevertheless, the fermentation rate in a fed-batch system is limited by the feed rate, where the feed rate is limited by the concentration of the cell mass (Chandel et al. 2013).

Continuous fermentation is carried out in stirred tank or plug flow reactors in which the feed is added continuously. The feed usually contains substrate, culture medium, and essential nutrients, which are pumped into an agitated vessel where the cells are in an active stage (Balat 2011). This type of fermentation shows a higher productivity than the batch or feed-batch modes. However, the maximum productivity is obtained only from a low dilution rate of feed.

During fermentation, microbial conversion of pentose (e.g., xylose and arabinose), hexoses (e.g., glucose), and disaccharides (e.g., xylobiose and cellobiose) into ethanol occurs according to the reactions shown in Equations 6.6 through 6.9 described by Tran et al. (2013). As per Equations 6.6 and 6.7, the ethanol yield from monosaccharides (hexoses and pentoses) is 510 g/kg, whereas the disaccharides (C10 and C12) produce 540 g ethanol per kg disaccharide. In relation to the fermentation conditions, the ethanol yield, volumetric productivity, and theoretical ethanol yield can be calculated using Equations 6.10 through 6.12 as reported by Wyman et al. (2005) and Tran et al. (2013).

$$C_6H_{12}O_6 \text{ (Glucose)} = 2C_2H_5OH + 2CO_2 \tag{6.6}$$

$$3C_5H_{10}O_5 \text{ (Xylose, arabinose)} = 5C_2H_5OH + 5CO_2 \tag{6.7}$$

$$C_{12}H_{22}O_{11} \text{ (Cellubiose)} + H_2O = 5C_2H_5OH + 5CO_2 \tag{6.8}$$

$$3C_{10}H_{18}O_9 \text{ (Xylobiose)} + 3H_2O = 5C_2H_5OH + 5CO_2 \tag{6.9}$$

Ethanol yield is calculated as the amount of ethanol produced per unit of substrate utilized:

$$Y_{EtOH} = \frac{CV}{m} \tag{6.10}$$

where:
Y_{EtOH} is the ethanol yield (g/kg)
C is the ethanol concentration (g/L)
V is the initial volume of liquid medium (L)
m is the mass of the substrate (kg)

Ethanol productivity is estimated as the amount of ethanol produced per unit of substrate utilized per unit of time. It is typically determined when ethanol concentration is maximal. Ethanol productivity is calculated as follows:

$$Q_{EtOH} = \frac{CV}{mt} = \frac{1000Y_{EtOH}}{t} \tag{6.11}$$

where:
Q_{EtOH} is the ethanol productivity (mg/kg h)
t is the time when ethanol concentration produced is maximum (h)

The maximum theoretical ethanol yield can be calculated as follows:

$$Y_{max}(\%) = \frac{\text{Ethanol produced in reactor (g)}}{\text{Initial sugar in reactor (g)} \times 0.511} \times 100 \qquad (6.12)$$

Ethanol production during fermentation significantly varies depending on the nature and initial concentration of the substrate, fermentation mode and conditions, and microorganisms used for fermentation. Based on the investigations for ethanol production from some known LCBs, it can be noted that ethanol concentration in the fermentation broth can vary from 8.8 to 37.0 g/L if the initial substrate concentrations range from 50 to 120 g/L (Table 6.7).

6.5.5 Process Integration

As stated earlier, the conversion of LCB into ethanol includes four steps, where the pretreatment, hydrolysis, and fermentation are the major processing steps. Accomplishing these process steps require high energy and time that make the overall process complicated and economically noncompetitive. Recent research efforts have been made on the integration of these three key steps to enhance ethanol yield and productivities as well as decrease production costs (Lynd et al. 2008; Margeot et al. 2009). The schematic diagrams of these process integrations are shown in Figure 6.4a–e.

Typically, enzymatic hydrolysis of carbohydrate polymers and fermentation of soluble sugars are done separately, which is known as separate hydrolysis and fermentation (SHF). In a commercial SHF system, the joint liquid flow from two hydrolysis reactors (hemicellulose and cellulose) first enters the glucose (hexose) fermentation reactor. Once glucose fermentation is completed, the broth of the reactor is distilled to recover ethanol, leaving unconverted xylose and other pentoses behind. Subsequently, pentose fermentation is carried out in the second reactor, followed by a further distillation of the fermented broth (Hamelinck et al. 2005). With a slight modification in SHF, separate hydrolysis and cofermentation (SHCF) is a process in which both pentose and hexose sugars are produced in a separate hydrolysis of hemicellulose and cellulose, respectively, which are subsequently fermented together in the same reactor (Gírio et al. 2010). The performance of cellulose hydrolysis and fermentation of hexose sugars at a time is called simultaneous saccharification and fermentation (SSF). When SSF includes the cofermentation of hexose and pentose sugars, it is called simultaneous saccharification and cofermentation (SSCF). The consolidated bioprocessing (CBP) is another process integration that not only includes hydrolysis of carbohydrate polymers and fermentation of all kinds of soluble sugars but also produces hydrolyzing enzymes during the process, thereby conducting enzyme production, hydrolysis, and fermentation in a single step. Table 6.8 summarizes these process integrations along with their advantages and disadvantages.

6.5.6 Product Recovery

Ethanol produced during fermentation remains dissolved in the fermentation broth that can be recovered and purified through distillation (Mosier et al. 2005). Distillation is a mature process for ethanol recovery on both laboratory and industrial scales. The first step in distillation results in a separation of the ethanol fraction from the fermentation broth through a simultaneous evaporation and condensation of ethanol. Finally, ethanol at a concentration of 95% is obtained. The bottom fraction of the broth left after distillation contains mainly lignin, unreacted carbohydrates, ash, protein, residual enzymes, and

TABLE 6.7

Ethanol Production from Several LCBs during Fermentation

Feedstocks	Solid Load (%, w/v)	Microorganisms	Fermentation Mode	Temperature (°C)	Ethanol Concentration (g/L)	Ethanol Yield	References
Corn stover	10.0	*Escherichia coli* strain FBR 5	Batch	37	20.9	0.49[a]	Saha et al. 2013
Rice straw	5.0	*Saccharomyces cerevisiae* MTCC 174	Batch	30	19.2	0.5[a]	Singh and Bishnoi 2012
Salix	10.0	Baker's yeast	Batch	37	32.0	76.0[c]	Sassner et al. 2006
Salix	11.0	Baker's yeast	Fed–batch	37	33.0	62.0[c]	Sassner et al. 2006
Wheat straw	8.6	*E. coli* FBR5	Batch	35	17.4–24.9	0.2–0.3[b]	Saha et al. 2011
Wheat straw	8.6	*E. coli* FBR5	Fed–batch	35	26.7	0.31[b]	Saha et al. 2011
Water hyacinth	7.8	*Picia stipites* NCIM–3497	Batch	30	—	0.43[a]	Kumar et al. 2009
Wheat straw	10.0–12.0	*Kluyveromyces marxianus* CECT 10875	Fed–batch	42	36.2	0.25–0.33[a]	Tomás-Pejó et al. 2009
Barley straw	7.5	Baker's yeast	Fed–batch	35	22.4	80.0[a]	Linde et al. 2007
Poplar	10.0	*K. marxianus* CECT 10875	Batch	42	19.0	71.2[c]	Ballesteros et al. 2004
Eucalyptus	10.0	*K. marxianus* CECT 10875	Batch	42	17.0	62.5[c]	Ballesteros et al. 2004
Sweet sorghum bagasse	10.0	*K. marxianus* CECT 10875	Batch	42	16.2	60.9[c]	Ballesteros et al. 2004
Paper sludge	10.0	*Kluyveromyces fragilis*	Batch	40	8.8	63.7[c]	Olofsson et al. 2008
Sugarcane leaves	10.0	*S. cerevisiae*	Batch	40	22.0	—	Olofsson et al. 2008
Switchgrass	7.5	*S. cerevisiae* and *Brettanomyces claussenii* mixed culture	Batch	37	37.0	87[c]	Olofsson et al. 2008

[a] Ethanol yield is expressed as g/g of sugar.

[b] Ethanol yield is expressed as g/g of biomass.

[c] Ethanol yield is expressed as % of theoretical yield.

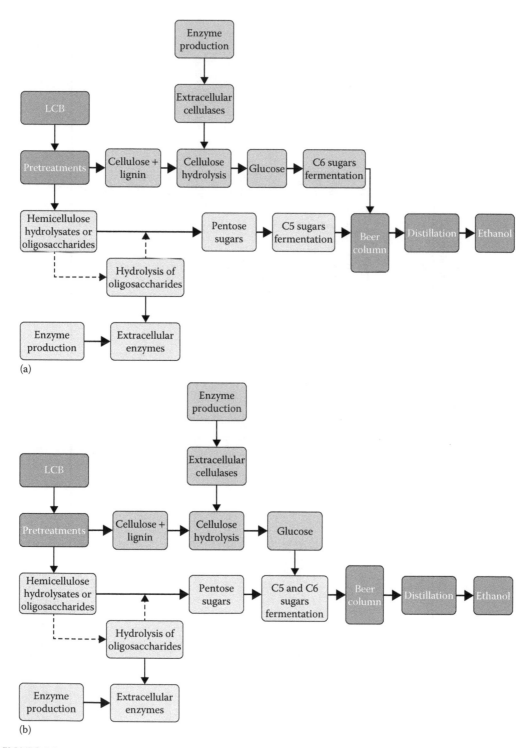

FIGURE 6.4
Schematic diagram of process integration strategies in the bioethanol production from LCB. General strategy for bioethanol production using (a) SHF, (b) SHCF. *(Continued)*

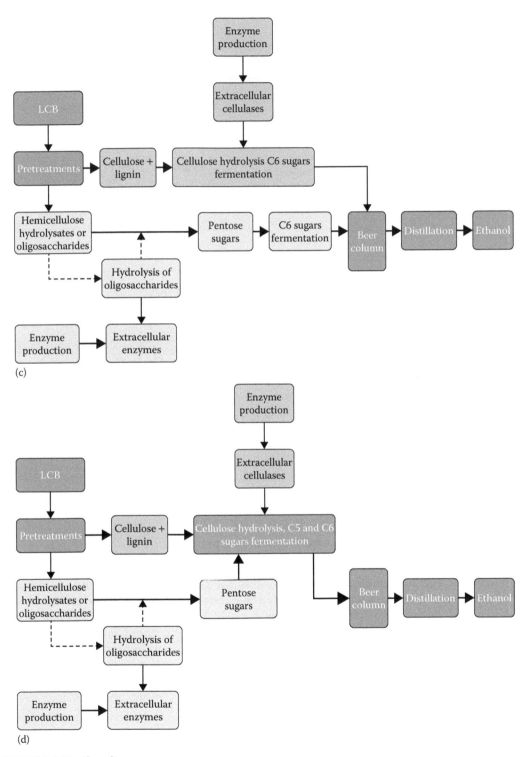

FIGURE 6.4 (Continued)
Schematic diagram of process integration strategies in the bioethanol production from LCB. General strategy
for bioethanol production using (c) SSF, (d) SSCF.   (*Continued*)

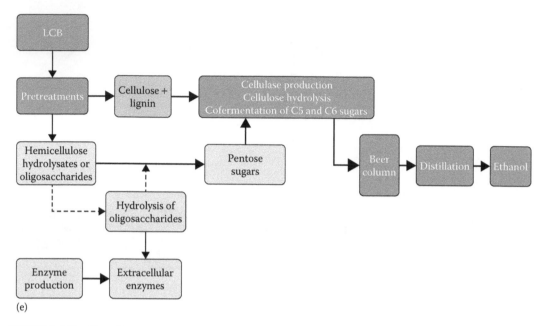

(e)

FIGURE 6.4 (Continued)
Schematic diagram of process integration strategies in the bioethanol production from LCB. General strategy for bioethanol production using (e) CBP. (Adapted from Hamelinck, C.N., G. Van Hooijdonk and A.P. Faaij, *Biomass Bioenergy*, 28(4), 384–410, 2005; Margeot, A., B. Hahn-Hagerdal, M. Edlund, R. Slade and F. Monot, *Curr. Opin. Biotechnol.*, 20(3), 372–380, 2009.)

microbial cells that are treated further in a waste water treatment step to separate different components and stillage (Zhang et al. 2009).

6.6 Conclusion: Challenges and Future Perspectives

LCBs have been proved to be one of the major renewable feedstocks for sustainable and economically viable ethanol production, because they are available with low prices and rich in carbohydrate. Many efforts have been made in recent years on the technical and technological aspects of ethanol production from LCBs. Nevertheless, some challenges still need to be addressed for a sustainable lignocellulosic ethanol production process, which can be described from the perspectives of biomass, pretreatment, hydrolysis, and fermentation configurations (Sarkar et al. 2012):

- Regarding feedstock, major hindrances are supply, harvesting, and handling.
- In consideration with the pretreatment process, the major challenge is to develop cost-effective and feasible methods, where high energy consumption is still a key issue. Furthermore, capital equipment needed for demonstrating some technologies on a large scale, such as steam explosion, does not exist. The recovery of chemicals used during chemical pretreatment as well as the treatment of wastewater produced are also important issues to be addressed.

TABLE 6.8

A Comparative Summary of the Process Integration in Hydrolysis and Fermentation

Technology	Process Description	Conversion Efficiency (%)	Yield (%)	Advantages	Disadvantages
SHF	• Hydrolysis and fermentation are done separately. • Enzymatic hydrolysis is done at 45°C–50°C, and fermentation at about 30°C.	75[a]	75–85[f] 85–90[g]	• Each step can be done under the optimum conditions. • Fermentation can be run in the continuous mode with cell recycling.	• Risk for end-product inhibition of the enzymes. • Risk for osmotic stress on yeast at high sugar level.
SHCF	Hydrolysis of cellulose and hemicellulose is done separately prior to conducting fermentation of the resultant five and six carbon sugars together.	—	—	Simultaneous conversion of hexose and pentose can be achieved in a single step.	The most commonly used yeast cannot be used, requiring recombinant strains.
SSF	Cellulose hydrolysis and fermentation take place at a time.	75–90[b] 70–80[a] 90–95[c] 80–92[d]	80–92.5[g]	• Utilization of high concentration of dry matter. • End product inhibition of the enzymes can be avoided. • High ethanol yield. • Low process time. • Requiring low amounts of enzyme. • Reduced chance for microbial contaminations.	• High viscosity may result in more power consumption. • Low mixing and heat transfer efficiency. • Requiring a compromise between the optimum conditions of the enzymatic reaction and the fermenting organism. • Cells cannot be recycled as it is mixed with the biomass.
SSCF	Simultaneous saccharification of both cellulose and hemicellulose, and cofermentation of both glucose and xylose using recombinant strains.	45–65[b] 90–96[a] 80–90[e] 90–95[c] 90[d]	85–92[g]	Simultaneous conversion of hexose and pentose sugars. • High solid load. • Short processing time. • High ethanol yield. • Low production costs.	• The most commonly used yeast cannot be used, requiring recombinant strains. • Low rate of cellulose hydrolysis.

(Continued)

TABLE 6.8 (Continued)

A Comparative Summary of the Process Integration in Hydrolysis and Fermentation

Technology	Process Description	Conversion Efficiency (%)	Yield (%)	Advantages	Disadvantages
CBP	Bioconversion of cellulose and hemicellulose into ethanol without adding any enzymes.	88–98[b] 97–99[a] 92–95[e] 92–95[c] 92–95[d]	85–95[f] 92–95[g]	• Low cost. • Desired yield. • Reduced number of reactors. • Simplification in the operation. • Reduction in the costs of chemicals. • No operating cost or capital investments for extracellular enzymes.	• Instability of the recombinant microorganisms. • Low ethanol yield. • Formation of the by-product (acetic acid, glycerol, and lactic acid). • Limited tolerance of the recombinant organism to ethanol (3.5% w/v).

Sources: Balat, M., *Energy Convers. Manage.,* 52, 858–875, 2011; Chandel, A.K., E. Chan, R. Rudravaram, M.L. Narasu, L.V. Rao and P. Ravindra, *Biotechnol. Molecul. Biol. Rev.,* 2, 14–32, 2007; Girio, F., C. Fonseca, F. Carvalheiro, L. Duarte, S. Marques and R. Bogel-Łukasik, *Bioresour. Technol.,* 101, 4775–4800, 2010; Hamelinck, C.N., G. Van Hooijdonk and A.P. Faaij, *Biomass Bioenergy,* 28(4), 384–410, 2005; Mosier, N., C. Wyman, B. Dale, R. Elander, Y. Lee, M. Holtzapple and M. Ladisch, *Bioresour. Technol.,* 96, 673–686, 2005.)

[a] Conversion efficiency of cellulose into fermentable sugars.
[b] Conversion efficiency of hemicellulose into fermentable sugars.
[c] Conversion efficiency of glucose into ethanol.
[d] Conversion efficiency of fermentable sugars (other than glucose and xylose) into ethanol.
[e] Conversion efficiency of xylose into ethanol.
[f] Sugar yield.
[g] Ethanol yield.

- With regard to the hydrolysis process, the major bottleneck is to achieve an efficient and cost-effective enzyme system to hydrolyze cellulose and hemicellulose with high yields.
- In case of fermentation configuration, the difficulty is the cofermentation of xylose and glucose. Even though recombinant microorganisms have been developed and suggested to address this issue, their stability and maintenance in the commercial facility poses additional challenges.

A further research efforts should be kept continue on the following aspects as attempts to overcome these challenges:

- Implementation of the process integration on pilot, demonstration, and commercial scales incorporating pretreatment, enzymatic hydrolysis, and fermentation (Hahn-Hägerdal et al. 2006).
- Improvement in enzymatic hydrolysis through developing potential enzyme systems, which will require low production costs and have potential for working in high solid load (Galbe and Zacchi 2002; Hahn-Hägerdal et al. 2006).
- Screening for naturally occurring microorganisms having the ability to ferment all kinds of sugars (polymers, oligosaccharides, and monosaccharides), efficient fermentation, and tolerance to inhibitors (Gírio et al. 2010).
- Development of promising recombinant microorganisms, strain improvement, and successful exploitation of these microorganisms into commercial ethanol production.

References

ABE. 2015. Abengoa Bioenergy. http://www.abengoabioenergy.com/web/es/index.html (accessed August 19, 2015).

Agarwal, A.K. 2007. Biofuels (alcohols and biodiesel) applications as fuels for internal combustion engines. *Progress in Energy and Combustion Science* 33(3): 233–271.

Aggarwal, N., P. Niga, D. Singh and B. Yadav. 2001. Process optimization for the production of sugar for the bioethanol industry from Tapioca, a non-conventional source of starch. *World Journal of Microbiology and Biotechnology* 17(8): 783–787.

Alriksson, B., A. Cavka and L.J. Jönsson. 2011. Improving the fermentability of enzymatic hydrolysates of lignocellulose through chemical in-situ detoxification with reducing agents. *Bioresource Technology* 102(2): 1254–1263.

Alvira, P., E. Tomás-Pejó, M. Ballesteros and M. Negro. 2010. Pretreatment technologies for an efficient bioethanol production process based on enzymatic hydrolysis: A review. *Bioresource Technology* 101(13): 4851–4861.

Amiri, H., K. Karimi and H. Zilouei. 2014. Organosolv pretreatment of rice straw for efficient acetone, butanol, and ethanol production. *Bioresource Technology* 152: 450–456.

Anonymous. 2014. U.S. Cellulosic Ethanol Plants, New or Under Construction. http://www.grainnet.com/pdf/cellulosemap.pdf (accessed February 7, 2014).

Aswathy, U., R.K. Sukumaran, G.L. Devi, K. Rajasree, R.R. Singhania and A. Pandey. 2010. Bioethanol from water hyacinth biomass: An evaluation of enzymatic saccharification strategy. *Bioresource Technology* 101(3): 925–930.

Avci, A., B.C. Saha, G.J. Kennedy and M.A. Cotta. 2013. Dilute sulfuric acid pretreatment of corn stover for enzymatic hydrolysis and efficient ethanol production by recombinant Escherichia coli FBR5 without detoxification. *Bioresource Technology* 142: 312–319.

Bajaj, B.K., H. Pangotra, M.A. Wani, P. Sharma and A. Sharma. 2009. Partial purification and characterization of a highly thermostable and pH stable endoglucanase from a newly isolated Bacillus strain M-9. *Indian Journal of Chemical Technology* 16(16): 382–387.

Balat, M. 2007. An overview of biofuels and policies in the European Union. *Energy Sources, Part B* 2(2): 167–181.

Balat, M. 2011. Production of bioethanol from lignocellulosic materials via the biochemical pathway: A review. *Energy Conversion and Management* 52(2): 858–875.

Balat, M. and H. Balat. 2009. Recent trends in global production and utilization of bio-ethanol fuel. *Applied Energy* 86(11): 2273–2282.

Ballesteros, M., J. Oliva, M. Negro, P. Manzanares and I. Ballesteros. 2004. Ethanol from lignocellulosic materials by a simultaneous saccharification and fermentation process (SFS) with Kluyveromyces marxianus CECT 10875. *Process Biochemistry* 39(12): 1843–1848.

Bell, P.J. and P.V. Attfield. 2009. Breakthrough in yeast for making bio-ethanol from lignocellulosics. *Disponível em:* Acesso em, 15. http://www.eri.ucr.edu/ISAFXVCD/ISAFXVAF/YMBL.pdf.

Bhatti, H.N., M.A. Hanif and M. Qasim. 2008. Biodiesel production from waste tallow. *Fuel* 87(13): 2961–2966.

BluFire. 2015. http://bfreinc.com/docs/IZUMI_Status_2004_for_BlueFire_051606.pdf.

Brodeur, G., E. Yau, K. Badal, J. Collier, K. Ramachandran and S. Ramakrishnan. 2011. Chemical and physicochemical pretreatment of lignocellulosic biomass: A review. *Enzyme Research* 2011: 17, Article ID 787532.

Brosse, N., R.E. Hage, P. Sannigrahi and A. Ragauskas. 2010. Dilute sulfuric acid and ethanol organosolv pretreatment of Miscanthus x Giganteus. *Cellulose Chemistry and Technology* 44(1–3): 71–78.

Buah, W., A. Cunliffe and P. Williams. 2007. Characterization of products from the pyrolysis of municipal solid waste. *Process Safety and Environmental Protection* 85(5): 450–457.

Carioca, J., P. Pannirselvam, E. Horta and H. Arora. 1985. Lignocellulosic biomass fractionation: I-solvent extraction in a novel reactor. *Biotechnology Letters* 7(3): 213–216.

Cavka A. and L.J. Jönsson. 2013. Detoxification of lignocellulosic hydrolysates using sodium borohydride. *Bioresource Technology* 136: 368–376.

Chandel, A.K., E. Chan, R. Rudravaram, M.L. Narasu, L.V. Rao and P. Ravindra. 2007. Economics and environmental impact of bioethanol production technologies: An appraisal. *Biotechnology and Molecular Biology Review* 2(1): 14–32.

Chandel, A.K., S.S. Da Silva and O.V. Singh. 2013. Detoxification of lignocellulose hydrolysates: Biochemical and metabolic engineering toward white biotechnology. *BioEnergy Research* 6(1): 388–401.

Chang, V.S. and M.T. Holtzapple. 2000. Fundamental factors affecting biomass enzymatic reactivity. *Twenty-First Symposium on Biotechnology for Fuels and Chemicals*. Springer, Totowa, NJ. pp. 5–37.

Claassen, P., J. Van Lier, A.L. Contreras, E. Van Niel, L. Sijtsma, A. Stams, S. De Vries and R. Weusthuis. 1999. Utilisation of biomass for the supply of energy carriers. *Applied Microbiology and Biotechnology* 52(6): 741–755.

De Bari, I., F. Nanna and G. Braccio. 2007. SO_2-catalyzed steam fractionation of aspen chips for bioethanol production: Optimization of the catalyst impregnation. *Industrial & Engineering Chemistry Research* 46(23): 7711–7720.

De Oliveira, M.E.D., B.E. Vaughan and E.J. Rykiel Jr. 2005. Ethanol as fuel: Energy, carbon dioxide balances, and ecological footprint. *BioScience* 55(7): 593–602.

De Vrije, T., G. De Haas, G. Tan, E. Keijsers and P. Claassen. 2002. Pretreatment of Miscanthus for hydrogen production by Thermotoga elfii. *International Journal of Hydrogen Energy* 27(11): 1381–1390.

Demirbas, A. 2008. Biofuels sources, biofuel policy, biofuel economy and global biofuel projections. *Energy Conversion and Management* 49(8): 2106–2116.

Demirbas, A. 2009. Political, economic and environmental impacts of biofuels: A review. *Applied Energy* 86: S108–S117.

Doornbosch, R. and R. Steenblik. 2008. Biofuels: Is the cure worse than the disease? Organisation for Economic Co-operation and Development. http://www.oecd.org/sd-roundtable/39411732.pdf.

Drummond, A.-R.F. and I.W. Drummond. 1996. Pyrolysis of sugar cane bagasse in a wire-mesh reactor. *Industrial & Engineering Chemistry Research* 35(4): 1263–1268.

Duff, S.J. and W.D. Murray. 1996. Bioconversion of forest products industry waste cellulosics to fuel ethanol: A review. *Bioresource Technology* 55(1): 1–33.

EBTP-SABS. 2014. Cellulosic ethanol (CE). http://www.biofuelstp.eu/cellulosic-ethanol.html#crescentino.

Escobar, J.C., E.S. Lora, O.J. Venturini, E.F. Yáñez, E.F. Castillo and Almazan, O. 2009. Biofuels: Environment, technology and food security. *Renewable and Sustainable Energy Reviews* 13(6): 1275–1287.

Esteghlalian, A., A.G. Hashimoto, J.J. Fenske and M.H. Penner. 1997. Modeling and optimization of the dilute-sulfuric-acid pretreatment of corn stover, poplar and switchgrass. *Bioresource Technology* 59(2): 129–136.

Ethtec. 2014. http://www.ethtec.com.au/default.asp?ID=43 (accessed March 27, 2014).

Ewanick, S.M., R. Bura and J.N. Saddler. 2007. Acid-catalyzed steam pretreatment of lodgepole pine and subsequent enzymatic hydrolysis and fermentation to ethanol. *Biotechnology and Bioengineering* 98(4): 737–746.

Fengel, D. and G. Wegener. 1983. *Wood: Chemistry, Ultrastructure, Reactions*. Walter de Gruyter, Berlin, Germany.

Fockink, D.H., M.A.C. Maceno and L.P. Ramos. 2015. Production of cellulosic ethanol from cotton processing residues after pretreatment with dilute sodium hydroxide and enzymatic hydrolysis. *Bioresource Technology* 187: 91–96.

Fuels-America. 2014. Four commercial scale cellulosic ethanol biorefineries to enter production this year. http://www.fuelsamerica.org/blog/entry/four-commercial-scale-cellulosic-ethanol–biorefineries-to-enter-production.

Galbe, M. and G. Zacchi. 2002. A review of the production of ethanol from softwood. *Applied Microbiology and Biotechnology* 59(6): 618–628.

Galbe, M. and G. Zacchi. 2007. Pretreatment of lignocellulosic materials for efficient bioethanol production. In: Olsson (Ed.), *Biofuels*, Springer, Berlin, Heidelberg, pp. 41–65.

Gírio, F., C. Fonseca, F. Carvalheiro, L. Duarte, S. Marques and R. Bogel-Łukasik. 2010. Hemicelluloses for fuel ethanol: A review. *Bioresource Technology* 101(13): 4775–4800.

Gonzalez, R.W., T. Treasure, R.B. Phillips, H. Jameel and D. Saloni. 2011. Economics of cellulosic ethanol production: Green liquor pretreatment for softwood and hardwood, greenfield and repurpose scenarios. *Bioresources* 6(3): 2551–2567.

Gouvea, B., C. Torres, A. Franca, L. Oliveira and E. Oliveira. 2009. Feasibility of ethanol production from coffee husks. *Biotechnology Letters* 31(9): 1315–1319.

Greer, D. 2005. Creating Cellulosic Ethanol, Spinning Straw Into Fuel, Magazine, *Biocycle*, April, pp. 61–67.

Green, K.R. and W.A. Lowenbach. 2001. MTBE Contamination: Environmental, legal, and public policy challenges, guest editorial. *Environmental Forensics* 2(1): 3–6.

Guimarães, J., E. Frollini, C. Da Silva, F. Wypych and K. Satyanarayana. 2009. Characterization of banana, sugarcane bagasse and sponge gourd fibers of Brazil. *Industrial Crops and Products* 30(3): 407–415.

Hadar, Y. 2013. Sources for lignocellulosic raw materials for the production of ethanol. In: V. Faraco (Ed.), *Lignocellulose Conversion*, Springer, Berlin, Heidelberg, pp. 21–38.

Hahn-Hägerdal, B., M. Galbe, M.-F. Gorwa-Grauslund, G. Lidén and G. Zacchi. 2006. Bio-ethanol–the fuel of tomorrow from the residues of today. *Trends in Biotechnology* 24(12): 549–556.

Hamelinck, C.N., G. Van Hooijdonk and A.P. Faaij. 2005. Ethanol from lignocellulosic biomass: Techno-economic performance in short-, middle-and long-term. *Biomass and Bioenergy* 28(4): 384–410.

Hayes, D.J. 2009. An examination of biorefining processes, catalysts and challenges. *Catalysis Today* 145(1): 138–151.

Howard, R., E. Abotsi, E.J. Van Rensburg and S. Howard. 2004. Lignocellulose biotechnology: Issues of bioconversion and enzyme production. *African Journal of Biotechnology* 2(12): 602–619.

Inbicon. 2015. http://www.inbicon.com (accessed January 21, 2015).

Iogen. 2014. http://www.iogen.ca/ (accessed January 23, 2014).

Jönsson, L.J., B. Alriksson and N.-O. Nilvebrant. 2013. Bioconversion of lignocellulose: Inhibitors and detoxification. *Biotechnology for Biofuels* 6(1): 16.

Kallioinen, A., J. Uusitalo, K. Pahkala, M. Kontturi, L. Viikari, N. von Weymarn and M. Siika-aho. 2012. Reed canary grass as a feedstock for 2nd generation bioethanol production. *Bioresource Technology* 123: 669–672.

Kar, Y. and H. Deveci. 2006. Importance of P-series fuels for flexible-fuel vehicles (FFVs) and alternative fuels. *Energy Sources, Part A* 28(10): 909–921.

Kim, S. and B.E. Dale. 2004. Global potential bioethanol production from wasted crops and crop residues. *Biomass and Bioenergy* 26(4): 361–375.

Kim, M. and D.F. Day. 2011. Composition of sugar cane, energy cane, and sweet sorghum suitable for ethanol production at Louisiana sugar mills. *Journal of Industrial Microbiology & Biotechnology* 38(7): 803–807.

Kourkoutas, Y., A. Bekatorou, I.M. Banat, R. Marchant and A. Koutinas. 2004. Immobilization technologies and support materials suitable in alcohol beverages production: A review. *Food Microbiology* 21(4): 377–397.

Kreith, F. and S. Krumdieck. 2013. *Principles of Sustainable Energy Systems*. CRC Press, Boca Raton, FL.

Kuhad, R.C. and A. Singh. 1993. Lignocellulose biotechnology: Current and future prospects. *Critical Reviews in Biotechnology* 13(2): 151–172.

Kumar, A., L. Singh and S. Ghosh. 2009. Bioconversion of lignocellulosic fraction of water-hyacinth (Eichhornia crassipes) hemicellulose acid hydrolysate to ethanol by Pichia stipitis. *Bioresource Technology* 100(13): 3293–3297.

Lal, R. 2005. World crop residues production and implications of its use as a biofuel. *Environment International* 31(4): 575–584.

Lamsal, B., J. Yoo, K. Brijwani and S. Alavi. 2010. Extrusion as a thermo-mechanical pre-treatment for lignocellulosic ethanol. *Biomass and Bioenergy* 34(12): 1703–1710.

Larson, E.D. 2008. Biofuel production technologies: Status, prospects and implications for trade and development. United Nations Conference on Trade and Development, New York and Geneva.

Larsson, S., A. Reimann, N.O. Nilvebrant and L.J. Jönsson. 1999. Comparison of different methods for the detoxification of lignocellulose hydrolyzates of spruce. *Applied biochemistry and biotechnology* 77(1–3): 91–103.

Lavarack, B., G. Griffin and D. Rodman. 2002. The acid hydrolysis of sugarcane bagasse hemicellulose to produce xylose, arabinose, glucose and other products. *Biomass and Bioenergy* 23(5): 367–380.

Lee, J. 1997. Biological conversion of lignocellulosic biomass to ethanol. *Journal of Biotechnology* 56(1): 1–24.

Lee, Y., P. Iyer and R.W. Torget. 1999. Dilute-acid hydrolysis of lignocellulosic biomass. In: G. T. Tsao et al. (eds.), *Recent progress in bioconversion of lignocellulosics*, Springer, Berlin, Heidelberg, pp. 93–115.

Leonard, R.H. and G.J. Hajny. 1945. Fermentation of wood sugars to ethyl alcohol. *Industrial & Engineering Chemistry* 37(4): 390–395.

Lewandowski, I., J.M. Scurlock, E. Lindvall and M. Christou. 2003. The development and current status of perennial rhizomatous grasses as energy crops in the US and Europe. *Biomass and Bioenergy* 25(4): 335–361.

Li, S., X. Zhang and J.M. Andresen. 2012. Production of fermentable sugars from enzymatic hydrolysis of pretreated municipal solid waste after autoclave process. *Fuel* 92(1): 84–88.

Limayem, A. and S.C. Ricke. 2012. Lignocellulosic biomass for bioethanol production: Current perspectives, potential issues and future prospects. *Progress in Energy and Combustion Science* 38(4): 449–467.

Linde, M., M. Galbe and G. Zacchi. 2007. Simultaneous saccharification and fermentation of steam-pretreated barley straw at low enzyme loadings and low yeast concentration. *Enzyme and Microbial Tchnology* 40(5): 1100–1107.

Ludueña, L., D. Fasce, V.A. Alvarez and P.M. Stefani. 2011. Nanocellulose from rice husk following alkaline treatment to remove silica. *Bioresources* 6(2): 1440–1453.

Lynd, L.R., J.H. Cushman, R.J. Nichols and C.E. Wyman. 1991. Fuel ethanol from cellulosic biomass. *Science (Washington)* 251(4999): 1318–1323.

Lynd, L.R., R.T. Elamder and C.E. Wyman. 1996. Likely features and costs of mature biomass ethanol technology. *Applied Biochemistry and Biotechnology* 57(1): 741–761.

Lynd, L.R., M.S. Laser, D. Bransby, B.E. Dale, B. Davison, R. Hamilton, M. Himmel, M. Keller, J.D. McMillan and J. Sheehan. 2008. How biotech can transform biofuels. *Nature Biotechnology* 26(2): 169–172.

Maitan-Alfenas, G.P., E.M. Visser and V.M. Guimarães. 2015. Enzymatic hydrolysis of lignocellulosic biomass: Converting food waste in valuable products. *Current Opinion in Food Science* 1: 44–49.

Malça, J. and F. Freire. 2006. Renewability and life-cycle energy efficiency of bioethanol and bio-ethyl tertiary butyl ether (bioETBE): Assessing the implications of allocation. *Energy* 31(15): 3362–3380.

Margeot, A., B. Hahn-Hagerdal, M. Edlund, R. Slade and F. Monot. 2009. New improvements for lignocellulosic ethanol. *Current Opinion in Biotechnology* 20(3): 372–380.

McCarthy, J.E. and M. Tiemann. 2006. MTBE in gasoline: Clean air and drinking water issues. *Congressional Research Service Reports*. Washington, DC.

McKendry, P. 2002. Energy production from biomass (part 1): Overview of biomass. *Bioresource Technology* 83(1): 37–46.

Medina de Perez, V.I. and A.A. Ruiz Colorado. 2006. Ethanol producction of banana shell and cassava starch. *Dyna* 73(150): 21–27.

Menon, V. and M. Rao. 2012. Trends in bioconversion of lignocellulose: Biofuels, platform chemicals & biorefinery concept. *Progress in Energy and Combustion Science* 38(4): 522–550.

Monavari, S., M. Galbe and G. Zacchi. 2009. Impact of impregnation time and chip size on sugar yield in pretreatment of softwood for ethanol production. *Bioresource Technology* 100(24): 6312–6316.

Mosier, N., C. Wyman, B. Dale, R. Elander, Y. Lee, M. Holtzapple and M. Ladisch. 2005. Features of promising technologies for pretreatment of lignocellulosic biomass. *Bioresource Technology* 96(6): 673–686.

Mussatto, S.I., G. Dragone, P.M. Guimarães, J.P.A. Silva, L.M. Carneiro, I.C .Roberto, A. Vicente, L. Domingues and J.A. Teixeira. 2010. Technological trends, global market, and challenges of bio-ethanol production. *Biotechnology Advances* 28(6): 817–830.

Naik, S., V.V. Goud, P.K. Rout and A.K. Dalai. 2010. Production of first and second generation biofuels: A comprehensive review. *Renewable and Sustainable Energy Reviews* 14(2): 578–597.

Nigam, J. 2001. Ethanol production from wheat straw hemicellulose hydrolysate by *Pichia stipitis*. *Journal of Biotechnology* 87(1): 17–27.

Nigam, J. 2002. Bioconversion of water-hyacinth (Eichhornia crassipes) hemicellulose acid hydroly-sate to motor fuel ethanol by xylose–fermenting yeast. *Journal of Biotechnology* 97(2): 107–116.

Nigam, P.S. and A. Singh. 2011. Production of liquid biofuels from renewable resources. *Progress in Energy and Combustion Science* 37(1): 52–68.

Nikolić, S., L. Mojović, D. Pejin, M. Rakin and M. Vukašinović. 2010. Production of bioethanol from corn meal hydrolyzates by free and immobilized cells of *Saccharomyces cerevisiae* var. ellipsoi-deus. *Biomass and Bioenergy* 34(10): 1449–1456.

Obama, P., G. Ricochon, L. Muniglia and N. Brosse. 2012. Combination of enzymatic hydrolysis and ethanol organosolv pretreatments: Effect on lignin structures, delignification yields and cellulose-to-glucose conversion. *Bioresource Technology* 112: 156–163.

Ogbonna, J.C., H. Mashima and H. Tanaka. 2001. Scale up of fuel ethanol production from sugar beet juice using loofa sponge immobilized bioreactor. *Bioresource Technology* 76(1): 1–8.

Olofsson, K., M. Bertilsson and G. Lidén. 2008. A short review on SSF-an interesting process option for ethanol production from lignocellulosic feedstocks. *Biotechnology for Biofuels* 1(7): 1–14.

Olsson, L. and B. Hahn-Hägerdal. 1996. Fermentation of lignocellulosic hydrolysates for ethanol production. *Enzyme and Microbial Technology* 18(5): 312–331.

Olsson L., B. Hahn-Hägerdal and G. Zacchi. 1995. Kinetics of ethanol production by recombinant *Escherichia coli* KO11. *Biotechnology and Bioengineering* 45(4): 356–365.

Palmqvist, E. and B. Hahn-Hägerdal. 2000. Fermentation of lignocellulosic hydrolysates. II: Inhibitors and mechanisms of inhibition. *Bioresource Technology* 74(1): 25–33.

Park, N., H.-Y. Kim, B.-W. Koo, H. Yeo and I.-G. Choi. 2010. Organosolv pretreatment with various catalysts for enhancing enzymatic hydrolysis of pitch pine (*Pinus rigida*). *Bioresource Technology* 101(18): 7046–7053.

Paulová, L., K. Melzoch, M. Rychtera and P. Patáková. 2013. *Production of 2nd Generation of Liquid Biofuels*. INTECH Open Access Publisher, Rijeka, Croatia.

Perlack, R.D., L.L. Wright, A.F. Turhollow, R.L. Graham, B.J. Stokes and D.C. Erbach. 2005. Biomass as feedstock for a bioenergy and bioproducts industry: The technical feasibility of a billion-ton annual supply. DTIC Document.

Pickett, J., D. Anderson, D. Bowles, T. Bridgwater, P. Jarvis, N. Mortimer, M. Poliakoff and J. Woods. 2008. *Sustainable Biofuels: Prospects and Challenges*. The Royal Society, London, UK.

Prasad, S., A. Singh and H. Joshi. 2007. Ethanol as an alternative fuel from agricultural, industrial and urban residues. *Resources, Conservation and Recycling* 50(1): 1–39.

Reczey, K., Z. Szengyel, R. Eklund and G. Zacchi. 1996. Cellulase production by T. reesei. *Bioresource Technology* 57(1): 25–30.

Reddy, N. and Y. Yang. 2005. Biofibers from agricultural by-products for industrial applications. *Trends in Biotechnology* 23(1): 22–27.

Rengsirikul, K., Y. Ishii, K. Kangvansaichol, P. Sripichitt, V. Punsuvon, P. Vaithanomsat, G. Nakamanee and S. Tudsri. 2013. Biomass yield, chemical composition and potential ethanol yields of 8 cultivars of napiergrass (*Pennisetum purpureum Schumach.*) harvested 3-monthly in central Thailand. *Journal of Sustainable Bioenergy Systems* 3(2): 107.

Rodríguez-Chong, A., J.A. Ramírez, G. Garrote and M. Vázquez. 2004. Hydrolysis of sugar cane bagasse using nitric acid: A kinetic assessment. *Journal of Food Engineering* 61(2): 143–152.

Ruiz, E., C. Cara, P. Manzanares, M. Ballesteros and E. Castro. 2008. Evaluation of steam explosion pre-treatment for enzymatic hydrolysis of sunflower stalks. *Enzyme and Microbial Technology* 42(2): 160–166.

Saarela, U., K. Leiviskä and E. Juuso. 2003. Modelling of a fed-batch fermentation process. Control Engineering Laboratory, Department of Process and Environmental Engineering, University of Oulu, Report A No. 21, Finland.

Saha, B.C. 2003. Hemicellulose bioconversion. *Journal of Industrial Microbiology and Biotechnology* 30(5): 279–291.

Saha, B.C. 2004. Lignocellulose biodegradation and applications in biotechnology. *ACS Symposium Series*. American Chemical Society, Washington, DC, 1999. pp. 2–35.

Saha, B.C., N.N. Nichols, N. Qureshi and M.A. Cotta. 2011. Comparison of separate hydrolysis and fermentation and simultaneous saccharification and fermentation processes for ethanol production from wheat straw by recombinant Escherichia coli strain FBR5. *Applied Microbiology and Biotechnology* 92(4): 865–874.

Saha, B.C., T. Yoshida, M.A. Cotta and K. Sonomoto. 2013. Hydrothermal pretreatment and enzymatic saccharification of corn stover for efficient ethanol production. *Industrial Crops and Products* 44: 367–372.

Saini, J.K., R. Saini and L. Tewari. 2015. Lignocellulosic agriculture wastes as biomass feedstocks for second-generation bioethanol production: Concepts and recent developments. *3 Biotech* 5(4): 337–353.

Sánchez, C. 2009. Lignocellulosic residues: Biodegradation and bioconversion by fungi. *Biotechnology Advances* 27(2): 185–194.

Sanchez, O.J. and C.A. Cardona. 2008. Trends in biotechnological production of fuel ethanol from different feedstocks. *Bioresource Technology* 99(13): 5270–5295.

Sarkar, N., S.K. Ghosh, S. Bannerjee and K. Aikat. 2012. Bioethanol production from agricultural wastes: An overview. *Renewable Energy* 37(1): 19–27.

Sassner, P., M. Galbe and G. Zacchi. 2006. Bioethanol production based on simultaneous saccharification and fermentation of steam-pretreated Salix at high dry-matter content. *Enzyme and Microbial Technology* 39(4): 756–762.

Sassner, P., C.-G. Mårtensson, M. Galbe and G. Zacchi. 2008. Steam pretreatment of $H_2SO_4^-$ impregnated Salix for the production of bioethanol. *Bioresource Technology* 99(1): 137–145.

Saxena, R., D. Adhikari and H. Goyal. 2009. Biomass-based energy fuel through biochemical routes: A review. *Renewable and Sustainable Energy Reviews* 13(1): 167–178.

Schmitt, E., R. Bura, R. Gustafson, J. Cooper and A. Vajzovic. 2012. Converting lignocellulosic solid waste into ethanol for the State of Washington: An investigation of treatment technologies and environmental impacts. *Bioresource Technology* 104: 400–409.

Searchinger, T., R. Heimlich, R.A. Houghton, F. Dong, A. Elobeid, J. Fabiosa, S. Tokgoz, D. Hayes and T.-H. Yu. 2008. Use of US croplands for biofuels increases greenhouse gases through emissions from land-use change. *Science* 319(5867): 1238–1240.

Sims, R.E., W. Mabee, J.N. Saddler and M. Taylor. 2010. An overview of second generation biofuel technologies. *Bioresource Technology* 101(6): 1570–1580.

Singh, A. and N.R. Bishnoi. 2012. Optimization of ethanol production from microwave alkali pretreated rice straw using statistical experimental designs by *Saccharomyces cerevisiae*. *Industrial Crops and Products* 37(1): 334–341.

Singh, A., D. Pant, N.E. Korres, A.-S. Nizami, S. Prasad and J.D. Murphy. 2010. Key issues in life cycle assessment of ethanol production from lignocellulosic biomass: Challenges and perspectives. *Bioresource Technology* 101(13): 5003–5012.

Söderström, J., M. Galbe and G. Zacchi. 2004. Effect of washing on yield in one-and two-step steam pretreatment of softwood for production of ethanol. *Biotechnology Progress* 20(3): 744–749.

Sørensen, A., P.J. Teller, T. Hilstrøm and B.K. Ahring. 2008. Hydrolysis of Miscanthus for bioethanol production using dilute acid presoaking combined with wet explosion pre-treatment and enzymatic treatment. *Bioresource Technology* 99(14): 6602–6607.

Stevens, D., M. Wörgetter and J.N. Saddler. 2004. Biofuels for transportation: An examination of policy and technical issues. *IEA Bioenergy Task*, 2001–2003.

Sun, Y. and J. Cheng. 2002. Hydrolysis of lignocellulosic materials for ethanol production: A review. *Bioresource Technology* 83(1): 1–11.

Szczodrak, J. and J. Fiedurek. 1996. Technology for conversion of lignocellulosic biomass to ethanol. *Biomass and Bioenergy* 10(5): 367–375.

Szymanowska-Powałowska, D., G. Lewandowicz, P. Kubiak and W. Błaszczak. 2014. Stability of the process of simultaneous saccharification and fermentation of corn flour. The effect of structural changes of starch by stillage recycling and scaling up of the process. *Fuel* 119: 328–334.

Taherzadeh, M.J. and K. Karimi. 2008. Pretreatment of lignocellulosic wastes to improve ethanol and biogas production: A review. *International Journal of Molecular Sciences* 9(9): 1621–1651.

Tian, S., J. Zang, Y. Pan, J. Liu, Z. Yuan, Y. Yan and X. Yang. 2008. Construction of a recombinant yeast strain converting xylose and glucose to ethanol. *Frontiers of Biology in China* 3(2): 165–169.

Tomás-Pejó, E., J. Oliva, A. González, I. Ballesteros and M. Ballesteros. 2009. Bioethanol production from wheat straw by the thermotolerant yeast Kluyveromyces marxianus CECT 10875 in a simultaneous saccharification and fermentation fed-batch process. *Fuel* 88(11): 2142–2147.

Tran, D.-T., I. Yet-Pole and C.-W. Lin. 2013. Developing co-culture system of dominant cellulolytic Bacillus sp. THLA0409 and dominant ethanolic Klebsiella oxytoca THLC0409 for enhancing ethanol production from lignocellulosic materials. *Journal of the Taiwan Institute of Chemical Engineers* 44(5): 762–769.

van der Weijde, T., C.L.A. Kamei, A.F. Torres, W. Vermerris, O. Dolstra, R.G. Visser and L.M. Trindade. 2013. The potential of C4 grasses for cellulosic biofuel production. *Frontiers in Plant Science* 4: 107. doi:10.3389/fpls.2013.00107.

Van Zyl C., B.A. Prior, J.C. Du Preez. 1988. Production of ethanol from sugar cane bagasse hemicellulose hydrolyzate by *Pichia stipitis*. *Applied Biochemistry and Biotechnology* 17(1–3): 357–369.

Vohra, M., J. Manwar, R. Manmode, S. Padgilwar and S. Patil. 2014. Bioethanol production: Feedstock and current technologies. *Journal of Environmental Chemical Engineering* 2(1) 573–584.

Wang, X., H. Feng and Z. Li. 2012. Ethanol production from corn stover pretreated by electrolyzed water and a two-step pretreatment method. *Chinese Science Bulletin* 57(15): 1796–1802.

Wildschut, J., A.T. Smit, J.H. Reith and W.J. Huijgen. 2013. Ethanol-based organosolv fractionation of wheat straw for the production of lignin and enzymatically digestible cellulose. *Bioresource Technology* 135: 58–66.

Wyman, C.E. 1999. Biomass ethanol: Technical progress, opportunities, and commercial challenges. *Annual Review of Energy and the Environment* 24(1): 189–226.

Wyman, C.E., B.E. Dale, R.T. Elander, M. Holtzapple, M.R. Ladisch and Y. Lee. 2005. Coordinated development of leading biomass pretreatment technologies. *Bioresource Technology* 96(18): 1959–1966.

Wyman, C.E., B.E. Dale, R.T. Elander, M. Holtzapple, M.R. Ladisch, Y. Lee, C. Mitchinson and J.N. Saddler. 2009. Comparative sugar recovery and fermentation data following pretreatment of poplar wood by leading technologies. *Biotechnology Progress* 25(2): 333–339.

Yang, B. and C.E. Wyman. 2008. Pretreatment: The key to unlocking low-cost cellulosic ethanol. *Biofuels, Bioproducts and Biorefining* 2(1): 26–40.

Yao, C., X. Yang, R. Roy Raine, C. Cheng, Z. Tian and Y. Li. 2009. The effects of MTBE/ethanol additives on toxic species concentration in gasoline flame. *Energy & Fuels* 23(7): 3543–3548.

Zabed, H., G. Faruq, J.N. Sahu, M.S. Azirun, R. Hashim and A.N. Boyce. 2014. Bioethanol production from fermentable sugar juice. *The Scientific World Journal* 2014: 11, 957102.

Zabed, H., G. Faruq, J.N. Sahu, A.N. Boyce and P. Ganesan. 2016a. A comparative study on normal and high sugary corn genotypes for evaluating enzyme consumption during dry-grind ethanol production. *Chemical Engineering Journal* 287: 691–703.

Zabed, H., G. Faruq, A.N. Boyce, J.N. Sahu and P. Ganesan. 2016b. Evaluation of high sugar containing corn genotypes as viable feedstocks for decreasing enzyme consumption during dry-grind ethanol production. *Journal of the Taiwan Institute of Chemical Engineers* 58: 467–475.

Zakaria, M.R., S. Hirata, S. Fujimoto and M.A. Hassan. 2015. Combined pretreatment with hot compressed water and wet disk milling opened up oil palm biomass structure resulting in enhanced enzymatic digestibility. *Bioresource Technology* 193: 128–134.

Zhang, K., Z. Pei and D. Wang. 2016. Organic solvent pretreatment of lignocellulosic biomass for biofuels and biochemicals: A review. *Bioresource Technology* 199: 21–33.

Zhang, M., F. Wang, R. Su, W. Qi and Z. He. 2010. Ethanol production from high dry matter corncob using fed-batch simultaneous saccharification and fermentation after combined pretreatment. *Bioresource Technology* 101(13): 4959–4964.

Zhang, S., F. Maréchal, M. Gassner, Z. Périn-Levasseur, W. Qi, Z. Ren, Y. Yan and D. Favrat. 2009. Process modeling and integration of fuel ethanol production from lignocellulosic biomass based on double acid hydrolysis. *Energy & fuels* 23(3): 1759–1765.

Zhang, Y.-H.P., S.-Y. Ding, J.R. Mielenz, J.-B. Cui, R.T. Elander, M. Laser, M.E. Himmel, J.R. McMillan and L.R. Lynd. 2007. Fractionating recalcitrant lignocellulose at modest reaction conditions. *Biotechnology and Bioengineering* 97(2): 214–223.

Zhao, J. and L. Xia. 2010. Ethanol production from corn stover hemicellulosic hydrolysate using immobilized recombinant yeast cells. *Biochemical Engineering Journal* 49(1): 28–32.

Zheng, Y., Z. Pan and R. Zhang. 2009. Overview of biomass pretreatment for cellulosic ethanol production. *International Journal of Agricultural and Biological Engineering* 2(3): 51–68.

Zhu, J. and X. Pan. 2010. Woody biomass pretreatment for cellulosic ethanol production: Technology and energy consumption evaluation. *Bioresource Technology* 101(13): 4992–5002.

Zhu, L., J.P. O'Dwyer, V.S. Chang, C.B. Granda and M.T. Holtzapple. 2008. Structural features affecting biomass enzymatic digestibility. *Bioresource Technology* 99(9): 3817–3828.

Zhu, Y., Y. Lee and R.T. Elander. 2005. Optimization of dilute-acid pretreatment of corn stover using a high-solids percolation reactor. *Twenty-Sixth Symposium on Biotechnology for Fuels and Chemicals*. Springer. pp. 1045–1054.

7

Butanol from Renewable Biomass: Highlights of Downstream Processing and Recovery Techniques

Sonil Nanda, Ajay K. Dalai, and Janusz A. Kozinski

CONTENTS

ABSTRACT The concerns relating to global warming, climate change, and increasing energy demands have led to significant research in the development of alternative energy to fossil-based sources worldwide. The prime focus has been on the emerging technologies to recover the chemical energy from plant-based biomass and other organic wastes in the form of liquid and gaseous biofuels. For realizing the tangible potential of bioenergy, there is an immediate need for breakthrough research in understanding the physicochemical properties of biomass, conversion technologies, and advanced engineering for larger scale biomass processing. This chapter is focused on biobutanol as an advanced next-generation biofuel. It is dedicated to advancing new developments and approaches in acetone–butanol–ethanol fermentation with a special emphasis on the butanol recovery techniques. The recovery techniques vividly discussed here include distillation, adsorption, liquid–liquid extraction, pervaporation, perstraction, gas stripping, and supercritical fluid extraction. The mechanisms for each recovery technology have been thoroughly discussed along with their benefits and limitations. Butanol is a progressive biofuel with fuel properties fairly compatible with gasoline; hence, advancements in its recovery technologies can mostly aid in its cost-effective production from lignocellulosic biomass and other waste resources.

7.1 Introduction

The interest in the production of renewable transportation fuels has been escalating momentarily. A few reasons that are contributing to this shift include, but are not limited to, the rising fuel prices, skyrocketing demand for fossil fuels, greenhouse gas (GHG) emissions, and global warming. The worldwide consumption of petroleum and other fossil-based liquid fuels, being 86 million barrels per day (BPD) in 2008, is predicted for an increase to 98 million BPD in 2020 and 112 million BPD in 2035 (USEIA, 2011). The steady search for alternative fuels to replace gasoline has brought two biofuels into the limelight, namely, ethanol and butanol. The chief attention has been on developing technologies to recover the chemical energy in the form of liquid fuels from plant-based residues and other organic wastes. The waste biomasses that have a tremendous potential to supplement the production of such fuels include lignocellulosic feedstocks (e.g., agricultural and forestry residues), municipal solid wastes, and sewage sludge.

The lignocellulosic biomasses have a tendency to achieve energy security and reduce GHG emissions (Nanda et al., 2015b, 2014b). The waste plant biomass is nonedible and lignocellulosic in nature. Lignocellulosic materials incorporate the agricultural residues, dedicated energy crops (e.g., temperate grasses), wood residues (e.g., sawmill and paper mill discards), and municipal paper waste. Lignocellulosic biomass is an inexpensive resource found abundantly on a global scale and can support the production of alternative liquid fuels. About 40 million tons of lignocellulosic biomass is generated globally (Sanderson, 2011). The renewable resources such as sugarcane bagasse, corn stover, corn fiber, wheat straw, and other agricultural by-products are recognized as potential feedstocks for the production of these alternative fuels. In addition, the life cycle assessment studies used for the production of bio-oils and bioethanol indicate the carbon-neutral nature of lignocellulosic biomass. In addition, as lignocellulosic materials are nonedible, they possess a negligible threat to the domestic and international food security unlike food-based fuel feedstocks (e.g., corn, wheat, potato, cassava, etc.) (Nanda et al., 2015b). As lignocellulosic biomass do not complete with the food supply, they have broad potential in bioethanol and biobutanol refineries.

The dominant source of today's bioethanol is corn and sugarcane. Because these raw materials are food-based crops, bioethanol is often surmounted by the food-versus-fuel debate (Graham-Rowe, 2011). In addition to this exploitation of food crops for fuel, the resulting inflation in the food prices is directing a global interest in exercising waste biomass for fuel production. Although there have been many advance researches on bioethanol production, its increased dependency on food crops has raised concerns inclining the attention toward compromised food security. Although lignocellulosic biomasses are being deployed as next-generation feedstocks for bioethanol production, the development of innovative methods for their efficient bioconversion is still a significant challenge. Hence, there has always been inquisitiveness to seek a superior fuel with flexibility in raw material selection, better bioconversion, and attractive fuel features than ethanol.

Butanol is one such fuel that can be derived from lignocellulosic feedstocks using *Clostridium* spp., especially *Clostridium acetobutylicum* and *Clostridium beijerinckii*. As a fuel, butanol has 30% higher energy content than ethanol and is nonsensitive to water with less corrosiveness, lower volatility, less flammability, and reduced hazardousness to handle (Nanda et al., 2014a). Its low vapor pressure facilitates its usage in existing gasoline supply pipelines, and it has a tendency to be mixed with gasoline in flexible proportions (Qureshi and Ezeji, 2008). In addition, the flexible gasoline blends allow butanol to be used in existing vehicular engines without any modification. Biobutanol is also an excellent fuel extender and a cleaner fuel as it contains 22% oxygen. These appealing features make butanol an attractive transportation fuel derived from lignocellulosic biomass and other biogenic wastes.

7.2 The History of Butanol Fermentation: Primitive Origin to Recent Development

The description of historical events in this section is an abridged version of the information presented by Jones and Woods (1986) as well as Köpke and Dürre (2011). The indication of butanol production through bioconversion is not utterly contemporary. It dates back to 1861 when the French chemist and renowned microbiologist Louis Pasteur reported the production of butanol through microbial fermentation. In his initial reports, Pasteur indicated butanol as a product of *Vibrion butyrique*, which was presumably *C. butyricum* or *C. acetobutylicum* (Sauer et al., 1993). Likewise, acetone production via fermentation was reported for the first time in 1905 by the Austrian biochemist Franz Schardinger.

It all began in the initial twentieth century when the shortage of natural rubber stimulated interest in producing synthetic rubber (Jones and Woods, 1986). The two routes extensively proposed for synthesizing artificial rubber were: (1) polymerizing isoprene produced from isoamyl alcohol, or (2) polymerizing butadiene produced from butanol. With this agenda, in the United Kingdom (*c.* 1910), a company named Strange and Graham Ltd. employed a group of well-known researchers, namely, Auguste Fernbach and Moïse Schoen (from Institut Pasteur, Paris) as well as William Perkins and Charles Weizmann (from Manchester University, Manchester). In 1911, Fernbach isolated a bacterial culture that was able to ferment potato starch to butanol with lacking abilities to convert maize starch. In the following year (*c.* 1912), Weizmann left Strange and Graham Ltd. but continued working at Manchester University in the same field of research. By 1914, he had isolated a culture (later to be called as *C. acetobutylicum*), which was able to produce butanol from starchy materials (both potato starch and maize starch) much efficiently than Fernbach's original culture (Jones and Woods, 1986). In 1915, Weizmann applied for a patent for his acetone–butanol producing bacterial culture. Much before realizing the final objective of producing synthetic rubber, many ventures in Asia started producing natural rubber during those years through mass plantation that also decreased its price subsequently (Dürre, 2007). Meanwhile (*c.* 1913), Strange and Graham Ltd. established a plant in Rainham, London, for producing acetone and butanol from potatoes that later transferred its operations to a new location in King's Lynn, Norfolk (Köpke and Dürre, 2011).

Soon after in 1914, the development in acetone and butanol bioproduction was significantly transformed with the outbreak of World War I (WWI). During WWI, large quantities of cordite were required by the British army to manufacture ammunitions (Dürre, 2007). Cordite is a smokeless military propellant developed and produced in the United Kingdom in 1889 as a substitute for gunpowder. The mass production of cordite necessitated vast quantities of acetone for use as a solvent. There was a tremendous shortage of acetone in the United Kingdom during WWI, as initially it was produced through chemical routes from calcium acetate imported from Germany, Austria, and the United States. The terminated supplies of calcium acetate during WWI and the high demands for acetone triggered the British government to contract Strange and Graham Ltd. in 1915 for supplying biologically produced acetone. Using Fernbach's culture, Strange and Graham Ltd. produced around 970 lb of acetone per week using potato starch (Jones and Woods, 1986). With the demand for acetone still at elevation, the British government soon directed Strange and Graham Ltd. to replace Fernbach's process with Weizmann's highly efficient *C. acetobutylicum* followed by adapting six new distilleries for the same. The new process replaced the feedstock from potato starch with maize starch, resulting in acetone production mounting up to 2200 lb per week (Köpke and Dürre, 2011). Weizmann's process guaranteed a relentless supply of acetone in the United Kingdom, Canada, and the United States during WWI.

As a war consequence, the supply of grains and other food crops had crippled, thus adversely impacting Weizmann's fermentation process with compromised acetone production. In 1916, the British Empire transferred all its operations on acetone fermentation to Canada, notably at the Gooderham and Worts Distillery in Toronto, Ontario (Jones and Woods, 1986). It has been documented that about 3000 tons of acetone and 6000 tons of butanol were produced by the Canadian allies in Toronto until 1918. By 1917, France, the United States, and British India started adapting to the acetone–butanol fermentation to supply acetone for cordite manufacturing. However, all the distilleries terminated acetone–butanol fermentation after WWI ceasefire in 1918 as the acetone was no longer required for ammunition.

In the hiatus, butanol had accumulated in massive quantities as a by-product of acetone production through Weizmann's fermentation process. Being considered as a redundant product, butanol and its esters (butyl acetate) were tested as solvents for nitrocellulose lacquers by the then booming automobile industries. The lacquers for automobile industries were produced using fusel oil and amyl acetate obtained from amyl alcohol, which is in turn a by-product of ethanol fermentation. Butanol and its esters turned out to be ideal solvents to replace fusel oil and amyl acetate, which was a turning point in butanol research making it lucrative than redundant. In 1919, Commercial Solvents Corporation, Maryland, acquired the Weizmann's fermentation process and the Terre Haute plant in Indiana. The Commercial Solvents Corporation started its operations by 1920 with installing plants in the United States (Terre Haute in Indiana and Peoria in Illinois) and Liverpool in the United Kingdom. The Peoria plant used molasses as a feedstock and was the largest distillery housing 32 fermentors, each having a butanol production capacity of 50,000 gallons (Jones and Woods, 1986). Soon after Weizmann's patent expired in 1936, many new distilleries opened in several locations in the United States, the United Kingdom, Australia, former USSR, Brazil, Puerto Rico, South Africa, Egypt, Japan, and today's Taiwan (Köpke and Dürre, 2011). With the advent of numerous distilleries between 1935 and 1941, several butanol-producing strains were also developed and around 18 patents were issued to different companies (Jones and Woods, 1986).

The acetone–butanol fermentation saw a renaissance with the beginning of World War II (WWII) in 1939. The existing distilleries in the United Kingdom and allied nations underwent technical rehabilitation to increase the acetone production and supply for ammunition (cordite) manufacturing. For instance, many batch distillation units, multiple-column continuous distillation systems, and semicontinuous distillation systems were installed in numerous distilleries (Jones and Woods, 1986). Many new developments in Weizmann's fermentation process were made to increase butanol production rather than acetone after WWI. One of such alterations included a change in feedstock selection from the conventional starch (potato or maize) to molasses. The supply of molasses in the United Kingdom dwindled due to the hampered wartime cross-border trading impact. The solvent-producing distilleries established in the United States post-WWI quenched the demand of acetone by exporting the solvent supplies to the United Kingdom and other allied countries. With the shortage of molasses also experienced in the United States during WWII, some distilleries switched back to maize starch as the feedstock. To increase the acetone supply, many new refineries were also established in Japan, British India, Australia, and South Africa. Most of these new distilleries continued the production of acetone and butanol during and post-WWII using molasses and starchy materials.

As WWII came to an end in 1945, copious tons of acetone and butanol remained underutilized. The interest in acetone–butanol fermentation gradually declined by 1950 and the operations in the respective distilleries in the United States and the United Kingdom were entirely terminated by 1960. The revolutionary growth in petrochemical industries in the 1950s also made the conditions acute for the existence of acetone–butanol distilleries (Qureshi and Blascheck, 2005). Additionally, molasses became scarce due to drought (Qureshi et al., 2007), and the limited available stock was used as cattle feed. The subsidized cost of molasses created severe economic challenges for their utilization as raw materials in fermentation. Most of the acetone–butanol distilleries established worldwide to supply acetone during WWII gradually stopped functioning. The reasons behind this failure were as follows: (1) relatively lower solvent production cost via petrochemical routes, (2) moderately lower production rates in fermentation than petrochemical routes, and (3) higher cost of molasses and other raw materials for fermentation in the post-WWII scenario.

It has been documented that the National Chemical Products Company in South Africa continued acetone–butanol fermentation until 1983 using molasses as substrates (Köpke and Dürre, 2011). The company used *C. beijerinckii* and *C. saccharobutylicum* as the solvent-producing strains and employed 12 fermentors, each having a capacity of 90,000 L (Jones, 2001; Keis et al., 2001). The former USSR continued acetone–butanol–ethanol (ABE) fermentation until the late 1980s operating at least eight distilleries with the largest one in Dokshukino, Russia (Zverlov et al., 2006). These distilleries used *C. acetobutylicum* to ferment the processing wastes from barley, rye, wheat, and potato. Surprisingly, in China, more than 30 acetone–butanol refineries were active by the 1980s, producing over 170,000 tons of solvent (Chiao and Sun, 2007). The Chinese distilleries used *C. acetobutylicum* to ferment the substrates such as corn, cassava, potato, and sweet potato. This large group of Chinese bioproduction enterprise came to a closure during the late 1990s due to the rapid increase in petrochemical infrastructures.

Until around 2005, butanol was considered to be an industrial precursor for the production of chemicals such as acrylate, amino resins, butyl acetate, butylamine, glycol ether, and methacrylate ester. The applications of these compounds are manifold, a few of which can be found in adhesives, alkaloids, antibiotics, beauty cosmetics, camphor, deicing fluid, detergents, elastomers, emulsifiers, flavoring agents, flocculants, flotation aids, hydraulic and brake fluids, paint thinners, perfumes, pesticides, printing ink, safety glass, superabsorbents, textiles, and so on (Dürre, 2007). Until recently, the acetone–butanol fermentation (now popularly called as acetone–butanol–ethanol or ABE fermentation) has seen a renaissance with high prospective and improvements based on scientific investigations.

In 2006, approximately 10 ABE refineries in China had reinstated operations. The solvent production capacity from these distilleries being around 210,000 tons is expected to increase up to 1,000,000 tons (Ni and Sun, 2009). Recently, GranBio (Brazilian Biotech Company in Alagoas) and Rhodia (Solvay International Chemical Group in Belgium) came in partnership to produce biobutanol from sugarcane bagasse in an upcoming refinery in Brazil (Lane, 2013). The plant, expected to be functional soon, envisages producing 100 kilotons of solvents per year. Butamax™ Advanced Biofuels is a joint venture of British

Petroleum (London) and DuPont (Delaware) to develop biobutanol production technology by offering an inexpensive, high-value, and drop-in biofuel for the global transportation sector. Butamax technology is designed for converting sugars from various biomass feedstocks, including corn and sugarcane, into butanol using existing biofuel production facilities (Butamax™, 2015). In summary, some global companies that have recently announced their ventures in commercializing butanol as a next-generation biofuel include Butamax, GreenBiologics (Oxon), Cobalt Biofuels (California), Tetravitae Bioscience, Inc. (Illinois), Gevo (Colorado), METabolic EXplorer (Clermont-Ferrand, France), Butalco (Fürigen, Switzerland), and Cathay Industrial Biotech (Shanghai, China).

7.2.1 Biomass Pretreatment and Hydrolysis

Lignocellulosic biomass represents a widely available and largely unexploited source of raw material for the generation of liquid and gaseous fuels. Prior to the biomass conversion, a fundamental characterization is indispensable in considering it as a feedstock for biofuel production because of its various physiochemical, thermochemical, and biochemical characteristics. In all cases, the intrinsic properties of biomass determine both the choice of the conversion process and any subsequent processing complexities that may arise (McKendry, 2002). The type of biomass (woody or herbaceous) equally explains the amount of energy stored in it. Furthermore, the sugar (pentose, C_5, and hexose, C_6) composition in the biomass decides the theoretical yield of fuel alcohols through fermentation and can therefore have a considerable impact on the process economics and productivity. Hence, it is the interaction between all these aspects that enables the deployment of waste biomass as an economical and efficient energy source.

It is evident that the cost of feedstock is a prime factor in influencing the production cost of biofuels (Festel, 2008; Qureshi and Blaschek, 2000). Therefore, economically viable and abundantly available lignocellulosic materials such as corn stover, corn fiber, wheat straw, rice straw, sugarcane bagasse, and perennial grasses could be potentially used to produce these biofuels. Moreover, lignocellulosic biomasses are non-contenders to the food grains conventionally used as biofuel substrates. Chemically, lignocellulosic residues are composed of 35%–55% cellulose, 20%–40% hemicellulose, and 10%–25% lignin by weight (Nanda et al., 2013). Cellulose is a glucose polymer comprising 1,4-β-linked D-glucose subunits. Hemicellulose is a mixture of polysaccharides composed of pentose and hexose sugars such as glucose, mannose, xylose, and arabinose. Lignin is a phenylpropane polymer linked by ester bonds that act as the glue and tightly binds with cellulose and hemicellulose. The cellulose fibers are sheltered by a firm matrix of hemicellulose and lignin as a result of which their tough interchain hydrogen-bonding network and the higher order structure in plants contribute to biomass recalcitrance (Nanda et al., 2015a). Hemicelluloses are covalently attached to lignin to form an extremely complex structure with cellulose. Therefore, a pretreatment process is necessary for the biomass to break down its internal polymeric framework and release the sugars that can then be fermented to alcohols.

The pretreatment procedure for lignocellulosic biomass usually involves the use of acids and alkali followed by hydrolytic enzymes. The conversion of lignocellulosic materials to alcohols requires the following: (1) delignification to liberate cellulose and hemicellulose from the complex with lignin, (2) depolymerization of the biopolymers to produce fermentable sugars, and (3) fermentation of mixed C_5 and C_6 sugars to ethanol or butanol. The dilute sulfuric acid hydrolysis is the most widely used pretreatment technology in lignocellulosic biofuel refineries. Some of the feedstocks reported to have undergone dilute acid pretreatment include Bermuda grass (Sun and Cheng, 2005), corn stover (Lloyd and Wyman, 2005), corn fiber (Qureshi et al., 2008b), rice straw (Karimi et al., 2006; Roberto et al., 2003), pinewood (Marzialetti et al., 2008; Nanda et al., 2014a; Zhu et al., 2010), reed canary grass (Dien et al., 2006), silver grass (Guo et al., 2008), sugarcane bagasse (Lavarack et al., 2002; Rodriguez-Chong et al., 2004), switch grass (Dien et al., 2006), timothy grass (Nanda et al., 2014a), and wheat straw (Nanda et al., 2014a; Qureshi et al., 2007; Saha et al., 2005).

In most of the dilute acid pretreatments, there is an efficient recovery of hemicelluloses without the removal of lignin. This is because lignin is insoluble in any acid or solvent (Sluiter, 2011). In contrast to herbaceous materials, woody biomass contains a higher degree of lignin approaching 20%–30% (McKendry, 2002). Only 0.2%–0.5% of lignin in softwoods and 3%–5% lignin in hardwoods are acid soluble (TAPPI, 2011). Because lignin is soluble in alkali, an alkaline treatment of biomass prior to its acidic treatment is found to delignify lignocellulosic biomass. In a study for obtaining cellulose nanofibers from curaua leaves, a NaOH treatment (5%–17.5%) at 70°C for 1 h was employed prior to dilute acid treatment to check the cellulose and hemicellulose yields (Correa et al., 2010). An increase in the concentration of NaOH resulted in higher cellulose recovery due to enhanced removal of noncellulosic components, especially lignin and hemicellulose. The percentage recovery of cellulose, hemicellulose, and lignin in 5% NaOH treatment was 83.1, 8.1, and 8.8, whereas in 17.5% NaOH, it was 91, 0.2, and 4.2, respectively. Another study suggests an acidified NaCl treatment at 75°C for 1 h to remove lignin from woody biomass (Chen et al., 2011). Nanda et al. (2015a) showed the distribution of cellulose, hemicellulose, and lignin in spectroscopic maps generated from pinewood, timothy grass, and wheat straw treated with acidified sodium chlorite.

Despite delignification and acid hydrolysis, the hydrolysates often result in the generation of some chemical by-products that inhibit cell growth and fermentation. The inhibitors include furfural, hydroxymethylfurfural, acetic, ferulic, glucuronic, and coumaric acids along with phenolic compounds (Ezeji et al., 2007a). Hence, for a successful production of butanol, it is very crucial to remove these inhibitors from the substrate solution. Qureshi et al. (2010a) suggested that over-liming the acidic hydrolysate can help in reducing the toxicity during the fermentation process. Over-liming involves increasing the initial pH of the hydrolysate to 10 with $Ca(OH)_2$ and the addition of 1 g/L Na_2SO_3. Thereafter, the solution is boiled at 90°C for 30 min followed by bringing back the final pH to 7 with H_2SO_4. In contrast to other detoxification methods such as the use of resin columns, $Ca(OH)_2$ aids in the maximum utilization of sugars in the medium. In a study by Ezeji et al. (2007b), $Ca(OH)_2$-treated corn fiber hydrolysate left behind 17 g/L of total residual sugars after fermentation from an initial sugar concentration of 56.9 g/L in untreated hydrolysate, whereas resin (XAD4)-treated hydrolysate presented 27.6 g/L of residual sugars.

7.2.2 ABE Fermentation

Butanol production by fermentation, widely known as acetone–butanol–ethanol or ABE fermentation, is one of the oldest fermentation processes as discussed earlier. Butanol is a major coproduct of ABE fermentation as the typical proportions of acetone, butanol, and ethanol are 3:6:1. *Clostridium* spp., particularly *C. acetobutylicum* or *C. beijerinckii*, are capable of converting broad-ranging carbon sources (e.g., glucose, galactose, cellobiose, mannose, xylose, and arabinose) to fuels and chemicals such as butanol, acetone, and ethanol (Ezeji et al., 2007a). Recently, research has been directed at exploring the area of butanol production from lignocellulosic materials. Attempts have been made to obtain butanol from many starch-based and cellulosic substrates. Table 7.1 summarizes a few lignocellulosic

TABLE 7.1

Lignocellulosic Biomasses and *Clostridium* spp. Used for Butanol Production

Feedstock	Bacteria	Total ABE Concentration (g/L)	References
Aspen wood	*C. acetobutylicum* P262	20.1–24.6	Parekh et al., 1988
Bagasse	*C. saccharoperbutylacetonicum* ATCC 27022	18.1	Soni et al., 1982
Barley straw	*C. beijerinckii* P260	26.6	Qureshi et al., 2010a
Corn fiber	*C. beijerinckii* BA101	9.3	Qureshi et al., 2008b
Corn stover	*C. acetobutylicum* P262	25.8	Parekh et al., 1988
Pinewood	*C. acetobutylicum* P262	17.6	Parekh et al., 1988
Pinewood	*C. beijerinckii* B-592	18.5	Nanda et al., 2014a
Rice straw	*C. saccharoperbutylacetonicum* ATCC 27022	13.0	Soni et al., 1982
Wheat straw	*C. beijerinckii* B-592	17.9	Nanda et al., 2014a
Wheat straw	*C. acetobutylicum* IFP 921	17.7	Marchal et al., 1984
Wheat straw	*C. beijerinckii* P260	21.4	Qureshi et al., 2008a
Switchgrass	*C. beijerinckii* P260	14.6	Qureshi et al., 2010b
Timothy grass	*C. beijerinckii* B-592	17.4	Nanda et al., 2014a

biomasses used for butanol production. Some of the traditional feedstocks for butanol are corn, millet, wheat, rice, tapioca, whey permeate, molasses, and potatoes. Because these substrates are starch based, the strong amylolytic activity of clostridia restricts the requirement of biomass pretreatment or hydrolysis (Qureshi and Blaschek, 2005). The high cost of starchy substrates has shifted the focus toward some of the agricultural residues such as corn stover, corn fiber, corn cobs, wheat straw, barley straw, rice straw, and switchgrass (Qureshi and Ezeji, 2008). Hemp waste and sunflower shells have also been investigated for butanol production (Zverlov et al., 2006).

The ABE fermentation carried out using anaerobic bacterium such as *C. acetobutylicum* or *C. beijerinckii* is biphasic involving acidogenesis and solventogenesis. Although the acidogenic phase involves the production of acids (e.g., acetic and butyric acid), the solventogenic phase is related to the production of solvents (e.g., acetone, butanol, and ethanol). The ABE-producing bacteria can utilize both starchy and lignocellulosic substrates; however, the later must be hydrolyzed using a suitable pretreatment method (i.e., dilute acid and enzymatic hydrolysis). The butanol-producing clostridia can metabolize both pentose and hexose sugars in the biomass hydrolysates. This property of clostridia is advantageous over the widely used ethanol-producing yeast *Saccharomyces cerevisiae* that is inefficient in metabolizing pentose sugars, for example, xylose (Ha et al., 2011).

A typical butanol production can be attained in batch reactors under the anaerobic environment, and the product is recovered using distillation. During the fermentation, which lasts for 36–72 h, the total ABE up to 20–25 g/L can be obtained (Qureshi and Ezeji, 2008). However, butanol toxicity to the fermenting microorganisms is a major barrier for the commercial bioproduction of the alcohol that limits its concentration in the fermentation broth (Ezeji et al., 2007c). Butanol inhibition can be noticed at concentrations of 5–10 g/L as the butanol-producing bacterial cultures can rarely tolerate more than 2% butanol (Liu and Qureshi, 2009; Qureshi and Ezeji, 2008). During the past decade, the application of molecular techniques to the solventogenic clostridia has resulted in the development of hyper-butanol-producing strains such as *C. beijerinckii* P260 (Qureshi et al., 2007) and *C. beijerinckii* BA101 (Ezeji et al., 2007c).

The final concentration of total solvents (acetone, butanol, and ethanol) produced ranges from 12 to 20 g/L in a typical batch fermentation process (Lee et al., 2008). The fed-batch and continuous fermentation do not seem to be economically feasible because of butanol toxicity problems and the biphasic nature of ABE fermentation. To overcome these technical barriers, fed-batch fermentation systems have been coupled with an *in situ* recovery process such as gas stripping (Ezeji et al., 2004b). Several multistage continuous fermentation systems have also been investigated (Godin and Engasser, 1990). The pH of the fermentation medium is very crucial to the biphasic ABE fermentation. In the acidogenic phase, the formation of acetic acid and butyric acid lowers the pH. The solventogenic phase starts when the pH reaches a critical point beyond which the acids are reassimilated, thereby generating acetone and butanol (Lee et al., 2008). Although low pH is preferable for solvent production, the pH below 4.5 results in enough acid formation making the solventogenic phase brief and incompetent. A solution for enhancing clostridial growth, carbohydrate utilization, and butanol production is by increasing the buffering capacity of the fermentation medium (Bryant and Blaschek, 1988).

As a next-generation alcohol-based fuel, butanol can be used in pure form or blended with gasoline at higher concentrations, unlike ethanol that can be blended only up to 85% (Dürre, 2007). Butanol can be blended with gasoline at higher concentrations providing twice the renewable carbon content in every gallon. The United States regulations allow butanol to be blended at 16% by volume compared to 10% by volume blending as suitable for ethanol (Butamax™, 2015). Biobutanol is also referred to as a drop-in fuel that can be used in existing motor engines and vehicular infrastructures. Moreover, the energy content of butanol is 29.2 MJ/L, that is, 30% higher than that of ethanol (21.2 MJ/L) and much closer to gasoline (32.5 MJ/L) (Nanda et al., 2014a). Butanol's low vapor pressure, nonhygroscopic nature, less volatility, and less flammability also facilitate its blending and supply in existing gasoline channels and pipelines (Qureshi and Ezeji, 2008). Although with many advantages, the production of butanol has some drawbacks such as (1) lower yields, (2) butanol toxicity due to lower solvent tolerance by clostridia, and (3) expensive recovery technologies (Zheng et al., 2009).

Owing to the partial miscibility of butanol in water, some technical issues are encountered during its recovery. The solubility of biobutanol (*n*-butanol) in water is 7.7% (77 g/L) at 20°C, 7.3% (73 g/L) at 25°C, and 7.1% (71 g/L) at 30°C. *sec*-Butanol (2-butanol) has a substantially greater solubility, and *tert*-butanol is fully miscible in water, whereas *n*-butanol (1-butanol) and isobutanol are sparingly water soluble. All alcohols have a hydroxyl group that makes them polar and tends to promote solubility in water. With the increase in the carbon chain of the alcohol, its solubility in water is lessened. Although methanol (CH_3OH), ethanol (C_2H_5OH), and propanol (C_3H_7OH) are completely miscible with water, *n*-butanol (C_4H_9OH) is moderately soluble because of the balance between two opposing solubility trends.

Figure 7.1 represents the steps involved in butanol fermentation. Although the upstream processing includes feedstock treatment, the downstream processing is related to alcohol recovery and purification. The starch-based feedstock can readily undergo ABE fermentation due to the amylolytic activity of clostridia, whereas lignocellulosic biomasses are pretreated using dilute acids and cellulolytic enzymes. The partial solubility of butanol in water creates impediments in its downstream processing and extraction from the fermentation broth. The high energy requirements subsequently add to the process expenditures making the overall butanol production expensive. To make ABE fermentation profitable, many *in situ* product recovery systems have been developed, a few of which include adsorption, liquid-liquid extraction, pervaporation, perstraction, gas stripping,

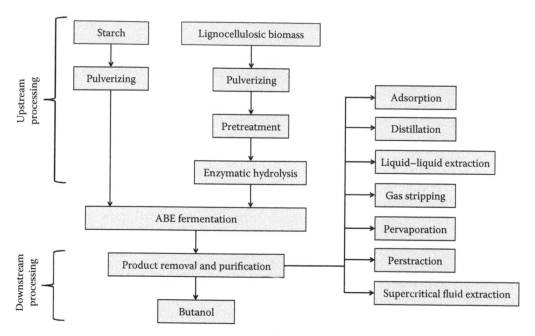

FIGURE 7.1
Typical biological route of biomass conversion to butanol.

and supercritical fluid extraction. The research on these butanol recovery techniques is still at its infancy stage, and none of these technologies have been implemented at the commercial level. Section 7.3 comprehensively discusses each of these extraction technologies with advantages, challenges, and recent developments. Table 7.2 summarizes the advantages and disadvantages of each of the butanol extraction techniques with highlights on butanol selectivity and energy requirement.

7.3 Butanol Recovery and Extraction Techniques

7.3.1 Adsorption

Adsorption is a promising butanol extraction technique and regarded as one of the most energy-efficient recovery processes compared to distillation, gas stripping, and pervaporation. The process involves the adsorption of butanol onto a packed column from the dilute solution or fermentation broth. In the next step, butanol is desorbed by heat treatment or displacer to achieve a concentrated solution as well as adsorbent regeneration (Vane, 2008). In adsorption, the alcohol is transferred from the feed mixture to a solid adsorbent material (Figure 7.2). The extractant is usually contained in a packed column contactor serving as both the adsorber and the desorber. The process usually involves the loading and unloading of the adsorbent. Similar to liquid extractants, the solid adsorbent should ideally demonstrate a particular sorption selectivity toward the alcohol compared to water. A greater separation factor between water and alcohol as well as high distribution coefficient are desirable for a model adsorbent. Furthermore, the adsorbent should be stable and have little or no solubility in water and alcohol.

TABLE 7.2

Advantages and Limitations of Butanol Recovery Technologies

Recovery Technique	Advantages	Limitations	Selectivity	Energy Requirement (MJ/kg)	References
Adsorption	• Easy operation • Less energy intensive	• Low selectivity • High material cost • High adsorbent regeneration cost	130–630	1.3–33	Oudshoorn et al., 2009; Xue et al. 2014b
Distillation	• Traditionally used • Less infrastructure requirement	• High operational cost due to low concentration of butanol • Highly energy intensive	72	—	Ezeji et al., 2004a; Lee et al., 2008; Oudshoorn et al., 2009
Gas stripping	• High efficiency • High sugar utilization • Reduced butanol toxicity • Easy operation • No fouling	• Low selectivity • Low efficiency	4–22	14–31	Ezeji et al., 2004b; Oudshoorn et al., 2009; Xue et al., 2014b
Liquid–liquid extraction	• Less energy intensive • High selectivity • High efficiency	• High operational cost • Toxicity by the extractant to the clostridia • Emulsion formation by the extractant	1.2–4100	7.7–26	Lee et al., 2008; Oudshoorn et al., 2009; Xue et al., 2014b
Perstraction	• High selectivity • Low toxicity to clostridia	• Fouling issues • High material cost	1.2–4100	7.7	Xue et al., 2014b
Pervaporation	• Selective removal of volatile liquids and solvents • Less energy intensive • High flux and greater allowable permeate pressure • Low permeate condensation cost	• High membrane cost • High-temperature requirement • Vulnerable to temperature-sensitive compounds and cells • Risk for material failure	2–209	2–145	Lee et al., 2008; Oudshoorn et al., 2009; Vane, 2005; Xue et al., 2014b
Supercritical fluid extraction	• Cost-effective • Recyclable • Feasible for large-scale processes	• Needs more research	550	—	Laitinen and Kaunisto, 1999; Oudshoorn et al., 2009; Xue et al., 2014b

In recent years, broad investigations have been done on adsorbent selection, enhancing adsorptive capabilities and regenerating the adsorbent for an energy and cost-efficient butanol recovery. Adsorbent materials such as activated carbon, zeolite, and polymeric resins have been explored as adsorbents for butanol recovery from the model solution or fermentation broth. Silicalite, adsorbent resins, Amberlite (XAD2, XAD4, XAD7, XAD8, and

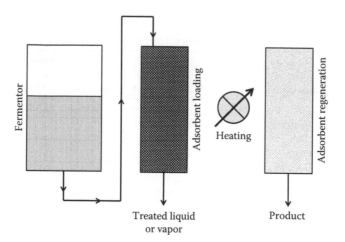

FIGURE 7.2
Representation of butanol recovery by adsorption.

XAD16), bonopore, polyvinyl pyridine, and so on have been thoroughly studied for butanol recovery (Qureshi et al., 2005). Although without the highest adsorption ability, silicalite is reported as an attractive adsorbent for its potential to concentrate butanol from dilute solutions up to 790–810 g/L (Qureshi et al., 2005). This high concentration could result in complete desorption and smooth silicalite regeneration by heat treatment. The resin-based adsorbents such as Amberlite (XAD2, XAD4, and XAD8) and bonopore have an adsorption capacity of less than 0.1 g butanol per gram of adsorbent (Qureshi et al., 2005).

Apart from the adsorption capacity, the biocompatibility of adsorbents also needs to be evaluated during the *in situ* product recovery process. Owing to the high water concentrations in the fermentation broth and low butanol levels, hydrophobic adsorbent materials are highly desirable. The overall selectivity of hydrophobic adsorbents (e.g., zeolite) is estimated to be between 130 and 630 (Oudshoorn et al., 2009). However, during the recovery process, there are possibilities that acetone and ethanol may compete with butanol for the adsorption sites on the adsorbent material as all the three solvents occur in the fermentation broth at the same time (Xue et al., 2014b). Similarly, adsorption of acids (i.e., acetate and butyrate) and nutrients from the fermentation broth could also block the adsorption sites, thereby lowering the recovery efficiency. In the absence of alcohol, silicalite-1 adsorbs about 40 mg water per gram of silicalite at room temperature (Vane, 2008). Upon addition of alcohol into the water, it is preferentially adsorbed causing a reduction in water adsorption.

The regeneration of adsorbent also needs more research to make the recovery process cost-effective. The regeneration of activated carbon is tedious, as its stability and homogeneity are less favorable compared to silica-based adsorbents. Some studies suggest that gradual heating allows the desorption from silicalite-1, thereby enabling the recovery of butanol fraction (Holtzapple and Brown, 1994). Although promising, more research on adsorbent materials could help understand their application in an integrated fermentation and butanol recovery process.

7.3.2 Distillation

Distillation is a traditional method for butanol recovery from the aqueous fermentation broth. Most of the energy consumption during distillation occurs from the evaporation of water from the broth. Distillation requires high energy as butanol has a boiling point

(117.7°C) greater than that of water. The performance of distillation is related to the applied energy integration because the energy consumption determines the process economics (Oudshoorn et al., 2009). Another reason for the high cost of distillation is that the concentration of butanol in the fermentation broth is low due to product inhibition.

The typical concentration of total solvents (acetone, butanol, and ethanol) in the fermentation broth is 18–33 g/L, of which butanol is only about 13–18 g/L (Ezeji et al., 2004a). This makes the distillation process energy intensive. The energy savings by distillation can be achieved if the concentration of butanol is increased from 10 to 40 g/L (Philips and Humphrey, 1985). In a binary system containing butanol and water, the energy required to recover butanol from a 0.5 wt.% solution to a concentrated 99.9 wt.% butanol is estimated to be around 79.5 MJ/kg (Matsumura et al., 1988). This is much higher than the energy density of butanol itself, that is, 36 MJ/kg. The energy consumption could potentially reduce to nearly 36, 6, and 3 MJ/kg upon the increase in butanol concentrations up to 1, 10, and 40 wt.%, respectively (Xue et al., 2014b).

A heterogeneous distillation, especially involving an azeotrope, is used to separate butanol from water mixture. The typical mechanism of distillation is represented in Figure 7.3. Heterogeneous distillation is employed in the scenario where the liquid phase of the mixture is immiscible as in the case of butanol–water mixture. A binary azeotrope is obtained at 92.7°C. The conversion of a feed mixture containing 20 g/L butanol into an azeotropic mixture at 0.1 MPa leads to a selectivity of 72 (Oudshoorn et al., 2009). As butanol is only soluble in water up to approximately 7.7 wt.%, and because the azeotrope occurs above the solubility limit, two liquid phases are formed. The upper phase contains 79.9 wt.% butanol, whereas the lower phase contains 7.7 wt.% butanol (Vane, 2008). The azeotrope is broken down by introducing a ternary compound into the mixture or by altering the pressure. The azeotropic phase separation together with the relative volatility behavior could be advantageous to design and separate butanol in newer systems with multicolumn distillation units and integrated phase separation. In contrast to ethanol distillation, butanol can be separated from its dilute mixture using a combination of multicolumn distillation system and phase separation units.

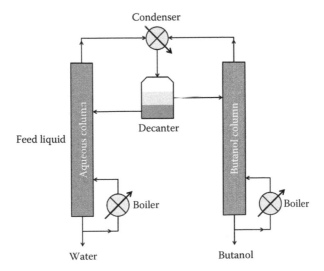

FIGURE 7.3
Representation of butanol recovery by distillation. (Adapted from Vane, L. M., *Biofuels Bioprod. Bioref.*, 2, 553–588, 2008.)

7.3.3 Gas Stripping

Gas stripping is a promising recovery technique that allows the selective removal of alcohols from the fermentation broth. The process occurs continuously with solvent removal from the vapor phase by condensation in a cold trap or via a molecular sieve. In principle, gas can be purged into the fermentor through a rotating shaft, and the volatiles are condensed and recovered from the condenser (Figure 7.4). As the butanol fermentation is anaerobic, extreme care should be taken to avoid the entry of O_2 into the fermentation medium through gas stripping. In the first step, usually N_2 or the fermentation gases (i.e., CO_2 and H_2) are purged through the fermentation broth. CO_2 and H_2 are produced during the biphasic ABE fermentation; hence, they do not interfere with the clostridial metabolism. Upon purging, the gas is bubbled through the fermentor that captures the solvents, including butanol. The gas–solvent (vapors) mixture passes through a condenser before being collected in the liquid state. Once the solvents are condensed, the gas is recycled back into the fermentor to capture more solvents in a continual mode. This process continues until all the fermentable sugars in the bioreactor are utilized by the culture (Ezeji et al., 2004b).

This process is beneficial in its simplicity, cost-effectiveness, nontoxicity, no fouling, or clogging due to the presence of biomass or cells (Ezeji et al., 2003). However, the mass transfer limitation is determined by the gas/liquid interface, which is a function of the gas bubble size (Oudshoorn et al., 2009). Gas stripping has been tested for butanol recovery in batch (Ezeji et al., 2003, 2005, 2007b), fed-batch (Ezeji et al., 2007b, 2004b; Lu et al., 2012), and continuous scale (Groot et al., 1989) reactors as a recovery technique. In all cases, the integration of gas stripping to ABE fermentation significantly increased the butanol productivity, yield, and most importantly, the substrate conversion rate. Gas stripping is also found to eliminate the substrate (sugar) and product (butanol) inhibition that normally restricts ABE production and sugar utilization to less than 20 and 60 g/L, respectively. Ezeji et al. (2004b) integrated gas stripping to a fed-batch ABE fermentation system and found the solvent productivity improvement by 400% than the control batch fermentation productivity. Although the control batch reactor resulted in the solvent yield of 0.39 g/g

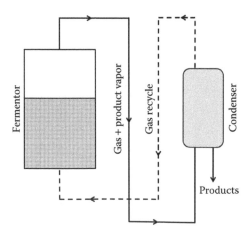

FIGURE 7.4
Representation of butanol recovery by gas stripping. (Adapted from Ezeji, T. C. et al., *Chem. Rec.*, 4, 305–314, 2004a.)

and the productivity of 0.29 g/L/h, gas stripping enhanced the solvent yield and productivity by 0.47 g/g and 1.16 g/L/h, respectively.

The selectivity of butanol and its removal rate by gas stripping depends on the parameters such as gas flow rate, butanol concentration in the fermentation broth, and condensation rate (Xue et al., 2014a). As a result of butanol toxicity and low final titer levels, the condensate obtained after gas stripping often contains less than 70 g/L butanol, which is below its critical concentration for phase separation (Xue et al., 2014a). With a low butanol selectivity of 4–22, gas stripping usually removes substantial amounts of water necessitating high energy input for its vaporization and condensation (Oudshoorn et al., 2009). The energy requirement by gas stripping and subsequent distillation or steam evaporation is estimated to be between 14 and 31 MJ/kg (Xue et al., 2014a).

7.3.4 Liquid–Liquid Extraction

In this process, an extraction solvent is mixed with the fermentation broth. All the three solvents, namely, acetone, butanol, and ethanol, are separated by an extracting solvent and further recovered through back-extraction by another extracting solvent or via distillation (Ezeji et al., 2004a). A simplified representation of liquid–liquid extraction is shown in Figure 7.5. This process relies on the differences between the distribution coefficients of the solvents. As butanol is more soluble in the organic phase (i.e., extractant) compared to the aqueous phase (i.e., fermentation broth), it is selectively concentrated in the extractant. An efficient butanol recovery by liquid–liquid extraction should have the following traits: (1) selection of an ideal extractant with an optimal distribution coefficient, (2) optimization of the broth/extractant ratio, and (3) short extraction time (Xue et al., 2014b).

There are some considerations for an ideal extracting solvent such as: (1) cost-effectiveness, (2) nontoxic nature toward cell culture, (3) immiscible and no emulsion formation in the fermentation broth, (4) high partition coefficient for fermentation products, and (5) stability during sterilization and recovery. The extractants explored so far for butanol recovery include corn oil (Wang et al., 1979), polyoxyalkylene ethers (Griffith et al., 1983), vegetable oil (Griffith et al., 1984), hexanol (Traxler et al., 1985), aromatic hydrocarbons, ketones,

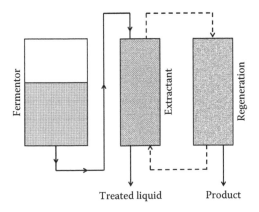

FIGURE 7.5
Representation of butanol recovery by liquid–liquid extraction and/or perstraction. (Adapted from Vane, L. M., *Biofuels Bioprod. Bioref.*, 2, 553–588, 2008.)

alkanes (Roffler et al., 1987), esters (Roffler et al., 1987), dibutyl phthalate (Wayman and Parekh, 1987), decanol and oleyl alcohol (Evans and Wang, 1988), crude palm oil (Ishizak et al., 1999), and biodiesel (Li et al., 2010).

Among all the extractants, oleyl alcohol and its mixtures are widely used due to their high extraction abilities. Furthermore, the ABE-enriched biodiesel obtained as a result of liquid–liquid extraction demonstrated improved fuel properties compared to petroleum-based diesel with higher octane number and lower cold filter plugging point (Xue et al., 2014b). Although alkanes are poor extractants, they are nontoxic and separate easily from the fermentation broth (Maddox, 1989). However, upon substitution with halogen groups, their partition coefficient is improved. In contrast, esters are good extractants and nontoxic but pose difficulties during the phase separation. The most widely used extractants, for example, alcohols, are easy to separate but suffer from toxicity problems. High butanol selectivity is attained with nonpolar extractants (Oudshoorn et al., 2009). As water does not readily dissolve in nonpolar solvents, an increase in hydrophobicity negatively impacts the solubility of butanol in the solvent.

Recently, ionic liquids are also gaining attention as an extractant for biobutanol recovery. Ionic liquids are salts in the liquid state and composed entirely of ions that could potentially substitute the organic solvents used in liquid–liquid extraction. The hydrophobicity and polarity of ionic liquids make them attractive to enhance selectivity, extraction efficiency, and distribution coefficient of butanol. The insignificant vapor pressure of ionic liquids also allows the extracted solvents to be separated by conventional low-pressure distillation with less energy input (Dai et al. 1999). A few imidazolium-based ionic liquids (e.g., [Tf2N] and [Omim][Tf2N]) have been studied in butanol recovery. The ionic liquid [Omim][Tf2N] showed a high butanol distribution coefficient of 1.94 and a selectivity of 132 (Ha et al., 2010).

The problems encountered with liquid–liquid extraction include cell toxicity imparted by the extractant and subsequent emulsion formation. To overcome these issues, a membrane could be used to separate the fermentation broth from the extractant. This could provide more surface area for the exchange of butanol between the two immiscible phases (also known as perstraction, discussed in the following Section 7.3.5). Although liquid–liquid extraction needs less energy input compared to other recovery techniques, it is potentially cumbersome for large-scale ABE fermentation processes. The larger scale fermentation requires enormous amounts of extractant that could significantly increase the capital investment. Higher levels of extractant could also result in: (1) cell inactivation, (2) loss due to incomplete phase separation, and (3) formation of an emulsion with difficulties in separation (Xue et al., 2014b).

7.3.5 Perstraction

Among all butanol recovery techniques, liquid–liquid extraction is considered as an economical downstream technique. However, as discussed earlier, many extractants used in liquid–liquid extraction turn out to be toxic to the bacterial cells, thereby inhibiting their grown and fermentative abilities. As aforementioned, liquid–liquid extraction has been reported to cause: (1) accumulation and inactivation of bacterial cells at the interface, (2) loss of extractant as a result of incomplete phase separation, (3) extraction of reaction intermediates along with butanol, and (4) formation of emulsions difficult to separate (Maddox, 1989; Qureshi et al., 1992). The use of butanol-permeable membranes between the fermentation broth and the extraction solvents has been found to overcome these problems to the most extent. This reformed version of liquid–liquid extraction is called as perstraction.

In perstraction, the diffusion of fermentation products through the permeable membrane is the rate-controlling process. This is because of the vapor pressure difference between the diffusing components at the feed phase and the extractant phase. After diffusion through the permeable membrane, the products are rapidly dissolved in the extractant. The membranes separate the organic and aqueous phases, thus preventing a settling compartment for the two liquid phases (Oudshoorn et al., 2009). This avoids any possible contamination of the organic phase by the bacterial cells, thereby reducing the chances of toxicity by the organic phase (either butanol or extractant) to the bacterial cells. In order to make the process efficient, it is highly desirable that the diffusion rates of the fermentation products exceed their production rate in the fermentor (Qureshi and Maddox, 2005).

It should also be noted that membranes can introduce an additional mass transfer limitation. Until the extractant diffuses through the membrane and the saturation of aqueous phase is attained, the fermentation may miscarry because of its toxicity to cells (Xue et al., 2014b). Qureshi et al. (1992) used silicone membrane and oleyl alcohol as the extraction solvent in a perstraction-based ABE fermentation of whey permeate. The membrane allowed butanol diffusion into the extractant, thus limiting the diffusion of acetone and acids, which resulted in a higher solvent yield of 0.37 g/g. About 58 g of total solvents (i.e., acetone, butanol, and ethanol) with a productivity of 0.24 g/L/h were reported from the study. In a similar study, Qureshi and Maddox (2005) also reported a high total solvent recovery of 137 g/L with a productivity of 0.21 g/L/h using perstraction-integrated ABE fermentation of whey permeate/lactose. The developments of advanced membranes are under way that would feature higher butanol selectivity and acetone–butanol–ethanol flux.

7.3.6 Pervaporation

Pervaporation is a solvent recovery technique that applies a combination of membrane permeation and evaporation (Wijmans and Baker, 1993). It is commercially used for the dehydration of organic solvents (Wee et al., 2008). It is a membrane-based recovery technique that employs a nonporous or molecularly porous membrane. One side of the membrane is brought into contact with the fermentation broth, whereas on the other side, vacuum or gas purge is applied to create permeate vapor stream. The components from the feed liquid diffuse through the membrane and evaporate into the permeate vapor phase (Figure 7.6). The target components diffuse through the membrane, desorb as vapor into the permeate under vacuum or low pressure, and finally condense at low temperatures. The use of pervaporation for butanol extraction is based on the high selective permeation and hydrophobicity of butanol through the hydrophobic membrane (Xue et al., 2014b).

A few hydrophobic polymeric membranes studied for butanol recovery include oleyl alcohol liquid membrane, poly[1-(trimethylsilyl)-1-propyne], polydimethylsiloxane (PDMS), polyether block amide, polypropylene, polysiloxane, polytetrafluoroethylene, polyurethane, polyvinylidene fluoride, silicalite-filled PDMS, silicalite-silicone, silicone, zeolite-filled PDMS, and zeolite membrane Ge-ZSM-5 (Oudshoorn et al., 2009; Qureshi and Blaschek, 1999; Xue et al., 2014b). PDMS composite membranes have shown an excellent recovery activity due to their highly hydrophobic nature along with thermal, chemical, and mechanical stability. Hydrophobicity of the membrane has a strong effect on selectivity because it limits the water flux. In case of alcohol recovery from a dilute solution, the use of hydrophobic alcohol-selective membranes could result in a permeate enriched with alcohol (Vane, 2008).

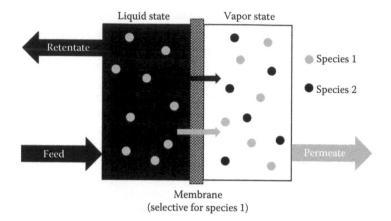

FIGURE 7.6
Representation of butanol recovery by pervaporation. (Adapted from Vane, L. M., *J. Chem. Technol. Biotechnol.*, 80, 603–29, 2005.)

A low vapor pressure or vacuum aids in increasing the transport flux and selectivity. The product flux through the membrane and the selectivity of the target compound are influenced by the aqueous- and gas-phase compositions, membrane area, temperature, and pressure (Oudshoorn et al., 2009). A higher membrane thickness tends to decrease the product transport flux. The flux limitations can be prevented by altering the process kinetics, although the kinetics is dependent on the temperature and product phase concentration. An important consideration is to enrich the permeate relative to the feed components because different compounds exhibit variable sorption and diffusion behavior with the membrane material (Vane, 2008). As liquid diffuses through the membrane, the concentration of the permeating compound is alleviated. The butanol concentration in the product phase can be lowered by: (1) increasing the volume using the vacuum or (2) applying a stripping gas that dilutes the system but increases the driving force for separation (Oudshoorn et al., 2009).

Pervaporation usually requires relatively higher temperatures to subordinate the operating cost. Pervaporation at 35°C requires more energy-intensive condensation temperatures, lower permeate pressures, and larger membrane area (Vane, 2008). The membrane area required for alcohol separation decreases as the temperature increases due to higher driving forces at higher process temperatures. This means that the optimum temperature for pervaporation could be inhibitory toward the microbial cells and deactivate many enzymes, proteins, and metabolites.

Vane (2005) proposed a few considerations for the improvement of pervaporation as an efficiently integrated fermentation–extraction process, a few of which are discussed here. The energy efficiency of pervaporation can be increased with an improved alcohol–water separation factor and reuse of the process heat. The operational costs can be lowered by increasing the membrane flux to reduce the area that consequently minimizes the membrane/module cost per unit area. Microfiltration and ultrafiltration techniques could help increase the cell density in the fermentor and allow higher pervaporation temperature. Dephlegmator fractional condensation technology could also bring synergy to butanol recovery along with the dehydration of organic components, thereby mitigating the capital cost. More research is required in this area to better understand the membrane and module stability and potential fouling behavior.

7.3.7 Supercritical Fluid Extraction

Supercritical fluids are new contenders in the list of extractants used in liquid–liquid extraction. A supercritical fluid is any compound at a temperature and pressure above its critical point where distinctive gas and liquid phases do not exist. Supercritical fluids, having both liquid-like and gas-like properties, can dissolve organic materials efficiently. The breakthrough research on supercritical fluids is now making them an attractive substitute for organic solvents in a broad range of applications. The most commonly explored supercritical fluids include supercritical water (SCW) and supercritical carbon dioxide (SCCO$_2$). Water above its critical temperature ($T_c > 374°C$) and critical pressure ($P_c > 22$ MPa) is called SCW. Similarly, CO$_2$ above its critical temperature ($T_c > 30.95°C$) and critical pressure ($P_c > 7.38$ MPa) is known as SCCO$_2$.

SCW is widely used in biomass gasification eliminating the need for biomass drying to produce H$_2$-rich synthesis gas from a broad range of waste biomass resources (Kruse, 2008; Reddy et al., 2014) and model compounds (Fang et al., 2008a, 2008b; Nanda et al., 2015c, 2015d). The major product as a result of SCW gasification of organics is synthesis gas that is primarily a mixture of H$_2$ and CO. Other gases that are also generated as a result of SCW gasification include CO$_2$, CH$_4$, and C$_2$H$_6$. However, their composition and yield are influenced by the gasification temperature, pressure, feed concentration, residence time, and use of catalysts (Reddy et al., 2014). The generation of gas products as a result of SCW gasification is dependent on the holistic interactions between a series of reactions such as steam reforming reaction, water–gas shift reaction, methanation reaction, and hydrogenation reaction.

SCCO$_2$ is used for the fractionation of organic components and extraction of bioactive compounds (Naik et al., 2010; Rout et al., 2009). The extraction process by supercritical fluids compared to other organic solvents is relatively faster because of their low viscosities and high diffusivities. The selectivity of supercritical fluid extraction is controlled by the density of the medium (Oudshoorn et al., 2009). The extracted compound is recovered by depressurizing the system and reducing the temperature, thereby allowing the supercritical fluid to return to the gaseous state. Figure 7.7 illustrates a simplified schematic of

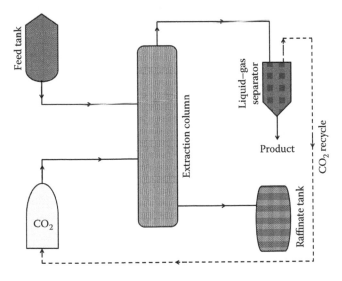

FIGURE 7.7
Representation of supercritical fluid (SCCO$_2$) extraction for butanol. (Adapted from Laitinen, A., and J. Kaunisto, *J. Supercrit. Fluids*, 15, 245–252, 1999.)

SCCO$_2$ extraction of butanol. SCCO$_2$, the most widely studied supercritical solvent, has been investigated for the decaffeination of coffee beans (Peker et al., 1992), extraction of hops for beer production (Guo-qing et al., 2005), fractionation of bio-oil from lignocellulosic biomass (Naik et al., 2010; Rout et al., 2009), and derivation of essential oils and pharmaceutical products from plants residues.

CO$_2$ is a suitable solvent for supercritical fluid extraction because it is recyclable, nonflammable, nontoxic, and inexpensive (Laitinen and Kaunisto, 1999). Owing to its reaction conditions, that are, 31°C and 7.4 MPa, its application is technically feasible even for large-scale extraction processes. The recovery of butanol using supercritical fluids relies on the partition coefficient of butanol and water as well as the solubility of the organic solvent in the aqueous phase (Oudshoorn et al., 2009). After butanol extraction using SCCO$_2$, the CO$_2$ can be separated by reducing the pressure and is easily removed for recycling from the extracted products. There is a lack of accessible information on the use of supercritical fluids for the downstream recovery of butanol; hence, future research could address their potential use and commercial applications.

7.4 Conclusions and Perspectives

Biobutanol is an attractive fuel over bioethanol as it is produced from waste biomass and has better fuel properties comparable to gasoline. The major issues surfacing the bioproduction of butanol are product toxicity toward *Clostridium* sp. that results in low butanol yields. Another technical challenge is the recovery of butanol from the fermentation broth, which is due to its hydrophobic nature and low titer levels. The low titer levels make the final recovery process expensive and energy intensive. The surfacing of new technologies for the recovery of butanol could allow it to emerge as a candidate fuel to tackle the global energy crisis.

The recovery methods discussed in this chapter include the conventional distillation, adsorption, gas stripping, liquid–liquid extraction, perstraction, pervaporation, and supercritical fluid extraction. The adsorption using molecular sieves and resins is attractive as the energy requirement is less compared to gas stripping and pervaporation. However, low selectivity, the high price of resins, and the cost for their regeneration after extensive operations are the factors for consideration. However, gas stripping has been found to be simple, highly efficient with reduced cell toxicity, and high substrate utilization. The condensation is performed prior to recovering the products and recycling the carrier gas. Moreover, the carrier gases used include CO$_2$ and H$_2$, which are produced during the biphasic ABE fermentation; hence, they do not interfere with the butanol-producing clostridia. Liquid–liquid extraction uses an extractant and separates the products based on the differences between the distribution coefficients of the solvents. However, several drawbacks include the toxicity of the extractant toward the bacterial cells and emulsion formation leading to separation problems. Nevertheless, liquid–liquid extraction is highly selective to butanol and requires less energy input.

Perstraction is a variation of liquid–liquid extraction in the way that a permeable membrane separates the cell culture from the extracting solvent. This prevents the issues of toxicity and emulsion formation. However, the membranes are usually expensive and often suffer from clogging and fouling due to the bacterial cells. Pervaporation uses the principle of membrane permeation using nonporous membrane followed by evaporation to dehydrate the organic solvent and separate the products. It is less energy intensive and moderately selective compared to liquid–liquid extraction and adsorption. Supercritical fluid, especially $SCCO_2$, has been primordially tested and found to have high butanol selectivity. Although in its infancy, the technology is promising due to its cost-effectiveness, recyclability, and feasibility for larger scale recovery techniques. The advancements in butanol recovery technologies could expedite the development of butanol as a next-generation biofuel for flexible use in the existing vehicular engines and power generation systems.

Acknowledgment

The authors thank the funding provided by the Natural Sciences and Engineering Research Council of Canada and Canada Research Chair program for conducting this bioenergy research.

References

Bryant, D. L., and H. P. Blaschek. 1988. Buffering as a means for increasing growth and butanol production by *Clostridium acetobutylicum*. *J. Ind. Microbiol.* 3:49–55.

Butamax™. 2015. A joint venture between BP and DuPont. http://www.butamax.com (accessed September 29, 2015).

Chen, W., H. Yu, Y. Liu, P. Chen, M. Zhang, and Y. Hai. 2011. Individualization of cellulose nanofibers from wood using high-intensity ultrasonication combined with chemical pretreatments. *Carbohydr. Polymer.* 83:1804–1811.

Chiao, J.-S., and Z.-H. Sun. 2007. History of the acetone-butanol-ethanol fermentation industry in China: Development of continuous production technology. *J. Mol. Microbiol. Biotechnol.* 13:12–14.

Correa, A. C., E. de Morais Teixeira, L. A. Pessan, and L. H. C. Mattoso. 2010. Cellulose nanofibers from curaua fibers. *Cellulose.* 17:1183–1192.

Dai, S., Y. H. Ju, and C. E. Barnes. 1999. Solvent extraction of strontium nitrate by a crown ether using room-temperature ionic liquids. *J. Chem. Soc., Dalton Trans.* 8:1201–1202.

Dien, B. S., H. J. G. Jung, K. P. Vogel, M. D. Casler, J. F. S. Lamb, L. Iten, R. B. Mitchell, and G. Sarath. 2006. Chemical composition and response to dilute-acid pretreatment and enzymatic saccharification of alfalfa, reed canarygrass, and switchgrass. *Biomass Bioenerg.* 30:880–891.

Dürre, P. 2007. Biobutanol: An attractive biofuel. *Biotechnol. J.* 2:1525–1534.

Evans, P. J., and H. Y. Wang. 1988. Enhancement of butanol formation by *Clostridium acetobutylicum* in the presence of decanol-oleyl alcohol mixed extractants. *Appl. Environ. Microbiol.* 54:1662–1667.

Ezeji, T. C., P. M. Karcher, N. Qureshi, and H. P. Blaschek. 2005. Improving performance of a gas stripping-based recovery system to remove butanol from *Clostridium beijerinckii* fermentation. *Bioprocess Biosyst. Eng.* 27:207–214.

Ezeji, T. C., N. Qureshi, and H. P. Blaschek. 2003. Production of acetone butanol ethanol by *Clostridium beijerinckii* BA101 and *in situ* recovery by gas stripping. *World J. Microb. Biot.* 19:595–603.

Ezeji, T. C., N. Qureshi, and H. P. Blaschek. 2004a. Butanol fermentation research: Upstream and downstream manipulations. *Chem. Rec.* 4:305–314.

Ezeji, T. C., N. Qureshi, and H. P. Blaschek. 2004b. Acetone butanol ethanol (ABE) production from concentrated substrate: Reduction in substrate inhibition by fed-batch technique and product inhibition by gas stripping. *Appl. Microbiol. Biotechnol.* 63:653–658.

Ezeji, T., N. Qureshi, and H. P. Blaschek. 2007a. Butanol production from agricultural residues: Impact of degradation products on *Clostridium beijerinckii* growth and butanol fermentation. *Biotechnol. Bioeng.* 97:1460–1469.

Ezeji, T. C., N. Qureshi, and H. P. Blaschek. 2007b. Production of acetone butanol (AB) from liqueWed corn starch, a commercial substrate, using *Clostridium beijerinckii* coupled with product recovery by gas stripping. *J. Ind. Microbiol. Biotechnol.* 34:771–777.

Ezeji, T. C., N. Qureshi, and H. P. Blaschek. 2007c. Bioproduction of butanol from biomass: From genes to bioreactors. *Curr. Opin. Biotechnol.* 18:220–227.

Fang, Z., T. Minowa, C. Fang, R. L. Smith Jr., H. Inomata, and J. A. Kozinski. 2008a. Catalytic hydrothermal gasification of cellulose and glucose. *Int. J. Hydrogen Energ.* 33:981–990.

Fang, Z., T. Sato, R. L. Smith Jr., H. Inomata, K. Arai, J. A. Kozinski. 2008b. Reaction chemistry and phase behavior of lignin in high-temperature and supercritical water. *Bioresour. Technol.* 99:3424–3430.

Festel, G. W. 2008. Biofuels—Economic aspects. *Chem. Eng. Technol.* 31:715–720.

Godin, C., and J. M. Engasser. 1990. Two-stage continuous fermentation of *Clostridium acetobutylicum*: Effects of pH and dilution rate. *Appl. Microbiol. Biotechnol.* 33:269–273.

Graham-Rowe, D. 2011. Agriculture: Beyond food versus fuel. *Nature* 474:S6–S8.

Griffith, W. L., A. L. Compere, and J. M. Googin. 1983. Novel neutral solvents fermentation. *Dev. Ind. Microbiol.* 24:347–352.

Griffith, W. L., A. L. Compere, and J. M. Googin. 1984. 1-Butanol extraction with vegetable oil fatty acid esters. *Dev. Ind. Microbiol.* 25:795–800.

Groot, W. J., R. G. J. M. van der Lans, and K. Ch. A. M. Luyben. 1989. Batch and continuous butanol fermentations with free cells: Integration with product recovery by gas-stripping. *Appl. Microbiol. Biotechnol.* 32:305–318.

Guo, G. L., W. H. Chen, W. H. Chen, L. C. Men, and W. S. Hwang. 2008. Characterization of dilute acid pretreatment of silvergrass for ethanol production. *Bioresour. Technol.* 99:6046–6053.

Guo-qing, H., X. Hao-ping, C. Qi-he, and R. Hui. 2005. Optimization of conditions for supercritical fluid extraction of flavonoids from hops (*Humulus lupulus* L.). *J. Zhejiang Univ. Sci.* 6B:999–1004.

Ha, S. H., N. L. Mai, and Y. M. Koo. 2010. Butanol recovery from aqueous solution into ionic liquids by liquid–liquid extraction. *Process Biochem.* 45:1899–1903.

Ha, S. J., J. M. Galazka, S. R. Kim, J. H. Choi, X. Yang, J. H. Seo, N. L. Glass, J. H. D. Cate, and Y. S. Jin. 2011. Engineered *Saccharomyces cerevisiae* capable of simultaneous cellobiose and xylose fermentation. *PNAS* 108:504–509.

Holtzapple, M. T., and R. F. Brown. 1994. Conceptual design for a process to recover volatile solutes from aqueous-solutions using silicalite. *Sep. Technol.* 4:213–229.

Ishizak, A., S. Michiwaki, E. Crabbe, G. Kobayashi, K. Sonomoto, and S. Yoshino. 1999. Extractive acetone-butanol-ethanol fermentation using methylated crude palm oil as extractant in batch culture of *Clostridium saccaroperbutylacetonicum* N1-4 (ATCC 13564). *J. Biosci. Bioeng.* 87:352–356.

Jones, D. T. 2001. Applied acetone-butanol fermentation. In *Clostridia. Biotechnology and Medical Applications*, ed. H. Bahl, and P. Durre, 125–168. Weingheim, Germany: Wiley-VCH Verlag.

Jones, D. T., and D. R. Woods. 1986. Acetone-butanol fermentation revisited. *Microbiol. Rev.* 50:484–524.

Karimi, K., S. Kheradmandinia, and M. J. Taherzadeh. 2006. Conversion of rice straw to sugars by dilute-acid hydrolysis. *Biomass Bioenerg.* 30:247–253.

Keis, S., R. Shaheen, and D. T. Jones. 2001. Emended descriptions of *Clostridium acetobutylicum* and *Clostridium beijerinckii*, and descriptions of *Clostridium saccharoperbutylacetonicum* sp. nov. and *Clostridium saccharobutylicum* sp. nov. *Int. J. Syst. Evol. Microbiol.* 51:2095–2103.

Köpke, M., and P. Dürre. 2011. Biochemical production of biobutanol. In *Handbook of Biofuels Production: Processes and Technologies*, ed. R. Luque, J. Campelo, and J. Clark, 221–257. Cambridge: Woodhead Publishing Limited.

Kruse, A. 2008. Supercritical water gasification. *Biofuels, Bioprod. Bioref.* 2:415–437.

Laitinen, A., and J. Kaunisto. 1999. Supercritical fluid extraction of 1-butanol from aqueous solutions. *J. Supercrit. Fluids* 15:245–252.

Lane, J. 2013. GranBio and Rhodia ink pact for biobased n-butanol, in Brazil, from bagasse. Biofuel Digest. http://www.biofuelsdigest.com/bdigest/2013/08/12/granbio-and-rhodia-ink-pact-for-biobased-n-butanol-in-brazil-from-bagasse (accessed September 29, 2015).

Lavarack, B. P., G. J. Griffin, and D. Rodman. 2002. The acid hydrolysis of sugarcane bagasse hemicellulose to produce xylose, arabinose, glucose and other products. *Biomass Bioenerg.* 23:367–380.

Lee, S. Y., J. H. Park, S. H. Jang, L. K. Nielsen, J. Kim, and K. S. Jung. 2008. Fermentative butanol production by Clostridia. *Biotechnol. Bioeng.* 101:209–228.

Li, Q., H. Cai, B. Hao, C. L. Zhang, Z. N. Yu, S. D. Zhou, and C. J. Liu. 2010. Enhancing clostridial acetone–butanol–ethanol (ABE) production and improving fuel properties of ABE-enriched biodiesel by extractive fermentation with biodiesel. *Appl. Biochem. Biotechnol.* 162:2381–2386.

Liu, S., and N. Qureshi. 2009. How microbes tolerate ethanol and butanol? *New Biotechnol.* 26:117–121.

Lloyd, T. A., and C. E. Wyman. 2005. Combined sugar yields for dilute sulfuric acid pretreatment of corn stover followed by enzymatic hydrolysis of the remaining solids. *Bioresour. Technol.* 96:1967–1977.

Lu, C., J. Zhao, S.-T. Yang, and D. Wei. 2012. Fed-batch fermentation for *n*-butanol production from cassava bagasse hydrolysate in a fibrous bed bioreactor with continuous gas stripping. *Bioresour. Technol.* 104:380–387.

Maddox, I. S. 1989. The acetone-butanol-ethanol fermentation: Recent progress in technology. *Biotechnol. Genet. Eng. Rev.* 7:189–220.

Marchal, R., M. Rebeller, and J. P. Vandecasteele. 1984. Direct bioconversion of alkali-pretreated straw using simultaneous enzymatic hydrolysis and acetone butanol production. *Biotechnol. Lett.* 6:523–528.

Marzialetti, T., M. B. V. Olarte, C. Sievers, T. J. C. Hoskins, P. K. Agrawal, and C. W. Jones. 2008. Dilute acid hydrolysis of Loblolly pine: A comprehensive approach. *Ind. Eng. Chem. Res.* 47:7131–7140.

Matsumura, M., H. Kataoka, M. Sueki, and K. Araki. 1988. Energy saving effect of pervaporation using oleyl alcohol liquid membrane in butanol purification. *Bioprocess Eng.* 3:93–100.

McKendry, P. 2002. Energy production from biomass (part 1): Overview of biomass. *Bioresour. Technol.* 83:37–46.

Naik, S., V. V. Goud, P. K. Rout, and A. K. Dalai. 2010. Supercritical CO_2 fractionation of bio-oil produced from wheat-hemlock biomass. *Bioresour. Technol.* 101:7605–7613.

Nanda, S., R. Azargohar, A. K. Dalai, and J. A. Kozinski. 2015b. An assessment on the sustainability of lignocellulosic biomass for biorefining. *Renew. Sust. Energ. Rev.* 50:925–941.

Nanda, S., A. K. Dalai, and J. A. Kozinski. 2014a. Butanol and ethanol production from lignocellulosic feedstock: Biomass pretreatment and bioconversion. *Energ. Sci. Eng.* 2:138–148.

Nanda, S., J. Maley, J. A. Kozinski, and A. K. Dalai. 2015a. Physico-chemical evolution in lignocellulosic feedstocks during hydrothermal pretreatment and delignification. *J. Biobased Mater. Bioenerg.* 9:295–308.

Nanda, S., J. Mohammad, S. N. Reddy, J. A. Kozinski, and A. K. Dalai. 2014b. Pathways of lignocellulosic biomass conversion to renewable fuels. *Biomass Conv. Bioref.* 4:157–191.

Nanda, S., P. Mohanty, K. K. Pant, S. Naik, J. A. Kozinski, and A. K. Dalai. 2013. Characterization of North American lignocellulosic biomass and biochars in terms of their candidacy for alternate renewable fuels. *Bioenerg. Res.* 6:663–677.

Nanda, S., S. N. Reddy, H. N. Hunter, I. S. Butler, and J. A. Kozinski. 2015d. Supercritical water gasification of lactose as a model compound for valorization of dairy industry effluents. *Ind. Eng. Chem. Res.* 54:9296–9306.

Nanda, S., S. N. Reddy, H. N. Hunter, A. K. Dalai, and J. A. Kozinski. 2015c. Supercritical water gasification of fructose as a model compound for waste fruits and vegetables. *J. Supercrit. Fluids* 104:112–121.

Ni, Y., and Z. Sun. 2009. Recent progress on industrial fermentative production of acetone–butanol–ethanol by *Clostridium acetobutylicum* in China. *Appl. Microbiol. Biotechnol.* 83:415–423.

Oudshoorn, A., L. A. M. van der Wielen, and A. J. J. Straathof. 2009. Assessment of options for selective 1-Butanol recovery from aqueous solution. *Ind. Eng. Chem. Res.* 48:7325–7336.

Parekh, S. R., R. S. Parekh, and M. Wayman. 1988. Ethanol and butanol production by fermentation of enzymatically saccharified SO_2-prehydrolysed lignocellulosics. *Enzyme Microbial. Technol.* 10:660–668.

Peker, H., M. P. Srinivasan, J. M. Smith, and B. J. McCoy. 1992. Caffeine extraction rates from coffee beans with supercritical carbon dioxide. *AIChE J.* 38:761–770.

Philips, J. A., and A. E. Humphrey. 1985. Microbial production of energy: Liquid fuels. In *Biotechnology and Bioprocess Engineering*, ed. T. K. Ghose, 157–186. New Delhi, India: United India Press.

Qureshi, N., and H. P. Blaschek. 1999. Butanol recovery from model solution/fermentation broth by pervaporation: evaluation of membrane performance. *Biomass Bioenerg.* 17:175–184.

Qureshi, N., and H. P. Blaschek. 2000. Economics of butanol fermentation using hyper-butanol producing *Clostridium beijerinckii* BA101. *T. I. Chem. Eng.* 78:139–144.

Qureshi, N., and H. P. Blascheck. 2005. Butanol production from agricultural biomass. In *Food Biotechnology*, ed. A. Pometto, K. Shetty, G. Paliyath, and R. E. Levin, 525–549. Boca Raton, FL: CRC Press.

Qureshi, N., and T. C. Ezeji. 2008. Butanol, "a superior biofuel" production from agricultural residues (renewable biomass): Recent progress in technology. *Biofuels, Bioprod. Bioref.* 2:319–330.

Qureshi, N., T. C. Ezeji, J. Ebener, B. S. Dien, M. A. Cotta, and H. P. Blaschek. 2008b. Butanol production by *Clostridium beijerinckii*. Part I: Use of acid and enzyme hydrolyzed corn fiber. *Bioresour. Technol.* 99:5915–5922.

Qureshi, N., S. Hughes, I. S. Maddox, and M. A. Cotta. 2005. Energy-efficient recovery of butanol from model solutions and fermentation broth by adsorption. *Bioprocess Biosyst. Eng.* 27:215–222.

Qureshi, N., and I. S. Maddox. 2005. Reduction in butanol inhibition by perstraction: Utilization of concentrated lactose/whey permeate by *Clostridium acetobutylicum* to enhance butanol fermentation economics. *Food Bioprod. Process.* 83:43–52.

Qureshi, N., I. S. Maddox, and A. Friedl. 1992. Application of continuous substrate feeding to the ABE fermentation: Relief of product inhibition using extraction, perstraction, stripping and pervaporation. *Biotechnol. Prog.* 8:382–390.

Qureshi, N., B. C. Saha, and M. A. Cotta. 2007. Butanol production from wheat straw hydrolysate using *Clostridium beijerinckii*. *Bioprocess Biosyst. Eng.* 30:419–427.

Qureshi, N., B. C. Saha, B. Dien, R. E. Hector, and M. A. Cotta. 2010a. Production of butanol (a biofuel) from agricultural residues: Part I—Use of barley straw hydrolysate. *Biomass Bioenerg.* 34:559–565.

Qureshi, N., B. C. Saha, R. E. Hector, B. Dien, S. Hughes, S. Liu, L. Iten, M. J. Bowman, G. Sarath, and M. A. Cotta. 2010b. Production of butanol (a biofuel) from agricultural residues: Part II – Use of corn stover and switchgrass hydrolysates. *Biomass Bioenerg.* 34:566–571.

Qureshi, N., B. C. Saha, R. E. Hector, S. R. Hughes, and M. A. Cotta. 2008a. Butanol production from wheat straw by simultaneous saccharification and fermentation using *Clostridium beijerinckii*: Part I—Batch fermentation. *Biomass Bioenerg.* 32:168–75.

Reddy, S. N., S. Nanda, A. K. Dalai, and J. A. Kozinski. 2014. Supercritical water gasification of biomass for hydrogen production. *Int. J. Hydrogen Energ.* 39:6912–6926.

Roberto I. C., S. I. Mussatto, and R. C. L. B. Rodrigues. 2003. Dilute-acid hydrolysis for optimization of xylose recovery from rice straw in a semi-pilot reactor. *Ind. Crop. Prod.* 17:171–176.

Rodriguez-Chong, A., J. A. Ramirez, G. Garrote, and M. Vazquez. 2004. Hydrolysis of sugar cane bagasse using nitric acid. *J. Food Eng.* 61:143–152.

Roffler, S. R., H. W. Blanch, and C. R. Wilke. 1987. *In-situ* recovery of butanol during fermentation. Part I: Batch extractive fermentation. *Bioprocess Eng.* 2:1–12.

Rout, P. K., M. K. Naik, S. N. Naik, V. V. Goud, L. M. Das, and A. K. Dalai. 2009. Supercritical CO_2 fractionation of bio-oil produced from mixed biomass of wheat and wood sawdust. *Energ. Fuels* 23:6181–6188.

Saha, B. C., L. B. Iten, M. A. Cotta, and Y. V. Wu. 2005. Dilute acid pretreatment, enzymatic saccharification and fermentation of wheat straw to ethanol. *Process Biochem.* 40:3693–3700.

Sanderson, K. 2011. Lignocellulose: A chewy problem. *Nature* 474:S12–S14.

Sauer, U., R. Fischer, and P. Dürre, 1993. Solvent formation and its regulation in strictly anaerobic bacteria. *Curr. Top. Mol. Genet.* 1:337–351.

Sluiter, A., B. Hames, R. Ruiz, C. Scarlata, J. Sluiter, D. Templeton, and D. Crocker. 2011. Determination of structural carbohydrates and lignin in biomass. Technical Report NREL/TP-510-42618. National Renewable Energy Laboratory (NREL), Golden, CO.

Soni, B. K., K. Das, and T. K. Ghose. 1982. Bioconversion of agro-wastes into acetone butanol. *Biotechnol. Lett.* 4:19–22.

Sun, Y., and J. J. Cheng. 2005. Dilute acid pretreatment of rye straw and bermudagrass for ethanol production. *Bioresour. Technol.* 96:1599–1606.

Technical Association of the Pulp and Paper Industry (TAPPI). 2011. TAPPI Test Method T222 om-88. Acid-insoluble lignin in wood and pulp. TAPPI, Atlanta, GA.

Traxler, R. W., E. M. Wood, J. Mayer, and M. P. Wilson. 1985. Extractive fermentation for the production of butanol. *Dev. Ind. Microbiol.* 26:519–525.

United States Energy Information Administration (USEIA). 2011. International Energy Outlook 2011. http://www.eia.gov/forecasts/ieo/pdf/0484 (2011).pdf (accessed January 3, 2012).

Vane, L. M. 2005. A review of pervaporation for product recovery from biomass fermentation processes. *J. Chem. Technol. Biotechnol.* 80:603–629.

Vane, L. M. 2008. Separation technologies for the recovery and dehydration of alcohols from fermentation broths. *Biofuels, Bioprod. Bioref.* 2:553–588.

Wang, D. I. C., C. L. Cooney, A. L. Demain, R. F. Gomez, and A. J. Sinskey. 1979. Production of acetone and butanol by fermentation. MIT Quarterly Report to the US Department of Energy, COG-4198-9, 141–149.

Wayman, N., and R. Parekh. 1987. Production of acetone-butanol by extractive fermentation using dibutylphthalate as extractant. *J. Ferment. Technol.* 65:295–300.

Wee, S. L., C. T. Tye, and S. Bhatia. 2008. Membrane separation process—Pervaporation through zeolite membrane. *Sep. Purif. Technol.* 63:500–516.

Wijmans, J. G., and R. W. A. Baker. 1993. Simple Predictive treatment of the permeation process in pervaporation. *J. Membr. Sci.* 79:101–113.

Xue, C., G. Q. Du, J. X. Sun, L. J. Chen, S. S. Gao, M. L. Yu, S. T. Yang, and F. W. Bai. 2014a. Characterization of gas stripping and its integration with acetone–butanol–ethanol fermentation for high-efficient butanol production and recovery. *Biochem. Eng. J.* 83:55–61.

Xue, C., J.-B. Zhao, L.-J. Chen, F.-W. Bai, S.-T. Yang, and J.-X. Sun. 2014b. Integrated butanol recovery for an advanced biofuel: Current state and prospects. *Appl. Microbiol. Biotechnol.* 98:3463 3474.

Zheng, Y. N., L. Z. Li, M. Xian, Y. J. Ma, J. M. Yang, X. Xu, and D. Z. He. 2009. Problems with the microbial production of butanol. *J. Ind. Microbiol. Biotechnol.* 36:1127–1138.

Zhu, J. Y., X. Pan, and R. S. Zalesny Jr. 2010. Pretreatment of woody biomass for biofuel production: energy efficiency, technologies, and recalcitrance. *Appl. Microbiol. Biotechnol.* 87:847–857.

Zverlov, V. V., O. Berezina, G. A. Velikodvorskaya, and W. H. Schwartz. 2006. Bacterial acetone and butanol production by industrial fermentation in the Soviet Union: Use of hydrolyzed agricultural waste for biorefinery. *Appl. Microbiol. Biotechnol.* 71:587–597.

8

Oil from Algae

Prasenjit Mondal, Preety Kumari, Jyoti Singh, Shobhit Verma,
Amit Kumar Chaurasia, and Rajesh P. Singh

CONTENTS

ABSTRACT   Utilization of algal biomass for the production of various types of oils such as biodiesel and biojet fuel has attracted research attention around the world. This chapter provides a comprehensive review on the algal types and their suitability for oil production, algal biochemistry, and monitoring of the cellular components of microalgae, different techniques on cultivation and harvesting of algal biomass, oil extraction, and its upgradation techniques, standards for biodiesel and its stability issue, world scenario on the biodiesel production form algal biomass. It also reviews the biojet fuel production from algal biomass.

8.1 Introduction

Increasing crude oil prices, energy crisis, accelerated global warming, and so on, have led into renewed interest in exploring renewable energy resources such as solar energy, wind, biomass, and wastes, around the world to produce sustainable energy (Boussiba et al. 2012). It is estimated that ~50% of the energy demand of the world is likely to be met by renewable energy by 2050 (Mata et al. 2010). Amongst various renewable energy sources, the biomass-based fuel is widely used to meet the domestic energy requirement in most of the developing countries. Recently, it has been reported that biomass is used as the main energy resource by about 1.5 billion of the world population (Chisti 2007). It is also assumed that both European Union and USA will have higher biomass consumption respectively to about 700% and ~300%, respectively, by the year 2020 compared to its usage in the year 2010 (Kheira and Atta 2009).

Different types of biomass sources such as sugarcane, corn, soybeans, jatropha seeds, edible, and nonedible oils, lignocellulosic biomass, macro-, and microalgae have been used for the production of biofuel/oil. Amongst these feedstocks, the most promising resource for oil production is algae, because it has the highest oil content and productivity (Mata et al. 2010) (Table 8.1). Recently, microalgae have been recognized as a potential resource for the production of bioethanol, biodiesel, green gasoline, methanol, bio hydrogen, and

TABLE 8.1

Comparison of Microalgae with Other Biodiesel Feedstocks

Plant Source	Seed Oil Content (% oil w/w in biomass)	Oil Volumetric Productivity (L oil/ha year)	Land Use (m² year/kg biodiesel)	Biodiesel Productivity (kg biodiesel/ha year)
Soybean (*Glycine max* L.)	18	636	18	562
Jatropha (*Jatropha curcas* L.)	28	741	15	656
Microalgae (low oil content)	30	58,700	0.2	51,927
Hemp (*Cannabis sativa* L.)	33	363	31	321
Palm oil (*Elaeis guineensis*)	36	5366	2	4747
Sunflower (*Helianthus annuus* L.)	40	1070	11	946
Canola/Rapeseed (*Brassica napus* L.)	41	974	12	862
Physic nut	41–59	741	—	656
Camelina (*Camelina sativa* L.)	42	915	12	809
Corn/Maize (*Zea mays* L.)	44	172	66	152
Castor (*Ricinus communis*)	48	1307	9	1156
Microalgae (medium oil content)	50	97,800	0.1	86,515
Microalgae (high oil content)	70	136,900	0.1	121,104

Source: Data from Mata, T.M. et al., *Renew. Sustain. Energy Rev.*, 14, 217–232, 2010.

so on, because microalgae can be productively cultivated as well as it can transform solar energy to chemical energy through carbon dioxide fixation (Wang et al. 2014). Biodiesel as compared to fossil fuel have been found to be biodegradable, nontoxic, and hence eco-friendly compared to the fossil fuel.

There is similarity between lower plants and algae because both of these are simple organisms, but lack of several distinct organs (stomata, xylem, and phloem) that are present in higher plants and typically produce their food through photosynthesis. Algae can be the macroalgae, seaweeds, and microalgae. They can be unicellular such as *Chlorella* as well as multicellular such as brown algae. Seaweeds are the most complex and largest marine algae. It has been shown that organic carbon absorption is possible by some algae through phagotrophy, myzotrophy, or osmotrophy; thus, these solely do not depend on photosynthesis. The growth of the microalgae can be autotrophic through photosynthesis, heterotrophic through growth in the dark using sugar/starch, or mixotrophic through both of these growth modes. Algae are classified into a number of subgroups, based on the types of pigments they use for photosynthesis.

Rhodophyta (red algae): Phyllum of kingdom protista and generally called red algae because of the purplish color or characteristic red due to the presence of accessory pigments called phycobilins. The photosynthetic pigments called *a* and *d* chlorophylls as well as their pigment (accessory) are phycobilins, xanthophyll, and carotenoids. The example includes (of red algae) *Porolithon* sp., also called as coralline algae, a filamentous species such as *Pleonosporum* sp., and thalloid species such as *Chondruscrispus* (Irish moss).

Phaeophyta (brown algae): Phylum of kingdom protista and commonly known as brown algae. They contain carotenoid pigments called fucoxanthin, in the cell chloroplasts, which imparts the characteristic brown color. The photosynthetic pigments of brown algae are chlorophylls *a* and *c*, and their accessory pigments are xanthophylls and carotenoids, including fucoxanthin. Many types of seaweed such as sargassum weed (*Sargassum* sp.), kelps (*Laminaria* sp.), and giant kelps (*Marcocystis* sp. and *Nereocystis* sp.) belong to phaeophyta.

Chlorophyta (green algae): Division of the kingdom protista. They contain chlorophyll *a* and *b*, as photosynthetic pigments and carotenoids and xanthophyll as their accessory pigments are. Unicellular genera *Chlamydomonas* and *Chlorella* are the best some examples of green algae, complex green algae genera such as *Gonium* and *Volvox*, and filamentous species such as *Cladophora* and *Codium magnum*.

Cyanobacteria (blue-green algae): Phylum of prokaryotic aquatic bacteria that can perform photosynthesis and is found in moist terrestrial and aquatic habitats. They are referred to as blue-green algae, although they are not related to the algal groups. Most cyanobacteria contain chlorophyll *a*, which together with phycobilins give the cells a characteristic blue-green to grayish-brown color (e.g., *Cyanophyceae*, *Synechococcales*, *Prochlorophyta*). A few genera, however, lack phycobilins and have chlorophyll *b* instead giving them a bright green color (e.g., *Prochloron*, *Prochlorococcus*, *Prochlorothrix*).

Algal cells of the above-mentioned types possess different composition in terms of protein, fat(lipid)/oil and carbohydrates. Algal cells containing more lipids are more suitable for bio-oil/biodiesel production. Out of the above-mentioned types, the blue-green alga contains more lipid content and hence is more suitable for oil production than other types (Sakthivel 2011). Some advantages of microalgae as a potential biomass resource include:

- Microalgae are very efficient biological system for the production of biofuels and other organic compounds by using solar energy.
- Compared with terrestrial plants, microalgae can grow 10–50 times faster, resulting in an extremely high CO_2 fixation rate and are thus now considered as one of the most capable sources of oil for making biodiesel.
- Many algal strains have lipid content ranging of 15%–35% dry weight of biomass.
- Microalgae are able to grow year round in diverse conditions. Many species of microalgae thrive in salt water.
- Microalgae grow in aqueous nutrient media, but require less water than terrestrial crops, therefore reduce the load on fresh water resources.
- Microalgae can also be cultivated using seawater or brackish water and therefore, do not require extensive land resources, reducing associated impacts also, whereas not compromising the production of fodder, food, and other products derived from the terrestrial crops.
- Microalgal cultivation does not require pesticides or herbicides application.
- Microalgae could also be induced to produce valuable co-products such as protein, carbohydrates, residual biomass, and pigments, which can be used as a feed for other organism or fertilizer.
- Microalgal biomass production system can easily be adapted to various levels of operational and technological skills.

In the presence of light, algae utilize carbon dioxide and convert it into carbohydrates, which produce proteins, fats, and oils through various pathways. The lipid content of algal cell, the main component for oil production, varies from species to species. Nutrient (C, N, etc.) concentration, temperature, pH, light intensity, and CO_2 concentration in the medium also influence the lipid content in a cell. Further, the nature of fatty acids in algal biomass also varies depending on the growth conditions and type of nutrients. The algal species having high lipid content live in both saline and river water and grow naturally. Lipid productivity and lipid content of some microalgal species are provided in Table 8.2.

Use of microalgae for biofuel production has its pros and cons, as the production cost are still not competitive with fossil fuel therefore microalgal strain development is one of the most significant aspects of biofuel development simultaneously with other efforts including dewatering, harvesting, and oil extraction approaches (Marjakangas et al. 2015). Strain development typically includes metabolic engineering and genetic engineering along with the focusing on other parameters such as increased biomass productivity, fast growth, and increased lipid productivity (Ahmad et al. 2011). Recently, novel strategies have been reported to overcome the challenges therein the conventional approaches and to achieve maximum possible outcomes in terms of lipid yields, sustainability, and cost-effectiveness. These strategies include a combination of stress factors, co-culturing with other microorganisms, addition of phytohormones, and chemical additives.

This chapter mainly enumerates on microalgae and how these could be a potential bioresource for biodiesel production. In this, we have elaborated upon the entire biodiesel production process, ranging from the cultivation and harvesting of microalgal biomass to the extraction and the biodiesel production along with the approaches to increase biodiesel

TABLE 8.2

Biomass Productivity, Lipid Content, and Lipid Productivity of the Microalgae Strains

Algal Group	Microalgal Strains	Habitat	Biomass Productivity (g/L/day)	Lipid Content (% biomass)	Lipid Productivity (mg/L/day)	References
Diatoms	*Chaetoceros muelleri F&M-M43*	Marine	0.07	33.6	21.8	Sakthivel 2011
	Chaetoceros calcitrans CS 178	Marine	0.04	39.8	17.6	Sakthivel 2011
	P. tricornutum F&M-M40	Marine	0.24	18.7	44.8	Sakthivel 2011
	Skeletonomacostatum CS 181	Marine	0.08	21.0	17.4	Sakthivel 2011
	Thalassioira pseudonana CS 173	Marine	0.08	20.6	17.4	Sakthivel 2011
	Chlorella vulgaris F&M-M49	Freshwater	0.20	18.4	36.9	Sakthivel 2011
	Chlorella sorokiniana IAM-212	Freshwater	0.23	19.3	44.7	Sakthivel 2011
	Chlorella sp. F&M-M48	Freshwater	0.23	18.6	42.1	Sakthivel 2011
	Chlorella vulgaris CCAP 211/11b	Freshwater	0.17	19.2	32.6	Sakthivel 2011
Green Algae	*Chlorococcum sp. UMACC 112*	Freshwater	0.28	19.3	53.7	Sakthivel 2011
	Scenedemus quadricauda	Freshwater	0.19	18.4	35.1	Sakthivel 2011
	Scenedemus F&M-M19	Freshwater	0.21	19.6	40.8	Sakthivel 2011
	Scenedemus sp. DM	Freshwater	0.26	21.1	53.9	Sakthivel 2011
	T. suecica F&M-M33	Marine	0.32	8.5	27.0	Sakthivel 2011
	Tetraselmis sp. F&M-M34	Marine	0.30	14.7	43.4	Sakthivel 2011
	Monodus subterraneus UTEX 151	Freshwater	0.19	16.1	30.4	Sakthivel 2011
	T. suecica F&M-M35	Marine	0.28	12.9	36.4	Mata et al. 2010
	Tetraselmis suecica	Marine	0.12–0.32	8.5–23.0	27.0–36.4	
	Ellipsoidion sp. F&M-M31	Marine	0.17	27.4	47.3	Sakthivel 2011
	Nannochloropsis sp. CS 246	Marine	0.17	29.2	49.7	Mata et al. 2010
	Nannochloropsissp	Freshwater	0.71–1.43	12.0–53.0	37.6–90.0	Talebi et al. 2015
	Dunaliella salina	Freshwater	0.05	18.9	10.26	Talebi et al. 2015
	D. salina (UTEX)	Freshwater	0.15	24.0	36.48	Mata et al. 2010
	Chlorella emersonii	Freshwater	0.036–0.041	25.0–63.0	10.3–50.0	Sakthivel 2011
	Chlorella vulgaris	Freshwater	0.02–0.2	5.0–58.0	11.2–40.0	Sakthivel 2011
Eustigmatophytes	*Nannochloropsis sp. F&M-M26*	Marine	0.21	29.6	61.9	Sakthivel 2011
	Isochrysis sp. F&M-M37	Marine	0.14	27.4	37.8	Sakthivel 2011
Prymnesiophytes	*Pavlova salina CS 49*	Marine	0.16	30.9	49.4	Sakthivel 2011
	Pavlova lutheri CS 182	Marine	0.14	35.9	50.2	Sakthivel 2011
Red algae	*Porphyridium cruentum*	Marine	0.37	9.5	34.8	Sakthivel 2011

Source: Talebi, A.F. et al., *BioMed Res. Int.*, 2015, 597198, 2015; Mata, T.M. et al., *Renew. Sustain. Energy Rev.*, 14, 217–232, 2010; Sakthivel et al. *Journal of Experimental Sciences* 2, 29–49, 2011.

production in microalgae. The advantages of microalgae compared with other available feedstocks, presence of different fatty acids, lipid content, lipid productivity, and lipid biosynthesis in microalgae had been critically reviewed. World scenario on the production of biofuel from algal biomass has also been enumerated.

8.2 Algal Biochemistry

Fatty acids synthesized by algae are building blocks for different of lipids, which sequentially produces biodiesel. Similar to higher plants, the most commonly synthesized fatty acids have chain lengths ranging from C16 to C18 (Table 8.3) and an insight into algal biochemistry would be relevant for modulating the production of these lipids from the microalgae (Lv et al. 2013). Although the understanding of lipid metabolism in algae have progressed but lipid biochemistry of algae have not been fully understood. Therefore, algal lipid biochemistry needs a critical review to understand the lipid metabolism and regulation.

In algae, photosynthesis is the key and fundamental driving force that supports all biofuel synthesis processes, converting atmospheric CO_2 into biomass, carbon storage products (e.g., carbohydrates and lipids), and H_2 (Beer et al. 2009). Algal cells have subcellular compartments such as mitochondria, chloroplast, and cytoplasm. Lipid metabolism in algae occurs in chloroplast, where the light energy is incorporated into the cell in the form of light quanta, which is absorbed by the pigments to drive the photosynthetic

TABLE 8.3

Major Fatty Acids Present in Algal Strains Are Listed

Algal Class	Major Fatty Acids	Major Polyunsaturated Fatty Acids (PUFAs)
Bacillariophyceae	C14:0, C16:0, C16:1 and C18:1	C20:5 and C22:6
Chlorophyceae	C14:0, C16:0, C18:0 and C18:1	C18:2, C18:3 and C18:4
Euglenophyceae	C16:0 and C18:1	C18:2 and C18:3
Chryotophyceae	C16:0 and C20:1	C18:3, C18:4 and C20:5
Chrysophyceae	C16:0 and C16:1 and C18:1	C20:5, C22:5 and C22:6
Prasinophyceae	C14:0, C16:0 and C18:1	C18:2, C18:3, C20:4 and C20:5
Dinophyceae	C14:0, C16:0	C18:4, C18:5, C20:4 and C22:6
Prymnesiophyceae	C14:0, C16:0, C16:1, C18:0 and C18:1	C18:2, C18:3, C18:4 and C22:6
Phaeophyceae	C14:0, C16:0 and C18:1	C18:2, C18:3 and C20:4
Rhodophyceae	C16:0, C16:1 and C18:1	C18:2, C20:4 and C20:5
Xanthophyceae	C14:0, C16:0 and C16:1	C16:3 and C20:5
Cyanobacteria	C16:0 and C16:1 and C18:1	C18:2 and C18:3
Haptophyceae	C14:0, C16:0, C16:1 and C18:1	C20:5 and C22:6
Raphidophyceae	C14:0, C16:0 and C16:1	C18:2, C18:3, C18:4 and C20:5

Source: Data from Bigogno, C. et al., *Phytochemistry* 60, 497–503, 2002; Cagliari, A. et al., *Int. J. Plant Biol.*, 2, 10, 2011; Lang, I. et al., *BMC Plant Biol.*, 11, 1, 2011, all modified.
C22:6, Docosahexaenoic acid; C22:5, C20:5, Eicosapentaenoic acid; C20:4, Arachidonic acid; Docosapentaenoic acid; C20:3, Dihomo-gamma-linolenic acid; C20:1, Eicosenoic acid; C18:5, Octadecatetraenoic acid; C18:4, Parinaric acid; C18:3, Linolenic acid; C18:2, Linoleic acid; C18:1, Oleic acid; C16:3, Hexadecatrienoic acid; C:16:1, Palmitoleic acid; C16:0, Palmitic acid; C14, Myristic acid.

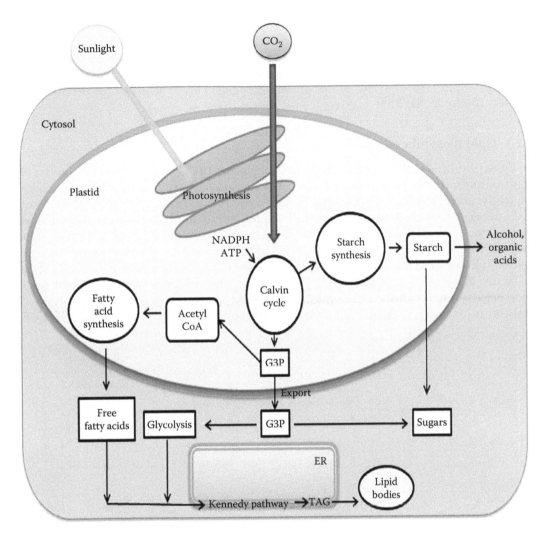

FIGURE 8.1
Metabolic pathways of algae rendered for biofuel production. (Modified from Radakovits, R. et al., *Metabol. Eng.*, 13, 89–95, 2011; Beer, L.L. et al., *Curr. Opin. Biotechnol.*, 20, 264–271, 2009.)

electron transport. During electron transport in the outer layers of thallakoids in chloroplast, NADPH is used to produce adenosine triphosphate (ATP) in the respiratory chain. The ATP generated in light reaction is used in the Calvin cycle, which is the primary pathway for CO_2 fixation as shown in Figure 8.1. The first step in the Calvin cycle is catalyzed by RubisCO enzyme where ATP and NADPH are used for conversion of CO_2 into 3-carbon molecule glycerate 3-phosphate (G3P). The G3P produced from the Calvin cycle is exported from the chloroplast to cytoplasm, where the carbon flow is basically divided into the sugar synthesis pathway and the glycolysis pathway for the formation of pyruvate that may also lead to Kennedy pathway occur in ER (endoplasmic reticulum) for triacylglycerol (TAG) biosynthesis (Sarkar and Shimizu 2015). The products generated from the Calvin cycle further goes into starch synthesis and fatty acid synthesis. The starch, sucrose, or other carbohydrates produced by the algal cells can be extracted and fermented to produce alcohol, organic acids, and other useful products. The conversion of G3P to pyruvate and after that to acetyl-CoA, via a reaction catalyzed by the pyruvate dehydrogenase complex (PDC), initiates the fatty acid biosynthetic pathway where free fatty acids (FFA) are synthesized and subsequently exported to ER where they are converted into TAGs by the action of ER enzymes. These TAG molecules get accumulated in to lipid bodies outside of ER. The numerous lipid bodies in the cytoplasm of algal cells can be utilized for production of biofuels (Bellou et al. 2014).

8.2.1 Lipid Biosynthesis

In algae, the lipids are basically classified into polar lipids, that is, phospholipids, and neutral lipids, that is, triglycerides. Polar lipids are used as structural component whereas the triglycerides are the main material in the production of biodiesel. The synthesis routes of triglycerides in algae mainly consist of the following steps: (i) the formation of acetyl coenzyme A (acetyl-CoA) in the cytoplasm; (ii) the elongation and de-saturation of carbon chain of fatty acids; and (c) the biosynthesis of triglycerides in the microalgal biological system (Huang et al. 2010). A simplified overview of lipid biosynthetic pathway is shown in Figure 8.2. Acetyl-CoA provided by photosynthetic reactions acts as a precursor for fatty acid synthesis in the chloroplast. Acetyl-CoA carboxylase (ACCase) catalyzes the first committed step in fatty acid biosynthesis by converting acetyl-CoA to malonyl-CoA.

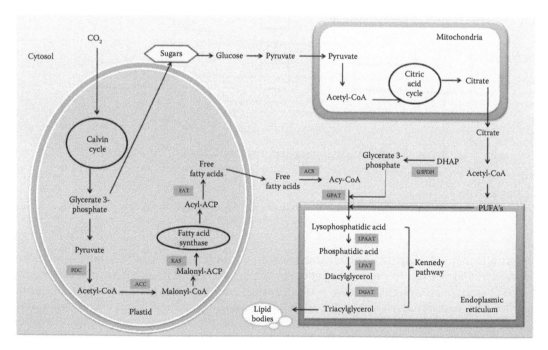

FIGURE 8.2
A simplified scheme showing lipid synthesis in microalgae. Abbreviations: ACC, acetyl-CoA carboxylase; ACP, acyl carrier protein; ACS, acyl-CoA synthetase; CoA, coenzyme A; DGAT, diacylglycerol acyltransferase; FAT, fatty acyl-ACP thioesterase; GPAT, glycerol-3-phosphate acyltransferase; G3PDH, glycerol-3-phosphate dehydrogenase; KAS, 3-ketoacyl-ACP synthase; LPAAT, lyso-phosphatidic acid acyltransferase; LPAT, lyso-phosphatidylcholine acyltransferase; PDC, pyruvate dehydrogenase complex. (Modified from Bellou, S., and G. Aggelis, *J. Biotechnol.*, 164, 318–329, 2013; Chen, J.E., and A.G. Smith, *J. Biotechnol.*, 162, 28–39, 2012; Radakovits, R. et al., *Metabol. Eng.*, 13, 89–95, 2011.)

Two types of ACCase are identified in algae, the heteromeric and homomeric. The heteromeric form is present in plastids whereas the homomeric form in cytosol. Further, the malonyl-CoA is transferred to the acyl carrier protein (ACP) by malonyl-CoA:ACP transacetylase, which then introduced in fatty acid synthesis cycle during the 3-ketoacyl-ACP synthase (KAS). The formation of 16- or 18-carbon chain of fatty acids is principally dependent on the reaction of two enzyme systems together with acetyl-CoA carboxylic enzyme along with fatty acid synthase in many algae. Fatty acid elongation occurs in ER and their synthesis requires specific classes of desaturases and enolases enzymes. The desaturates enzymes introduce double bonds at specific carbon atoms in the fatty acid chain in addition to the fatty acid elongation system that elongates the precursors in two-carbon increments (Gong and Jiang 2011). TAG biosynthesis and assembly is a complex process in algae can occur via the direct glycerol pathway, that is, Kennedy pathway where the acyltransferases residing in the ER catalyze sequential transport of the acyl group from the acetyl-CoA to glycerol 3-phosphate backbone and determine the final content of TAG as shown in Figure 8.2. TAG is then deposited into the cytosol as ER-derived lipid droplets (De Bhowmick et al. 2015).

8.3 Approaches for Enhanced Lipid Accumulation

Under ideal growth conditions, an alga produces larger amounts of biomass but with reasonably small lipid contents, which is around 5%–20% of their dry cell weight (DCW). Microalgal biomass and TAG compete for the energy fixed during photosynthesis; therefore, reprogramming of physiological pathways is required to divert the metabolic flux to encourage lipid production. Under stress conditions or unfavorable environmental conditions, many of the microalgae shift lipid biosynthetic pathways in the direction of the formation and amassing of neutral lipids (20%–50% DCW), primarily in the form of TAG (Hu et al. 2008), enabling microalgae to sustain these adversarial conditions. Use of nutrients stress including phosphorus and/or nitrogen starvation; varied pH, light intensity, temperature, irradiation; as well as the presence of heavy metals and other chemicals have been investigated to increase the lipid yield.

Nitrogen is the most abundant element of intracellular components as well as the primary constituent of most macromolecules. It has been noticed that when nitrogen levels are limited in growth media, protein biosynthesis cascade during growth is affected and excess carbons from photosynthesis are normally channeled to storage molecules such as TAGs. Thus, nitrogen starvation affects the lipid metabolism and alters the physiological pathways in algae that lead into increased lipid content, more specifically triacylglycerides for many microalgal species (Yao et al. 2014b).

Further, genetic engineering can also contribute as tool to enhance lipid accumulation in the microalgae (Sharma et al. 2012). Several key genes of fatty acid metabolism, such as G3PDH, ACCase, and DAGAT, have been found to take part in the enhanced lipid accumulation in the microalgal cells when exposed to stress or certain stimuli (Fan et al. 2014). Transforming microalgae for overexpression of such genes participating in the TAG biosynthesis or inhibiting the expression of the genes of other metabolic routes such as citric acid cycle have been found to increase the total neutral lipid accumulation in algae (Yao et al. 2014). Some important approaches for increasing lipid accumulation in microalgae are depicted in Figure 8.3.

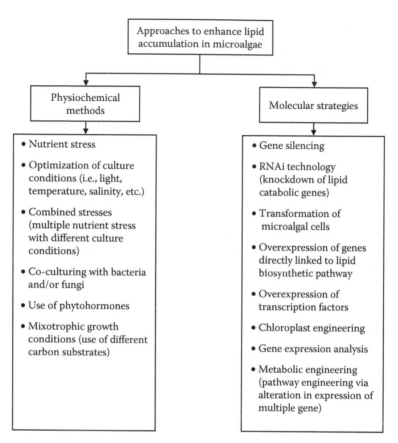

FIGURE 8.3
Strategies for enhanced lipid accumulation in microalgae.

8.3.1 Physiochemical Methods or Conventional Methods

Physiochemical parameters such as nutrients and culture conditions hold significant impact on the lipid accumulation inside microalgal cells. Some important observations on the lipid accumulation through conventional routes are summarized in Table 8.4. Physiochemical methods are easy to use and are fast delivering methods but these have certain drawbacks as well. In these methods, cellular growth is hampered and lipid productivity versus biomass proportion remains almost the same. Further, similar stress condition may not be fruitful for all kinds of microalgae and these methods often fail to highlight the reasons for lipid accumulation or better lipid productivity.

8.3.2 Molecular Methods

Modification in nutrients and conditions can work toward increasing lipid accumulation but only up to a certain limit. Attempts have been made to decipher the mechanism lying behind the increase in the lipid content at the molecular level and denoted that several key genes could be implicated for changes in lipid accumulation. Most of these genes are mainly associated with fatty acid biosynthetic pathway such as ACCase, glyceraldehyde 3-phosphate dehydrogenase (G3PDH), KAS, diacylglyecerol acyltransferase (DAGAT), and so on.

TABLE 8.4

Impact of Physiochemical Factors on Lipid Accumulation in Different Species of Microalgae

Species	Factors	Effect	Reference
Chlamydomonas reinhardtii	Nitrogen starvation	Nearly fivefold increase in neutral lipid content in low N condition in comparison with high N condition	Dean et al. 2010
Anacystis nidulans	Nitrogen starvation	Accumulation of saturated, longer-chain (C18) fatty acids	Gombos et al. 1987
Nannochloropsis oculata	Nitrogen starvation	Total lipid increased by 15.31%	Converti et al. 2009
Chlorella vulgaris	Nitrogen starvation	Total lipid increased by 16.41%	Converti et al. 2009
Chlorella vulgaris	Nitrogen starvation	Lipid productivity of 78 mg/L d achieved, nearly 1.5-fold of normal productivity	Yeh and Chang 2011
Scenedesmus sp.	Nitrogen & phosphorous starvation	~30% lipid content at N 2.5 mg/L compared with ~24% at N 25 mg/L	Xin et al. 2010
Chlorella sp.	Phosphate, potassium, iron starvation	Lipid productivity in N limitation went from 40.27 to 53.97 mg/L.d and in combined stress to 49.16 mg/L.d	Praveenkumar et al. 2012
Phaeodactylum tricornutum	Phosphorus starvation	Increase in total lipids with higher relative content of 16:0 and 18:1	Reitan et al. 1994
Chlamydomonas reinhardtii	Sulfur starvation	PG was increased by 2-fold	Sato et al. 2000
Chlorella vulgaris	Iron starvation	High levels of chelated Fe^{3+} results in high levels of lipid accumulation (3–7 fold increase)	Liu et al. 2008
Cyclotella cryptica	Silica starvation	Increased in total lipids from 27.6% to 54.1%	Roessler 1998
Nannochloropsis gaditana	Nitrogen starvation	5.7-fold increase in NgMCAT expression	Tian et al. 2013
Isochrysis galbana	Nitrogen starvation	Desaturation of fatty acids	Huerlimann et al. 2010
Chlorella vulgaris	Nitrogen starvation	Increased lipid accumulation in shorter time period	Mujtaba et al. 2012
Micractinium pusillum	Nitrogen starvation	Genes linked to pyruvate and acetyl-CoA synthesis were radically increased over 100 times after nitrogen starvation	Li et al. 2012
Dunaliella sp.	High salinity	High intracellular lipid content and high percentage of triacylglyceride in the lipid	Takagi and Yoshida 2006
Chlorella vulgaris	CO_2 concentration	Maximum lipid productivity of 29.5 mg. $L^{-1}.day^{-1}$ at 8% (v/v) CO_2 in which higher saturated fatty acids were obtained	Sibi et al. 2015
	Light intensity	Higher light intensity reduces total polar lipid content with a simultaneous increase in the amount of neutral storage lipids, mostly TAGs	Harwood 1998

The microalgal DNA sequences available in the genomic databases and accessibility of potent bioinformatics tools in public domain have empowered us to use microalgae as a valuable host for genetic modifications. Till date, at least 20 algal genomes have been sequenced, including the green algae—*Chlamydomonas reinhardtii, Coccomyxa* sp. C-169, *Micromonas CCMP 1545, Ostreococcus lucimarinus CCE9901, Ostreococcus tauri,* and *Volvox carteri*; red alga—*Cyanidioschyzon merolae*; brown alga—*Ectocarpus siliculosus*; diatoms—*Phaeodactylum tricornutum* and *Thalassiosira pseudonana*. There are several advantages of using molecular techniques over conventional approaches; mainly, it gives a molecular knowhow of lipid metabolism, provides better understanding of techniques and strategies to enhance lipid productivity, and it does not affect cellular growth, hence biomass yield, and desirable quality and quantity of oil can be produced. Moreover, it provides detailed diversified methods for increasing lipid content (Khozin-Goldberg and Cohen 2011).

Approaches that are mainly employed to enhance lipid accumulation in the microalgae include, first upregulation of the genes involved in lipid biosynthesis and second the downregulation or blocking of genes of lipid catabolism. Both of these strategies normally turn the metabolic flux toward fatty acid biosynthesis and are found promising.

8.3.2.1 Overexpression of Enzymes Participating in Biosynthesis of Fatty Acids

Overexpression of certain genes impacts into increased flux of the pathway and may also competitively block the other enzymes of the similar substrate, eventually increasing the lipid accumulation. Efforts have been made to analyze the effect of upregulation of different genes on lipid accumulation in microalgae, few of which are mentioned in Table 8.5. Further, besides overexpression of key genes of the pathway, enhanced expression of transcription factors for the lipid biosynthetic pathways genes have also been accounted for increase in lipid content in the microalgal cells (Kang et al. 2015; Zhang et al. 2014).

8.3.2.2 Thwarting Accessory Pathways to Enhance Lipid Biosynthesis

Numerous key metabolites are common in several different metabolic pathways and blocking enzymes of one pathway can lead to increased availability of the metabolite for the pathway of interest. For example, pyruvate, a product of glycolysis, can enter a TCA cycle for energy synthesis or move toward fatty acid biosynthesis. Thus, to enhance the flux of pyruvate toward fatty acid biosynthesis, enzymes of TCA can be blocked or downregulated; this will lead to a surge in the fatty acid production, ultimately enhancing the lipid accumulation in the microalgal cell. Besides this, blocking the enzymes participating in lipid catabolism such as lipases can also escalate the lipid accumulation. Majority of approaches working toward downregulation of certain genes utilize RNA interference as the tool to achieve their goal (Deng et al. 2013; Trentacoste et al. 2013).

Whether overexpressing the genes directly involved in lipid biosynthetic pathway or indirectly increasing their expression via transcription factors or even blocking the expression of genes dispensing the energy toward products other than TAG, the motive behind every metabolic manipulation has to be the increased metabolic flux toward accumulation of lipid inside the microalgal cells. Combinatorial approaches utilizing both the aforementioned approaches to engineer the lipid content have also been employed to maximize the output (Yao et al. 2014).

TABLE 8.5

Enhancement of Lipids in Microalgae via Molecular Approaches

Species	Gene	Approach	Effect	Reference
Chlamydomonas reinhardtii	Dof-type transcription factor	Overexpression	Twofold increase in total lipid	Ibáñez-Salazar et al. 2014
Phaeodactylum tricornutum	G3PDH	Overexpression	60% increase in neutral lipid content reaching up to 39.70% of total dry cell weight	Yao et al. 2014b
Chlamydomonas reinhardtii	Citrate synthase	Downregulation	TAG levels increased by 169.5% in response to 37.7% decrease in CrCIS activities	Deng et al. 2013
Schizochytrium sp.	Acetyl-CoA synthetase	Overexpression	The biomass and fatty acid proportion increased by 29.9% and 11.3%, respectively	Yan et al. 2013
Thalassiosira pseudonana	Lipase (Thaps3-264297)	Downregulation	Strains 1A6 and 1B1, respectively, contained 2.4- and 3.3-fold higher lipid content than wild-type during exponential growth	Trentacoste et al. 2013
Phaeodactylum tricornutum	Malic enzyme (PtME)	Overexpression	Total lipid content in transgenic cells markedly increased by 2.5-fold and reached a record 57.8% of dry cell weight	Xue et al. 2015
Chlamydomonas reinhardtii	DGTT4	Overexpression	Enhanced TAG accumulation	Iwai et al. 2014
Nannochloropsis salina	bHLH transcription factor	Overexpression	Increase in biomass production by 36% and FAME productivity by 33% under nitrogen stress	Kang et al. 2015
Synechocystis sp.	*maqu_2220* (acyl-CoA reductase), aldehyde-deformylatingoxygenase gene (*sll0208*)	Combinatorial Approach	Overexpression of *maqu_2220* and knockout of *sll0208* resulted in higher production of fatty alcohols	Yao et al. 2014a
Chlorella ellipsoidea	GmDof4 (Transcription factor)	Overexpression	Heterologous-expression of GmDof4 gene from soybean can significantly increase the lipid content but does not affect the growth rate	Zhang et al. 2014

8.4 Cultivation and Harvesting

Cultivation of microalgae could also be done in a photobioreactor (PBR), which could be of various types, an open production system such as raceway pond or closed systems. The operating cost of raceway ponds is significantly lower than that of the closed system PBRs. However, the risk of water evaporation, pH and temperature variation, cells illumination, and contamination by other organisms are higher, which more often reduces algal growth. In open ponds, biomass concentration is approximately 0.5 g/L while in PBR it is of the order of 1–12 g/L (Davis et al. 2011; Tredici 2004). Some commonly used closed system PBRs are tubular PBR, flat plate PBR, and column PBR. In these PBRs, the operating conditions are controlled more preciously than in raceway ponds and can produce more biomass containing good quality lipids in higher extent with respect to open pond systems. PBRs can be operated in batch, semicontinuous, or continuous mode. Advantages and limitations of different cultivation systems such as open pond, tubular PBR, flat plate PBR, and column PBR are summarized in Table 8.6.

TABLE 8.6

Comparison of Different Microalgae Production Systems

Culture System	Advantages	Limitations
Open Pond	• Low cost of manufacturing and operation. • Easy cleaning after each culture and easy maintenance.	• More chances of culture contamination (Ugwu et al. 2008). • Larger area of land is required and limited for some specific strain of algae (Ugwu et al. 2008). • Diffusion of CO_2 to the atmosphere and water evaporation is a challenge (Vree et al. 2015). • Low biomass productivity of approximately 25 g/m²-day to 72.5 g/m²-day in compared an inclined tubular PBR made with tube of 2.5 cm internal diameter (Singh et al. 2011).
Tubular Photobioreactor	• Offers larger illumination surface area compared to flat and column PBR. • Suitable for outdoor cultures and can be used with indoor cultures by providing artificial illumination such as fluorescent lamp. • Compared to flat PBR it is relatively cheap. • Good mixing mechanism with good biomass productivities.	• Accumulation of nondissolved CO_2 and O_2 along the tubes resulting in inhibition of culture growth (Richmond 2000). • Thin cell layer at the surface of wall results in wall growth that reduces the light exposure and hence growth is affected (Ugwu et al. 2008). • Mass transfer between nutrient in medium and algal cell is poor (Ugwu et al. 2008).
Flat PBR	• Power consumption is low as compared to other culture system. • Easy sterilization and good light path. • Oxygen build up is low. • Illumination surface area is larger.	• Scale up and temperature control is difficult (Ugwu et al. 2008). • Because of some wall, growth light exposure to the cells is reduced which consequently reduces photosynthetic efficiency (Azgin et al. 2014).

(Continued)

TABLE 8.6 (*Continued*)

Comparison of Different Microalgae Production Systems

Culture System	Advantages	Limitations
Column PBR	• Yield of algal biomass is higher than in open ponds and offers less space compared to open pond, flat PBR, column PBR	• As compared to tubular PBR smaller, surface area is exposing to light in column PBR (Ugwu et al. 2008). • Manufacturing material includes glass, acrylics which increases the production cost (Chiu et al. 2009). • When column PBR is scaled up, there is a decrease in the ratio of illumination surface area to the volume of reactor (Ugwu et al. 2008).

Tubular PBR may be of horizontal type, inclined, or vertical coil types. Yield of microbial biomass is influenced by the type of PBR used for cultivation because of variation in accumulation of undissolved CO_2 and O_2, mixing of nutrients, surface area exposed to light, and so on, in different PBR types (Meng et al. 2009; Ugwu et al. 2008). Figure 8.4 shows the comparison of biomass productivity in different types of PBR (Azgin et al. 2014; Vree et al. 2015).

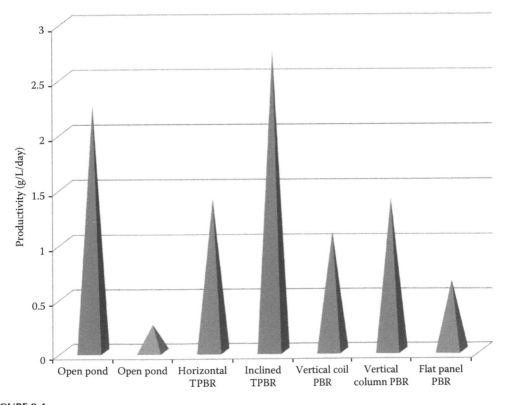

FIGURE 8.4
Production capacity comparison. (Adapted from Vree, J.H. et al., *Biotechnol. Biofuels.*, 8, 1, 2015; Azgın, C. et al., *J. Biol. Environ. Sci.*, 8, 24, 2014.)

8.4.1 Harvesting of Algal Biomass

Harvesting is the process of separating or concentrating the algal biomass from dilute microalgal culture. After biomass growth and lipid accumulation in the cell, the biomass needs to be harvested. For further conversion of algal biomass into useful biofuel products such as biodiesel, bioethanol, bio-oil, biogas, and so on, water removal or dewatering is required. For the production of value-added products on a large industrial scale, the main problem or hindrance associated with the use of microalgae is the harvesting or dewatering step because microalgae exist in suspension form containing about 0.1–2.0 g of dried algal biomass per liter of algal slurry... Approximately 20%–30% of the total production cost is associated with various harvesting methods including cultivation and processing of microalgae. The further steps for biofuel production such as lipid extraction and purification are interrelated with harvesting process. The more concentrated the biomass obtained after harvesting, the less will be the cost associated with extraction and purification. Therefore, selection of efficient and appropriate harvesting technique is of great importance for economically viable biofuel production. Microalgae harvesting consists of two steps such as bulk harvesting and thickening, as shown in Figure 8.5. In the bulk, harvesting step, microalgal biomass is separated from the bulk suspension of microalgae culture through dewatering to produce concentrated biomass slurry (~2%–7% dry weight). Some important techniques used in this step are sedimentation, flotation, and flocculation. In the second step, that is, thickening step, the microalgal slurry is further concentrated to separate algal biomass. In this process, agglomerations of algae into macroscopic mass are done to facilitate the dewatering. Some important techniques employed in this step are centrifugation, ultrasonic agitation, and filtration. Salient features of different harvesting techniques described below.

8.4.1.1 Gravity Sedimentation

Sedimentation is a physical process of separation of suspended particles from fluid in which suspended particles settle down under gravity at the bottom and form concentrated slurry and at the top a layer of clear liquid is formed. Large-size and high-density microalgae such as *Spirulina* settle very rapidly and can be properly harvested by the sedimentation method. This is a very energy-efficient method. One disadvantage of this method is

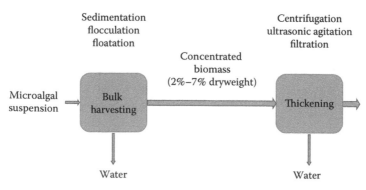

FIGURE 8.5
Harvesting of algal biomass.

that small-size and less-concentrated particles take much time for settling. Therefore, it is a slow process. Lamella plate settlers and sedimentation tank can be used to increase the microalgal harvesting efficiency by sedimentation. In wastewater treatment, harvesting of algal biomass by the sedimentation method is the most common one because large amount of volume can be treated. This method is less costly but not reliable without the use of flocculants.

8.4.1.2 Flocculation

In this method, microalgae cells are aggregated to increase the effective particle size. Flocculation works on the principle of a charge dispersion mechanism. Because of the adsorption of ions generating from organic matter microalgae cells carry a negative charge and because this common negative charge would not be able to self-aggregate so chemicals called flocculants are added to displace the negative charge on the microalgae surface and allow the agglomeration of microalgae cells. Both inorganic and organic compounds have been used as flocculants. The commonly used inorganic flocculants are aluminum chloride, aluminum sulfate, ferric sulfate, and ferric chloride. However, out of these, the aluminum sulfate has proved to be the most effective (Oh et al. 2001). Organic flocculants are polyelectrolytes or cationic polymers and are relatively more costlier. These are also not suitable for the harvesting of algal biomass from saline solution (Uduman et al. 2010), which is normally done by using combination of inorganic flocculants with organic flocculants or ozone oxidation after that adding of flocculants.

8.4.1.3 Flotation

Flotation is a physiochemical gravity separation process, in which air or gas bubble is allowed to pass through algal slurry and algae attach or adhere to the gas bubble and float to the liquid surface (Chen et al. 2011). The efficiency of flotation depends on the particle size, and the air/gas particle contact, and so on. The lesser the particle size and higher the air particle contact more the microalgae will be lifted to the top of the liquid surface. Based on the bubble size, the flotation can be grouped into three types such as dispersed air flotation, dissolved air flotation, and electrolytic flotation. A high-speed mechanical agitator and an air injection are required to form the bubbles in the range of 700–1500 μm in dispersed air flotation. Gas is introduced at the top of the vessel and then it mixes with liquid and allowed to pass through a disperser to produce the bubbles. Because algal cell surface carry, negative charge, bubbles act by interacting with them. The efficiency of this method can be enhanced by addition of surfactants or chemicals that give a net positive charge, increase the integrity of bubble, and avoid rupturing (Chen et al. 1998). Dissolved air flotation requires the reduction of the pressure of the water stream that is presaturated with air. Diffuser nozzles generate air bubble, adhere/attach to the microalgal cells, rise through the liquid, and make the algal cells to float on the surface. Biomass accumulated at the surface can be skimmed and collected. In electrolytic flotation, cathode made from an inactive metal that is electrochemically nondepositing. At cathode, inactive metal forms hydrogen bubbles from the electrolysis of water. The algal cells in the suspension attach to the gaseous bubbles and lift to the surface of the vessel. Although for efficient microalgae, harvesting flotation is much faster than sedimentation but one drawback of this method is that it is effective on bench scale only and not suitable for large-scale harvesting because of high operational cost and high energy required for bubble generation.

8.4.1.4 Centrifugation

This method involves the use of the centrifugal force for the sedimentation of heterogeneous mixtures by a centrifuge to separate algal suspension into regions of lesser and greater densities. After that by draining the culture medium, algae and water are separated out. This harvesting method recovers algae in excess of 90% and the biomass recovery efficiency depends on some factors such as the settling characteristics of the microalgae cell, depth of settling, and residence time of slurry. This method is one of the most common and widely used techniques due to its rapid and efficient nature. However, it is energy intensive and it requires high maintenance due to freely moving parts. To achieve high harvesting efficiency, longer residence time and hence more energy is required, which restrict it to become the cost-effective algae harvesting method with high efficiency.

8.4.1.5 Filtration

It is the physical or mechanical operation, which is used for the separation of suspended solid particles from liquid by passing the liquid through a medium that is permeable so that it can retain the algal biomass. In this method, pressure difference across the system is required which can be driven by vacuum, pressure, or gravity. The harvesting efficiency of this method depend ranges from 20% to 49%. Some important filtration methods are pressure filtration, vacuum filtration, ultrafiltration, microfiltration, dead-end filtration, and tangential flow filtration (TFF). Liquid is pulled through by suction from the filter and the microalgal biomass is captured on the filter in vacuum filtration. Microfiltration membrane is suitable for microalgae range in size from 2 to 30 µm for vacuum filtration (Fajardo et al. 2007; Brennan and Owende 2010). According to Milledge and Heaven (2013) for large microalgal cells, macrofiltration membranes are used. The efficiency of vacuum filtration depends on the size of membrane and microalgae cell. Pressure filtration also works on the principle of pressure difference by raising the pressure above the atmospheric pressure across the filter so that liquid can pass through the filter leaving the algal biomass deposit. The advantage of this process is that solid algal biomass deposited has low moisture content. The disadvantage is the difficulty in washing. In TFF, algal slurry is allowed to flow tangentially across a membrane and the algal cell size larger than membrane pore sizes are captured on the filter medium. This method recovers about 70%–89% algae. The other advantage of this method is that structure and properties of the filtered microalgae remain unchanged. However, pumping cost and membrane replacement restrict this method for large-scale harvesting. Large microalgae can be removed by vacuum filtration and ultra- and microfiltration recovers smaller size algae. Simple filtration methods such as dead-end filtration are not suitable for harvesting microalgae because of the problem associated such as back mixing. Comparisons of some harvesting methods are provided in Table 8.7.

8.5 Extraction of Lipid

Algal biomass is harvested after the lipid (bio-oil) is extracted from the algal biomass. Extraction of lipid from algae is very costly and the sustainability of the algae-based biofuel depends on it. Conventionally bio-oil is extracted from algal biomass through the

TABLE 8.7

Common Harvesting Methods and Their Effectiveness in Recovery of Biomass from Their Culture Medium

Microalgal Species	Method	Effectiveness	Condition	Reference
Spirulina Platensis	Vacuum filtration	—	Vacuum filtration, using cellulose membrane with 0.45 μm pore size	Stucki et al. 2009
Stephanodiscus hantzschii, S. astraea and *Cyclotella* sp.	Tangential flow filtration	70%–89% recovery	Cross filtration, 0.45 μm membrane pore	Petrusevski et al. 1995
Tetraselmis spp. and *Chaetoceros calcitrans*	Centrifugation	Greater than 95% recovery	—	Heasman et al. 2000
Chlorella+Oocytis	Centrifugation	90% recovery	—	Sim et al. 1988
B. braunii	Dispersed air flotation	93.6% recovery	—	Xu et al. 2010
Chlorella minutissima	Flocculation followed by sedimentation	60% Recovery efficiency	1g/L of $Al_2(SO_4)_3$ and $ZnCl_2$ and took 1.5 h and 6 h, respectively	Papazi et al. 2010
Dunaliella salina	Micro bubble flotation	99.2% recovery	Formation of microbubble, pH 5, (150 mg/L) ferric chloride as coagulant	Hanotu et al. 2012
Chlorella vulgaris	Gravity sedimentation	60% of biomass was recovered	Biomass with density variation between 0.620 and 0.820 OD at 685 nm are settled in 1 h	Ras et al. 2011
Chlorella vulgaris	Flocculation	85%–95% of biomass was recovered	Above 25 m mol/L aluminum sulfate was used as flocculant	Ras et al. 2011
		95% of biomass was recovered	Sodium hydroxide was used as flocculent at pH between 11 and 12	Ras et al. 2011

mechanical route (Expression/Expeller press) as well as the chemical method (solvent extraction). Ultrasonic-assisted extraction and supercritical fluid extraction are two developments in the chemical route.

8.5.1 Expeller Press

In an expeller, raw material is squeezed or pressed with an oil press under high pressure in a single step for the extraction of oil. Friction or continuous pressure is employed to move or compress the algal cells, which breaks their cell wall. This method can extract around 75% of oil present in biomass. To ensure good efficiency, algae must be dried before the extraction of oil. No purification is required as it does not require any solvent (Popoola and Yangomodou 2006). Because there is a variety in physical attributes of different algae strains so various press configurations (screw, expeller, piston, etc.) are used for specific algae types. A limitation of this method is that it needs algae in dried form, which is energy intensive. It is also a slow process.

8.5.2 Solvent Extraction

In solvent extraction method, a solvent is used for the extraction of algal oil. The principle of this method is based on the concept, that is, "like dissolves like." This mechanism of solvent extraction can be categorized into five steps.

- When a microalgae cell is exposed to a polar or nonpolar organic solvent, then the solvent permeates through the cell membrane into the cytoplasm.
- It then interacts with the lipid complex. The nonpolar organic solvent forms van der walls bonds with the neutral lipid in the complex, although the polar organic solvent forms the hydrogen bonding with the polar lipid in the complex.
- An organic lipid solvent complex can form through either of the above processes.
- The solvent lipid complex is formed and then diffuses across the cell membrane due to the concentration gradient, as shown in Figure 8.6.
- Then it diffuses across the static organic solvent film that surrounds the microalgal cell and enters into the bulk organic solvent, as shown in Figure 8.6 (Halim et al. 2012).

The solvent used in this method is inexpensive and can be recovered. This method has some drawbacks such as requirement of large quantity of solvent, expensive solvent recovery as well as toxicity, and flammability of some of the solvents. Extraction of bio-oil through solvent extraction is conventionally carried out in a soxhlet apparatus. This method can be either used alone or can be used together with the oil press/expeller method. After the oil extraction from algal biomass with expeller, the remaining biomass is mixed with cyclohexane to further extract the remaining oil from the cake. Oil is dissolved into cyclohexane and the residual biomass is filtered out. Then the oil is separated from the cyclohexane by distillation. If these two methods are used together, it can extract

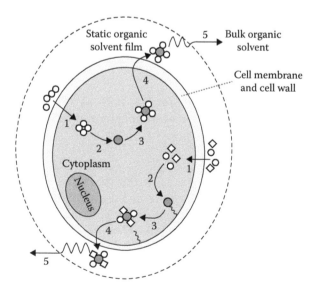

FIGURE 8.6
Schematic diagram of the solvent extraction mechanism. (Adapted from Halim, R. et al., *Biotechnol. Adv.*, 30, 709–732, 2012. With permission.)

more than 95% of the total oil present in algae. Further, the application of a two-step extraction process (extraction of the lipids from algal biomass by ethanol in the first step and the purification of the extracted lipids in second step by hexane) can increase the lipid recovery (Fajardo et al. 2007).

One important solvent extraction method is the Bligh and Dyer Method. This method can extract more than 95% of total lipid and can even be used for cells having water content up to 80%. The Bligh and Dyer method can be used for both dry and wet route (Chen et al. 2011). In this method, the solvent and culture are mixed in a ratio of 3:1 and ratio of methanol, chloroform, and water should be 2:1:1.8 (Chen et al. 2011; Pragya et al. 2013). Different types of solvents have been tested in the literature for the extraction of bio-oil from algal biomass; some of these processes are summarized in Table 8.8.

The use of solvents makes the mixture system to get homogenized and reduce the viscosity of the bio-oil and at the aging rate. It also increases the heating value and volatility of the bio-oils and reduces the acidity. From Table 8.8, it seems that the conventional solvents are used with the dry or concentrated biomass of algae for the extraction of lipid. Some commonly used solvents are hexane, isopropanol, methanol, and chloroform. Recently, oil extraction from algal biomass in wet phase has been investigated by many researchers (Fajardo et al. 2007; Halim et al. 2012; Lee et al. 1998). Some solvents such as piperylene

TABLE 8.8

Extraction of Microalgal Lipids Using Organic Solvents

Microalgae Species and Biomass Characteristics	Operating Conditions/Solvent	Oil/Lipid Yield (wt.% of dried microalgal biomass)	References
Chlorella sp. (0.1 g dried powder by lyophilization)	Soxhlet extraction: methylene chloride/methanol (2:1 v/v); 1000 mL organic solvent mixture/g dried microalgal biomass; Time; 180 min, Temp.: Not specified	11.9	Guckert et al. 1988
	Batch extraction: chloroform/ methanol/50 mM phosphate buffer (35:70:28 v/v/v/); Time: 2160 min, Temp.: not specified		
Botryococcus braunii (0.12 g in concentrated form)	Chloroform/methanol (2:1 v/v); 250 mL organic solvent mixture/g dried microalgal biomass; Time: 50 min; Degree of agitation: High but not specified; Temp: Not specified	28.6	Lee et al. 1998
Phaeodactylum tricornutum (10 g dried powder by lyophilization)	Ethanol, 5 mL organic solvent mixture/g dried microalgal biomass; Time: 1440 min; Degree of agitation: 500 rpm; Temp: 25°C	6.3	Fajardo et al. 2007
Chlorococcum sp. (4 g in concentrated form, residual water Content = 70 wt.%) or dried powder via thermal drying	Hexane, hexane/isopropanol (3:2 v/v); 75 mL organic solvent mixture/g dried microalgal biomass; Time: 450 min; Degree of agitation: 800 rpm; Temp: 25°C	6.8	Halim et al. 2012

sulfone, recycled DMSO substitute, and two components reversible ionic liquids, which have the capacity to convert from nonionic to ionic forms by CO_2, bubbling and back to nonionic form by N_2 bubbling, have been used for wet phase oil extraction. These solvents are also termed as switchable solvents. In a switchable solvent system the nonpolar lipid, captured by the nonpolar solvent, is separated from it when the solvent is converted to polar form by CO_2 bubbling. This method is energy efficient with respect to conventional dry phase oil extraction using solvents.

8.5.3 Ultrasonic Assisted Solvent Extraction

In ultrasonic-assisted extraction, the liquid is sonicated at elevated intensity sound waves, which propagate into the liquid nutrient media and result in flashing high-pressure (compression) and low-pressure (rarefaction) cycles. Throughout the low-pressure cycle, high-intensity ultrasonic waves create small vacuum bubbles or voids in the liquid. When the bubbles achieve a volume at which they are able to no longer absorb energy, they collapse aggressively during a high-pressure cycle. This phenomenon is termed cavitation. The resulting shear forces break the cell envelope mechanically and improve material transfer (Balakrishnan et al. 2015).

8.5.4 Supercritical Carbon-Dioxide Extraction

A fluid having temperature and pressure above its critical values is called a supercritical fluid, which has good solvent properties and can be used as an extracting solvent for the recovery of lipid from micro algal biomass. The high temperature and pressure rupture the cells and within a shorter extraction time higher total lipid yield is obtained (Canela et al. 2002; Macıas-Sánchez et al. 2005). Recently, it has been reported that if supercritical carbon dioxide ($SCCO_2$) is used as a fluid for the extraction of lipid from algal cells in packed bed, the ($SCCO_2$) comes on the surface of the packed mixture and lipids are desorbed from the microalgal biomass. Immediately upon dissolution, released lipids are solubilized by $SCCO_2$ and then form an $SCCO_2$–lipids complex. Because of the concentration gradient, this lipid complex diffuses across the static $SCCO_2$ film and enters the bulk $SCCO_2$ flow as shown in Figure 8.7 (Halim et al. 2012).

The solvent power of a supercritical fluid can be adjusted so that it can interact primarily with neutral lipids by changing the extraction temperature and pressure. A drawback

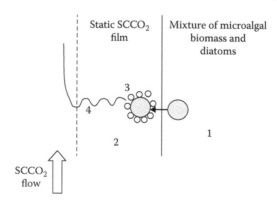

FIGURE 8.7
Schematic diagram of the supercritical carbon dioxide extraction mechanism. (Adapted from Halim, R. et al., *Biotechnol. Adv.*, 30, 709–732, 2012. With permission.)

TABLE 8.9

Some Studies on the Extraction of Microalgal Lipids by Supercritical Carbon Dioxide

Microalgal Species	Operating Conditions	Lipid Yield (wt.% of dried microalgal biomass)	Remarks	References
Chlorella vulgaris	Extraction pressure (bar): 200, 350; Extraction temp (°C): 40, 55; $SCCO_2$ flow rate: 0.41/min; Time: 500 min	13.0	Total lipid yield increased with P at constant T It decreased with T at constant lower P (200 bar) and increased with T at constant high P (350 bar). Optimum condition: 350 bar and 55°C	Mendes et al. 1995
Spirulina maxima	Extraction pressure (bar):- 100, 250, 350; Temp (°C): 50, 60; $SCCO_2$ flow rate: not specified; polar modifier, amount: ethanol, 10 mol% of CO_2	3.1	Similar effect of temperature and pressure on total lipid yield such as *Chlorella vulgaris*. Addition of polar modifier increased total lipid yield at constant T and P. Optimum condition: 350 bar, 60°C with ethanol addition (10 mol%).	Mendes et al. 2003
Nannochloropsis sp.	Extraction pressure (bar): 400, 550, 700; Temp (°C): 40, 55; $SCCO_2$ flow rate: 0.17 kg/min; Time: 360 min; polar modifier, amount: None	25.0	Enhancement in lipid extraction rate with pressure at constant temperature Slight increment in lipid extraction rate with temperature at constant pressure.	Andrich et al. 2005
Spirulina platensis	Extraction pressure (bar): 316, 350, 400, 450, 484; Temp (°C): 40, 55; $SCCO_2$ flow rate: 0.7 L/min; Time: 26.4, 40, 60, 80, 94 min; polar modifier (amount): Ethanol; 9.64, 11, 13, 15, 16.36 mL	8.6	Optimum condition: 400 bar, 13.7 mL ethanol, and 60 min.	Sajilata et al. 2008b

of this method is that the operational and infrastructure cost associated with this process makes it very expensive. Salient features of some lipid extraction methods using supercritical carbon dioxide are summarized in Table 8.9.

8.5.5 Ionic Liquids

Ionic liquids (ILs) are nonaqueous solutions that maintain at 0°C–140°C comprising asymmetric organic cation and inorganic/organic anion (Cooney et al. 2009). Its conductivity, polarity, hydrophobicity, and so on can be monitored by adjusting its composition and can be used for extracting polar and nonpolar lipids from algal biomass. The advantage of this process is that it avoids the application of toxic solvents for algal oil extraction. Oil extraction from scenedesmus, *Chlorella vulgaris* (Kim et al. 2012) biomass and algal lipids extraction studied have been in recent years. However, this method needs good amount of research and development activities for further development.

TABLE 8.10

Comparison of Various Lipid Extraction Processes

Process	Advantages	Disadvantages	References
Oil press	Simple to use No solvent required	Large biomass requirement Slow process	Mata et al. 2010
Solvent extraction	Solvent is expensive Solvent is recoverable/ reproducible	Most solvents are flammable and/or toxic Expensive solvent recovery Solvent requirement is large	Herrero et al. 2004
Supercritical fluid extraction	Solvent is non-toxic and nonflammable Operation is simple	Extraction of polar analyses from biomass is difficult Limited interaction between supercritical CO_2 and Biomass	Macias-Sánchez et al. 2005
Ultrasonic assisted extraction	Less extraction time Less solvent requirement Good penetration of solvent into cell, More release of cell contents into the medium	Scale up is not easy Power requirement is very high	Luque-Garcia et al. 2003

The selection of suitable oil extraction method is very important and depends on its speed, scalability, effectiveness, and tendency not to damage the extracted lipids. Comparison of different types of oil extraction processes is summarized in Table 8.10.

8.6 Biodiesel Production

The bio-oil produced from algal biomass through any of the methods discussed in the above section can be upgraded, blended, or directly used. However, the direct use of bio-oil in combustion engine is limited because it causes various problems such as filter clogging and plugging, engine knocking, coking of injectors on the head and piston of engine, excessive engine wear, and so on. The characteristics of the algal bio-oil due to which the problem arises are high viscosity, low cetane number, low flash point, presence of natural gums, and so on. Hence, algal bio-oil needs to be upgraded to biodiesel, which can meet all the international standards as shown in Table 8.11 and can be used in diesel engines solely.

Upgradation of bio-oil to biodiesel is done through transesterification, in which alcohol is reacted in the presence of a catalyst to alkyl esters and glycerol as a byproduct, that is, triglyceride is converted to fatty acids methyl esters (FAME) if methanol is used (Guo et al. 2012). Catalysts used for transesterification reaction can be acidic, basic, or enzyme based.

The most important characteristics of biodiesel include mainly oxidative stability, ignition quality, and cold flow properties. The structure of FAME determines the properties of biodiesel (Knothe et al. 2005). Although upgradation of bio-oil derived from edible and nonedible oil is widely reported, there is relatively less literature on the upgradation of algal oil to biodiesel as this is under development. Some works available on the upgradation of algal oil to biodiesel are summarized in Table 8.12.

Catalyst enhances the rate of transesterification reaction and hence the biodiesel yield. As evident from Table 8.13, the catalysts can be divided into three categories, that is, base catalysts, acid catalysts, and enzyme catalysts. Although overall conversion remains

TABLE 8.11

Physico-Chemical Properties of Biodiesel

Properties	Biodiesel
Chemical name of biodiesel	FAME (fatty acid methyl esters)
Chemical formula range of biodiesel	C14–C24
Boiling point range (°C) and flash point range (°C)	>202 157–182
Density range of biodiesel	860–894
Distillation range (k) of biodiesel	470–600
Vapor pressure (mm Hg at 22°C)	Less than 5
Physical appearance of biodiesel	Light to dark yellow transparent liquid
Solubility in water of biodiesel	Not soluble
Kinematic viscosity range (mm^2/s)	3,3–5,2
Biodegradability	More biodegradable than conventional diesel fuel
Reactivity of biodiesel	avoid strong oxidize agents, it is stable
Odor of biodiesel	Light soapy and oily odor

Source: Adopted from Demirbas et al., *Biodiesel*, 111–119, 2008.

TABLE 8.12

Upgradation of Algal Oil to Biodiesel

Microalgae Species (dry orwet)	Oil Extraction Method	Catalyst	Lipid Content (%)	Biodiesel Conversion (%)	References
Nannochlorop sis gaditana (dry biomass)	Stirring extraction	HCl	13.1	14.8	Rios et al. 2013
Chlorella vulgaris ESP-31 (dry biomass)	Ultrasonic assisted extraction	Lipase (*Burkholderia* sp.)	63.2	72.1	Tran et al. 2012
Chlorella vulgaris (dry biomass)	Stirring extraction	H$_2$SO$_4$	–	95.0	Lam et al. 2014
Chlorellaprotothecoides (dry biomass)	Soxhlet extraction	Lipase (*Candidia* sp.)	44–48.7	98.1	Li et al. 2007
Chlorella vulgaris (dry biomass)	Soxhlet extraction	lipase (*P. expansum*)	40.7	90.7	Lai et al. 2002
Chlorella sp. KR-1 (dry biomass)	Stirring extraction	Lipase (Novozyme435)	38.9	75.5	Lee et al. 2013
Tribonema minus (wet biomass)	Subcritical ethanol extraction	H$_2$SO$_4$	20.2	96.5	Wang et al. 2013
Nannochloropsis sp. (80%moisture)	Microwave assisted transesterification	NaOH	38.3	86.4	Wahidin et al. 2014
Chlorella sp. KR-1 (drybiomass)	–	–	40.9	89.7	Jo et al. 2014
Chlorella protothecoides (dry biomass)	–	–	90.8	87.8	Nan et al. 2015
Tetraselmis suecica (dry biomass) *Chlorella sorokiniana* (dry biomass)	–	H$_2$SO$_4$	23.0	78.0	Wahidin et al. 2014
Nannochloropsis oceanica (65% moisture)	In situ transesterification in water bath	H$_2$SO$_4$	19.1	90.6	Lam et al. 2014
Chlorella pyrenoidosa (80% moisture)	–	H$_2$SO$_4$	19.0	55.2	Chen et al. 2013

TABLE 8.13

Comparison of Processes for Biodiesel Production from Algal Oil Using Different Types of Catalyst

Parameters	Base Catalysis	Acid Catalysis	Enzyme Catalysis
Scale of application	Most widely used process. Currently, practically 100% of biodiesel is produced by the alkaline process	Acid catalysts are rarely used on the industrial scale because of their corrosive nature	Seems to be starting its application in industrial scale. For large-scale production this may not be economically viable due to high enzyme production costs.
Rate of reaction	It is a faster reaction	The reaction is slow. Speeding up the acid-catalyzed reaction requires an increase in temperature and pressure making it prohibitively expensive at large scale	Reactions are slower than the alkaline catalyst and the risk of enzyme inactivation due to methanol moderate reaction conditions thus less energy intensive and does not even run the reaction to completeness.
Effect of alcohol	It is reversible reaction, the rate of forward reaction increases with addition of more alcohol	It is reversible reaction, the rate of forward reaction increases with addition of more alcohol	If methanol is in a relatively high amount with respect to oil, it may inhibit and deactivate a large proportion of lipase.
Effect of FFA content	For alkali transesterification its amount should not exceed a certain limit. Alkaline catalysis is preferred over acid catalysis for oil samples containing FFA below 2.0%	It is suitable for transesterification of oils containing high levels of free fatty acids because it is more corrosive, its yield is lower in comparison to base catalyst	It is a viable method for parent oils containing high levels of free fatty acids as they can also be converted to alkyl esters.
Downstream recovery	Downstream recovery difficult due to soap formation	No soaps are formed, if the reagents are moisture free	Easier product recovery. If lipase is immobilized, it can easily be separated from the reaction mixture by filtration, or when the lipase is in a packed bed photobioreactor (PBR), no separation is necessary after transesterification.
Effect on environment	The alkaline process is not so environmental friendly. glycerol, which is a byproduct formed during the reaction, is usually contaminated with alkaline catalysts, thus, its purification to provide an added value to the alkaline process is not easy	Its environmental effects are similar to base catalyst	The subsequent separation and purification of biodiesel is easier than with alkali ne catalysts Immobilizational so increases the stability of the lipases and has potential for repeated use.

Source: Pragya, N. et al. *Renew. Sustain. Energy Rev.,* 24, 159–171, 2013.

similar for all the cases, the mechanism varies in the presence of different types of catalysts (Antunes et al. 2008). The mechanism of alkali-catalyzed transesterification can be explained with the help of different reaction steps, as shown in Figure 8.8.

Initially, alkoxide ion is formed by the reaction of alkali and alcohol (Prestep), and the conversion process starts with the attack by the alkoxide ion on the carbonyl carbon of the triglyceride molecule, which favors the formation of a tetrahedral intermediate (Step 1).

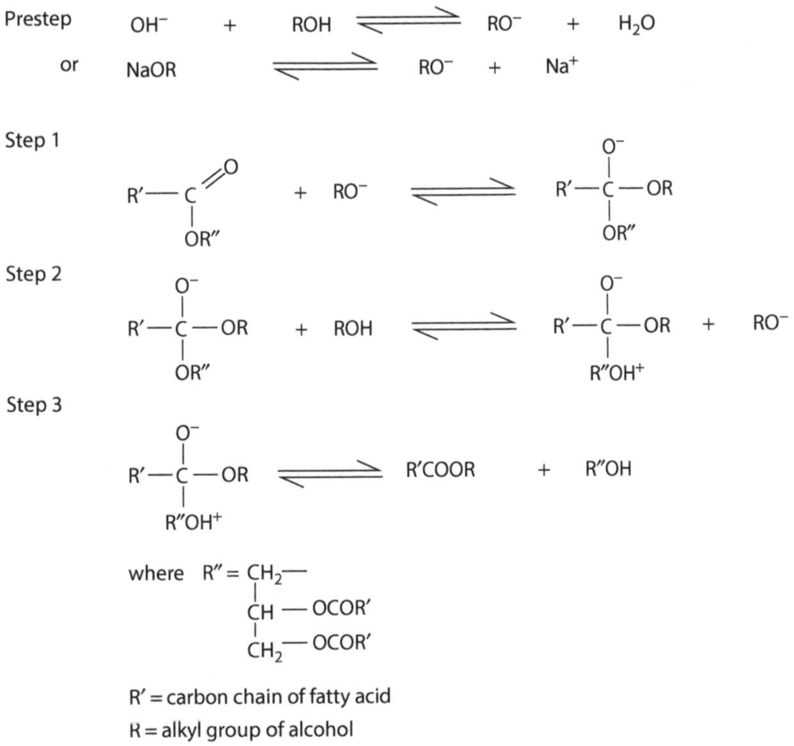

Prestep: $OH^- + ROH \rightleftharpoons RO^- + H_2O$

or $NaOR \rightleftharpoons RO^- + Na^+$

Step 1

Step 2

Step 3

where $R'' = CH_2-$
$CH-OCOR'$
CH_2-OCOR'

$R' = $ carbon chain of fatty acid

$R = $ alkyl group of alcohol

FIGURE 8.8
Mechanism of base-catalyzed transesterification.

This tetrahedral intermediate now reacts with alcohol to produce the alkoxide ion (Step 2). The tetrahedral intermediate rearranges to give an ester and a diglyceride (Step 3) (Ma and Hanna 1999).

The mechanism of acid-catalyzed transesterification is described in Figure 8.9. In the presence of acid catalysts, initially the protonation of carbonyl group of the ester takes place

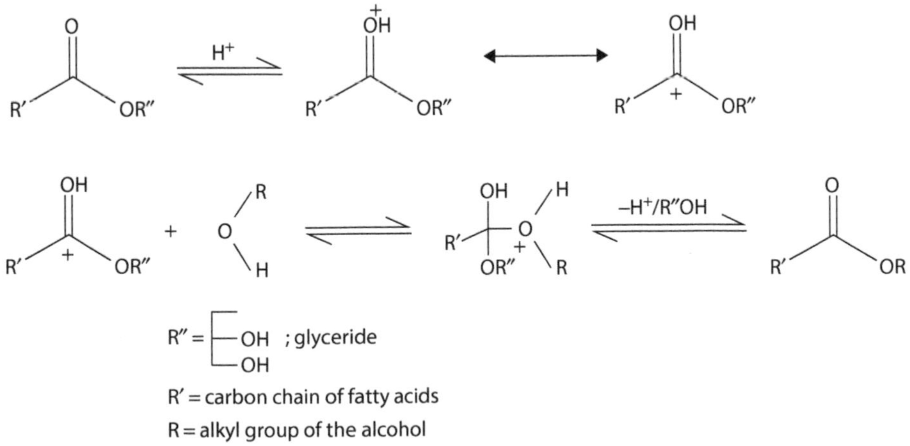

$R'' = \begin{cases} -OH \\ -OH \end{cases}$; glyceride

$R' = $ carbon chain of fatty acids

$R = $ alkyl group of the alcohol

FIGURE 8.9
Mechanism of acid-catalyzed transesterification.

and carbocation is formed, which after by the nucleophilic attack of the alcohol produces a tetrahedral intermediate. The glycerol eliminates by this intermediate to form a new ester and to regenerate the catalyst. However, it can be extended to di- and triglycerides.

Acid and base catalysts (homogeneous catalyst) act in the identical liquid phase as the reaction mixture (Meher et al. 2006). Homogeneous catalyst transesterification results in water and more soap formation and hence consumes more catalyst and reduces biodiesel yield. Moreover, if a homogeneous catalyst is used for biodiesel production at a large scale it requires much water for the separation and washing of equipment and wastewater treatment of effluent increases the overall cost.

In a homogeneous catalytic transesterification process, oil (algal or vegetable oil) with higher fatty acid leads to the formation of soap and water, separation problem, oil losses, and an increase in the production cost of biodiesel. Further, the rate of acid catalyzed transesterification process is very slow, hence, enormous efforts have been made to investigate the enzymatically catalyzed transesterification process. The enzymatic transesterification process avoids soap formation, performs at neutral pH, and lower temperature, so it becomes economical. The reusability of enzymes by immobilizing the solid support has created a new window of opportunity. Different methods of enzymatic immobilization such as cross-linking, covalent bonding, and microencapsulation have been reported. Lipase is the key enzyme used for the transesterification process, as this is cheap as well as able to catalyze both transesterification and hydrolysis of triglycerides at very mild conditions, so this is considered for biodiesel production. A transesterification reaction using a biocatalyst (enzyme) eliminates the disadvantages of the alkali-catalyzed process by producing product of very high purity with lesser or no downstream operations (Fukuda et al. 2001). However, this method has not been carried out mostly on an industrial scale because of few problems such as enzyme inhibition due to methanol, exhaustion of enzyme activity, and high cost of enzymes. Comparison of the processes for biodiesel production from algal oil using different types of catalyst is summarized in Table 8.13.

8.7 Quality Assessment of Biodiesel

There are few reports on biodiesel production from algal bio-oil but biodiesel from other feedstocks are commercially used in many countries, which has been given or conforms to recognized legislation/standard specifications such as ASTM D6751 (15C), EN14214, ON C1191, and DIN 51606 in the U.S., EU, Austria, and Germany, respectively. Algal biodiesel should also comply with these standards. Several analytical methods are employed in assessing the biofuel quality but the common ones used are chromatographic and spectroscopic methods (Knothe 2001). Quality parameters of biodiesel as per ASTM D6751 (15C) standard along with different test methods are summarized in Table 8.14 and the importance of some properties are described below:

> *Cetane Number (CN)*: It is an indicator of the ignition quality; the CN is a prime indicator of the fuel quality in the realm of diesel engines and gives a relative measurement of the interval between the beginning of injection and auto ignition of the fuel. Higher the cetane number the shorter the delay interval and the greater the combustibility of biodiesel.

TABLE 8.14

Biodiesel Standard

Property	Test Method	Limits	Units
Flash point (closed cup)	D 93	93 min	°C
Alcohol control. One of the following must be met:			
1. Methanol content	EN 14110	0.2 max	% volume
2. Flash point	D 93	130 min	130 min
Water and sediment	D 2709	0.050 max	% volume
Kinematic viscosity, 40°C	D 445	1.9–6.0	mm²/s
Sulfated ash	D 874	0.020 max	% mass
Sulfur	D5453	0.05 or 0.0015 max	% mass
Copper strip corrosion	D 130	No. 3 max	
Cetane number	D 613	47	min
Cloud point	D 2500		°C
Carbon residue	D 4530	0.050 max	% mass
Acid number	D 664	0.50 max	mg KOH/g
Free glycerin	D 6584	0.020	% mass
Total glycerin	D 6584	0.240	% mass
Phosphorus content	D 4951	0.001 max	% mass
Distillation temperature	D 1160	360	max °C
Atmospheric equivalent temperature, 90% recovered			
Sodium and potassium, combined	EN 14538	5 max	ppm (µg/g)
Calcium and magnesium, combined	EN 14538	5 max	ppm (µg/g)
Oxidation stability	EN 15751	3 min	Hours
Cold soak filterability	D7501	360 max	sec

Source: ASTM Standard D6751 (15C) http://www.astm.org/Standards/D6751.htm, accessed March 7, 2016.

Cloud Point (CP): It is also an indicator of ignition quality; the CP is a prime indicator of fuel quality in the realm of diesel engines and is the temperature at which oil starts to solidify. Even operating an engine at temperatures below oil's cloud point, to avoid waxing of the fuel heating will be necessary. Melting points of fatty acid esters mainly depend on unsaturation and chain length.

Flash Point (FP): It is the minimum temperature of a fuel at which it produces sufficient vapor for momentary flush in the application of an ignition source. Fuel's volatility varies inversely with the FP. For the proper safety and handling of fuel minimum FP temperatures are required.

Iodine Value (IV): It is the amount of iodine, measured in grams, absorbed by 100 mL of given oil. The IV of the oil is indicated by the degree of saturation.

Sulfur Percentage: It is the percentage, measured by weight, of sulfur present in the fuel. The sulfur content in diesel fuel used in on-road applications is limited by law to very small percentages.

Viscosity: It refers to the thickness of the oil, and is determined by measuring the amount of time taken for a given measure of oil to pass through an orifice of a specified size. Chain length and increasing saturation may lead to increase in viscosity.

Oxidative Stability: It is one of the most important technological challenges faced by biodiesel. It is affected by the presence of air, light, temperature, extraneous materials, container material, and headspace volume.

Peroxide Value: This peroxide value is not as much suitable for monitoring the oxidation as it tends to increase and then decrease due to its further oxidation, which leads to the formation of the secondary oxidation products. When the peroxide value reaches a plateau of about 350 mg equiv./kg ester during the biodiesel (SME) oxidation, viscosity and acid value continue to increase monotonically. Besides viscosity, the acid value has a suitable potential as a parameter for monitoring biodiesel quality during storage.

8.7.1 Stability Issue

Biodiesel produced from vegetable oils and other feedstocks is more prone to oxidation than a typical petroleum diesel and loses the stability in composition and properties unless it is modified or treated with additives to improve the stability. The degree of unsaturation present in biodiesel fuels makes them susceptible to oxidative and thermal polymerization, which may lead to the formation of insoluble products (loss of stability) that create problems inside the fuel system, particularly in the injection pump.

When biodiesel is exposed to light, heat and stress, or trace metals, it leads to the formation of free radical (initiator) or an electron imbalance on the site of double bond of a fatty acid. It is most probable to happen on a definite type of unsaturated fatty acid with two or more double bonds. The form radical (initiator) is highly reactive in nature and combines with any oxygen present. Once the free radical combines with oxygen, peroxide radical is formed in the fatty acid. When this peroxide radical bumps into another unsaturated fatty acid, it makes the new fatty acid into a radical and changes the peroxide into a hydroperoxide and then back into a free radical (propagation). This conversion of fatty acids into peroxides continues until the peroxide bumps into a radical rather than into an unconverted fatty acid (termination). The whole process can be presented through Equations 8.1 through 8.5.

Initiation:
$$RH + I \rightarrow R\bullet + IH \tag{8.1}$$

Propagation:
$$R\bullet + O_2 \rightarrow ROO \tag{8.2}$$

$$ROO\bullet + RH \rightarrow ROOH + F \tag{8.3}$$

Termination:
$$R\bullet + R\bullet \rightarrow R-R \tag{8.4}$$

$$ROO\bullet + ROO\bullet \rightarrow \text{Stable products} \tag{8.5}$$

Peroxide value increases dramatically in this process. Up to 100 new radicals are formed quickly from one single radical, which signifies that decomposition occurs at an exponentially speedy rate, and the oil spoils and becomes rancid very quickly.

The auto-oxidation process stops making further radicals when it bumps into another radical rather than an unradicalized fatty acid. When two radicals bump into each other, the fatty acids are damaged by either connecting the two fatty acids together at the radical (secondary oxidation) or splitting the fatty acid at the double bond, which form aldehydes, alcohols, and carbonic acids.

There are three suitable approaches to increase the resistance to the process of auto-oxidation. The first is to starve the biodiesel intended for oxygen. It is not an easy process because biodiesel absorbs four times as much oxygen as water; displacing the tank head-space with nitrogen can help to some extent. The second approach is to prevent the contact with substances and materials that promote or catalyze the reaction of oxidation such as pro-oxidants, higher temperature, trace metals, light, and so on. And the third approach is to use antioxidants. The three suitable and main ways anti-oxidants work are to attack the radicals to make them less reactive, interfere with the chemical reactions that form the radicals, and chemically remove the pro-oxidants similar to trace metals.

Both natural and synthetic antioxidants are available for increasing the stability of bio-diesel. Tocopherol (vitamin E) is an example of natural antioxidant. Some synthetic antioxidants are pyrogallol, propyl gallate, butylated hydroxyanisole, butylated hydroxytoluene, tert-butyl hydroquinone, and so on; the synthetic antioxidants are more efficient than the natural antioxidants.

Biodiesel fuels are monitored throughout the complete storage period by analysis of several (15) different properties. Several other properties do not show any important change during storage, even as others such as viscosity, peroxide value, and more dramatically, Rancimat Induction Period demonstrated changes associated with the nature of the starting product. The peroxide number and neutralization number of the biodiesel must be analyzed for long-term storage.

8.8 Jet Fuel from Microalgae

It is typical type of jet fuel, specially designed for use in aircrafts driven by gas-turbine engines. The most frequently used fuels for commercial flying are Jet A and Jet A-1, which are produced by a standardized international specification. Jet B is the only other jet fuel commonly used in civilian turbine-engine-powered aviation, which is used due to its enhanced cold-weather performance. Microalgae are generally used to produce Jet B fuel. Jet fuels are a mixture of a variety of different hydrocarbons. The range of their molecular weights or carbon numbers is controlled by the necessities for the product, for example, the freezing point or smoke point. Kerosene-type jet fuels (including Jet A and Jet A-1) have carbon numbers between about 8 and 16 (carbon atoms per molecule); naphtha-type jet fuel (including Jet B), between about 5 and 15.

Table 8.15 shows the classification of different types of commercial and military jet fuels. Light jet fuels (kerosene) are produced by refining the crude oil by distillation in the presence of a catalyst. Jet fuels used in commercial aviation are mainly of two grades: Jet A-1 and Jet A, both are kerosene-type jet fuels. Jet B is a mixture of gasoline with kerosene (a wide cut kerosene) but it is not usually used except in very cold climates. Sometimes fuels are chemically enhanced with antioxidants, dispersants and corrosion inhibitors for specific requirements such as JP-5 and JP-8.

Few groups have shown that jet fuels can also be produced from microalgae, other than fossil fuel crude oil (Campanella et al. 2012). In the recent past a new approach toward thermolysis of microalgal biomass has been developed; by using moderate temperatures (around 300°C) biomass is converted into bioleum, which can be further processed to produce various fuels (Figure 8.10). Using a similar approach, Wang et al. worked toward developing Jet B fuel from bioleum produced from microalgal biomass, and further

TABLE 8.15

Classification of Commercial and Military Jet Fuel

Civil/Commercial Jet Fuel	Military Jet Fuel
Type1: Jet A-1: It is based on kerosene grade fuels, which are suitable for mostly turbine engine aircraft. Flash point above 38°C (100°F). Freezing point maximum of (−47°C). Net heat of combustion minimum of (43.15 MJ/kg). Density at 15°C is 804 kg/m³.	**Type1: JP-4:** This is the military equivalent grade fuel of Jet B with the addition of corrosion inhibitor and anti-icing additives have high efficiency.
Type2: JET A: It is another kerosene-based fuel. Flash point above 38°C (100°F) Freezing point maximum (−40°C).	**Type2: JP-5:** These types of fuel possess high flash point type kerosene-based fuel.
Type3: TS-1: Kerosene-type BP Jet TS-1 with higher volatility (for lower temperatures and flash point is 28°C minimum), Net heat of combustion is 43.2 MJ/kg, density at 15°C is 787 kg/m³.	
Type4: JET B: It is distillate product covering the naphtha and kerosene fractions component. Freezing point maximum of (−50°C). Minimum of (42.8 MJ/kg) Net heat of combustion. Very high flammability. Significant demand in high cold climates and such climatic region.	**Type3: JP-8:** This is military equivalent of Jet A-1 with the addition of corrosion inhibitor with anti-icing additives.

Source: Modified from Elmoraghy, M., and I. Farag, *Int. J. Eng. Sci.*, 1, 22–30, 2012; Shell Global, Civil Jet Fuel Shell Global, http://www.shell.com/business-customers/aviation/aviation-fuel/civil-jet-fuel-grades. html.

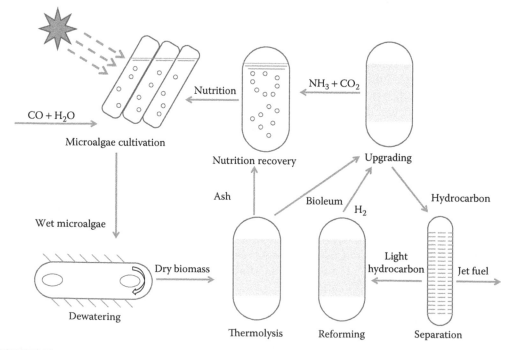

FIGURE 8.10

Concept flow diagram for microalgae to jet fuel. (From Wang, H.Y. et al., *Bioresour. Technol.*, 167, 349–357, 2014. With permission.)

TABLE 8.16

Properties of Bio-Jet B Compared to Industrial Standards for Jet B Product

Properties	Bio-Jet B	Industrial Jet B
Heating value (MJ/kg)	43.7	42.8 (min)
Freezing point (°C)	−60.1	−51 (max)
Flash point (°C)	15.8	—
Smoke point (mm)	30.3	25 (min)
Specific gravity	0.764	0.750–0.800
50% recovery point (°C)	182.3	190 (max)
End point (°C)	244.2	270 (max)
Production rate (kg/h)	826.8	—

Source: Data from Wang, H.Y. et al., *Bioresour. Technol.*, 167, 349–357, 2014.

upgraded it by hydrotreating and hydrocracking. From the physiochemical properties of biomolecules to reaction thermodynamics, all the parameters were simulated and optimized (Wang et al. 2014) (Table 8.16).

There is need for developing efficient methods for producing the Jet B aviation fuel from the microalgae; however, the dependence on the fossil fuel can also be decreased by blending the bio-oil produced from microalgae with jet fuel (Figure 8.11). Blending involves algae growth, harvesting, oil extraction, and its transesterification into biodiesel. The biodiesel can be blended with Jet A (or JP5) to produce biojet fuel blend that can be used in jet engines. Bio-JP-5 as an alternative to JP-8 fuel has already been accepted by the U.S. navy. Biodiesel is relatively nontoxic, contains no aromatics or sulfur, and is further biodegradable and safer to handle compared to diesel or jet fuels. Biodiesel also adds lubricity to the fuel. The main advantage of using biodiesel is the reduction in the emission of particulate matter (PM) (Elmoraghy and Farag 2012).

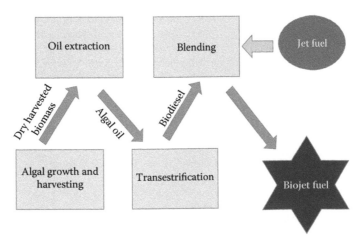

FIGURE 8.11
Biojet fuel production by the blending process.

8.9 World Scenario

8.9.1 Current Research in Microalgal Biofuels

It would not be farfetched to state that research in the algal biology for biofuel production has become a global phenomenon. Various institutions and universities, working in the field of renewable energy and biofuel in every corner of the world, are looking for opportunities promised by the microalgae in the development of a renewable energy source.

A number of institutes all over the world are working to make microalgae a viable and efficient energy source. Both industries and institutes are spending time and capital on the development of technology for the production of algal fuel. The examples of few institutes spread all over the world and yet working in the same research field are mentioned in Table 8.17.

TABLE 8.17

Institutes and Industries Focusing on Research in the Field of Algal Fuel

Industry/Institute	Country	Research Area	Reference
The Bioeconomy Institute (BEI) Iowa State University	United States	Cost-effective harvesting of microalgal cells	IOWA State University 2016
Leibniz University, Hannover	Germany	Development of disposable photobioreactors	Leibniz University Hannover 2016
Murdoch University	Australia	Industrial production of oil and carotenoids from microalgae, efficient use of solar energy by microalgae, sponge–algae interaction	Murdoch University 2016
University of Texas, Austin	United States	Extraction of oil and other high-value products from algae, cost-effective bio-oil extraction	University of Texas 2016
CFTRI Mysore	India	Enhancing lipid accumulation in microalgae, isolation and characterization of native microalgal strains	CFTRI 2016
Korea Advanced Institute of Science and Technology (KAIST)	South Korea	Using genome editing techniques for increased biofuel production from microalgae	KAIST 2014
Indian Institute of Technology Roorkee	India	Development of genetically engineered microalgae	IIT Roorkee 2016
Monash University	Australia	Bioremediation and biofuel production by algae	Monash University 2016
Indian Institute of Technology Kharagpur	India	Increasing biofuel production from microalgae	IIT Kharagpur 2016
Cellana Inc.	United States	Developing biodiesel production system from microalgae	Cellana Inc. 2013
Algenol	United States	Developing a third-generation advanced process to convert sunlight, CO_2 and water directly into ethanol, cost-effective ethanol production from microalgae	Algenol 2015

TABLE 8.18

Industries Producing Algal Biofuel at Commercial Scale

Industry	Country	Product	Production Capacity	Reference
Algenol	United States	Ethanol, gasoline, jet fuel, and diesel	8000 gallons/acre/year	Algenol 2015
Solazyme	United States	Renewable biofuel	—	Solazyme 2016
BioFuel Systems	Spain	Bio-oil and ethanol	—	Planet Ark 2006
Cellana Inc.	United States	Byproducts of biofuels	—	Cellana Inc. 2013
Sapphire Energy	United States	Crude oil from algae	100 barrels/day in 2012 (5000 barrels/day by 2018)	Sapphire Energy 2013
Synthetic Genomics	United States	Ethanol and bio-hydrogen	—	Synthetic Genomics 2009
Reliance	India	Biofuel	66,800 barrels/day (projected)	BiofuelsDigest 2015

8.9.2 Industries in Commercial Production of Algal Biofuel

Several industries utilizing the technology developed so for are producing algal biofuel at commercial scale. Few of the noted companies are listed in the Table 8.18.

References

Ahmad, A.L., N.M. Yasin, C.J.C. Derek, and J.K. Lim. 2011. Microalgae as a Sustainable Energy Source for Biodiesel Production: A Review. *Renewable and Sustainable Energy Reviews* 15(1): 584–593.

Algenol. 2015. Ethanol Producer Magzeen Algenol. http://www.ethanolproducer.com/articles/12629/algenol-ethanol-plant-will-have-capacity-to-produce-18-mmgy, accessed March 7, 2016.

Andrich, G., U. Nesti, F. Venturi, A. Zinnai, and R. Fiorentini. 2005. Supercritical Fluid Extraction of Bioactive Lipids from the Microalga Nannochloropsis sp. *European Journal of Lipid Science and Technology* 107(6): 381–386.

Antunes, W.M., C. de Oliveira Veloso, and C.A. Henriques. 2008. Transesterification of Soybean Oil with Methanol Catalyzed by basic Solids. *Catalysis Today* 133: 548–554.

ASTM Standard D6751 (15C). http://www.astm.org/Standards/D6751.htm, accessed March 3, 2016.

Azgın, C., O. Işık, L. Uslu, and B. Ak. 2014. A Comparison the Biomass of Productivity, Protein and Lipid Content of Spirulina platensis Cultured in the Pond and Photobioreactor. *Journal of Biological and Environmental Sciences* 8: 24.

Balakrishnan, N., A. Patricia, and J. Tony Pembroke. 2015. Ultrasonic intensification as a tool for enhanced microbial biofuel yields. *Biotechnol Biofuels* 8: 140.

Beer, L.L., E.S. Boyd, J.W. Peters, and M.C. Posewitz. 2009. Engineering Algae for Biohydrogen and Biofuel Production. *Current Opinion in Biotechnology* 20(3): 264–271.

Bellou, S., and G. Aggelis. 2013. Biochemical Activities in Chlorella sp. and Nannochloropsis Salina during Lipid and Sugar Synthesis in a Lab-Scale Open Pond Simulating Reactor. *Journal of Biotechnology* 164(2): 318–329.

Bellou, S., M.N. Baeshen, A.M. Elazzazy, D. Aggeli, F.Sayegh, and G. Aggelis. 2014. Microalgal Lipids Biochemistry and Biotechnological perspectives. *Biotechnology Advances* 32(8): 1476–1493.

Bigogno, C., I. Khozin-Goldberg, S. Boussiba, A.Vonshak, and Z. Cohen. 2002. Lipid and Fatty Acid Composition of the Green Oleaginous Alga Parietochloris incisa, the Richest Plant Source of Arachidonic Acid. *Phytochemistry* 60(5): 497–503.

BiofuelsDigest. 2015. Reliance and Algenol World Largest Oil Refinery for algal Biofuels BiofuelsDigest News. http://www.biofuelsdigest.com/bdigest/2015/01/26/worlds-largest-oil-refinery-adds-algae-demonstration-project/, accessed March 7, 2016.

Boussiba, C.H., S. Cuello, J.L. Duke, C.S. Efroymson, R.A. Golden, S.S. Holmgren, J. Johnson, D.L. Jones, M.E. Smith, and V.H. Steger. 2012. *Sustainable Development of Algal Biofuels in the United States*. The National Academies Press, Washington, DC.

Brennan, L., and P. Owende. 2010. Biofuels from Microalgae—A Review of Technologies for Production, Processing, and Extractions of Biofuels and Co-Products. *Renewable and Sustainable Energy Reviews* 14(2): 557–577.

Cagliari, A., R. Margis, F. dos Santos Maraschin, A.C. Turchetto-Zolet, G. Loss, and M. Margis-Pinheiro. 2011. Biosynthesis of Triacylglycerols (TAGs) in Plants and Algae. *International Journal of Plant Biology* 2(1): 10. doi:10.4081/pb.2011.e10.

Campanella, A., R. Muncrief, M.P., Harold, D.C. Griffith, N.M. Whitton, and R.S. Weber. 2012. Thermolysis of Microalgae and Duckweed in a CO_2-swept Fixed-bed Reactor: Bio-oil Yield and Compositional Effects. *Bioresource Technology* 109: 154–162.

Canela, A.P.R., P.T. Rosa, M.O. Marques, and M.A.A. Meireles. 2002. Supercritical Fluid Extraction of Fatty Acids and Carotenoids from the Microalgae Spirulina Maxima. *Industrial & Engineering Chemistry Research* 41(12): 3012–3018.

Cellana Inc. 2013. Cellana and Neste Oil Agreement on Algae Oil Feedstock For Biofuels Cellana Inc. http://cellana.com/press-releases/cellana-and-neste-oil-enter-into-multi-year-commercial-scale-off-take-agreement-for-algae-oil-feedstock-for-biofuels/, accessed March 7, 2016.

CFTRI. 2016. Department of Plant Cell Biotechnology CFTRI Mysore. http://www.cftri.com/Rareas/PCBT/rd.php, accessed March 7, 2016.

Chen, C., Z. Lu, X. Ma, J. Long, Y. Peng, L. Hu, and Q. Lu. 2013. Oxy-Fuel Combustion Characteristics and Kinetics of Microalgae Chlorella Vulgaris by Thermogravimetric Analysis. *Bioresource Technology* 144: 563–571.

Chen, C.Y., K.L. Yeh, R. Aisyah, D.J. Lee, and J.S. Chang. 2011. Cultivation, Photobioreactor Design and Harvesting of Microalgae for Biodiesel Production: A Critical Review. *Bioresource Technology* 102(1): 71–81.

Chen, J.E., and A.G. Smith. 2012. A Look at Diacylglycerol Acyltransferases (DGATs) in Algae. *Journal of Biotechnology* 162(1): 28–39.

Chen, Y.M., J.C. Liu, and Y.H. Ju. 1998. Flotation removal of algae from water. Colloids and Surfaces B. *Biointerfaces* 12(1): 49–55.

Chisti, Y. 2007. Biodiesel from Microalgae. *Biotechnology Advances* 25(3): 294–306.

Chiu, S.Y., M.T. Tsai, C.Y. Kao, S.C. Ong, and C.S. Lin. 2009. The Air-lift Photobioreactors with Flow Patterning for High-Density Cultures of Microalgae and Carbon Dioxide Removal. *Engineering in Life Sciences* 9(3): 254–260.

Converti, A., A.A. Casazza, E.Y. Ortiz, P. Perego, and M. Del Borghi. 2009. Effect of Temperature and Nitrogen Concentration on the Growth and Lipid Content of Nannochloropsis Oculata and Chlorella Vulgaris for Biodiesel Production. *Chemical Engineering and Processing: Process Intensification* 48(6): 1146–1151.

Cooney, M., G. Young, and N. Nagle. 2009. Extraction of Bio-Oils from Microalgae. *Separation & Purification Reviews* 38(4): 291–325.

Davis, R., A. Aden, and P.T. Pienkos. 2011. Techno-Economic Analysis of Autotrophic Microalgae for Fuel Production. *Applied Energy* 88(10): 3524–3531.

De Bhowmick, G., L. Koduru, and R. Sen. 2015. Metabolic Pathway Engineering Towards Enhancing Microalgal Lipid Biosynthesis for Biofuel Application—A Review. *Renewable and Sustainable Energy Reviews* 50: 1239–1253.

Dean, A.P., D.C. Sigee, B. Estrada, and J.K. Pittman. 2010. Using FTIR Spectroscopy for Rapid Determination of Lipid Accumulation in Response to Nitrogen Limitation in Freshwater Microalgae. *Bioresource Technology* 101(12): 4499–4507.

Demirbas, A. 2008. *Biodiesel*. Springer, London, UK, pp. 111–119.

Deng, X., J. Cai, and X. Fei. 2013. Effect of the Expression and Knockdown of Citrate Synthase Gene on Carbon Flux During Triacylglycerol Biosynthesis by Green Algae Chlamydomonas Reinhardtii. *BMC Biochemistry* 14(1): 1.

Elmoraghy, M., and I. Farag. 2012. Bio-jet Fuel from Microalgae: Reducing Water and Energy Requirements for Algae Growth. *International Journal of Engineering and Sciences* 1: 22–30.

Fajardo, A.R., L.E. Cerdan, A.R. Medina, F.G.A. Fernandez, P.A.G. Moreno, and E.M. Grima. 2007. Lipid Extraction from the Microalga Phaeodactylum Tricornutum. *European Journal of Lipid Science and Technology* 109(2): 120–126.

Fan, J., Y. Cui, M. Wan, W. Wang, and Y. Li. 2014. Lipid Accumulation and Biosynthesis Genes Response of the Oleaginous Chlorella Pyrenoidosa Under Three Nutrition Stressors. *Biotechnology for Biofuels* 7(1): 1.

Fukuda, H., A. Kondo, and H. Noda. 2001. Biodiesel fuel Production by Transesterification of Oils. *Journal of Bioscience and Bioengineering* 92(5): 405–416.

Gombos, Z., M. Kis, T. Pali, and L. Vigh. 1987. Nitrate Starvation Induces Homeoviscous Regulation of Lipids in the Cell envelope of the Blue-Green Alga, Anacystis Nidulans. *European Journal of Biochemistry* 165: 465.

Gong, Y., and M. Jiang. 2011. Biodiesel Production with Microalgae as Feedstock: From Strains to Biodiesel. *Biotechnology Letters* 33(7): 1269–1284.

Guckert, J.B., K.E. Cooksey, and L.L. Jackson. 1988. Lipid Sovent Systems are not Equivalent for Analysis of Lipid Classes in the Microeukaryotic Green Alga, Chlorella. *Journal of Microbiological Methods* 8(3): 139–149.

Guo, J., E. Peltier, R.E. Carter, A.J. Krejci, S.M. Stagg-Williams, and C. Depcik. 2012. Waste Cooking Oil Biodiesel use in two off-road Diesel Engines. *ISRN Renewable Energy* 2012: 1–10.

Halim, R., M.K. Danquah, and P.A. Webley. 2012. Extraction of Oil from Microalgae for Biodiesel Production: A Review. *Biotechnology Advances* 30(3): 709–732.

Hanotu, J., H.C. Bandulasena, and W.B. Zimmerman. 2012. Microflotation Performance for Algal Separation. *Biotechnology and Bioengineering* 109(7): 1663–1673.

Harwood, J.L. 1998. Membrane Lipids in Algae. In *Lipids in Photosynthesis: Structure, Function and Genetics*. Springer Netherlands, pp. 53–64.

Heasman, M., J. Diemar, W. O'connor, T. Sushames, and L. Foulkes. 2000. Development of Extended Shelf-Life Microalgae Concentrate Diets Harvested by Centrifugation for Bivalve Molluscs–A Summary. *Aquaculture Research* 31(8–9): 637–659.

Herrero, M., E. Ibáñez, J. Señoráns, and A. Cifuentes. 2004. Pressurized Liquid Extracts from Spirulina Platensis Microalga: Determination of their Antioxidant Activity and Preliminary Analysis by Micellar Electrokinetic Chromatography. *Journal of Chromatography A* 1047(2): 195–203.

Hu, Q., M. Sommerfeld, E. Jarvis, M. Ghirardi, M. Posewitz, M. Seibert, and A. Darzins. 2008. Microalgal Triacylglycerols as Feedstocks for Biofuel Production: Perspectives and Advances. *The Plant Journal* 54(4): 621–639.

Huang, G., F. Chen, D. Wei, X. Zhang, and G. Chen. 2010. Biodiesel Production by Microalgal Biotechnology. *Applied Energy* 87(1): 38–46.

Huerlimann, R., R. De Nys, and K. Heimann. 2010. Growth, Lipid Content, Productivity, and Fatty Acid Composition of Tropical Microalgae for Scale-Up Production. *Biotechnology and Bioengineering* 107(2): 245–257.

Ibáñez-Salazar, A., S. Rosales-Mendoza, A. Rocha-Uribe, J.I. Ramírez-Alonso, I. Lara-Hernández, A. Hernández-Torres, L.M.T. Paz-Maldonado, A.S. Silva-Ramírez, B. Bañuelos-Hernández, J.L. Martínez-Salgado, and R.E. Soria-Guerra. 2014. Over-Expression of Dof-type Transcription Factor Increases Lipid Production in Chlamydomonas Reinhardtii. *Journal of Biotechnology* 184: 27–38.

IIT Kharagpur. 2016. Nirupma Mallick Research Finding IIT Kharagpur. http://www.iitkgp.ac.in/fac-profiles/showprofile.php?empcode=bbmbS, accessed March 7, 2016.

IIT Roorkee. 2016. Department of Biotchnology IIT Roorkee. http://www.iitr.ac.in/departments/BT/pages/Research+MajorGroupsAreas.html, accessed March 7, 2016.

IOWA State University. 2016. Algal Research Program IOWA State University. http://www.biorenew.iastate.edu/research/signature/algae/, accessed March 7, 2016.

Iwai, M., K. Ikeda, M. Shimojima, and H. Ohta. 2014. Enhancement of Extraplastidic Oil Synthesis in Chlamydomonas Reinhardtii using a Type-2 Diacylglycerol Acyltransferase with a Phosphorus Starvation–Inducible Promoter. *Plant Biotechnology Journal* 12(6): 808–819.

Jo, Y.J., O.K. Lee, and E.Y. Lee. 2014. Dimethyl Carbonate-Mediated Lipid Extraction and Lipase-catalyzed in situ Transesterification for Simultaneous Preparation of Fatty Acid Methyl Esters and Glycerol Carbonate from *Chlorella sp.* KR-1 biomass. *Bioresource Technology* 158: 105–110.

KAIST. 2014. ToolGen & KAIST to Collborate on Biofuel Gene Editing. http://www.asianscientist.com/2014/09/tech/toolgen-kaist-collaborate-biofuel-gene-editing/, accessed March 7, 2016.

Kang, N.K., S. Jeon, S. Kwon, H.G. Koh, S.E. Shin, B. Lee, G.G. Choi, J.W. Yang, B.R. Jeong, and Y.K. Chang. 2015. Effects of Overexpression of a bHLH Transcription Factor on Biomass and Lipid Production in Nannochloropsis Salina. *Biotechnology for Biofuels* 8(1): 1.

Kheira, A.A.A., and N.M. Atta. 2009. Response of Jatropha Curcas L. to Water Deficits: Yield, Water use Efficiency and Oilseed Characteristics. *Biomass and Bioenergy* 33(10): 1343–1350.

Khozin-Goldberg, I., and Z. Cohen. 2011. Unraveling Algal Lipid Metabolism: Recent Advances in Gene Identification. *Biochimie* 93(1): 91–100.

Kim, Y.H., Y.K. Choi, J. Park, S. Lee, Y.H. Yang, H.J. Kim, T.J. Park, Y.H. Kim, and S.H. Lee. 2012. Ionic Liquid-mediated Extraction of Lipids from Algal Biomass. *Bioresource Technology* 109: 312–315.

Knothe, G. 2001. Analytical Methods used in the Production and Fuel Quality Assessment of Biodiesel. *Transactions of the ASAE* 44(2): 193–200.

Knothe, G. 2005. Dependence of Biodiesel Fuel Properties on the Structure of Fatty Acid Alkyl Esters. *Fuel Processing Technology* 86(10): 1059–1070.

Lai, J.Q., Z.L. Hu, P.W. Wang, and Z. Yang. 2012. Enzymatic Production of Microalgal Biodiesel in Ionic Liquid [BMIm][PF 6]. *Fuel* 95: 329–333.

Lam, M.K., I.S. Tan, and K.T. Lee. 2014. Utilizing Lipid-Extracted Microalgae Biomass Residues for Maltodextrin Production. *Chemical Engineering Journal* 235: 224–230.

Lang, I., L. Hodac, T. Friedl, and I. Feussner. 2011. Fatty Acid Profiles and Their Distribution Patterns in Microalgae: A Comprehensive Analysis of more than 2000 Strains from the SAG Culture Collection. *BMC Plant Biology* 11(1): 1.

Lee, O.K., Y.H. Kim, J.G. Na, Y.K. Oh, and E.Y. Lee. 2013. Highly Efficient Extraction and Lipase-catalyzed Transesterification of Triglycerides from Chlorella sp. KR-1 for Production of Biodiesel. *Bioresource Technology* 147: 240–245.

Lee, S.J., B.D. Yoon, and H.M. Oh. 1998. Rapid Method for the Determination of Lipid from the Green Alga Botryococcus Braunii. *Biotechnology Techniques* 12(7): 553–556.

Leibniz University Hannover. 2016. Microalgae Biotechnology Research Group Leibniz University Hannover. http://www.bgt-hannover.de/algae/, accessed March 7, 2016.

Li, X., H. Xu, and Q. Wu. 2007. Large-scale Biodiesel Production from Microalga Chlorella Protothecoides Through Heterotrophic Cultivation in Bioreactors. *Biotechnology and Bioengineering* 98(4): 764–771.

Li, Y., X. Fei, and X. Deng. 2012. Novel Molecular Insights into Nitrogen Starvation-Induced Triacylglycerols Accumulation Revealed by Differential Gene Expression Analysis in Green Algae Micractinium Pusillum. *Biomass and Bioenergy* 42: 199–211.

Liu, Z.Y., G.C. Wang, and B.C. Zhou. 2008. Effect of Iron on Growth and Lipid Accumulation in Chlorella Vulgaris. *Bioresource Technology* 99(11): 4717–4722.

Luque-García, J.L., and M.L. De Castro. 2003. Ultrasound: A Powerful Tool for Leaching. *TrAC Trends in Analytical Chemistry* 22(1): 41–47.

Lv, H., G. Qu, X. Qi, L. Lu, C. Tian, and Y. Ma. 2013. Transcriptome Analysis of Chlamydomonas Reinhardtii During the Process of Lipid Accumulation. *Genomics* 101(4): 229–237.

Ma, F. and M.A. Hanna. 1999. Biodiesel Production: A Review. *Bioresource Technology* 70(1): 1–15.

Macıas-Sánchez, M.D., C. Mantell, M. Rodrıguez, E.M. de La Ossa, L.M. Lubián, and O. Montero. 2005. Supercritical Fluid Extraction of Carotenoids and Chlorophyll a from Nannochloropsis Gaditana. *Journal of Food Engineering* 66(2): 245–251.

Marjakangas, J.M., C.Y. Chen, A.M. Lakaniemi, J.A. Puhakka, L.M. Whang, and J.S. Chang. 2015. Selecting an Indigenous Microalgal Strain for Lipid Production in Anaerobically Treated Piggery Wastewater. *Bioresource Technology* 191: 369–376.

Mata, T.M., A.A. Martins, and N.S. Caetano. 2010. Microalgae for Biodiesel Production and other Applications: A Review. *Renewable and Sustainable Energy Reviews* 14(1): 217–232.

Meher, L.C., D.V. Sagar, and S.N. Naik, 2006b. Technical Aspects of Biodiesel Production by Transesterification—A Review. *Renewable and Sustainable Energy Reviews* 10(3): 248–268.

Mendes, R.L., J.P. Coelho, H.L. Fernandes, I.J. Marrucho, J.M. Cabral, J.L.M. Novais, and A.A.F. Palavra. 1995. Applications of Supercritical CO_2 Extraction to Microalgae and Plants. *Journal of Chemical Technology and Biotechnology* 62(1): 53–59.

Mendes, R.L., B.P. Nobre, M.T. Cardoso, A.P. Pereira, and A.F. Palavra. 2003. Supercritical Carbon Dioxide Extraction of Compounds with Pharmaceutical Importance from Microalgae. *Inorganica Chimica Acta* 356: 328–334.

Meng, X., J.Yang, X. Xu, L. Zhang, Q. Nie, and M. Xian. 2009. Biodiesel Production from Oleaginous Microorganisms. *Renewable Energy* 34(1): 1–5.

Milledge, J.J., and S. Heaven. 2013. A Review of the Harvesting of Micro-Algae for Biofuel Production. *Reviews in Environmental Science and Bio/Technology* 12(2): 165–178.

Monash University. 2016. John Beardall Research Finding Monash University. https://www.monash.edu/research/people/profiles/profile.html?sid=521&pid=2711, accessed March 7, 2016.

Morgan-Kiss, R.M., J.C. Priscu, T. Pocock, L. Gudynaite-Savitch, and N.P. Huner. 2006. Adaptation and Acclimation of Photosynthetic Microorganisms to Permanently Cold Environments. *Microbiology and Molecular Biology Reviews* 70(1): 222–252.

Mujtaba, G., W. Choi, C.G. Lee, and K. Lee. 2012. Lipid Production by Chlorella Vulgaris after a Shift from Nutrient-Rich to Nitrogen Starvation Conditions. *Bioresource Technology* 123: 279–283.

Murdoch University. 2016. Algae R&D Center Murdoch University. http://www.murdoch.edu.au/Research-capabilities/Algae-R-and-D-Centre/, accessed March 7, 2016.

Nan, Y., J. Liu, R. Lin, and L.L. Tavlarides. 2015. Production of Biodiesel from Microalgae Oil (*Chlorella protothecoides*) by Non-Catalytic Transesterification in Supercritical Methanol and Ethanol: Process Optimization. *The Journal of Supercritical Fluids* 97: 174–182.

Oh, H.M., S.J. Lee, M.H. Park, H.S. Kim, H.C. Kim, J.H. Yoon, G.S. Kwon, and B.D. Yoon. 2001. Harvesting of Chlorella Vulgaris using a Bioflocculant from Paenibacillus sp. AM49. *Biotechnology Letters* 23(15): 1229–1234.

Papazi, A., P. Makridis, and P. Divanach. 2010. Harvesting *Chlorella Minutissima* using Cell Coagulants. *Journal of Applied Phycology* 22(3): 349–355.

Petrusevski, B., G. Bolier, A.N. Van Breemen, and G.J. Alaerts. 1995. Tangential Flow Filtration: A Method to Concentrate Freshwater Algae. *Water Research* 29(5): 1419–1424.

Planet Ark. 2006. World Environment News Planet Ark. http://www.planetark.com/dailynewsstory.cfm/newsid/37354/story.htm, accessed March 7, 2016.

Popoola, T.O.S., and O.D. Yangomodou. 2006. Extraction, Properties and Utilization Potentials of Cassava Seed Oil. *Biotechnology* 5(1): 38–41.

Pragya, N., K.K. Pandey, and P.K. Sahoo. 2013. A Review on Harvesting, oil Extraction and Biofuels Production Technologies from Microalgae. *Renewable and Sustainable Energy Reviews* 24: 159–171.

Praveenkumar, R., K. Shameera, G. Mahalakshmi, M.A. Akbarsha, and N. Thajuddin. 2012. Influence of Nutrient Deprivations on Lipid Accumulation in a Dominant Indigenous Microalga Chlorella sp., BUM11008: Evaluation for Biodiesel Production. *Biomass and Bioenergy* 37: 60–66.

Radakovits, R., P.M. Eduafo, and M.C. Posewitz. 2011. Genetic Engineering of Fatty acid Chain Length in Phaeodactylum Tricornutum. *Metabolic Engineering* 13(1): 89–95.

Ras, M., L. Lardon, S. Bruno, N. Bernet, and J.P. Steyer. 2011. Experimental Study on a Coupled Process of Production and Anaerobic Digestion of Chlorella vulgaris. *Bioresource Technology* 102: 200–206.

Reitan, K.I., J.R. Rainuzzo, and Y. Olsen. 1994. Effect of Nutrient Limitation on Fatty Acid and Lipid Content of Marine Microalgae1. *Journal of Phycology* 30(6): 972–979.

Richmond, A. 2000. Microalgal biotechnology at the turn of the millennium: a personal view. *Journal of Applied Phycology* 12(3): 441–451.

Ríos, S.D., J. Castañeda, C. Torras, X. Farriol, and J. Salvadó. 2013. Lipid Extraction Methods from Microalgal Biomass Harvested by Two Different Paths: Screening Studies Toward Biodiesel Production. *Bioresource Technology* 133: 378–388.

Roessler, P.G. 1988. Effects of Silicon Deficiency on Lipid Composition and Metabolism in the Diatom Cyclotella Cryptical. *Journal of Phycology* 24(3): 394–400.

Sajilata, M.G., R.S. Singhal, and M.Y. Kamat. 2008a. Fractionation of Lipids and Purification of γ-Linolenic Acid (GLA) from Spirulinaplatensis. *Food Chemistry* 109(3): 580–586.

Sajilata, M.G., R.S. Singhal, and M.Y. Kamat. 2008b. Supercritical CO_2 Extraction of γ-linolenic Acid (GLA) from Spirulina Platensis ARM 740 Using Response Surface Methodology. *Journal of Food Engineering* 84(2): 321–326.

Sakthivel, R. 2011. Microalgae Lipid Research, Past, Present: A Critical Review for Biodiesel Production, in the Future. *Journal of Experimental Sciences* 2: 10.

Sapphire Energy. 2013. New Study on algal crude oil Sapphire Energy. http://www.sapphireenergy. com/news-media, accessed March 7, 2016.

Sarkar, D., and K. Shimizu. 2015. An Overview on Biofuel and Biochemical Production by Photosynthetic Microorganisms with Understanding of the Metabolism and by Metabolic Engineering Together with Efficient Cultivation and Downstream Processing. *Bioresources and Bioprocessing* 2(1): 1–19.

Sato, N., M. Hagio, H. Wadaand, and M. Tsuzuki. 2000. Environmental Effects on Acidic Lipids of Thylakoid Membranes. *Biochemical Society Transactions* 28(6): 912–914.

Sharma, K.K., H. Schuhmann, and P.M. Schenk. 2012. High Lipid Induction in Microalgae for Biodiesel Production. *Energies* 5(5): 1532–1553.

Shell Global. 2016. Civil Jet Fuel Shell Global. http://www.shell.com/business-customers/aviation/ aviation-fuel/civil-jet-fuel-grades.html, accessed March 7, 2016.

Sibi, G., V. Shetty, and K. Mokashi. 2015. Enhanced Lipid Productivity Approaches in Microalgae as an Alternate for Fossil Fuels–A Review. *Journal of the Energy Institute.* 89(3): 330–334.

Sim, T.S., A. Goh, and E.W. Becker. 1988. Comparison of Centrifugation, Dissolved Air Flotation and Drum Filtration Techniques for Harvesting Sewage-grown Algae. *Biomass* 16(1): 51–62.

Singh, A., P.S. Nigam, and J.D. Murphy. 2011. Mechanism and Challenges in Commercialisation of Algal Biofuels. *Bioresource Technology* 102(1): 26–34.

Solazyme. 2016. Producing Sustainable, High Performance Oils and Ingredients Derived From Microalgae Solazyme. http://solazyme.com/company/, accessed March 7, 2016.

Stucki, S., F. Vogel, C. Ludwig, A.G. Haiduc, and M. Brandenberger. 2009. Catalytic Gasification of Algae in Supercritical Water for Biofuel Production and Carbon Capture. *Energy & Environmental Science* 2(5): 535–541.

Synthetic Genomics. 2009. J. Craig Research Finding Ethanol and Bio-hydrogen production from Synthetic Genomes-Genetically Engineered Mic4robes. http://www.xconomy.com/san-diego/2009/06/04/craig-venter-has-algae-biofuel-in-synthetic-genomics-pipeline/, accessed March 3, 2016.

Takagi, M. and T. Yoshida. 2006. Effect of Salt Concentration on Intracellular Accumulation of Lipids and Triacylglyceride in Marine Microalgae Dunaliella Cells. *Journal of Bioscience and Bioengineering* 101(3): 223–226.

Talebi, A.F., M. Tohidfar, S.M. Mousavi Derazmahalleh, A. Sulaiman, A.S. Baharuddin, and M. Tabatabaei. 2015. Biochemical Modulation of Lipid Pathway in Microalgae Dunaliella sp. for Biodiesel Production. *BioMed Research International* 2015: 597198.

The University of Texas. 2016. Algal Program The University of Texas, Austin. https://www.utexas. edu/research/cem/algae_project.html, accessed March 7, 2016.

Tian, J., M. Zheng, G. Yang, L. Zheng, J. Chen, and B. Yang. 2013. Cloning and Stress-Responding Expression Analysis of Malonyl CoA-acyl Carrier Protein Transacylase Gene of Nannochloropsis Gaditana. *Gene* 530(1): 33–38.

Tran, D.T., K.L. Yeh, C.L. Chen, and J.S. Chang. 2012. Enzymatic Transesterification of Microalgal Oil from *Chlorella vulgaris* ESP-31 for Biodiesel Synthesis using Immobilized Burkholderia Lipase. *Bioresource Technology* 108: 119–127.

Tredici, M.R. 2004. Mass Production of Microalgae: Photobioreactors. *Handbook of Microalgal Culture: Biotechnology and Applied Phycology* 1: 178–214.

Trentacoste, E.M., R.P. Shrestha, S.R. Smith, C. Glé, A.C. Hartmann, M. Hildebrand, and W.H. Gerwick. 2013. Metabolic Engineering of Lipid Catabolism Increases Microalgal Lipid Accumulation without Compromising Growth. *Proceedings of the National Academy of Sciences* 110(49): 19748–19753.

Uduman, N., Y. Qi, M.K. Danquah, G.M. Forde, and A. Hoadley. 2010. Dewatering of Microalgal Cultures: A Major Bottleneck to Algae-Based Fuels. *Journal of Renewable and Sustainable Energy* 2(1): 012701.

Ugwu, C.U., H. Aoyagi, and H. Uchiyama. 2008. Photobioreactors for Mass Cultivation of Algae. *Bioresource Technology* 99(10): 4021–4028.

Vree, J.H., R. Bosma, M. Janssen, M.J. Barbosa, and R.H. Wijffels. 2015. Comparison of Four Outdoor Pilot-Scale Photobioreactors. *Biotechnology for Biofuels* 8(1): 1.

Wahidin, S., A. Idris, and S.R.M. Shaleh. 2014. Rapid Biodiesel Production Using Wet Microalgae via Microwave Irradiation. *Energy Conversion and Management* 84: 227–233.

Wang, H., L. Gao, L. Chen, F. Guo, and T. Liu. 2013. Integration Process of Biodiesel Production from Filamentous Oleaginous Microalgae Tribonema minus. *Bioresource Technology* 142: 39–44.

Wang, H.Y., D. Bluck, and B.J. Van Wie. 2014. Conversion of Microalgae to Jet Fuel: Process Design and Simulation. *Bioresource Technology* 167: 349–357.

Xin, L., H. Hong-ying, G. Ke, and S. Ying-xue. 2010. Effects of Different Nitrogen and Phosphorus Concentrations on the Growth, Nutrient Uptake, and Lipid Accumulation of a Freshwater Microalga Scenedesmus sp. *Bioresource Technology* 101(14): 5494–5500.

Xu, L., F. Wang, H.Z. Li, Z.M. Hu, C. Guo, and C.Z. Liu. 2010. Development of An Efficient Electroflocculation Technology Integrated with Dispersed-air Flotation for Harvesting Microalgae. *Journal of Chemical Technology and Biotechnology* 85(11): 1504–1507.

Xu, L., P.J. Weathers, X.R. Xiong, and C.Z. Liu. 2009. Microalgal Bioreactors: Challenges and Opportunities. *Engineering in Life Sciences* 9(3): 178–189.

Xue, J., Y.F. Niu, T. Huang, W.D. Yang, J.S. Liu, and H.Y. Li. 2015. Genetic Improvement of the Microalga Phaeodactylum Tricornutum for Boosting Neutral Lipid Accumulation. *Metabolic Engineering* 27: 1–9.

Yan, J., R. Cheng, X. Lin, S. You, K. Li, H. Rong, and Y. Ma. 2013. Overexpression of Acetyl-CoA Synthetase Increased the Biomass and Fatty acid Proportion in Microalga Schizochytrium. *Applied Microbiology and Biotechnology* 97(5): 1933–1939.

Yao, L., F. Qi, X. Tan, and X. Lu. 2014a. Improved Production of Fatty Alcohols in Cyanobacteria by Metabolic Engineering. *Biotechnology for Biofuels* 7(1): 1.

Yao, Y., Y. Lu, K.T. Peng, T. Huang, Y.F. Niu, W.H. Xie, W.D. Yang, J.S. Liu, and H.Y. Li. 2014b. Glycerol and Neutral Lipid Production in the Oleaginous Marine Diatom Phaeodactylum Tricornutum Promoted by Overexpression of Glycerol-3-Phosphate Dehydrogenase. *Biotechnology for Biofuels* 7(1): 1.

Yeh, K.L., and J.S. Chang. 2011. Nitrogen Starvation Strategies and Photobioreactor Design for Enhancing Lipid Content and Lipid Production of a Newly Isolated Microalga Chlorella Vulgaris ESP-31: Implications for Biofuels. *Biotechnology Journal* 6(11): 1358–1366.

Zhang, J., Q. Hao, L. Bai, J. Xu, W. Yin, L. Song, L. Xu, X. Guo, C. Fan, Y. Chen, and J. Ruan. 2014. Overexpression of the Soybean Transcription Factor GmDof4 Significantly Enhances the Lipid Content of Chlorella Ellipsoidea. *Biotechnology for Biofuels* 7(1): 1.

9

Biofertilizers and Biopesticides: A Holistic Approach for Sustainable Agriculture

P. Balasubramanian and P. Karthickumar

CONTENTS

ABSTRACT Among the major environmental concerns in the world today, contamination of mother's breast milk through the excessive and injudicious use of agrochemicals is a grave threat to humankind. It has occurred due to the paradigm shift in agricultural practices from conventional natural products to anthropogenic chemicals as fertilizers to sustain the food demand of a rising human population. Though chemical pesticides could contribute substantially to modern agricultural production systems, they alter the ecological balance and an unintended effect of that is irrevocable harm to humans and other species. Ensuring environmentally sound and sustainable crop production without causing detrimental effects to biodiversity, therefore, is the most significant challenge for humankind in this century. The potential of biopesticides and biofertilizers in promoting sustainable agriculture has been evidenced in recent years. The demand for organic farming products is expected to escalate globally in the near future, as they are a cost-efficient and renewable source for sustainable agriculture. Integrated pest management (IPM) and integrated nutrient management (INM) are two key driving forces for biopesticides and biofertilizers. This chapter deals with the harmful effects of chemical fertilizers to the environment as well as the scope and benefits of biopesticides and biofertilizers to attain food security for the growing population through enhanced quality and restoration of soil fertility for sustainable development.

9.1 Introduction

As per a United Nations (UN) report, global population is projected to stabilize at around 9.2 billion by the year 2050 and the global agricultural production rate is projected to be doubled to meet the increasing demands (Ray et al., 2013). On the contrary, the primary natural resources needed for agricultural productivity, such as land, water, and biodiversity, are deteriorating both qualitatively and quantitatively at an alarming rate. For instance, regarding water availability, 30% of crop production will be at severe risk by the year 2025 (Pimentel et al., 2004). Moreover, as per World Bank projections for the year 2050, climate change could dampen crop yields by 20% or more (Alexandratos and Bruinsma, 2012). Therefore, the critical situation is that we have to ensure huge crop production to meet the ever rising demand for finite natural and nonrenewable resources. Since agriculture is both the victim and contributor to greenhouse gas (GHG) emissions, a two-pronged approach is needed to develop adaptive measures that will enhance agricultural resilience (Tuteja and Singh, 2012).

9.1.1 Importance of Soil Nutrients for Agricultural Biodiversity

Several nutritious elements are required for proper plant growth, which directly impacts agricultural biodiversity. Only a minor portion of the nutrients from the soil reserves is released each year through biological activity and/or chemical processes, and most agricultural lands are deprived of some essential nutrient or the other (Feller et al., 2012). In specific, 17 plant food nutrients are essential for proper crop development as shown in Figure 9.1. Although all the nutrients are equally important to plant growth, each of them is required in vastly different amounts. Based on this property, the essential elements needed for the plant growth could be categorized into three classes—primary (macro) nutrients, secondary nutrients, and micronutrients.

Nitrogen, phosphorus, and potassium, which are required in high quantities, are the primary macronutrients. They are frequently supplied to plants in the form of fertilizers. Nitrogen is needed in huge amounts for the growth of plants, since it is the basic constituent of proteins and nucleic acids. Phosphorus is one of the essential primary elements for plant nutrition that plays a significant role in several processes of plant cell growth such as photosynthesis, respiration, energy storage and transfer, cell division, and enlargement. Phosphorous could be assimilated by plants only as soluble phosphates. Moreover, most of the phosphatic soil content in nature exists in the form of rock phosphate and organic phosphorus, both of which are poorly soluble in water. The ability to revitalize the phosphorus content in the soil is solely dependent on the complex chemistry of the soil system. An inadequate availability of phosphorous to the crops could reduce the seed size, seed number, and viability. Potassium is another vital component needed for effective functioning of nutrient absorption, respiration, transpiration, and enzyme activity. However, potassium remains in ionic form and does not become part of plant compounds.

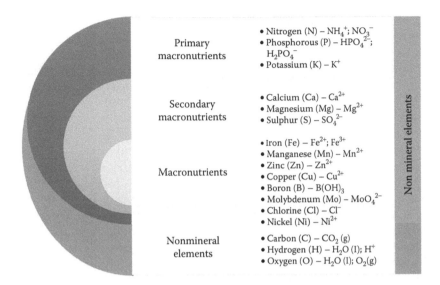

FIGURE 9.1
List of essential plant nutrient elements and their primary forms of availability.

The second class of nutrients such as sulfur, calcium, and magnesium are needed in slightly lesser amounts than primary nutrients for most crops. Often, the secondary nutrients are readily available and in adequate supply. However, these secondary nutrients are gaining significant attention in case of crop fertilization programs to overcome the shortcomings of unavailability as that could limit the overall plant growth even if all other primary nutrients are available unlimitedly in the soil. Micronutrients are elements required in very meager amounts for the overall development of plants and the cost-effective crop production. For instance, micronutrients such as boron, chlorine, copper, iron, manganese, molybdenum, and zinc act as activators and are responsible for the regulation of many functions in plants.

9.1.2 Problems with Existing Agricultural Systems

The land is a natural resource for human beings and a habitat for many living organisms. Population explosion over the past few centuries has raised the demand for greater agricultural productivity. The unprecedented increase in diversion of agricultural lands to nonagricultural purposes has mounted up the deficiency of arable lands. There has been a decrease in agriculture productivity due to several reasons such as increased soil nutrient deficiency, massive build-up of obnoxious weeds and pests, escalated production cost, and so on. As an outcome, the growth rate of agricultural productivity could not sustain the population growth rate. This situation mounted pressure on several developing countries and has led to the usage of chemical fertilizer and pesticides over natural fertilizers to increase productivity because of availability and affordability (Zendejas et al., 2015). Specifically, agriculture suffers from both constructive and harmful effects of the usage of chemical fertilizers and pesticides on the soil. Although a wide array of chemical pesticides have been applied to overcome the economic losses in agriculture, they have caused several environmental stresses on nontarget organisms present in the soil (Zahran, 1999). The use of chemical fertilizers and pesticides has resulted in land degradation and loss of soil fertility (Liu et al., 2009). Apart from being hazardous to the balance in the soil ecosystem, they cause adverse effects on human beings and the environment. Chemical pesticides such as atrazine, 2,4-dichlorophenoxyacetic acid, leave a sustained residue in food products like fruits and vegetables and are considered to be serious public health concern. Magner et al. (2015) investigated a pilot study of human exposure to pesticides from food and found higher concentrations of chemical residues in the urine samples of children. Lu et al. (2015) studied the impact of soil and water pollution on food safety and health risks in China and emphasized that a holistic approach is needed to tackle the problems of environmental pollution and food safety.

Another important problem caused by the usage of chemical fertilizer is contamination of groundwater and surface water along with nutrient imbalance of soil. For instance, inefficient uptake of nitrogen fertilizers by plants contributes to nitrate contamination of soils and groundwater, leading to health hazards and compromising agricultural sustainability (Santi et al., 2013). The primary nutrients present naturally in the soil are depleted due to the residual accumulation of chemical pesticides and fertilizers (Savci, 2012). Chemical fertilizers are highly resistant to degradation and therefore reduce soil fertility and decrease nitrogen fixation either through alteration of root nodules or reduction of the nodulation by soil bacteria. Further, the crops also develop resistance to the pesticide absorption (Zahran, 1999). In subsistence agricultural systems, crop yields are directly dependent on the inherent soil fertility and microbial processes that govern the mineralization and mobilization of nutrients required for plant growth (Choudhary et al., 2011). Since productivity is often limited by the availability of soil nutrients and relies on the interface of the rhizosphere between living roots and soils, any shift in the microbial community structure could lead to significant changes in agroecosystems. The benefits and drawbacks of using chemical fertilizers in agricultural activities are shown in Figure 9.2.

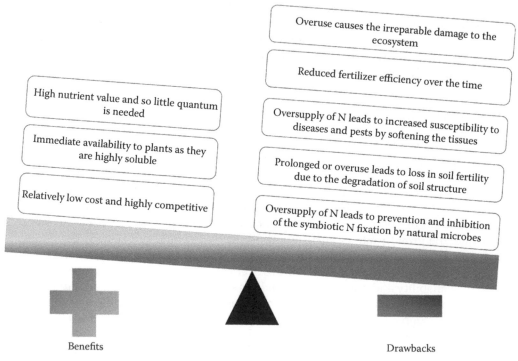

FIGURE 9.2
Benefits and drawbacks of using chemical fertilizers in agriculture.

There has been a growing concern in recent years about the excessive usage of chemical pesticides and its impact on nature and natural resources including humans (Pimentel, 1996; Margni et al., 2002). Over the years, the mounting evidence of the adverse effects of chemical pesticides has contributed to changing the conception of "pesticide as panacea." Further, it has strengthened the advocacy of biopesticides and led to intensive biopesticide research programs in various institutions across the world (Rossel et al., 2008). Moreover, particularly in the last three decades, this shift has resulted in an avalanche of reporting in scientific literature attempting to discover and develop newer and safer pesticides (Koul and Dhaliwal, 2001, 2002; Sinha, 2012).

The challenge of attaining sustainability that is inclusive of economic growth and food security has added another dimension due to the problems posed by climate change. The ever increasing emissions of GHGs have led to several environmental menaces like global warming, rise in sea level, disturbed rainfall pattern with frequent droughts and floods, and so on. These extreme environmental changes hinder the overall plant growth. For instance, in the recent decades, the yield of economically important crops has been drastically declining due to several abiotic stresses. Therefore, adverse impact of climate change on economic stability and overall agricultural productivity is causing increased vulnerability of resource-poor farmers (Tuteja and Singh, 2012). However, plants are subjected to overcome the environmental stresses through the development of tolerance, resistance, or avoidance mechanisms by adjusting to a gradual change in its environment that allow the plants to maintain performance across wide range of environmental conditions (Mohammad et al., 2007).

9.1.3 Need of Biofertilizers and Biopesticides for Sustainable Agriculture

Today's agricultural productivity is influenced by several biotic and abiotic factors such as high salinity, extreme temperature, and droughts and floods due to seasonal variations. Other biotic factors such as weeds and pests also significantly influence the qualitative and quantitative production of agricultural products. As stated above, the indiscriminate use of chemical fertilizers and pesticides has generated several environmental problems along with ill-effects of long-term health impacts on animals and humans. Some of these issues could be tackled effectively by promoting ecologically based management of nutrients and pests by the use of biofertilizers and biopesticides. Because they are natural, they ensure supply of essential nutrients, maintain soil structure, and establish a balance of ecosystems. The appropriate utilization of biofertilizers and biopesticides holds the sustainable approach to enhance agricultural productivity. Several countries such as Argentina, Australia, Brazil, Canada, India, the Philippines, South Africa, and the Unites States, among others have practiced these technologies for promoting sustainable agriculture. For instance, most of the nitrogen demand by the crops to date is supplied in the form of synthetic chemical fertilizers such as urea. Such chemical fertilizers impose a health hazard to agroecosystems and humans besides being quite expensive, leading to increase in the production cost (Tiwary et al., 1998). Encouraging the alternative means of soil fertilization solely relies on the natural/organic inputs to improve the nutrient supply and conserve the ecosystems (Araujo et al., 2008). The appropriate utilization of biofertilizers paves the way for the enrichment of micro- and macronutrients through fixation of nitrogen, solubilization or mineralization of phosphate and potassium, the release of plant growth–regulating substances, production of antibiotics, and biodegradation of organic matter in the soil (Sinha et al., 2014). Overall, it improves plant productivity by enhancing the soil fertility and promoting the soil ecosystem. Precisely, the narrow zone of soil surrounding plant roots named rhizosphere could comprise up to 10^{11} microbial cells per gram of root and above 30,000 prokaryotic species (Mendes et al., 2013; Bhardwaj et al., 2014).

It has been revealed that the effect of nitrogen fixation induced by nitrogen fixers is not only significant for legumes but also for nonlegumes. The symbiotic association of rhizosphere with nitrogen-fixing microbes and plants is a major driving force for the flourishment of the biosphere and adaption to a variety of environmental stresses (Santi et al., 2013). Therefore, the utilization of both chemical fertilizers and biofertilizers has its benefits and drawbacks in the context of nutrient supply, crop growth, soil fertility, and other overall environmental qualities (Chen, 2006). Therefore, the rewards need to be integrated and appropriate use of biofertilizers should be taken up to achieve balanced nutrient management for a dynamic crop growth. The benefits and drawbacks of using biological fertilizers in agricultural activities are shown in Figure 9.3.

Biological regulation is the process of maintaining particular microbial population at its desirable limit, beyond which it could act as a pest. To subdue pest densities while conserving and augmenting native agents, other living organisms are introduced to the environment by the process called importation. This is one of the three major ways to implement biological regulation. The other two processes are conservation/augmentation of natural enemies (disease-causing agents, predators, parasitoids) and application of microbial pesticides (Szewczyk et al., 2006).

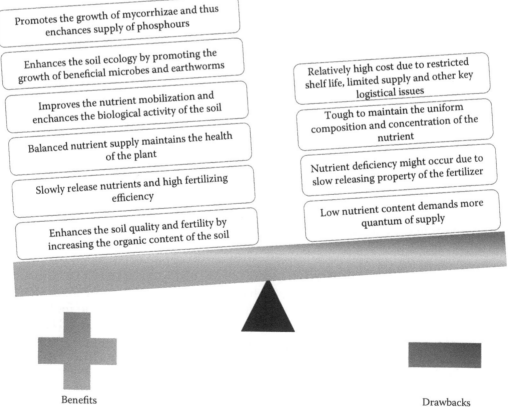

FIGURE 9.3
Benefits and drawbacks of using biofertilizers in agriculture.

9.2 Biofertilizers

Biofertilizers are defined as biologically effective products and/or microbe-based inoculants of bacteria, algae, and fungi either as individual agents or in combination that may help in biological nitrogen fixation or any other beneficial activity for the plants. It also includes organic fertilizers such as manures and other related products for the associated interaction of microbes with plants. Biofertilizers offer several beneficial functions to the plants such as nitrogen fixation, solubilizing, and stabilizing of phosphate in addition to aiding in the supply of other essential micronutrients (Bhardwaj et al., 2014). Based on the functions of biofertilizers, it could be classified into various categories, and the most significant ones are those that establish a symbiotic relationship where both the partners derive benefits from each other.

9.2.1 Types of Biofertilizers

Biofertilizers could be broadly classified as symbiotic nitrogen fixers (*Rhizobium* sp. for legumes); asymbiotic free nitrogen fixers (*Azotobacter, Azospirillum*, etc.); cyanobacteria (*Aulosria, Nostoc, Anabaena*, etc.); and algal biofertilizers (blue green algae in close association with/without Azolla) for wetland rice phosphate-solubilizing bacteria (PSB), mycorrhizae, organic fertilizers, and *Azotobacter/Azospirillum* (Zahran, 1999; Shridhar, 2012). This broad array of biofertilizers could act upon several crops and play a significant role in sustainable agriculture. Microbial species belonging to PSB are *Pseudomonas, Bacillus, Rhizobium, Agrobacterium, Burkholderia, Achromobacter, Micrococcus, Aerobacter, Enterobacter*, and *Flavobacterium*. These PSBs secrete organic acids and lower the pH to dissolute the bound phosphates present in the soil. The phosphatic biofertilizers such as the bacterial species including *Thiobacillus, Bacillus*, and others, along with the mycorrhizal fungus *Glomus* sp. help in increasing the solubility or availability of nutrient phosphate from its existing sparingly soluble forms. However, these microbes do not bring in phosphorus from outside, rather they deplete the soil phosphate reserves. While considering the low utilization efficiency of chemical phosphatic fertilizers, phosphate-solubilizing biofertilizers could play a significant role in improving the efficiency of utilization of phosphate residues left in the soil.

It is beyond the scope and context of this chapter to outline a complete and exhaustive literature survey of the list of microbes that could be utilized as biofertilizers. However, an attempt has been made to summarize the available literature briefly with special emphasis on the classifications of biofertilizers based on the existing literatures of nature and origin to convince the readers. Table 9.1 lists the various types of fertilizers with limited examples.

9.2.2 Mode of Action of Biofertilizers in the Soil

The appropriate use of biofertilizers aids in fixing the atmospheric nitrogen in the soil and root nodules of the legume crops and enhances the overall bioavailability of the nutrients to the plants without adversely affecting the soil and environment. Further, they scavenge the phosphate from soil layers and solubilize the insoluble forms of phosphates into available forms. Many times, the biofertilizers produce hormones for promoting the growth of root nodules and assists in soil mineralization through effective decomposition of the organic matter. Hence, the mode of operation of biofertilizers in the soil varies with the desired applications.

TABLE 9.1

Types of Biofertilizers with Examples

Categories	Subcategories	Examples
Nitrogen fixers	Free living	*Azotobacter: A. chroococcum, A. vinelandii, A. beijerinckii, A. insignis, A. macrocytogenes; Beijerinckia, Clostridium, Klebsiella, Anabaena, Nostoc, Trichodesmium*
	Symbiotic	*Rhizobium, Bradyrhizobium, Sinorhizobium, Azorhizobium, Mesorhizobium, Allorhizobium, Frankia, Anabaena azollae*
	Associative symbiotic	*Azoarcus* sp., Alcaligenes, Acetobacter diazotrophicus, *Azospirillium: A. amazonense, A. halopraeferens, A. brasilense, A. lipoferum,* Alcaligenes, Bacillus, Enterobacter, Herbaspirillum, Klebsiella, Pseudomonas, and Rhizobium
Phosphate solubilizers	Bacteria	*Arthrobacter* sp., *Bacillus* sp., *B. atrophaeus, B. circulans, B. licheniformis, B. amyloliquefaciens, B. firmus, B. megaterium, B. polymyxa, B. subtilis, Chryseomonas luteola,* Enterobacter *aerogenes, E. asburiae, E. taylorae, E. intermedium, Kluyvera cryocrescens, Micrococcus* sp., *Penibacillus macerans, Pseudomonas aerogenes, P. cepacia, P. straita, Vibrio proteolyticus, Xanthobacter agilis*
	Fungi and actinomycetes	*Aspergillus* sp., *A. amstelodemi, A. awamori, A. candidus, A. foetidus, A. fumigatus, A. aponicas, A. niger, A. tamari, A. terricola, A. flavus, Chaetomium nigricoler, Penicillium* sp., *P. digitatum, P. simplicissimum, P. bilaji, P. canescens, P. radicum, P. rugulosum, P. variable, Scwaniomyces occidentalis, Streptomyces* sp.
Phosphate stabilizers		*Pseudomonas, Bacillus, Rhizobium, Burkholderia, Achromobacter, Agrobacterium, Microccocus, Aereobacter, Flavobacterium* and *Erwinia, Pseudomonas putida, Pantoea agglomerans*
Phosphate stabilizers by mycorrhiza	Arbuscular mycorrhiza	*Acaulospora* sp., *Gigaspora* sp., *Glomus* sp., *Scutellospora* sp., *Sclerocystis* sp.
	Ectomycorrhiza	*Laccaria* sp., *Pisolithus* sp., *Boletus* sp., *Amanita* sp.
	Ericoid mycorrhizae	*Pezizella ericae*
	Orchid mycorrhiza	*Rhizoctonia solani*
Potassium solubilizers		*Bacillus mucilaginous*
Plant hormones producing bacteria	Gibberellins	*Bacillus pumilus* and *B. licheniformis,*
	Cytokinins	*Paenibacillus polymyxa*
Plant growth–promoting rhizobacteria		*Achromobacter, Pseudomonas fluorescens*
Biofertilizers for micronutrients	Zinc solubilizers	*Bacillus* sp., *B. subtilis, B erythropolis, B. pumilus, P. rubiacearum, Thiobacillus thioxidans,* and *Saccharomyces* sp.
	Silicate solubilizers	*Bacillus* sp.

Biofertilizers are mainly prepared in a liquid medium, and there are several ways of applying them to the crops. Three ways of using liquid biofertilizers are seed treatment, root dipping, and soil application. Seed treatment is the most commonly used method due to its high effectiveness and relatively low cost. In this method, the seedlings are soaked in the biofertilizer solution followed by their transplantation. In the case of root dipping, roots are dipped in the synthesized biofertilizer solution before transplantation. This method is quite useful for most of the vegetables and crops. Soil application is carried out during the preparatory phase and leveling of the soil before seeding.

The various methods of application of biofertilizers have been summarized in Table 9.2. Bacteria belonging to different genera including *Achromobacter, Azospirillum,*

TABLE 9.2

Method of Application of Biofertilizers and Its Suitable Crops

Method of Application	Efficacy in Crops	References
Soil treatment: Soil is mixed with biofertilizers before planting or sowing. One part of biofertilizers is to be added for four parts of carrier	Maize and wheat	Schoebitz et al. 2014
Seed treatment: One part of biofertilizers is to be added for 50 part of seeds in the prepared slurry (water to be added in the ratio of 1:2)	Pulses, oilseeds, and fodder crops	Singh et al. 2013b
Seedling treatment: Seedlings are dipped in solution (one part of biofertilizers is to be added for 10 part of water) for 40 minutes	Paddy, tomatoes, potato, onion, marigold, and jasmine	Andrade et al. 2013
Set treatment: Sets are immersed in the biofertilizers solution for 30 minutes	Sugarcane, banana, and grapes	Shen et al. 2013
Fertigation: Biofertilizers are added in water with required micronutrients	Orchard crops like mango, lemon, custard apple, vine, and peaches	Singh et al. 2013a

Bacillus, Burkholderia, Enterobacter, Microbacterium, Methylobacterium, Pseudomonas, Pantoea, Paenibacillus, Rhizobium, and *Variovorax* provide tolerance to host plants under different abiotic stress environments (Egamberdieva and Kucharova, 2009). Free-living nitrogen-fixing bacteria such as *Azotobacter chroococcum* and *Azospirillum lipoferum* have been found to have the ability to release phytohormones similar to gibberellic acid (GA) and indole-acetic acid (IAA) that could stimulate plant growth, absorption of nutrients, and photosynthesis in addition to fixing nitrogen (Fayez et al., 1985; Mahfouz and Sharaf-Eldin, 2007).

9.3 Biopesticides

Biopesticides are pesticides derived from either animals, plants, or microorganisms such as bacteria and viruses. These biobased control agents are advantageous because of their inherently less harmful nature in comparison to chemical pesticides and are thus considered to be ecofriendly and safe. Further, these biopesticides are more target-specific than chemical pesticides, that is, they affect only the target pests and their close relatives. On the contrary, chemical pesticides often destroy beneficial insects as well and thereby harm the ecosystem. Due to their high target specificity and nonpersistent and rapid decomposition nature, the development of resistance in the particular pest is also reduced to a great extent. The frequency and quantum of biopesticides required on the farm are relatively small. Moreover, the yield and quality are enriched along with better social acceptability. They are also highly suited for economically deprived rural areas. This has led to an upsurge in economic benefits of these farm products because of the enhancement of product quality and, thereby, increase in export businesses and related activities.

Biopesticides have been accepted globally, of late, and the net market witnessed the remarkable growth rate of 10%–15% per annum (Wahab, 2009). To date, biopesticides share a small pie of global pesticide market, yet the growth of biopesticides is faster due to rising demand for organic agricultural products from Western countries (Hassan and Ayhan, 2014). Though hundreds of naturally occurring insect-specific (entomopathogenic) bacteria have been isolated from insects, plants, and the soil, very few have been studied intensively. Much attention has been given to *Bacillus thuringiensis* (*Bt*), the most promising biological control agent for pest and insect management, as many of its strains are toxic to particular groups of insects. As an outcome, *Bt* has been commercialized and sold under various trade names over the past several decades. A wide array of these natural bacteria isolated from the soil and plants could produce a crystal protein, which is toxic to broad groups of insects.

9.3.1 Types of Biopesticides

Biopesticides can be categorized into three types: microbial pesticides, biochemical pesticides, and plant-incorporated protectants. Besides these, biopesticides also include entomopathogenic nematodes, parasites and predators, plant extracts and secondary metabolites, and so on. Biopesticides may also be classified according to the target organism: bioinsecticides, bioherbicides, and biofungicides. Approximately 75% of biopesticides used currently consists of *Bt*-based products. The live microbe form is an effective microbial pesticide; purified toxin from this strain is the world's most widely used biochemical biopesticide, and the genetic material encoding the *Bt* toxin makes powerful plant-incorporated protectants as well (Olson, 2015).

Microbial pesticides are obtained from naturally occurring or genetically altered bacteria, fungi, algae, viruses, or protozoans. They suppress pests either by producing a toxin specific to the pest and thereby causing disease, or by preventing the establishment of other microorganisms through competition, or by various other mechanisms of action. Biochemical pesticides are closely related to the conventional chemical pesticides. However, they are distinguished from conventional pesticides by their nontoxic mode of action of target organisms (as they are usually species-specific) and their natural occurrence. In the case of biopesticide formulations, it might be a registered biochemical or microbial pesticide that contains one or more active ingredients from the categories described above. The active ingredient(s) is primarily responsible for the pesticidal activity. In addition to the active ingredient, the product formulation contains one to several other inert ingredients. Plant-incorporated protectants are genetically modified plants that synthesize the pesticidal substances for the desired activity. The current advancements in recombinant DNA technologies have contributed to modern agriculture through the development of plant-incorporated protectants. However, the potential risks of genetically engineered plant-incorporated protectants have to be examined in view of numerous factors, such as risks to human health, nontarget organisms, the environment, potential for gene flow, and the need for insect resistance management plans (EPA, 2015). It is beyond the scope and context of this chapter to outline the genetically engineered plant-incorporated protectants here due to the strong advocacy for natural indigenous resources as biofertilizers. The types of biopesticides and their characteristics have been tabulated in Table 9.3.

9.3.2 Mode of Action of Biopesticides in the Soil

The major feature that differentiates biopesticides from synthetic chemical pesticides is their mode of action. Many biopesticides have various modes of action such as disruption of mating, antifeeding, suffocation, and desiccation, while most of the chemical insecticides are neurotoxic to pests (Olson, 2015). The most commonly used biopesticides are *Bt*, baculoviruses, and neem. In addition to these, *Trichoderma* and *Trichogramma* are also frequently used. *Trichoderma* is a fungicide and *Trichogramma* is a biobased control agent

TABLE 9.3

Types of Biopesticides and Their Characteristics with Examples

Types and Subtypes		Characteristics
Microbial pesticides		
Bacterial	Insecticide, e.g., *Bacillus thuringiensis* (Bt)	Target pests are moths and butterflies specifically
	Bactericide, e.g., *Bacillus subtilis* (Bs)	Target pests are bacterial and fungal pathogens (*Rhizoctonia, Fusarium, Aspergillus*, and others)
	Fungicide/bactericide *Pseudomonas fluorescens*	Used to control many fungal, viral, and bacterial diseases
Fungal	Insecticide, e.g., *Beauveria bassiana*	Used to control foliar feeded insects
	Fungicide, e.g., *Trichoderma viride/ harzianum*	Used to control soil-borne fungal diseases
	Fumigant, e.g., *Muscodor albus*	Used to control bacterial and soil-borne diseases
Viral	Insecticide, e.g., Nucleopolyhedrosis virus	Used to control specific species of Lepidoptera (88%), Hymenoptera (6%), and Diptera (5%)
	Insecticide, e.g., Granulosis Virus	Used to control specific species of Lepidoptera
Others	Biological controls	Microscopic and macroscopic predators are used in integrated pest management systems
Biochemical pesticides		
Insect pheromones	Production of message-bearing substances from plant or animal sources	It does not kill the target pest. Mainly used to attract insect toward trap or disturb mating
Plant extracts and oils	Insecticides/herbicides, e.g., pyrethrum	Used to kill insect or paralyzes quickly by altering the electrical impulse
Plant growth regulators	Auxins, e.g., Indole-3-butyric acid	Used to shoot elongation for fruit trees
	Gibberellins, e.g., gibberellic acid	Used for stimulating the cell division and elongation in fruit trees
	Cytokinins, e.g., kenetin	Used for stimulating cell division especially in bud initiation
	Ethylene and ethylene generators	Used for increase the ripening activity
	Growth inhibitors and retardants, e.g., abscisic acid	Used to shortening the internodes specifically flower production
Insect growth regulators	Juvenile hormone-based insecticides	Disturbs the emergence of adult and formation of immature
	Chitin synthesis inhibitors	To control the production of new exoskeleton after molting
Biopesticides formulations		
Inerts	Inert ingredients are grouped into four types based on toxicological concern, priority for testing, new toxics, and least toxic concern	Effectiveness is based on target pest attachment or absorption of active ingredient or died after eaten the inerts Main reasons behind the adding inerts in product formulations are increased performance, ease mode of apply, spreadability or stickiness in plant leaves and soil, and solubility

for parasites, as they are the predators of pests and their eggs. In general, biopesticides destroy pests either by producing a toxin or causing the disease, or competing with others to predominantly flourish. The specific mode of action of these biopesticides and bio-control agents are briefly described below: *Bt* is mainly used against the pathogen of lepi-dopterous pests that includes American bollworm in the case of cotton and stem borers in rice crops (Gupta and Dikshit, 2010). The mechanism of action of a *Bt*-based biopesticide depends on the release of toxins after it has been ingested by the pest larvae. The tox-ins damage the midgut of the larva, finally killing the pest and protecting the crop from attack. However, it is important that the pH of the midgut should be alkaline for the appro-priate action of biopesticides and the specific gut membranes bind strongly with the toxin.

Baculoviruses are target-specific viruses that infect and destroy numerous import plant pests under the category of lepidopterous pests of cotton, rice, and vegetables. They are widely used as insect pest control agents as well as protein expression vectors. Baculoviruses do not replicate in vertebrates, plants, and microorganisms. Their host range is generally limited to arthropods and most particularly insects. To date, most of the registered bacu-lovirus products are for control of lepidopteran and sawfly forest pests (Rohrmann, 2013). The mode of operation of baculoviruses follows a complex replicative cycle for virus pro-duction through effective protein synthesis by seizing and diverting the metabolic mecha-nism of the host cell (Hubbard et al., 2014). The benefits and drawbacks of baculoviruses as insecticides and the recent progress made in genetic enhancement of baculoviruses for improved insecticidal efficacy have been reviewed in detail elsewhere (Popham et al., 2015).

Neem tree (*Azadirachta indica*) is a natural source of numerous chemicals that exhibit antipest properties. For instance, the popularity of "azadirachtin" that affects the repro-ductive and digestive systems of some important pests has led to the development of effective formulations of neem that are commercially produced. Due to their nontoxic and noncarcinogenic nature, there is an enormous upsurge in the demand for the neem-based biopesticides, which has further led to its intensive industrialization.

Trichoderma is an effective fungicide for dryland crops such as groundnut, black gram, green gram, and chickpea, which are susceptible to soil-borne diseases such as root rot (Chandler et al., 2011). The principal mode of action of *Trichoderma* was presumed to exhibit either competition for space and resources or antibiosis and mycoparasitic (Harman, 2006). However, these fungi exhibit localized resistance through the released bioactive molecules by colonizing the root epidermis and outer cortical layers of the plant. Likewise, the fungi could act as biocontrol agents through various modes of a mechanism such as enzyme inhibition, competition for nutrients, and opportunistic plant symbionts. *Trichogramma* is effective against pests of vegetables and fruits and lepidopteran pests like the sugarcane internode borer, pink bollworm, and sooted bollworms in cotton and stem borers in rice. The mode of action of *Trichogramma* is through laying of eggs on various lepidopteran pests and the larvae then feed on and destroy the host eggs.

9.4 Scope of Biofertilizers and Biopesticides

The recent decades have witnessed a slow but steady rise in the industrial activities asso-ciated with biofertilizers and biopesticides as a potential supplementary and ecofriendly input to their chemical counterparts. In order to nullify (or respond to) the damages caused by the insect pests, the plants have started synthesizing specific secondary plant chemicals.

It could be highly possible as the plants and insects have simultaneously evolved in the biosphere for millions of years. Hence, the development and implication of botanical pesticides (originating from the plants) is one of the easiest and viable alternatives to the intensified use of chemical fertilizers as they are ecofriendly, economical, and effective against pests. Those bioactive chemicals include a wide array of types such as insecticides, antifeedants, insect growth regulators (IGRs), hormones, repellents, attractants, and so on. Therefore, natural pesticides from plants should be treated as an important alternative source for chemical pesticides. The acknowledgment of scope of bioactive plant species rekindled the quest for search of new bioactive plant species. To date, the selection of bioactive plant species was based on random screening, phytochemical targeting, ethnobotanical survey, chemotaxonomic approach, and targeted screening approach (Charles, 2011). Over the years, more than 6000 species of plants have been screened and nearly 2400 plants belonging to 235 families have been found to possess significant biological activity against insect pests (Saxena, 1998; Koul and Walia, 2009). Till date, around 2500 bioactive plant species, over 1000 protozoa pathogenic organisms to prevent insect attack, 750 fungal species to avoid attack of terrestrial and aquatic arthropods, baculovirus that can infect over 700 species of invertebrates, and a wide array of other microbial agents and macrobiological agents have been documented (Raj Paroda, 2009). These bioagents and their role in sustainable agriculture should be highlighted further in order to encourage future generations to adopt agricultural economy in a sustainable way. There are approximately 400 registered biopesticide active ingredients and over 1250 actively registered biopesticide products in the market (Raja, 2013). Table 9.4 lists the limited commercially explored biocontrol agents and their uses with examples.

Surprisingly, there was limited number of research and development as well scientific publications on biopesticides until the early 1990s. However, globally, biopesticides witnessed a significant growth in research and commercialization after 2000 due to serious recognitions of the long-term ill effects of unprecedented utilization of chemical fertilizers and pesticides as well as due to a rising demand for organic agricultural products from Western countries. In the last decade, the focus has shifted toward development of newer and safer pesticides for agricultural uses. Research on biopesticides has been dominated by microbial pesticides, particularly *Bt* research (Sinha, 2012). For all crop types, bacterial, fungal, viral, predator, and other biopesticides claim about 74, 10, 5, 8, and 3% of the market, respectively (Thakore, 2006). Moreover, there is an enormous upsurge in reverting to traditional indigenous systems in farming activities due to the realization of adverse impacts of chemical fertilizers and its injudicious use. As an outcome, intensive research activities are being carried out, and several products have also been made available in the global market.

9.4.1 Global Biofertilizer and Biopesticide Market

Recently, the market potential of biofertilizer- and biopesticide-based industries has been accelerating due to intensified utilization in current agricultural practices and increased demand for organic produce globally. Based on the report by Markets and Markets (2016), the worldwide biofertilizers market is expected to reach USD 1.88 billion by 2020 at a compound annual growth rate (CAGR) of 14.0% from 2015 to 2020. Based on the biofertilizer product types, the market has been segmented as nitrogen-fixing, phosphate-solubilizing, potash-mobilizing, and others (zinc, boron, and sulfur-solubilizing biofertilizers). The market has been fragmented based on the crop types, as cereals and grains, pulses and oilseeds, fruits and vegetables, and others. The minor categories such as turf and ornamentals, plantation crops, fiber crops, spices, silage, and forage crops were deliberated in the *Others*

TABLE 9.4

List of Commercially Explored Biocontrol Agents and Their Uses with Examples

Biocontrol Agents	Pests	Crops
Bacteria		
Bacillus popilliae	Japanese beetle, (Popillia japonica), milky diseases of beetles (coleoptera)	
Bacteriophages of *Xanthomonas* sp. and *Pseudomonas syringae*	Bacterial spot in pepper and tomatoes and bacterial speck in tomatoes	Tomatoes and pepper
Pseudomonas syringae strain ESC 10	Ice inducing bacteria and biological decay	Apples, pears, lemons, oranges, or grapefruit after the fruit is harvested
Pantoea agglomerans strain E325	Fireblight (*Erwinia amylovora*)	Apples and pears
Virus		
Coding moth granulosis virus	Coding moth	Apple, pear, walnut, and plum
Cabbage army worm nuclear polyhedrosis virus	Cabbage moth, American bollworm, diamondback moth, potato tuber moth	Cabbage, tomatoes, cotton
Spodoptera littoralis	*Spodoptera littoralis*	Cotton, corn, tomatoes
Helicoverpa zea	*Helicoverpa zea*, and cotton bollworm, *Heliothis virescens*, tobacco budworm	Cotton and vegetables
Spodoptera exigua	Beet armyworm (*Spodoptera exigua*)	Vegetable crops, greenhouse flowers
Anagrapha falcifera	Celery looper (*Anagrapha falcifera*)	Vegetables
Autographa californica	Alfala looper (*Autographa californica*)	Alfalfa and other crops
Orgyia psuedotsugata	Douglas fir tussock moth (*Orgyia psuedotsugata*)	Forest habitat, lumber
Lymantria dispar	Gypsy moth (*Lymantria dispar*)	Forest habitat, lumber
Fungi		
Bacillus pumilus QST 2808	Rust, powdery mildew, cercospora, and brown spot	Soybeans, cereal crops, and potatoes
Coniothyrium minitans strain CON/M/91-08	*Sclerotinia sclerotiorum, Sclerotinia minor*	Agricultural soils
Bacillus subtilis GB03	*Rhizoctonia, Fusarium, Alternaria, Aspergillus,* and others that attack the root systems of plants	Cotton, peanuts, soybeans, wheat, barley, peas, and beans
Trichoderma harzianum Rifai strain KRL-AG2	*Fusarium, Pythium,* and *Rhizoctonia*	Flowers, ornamentals, fruiting vegetables, herbs, and spices, hydroponic crops, leafy vegetables, cole crops, pome fruits
Bacillus subtilis strain QST 713	Fire blight, botrytis, sour rot, rust, bacterial spot, and white mold	Vegetables, fruit, nut, and vine crops
Bacillus pumilus QST 2808	Fungal pests such as molds, mildews, blights, and rusts	Many food and nonfood crops, including trees susceptible to sudden oak death syndrome

category. On the basis of microorganisms, the biofertilizers market has been segmented as *Rhizobium, Azotobacter, Azospirillum, Cyanobacteria,* PSB, and others (potash-mobilizing, zinc biofertilizers, sulfur-solubilizing biofertilizers, and manganese solubilizers). Based on the application type, the biofertilizers market has been segmented as seed, soil, and others (set treatment, foliar treatment, root dipping, and seedling root treatment). On the basis of region, the biofertilizers market has been segmented as North America, Europe, Asia-Pacific, Latin America, and rest of the world (Markets and Markets, 2016).

Currently, the nitrogen-fixing biofertilizer category holds the largest market share in global biofertilizer market and is expected to grow at a CAGR of 13.25% till 2020. The second largest market share is from phosphate-solubilizing biofertilizer segment and expected to rise at a CAGR of 20.75% till 2020. Further, these trends were expected to follow till the projected time. In case of biofertilizer marketing potential by microorganism type, Azospirillum-based biofertilizers currently control the largest market share in the global biofertilizer market. The global Azospirillum biofertilizer market is expected to grow at a CAGR of 17% till 2020. The second largest market share in this category is from *Azotobacter* type and expected to grow at a CAGR of 11% till 2020. In terms of biofertilizer marketing potential by crop type, cereals and grains currently control the largest market share in the global biofertilizer market in terms of consumption. The global cereals and grains biofertilizer market is expected to grow at a CAGR of 13% till 2020. Based on the terms of geographical locations, the North American biofertilizer market controls the largest market share of the global biofertilizer market. The market share is expected to grow at a CAGR of 16.65% till 2020. Next to North American market share, European biofertilizer market is expected to grow at a CAGR of 14.90% till 2020. Globally, the third largest market share is the Asia-Pacific biofertilizer market and expected to register second largest growth rate of 11.40% till 2020 (Business Wire, 2016).

Based on the report by Markets and Markets (2016), the biopesticide market is projected to reach USD 6.6 billion by 2020 and is expected to grow at a CAGR of 18.8% from 2015 to 2020. Based on the biopesticide product types, the global market has been segmented as bioinsecticides, biofungicides, bioherbicides, bionematicides, and others (sulfur, oil, insect repellent, moth control, and other biochemicals). The market has been fragmented based on the crop types, as cereals and grains, pulses and oilseeds, fruits and vegetables, and others. The minor categories such as turf and ornamentals, plantation crops, fiber crops, spices, silage, and forage crops were deliberated in the *Others* category. On the basis of origin, the biopesticide market has been segmented as beneficial insects, microbial pesticides, and biochemicals. The biopesticide market has been classified as liquid formulation and dry formulations on the basis of formulation type. Based on the application type, the biopesticide market has been segmented as foliar spray, seed treatment, soil treatment, and postharvest category. On the basis of region, the biopesticides market has been segmented as North America, Europe, Asia-Pacific, and rest of the world such as Brazil, Argentina, and South Africa (Markets and Markets, 2016). Currently, the major share of the global biopesticide market is from North America, but Europe is expected to show the highest growth rate during the forecast period. However, developing regions such as Asia-Pacific and South America are also expected to grow significantly during the forecast period. The biopesticides market in Asia-Pacific is expected to increase globally by CAGR of 17.8% during 2015–2020 (Research and Markets, 2016).

9.4.2 Production of Biofertilizers and Biopesticides

Increase in the demand for the biofertilizers in the recent decades have paved the way for new entrepreneurs into biofertilizer production. The strategy involves choosing active microbes, isolation, and selection of target microbes, selection of propagation methods and appropriate carrier materials, prototype testing, and large-scale testing. The first step involves the selection of active microbial species based on the purpose of fertilizing property and crop requirements. For instance, one should decide on whether to use organic acid-secreting bacteria or nitrogen fixer or a combination of few organisms. Then, isolation of the target microbes from their actual habitation is carried out. Figure 9.4 outlines the strategies in synthesizing biofertilizers.

In general, microorganisms are usually isolated from plant roots or are trapped using other organic materials. Further, the isolated organisms would be subjected to the general bioprocess development that involves a series of steps. The isolated microbes will be grown first on Petri plates, then in a shaker flask, and finally on a laboratory-scale fermenter before proceeding to actual field-level implementation to select the best organisms and optimize influencing parameters. Growth conditions could be experimented using shaker flask experiments to optimize the medium, while fermentor-based experiments are used to optimize operating conditions. Since bacteria are extremely perishable and sensitive to

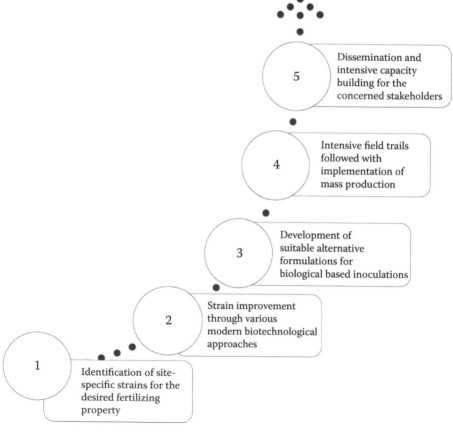

FIGURE 9.4
Strategies in synthesizing biofertilizers.

environmental factors, there is always a need for developing efficient and resistant strains that can withstand local ecological conditions, replenish soil fertility, and improve nutrient uptake by plants. At this stage, it is quite imperative to decide the form of the biofertilizer product wisely so that the right carrier material could be recommended. For instance, if biofertilizer in powder form is desired, then organic materials such as tapioca flour or peat could be used as carriers. Incorporation of microbes in the carrier material enables easy handling, long-term storage, and high effectiveness of biofertilizers. Sterilization of carrier material is crucial for maintaining high concentrations of inoculant bacteria on the carriers for long storage periods. Gamma irradiation or autoclaving can be used as a method of sterilization.

While selecting the propagation method, it is equally important to find out the necessary optimum conditions for the growth of the isolated organisms. It could be achieved by obtaining a growth profile by varying several influencing parameters and all other environmental conditions. Finally, a prototype (usually in different forms) is made and tested before proceeding with large-scale testing at various environmental setups to scrutinize effectiveness and limitability. Although many microbial species can be used beneficially as biofertilizers, the technique of mass production is standardized for very few particular strains such as *Rhizobium, Azospirillum, Azotobacter*, and *Phosphobacteria*. However, the survival of the inoculant strains depends on various influencing factors such as soil composition, physiological status, temperature, pH, and moisture content. Several biotic factors contribute to competition; predation and root growth also affect the survival of the inoculated strains.

Most of the biocontrol agents are quite stable in dry form, as they are free from water and extremes of heat. Basically, liquid biofertilizers have been classified into two types: dry products (dust, granules, and briquettes) and liquid suspensions (oil or water-based and emulsions). Liquid formulations are a sprayable form of the microbial agents suspended in suitable liquid medium to maintain their viability for the stipulated time that enhances the biological activity of the target site. In the case of flowable liquid formulations, the products are quite sensitive to heat and must be protected during storage to avoid settlability of active ingredients. Hence, the shelf-life of flowable formulations is much shorter than that of dry formulations. The formulation should have the following basic characteristics:

- Stabilization of the microorganisms during production, distribution, and storage of the product.
- Easy deliverable to the target site (field) on time.
- Exhibiting perseverance at the target site (field) through the protection of the microorganisms from adverse environmental factors.
- Enhancement of the microorganism's activity at the target site by increasing its viability, reproduction, contact, and interaction with the target crops (Pindi and Satyanarayana, 2012).

9.4.3 Criteria for Ensuring the Quality of Biofertilizers and Biopesticides

Biofertilizers could be formulated either as a carrier- or noncarrier-based microbial inoculants. In addition to the active microbes, several other amendments such as carriers, fillers, and extenders have been utilized in experimental and commercial formulations of various biocontrol agents. Amendments are specifically added to

TABLE 9.5

List of the Amendments with Examples

Purpose of the Amendments	Examples
Binders	Gum arabic, carboxymethylcellulose
Desiccants	Silica gel, anhydrous salts
Nutrients	Molasses, peptone
Stabilizers	Lactose, sodium benzoate
Surfactants	Tween 80
Thickeners	Xanthan gum
Carriers	
Liquid carriers	Vegetable oils
Mineral carriers	Kaolinite clay, diatomaceous earth
Organic carriers	Grain flours
UV protectants	
Dispersants	Microcrystalline cellulose
Light blockers	Lignin
Optical brighteners	Blankophor
Stickers	Pregelatinized corn flour
Sunscreens	Oxybenzone

improve the physicochemical and nutritional properties of the formulated biocontrol agents. A few amendments along with limited examples are tabulated in Table 9.5.

Most of the times, carrier-based preparations are typically followed in the case of microbial inoculants for nitrogen-fixing or phosphate-solubilizing activities, since their prolonged persistence in soils is vital. A good quality carrier material for microbial inoculants should consist of carbon, nitrogen, and vitamin sources, which can promote growth and survival of bacteria (Bashan, 1998). Compost is one such carrier material that has these characteristics. Organic carrier materials are the most cost-efficient carriers for the preparation of microbial inoculants due to their widespread availability. The solid inoculants carry a number of microbial cells and support the survival of these cells for longer periods of time. The mass production of carrier-based microbial biofertilizers involves three stages: culturing of the desired microorganisms, processing of the carrier material, mixing the carrier and the broth culture, and, finally, packing in an appropriate container for the desired purpose. However, before proceeding with the mass production of carrier-based microbial biofertilizers, the selection of carrier materials should be ensured for the following criteria: nontoxic to the plant as well as microbial inoculant strain(s), cost-effective and ubiquitous, easy to process, and sterilization. In addition, the carrier materials should have good moisture absorption and pH buffering capacity. Likewise, the selection of microbial inoculants for biofertilizer applications should satisfy the following criteria:

- ability to fix atmospheric N_2 over a broad range of varying environmental conditions of the soil
- ability to compete with other potent microbial strains
- ability to multiply in a liquid culture medium and survive in dormant stage in the carrier as well as in seed pellets
- ability to persist in soil and to form nodules

- ability to continue to fix atmospheric N_2 in the presence of soil nitrogen
- ability to migrate beneath the soil and to colonize the microbes present in the roots of the plant
- ability to be stable during storage

These criteria have to be fulfilled before proceeding with the mass production and application in the field level. Similar kind of criteria for selecting the appropriate microbial consortium cultures in view of biopesticide applications are as follows: ability to compete with other microbial strains present in the soil; ability to control pests; ability to maintain ecological balance; ability to sustain the productivity of the soil; and so on.

9.5 Limitations of Using Biofertilizers and Biopesticides

If given the right impetus, biopesticides and biofertilizers could pave the way for sustainable agriculture. However, their use has not reached significant levels due to various natural and odd reasons. The market share of most of the biocontrol agents is very meager. However, in the case of seedling disease, root rot, and postharvest disease control, the market share are competing with synthetic fungicides. Yet, effective biocontrol agents have not been identified for the control of serious foliar diseases such as grape downy mildew, potato late blight, wheat powdery mildew, and apple scab (Froyd, 1997). Therefore, in addition to routine bioprospecting aspects of novel biocontrol agents, several strategies have to be adapted for scale-up of the demonstrated biocontrol agents. A thorough investigation is needed at regional and global level to discuss the issues that have to be resolved to promote their use for sustainable agriculture.

Biofertilizers that contain useful enzymes and microbes could increase plant growth and quality of crops, as well as reduce the cost of fertilizer and pesticide applications (Chen, 2006; Zarei et al., 2012). Since the production technology of biofertilizers is relatively simple, the installation capital cost is very low compared with that of the chemical fertilizer plants (Chen, 2006; Wong et al., 2015). Biofertilizers are low-cost renewable source of nutrient that supplements the chemical fertilizer and gaining attention from small and marginal farmers (Bhattacharjee and Dey, 2014). The cost-effectiveness of biofertilizers and biopesticides invites lot of debates in view of several dimensions. This could be due to the fact that several factors on environmental economics are not being considered in the techno-economical analysis. Recently, several studies have reported the cost–benefit analysis of commercially available biofertilizers and biopesticides. For instance, Jefwa et al. (2014) evaluated over 80 microbial inoculant products on major legumes, cereals, and banana crops across diverse agro-ecological conditions in Ethiopia, Nigeria, and Kenya at both laboratory and field conditions. The study revealed around 30% increase in the yield and benefit–cost ratio of almost five. The efficient utilization of locally available indigenous natural resources for the fertilizing activities should be promoted for the sustainable use of biofertilizers and biopesticides. For instance, if a farmer utilizes his locally available natural bioresources for fertilizing and pesticiding activities, the manufacturing cost will be low. Further, it minimizes the overall cost of the product due to less money spent on managing the costs such as transport, storage, dealership, and so on. Even in the case of synthetic formulations of microbial inoculants, the cost-effective organic carrier materials are most efficiently utilized due to the widespread availability of natural renewable resources. Since the key ingredients (inputs) for biofertilizers were demonstrated to be cost effective, the ease availability and standard quality of biofertilizers could resulted in intensive usage at grass root levels. Two reasons were reported to be important for the discontinuation of use of biofertilizers in India: availability and quality. Both of these issues could be addressed by promoting the utilization of regionally available natural bioresources at grassroot levels. The cost-effectiveness of biofertilizers and biopesticides could be comprehended well with the few following cited examples, but are not limited to

1. Though the direct application of raw organic wastes such as solid urban waste, food factory waste, sewage sludge, agricultural residues, and domestic waste are inappropriate for land and agricultural production (Ahmad et al., 2007), their utilization in land application has emerged as an attractive and cost-effective strategy in recent decades due to its richness in organic content for composting. These wastes have been demonstrated to supply plant nutrients and organic matter to the soil for efficient crop production. Unlike fast-release chemical fertilizers, these organic wastes minimize the cost of purchasing nonorganic fertilizers and the environmental impact associated with fertilizer production and use (Vakili et al., 2015; Zendejas et al., 2015).

2. The application of PSB such as *Bacillus megatherium* var. *phosphaticum* enhanced the phosphate availability of the soil by increasing the PSB population in the rhizosphere. While used in conjunction with phosphate fertilizers, PSB reduced the required phosphate dosage by 25%. In addition, 50% of costly superphosphate could be replaced by a cheap rock phosphate, when applied in combination with PSB (Sundara et al., 2002; Chen, 2006).

Yet, the biopesticides are largely regulated by the protocols developed for chemical pesticides. Several technological and policy gaps have to be addressed to avoid the imposition of burdensome costs on the biopesticide industry under this situation. A proper regulatory system should be developed to balance the broadly defined costs and benefits of biopesticides compared with synthetic pesticides (Kumar and Singh, 2014). To summarize, several optimistic studies on paradigm shift for the rejuvenation of biofertilizers and biopesticides for sustainable agriculture were reported with the cost—benefit analysis. Though biofertilizers seem to be a cost-efficient and ecofriendly technology, several constraints limit their application, and they are highlighted in Table 9.6.

Biofertilizers could not completely replace but complement conventional chemical fertilizers. Due to the low nutrient density of biofertilizers, the quantum, frequency, and application procedures might differ based on site-specific conditions. Usually, they require long-term storage and, therefore, careful maintenance of the activeness of the inocula is essential. Also, extreme soil conditions like too hot or dry, acidic or alkaline, might influence the efficacy of biopesticides if other microorganisms contaminate the carrier medium. Moreover, biological control agents are less effective if the soil contains an excess of their natural microbiological predators.

While implementing the biofertilizer technology, various constraints regarding environmental, technological, infrastructural, financial, human resources, awareness, quality, and marketing are being discussed in public domains. These numerous constraints might affect the production techniques severely and trigger the alteration in marketing and/or their

TABLE 9.6

List of the Constraints for Using Biopesticides and Biofertilizers in Agriculture

Constraints	Purpose of the Constraints
Technology	• Use of improper, less-efficient strains for production • Lack of qualified technical personnel in production units • Unavailability of good-quality carrier material or use of different carrier materials by various producers without knowing the quality of the materials • Production of poor quality inoculants without understanding the basic microbiological techniques • Short shelf-life of inoculants
Infrastructure	• Nonavailability of suitable facilities for production • Lack of essential equipment, power supply, etc. • Space availability for laboratory, production, storage, etc. • Lack of facility for cold storage of inoculant packets
Finance	• Nonavailability of sufficient funds and problems in getting bank loans • Less return by sale of products in smaller production units
Environment	• Seasonal demand for biofertilizers • Simultaneous cropping operations and short span of sowing/planting in a particular locality • Soil characteristics like salinity, acidity, drought, water logging, etc.
Human resources	• Lack of technically qualified staff in the production units • Lack of suitable training on the production techniques • Ignorance on the quality of the product by the manufacturer • Nonavailability of quality specifications and quick quality control methods • No regulation or act on the quality of the products • Lack of awareness of the technology and its benefits • Problem in the adoption of the technology by the farmers due to different methods of inoculation • No visual difference in the crop growth immediately as that of inorganic fertilizers
Dissemination and capacity building	• Unawareness on the benefits of the technology • Problem in the adoption of the technology by the farmers due to lack in hands-on training on various methods of inoculation • Lack of technically qualified personnel to train the end users • Unawareness on the damages caused on the ecosystem by continuous application of inorganic fertilizer
Market	• Nonavailability of right inoculant at the right place in right time • Lack of retail outlets or the market network for the producers

Source: Entrepreneurial Training Manual, TNAU Agritech Portal.

further usage. Despite encouraging results, microbial inoculants-based biofertilizers have not got a widespread application in agriculture. This is mainly about the choice of bacterial strain used, the responses of plant species vary under unfavorable rhizosphere conditions. On the other hand, real competitive ability and high saprophytic competence are major factors determining the success of a bacterial strain as an inoculant. Therefore, studies to find out the competitiveness and persistence of specific microbial populations in complex environments, such as the rhizosphere, should be carried out to obtain efficient inoculants. The key critical factors responsible for effectiveness of biofertilizers discussed by Kalra and Khanuja (2005) are as follows:

• Host specificity and suitability of the species for the target crop.
• Identification of strains suited to the agroecosystem, particularly the soil pH and moisture conditions.

- Significant cell count of living organisms presents on the product, its purity, and its level of contamination.

- Conditions of the carrier material in which the culture is packed and the quality of the packing material, which determines the shelf-life.

- Appropriate packing, storage, and distribution of the product by stakeholders before application (Teng, 2007).

Further, several other factors also limit the extensibility of biofertilizers and biopesticides in real-time applications. Though several microbial strains have been shown to enhance the agricultural productivity on a laboratory scale, only few strains could be acclimatized for real-time scenarios. Precisely, the limited availability of active microbes with field-level testing and the appropriate growing nutrient medium are the two key issues that hinder the commercial availability of these biofertilizers in the market.

9.6 Integrated Nutrient Management and Integrated Pest Management

Integrated nutrient management (INM) and integrated pest management (IPM) are the two key driving forces for biopesticides and biofertilizers, as they are cost-efficient and renewable source for sustainable agriculture. Much attention is now being paid to develop an INM system that maintains or enhances soil productivity through the balanced use of all sources of nutrients—including chemical fertilizers, organic fertilizers, and biofertilizers. The basic concept underlying INM is the adjustment of soil fertility and plant nutrient supply to an optimum level for sustaining desired crop productivity through optimization of the benefits from all possible sources of plant nutrients in an integrated manner. Further, the implementation of IPM ensures a healthy ecosystem to enhance the agricultural productivity (Tilman et al., 2002). While we are striving hard to accomplish rapid advancements on agricultural technology and devising sophisticated approaches for crop protection, the IPM could offers an alternative to evade the resistance and non-targeted problems associate with the usage of conventional chemical fertilizers and pesticides. IPM is a systematic approach that combines different crop protection practices along with a cautious monitoring of pests and their natural predators. The aim is to manage pest population below the levels that cause economic damage. The main components of IPM are pesticides with high levels of selectivity and low-risk compounds, breed of crop cultivars with pest resistance, cultivation practices (crop rotation, intercropping, undersowing), physical methods (mechanical weeders), natural products (semi-chemical, biocidal plant extracts), biological control with natural predators and microbial pathogens, informing farmers when it is economically beneficial to apply pesticides (timing of pest activity and scouting), and other controls. Thus, IPM is an example of sustainable intensification, defined as "producing more output from the same area of land while reducing the negative environmental impacts and at the same time increasing contributions to natural capital and the flow of environmental services" (Pretty and Bharucha, 2014, 2015).

Organic farming practices aim to enhance biodiversity, biological cycles, and soil biological activity so as to achieve optimal natural systems that are ecologically and economically sustainable (Samman et al., 2008). Manure management is a decision-making process aimed at combining profitable agricultural production with minimal nutrient losses from manure. The selection of manure management and treatment options increasingly depends on environmental regulations for preventing pollution of land, water, and air (Karmakar et al., 2007).

In recent years, biofertilizers have emerged as a promising component of the INM system in agriculture. The entire structure of farming depends on microbial activities in many significant ways, and there appears to be a tremendous potential for making use of microorganisms in increasing crop production. Moreover, some microbes are helpful for plant growth in multiple ways, like *Azotobacter*, which help in both nitrogen fixation and stimulating root development. Microbiological fertilizers are an important component of ecofriendly sustainable agricultural practices (Bloemberg et al., 2000).

Soil microbes play an important role in many critical ecosystem processes, including nutrient cycling and homeostasis, decomposition of organic matter, and promoting plant health and plant growth through biofertilization (Han et al., 2007). Certain strains are referred to as plant growth-promoting rhizobacteria (PGPR), and these can be used as inoculant biofertilizers (Kennedy et al., 2004). They include species of *Azotobacter* and *Azospirillum*, both of which have direct and indirect effects on plant growth and pest resistance (Persello et al., 2003). Soil microorganisms such as *Azotobacter* and *Azospirillum* as N_2-fixing bacteria could be valuable sources to enhance plant growth and produce considerable amounts of biologically active substances that can promote the growth of the reproductive organs and increase crop productivity (Ebrahimi et al., 2007; Akbari et al., 2011; Singh et al., 2011). Biofertilizers enhance the productivity and sustainability of the soil, as they are low-cost, ecofriendly, and renewable source of plant nutrients. Therefore, they play a significant role in the sustainable agricultural system. In this context, biofertilizers can be considered as key components of INM (Mohammadi and Sohrabi, 2012).

Figure 9.5 outlines the several possible interactions between plants, pathogens, and the plants associated beneficial microbes. Competitive interactions between pathogens and beneficial microbes along with other significant interactions such as antibiosis (an antagonistic association between two microorganisms) for the prevention of growth or development of an organism by a substance or another organism play a crucial role in the overall health of the plant. However, plant growth and suppression of the plant-associated diseases could be ensured by integrated application of the biopesticides and biofertilizers. Since the plant-associated microbes support the essential macro- and micronutrients, integrating them along with hormonal stimulation enhances disease resistance. Thus, the implementation of INM and IPM would pave the way for sustainable agricultural practices along with environmental protection and human health. There are basically four types of IPM interventions: management of pesticide components, breeding of crop (or livestock), deployment of pheromones and/or release of parasites (or predators), and establishing the agro-ecological habitat (Pretty and Bharucha, 2015).

Based on scientific knowledge and a fundamental understanding of pest and plant nutrition, environmentally sound integrated biological pest control and INM systems, which emphasize on the use and integration of diverse methods such as cultural (crop rotation), physical (organic amendments, chemicals, microorganisms), and resistant hosts should be developed. Ideally, pest control and nutrient supply must be integrated into one system. Adequate research is crucial for the successful integration of biopesticides and biofertilizers as integrated crop management (ICM), where a practical approach is followed for crop production that includes IPM as a subentity. It is based on the understanding of complex balance between the environment and agriculture through the encapsulation of numerous basic components for management of crop, nutrients, pest, and lastly economics (Kumar and Singh, 2014).

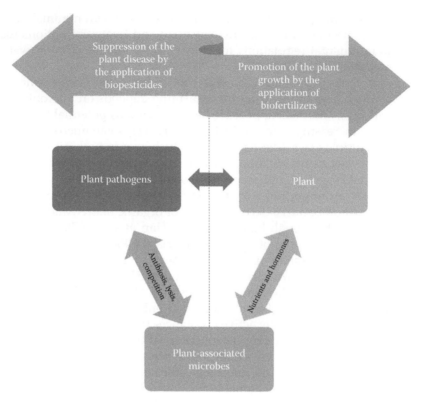

FIGURE 9.5
Role of biopesticides and biofertilizers in the several interactions among plants and microbes.

9.7 Role of Biofertilizers and Biopesticides in Sustainable Development

Since biopesticides emerged to address the major twin issues of crop protection while ensuring environmental sustainability, their role in overall sustainability is substantial. Enhancement in agricultural sustainability necessitates the optimal use and management of soil fertility and soil physical property and relies on soil biological processes and soil biodiversity. The integrated use of biofertilizers (for crop nutrition), biopesticides (for crop protection), and biostimulants (for crop enhancement) is the need of the hour for ensuring agricultural resilience. Hence, IPM should be treated as a subentity of sustainable agriculture, as the utilization of pest-control practices aids in enhancing the efficiency and economic viability of agricultural practices, environmental protection, and human health.

In general, sustainable development refers to the mode of human development through which resources are utilized wisely to meet the demands of both the current and future generations. Sustainable development tries to ensure sustainability in all three aspects of life: environment, economy, and socio-politics. More specifically, economic viability extends to economic growth across generations. While extending this paradigm to sustainable agriculture, it could be defined as the application of ecofriendly farming principles to enhance crop production without damaging the natural biotic and abiotic resources. Thus, the term sustainable agriculture represents an integrated system of site-specific plant and

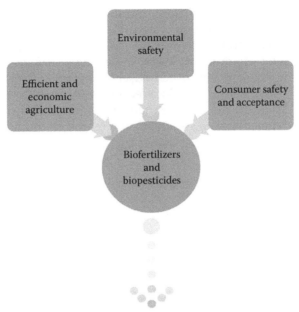

Sustainable agriculture and economic development

FIGURE 9.6
Role of biofertilizers and biopesticides in sustainable development.

animal production as well as manure management practices. The role of biofertilizers and biopesticides in promoting sustainable development is shown in Figure 9.6.

This concept of sustainable agriculture warrants sustainability in all three aspects across generations to fulfill human needs, enhance environmental quality, and sustain the economic viability of farm operations. For instance, instead of promoting economy at the cost of the environment, as with the current way of development, the implementation of sustainable agriculture would ensure the promotion of both economy and environment in a win–win situation. This could be achieved through well-planned management strategies that help the producer select hybrids and varieties through soil-conserving cultural practices, soil fertility programs, and pest management programs. The goal of sustainable agriculture is to minimize the adverse impacts while providing a sustained level of production and profit. Recently, the confluence of economy and environment through promoting sustainable agricultural activities has contributed to the rise of a new paradigm called sustainable agriculture economic development (SAED). Pallabi and Debiprasad (2014) demonstrated that the utilization of Azolla as biofertilizer is significantly greater than chemical fertilizer and showing better SAED.

The application of microbial inoculants-based biofertilizers is a natural and efficient way to enhance and promote the mineral economy of the soil. Biofertilizers could not be treated as a complete replacement for chemical fertilizers, rather they should be addressed as the next best alternative for the synthetic ones. The utilization of biopesticides lessens the utilization of chemical fertilizers and exploits the alternative for sustainable agriculture. In recent decades, there has been an expansion of research related to biofertilizers and biopesticides, since they act as natural stimulators of plant growth and development. Consequently, there has been considerable interest in the possible use of inoculants of effective microbes for the development of biofertilizers and biopesticides. Since the organic farming is one of the oldest forms of agricultural activities on the earth that still offers numerous benefits such as conservation of natural resources such as maintenance of soil fertility, water quality, prevention of topsoil erosion, and preservation of natural biodiversity, it could be advocated at grassroots levels. In view of social advantages, it aids in the generation of rural employment, improved household nutrition, and reduced dependence on external inputs. Further, it boosts the local economy through the high export potential of organic products. Thus, the practice of organic farming could also ensure the sustainability in all the three dimensions of society, economy, and environment.

9.8 Summary and Prospects

Agriculture has sustained human lives since the time of the ancient civilizations. With the rapid decline in agricultural farmlands due to the industrialization and rapid urbanization, modern agriculture inevitably continues to play a significant role in the survival of humanity. The diminishing quantity of finite natural resources to support agricultural activities are also under high stress, further fueling the need for adopting and implementing novel practices for sustainable agriculture. Efficient nutrition management should be ensured at all possible levels to enhance and sustain agricultural production while conserving the environment. Despite the sources of nutrient—chemical, organic, or microbial fertilizers—every type has its advantages and disadvantages. Hence, developing a suitable INM system based on these various kinds of fertilizers may be an extremely challenging task. On the other hand, one should not avoid the environmental consequences of pest management. Therefore, it is highly advisable to focus on IPM. INM and IPM should go hand in hand for a win–win situation to ensure the practice of sustainable agriculture. However, much research is still needed to address these challenges.

It is worth noting that biofertilizers are part of nature in the form of microorganisms that colonize the rhizosphere. Biofertilizers containing those microorganisms can play a significant role in crop improvement. Biofertilizers and biopesticides are a modern technology for global agriculture. It promises to overcome the limitations of conventional chemical-based technologies. In fact, it should be treated as a forgotten gift from nature to supplement modern agricultural activities. Biofertilizers and biopesticides are emerging products in the commercial market that are likely to sustain in the long run once sufficient scientific research and development has been done and the technological know-how has been disseminated to concerned stakeholders. Advocating the environmental paradigm of *think global and act local*, the utilization of biofertilizers and biopesticides at the regional level will not only have an impact on national agricultural productivity but also contribute to world economic development due to sustainable agriculture. There is also the undeniable fact that these holistic approaches contribute to the universal well-being of the world.

References

Ahmad, R., Shahzad, S.M., Khalid, A., Arshad, M., and M.H. Mahmood (2007). Growth and yield response of wheat (*Triticum aestivum* L.) and maize (*Zea mays* L.) to nitrogen and L-tryptophan enriched compost. *Pakistan Journal of Botany*, 39(2), 541–549.

Akbari, P., Ghalavand, A., Modarres Sanavy, A.M., and M.A. Alikhani (2011). The effect of biofertilizers, nitrogen fertilizer and farmyard manure on grain yield and seed quality of sunflower (*Helianthus annus* L.). *Journal of Agricultural Technology*, 7(1), 173–184.

Alexandratos, N., and J. Bruinsma (2012). World agriculture towards 2030/2050: The 2012 revision. ESA Working Paper No 12-03, pp. 1–147.

Andrade, M.M.M., Stamford, N.P., Santos, C.E.R.S., Freitas, A.D.S., Sousa, C.A., and M.A. Lira-Junior (2013). Effects of biofertilizer with diazotrophic bacteria and mycorrhizal fungi in soil attribute, cowpea nodulation yield and nutrient uptake in field conditions. *Science, Horticulture*, 162, 374–379.

Araujo, A.S.F., Santos, V.B., and R.T.R. Monteiro (2008). Responses of soil microbial biomass and activity for practices of organic and conventional farming systems in Piaui state, Brazil. *European Journal of Soil Biology*, 44, 225–230.

Bashan, Y. (1998). Inoculants of plant growth-promoting bacteria for use in agriculture. *Biotechnology Advances*, 16(4), 729–770.

Bhardwaj, D., Ansari, M.W., Sahoo, R.K., and N. Tuteja (2014). Biofertilizers function as a key player in sustainable agriculture by improving soil fertility, plant tolerance, and crop productivity. *Microbial Cell Factories*, 13(1), 66.

Bhattacharjee, R., and U. Dey (2014). Biofertilizer, a way towards organic agriculture: A review. *African Journal of Microbiology Research*, 8(24), 2332–2343.

Bloemberg, G.V., Wijfjes, A.H.M., Lamers, G.E.M., Stuurman, N., and B.J.J. Lugtenberg (2000). Simultaneous imaging of *Pseudomonas fluorescens* WCS3655 populations expressing three different autofluorescent proteins in the rhizosphere: New perspective for studying microbial communities. *Molecular Biology and Plant Microbial Interactions*, 13, 1170–1176.

Business Wire (2016). Global Biofertilizer Market 2016–2020—Research and Markets. http://www.businesswire.com/news/home/20160426005882/en/Global-Biofertilizer-Market-2016-2020—Main-Growth (accessed May, 2016).

Chandler, D., Bailey, A.S., Tatchell, G.M., Davidson, G., Greaves, J., and W.P. Grant (2011). The development, regulation and use of biopesticides for integrated pest management. *Philosophical Transactions of the Royal Society of London B: Biological Sciences*, 366(1573), 1987–1998.

Charles, K. (2011). Tanzanian botanical derivatives in the control of malaria vectors: Opportunities and challenges. *Journal of Applied Science and Environmental Management*, 15(1), 127–133.

Chen, J.H. (2006). The combined use of chemical and organic fertilizers and/or biofertilizer for crop growth and soil fertility. *Taipei Food Fertilizer Technology Bulletin*, 16, 1–10.

Choudhary, D.K., K.P. Sharma, and R.K. Gaur (2011). Biotechnological perspectives of microbes in agroecosystems. *Biotechnology Letters*, 33, 1905–1910.

Ebrahimi, S., H.I. Naehad, R. Shirani, A.H. Abbas, G. Akbari, R. Amiry, and S.A.M. Sanavy (2007). Effect of *Azotobacter chroococcum* application on quantity and quality forage of rapeseed cultivars. *Pakistan Journal of Biological Sciences*, 10(18), 3126–3130.

Egamberdieva, D., and Z. Kucharova (2009). Selection for root colonizing bacteria stimulating wheat growth in saline soils. *Biology and Fertility of Soil*, 45, 563–571.

EPA (2015). Plant Incorporated Protectants and their Regulations, https://www3.epa.gov/pesticides/chem_search/reg_actions/pip/index.htm (accessed November 15, 2015).

Fayez, M., Emam, N.F., and H.E. Makboul (1985). The possible use of nitrogen fixing *Azospirilum* as biofertilizer for wheat plants. *Egyptian Journal of Microbiology*, 20(2), 199–206.

Feller, C., Blanchart, E., Bernoux, M., Lal, R., and R. Manlay (2012). Soil fertility concepts over the past two centuries: The importance attributed to soil organic matter in developed and developing countries. *Archives of Agronomy and Soil Science*, 58(1), 3–21.

Froyd, J.D. (1997). Can synthetic pesticides be replaced with biologically-based alternatives?—An industry perspective. *Journal of Industrial Microbiology and Biotechnology*, 19(3): 192–195.

Gupta, S., and A.K. Dikshit (2010). Biopesticides: An eco-friendly approach for pest control. *Journal of Biopesticides*, 3(1), 186–188.

Han, X.M., Wang, R.Q., Liu, J., Wang, M.C., Zhou, J., and W.H. Guo (2007). Effects of vegetation type on soil microbial community structure and catabolic diversity assessed by polyphasic methods in North China. *Journal of Environmental Sciences*, 19, 1228–1234.

Harman, G.E. (2006). Overview of mechanisms and uses of *Trichoderma sp*.*Phytopathology*, 96(2): 190–194.

Hassan, E., and G. Ayhan (2014). Production and consumption of biopesticides. In *Advances in Plant Biopesticides*, edited by D. Singh, pp. 361–379.

Hubbard, M., R.K. Hynes, M. Erlandson, and K.L Bailey (2014). The biochemistry behind biopesticide efficacy. *Sustainable Chemical Processes*, 2, 18.

Jefwa, J.M., Pypers, P., Jemo, M., and M. Thuita et al. (2014). Do commercial biological and chemical products increase crop yields and economic returns under smallholder farmer conditions? In *Challenges and Opportunities for Agricultural Intensification of the Humid Highland Systems of Sub-Saharan Africa*, edited by B. Vanlauwe, P. van Asten, and G. Blomme, Springer International Publishing, Switzerland, pp. 81–96.

Kalra, A., and S.P.S. Khanuja (2007). Research and development priorities for biopesticide and biofertiliser products for sustainable agriculture in India. Business Potential for Agricultural Biotechnology, edited by P.S. Teng, Asian Productivity Organization, Tokyo, pp. 96–102.

Karmakar, S., Lague, C., Agnew, J., and H. Landry (2007). Integrated decision support system (DSS) for manure management. A review and perspective. *Computers and Electronics in Agriculture*, 57, 190–201.

Kennedy, I.R., Choudhury, A.T.M.A., and M.L. Kecskes (2004). Nonsymbiotic bacterial diazotrophs in crop farming systems: Can their potential for plant growth promotion be better exploited? *Soil Biology and Biochemistry*, 36(8), 1229–1244.

Koul, O., and G.S. Dhaliwal (2001). *Phytochemical Biopesticides: Advances in Biopesticide Research*. Taylor and Francis, London, p. 236.

Koul, O., and G.S. Dhaliwal (2002). *Microbial Biopesticides: Advances in Biopesticide Research*. Taylor and Francis, London, p. 352.

Koul, O., and S. Walia (2009) Comparing impacts of plant extracts and pure allelochemicals and implications for pest control. *CAB Reviews: Perspectives in Agriculture, Veterinary Science, Nutrition and Natural Resources*, 4, 1–30.

Kumar, S., and A. Singh, (2014) Biopesticides for integrated crop management: Environmental and regulatory aspects. *Journal of Biofertilizers and Biopesticides*, 5, e121.

Liu, Y., Zhang, J.B., and J. Du (2009). Factors affecting the reduction of fertilizer application by farmers: Empirical study with data from Jianghan plain in Hublei province. *Contributed Paper Prepared for Presentation at the International Association of Agricultural Economists Conference*, Beijing, China.

Lu, Y., S. Song, R. Wang, and Z. Liu et al. (2015). Impacts of soil and water pollution on food safety and health risks in China. *Environment International*, 77, 5–15.

Magner, J., P. Wallberg, J. Sandberg, and A.P. Cousins (2015). Human exposure to pesticides from food, IVL Report U, 5080.

Mahfouz, S.A., and M.A. Sharaf-Eldin (2007). Effect of mineral vs. biofertilizer on growth, yield, and essential oil content of fennel (*Foeniculum vulgare Mill.*). *International Agrophysics*, 21, 361–366.

Margni, M., Rossiser, D., Crettaz, P., and O. Jolliet (2002). Life cycle impact assessment of pesticides on human health and ecosystems. *Agriculture Ecosystem and Environment*, 93, 79–92.

Markets and Markets (2016). Biofertilizers market by type (nitrogen-fixing, phosphate-solubilizing & potash-mobilizing), Microorganisms (*Rhizobium, Azotobacter, Azospirillum, Cyanobacteria* & phosphate-solubilizing bacteria), application, crop type & by region—Global Forecast to 2020. http://www.marketsandmarkets.com/PressReleases/compound-biofertilizers-customized-fertilizers.asp (accessed May, 2016).

Mendes, R., Garbeva, P., and J.M. Raaijmakers (2013). The rhizosphere microbiome: Significance of plant beneficial, plant pathogenic, and human pathogenic microorganisms. *FEMS Microbiology Reviews*, 37(5), 634–663.

Mohammad, S.K., A. Zaidi, and P.A. Wani (2007). Role of phosphate solubilizing microorganisms in sustainable agriculture—A review. *Agronomy for Sustainable Development*, 27(1), 29–43.

Mohammadi, K., and Y. Sohrabi (2012). Bacterial biofertilizers for sustainable crop production: A review. *Journal of Agricultural and Biological Science*, 7, 307–316.

Olson, S. (2015). An analysis of the biopesticide market now and where it is going. *Outlooks on Pest Management*, 26(5), 203–206.

Pallabi, M., and D. Debiprasad (2014). Rejuvenation of biofertilizer for sustainable agriculture and economic development. *Consilience: The Journal of Sustainable Development*, 11(1), 41–61.

Paroda, R. (2009). Proceedings and Recommendations of Expert Consultation on Biopesticides and Biofertilizers for Sustainable Agriculture organized by Asia-Pacific Association of Agricultural Research Institutions (APAARI) Asia-Pacific Consortium on Agricultural Biotechnology (APCoAB) Council of Agriculture (COA).

Persello, C.F., Nussaume, L., and C. Robaglia (2003). Tales from the underground: Molecular plant–rhizobacteria interactions. *Plantcell Environment*, 26(2), 189–199.

Pimentel, D. (1996). Green revolution agriculture and chemical hazards. *Science of the Total Environment*, 188, 86–98.

Pimentel, D., Berger, B., Filiberto, D., Newton, M., Wolfe, B., Karabinakis, E., Clark, S., Poon, E., Abbett, E., and S. Nandagopal (2004). Water resources: Agricultural and environmental issues. *BioScience*, 54(10), 909–918.

Pindi, P.K., and S.D.V. Satyanarayana (2012). Liquid microbial consortium—A potential tool for sustainable soil health. *Journal of Biofertilizers and Biopesticides*, 3, 124.

Popham, J.R., T. Nusawardani, and B.C. Bonning (2016). Introduction to the use of baculoviruses as biological insecticides. In *Baculovirus and Insect Cell Expression Protocols of Methods in Molecular Biology*, edited by W. Murhammer, Springer, Newyork, 1350, pp. 383–392.

Pretty, J., and Z.P. Bharucha (2014). Sustainable intensification in agricultural systems. *Annals of Botany*, 114(8), 1571–1596.

Pretty, J., and Z.P. Bharucha (2015). Integrated pest management for sustainable intensification of agriculture in Asia and Africa. *Insects*, 6(1), 152–182.

Raja, N. (2013). Biopesticides and biofertilizers: Ecofriendly sources for sustainable agriculture. *Journal of Biofertilizers and Biopesticides*, 4(1), 1000e112.

Ray, D.K., Mueller, N.D., West, P.C., and J.A. Foley (2013). Yield trends are insufficient to double global crop production by 2050. *PLos One*, 8(6), e66428. doi:10.1371/journal.pone.0066428.

Research and Markets (2016). Asia Pacific biopesticides market—Growth, trends and forecasts (2015–2020). http://www.researchandmarkets.com/research/s765jq/asia_pacific (accessed May, 2016).

Rohrmann, G.F. (2013). *Baculovirus Molecular Biology*, 3rd ed. In National Library of Medicine (US). Bethesda (MD): National Center for Biotechnology Information.

Rossel, G., Quero, C., Coll, J., and A. Guerrero (2008). Biorational insecticides in pest management. *Journal of Pesticide Science*, 332, 103–121.

Samman, S., Chow, J.W.Y., Foster, M.J., Ahmad, Z.I., Phuyal, J.L., and P. Petocz (2008). Fatty acid composition of edible oils derived from certified organic and conventional agricultural methods. *Food Chemistry*, 109, 670–674.

Santi, C., Bogusz, D. and C. Franche (2013). Biological nitrogen fixation in non-legumes plants. *Annals of Botany*, 111, 743–767.

Savci, S. (2012). An agricultural pollutant: Chemical fertilizer. *International Journal of Environmental Science and Development*, 3(1), 77–80.

Saxena, R.C. (1998). Botanical pest control. In *Critical Issues in Insect Pest Management*, edited by G.S. Dhaliwal, and E.A. Heinrichs, Commonwealth Publishers, New Delhi, India, pp. 155–179.

Schoebitz, M., Mengual, C., and A. Roldan (2014). Combined effects of clay immobilized *Azospirillum brasilense* and *Pantoea dispersa* and organic olive residue on plant performance and soil properties in the revegetation of a semiarid area. *Science of the Total Environment*, 466, 67–73.

Shen, Z., Zhong, S., Wang, Y., Wang, B., Mei, X., and R. Li (2013). Induced soil microbial suppression of banana fusarium wilt disease using compost and biofertilizers to improve yield and quality. *European Journal of Soil Biology*, 57, 1–8.

Shridhar, B.S. (2012). Review: Nitrogen-fixing microorganisms. *International Journal of Microbiological Research*, 3(1), 46–52.

Singh, J.S., Pandey, V.C., and D.P. Singh (2011). Efficient soil microorganisms: A new dimension for sustainable agriculture and environmental development. *Agriculture, Ecosystems and Environment*, 140, 339–353.

Singh, Y.V., Singh, K.K., and S.K. Sharma (2013b). Influence of crop nutrition on grain yield, seed quality and water productivity under two rice cultivation systems. *Rice Science*, 20, 129–138.

Singh, R., Sonim, S.K., Patel, R.P., and A. Kalra (2013a). Technology for improving essential oil yield of *Ocimum basilicum* L. (sweet basil) by application of bioinoculant colonized seeds under organic field conditions. *Industrial Crops Production*, 45, 335–342.

Sinha, B. (2012). Global biopesticide research trends: A bibliometric assessment. *Indian Journal of Agricultural Sciences*, 82(2), 95–101.

Sinha, R.K., Valani, D., Chauhan, K., and S. Agarwal (2014). Embarking on a second green revolution for sustainable agriculture by vermiculture biotechnology using earthworms: Reviving the dreams of Sir Charles Darwin. *International Journal of Agriculture, Health and Safety*, 1, 50–64.

Sundara, B., Natarajan, V., and K. Hari (2002). Influence of phosphorus solubilizing bacteria on the change in soil available supply system on soil quality restoration in a red and laterite soil. *Archives of Agronomy and Soil Science*, 49, 631–637.

Szewczyk, B., Hoyos-Carvajal, L., Paluszek, M., Skrzecz, I., and M. Lobode Souza (2006). Baculoviruses—Re-emerging biopesticides. *Biotechnology Advances*, 24(2), 143–160.

Teng, P.S. (2007). Business potential for agricultural biotechnology products, Report of the Asian Productivity Organization, Tokyo, pp. 1–197.

Thakore, Y. (2006). The biopesticide market for global agricultural use. *Industrial Biotechnology*, 2(3), 192–208.

Tilman, D., Cassman, K.G., Matson, P.A., Naylor, R., and S. Polasky (2002). Agricultural sustainability and intensive production practices. *Nature*, 418(6898), 671–677.

Tiwary, D.K., Abuhasan, M.D., and P.K. Chattopadhyay (1998). Studies on the effect of inoculation with *Azotobacter* and *Azospirillum* on growth, yield, and quality of banana. *Indian Journal of Agriculture*, 42, 235–240.

TNAU (Tamilnadu Agricultural University), Entrepreneurial training manual, http://agritech.tnau.ac.in/org_farm/orgfarm_biofertilizertechnology.html (accessed November 10, 2015).

Tuteja, N., and S. Sarvajeet (2012). *Plant Acclimation to Environmental Stress*. Springer, New York.

Vakili, M., Rafatullah, M., Ibrahim, M.H., Salamatinia, B., Gholami, Z., and H.M. Zwain (2015). A review on composting of oil palm biomass. *Environment, Development and Sustainability*, 17, 691–709.

Wahab, S. (2009). Biotechnological approaches in the management of plant pests, diseases, and weeds for sustainable agriculture. *Journal of Biopesticides*, 2(2), 115–134.

Wong, W.S., Tan, S.N., Ge, L., Chen, X., and J.W.H. Yong (2015). The importance of phytohormones and microbes in biofertilizers. In *Bacterial Metabolites in Sustainable Agroecosystem*, edited by D.K. Maheshwari, Springer International Publishing, Switzerland, pp. 105–158.

Zahran, H.H. (1999). Rhizobium-legume symbiosis and nitrogen fixation under severe conditions and in an arid climate. *Microbiology and Molecular Biology Reviews*, 63(4), 968–989.

Zarei, I., Sohrabi, Y., Heidari, G.R. Jalilian, A., and K. Mohammadi (2012). Effects of biofertilizers on grain yield and protein content of two soybean *(Glycine max L.)* cultivars. *African Journal of Biotechnology*, 11(27), 7028–7037.

Zendejas, H.S.L., Oba, M.S., Oba, A.S., Gallegos, E.G., and J.L.G. Estrada (2015). A bioeconomic approach for the production of biofertilizers and their influence on Faba Bean (*Vicia faba L*) productivity. *Journal of Natural Sciences*, 3(2), 75–91.

10

Production of Life-Saving Drugs from Marine Sources

Rakesh Maurya and Sudhir Kumar

CONTENTS

ABSTRACT Natural-product-based drug discovery has imparted a renaissance, particularly, to the field of marine natural-product-based drug discovery in the past few decades. A large number of structurally diverse chemical compounds discovered from marine sources have beneficial health effects and reduce the risk of a number of life-threatening and debilitating diseases. Inspirational drug discovery processes by semisynthetic modification of natural leads have provided an additional factor of force to drive marine natural product pharmacognosy ahead. A number of marine natural products and synthetic analogs as well are being evaluated for indications of different diseases including cancer, neurodegenerative disorders, infectious disease, inflammation and protozoan diseases. Prosperous biodiversity of marine environment is the major source of new biologically active molecules. This chapter presents an account of 198 compounds of marine origin and 206 references directed toward the discovery of new anticancer, anti-infective, anti-inflammatory, anticoagulating and antidiabetic agents from marine sources particularly emphasizing tremendous opportunities involved in developing new pharmaceuticals from the sea, along with compounds that are in the market and in clinical trial.

10.1 Introduction to Marine Pharmacology

Rich chemical diversity of marine secondary metabolites has proven to be a great source of extremely potent bioactive compounds. Several of them have inspired the development of life-saving pharmaceuticals; others are used as food additives and beauty products. Although the numbers of marine natural products (NPs) in market are relatively few, they have the potential for the development of new life-saving drugs (Molinski et al. 2009). Approximately half of all life forms inhabit oceans and seas that cover 70% of the earth's surface. Estimated number of species represented by sponges, echinoderms, bryozoans, tunicates, shellfish, bacteria, fishes, seaweeds, and so on inhabiting the world's oceans may come close to 10 million (Mora et al. 2011). Extreme conditions of pressure and temperature have facilitated extensive biological and chemical diversification. To survive in these surroundings, marine organisms are armored with potent and highly active secondary metabolites to either protect the user of it from predation or to kill the prey.

Several bioactive compounds have been isolated from various marine invertebrates such as sponges, echinoderms, jellyfishes, mollusks, bryozoans, and a few others. Marine organisms are the ocean of immense bioactive substances of diverse structural classes, including alkaloids, polyketides, terpenes, steroids, and peptides, responsible for the development of more than 15,000 different products (Hu et al. 2015) and hundreds of new compounds being discovered every year.

Marine sponges are most primitive and diverse animals, comprising approximately 8000 sponge species (Van Soest et al. 2012). Early pharmaceuticals inspired from marine NPs were isolated from the marine sponge *Cryptotethya crypta*. The discovery of the nucleosides spongothymidine and spongouridine played a significant role in the synthesis of cytotoxic compounds. Sponges produce remarkable chemical diversity of toxins and other compounds to protect themselves from predators and to compete with other species. Marine sponges are currently one of the richest sources of new marine NPs reported yearly. Cnidarians are second after sponges in terms of new marine NPs, particularly terpenoid metabolites, reported annually (Blunt et al. 2005). Echinodermata, of which there are over 6000 species known worldwide, is one of the most distinct phyla among the marine invertebrates and includes sea urchins, sea cucumbers, sea stars, sea lilies, feather stars, brittle stars, sand dollars, and sea cucumbers. They are proven to be an abundant source of bioactive glycosylated metabolites showing antimicrobial, antitumor, anticoagulant, cardioprotective, and even antiviral activities. Approximately 500 species of bryozoans, prospering in marine environment ranging from shallow water to deep water of about 5000 m, are a considerable source of marine metabolites (Clarke and Lidgard 2000). Shallow-water ascidians, comprising over 2800 known species (Shenkar and Swalla 2011), are among the producers of biologically important metabolites.

Several bioactive secondary metabolites have successfully been approved for clinical trial and a large number of them have shown the potential to be developed as potential therapeutics for life-threatening diseases. Despite the undying problem of supply and low availability of biologically active compounds isolated from natural material, the development of advanced methods of sampling identification and competent techniques of isolation have boosted marine NP research. Synthetic chemists are taking up the challenge of drug development from ocean by developing methods of synthesis and artificially producing many marine NPs in gram scale (Morris 2013). There are several excellent reviews on marine NP and drug development from marine sources (Newman and Cragg 2004; Glaser and Mayer 2009; Molinski et al. 2009; Sashidhara et al. 2009; Gerwick and Moore 2012). This chapter describes research on natural marine pharmacological compounds isolated from a marine source. The selected compounds will be reviewed under the heading of each pharmacological activity.

10.2 Marine Natural Products and Inspired Drugs

The story of marine and marine inspired drugs begins with the discovery of spongothymidine (1) and spongouridine (2) from the Caribbean sponge *Tectitethya crypta* (also known as *Cryptotethya crypta* and *Tethya crypta*) in the 1950s by Bergmann (Bergmann and Feeney 1950). These compounds are glycosides of D-arabinose rather than D-ribose. The discovery of spongothymidine in marine sponges, a nucleoside with a modified sugar instead of a modified base, directed the design and synthesis of a new generation, sugar modified nucleoside (Hamann et al. 2007). Compounds derived from marine sponge *Cryptotethya crypta* have inspired the synthesis of several ara-nucleosides (where "ara" represents arabinose). Two ara-nucleosides inspired drugs that were able to reach clinics from labs are cytarabine (3) followed by vidarabine (4).

10.2.1 Life-Saving Drugs in Market: Discovery, Production, and Pharmacology

10.2.1.1 Cytarabine

Cytarabine (3) received FDA's approval as an anticancer drug in June 1969 (Schwartsmann et al. 2003). It (also known as ara-C or 1-β-D-arabinofuranosylcytosine) received the pleasure of being the first marine-derived drug used for the management of leukemia. Inspired by arabinose nucleosides, cytarabine or cytosine arabinoside was produced artificially by R. Walwick in 1959 and natural-sourced cytarabine was isolated from the fermentation broth of *Streptomyces griseus*. Subsequently, cytarabine was isolated from the gorgonian *Eunicella cavolini* (Cimino et al. 1984). Cytarabine is sold under the trade name Cytosar-U® or Depocyt® and is prescribed mainly for the treatment of non-Hodgkin lymphoma and acute myeloid leukemia (AML; Wang et al. 1997).

Cytarabine interferes with DNA synthesis and its mechanism. It is rapidly converted into cytosine arabinoside triphosphate in the body which replaces cytidine triphosphate during DNA synthesis leading to fabrication faulty DNA. Cytarabine also inhibits important enzymes of DNA synthesis mechanism, RNA polymerase and nucleotide reductase (Severin et al. 2003). Despite side effects such as cerebellar toxicity, granulocytopenia, thrombocytopenia, leukopenia, anemia, and gastrointestinal disturbance associated with high dosing, it is a drug of choice for myeloid leukemia and non-Hodgkin lymphoma because no better approaches are available.

10.2.1.2 Vidarabine

Vidarabine (4) or 9-β-D-arabinofuranosyladenine (ara-A) is a glycoside analog of adenosine where D-ribose is replaced with D-arabinose which is inspired by the knowledge gained from the chemical structure of spongouridine (2). The synthesis of vidarabine was first achieved in the laboratory of Bernard Randall Baker at the Stanford Research Institute (now known as SRI International) and finally natural vidarabine was obtained from the fermentation broth of *Streptomyces antibioticus*. It is the first systemic nucleoside analog antiviral agent that has been licensed for the management of herpes virus infection in human. Vidarabine interferes with the synthesis of viral DNA. Diphosphorylated and triphosphorylated vidarabine are active drugs produced in vivo that interfere with the synthesis of viral DNA. Diphosphorylated vidarabine prevents the reduction of nucleotide diphosphates by inhibiting ribonucleotide reductase enzyme, thereby inducing a decline of viral replication. Triphosphorylated vidarabine competitively inhibits dATP leading to the formation of "faulty" DNA. Vidarabine has some significant limitations such as rapid metabolism by adenosine deaminase, low intramuscular and intestinal absorption, and requirement of large fluid volumes for intravenous administration over prolonged periods (Cohen 1979; Whitley et al. 2012).

10.2.1.3 Ziconotide

Ziconotide (5), or Prialt®, also known as SNX-111, is an uncommon non-opioid, not addictive analgesic agent which is 1000 times more effective than morphine (de la Calle Gil et al. 2015). It is the synthetic form of a ω-conotoxin peptide derived from *Conus magus* (Cone Snail). The ω-conotoxins is a chemical group of structurally related polypeptide molecules discovered in the venom of predatory cone snails (genus *Conus*) of different species. ω-conotoxins are comprised of several different peptides such as ω-GVIA, ω-MVIIA, ω-MVIIC, and ω-CVID, each having potent physiologic capabilities. Ziconotide (ω-MVIIA) is a small peptide molecule made up of 25 amino acids sequence H-Cys-Lys-Gly-Lys-Gly-Ala-Lys-Cys-Ser-Arg-Leu-Met-Tyr-Asp-Cys-Cys-Thr-Gly-Ser-Cys-Arg-Ser-Gly-Lys-Cys-NH$_2$ in which six cysteine residues are linked in pairs by three disulphide bonds.

The powerful analgesic effect of ziconotide results from its intrathecal injection that interferes pain signaling at the spinal cord by selectively blocking neurotransmission at N-type calcium channels on nerve cells (Wang et al. 2016). Ziconotide was developed into an artificially manufactured drug by Elan Corporation. U.S. FDA approved for its sale under the name Prialt for the treatment of severe chronic pain in patients suffering from cancer, AIDS, or certain neurological disorders. Ziconotide is delivered directly into fluid surrounding the spinal cord through intrathecal route. It is effective and safer than morphine as no addiction or tolerance is developed on its prolonged administration.

10.2.1.4 Trabectedin

Trabectedin (6) is a complex tetrahydroisoquinoline alkaloid, first time isolated from Caribbean collection of a tunicate, *Ecteinascidia turbinata* (Valoti et al. 1998). E. J. Corey developed a method of total synthesis of ecteinascidin 743 to overcome the problem of low yields from the sea squirt (1 g of trabectedin is isolated from 1 t of sea squirt; Corey et al. 1996).

Although the initial clinical trials were performed with the NP, the subsequent developments were made of a semisynthetically produced trabectedin from the microbial product cyanosafracin B (7). The development of semisynthetic methods from cyanosafracin B, an antibiotic obtained by fermentation of the bacterium *Pseudomonas fluorescens*, has solved the all-time problem of supply with marine-sourced secondary metabolites (Molinski et al. 2009). Trabectedin was approved by EMEA for the treatment of patients with advanced soft-tissue sarcoma and was sold under the brand name Yondelis® by Zeltia and Johnson and Johnson. Trabectedin was categorized as a medicine used in rare diseases (orphan medicine) on May 30, 2001, for soft-tissue sarcoma and in October 2003 for ovarian cancer because these are considered rare (http://www.prnewswire.co.uk/news-releases/yondelisr-granted-orphan-drug-designation-by-the-us-fda-for-the-treatment-of-ovarian-cancer-154027485.html, retrieved on November 30, 2015).

10.2.1.5 Eribulin Mesylate

Eribulin (8) is a fully synthetic analog of a large naturally occurring polyether macrolide halichondrin B (9) that was purified from the marine sponge *Halichondria okadai*. Halichondrin B has shown potent anticancer activity against murine cancer cells in both *in vitro* and *in vivo* studies and indicated that it disrupts microtubules (Hirata and Uemura 1986). The efforts to combat limited supply of halichondrin B, the complete chemical synthesis was achieved by Yoshito Kishi and colleagues at Harvard University in 1992 (Aicher et al. 1992).

Several structurally simplified synthetic analogs facilitated the development and optimization of structurally simplified analogs of eribulin, such as E7389, ER-086526, and NSC-707389, that retained anticancer activity (Yu et al. 2005). The most potent analog among them, E7389, that ultimately became eribulin (Halaven®) was approved by the FDA in 2010 for the treatment of refractory metastatic breast cancer (mBC).

10.2.1.6 Brentuximab Vedotin

Brentuximab vedotin (Adcetris®) is an antibody–drug conjugate (ADC) with a potent microtubule inhibitor in which an anti-CD30 antibody conjugated an analog of dolostatin-10, that is, monomethyl auristatin E (MMAE). Although dolastatins were isolated for the first time from the sea hare *Dolabella auricularia* (Phylum Mollusca), later they were found to be produced by cyanobacteria (Luesch et al. 2001). Dolostatin-10 was withdrawn from phase II of clinical trials due to some toxic effect associated (Francisco et al. 2003). Dolostatin-10 analog was then conjugated with an antibody that targeted the CD30 antigen present on the surface of Hodgkin lymphoma cells to produce brentuximab vedotin (Armitage et al. 2015). The ADC brentuximab vedotin was launched in 2011 under the trade name Adcetris® by Seattle Genetics and indicated for the management of Hodgkin and systemic large cell lymphoma. The drug is proven to be less toxic than the parent compound, dolostatin-10. However, so far described adverse effects are neutropenia, peripheral sensory neuropathy, fatigue, and hyperglycemia (http://www.fda.gov/Drugs/InformationOnDrugs/ApprovedDrugs/ucm458815.htm, retrieved on November 30, 2015).

10.2.1.7 Omega-3-Acid Ethyl Esters

Omega-3-acid ethyl esters (12, 13) are lipid-regulating agents developed by Reliant Pharmaceuticals and sold under the brand name Lovaza® by GlaxoSmith Kline. Lovaza is an FDA-approved drug for the treatment of patients with hypertriglyceridemia (Weintraub 2014). The mechanism of action of Lovaza is not understood completely. But is it found that the drug produces hypotriglyceridic effect probably by the inhibition of acyl CoA, 1,2-diacylglycerol acyltransferase. It decreases lipogenesis in the liver and the VLDL-cholesterol level, and increases HDL-cholesterol, mitochondrial and peroxisomal β-oxidation in the liver, and plasma lipoprotein lipase activity.

10.2.1.8 Iota-Carrageenan

Carrageenan (14) is a family of linear sulphated polysaccharides, a common food additive, that are extracted from red edible seaweeds. Carrageenan finds its medical use in the form of iota-carrageenan. Iota-carrageenan produced mainly from *Eucheuma denticulatum* is sold as an over-the-counter (OTC) drug by the trade name of Carragelose®. Chemically, iota-carrageenan is linear sulphated polysaccharides having two sulphates per disaccharide (Ahmadi et al. 2015).

The use of carrageenan as food additive has many objections, as it causes gastrointestinal inflammation; intestinal lesions, ulcerations, and even malignant tumors. Whereas Marinomed Biotechnologie GmbH developed an effective nasal spray for the treatment of common cold containing antiviral iota-carrageenan (Carragelose). Carragelose creates a protective physical barrier in the nasal cavity which protects from viral attack. In addition, it has the potential to manage patients with viral conjunctivitis, an infection for which there is currently no approved etiological treatment (Figure 10.1).

10.2.2 Marine Natural Products and Their Synthetic Analogs in Clinical Trial

There are considerable numbers of interesting molecules that have either been isolated from marine sources or have been produced artificially based on the knowledge gained from a prototypical compound isolated from marine sources. Many of them are in phases II and III of clinical trials mainly for cancer and analgesia (Table 10.1).

Plitidepsin (15) or dehydrodidemnin B is a marine cyclic depsipeptide extracted from the ascidian *Aplidium albicans* (Newman and Cragg 2004). Plitidepsin is extremely potent to induce rapid p53-independent apoptosis with IC_{50} values in nanomolar range. The drug is marketed under the trade name Aplidin® by PharmaMar and is undergoing phase III of clinical trial for the treatment of multiple myeloma.

Lurbinectedin (16) is a dimeric isoquinoline alkaloid similar to trabectedin but has a tetrahydro-β-carboline moiety as a replacement of tetrahydroisoquinoline present in trabectedin. Currently, it is in phase III of clinical development for ovarian cancer. Lurbinectedin binds covalently to minor grooves of DNA inducing break formation in double-stranded DNA, thereby leading to cell apoptosis.

Panobinostat (17; Farydak®) is a drug developed by Novartis for the treatment of various cancers. Panobinostat induces apoptosis of malignant cells via inhibition of multiple histone deacetylase enzymes. It is in phase II/III of clinical trial for the treatment of chronic myeloid leukemia (Bailey et al. 2015).

Salinosporamide A (18) is a potent proteasome inhibitor in phase II of clinical trial for the treatment of multiple myeloma. This marine NP is produced by marine bacteria found

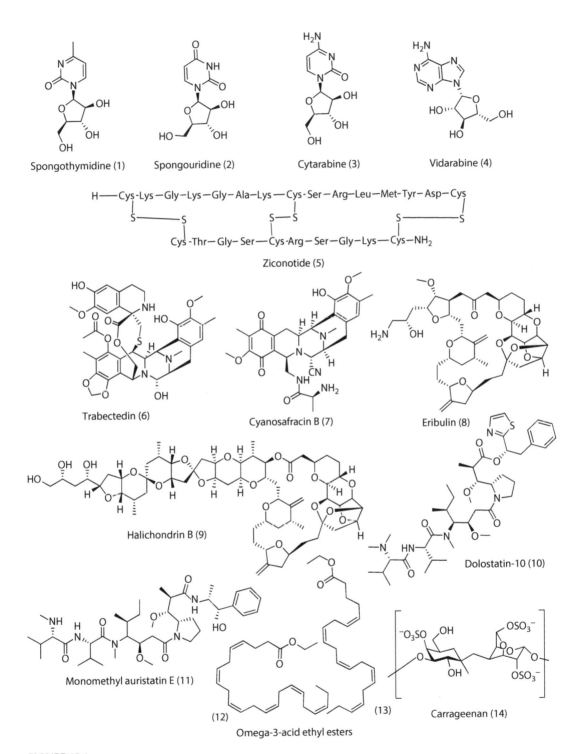

FIGURE 10.1
Chemical structures of marketed drugs and related marine compounds.

TABLE 10.1

Natural Marine and Derived Compounds in Market and Clinical Trial

S. No.	Name	Alternative Name	Source	Activity	Clinical Status
1	Cytarabine	Cytosar-U®	Spongothymidine derivative	Cancer	FDA approved
2	Vidarabine	Vira-A®	Spongouridine derivative	Anti-viral	FDA approved
3	Ziconotide	Prialt®	NP, synthetic version of ω-conotoxin MVIIA	Neuropahtic Pain	FDA approved
4	Trabectedin	Yondelis®	NP, Ecteinascidin 743	Cancer	EU approved
5	Eribulin mesylate	Halaven®	Halichondrin B inspired	Cancer	FDA approved Breast cancer
6	Brentuximab vedotin	Adcetris®	Dolastatin-10 synthetic analog monomethylauristatin E	Cancer	FDA approved
7	Omega-3-acid ethyl esters	Lovaza®	Derivative omega-3-fatty acids	Hypertriglyceridemia	FDA approved
8	Iota-carrageenan	Carragelose®	NP, *Eucheuma denticulatum*	Antiviral	FDA approved
9	Plitidepsin	Aplidine® Dehydrodidemnin B	Ascidian *Aplidium albicans*	Cancer	Phase III Multiple myeloma
10	Lurbinectedin	PM1183 Tryptamicidin	NP derivative	Cancer	Phase III Ovarian cancer;
11	Panobinostat	Farydak; LBH-589	NP-derived	Cancer	Phase II/III Chronic myeloid leukemia
12	Bryostatin-1	Bryostatin-1-Neurotrope	NP, *Bugula neritina*	Alzheimers disease	Phase II Alzheimers disease
13	Salinosporamide A	NPI-0052; Marizomib	NP, *Salinospora* strain CNB-392	Cancer	Phase II Multiple myeloma
14	PM 060184	PM060184	NP, *Lithoplocamia lithistoides*	Cancer	Phase I Solid tumors
15	Glembatumumab vedotin	CDX-011; CR 011 ADC	MMAE-based ADC	Cancer	Phase II Breast cancer; Malignant melanoma

in ocean sediment, *Salinispora tropica* and *Salinispora arenicola*. A significantly potent proteasome inhibitor, salinosporamide A showed proteasomal chymotrypsin-like proteolytic inhibition activity with an IC_{50} value of 1.3 nM (Feling et al. 2003). Furthermore, it exhibited significant in vitro cytotoxicity against HCT-116 human colon carcinoma (IC_{50} 11 ng/mL).

PM060184 (19) is a tubulin-binding agent (Martinez-Diez et al. 2014) originally isolated from the marine sponge *Lithoplocamia lithistoides*. Currently produced by total synthesis (Martin et al. 2013), it is under evaluation in clinical studies on patients with advanced cancer diseases (https://clinicaltrials.gov/ct2/show/NCT01299636).

Bryostatin 1 (20) is a cyclic macrolide originally isolated from the marine bryozoan *Bugula neritina* (Order *Cheilostomata*; Pettit et al. 1982), and later it was proposed to be produced by symbiotic bacteria of the bryozoans (Davidson et al. 2001). As in 2015, bryostatin 1 (20) is in phase II of clinical studies for the therapy of modest as well as severe Alzheimer's disease (Figure 10.2; https://clinicaltrials.gov/ct2/show/NCT02431468?term=Bryostatin-1&rank=12).

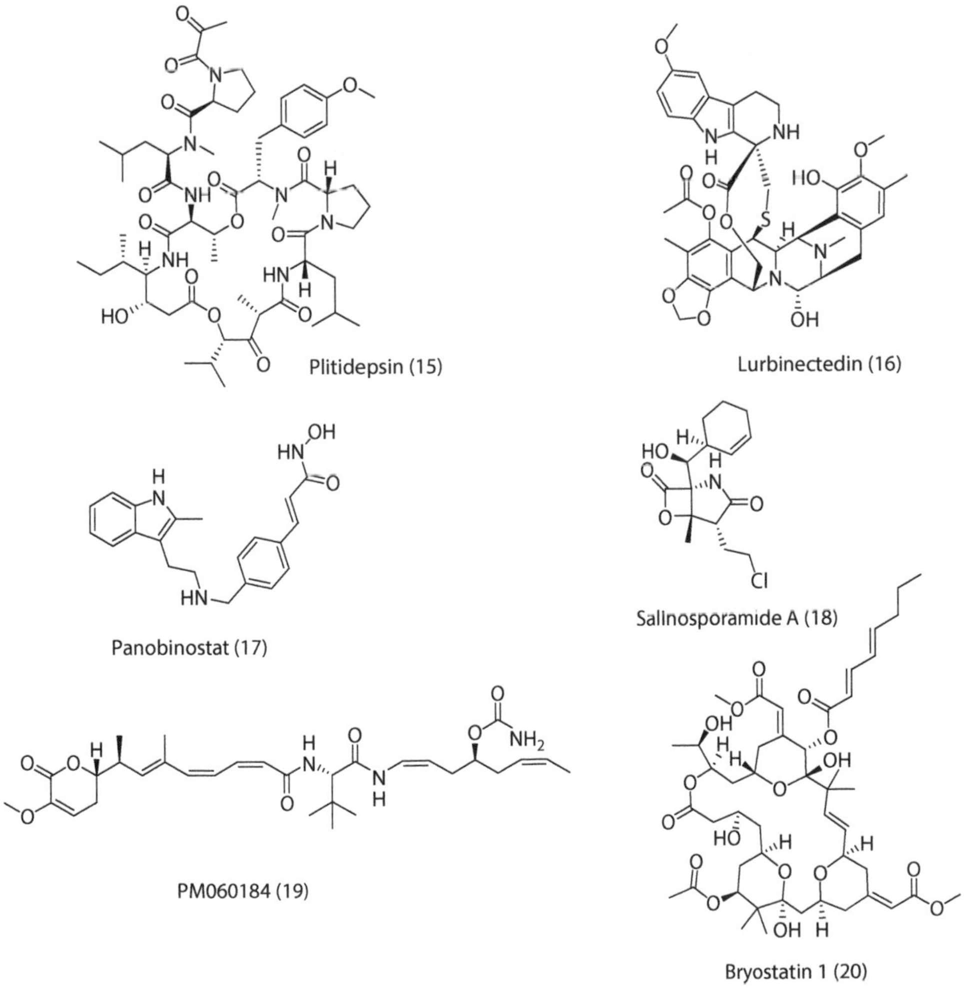

FIGURE 10.2
Chemical structures of marine and derived compounds in clinical trial.

10.3 Anticancer Marine Natural Products

Tremendous diversity of toxic metabolites produced by marine sources has been found active against cancer. Many of them have the potential to be developed as newer and safer drugs for the treatment of cancer. Considerable numbers of anticancer drugs used in the clinic are either NPs or derived from NPs. Several new antineoplastic compounds derived from marine NPs are now in the preclinical pipeline. Researchers have isolated hundreds of compounds each year and evaluated their anticancer potential. Even though many compounds could not reach the market, these have guided the clinical trial of other NPs or derived agents. These are crucial compounds in the understanding of the mechanism of drug action.

10.3.1 Microtubule Inhibitors

Antimicrotubule agents generally act by either inhibiting or stabilizing the polymerization of tubulin via binding on the tubulin surface. Cryptophycins exhibit potent antiproliferative and antimitotic effect by destabilizing microtubule dynamics as well as inducing hyperphosphorylation of the antiapoptotic protein BCl-2 (B-cell leukemia/lymphoma 2; Wagner et al. 1999). Cryptophycin-52 (21), an analog of the marine NP Cryptophycin 1, had undergone phase I of clinical trials for the treatment of nonsmall cell lung cancer and solid tumors. However, due to adverse effects, phase II of trial was discontinued. Kobayashi et al. (1994) isolated another related cyclic didepsipeptide, arenastatin-A (22), from marine sponge *Dysidea arenaria*. Later, it was found to be identical to cryptophycin-24 (22). Arenastatin-A (22) exhibited extremely potent cytotoxicity against KB cells with IC_{50} 5 pg/mL. Dolastatins are significant tubulin polymerization inhibitors isolated from marine sources. Dolastatin-10 (10), isolated for the first time from the Indian Ocean mollusk *Dolabella auricularia* (Luesch et al. 2001), is a tubulin interactive agent binding close to the vinca domain (Bai et al. 1990) which entered phase I of clinical trials in the 1990s. Cemadotin (23) is a water-soluble synthetic analog of dolastatin that exhibited potent antiproliferative and antitumor activities (Supko et al. 2000) through action on microtubules which blocks cells at mitosis (Jordan et al. 1998). Didemnin B (24) was isolated from extracts made of the tunicate *Trididemnum solidum* that demonstrated excellent cytotoxicity against P388 and L1210 murine leukemia cell lines (Rinehart et al. 1981). Didemnin B (24) was the first defined chemical compound directly from a marine source that was taken for preclinical and clinical trials under the financial support of the National Cancer Institute (NCI) in 1980s (Cain et al. 1992). However, the compound was withdrawn from clinical trial due to severe toxicity observed. Phenylhistin (Halimide; 25) is a NP produced by the fungus *Aspergillus ustus* (Kanoh et al. 1997) that belongs to a class of naturally occurring 2,5-diketopiperazines. Phenylhistin (25) exhibits cytotoxic effects against various types of tumor cell lines by interfering in the microtubule-formation mechanism (Kanoh et al. 1999).

Hemiasterlins are another group of potent microtubule-disrupting agents (Anderson et al. 1997). Tripeptide hemiasterlin (26) was isolated from the sponge *Hemiasterella minor* (Talpir et al. 1994). Taltobulin (27), a synthetic analog of hemiasterlin, inhibited proliferation of hepatic tumor cell lines (Vashist et al. 2006), bladder cancer (Matsui et al. 2009), and prostate cancer (Hadaschik et al. 2008) like its precursor inhibited microtubules formation (Lo et al. 2004; Ravi et al. 2005).

Dictyostatins are macrolactone isolated from the sponge *Spongia sp*. Dictyostatin-1 (28) induced tubulin polymerization by binding to a taxoid site against human ovarian carcinoma cells (Madiraju et al. 2005). Discodermolide (29) is a metabolite originally isolated from deep-sea sponge *Discodermia dissoluta* (Gunasekera et al. 1990) and is produced synthetically using many others (Arefolov and Panek 2002; Mickel et al. 2005; Mickel 2007). Discodermolide induces microtubule stabilization (ter et al. 1996). Marine sponge-derived microtubule-stabilizing agent Peloruside A (30), isolated from *Mycale hentscheli*, has potent paclitaxel-like microtubule-stabilizing activity at nanomolar concentration (Hood et al. 2002). A macrolide laulimalide (31), isolated from marine sponges *Hyattella sp.* (Corley et al. 1988) and the Okinawan sponge *Fasciospongia rimosa* (Jefford et al. 1996) promotes anomalous tubulin polymerization.

Avarol (32), a sesquiterpenoid hydroquinone, was first isolated from the marine sponge *Disidea avara* (Minale et al. 1974). At a concentration of 0.9 µM, avarol reduced the cell growth to 50% in an in vitro assay of L5178y mouse lymphoma cells. Avarol interfered with mitotic processes, preventing telophase formation (Müller et al. 1985; Figure 10.3).

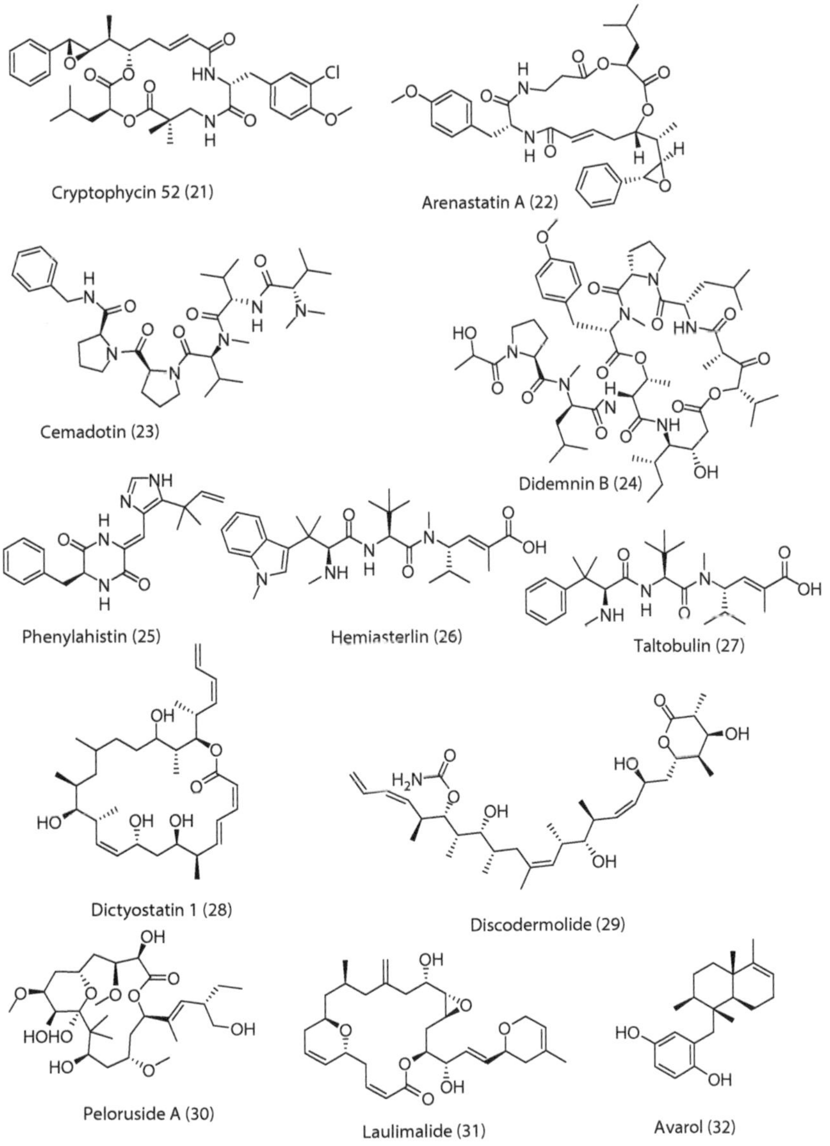

Cryptophycin 52 (21)

Arenastatin A (22)

Cemadotin (23)

Didemnin B (24)

Phenylahistin (25)

Hemiasterlin (26)

Taltobulin (27)

Dictyostatin 1 (28)

Discodermolide (29)

Peloruside A (30)

Laulimalide (31)

Avarol (32)

FIGURE 10.3
Chemical structure of tubulin inhibitor marine natural products.

10.3.2 Actin Filaments Inhibitors

Actin is an essential component of the cell's cytoskeleton which participates in the regulation of many motor functions in the cell, such as cell migration, cell division, and muscle contraction. Notable number of highly potent cytotoxic agents that interact with actin have been produced by marine organisms (Figure 10.4). Latrunculins were the first actin-binding substances isolated from a marine source. Ichthyotoxicity principle isolated from sponge *Latrunculia magnifica*, Latrunculin A (33), was later shown to be cytotoxic (Kashman et al. 1980). Latrunculin A (33), a well-known actin-binding macrolide, arrests polylysine-induced nucleation at the level of an antiparallel dimer (Bubb et al. 2002). Antitumor macrolide aplyronine A (34) inhibits the polymerization of globular actin to fibrous actin by forming 1:1 complex with globular actin. Aplyronine A (34) binds to actin molecule by intercalating its aliphatic tail into a hydrophobic cleft of the

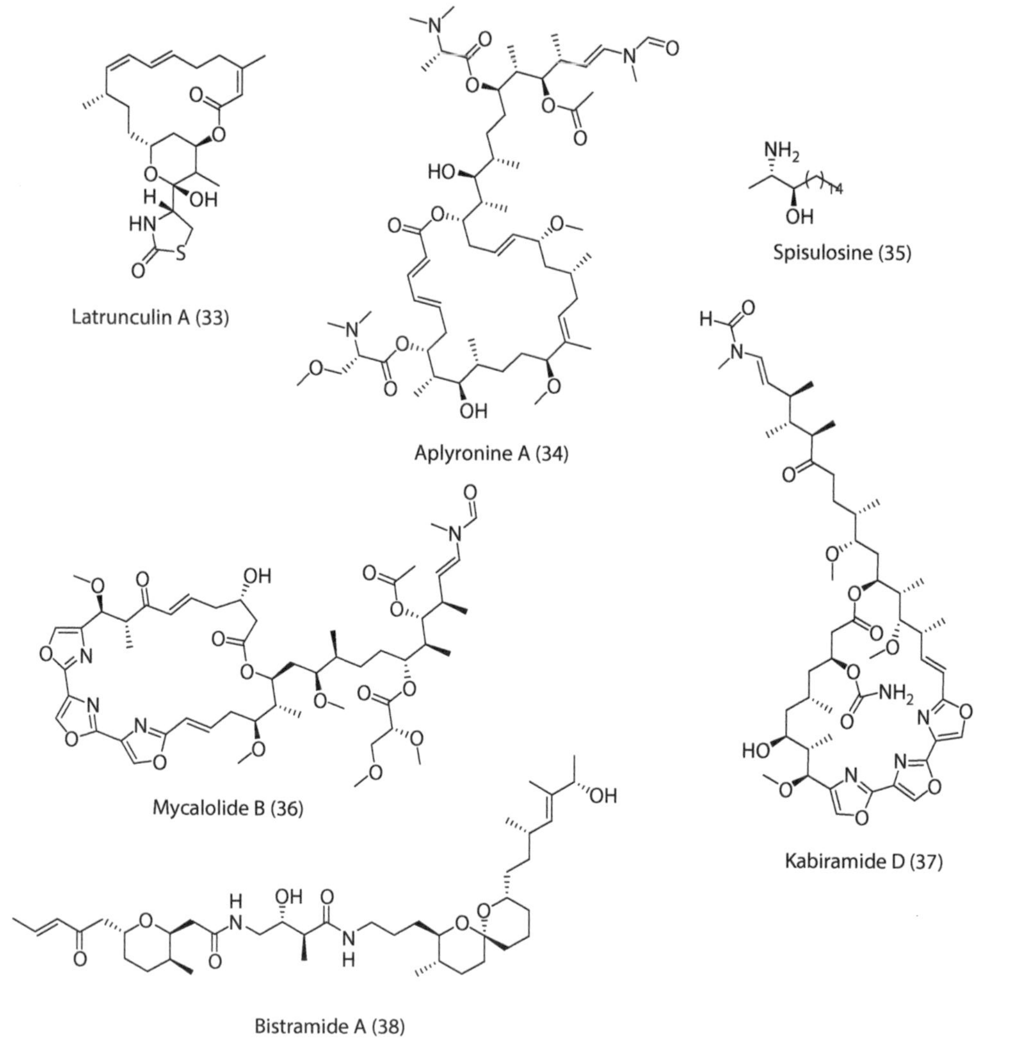

FIGURE 10.4
Cytotoxic marine natural products that interact with actin.

(*Continued*)

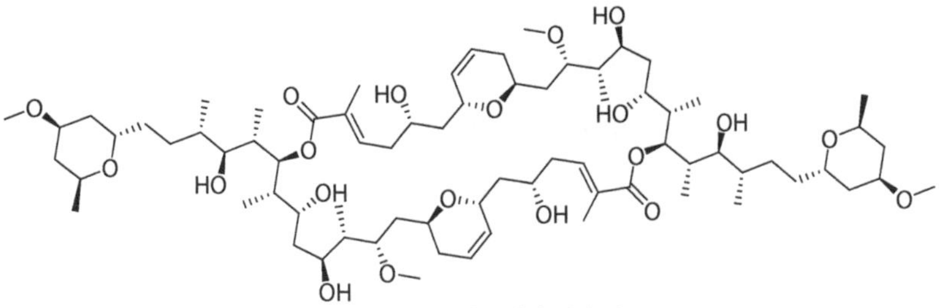

Misakinolide A (39)

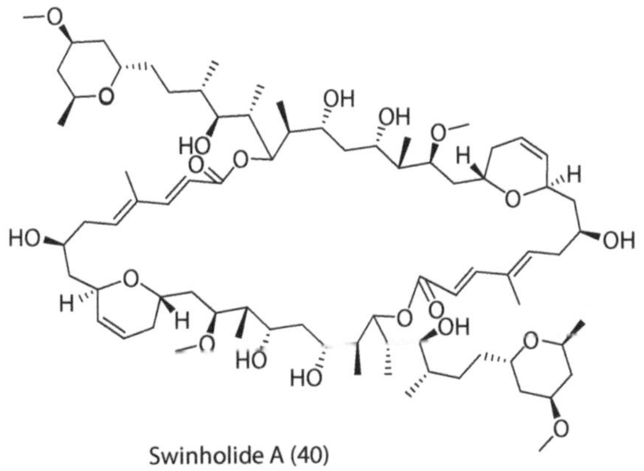

Swinholide A (40)

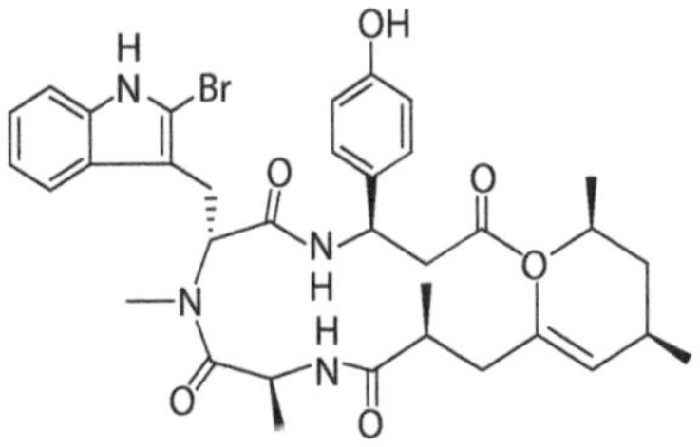

Jasplakinolide (41)

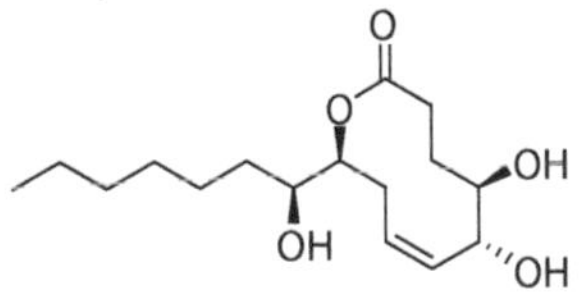

Microcarpalide (42)

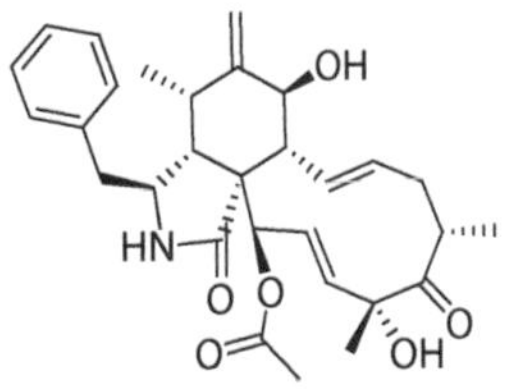

Cytochalasin D (43)

FIGURE 10.4 (Continued)
Cytotoxic marine natural products that interact with actin.

actin molecule (Hirata et al. 2006). Spisulosine (35; ES-285), isolated from the sea mollusc *Spisula polynyma*, is a novel antiproliferative (antitumoral) compound of marine origin. It inhibits cell proliferation by preventing assembly of actin stress fibers (Cuadros et al. 2000). Marine sponge *Mycale sp.* derived NP mycalolide B (36) and a related kabiramide D (37) are actin-depolymerizing cytotoxic compounds (Wada et al. 1998). A polyketide bistramide-A (38), isolated from the marine ascidian *Lissoclinum bistratum*, disrupts actin cytoskeleton both by depolymerizing F-actin and binding directly to monomeric actin in vitro (Rizvi et al. 2006; Statsuk et al. 2005). Dimeric macrolides, bistheonellide-A or misakinolide-A (39), isolated from an Okinawan *Theonella sp.* sponge (Terry et al. 1997) and swinholide A (40) isolated from *Symploca* and *Geitlerinema species* of cyanobacteria (Andrianasolo et al. 2005) bind with two molecules of actin per molecule of these cyto-toxic agents. Jasplakinolide (41), a cyclodepsipeptide isolated from marine sponge, *Jaspis johnstoni*, showed 50% growth inhibition of prostate carcinoma cell lines PC-3, LNCaP, and TSU-Pr1 at 65 nM, 41 nM, and 170 nM, respectively, on 48 h exposure (Senderowicz et al. 1995). Microcarpalide (42), isolated from culture broth of an endophytic fungus hosted by a *Ficus microcarpa L.* (Ratnayake et al. 2001), cause complete disruption of actin microfilament of NIH/3T3 fibroblasts incubated with microcarpalide (42; Furstner et al. 2007). Cytochalasin D (43) showed cytotoxicity at 100 nM against T cells (Suria et al. 1999).

10.3.3 Induction of Apoptosis in Cancer Cells

Cephalostatin-1 (44), a bis-steroidal marine natural product (MNP) obtained from the Indian Ocean collection of hemichordata *Cephalodiscus*, inactivated the antiapoptotic mito-chondrial protein Bcl-2 by hyperphosphorylation (Lopez-Anton et al. 2006). Curacin-A (45), isolated from the cyanobacterium *L. Majuscule* (Gerwick et al. 1994), is exquisitely potent but is effectively insoluble in any formulation (Wipf and Xu 1996). Aaptamine (46) is a prominent cancer cell growth inhibitory constituent isolated from marine sponge *Hymeniacidon sp.* (Pettit et al. 2004). It exerted an antiproliferative effect by inducing apop-tosis (Dyshlovoy et al. 2014). Dysidiolide (47), a secondary metabolite obtained from the Caribbean sponge *Dysidea etheria*, inhibited cdc25A protein phosphatase with an IC_{50} value of 9.4 μM. Moreover, dysidiolide (47) showed IC_{50} 4.7 μM in vitro inhibition of A-549 human lung carcinoma and P388 cells (IC_{50} 1.5 μM). Cycloprodigiosin hydrochloride (48), a member of the prodigiosin family of compounds, exhibited protein synthesis inhibition-induced apoptosis in PC12 cells (Kawauchi et al. 2008). Pycnidione (49), isolated from the fermented broth of *Theissenia rogersii* 92031201, exhibits antiproliferative activities on A549 cells with a 50% growth inhibition (GI_{50}) value of 9.3 nM at 48 h, triggering apoptosis in the A549 cells by enhancing PAI-1 production, and activated caspase-8 and -3 (Figure 10.5).

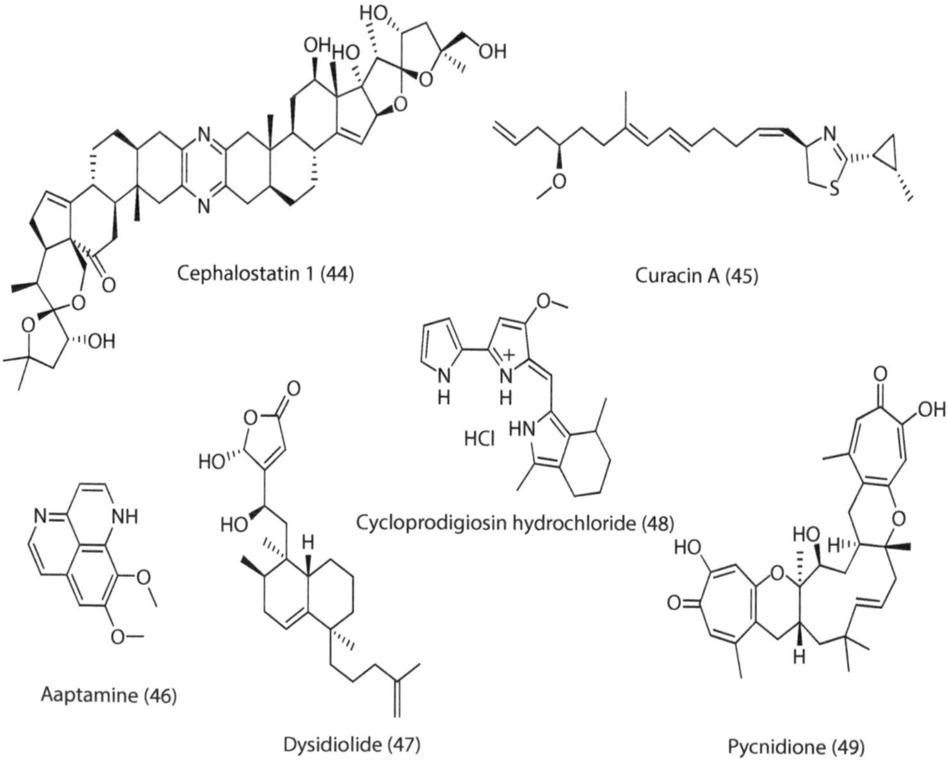

FIGURE 10.5
Chemical structures of apoptosis inducer marine natural products.

10.3.4 Anticancer Natural Product Antibody–Drug Conjugates

Antibody–drug conjugates or ADCs play an increasing role in cancer treatment as they are highly potent biopharmaceuticals designed to selectively target cancer cells. This combination of antibodies to cytotoxic drugs conveys the benefits of highly potent drugs. ADCs consist of covalently linked monoclonal antibodies aimed at certain tumor markers differentially overexpressed in tumor cells and at selectively delivering cytotoxic (anticancer) payload to tumor cells so that healthy cells are less severely affected (Kovtun and Goldmacher 2007).

Gemtuzumab ozogamicin (trade name Mylotarg®, Pfizer/Wyeth) was the first ADC to receive marketing approval in 2001 by U.S. FDA for the treatment of acute myelogenous leukemia. However, it was withdrawn from market in June 2010 by the sponsor due to the absence of significant benefit and safety reasons in post-approval phase III of clinical trial. Trastuzumab emtansine (Herceptin®, Genentech and Roche), an ADC composed of trastuzumab linked via a noncleavable linker to DM1, was approved in February 2013 for the treatment of HER2-positive refractory/relapsed mBC. Later, it was withdrawn from market. As a result, only one ADC Brentuximab vedotin (trade name: Adcetris, marketed by Seattle Genetics and Millennium/Takeda) is in the market.

Doxorubicin (50), calicheamicin (51), auristatins, and maytansine (52) are important cytotoxic payloads conjugated with antibodies. Calicheamicin is a DNA-alkylating agent produced by *Micromonospora echinospora ssp. calichensis*. Gemtuzumab ozogamicin (Wyeth) is a representative example of a calicheamicin-based ADC. Another calicheamicin-based ADC, Inotuzumab ozogamicin (CMC-544, Pfizer) derived via linking an antibody anti-CD22 IgG4 monoclonal antibody is in phase III of clinical trials.

Doxorubicin is an anticancer agent belonging to the anthracycline family, hemisynthesized from a NP of *Streptomyces actinobacteria* used in the treatment hematological malignancies, carcinomas, and soft-tissue sarcomas. Doxorubicin conjugated with the chimeric anti-LeY cBR96 monoclonal antibody through a hydrazone acid-labile linker is under evaluation.

MMAE (11) and F (53), collectively known as auristatins are structurally modified equipotent derivatives of dolastatin-10 (10). The dolastatins are pentapeptides, originally discovered as constituents of the sea hare *Dolabella auricularia*. These agents prevent the formation of the mitotic machinery and block α-tubulin polymerization, interacting with the vinca alkaloid binding site on α-tubulin. Antibody conjugation with MMAE and MMAF has provided a number of auristatin-based ADCs that have undergone clinical trials—one of them brentuximab vedotin is currently in market.

Other MMAE conjugated with antibodies, such as NaPi2b (RG7599; Roche), CD22 (RG7593; Roche), anti-MUC16 (RG7458; Roche), PSMA (Progenics Pharmaceuticals), and the tumor-linked glycoprotein NMB (CDX-011; Celldex Therapeutics), as well as MMAF conjugates (PF-06263507, a 5T4-targeted ADC; Pfizer) are under clinical development.

Cytotoxic agent maytansine (52), a benzoansamacrolide first isolated from the bark of the Ethiopian shrub *Maytenus ovatus*, leads to the formation of derivatives emtansine (54; also known as DM1) and ravtansine (55; also known as DM4) that bind to tubulin near the vinca alkaloid binding site. Trastuzumab emtansine (T-DM1; Roche in partnership with ImmunoGen) is an important ADC of emtansine to trastuzumab through a stable thioether linker that target human epidermal growth factor receptor (HER2; Figure 10.6).

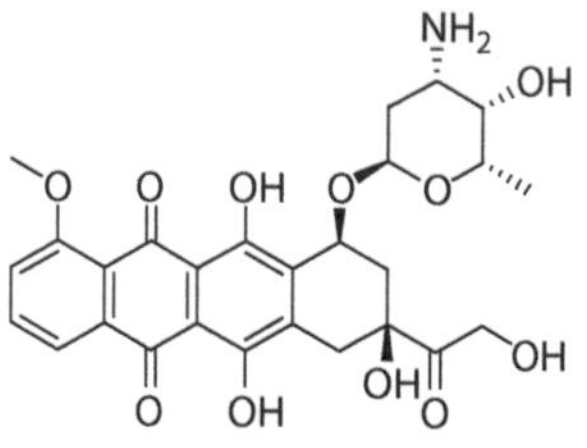

Doxorubicin (50)

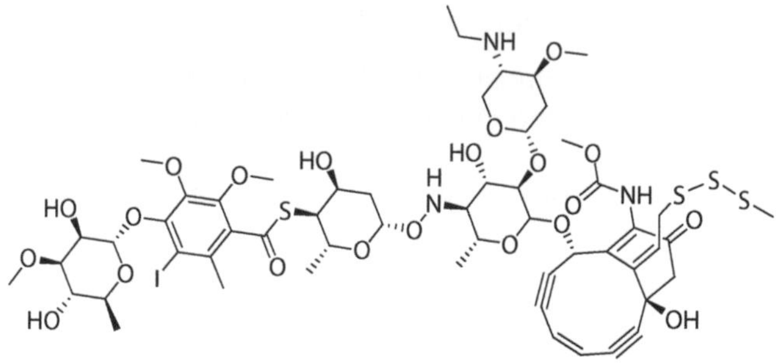

Calicheamicin (51)

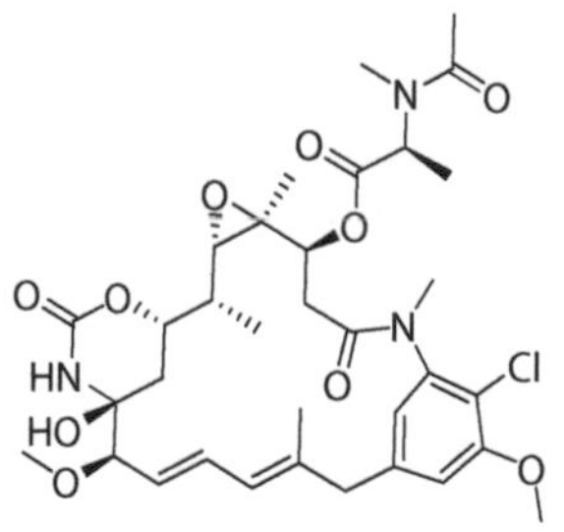

Maytansine (52)

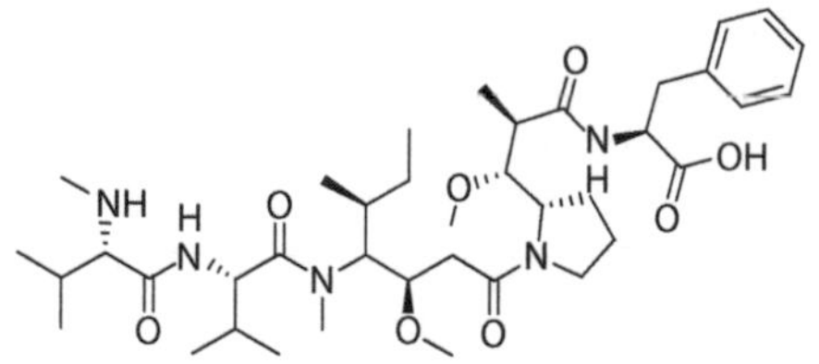

Monomethylauristatin F (53)

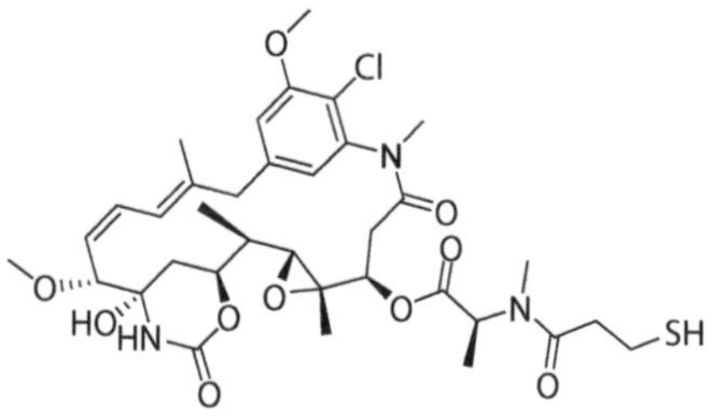

Emtansine (54)

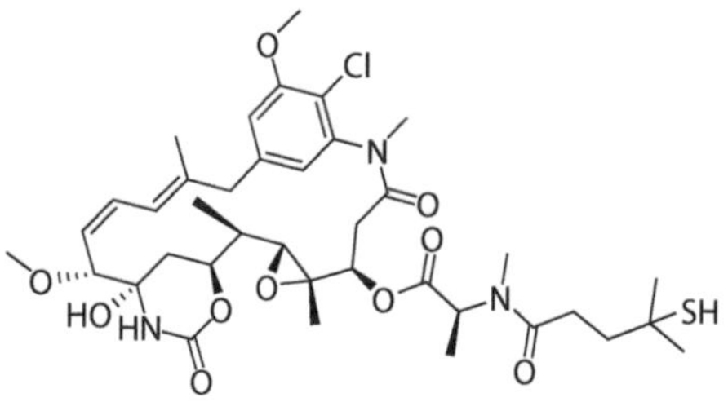

Ravtansine (55)

FIGURE 10.6
Chemical structures of ADC coupled and related cytotoxic compounds.

10.4 Anti-Infective Potential of Marine Natural Products

NPs have been the most important source of anti-infective compounds since the early days of anti-infective research. Marine organisms are potential producers of anti-infective secondary metabolite. In fact, marine organisms are naturally bestowed with an effective defense mechanism as an instrument of survival in marine conditions. At this time, any new antibiotic entering the clinical trials' pipeline needs to be advanced with the emergence of multidrug resistant (MDR) bacterial strains. There are a large number of compounds having anti-infective potential; some selected ones that have undergone extensive investigation or have shown significant activity are described further. Because of enormous diversity and several examples of marine NPs with promising antibacterial activity, the ocean continues to be a tremendous opportunity to discover new antibiotics. Coupled with synthetic analogs with enhanced pharmacological profiles, they are exceptionally attention-grabbing high-value materials for applications in benefit of human beings. Marine chemicals with potential antibacterial, antifungal, antiprotozoal, antituberculosis, and antiviral activities will be discussed in this section.

10.4.1 Antifungal Marine Natural Products

Antifungal screen of marine NPs has led to the identification of several interesting antifungal NPs, both by the way of chemical structure and biological activity. SCY-07876 and NP213 are two NP-derived compounds that could reach clinical trials against fungal infections.

Enfumafungin (56), an acidic triterpenoid isolated from the fungus *Hormonema*, showed antifungal activity by inhibiting 1,3-β-D-glucan synthase (Lamoth and Alexander 2015). A semisynthetic derivative of enfumafungin SCY-07876 (57; MK-3118) is in clinical trials against fungal infections (Pfaller et al. 2013). Another important cyclic cationic antifungal peptide NP213 (58; Novexatin®), a derived analog of natural cationic peptides with seven arginine residues (Li and Breaker 2012), is being developed by Nova Biotics Ltd. Clinical studies have established Novexatin as safe, well-tolerated and effective agent for the treatment of onychomycosis (nail infections).

Phidolopin (59), a purine derivative isolated from bryozoan *Phidolophora pacifica* (Ayer et al. 1984), has exhibited a broad spectrum antifungal activity against *Phythium ultimum*, *Rhizoctonia solani*, and *Helmithosporium sativum* with a minimum inhibitory concentration of 70 μg per 6 mm disc (Christophersen 1985). Meridine (60), a polycyclic alkaloid, is a cytotoxic agent (Guittat et al. 2005) isolated from South Australian marine ascidian *Amphicarpa meridian* (Schmitz et al. 1991) and from marine sponge *Corticum sp.* Meridine exhibited antifungal effect against *Candida albicans*, *Cryptococcus*, *Trichophyton mentagrophytes*, and *Epidermophyton floccosum* by the inhibition of nucleic acid biosynthesis (McCarthy et al. 1992).

Ptilomycalin A (61) is an antifungal spirocyclic guanidine alkaloid isolated from a marine sponge *Monanchora arbuscula* (Gallimore et al. 2005). It inhibits the production of melanin in *Cryptococcus neoformans* in vitro with an IC_{50} of 7.3 μM, through the inhibition of laccase biosynthesis in the melanin biosynthetic pathway (Dalisay et al. 2011). Melanin is deposited as a protective covering in the cell wall of *C. neoformans* and contributes toward cell wall's thickness and strength and protects the fungal cells from phagocytosis by macrophages. Additionally, melanin deposition decreases the susceptibility of *C. neoformans* to antifungal agents (Wang et al. 1995). Naamine-G (62) exhibits powerful activity against phytopathogenic fungal strain *Cladosporium herbarum* with 20 mm zone of inhibition in the agar plate diffusion assay (20 μg/disk; Hassan et al. 2004).

The bengazoles, a family of marine NPs, were first reported from sponge *Jaspis sp.* with anthelmintic activity against the nematode *Nippostrongylus brasiliensis* (Adamczeski et al. 1988). Afterward, bengazole-A (63) exhibited ergosterol-dependent in vitro antifungal activity against *C. albicans* with a potency similar to amphotericin-B (Richter et al. 2004).

Untenospongin-B (64), a marine NP isolated from sponge *Hippospongia communis*, was found to possess a broad and strong activity against bacteria and human pathogenic fungi. It showed antifungal activity against *Candida albicans* and *Aspergillus fumigatus* in an in vitro assay. Comparatively, it was found to be more active than amphotericin B against *C. tropicalis* (R2 CIP 1275.81) and *F. oxysporum* (CIP 108.74). In addition, it showed an activity profile similar to amphotericin-B against *C. albicans* (ATCC 10231) and *A. niger* (CIP 1082.74; Rifai et al. 2004). Untenospongin-B (64) showed growth inhibition diameters of 17 mm, 14 mm, 20 mm, and 13 mm against *Aspergillus fumigatus*, *Aspergillus niger*, *Arthoderma simii*, and *Trichophyton rubrum*, respectively. A novel marine antifungal NP (2S, 3R)-2-aminododecan-3-ol, isolated from the ascidian *Clavelina oblonga*, exhibited fungal growth with MIC values of 0.7 µg/mL and 30 µg/mL against *Candida albicans* ATCC 10231 and *Candida glabrata*, respectively (Kossuga et al. 2004).

Dysideasterol-A (65), a sterol isolated from a sponge, was found to reverse fluconazole resistance mediated by a *Candida albicans* MDR efflux pump. A combination of fluconazole with dysideasterol-A (3.8 µM) decreased IC_{50} of fluconazole from 300 to 8.5 µM (35-fold enhancement). Configuration at C-6 position of dysideasterol-A was revised by Jacob et al. (2003). Geodisterol-3-O-sulfite (66) and 29-demethylgeodisterol-3-O-sulfite (67) are two new sulfated sterols identified as active constituents through bioassay-guided fractionation of the extract of *Topsentia sp*. These sulfated sterols improved the activity of fluconazole in fluconazole-resistant *Candida albicans* and in wild *Saccharomyces cerevisiae* strain overexpressing the *Candida albicans* efflux pump, MDR1. The active constituents reversed the efflux pump-mediated fluconazole resistance (Digirolamo et al. 2009).

Tanikolide (68), an antifungal compound, was purified from the lipid extract of marine cyanobacterium *Lyngbya majuscula* collected from Madagascar. Tanikolide showed a 13-mm zone of inhibition at 100 µg/disk to *Candida albicans* (ATCC 14053; Singh et al. 1999). Red Sea sponge *Theonella swinhoei* afforded a new bicyclic glycopeptide theonellamide G (69) that exhibited remarkable antifungal activity toward amphotericin B resistant and nonresistant strains of *Candida albicans* with IC_{50} of 2.0 and 4.49 µM, respectively (Youssef et al. 2014; Figure 10.7).

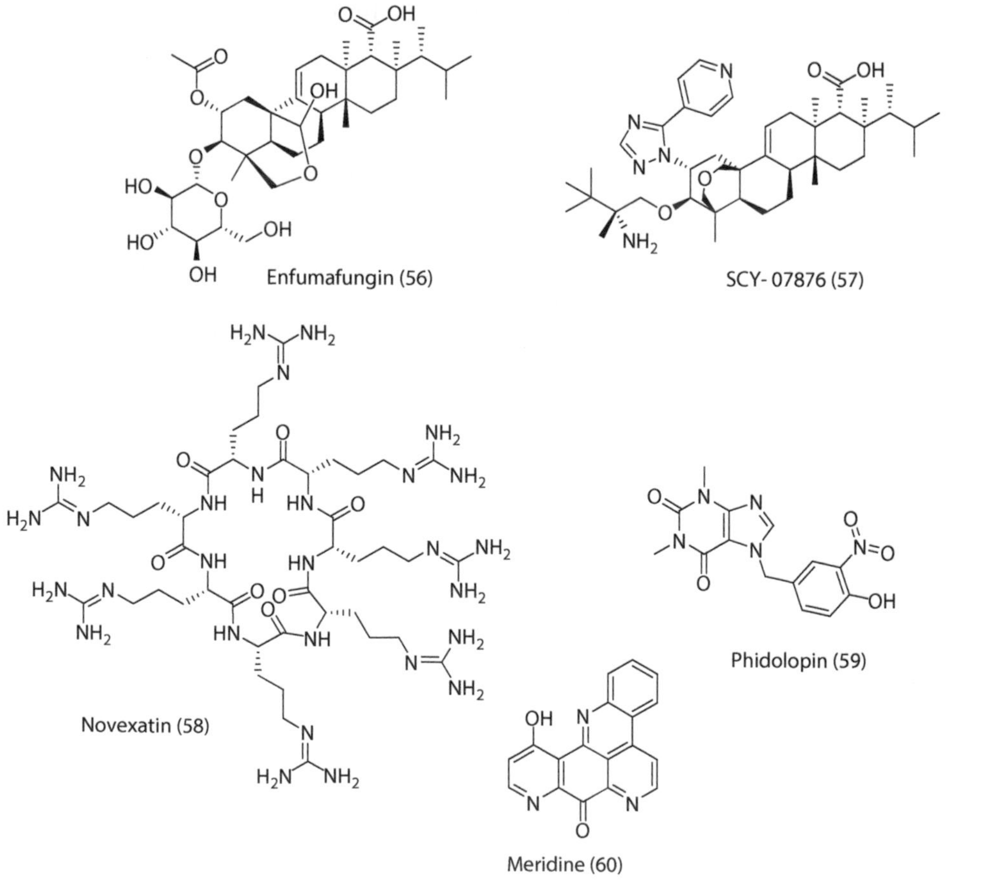

FIGURE 10.7
Chemical structures of antifungal marine natural products.
(*Continued*)

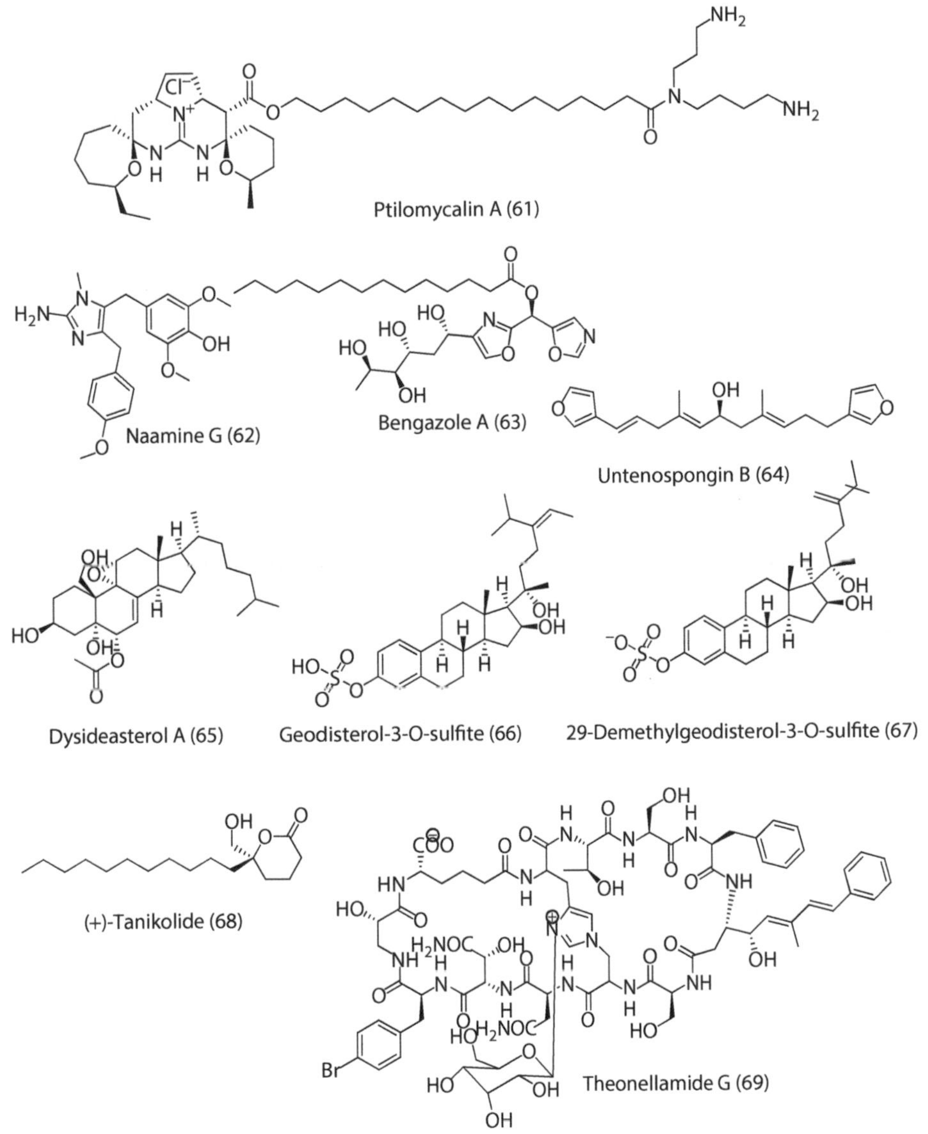

FIGURE 10.7 (Continued)
Chemical structures of antifungal marine natural products.

10.4.2 Antituberculosis Activity

Serrulatane-type diterpenes erogorgiaene (70) and 7-hydroxyerogorgiaene (71) were isolated from *Pseudopterogorgia elisabethae*. Compound 70 showed 96% growth inhibition of *Mycobacterium tuberculosis* (*M. tuberculosis*) H37Rv at a concentration of 12.5 μg/mL and compound 71 showed 77% mycobacterial growth inhibition at a concentration of 6.25 μg/mL.

Homopseudopteroxazole (72) inhibited 80% of the growth whereas pseudopteroxazole (73) induced 97% growth inhibition of Mycobacterium tuberculosis H37Rv at a concentration of 12.5 μg/mL (Rodriguez and Ramirez 2001; Rodriguez and Rodriguez 2003).

Ileabethoxazole (74) a perhydroacenaphthene-type diterpene alkaloid containing benzoxazole moiety showed MIC value of 61 µg/mL. It, at the concentration range of 128–164 µg/mL, showed 92% growth inhibition of *M. tuberculosis* (H37Rv). At concentrations 4, 8, 16, and 32 µg/mL, ileabethoxazole (74) exhibited 29%, 38%, 54%, and 73% inhibition, respectively (Rodríguez et al. 2006).

In an in vitro assay, Agelasine F (75), an *Agelas sp.* sponge-derived compound, inhibited drug resistant strains of *M. tuberculosis* at concentration 3.13 µg/mL (Mangalindan et al. 2000). Bipinnapterolide B (76) is an oxapolycyclic diterpene isolated from the Colombian gorgonian coral *Pseudopterogorgia bipinnata*. Bipinnapterolide B (76) caused 66% inhibition against *M. tuberculosis* H37Rv at 128 µg/mL (Ospina et al. 2007).

Manzamine-type polycyclic alkaloids have exhibited wide anti-infective activity against malaria, leishmania, tuberculosis, and HIV-1. In 1986, manzamine A (77) was isolated from a sponge harvested near the coast of Okinawa (Sakai et al. 1986). Compound 78 (8-hydroxymanzamine A) isolated from sponge *Pachypellina sp.* (Ichiba et al. 1994) as well as 6-hydroxymanzamine E (79), manzamine A (77), and manzamine F (80) isolated from an Indonesian *Acanthostrongylophora* sponge exhibited MIC values of 0.4, 0.9, 1.5, and 2.6 µg/mL, respectively, against *M. tuberculosis* (H37Rv; Rao et al. 2006). A tetracyclic bis-piperidine alkaloid neopetrosiamine A (81), isolated from the marine sponge *Neopetrosia proxima*, was tested in a microplate alamar blue assay exhibiting MIC value of 7.5 µg/mL against the pathogenic strain of *M. tuberculosis* (H37Rv; Wei et al. 2010).

Two novel $5(6{\rightarrow}7)$abeo-sterols, parguesterols A (82) and B (83), were isolated from sea sponge *Svenzea zeai*. Parguesterols A and B show antituberculosis activity against *M. tuberculosis* H37Rv and low toxicity against Vero cells (IC_{50} value 52 µg/mL; Wei et al. 2007). Puupehenone (84), a tetracyclic terpene, isolated from deep-water marine sponge *Strongylophora hartmani* (Kohmoto et al. 1987), exhibited 99% inhibition of *M. tuberculosis* H37Rv at 12.5 µg/ml (Inman et al. 2010).

Culture of *Trichoderma sp.*, a fungus growing on marine sponge, afforded Trichoderins A (85) and B (86) which showed potent inhibitory activity against the pathogenic strain *M. tuberculosis* H37Rv with MIC values of 0.12 and 0.13 µg/mL, respectively (Pruksakorn et al. 2010). Litosterol (87), a C-19 hydroxy steroids isolated from soft coral *Nephthea chabroli* (Rao et al. 1999) and soft coral *Litophyton viridis* (Iguchi et al. 1989), showed 90% growth inhibition of *M. tuberculosis* with MIC value of 3.13 µg/mL (El Sayed et al. 2000). Axisonitrile-3 (88), a sesquiterpene possessing a cyano group and purified from marine sponge *Acanthella klethra*, was found to possess potent inhibitory activity with MIC value of 2.0 µg/mL against *M. tuberculosis* (Konig et al. 2000; Figure 10.8).

10.4.3 Anthelmintic Activity

Helminthiasis contributes substantially to malnutrition, reduced food intake, weakened digestion, severe blood disorder like iron-deficient anemia, growth impairment, cognitive changes particularly in young children, and enormous economic losses in livestock animals. Although excellent commercial anthelmintics are available, growing resistance to current drugs necessitates the search for new anthelmintics. Although the literature is not very extensive, the structure of a variety of anthelmintic marine NP will be discussed.

Bioactivity guided fractionation and isolation of the ethanol extract of sponges *Amphimedon spp.* (LD_{50} 4.0 µg/mL and LD_{99} 130 µg/mL) against nematode *Haemonchus contortus* resulted in the isolation of macrocyclic lactone and lactams. Amphilactams A–D (89–92), exhibited in vitro LD_{99} activities at 7.5, 47, 8.5, and 0.39 µg/mL, respectively. These

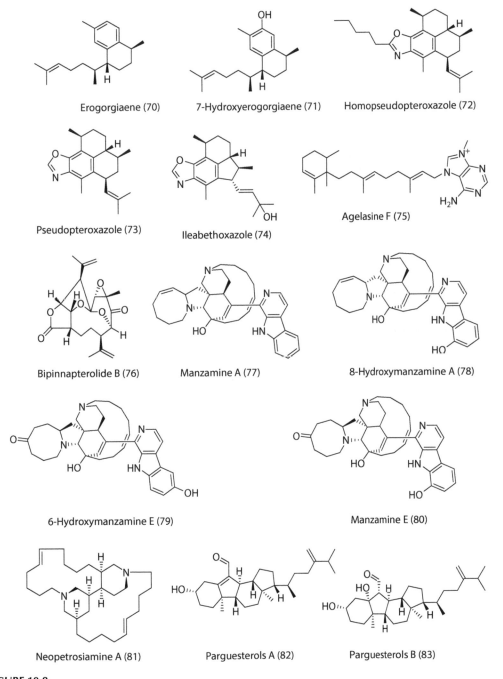

FIGURE 10.8

Chemical structures of antitubercular marine natural products.   *(Continued)*

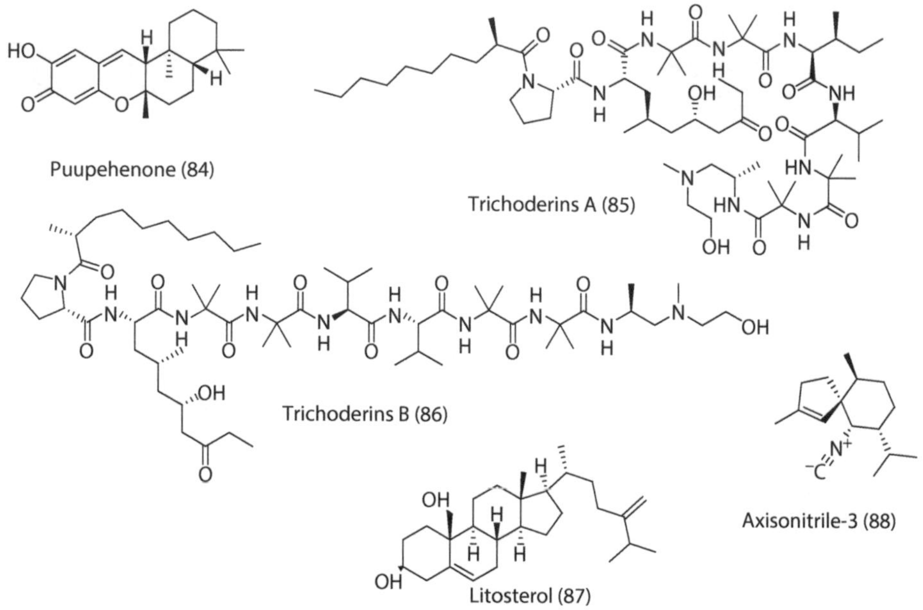

FIGURE 10.8 (Continued)
Chemical structures of antitubercular marine natural products.

compounds restrain larval development at the L1 stage of nematode *Haemonchus contortus* but no activity against nematode eggs (Ovenden et al. 1999). Geodin A Mg salt (93), chemically macrocyclic polyketide lactam tetramic acid, is a potent nematocide isolated from southern Australian marine sponge *Geodia*. Geodin A which exists as Mg salt in a natural source was found to be nematocidal to *Haemonchus contortus* with LD_{99} value of 1 µg/mL (Capon et al. 1999). Bislobane (94) obtained from red alga *Laurencia scoparia* displayed weak anthelmintic activity with an EC_{50} of 0.11 mM against the parasitant stage (L4) of a rat gastrointestinal parasite *Nippostrongylus brasiliensis*, which is similar to human parasite hookworms (Davyt et al. 2006). Nafuredin (95) is produced by a fungus *Aspergillus niger* FT-0554 isolated from a marine sponge, as well as cultured broth of the strain FT-0554. Nafuredin (95) inhibited NADH-fumarate reductase (complexes I+II) from adult *Ascaris suum* (pig roundworm) at IC_{50} of 12 nM. NADH-fumarate reductase is an important enzyme involved in the electron transport system of anaerobic metabolism found in many anaerobic organisms (Omura et al. 2001). Nafuredin (95) also showed anthelmintic activity against *Haemonchus contortus* (wireworm) in in vivo trials with sheep (Ui et al. 2001). Significant nematocidal (LD_{99} 5.2 µg/mL) activity against parasitic nematode *Haemonchus contortus* was exhibited by onnamide F (96), a marine NP isolated from marine sponge, *Trachycladus laevispirulifer*. Onnamide F (96) was also active against *S. cerevisiae* with LD_{99} value of 1.4 µg/mL (Vuong et al. 2001). (−)-echinobetaine A (97; Capon et al. 2005a) and (+)-echinobetaine B (98; Capon et al. 2005b) are betaine-type nematocidal agents present in a southern Australian marine sponge of the genus *Echinodictyum*. (−)-echinobetaine A (97) and (+)-echinobetaine B (98) are nematocidal (LD_{99} 83 and 8.3 µg/mL, respectively) to the parasite *Haemonchus contortus*.

The acyclic lipid thiocyanatins isolated from the Australian sponge *Oceanapia sp.* exhibits potent nematocidal activity. Thiocyanatin A (99) displayed powerful nematocidal activity against the animal parasite *Haemonchus contortus* (LD$_{99}$ 1.3 µg/mL). Although the mechanism of action is undetermined, 2-alcohol, SCN functionalities and chain length may have vital role nematocidal activity (Figure 10.9).

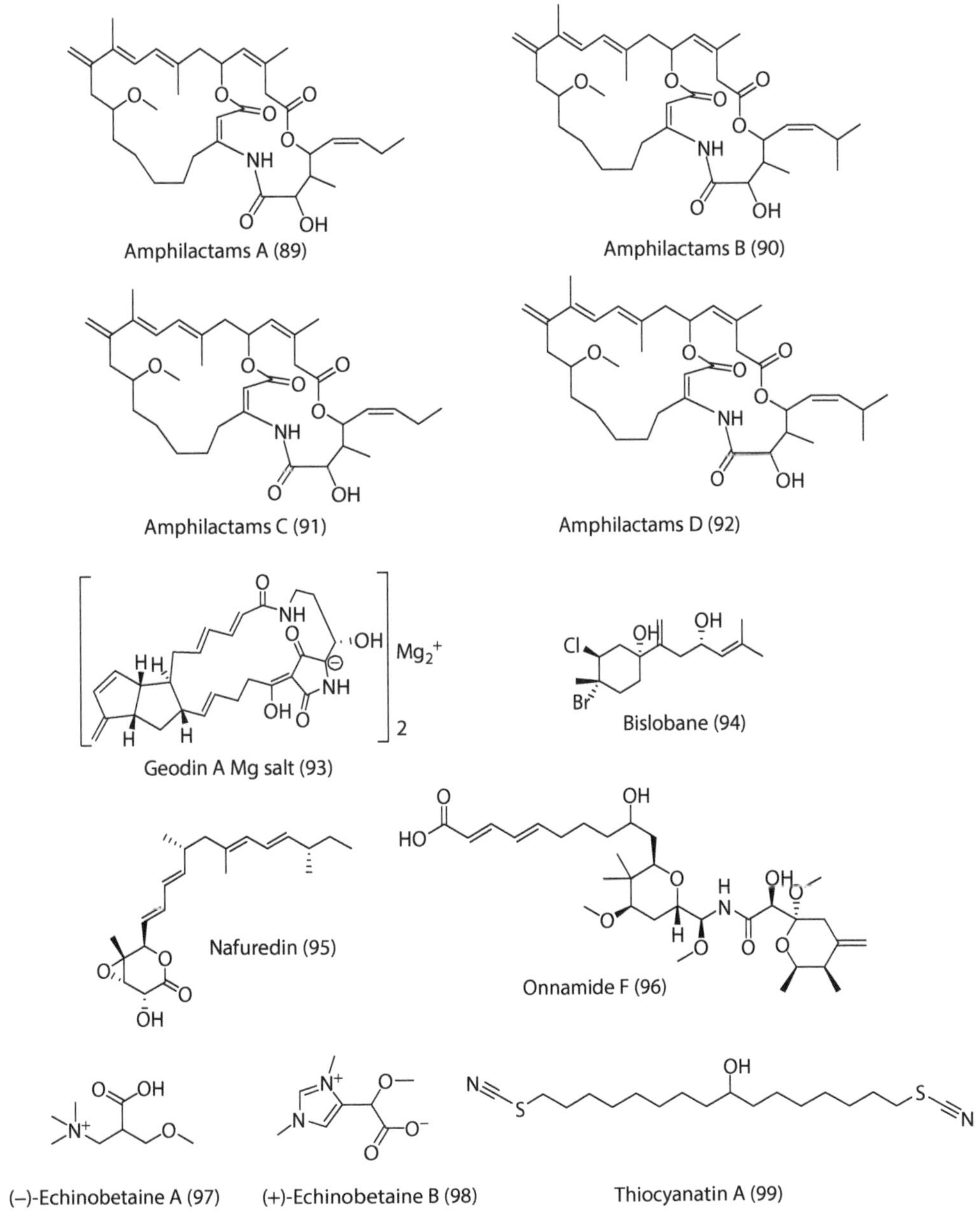

Amphilactams A (89)

Amphilactams B (90)

Amphilactams C (91)

Amphilactams D (92)

Geodin A Mg salt (93)

Bislobane (94)

Nafuredin (95)

Onnamide F (96)

(−)-Echinobetaine A (97)   (+)-Echinobetaine B (98)   Thiocyanatin A (99)

FIGURE 10.9
Chemical structures of anthelmintic marine natural products.

10.4.4 Antiprotozoal Activity

Parasitic protozoans that belong to the genus *Plasmodium* (*P. falciparum*, *P. ovale*, *P. vivax*, and *P. malariae*) are responsible for most severe diseases. Malaria is still a common cause of death, particularly in tropical countries. (+)-7-bromotrypargine (100), a β-carboline, originally isolated from the skin of African frog *Kassina senegalensis* was later isolated from the DCM/MeOH extract of Australian marine sponge *Ancorina sp*. Compound (+)-7-bromotrypargine (100) displays IC_{50} of 5.41 μM (Dd2) and 3.51 μM (3D7) against chloroquine-resistant (Dd2) and chloroquine-sensitive (3D7) strains of *Plasmodium falciparum* (Davis et al. 2010). (S)-Curcuphenol (101) and 15-hydroxycurcuphenol (102) was isolated from the Jamaican sponge *Didiscus oxeata*. Compounds (S)-curcuphenol (101) and 15-hydroxycurcuphenol (102) exhibited antimalarial activity with MIC values of 3.6 and 3.8 μg/mL, respectively, against *Palsmodium falciparium* (D6 clone; El Sayed et al. 2002). Ascosalipyrrolidinone A (103) is an alkaloid isolated from the obligate marine fungus *Ascochyta salicornia* which showed IC_{50} of 736 ng/mL against K1 (Thailand; resistant to chloroquine and pyrimethamine) and 378 ng/mL against NF 54 (an airport strain of unknown origin; susceptible to standard antimalarials; Osterhage et al. 2000). Bielschowskysin (104) is highly oxygenated diterpenoid isolated from the Caribbean gorgonian octocoral *Pseudopterogorgia kallos*. It exhibits antimalarial activity against *Plasmodium falciparum* (Marrero et al. 2004). Gracilioethers A–C (105–107) isolated from the marine sponge *Agelas gracilis* showed good antimalarial activities (IC_{50} 10 μg/mL, 0.5 μg/mL, and 10 μg/mL, respectively) against *Plasmodium falciparum*. In addition, gracilioether B (106) inhibited the growth of *Leishmania major* (68% at 10 μg/mL; Ueoka et al. 2009).

Marine fungus *Halorosellinia oceanic* derived cytochalasin Q (108), 5-carboxymellein (109), and halorosellinic acid (110; ophiobolane sesterterpene) which display antimalarial activity with IC_{50} of 17, 4, and 13 μg/mL, respectively (Chinworrungsee et al. 2001).

NP, heptyl prodigiosin (111), isolated from a culture of α-proteobacteria from a marine tunicate, exhibits antimalarial activity against the chloroquine-sensitive strain *Plasmodium falciparum* 3D7 similar to quinine (Lazaro et al. 2002). Lepadins are decahydroquinoline alkaloids showing significant and selective antiplasmodial and antitrypanosomal activities. Lepadin D (112) has an IC_{50} of 6.169 μg/mL, whereas lepadin E (113) and lepadin F (114) show IC_{50} 0.4 μg/mL and 208 ng/mL, respectively, against *P. falciparum*'s clone K1.

Plakortide F (115), antimalarial principle from Caribbean sponge *Plakorti ssp*., exhibited IC_{50} of 480 ng/mL against *P. falciparum* (D6 clone) and 390 ng/mL against *P. falciparum* (chloroquine-resistant W2 clone) in vitro (Gochfeld and Hamann 2001). Pycnidione (116) is a small tropolone first isolated from the fermentation of *Phoma sp*. (Harris et al. 1993) and from the fermented broth of *Theissenia rogersii* 92031201 (Hsiao et al. 2012). Pycnidione (116) exhibited activities against *Plasmodium falciparum* in the sub-micromolar (μM) range (Wright and Lang-Unnasch 2005). Marine cyanobacterium-derived NP, symplostatin-4 (117) possesses significant antimalarial activity (ED_{50} of 74 nM against *Plasmodium falciparum*, strain 3D7; Figure 10.10).

10.4.5 Antiviral Activities

Echrebsteroid C (118), a steroid obtained from South China sea gorgonian *Echinogorgia rebekka*, exhibits promising antiviral activity against respiratory syncytial virus with an IC_{50} of 0.19 μM (Cao et al. 2014). Marine NPs purpurquinone B (119) and purpurester A (120) are antiviral compounds from marine fungus. Purpurester A (120) has been isolated

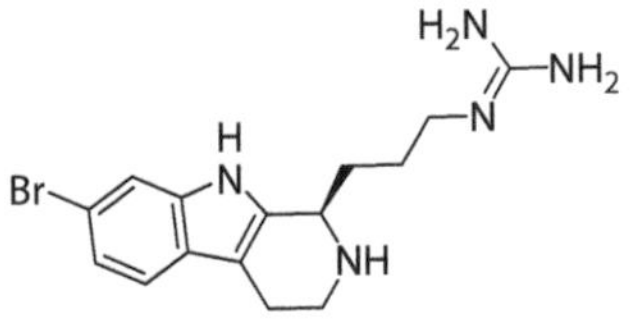

(+)-7-Bromotrypargine (100)

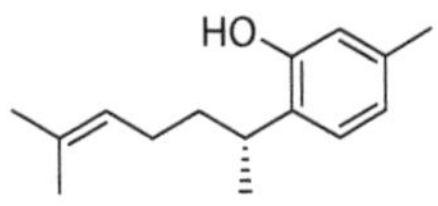

(S)-Curcuphenol (101)

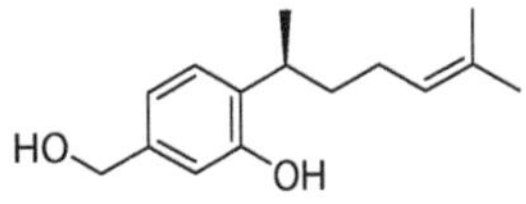

15-Hydroxycurcuphenol (102)

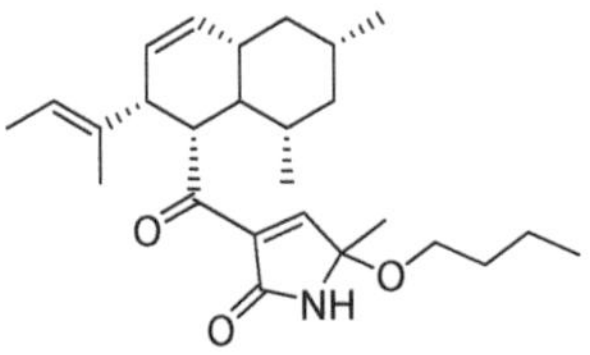

Ascosalipyrrolidinone A (103)

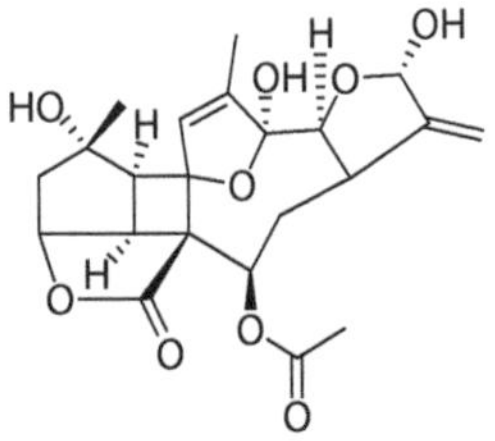

Bielschowskysin (104)

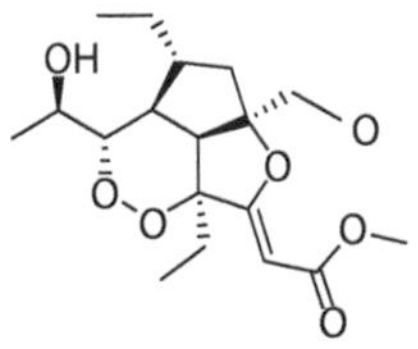

Gracilioether A (105)

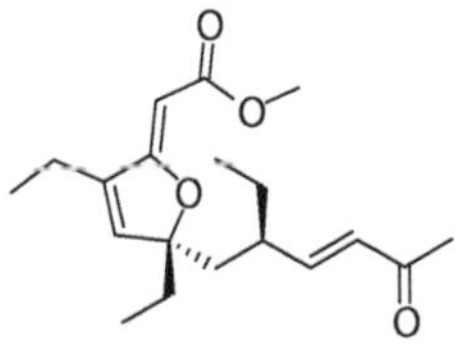

Gracilioether B (106)

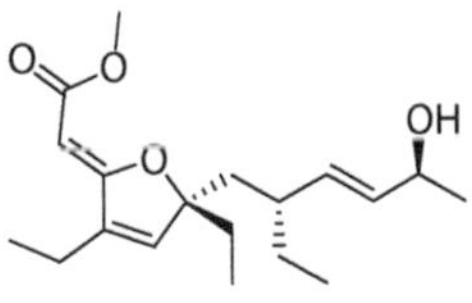

Gracilioether C (107)

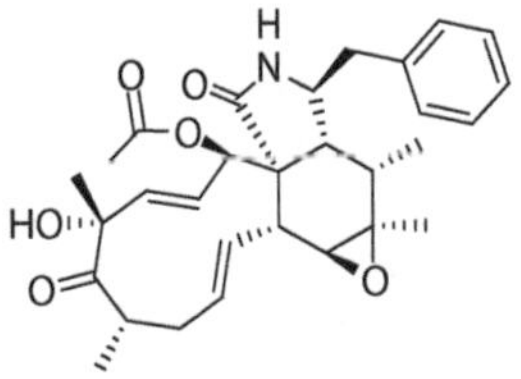

Cytochalasin Q (108)

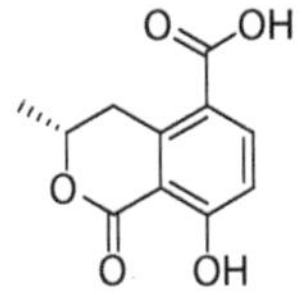

5-Carboxymellein (109)

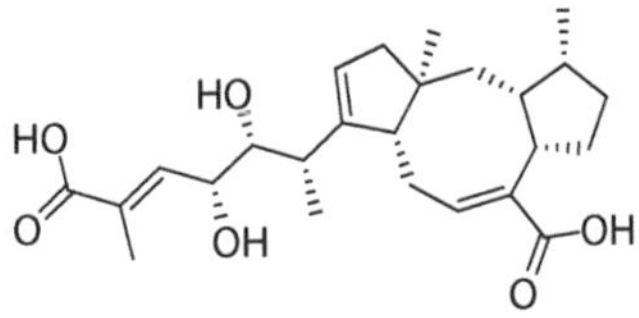

Halorosellinic acid (110)

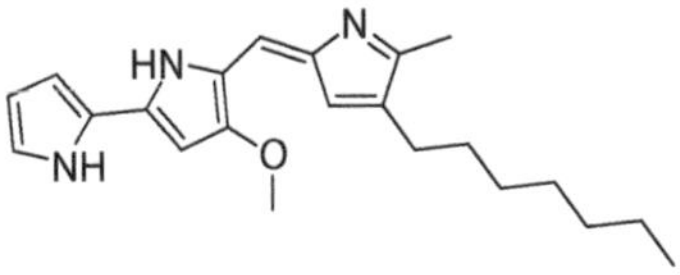

Heptyl prodigiosin (111)

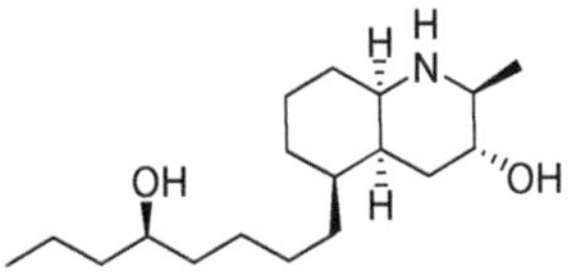

Lepadin D (112)

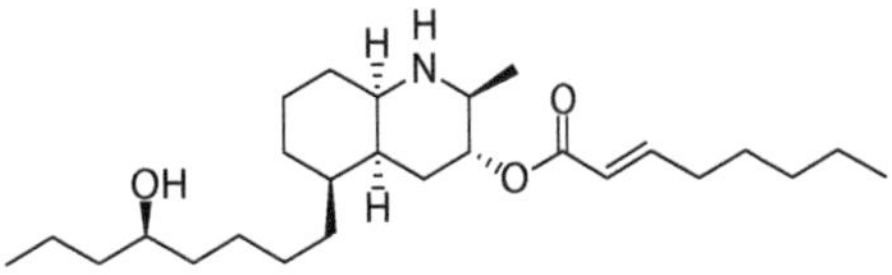

Lepadin E (113)

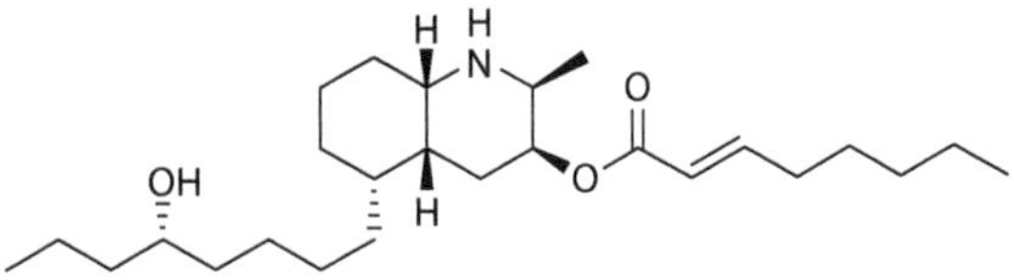

Lepadin F (114)

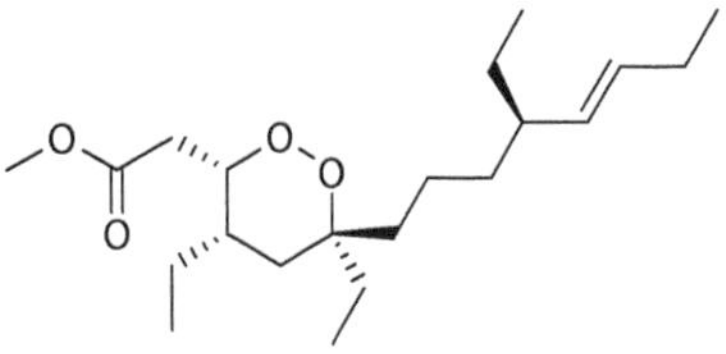

Plakortide F (115)

FIGURE 10.10
Chemical structures of antimalarial agents.

(Continued)

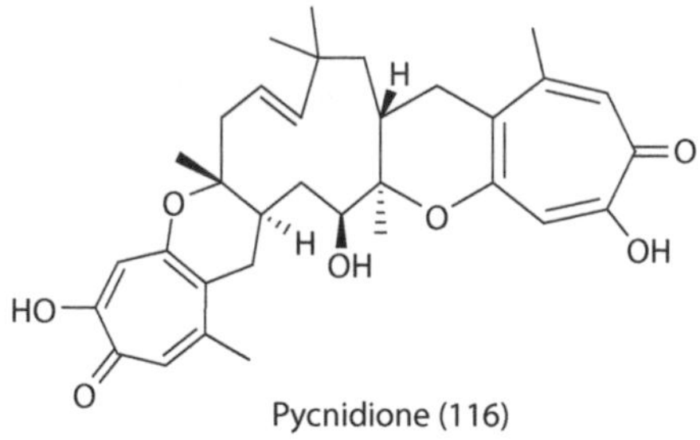

Pycnidione (116) 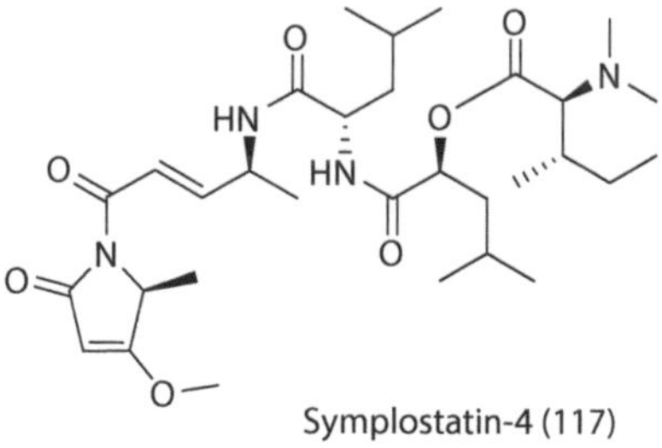   Symplostatin-4 (117)

FIGURE 10.10 (Continued)
Chemical structures of antimalarial agents.

from ethyl acetate extract of *Penicillium purpurogenum JS03-21*. Purpurquinone B (119) and purpurester A (120) demonstrate significant antiviral activity with IC_{50} values of 61.3, 64.0, and 85.3 μM, respectively, against influenza virus H1N1 (Wang et al. 2011). A hydroanthraquinone derivative, tetrahydroaltersolanol C (121), and an alterporriol-type anthranoid dimer, alterporriol Q (122), were purified from microbial culture as well as mycelia of *Alternaria sp.* ZJ-2008003, a fungus growing on a soft coral *Sarcophyton sp.* Tetrahydroaltersolanol C (121) and alterporriol Q (122) exhibited antiviral activity with an IC_{50} of 65 and 39 μM, respectively, against the porcine reproductive and respiratory syndrome virus (PRRSV; Zheng et al. 2012). Oxazole-containing alkaloid (–)-hennoxazole A (123) was isolated from *Polyfibrospongia sp.* sponge showed activity against herpes simplex virus type 1 (IC_{50} 0.6 μg/mL; Ichiba et al. 1991). Halovir A–E (123–127), isolated from the marine fungus *Scytidium sp.* sourced from the Caribbean seagrass *Halodule wrightii*, have shown potent antiviral activity against HSV-1 and HSV-2. Halovirs A, B, C, D, and E showed ED_{50} of 1.1, 3.5, 2.2, 2.0, and 3.1 μM, respectively, when added to cells infected with HSV-1 for 1 h (Rowley et al. 2003). Clathsterol (128) isolated from the Red Sea sponge *Clathria sp.* showed inhibitors of human immunodeficiency virus type 1 (HIV-1) reverse transcriptase (RT) at a concentration of 10 μM (Rudi et al. 2001). Microspinosamide (129) a cyclic depsipeptide was obtained from marine sponge *Sidonops microspinosa* collected from Indonesia. Compound 129 inhibited HIV-1 interactions with EC_{50} of 0.2 μg/mL in an in vitro XTT-based assay (Rashid et al. 2001). Antiretroviral agents thalassiolin A–C (130–132) were isolated from the Caribbean seagrass *Thalassia testudinum*, as that showed RT and protease inhibition. Thalassiolin A (130) was most active, inhibiting the integrase terminal cleavage (IC_{50} 2.1 mM) and strand transfer (IC_{50} 0.4 mM) activities (Rowley et al. 2002). Papuamide A (133) is marine derived cyclic depsipeptides having cytoprotective activity against HIV-induced T cell death (Andjelic et al. 2008). Polycitone A (134), an aromatic alkaloid isolated from the ascidian *Polycitor sp.*, displayed significant inhibitory activity against RT of HIV, retrovirus B, and retrovirus C (Loya et al. 1999). A new polycyclic bromoindole alkaloid dragmacidin F (135) was isolated by Cutignano and collaborators from the Mediterranean sponge *Halicortex sp.* (Cutignano et al. 2000). It shows weak inhibition of herpes simplex virus (HSV-1) infected cells from HSV-induced destruction (IC_{50} 95.8 μM) and furthermore delays syncytia formation by HIV-2 (IC_{50} 0.91 μM). Neamphamide A (136) was isolated as the principle active constituent against HIV-1 infection from marine sponge *Neamphius huxleyi*. Cytopathic effect of HIV-1 infection was inhibited efficiently by neamphamide A (136) with EC_{50} of 28 nM in an XTT-based cell viability assay using HIV-1RF infected human cell line CEM-SS (Oku et al. 2004; Figure 10.11).

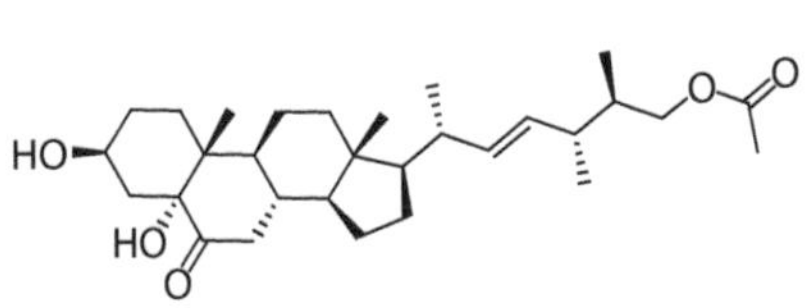

Echrebsteroid C (118)

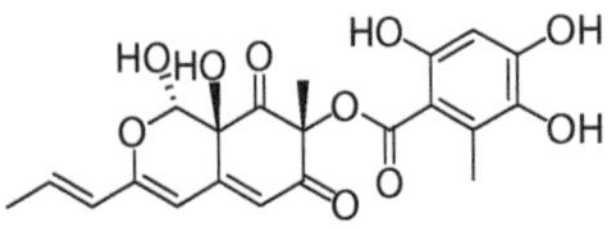

Purpurquinone B (119)

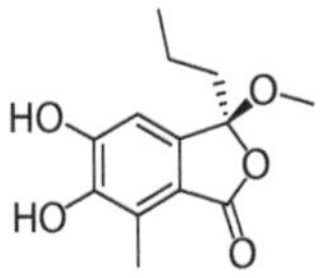

Purpurester A (120)

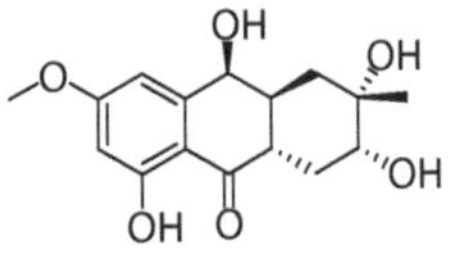

Tetrahydroaltersolanol C (121)

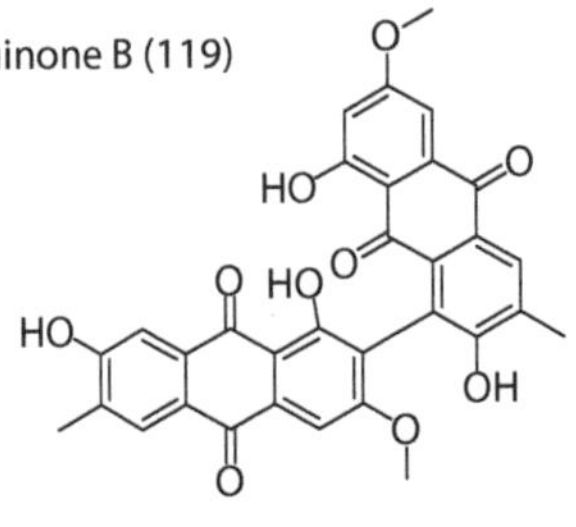

Alterporriol Q (122)

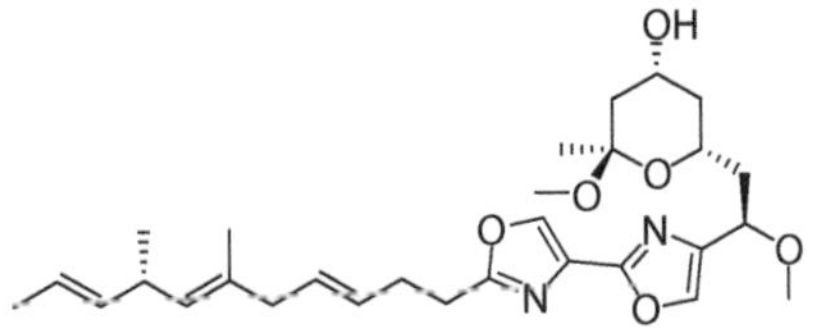

(–)-Hennoxazole A (123)

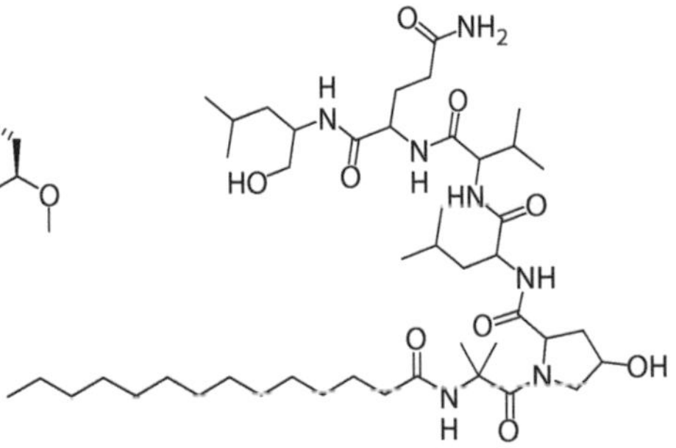

Halovir A (124)

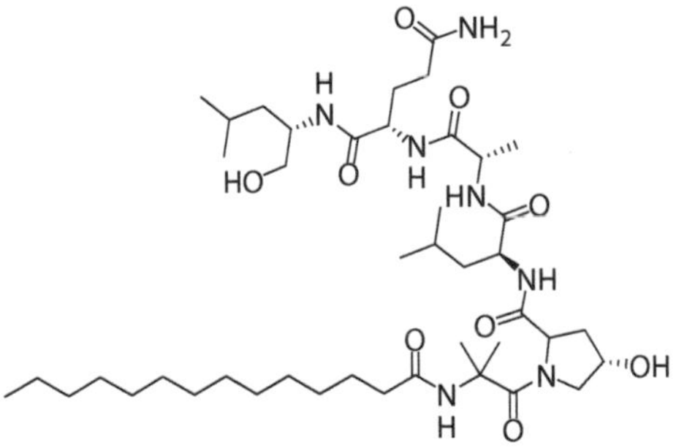

Halovir B (125)

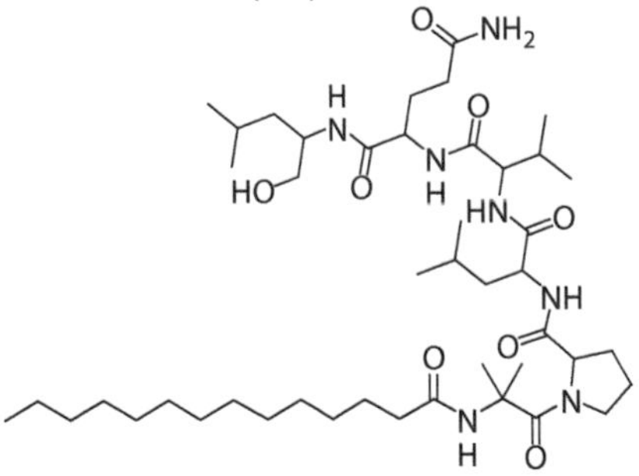

Halovir C (126)

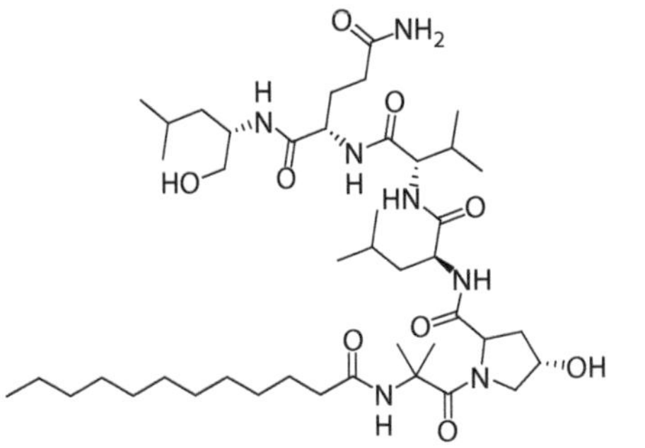

Halovir D (127)

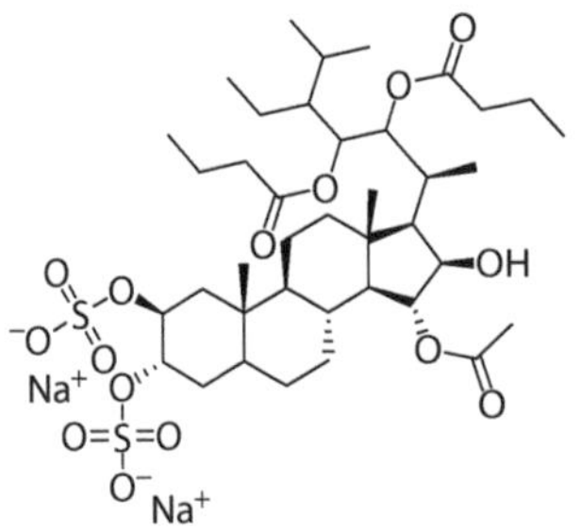

Clathsterol (128)

FIGURE 10.11
Chemical structures of natural antivirus agents from marine source.

(Continued)

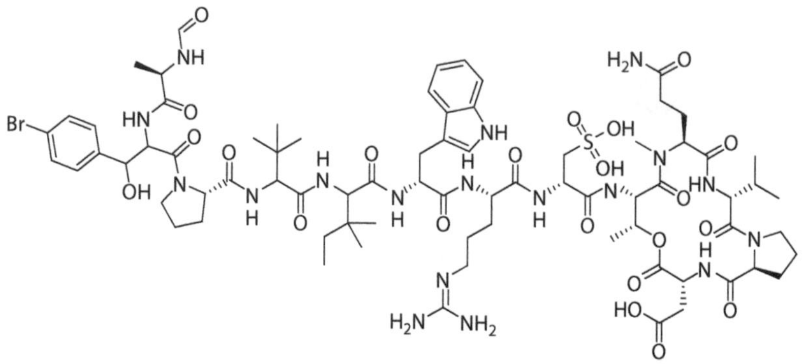

Microspinosamide (129)

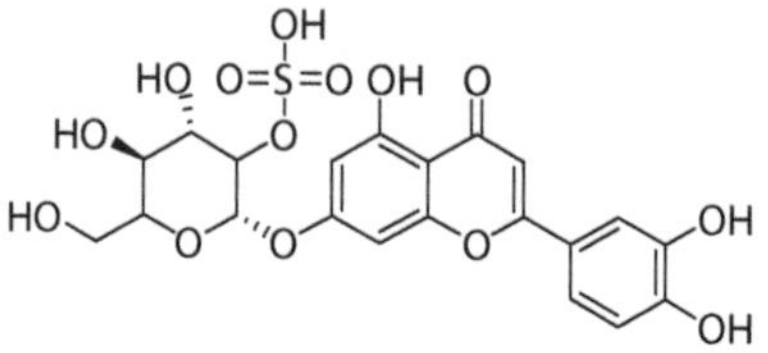

Thalassiolin A (130)

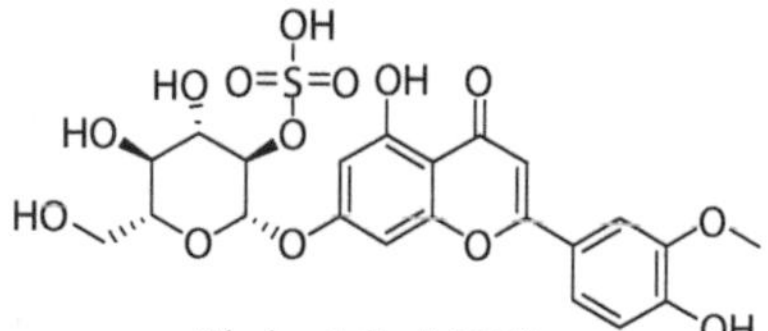

Thalassiolin B (131)

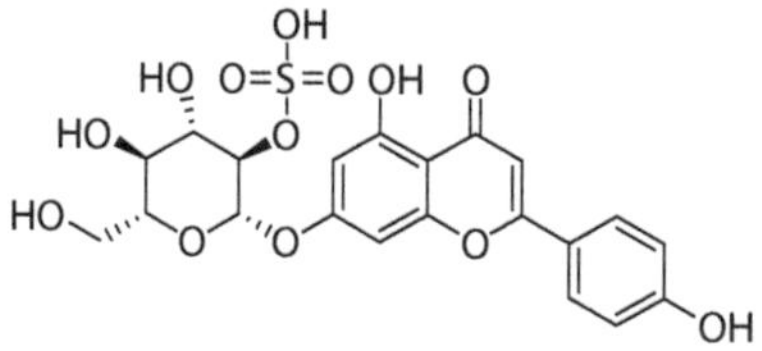

Thalassiolin C (132)

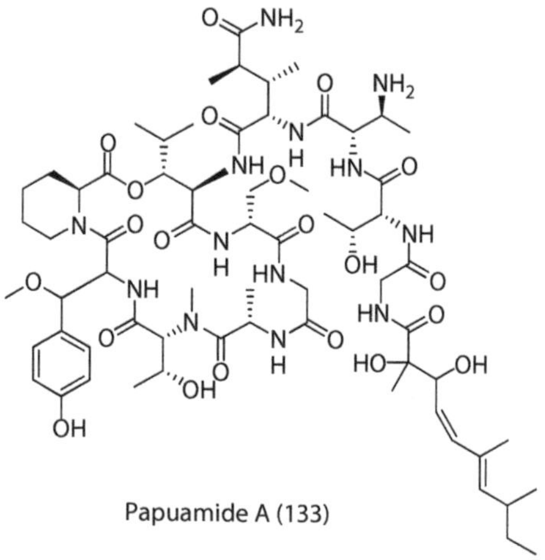

Papuamide A (133)

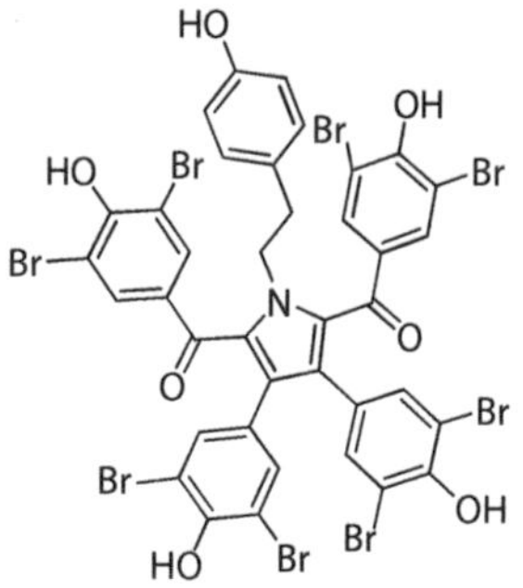

Polycitone A (134)

FIGURE 10.11 (Continued)
Chemical structures of natural antivirus agents from marine source.

(*Continued*)

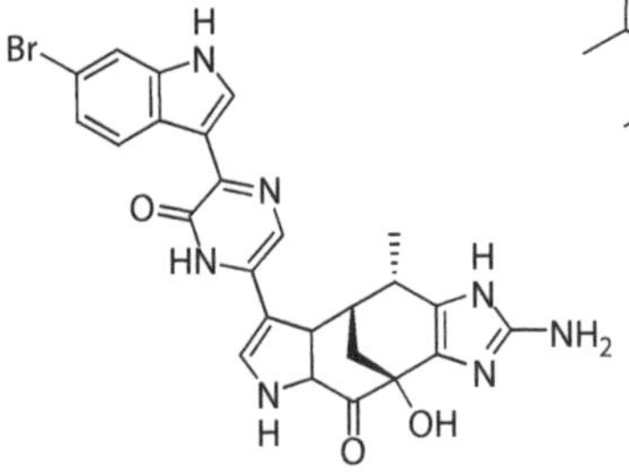

Dragmacidin F (135)

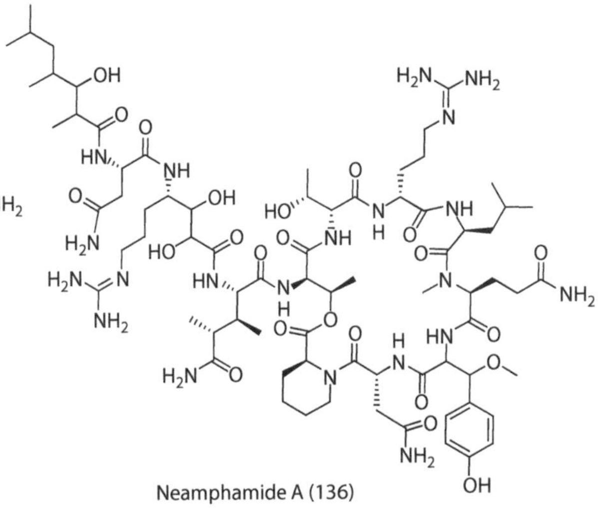

Neamphamide A (136)

FIGURE 10.11 (Continued)
Chemical structures of natural antivirus agents from marine source.

10.5 Anti-Inflammatory Agents

The management of inflammatory processes correlates with prostaglandin (PG), interleukin (IL), leukotrienes (LT), nitric oxide (NO), and tumor necrosis factor-α (TNF-α) inhibition. These mediators are released and play a crucial role in progression of inflammation. Inflammation is directly related to rheumatoid arthritis, septic shock, psoriasis, and asthma.

Inhibition of LPS (lipopolysaccharide), stimulated production of inducible nitric oxide synthase (iNOS) protein, and NO-2 (Nitrite) were observed in RAW264.7 cells and primary macrophages, preincubated with 11-oxoaerothionin (137). Inflammatory cytokines and PGE2 as well as expressions of NO_2 and iNOS were suppressed by 11-oxoaerothionin (Ivo de Medeiros et al. 2012). Coscinolactams A (138) and B (139) are two novel nitrogen-containing cheilanthane sesterterpenoids isolated from marine sponge *Coscinoderma matthewsi* that showed moderate inhibitory activity against PGE2 and NO production (De Marino et al. 2009). A new lactone, penicillinolide A (140), isolated from the organic extract of *Penicillium sp.* SF-5292, suppressed suppresses the expression of iNOS, COX-2, and TNF-α, as a result inhibition in the production of NO and PGE2 occurred (Lee et al. 2013). Anti-PLA2 NPs, scalaradial (141), a 1,4-dialdehyde marine terpenoid which was isolated from the sponge *Cacospongia mollior* (Monti et al. 2007), has selectively inhibited type II phospholipase A2 (PLA2; IC_{50} 0.07 µM; Marshall et al. 1994), as well as bee venom PLA2 with an IC_{50} of 0.07 µM (de and Jacobs 1991). Tanzawaic acid A (142) and D (143), isolated from *Penicillium sp.* (SF-6013), suppressed NO production with IC_{50} of 37.8 and 7.1 µM, respectively. Moreover, tanzawaic acid A (142) also inhibited LPS-stimulated NO production in murine macrophages (RAW264.7) with an IC_{50} of 27.0 µM. These inhibitory effects are associated with the inhibition of LPS-stimulated expression of iNOS and cyclooxygenase-2 (COX-2) in RAW264.7 and BV2 cells (Quang et al. 2014). Cyclic depsipeptides halipeptins A (144) and B (145) demonstrated significant reduction in in vivo anti-inflammatory activity via intraperitoneal dose of 0.3 mg/Kg in mice (60% reduction of carrageenan induced edema; Randazzo et al. 2001). Splenocin B (146) a nine-membered bis-lactone isolated from *Streptomyces sp.* (California) showed suppression of cytokine production with minimal mammalian cell cytotoxicity (Strangman et al. 2009). Pyrenocine A (147), derived from marine fungus *Penicillium paxilli* Ma(G)K, was able to inhibit inflammatory cytokinase, PGE2, and nitrite production consequently exhibiting deactivation of LPS-induced macrophage. Moreover, pyrenocine A (147) also inhibited the LPS-stimulated expression of genes related to NFκB-mediated signal transduction on macrophages (Toledo et al. 2014). Scytonemin (148) a yellow–green pigment isolated from the extracellular sheath of cyanobacteria *Stigonema spp.* (Proteau et al. 1993) showed remarkable inhibition of the anti-inflammatory drug target IκB kinase (Stevenson et al. 2002). *Octocorallia* (octocoral) derived eunicol (149) and fuscol (150; Kerr et al. 2014), and *P. foliascens* derived foliaspongin (151; Kikuchi et al. 1983) are anti-inflammatory terpenes.

Manoalide (152; Glaser and Jacobs 1986) and related marine sesterterpenoid cladocoran-A (153) imparted anti-inflammatory properties by inactivation of human group IIA PLA2 (Monti et al. 2011). Luffariellolide (154), a sesterterpene, first isolated in 1987 from Palauan sponge *Luffariella sp.* (Albizati et al. 1987) showed potent in vivo anti-inflammatory activity through partially reversible inhibition of PLA2. Luffariellolide (154) and related NPs, like manoalide, do not interact with the active site instead react at the surface of PLA2 with lysine residue, thereby destroying the ability that moves across the membrane. *Euryspongia n. sp.* collected in Vanuatu afforded two steroids, petrosterol (155) and 3β-Hydroxy-24-Norchol-5-En-23-Oic acid (156; Mandeau et al. 2005) which exhibited anti-inflammatory activity against 6-keto-PGF1 release in a human keratinocyte cell line HaCaT.

Two dendalones, 3-hydroxybutyrate (157) and flexibilide (158) or sinularin isolated from dictyoceratid sponge and soft coral, respectively (Buckle et al. 1980), and a cyclic heptapeptides, cyclomarin-A (158) isolated from cultured marine bacterium *Streptomyces sp.* (Renner et al. 1999) were active after oral administration in in vivo animal models of inflammation. Pseudopterosin A (159), diterpene marine NPs isolated from the marine gorgonian *Pseudopterogorgia elisabethae* inhibits phagocytosis (Moya and Jacobs 2006) and reduced phorbol myristate acetate (PMA) induced mouse ear edema (Figure 10.12).

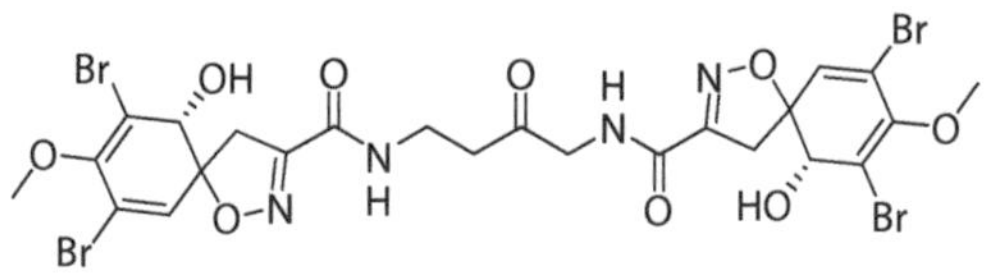

11-Oxoaerothionin (137)

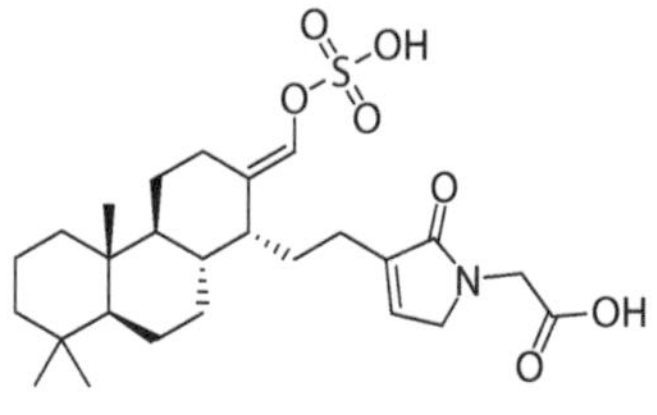

Coscinolactams A (138)

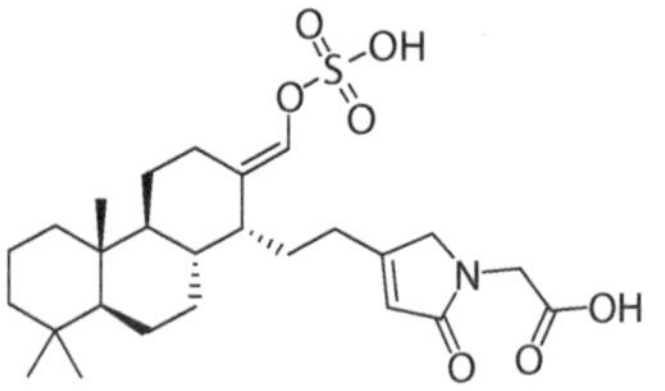

Coscinolactams B (139)

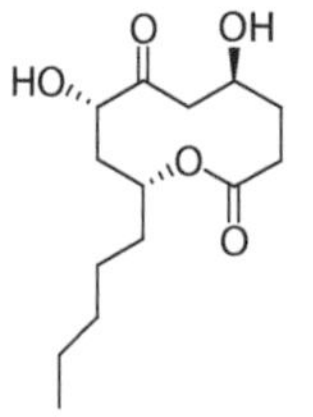

Penicillinolide A (140)

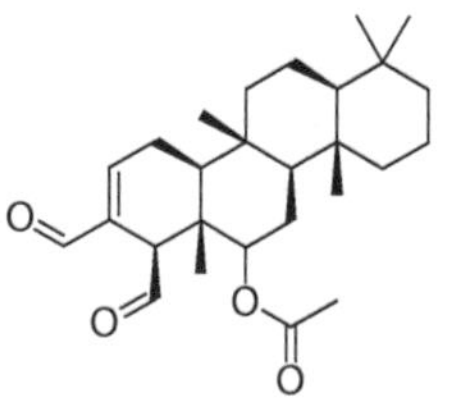

Scalaradial (141)

FIGURE 10.12
Chemical structures of anti-inflammatory marine natural products.

(Continued)

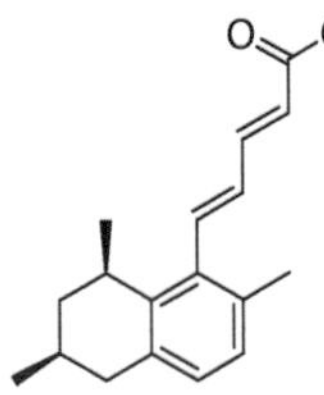

Tanzawaic acid A (143)

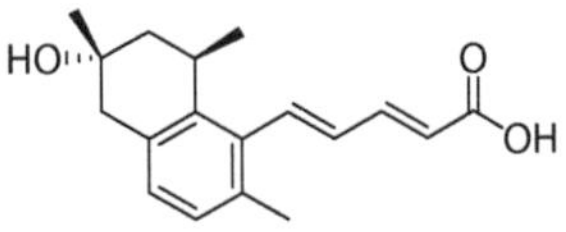

Tanzawaic acid D (143)

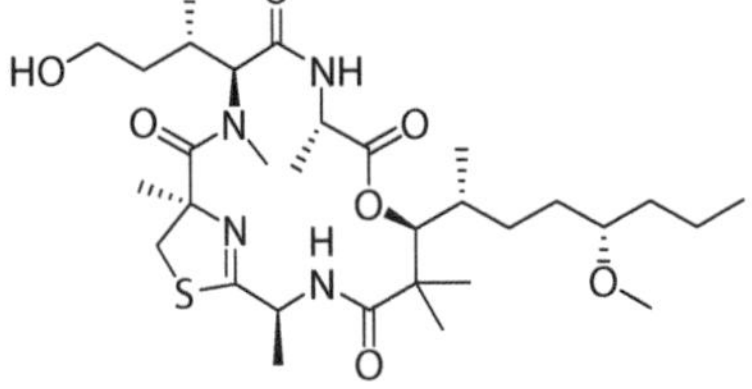

Halipeptin A (144)

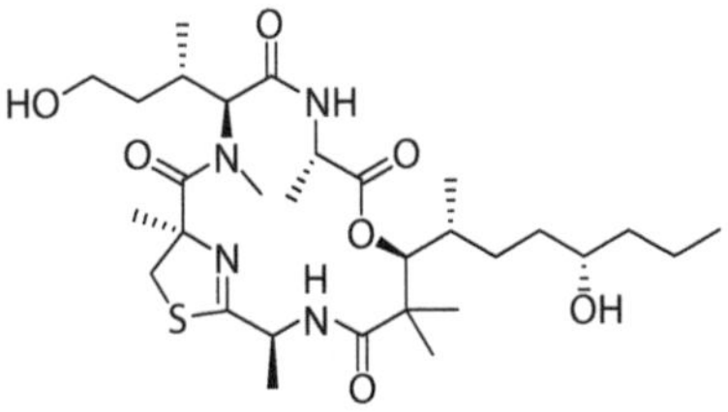

Halipeptin B (145)

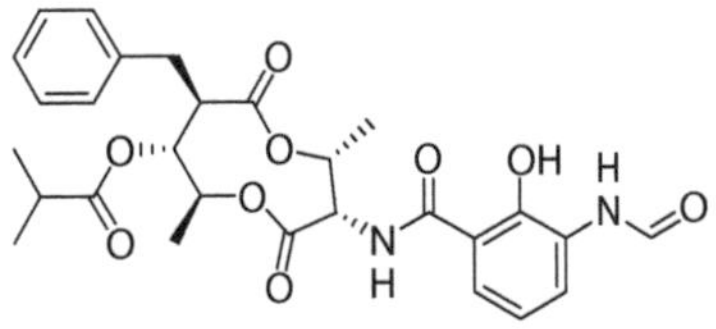

Splenocin B (146)

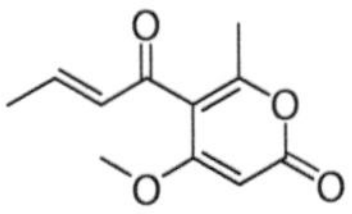

Pyrenocine A (147)

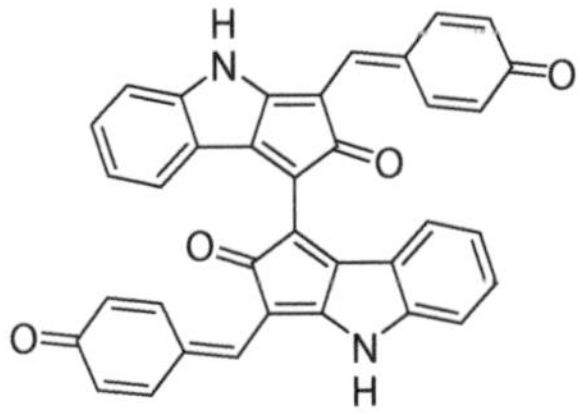

Scytonemin (148)

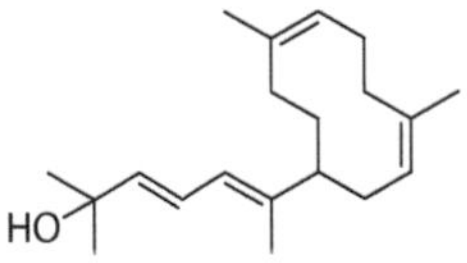

Eunicol (149)

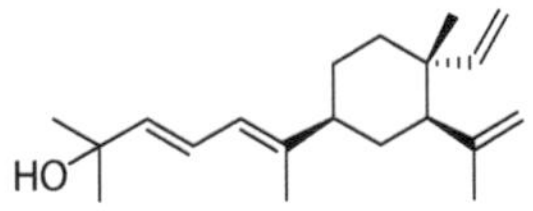

Fuscol (150)

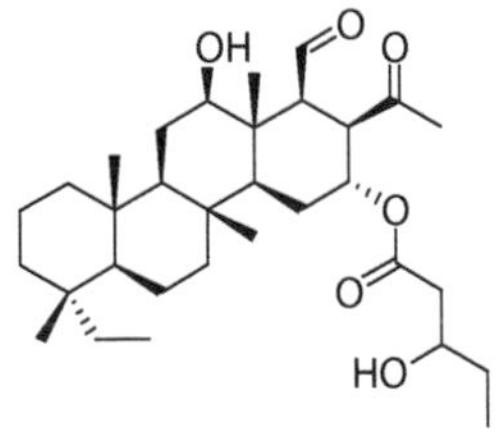

Foliaspongin (151)

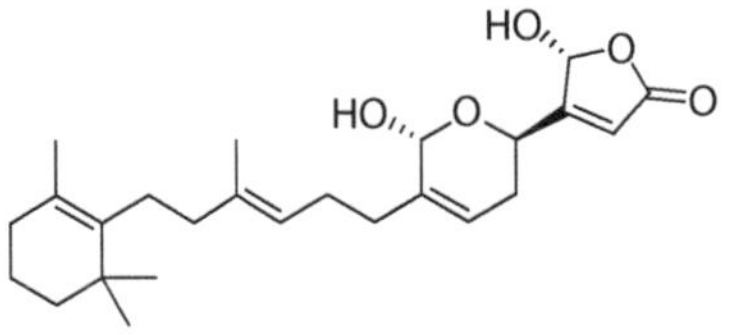

Manoalide (152)

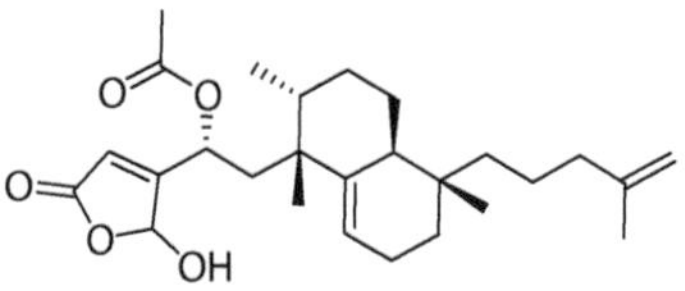

Cladocoran A (153)

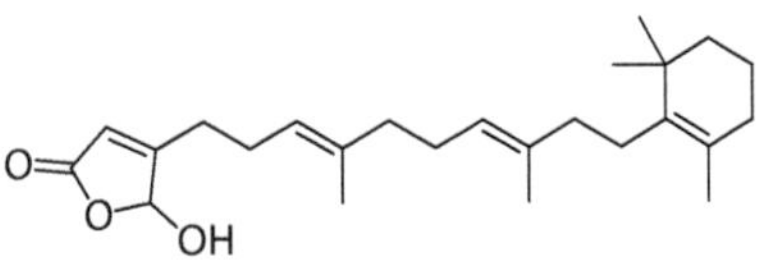

Luffariellolide (154)

FIGURE 10.12 (Continued)
Chemical structures of anti-inflammatory marine natural products. *(Continued)*

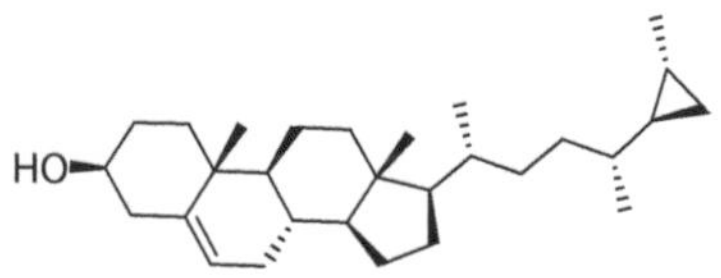

Petrosterol (155)

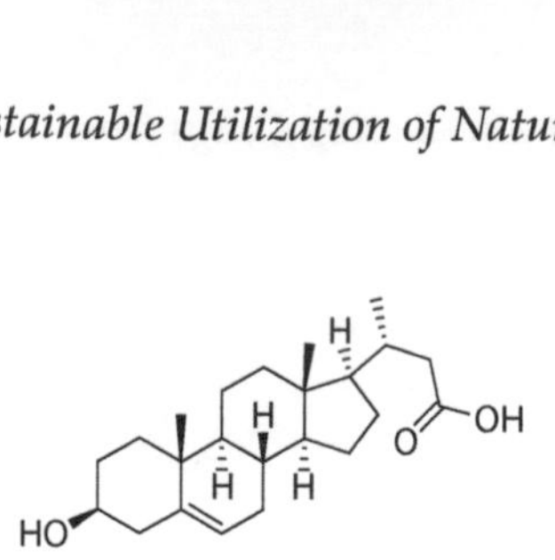

3-β-Hydroxy-24-norchol-5-en-23-oic acid (156)

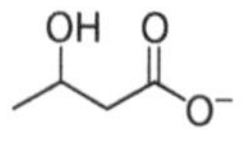

3-Hydroxybutyrate (157)

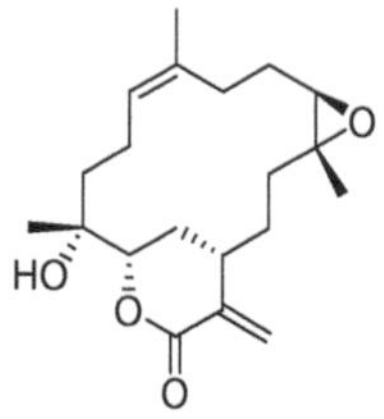

Flexibilide (158)

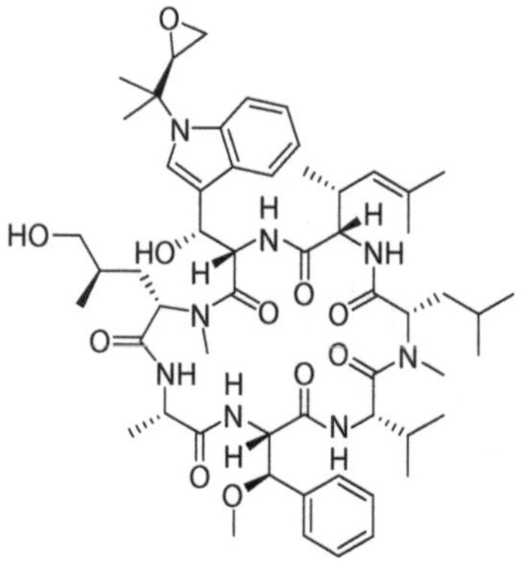

Cyclomarin A (158)

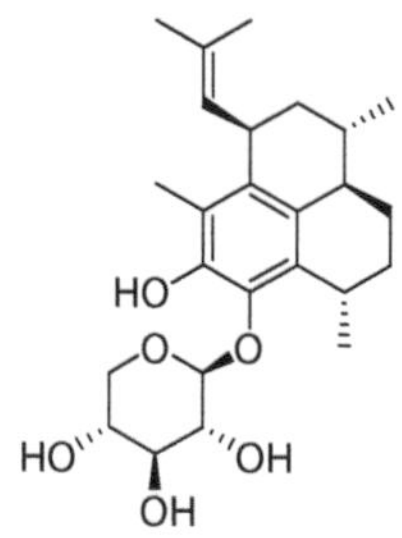

Pseudopterosin A (159)

FIGURE 10.12 (Continued)
Chemical structures of anti-inflammatory marine natural products.

10.6 Anticoagulants

Heparin is a sulfated glycosaminoglycan (GAG) widely used in anticoagulant therapy. GAGs have the potential to be an alternative source of heparin. Crude and purified samples GAGs isolated from marine bivalve *Donax faba* of 58 and 114 USP units/mg correspondingly in *D. faba* exhibited anticoagulant activity (Periyasamy et al. 2013). GAGs from marine polychaete (*Nereis sp.*) showed the anticoagulant activity of the 58 USP units/mg from crude, whereas purified samples showed activity of 114 USP units/mg (Singh et al. 2013). Sargahydroquinoic acid (SHQA; 160) and sargaquinoic acid (SQA; 161) demonstrated strong inhibition of collagen-induced platelet aggregation in both in vitro and in vivo studies (Park et al. 2013). Marine antithrombotic peptides dysinosin A, B, C, and D (162–165), isolated from sponge *Lamellodysidea chlorea*, were found to suppress blood coagulation cascade serine proteases factor VIIa and thrombin by binding to factor VIIa and thrombin proteases (Carroll et al. 2002). Oscillarin (166) was isolated from the algal cultures of *Oscillatoria agardhii* (strain B2 83), its previously assigned structure was later corrected by Hanessian et al (2004). It interacted with human α-thrombin producing antithrombotic effect (Wu et al. 2009).

New marine NPs clavatadines A (167) and B (168) were isolated from sponge *Suberea clavata*. Compounds 167 and 168 inhibited serine protease Factor XIa (FXIa) with IC_{50} of 1.3 and 27 µM, respectively (Buchanan et al. 2008).

Diterpenes pachydictyol-A (169), isopachydictyol-A (170), and dichotomanol (171) are promising anticoagulant and platelet aggregation molecules isolated from the Brazilian marine alga *Dictyota menstrualis*. Dichotomanol (171) inhibited collagen and ADP-induced aggregation in platelet-rich plasma with IC_{50} of 0.31 mM and 1.06 mM, respectively, whereas pachydictyol-A/isopachydictyol-A (0.18–0.7 mM) inhibited collagen (IC_{50} 0.12 mM) and thrombin (IC_{50} 0.25 mM) induced aggregation on washed platelets (de Andrade Moura et al. 2014; Figure 10.13).

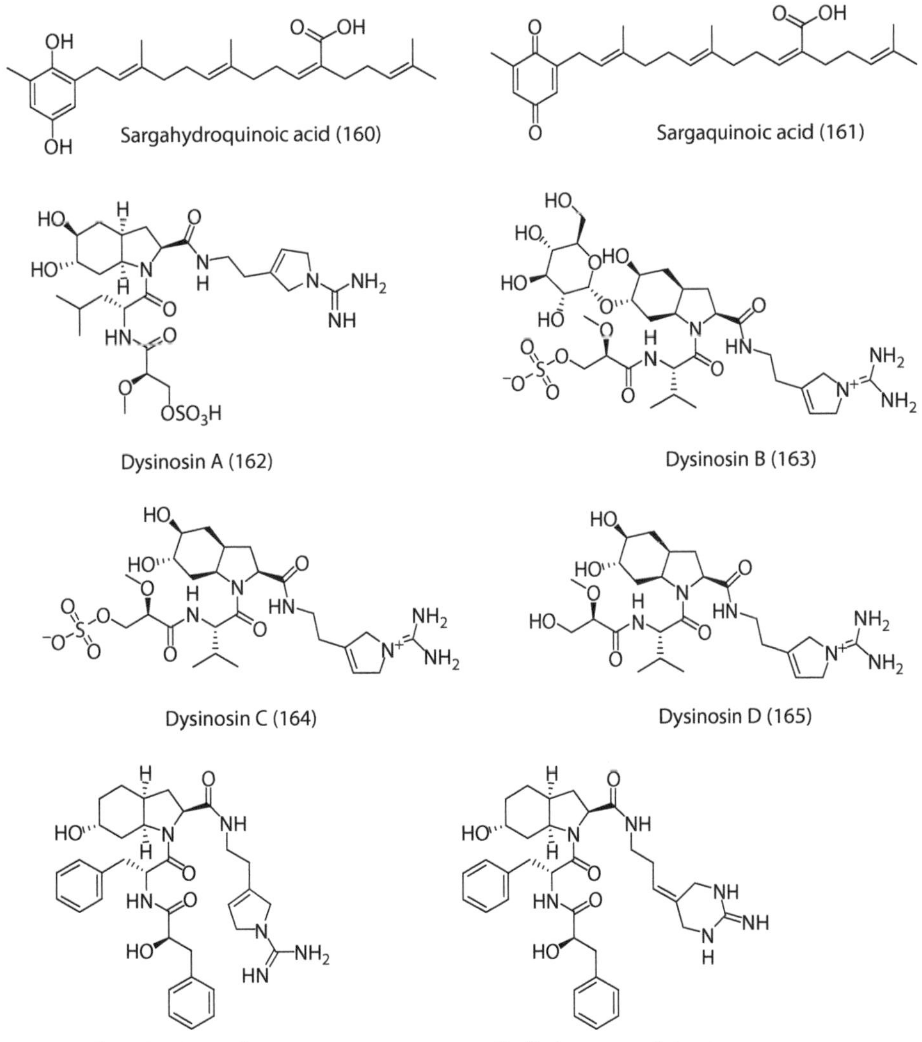

Sargahydroquinoic acid (160)

Sargaquinoic acid (161)

Dysinosin A (162)

Dysinosin B (163)

Dysinosin C (164)

Dysinosin D (165)

Oscillarin (166) actual structure

Oscillarin proposed structure

FIGURE 10.13
Chemical structures of marine anticoagulant compounds.

(Continued)

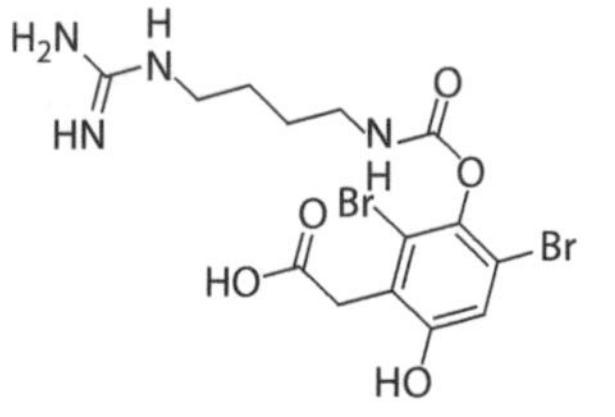

Clavatadines A (167) 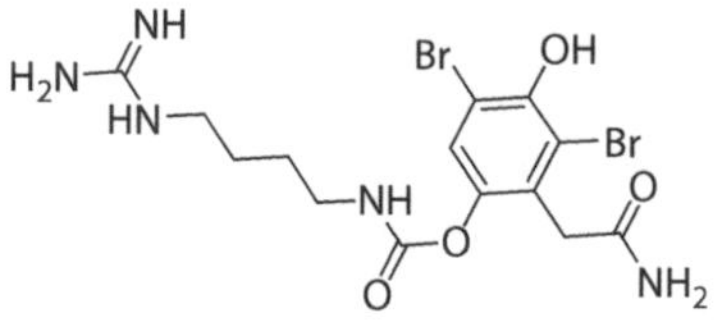   Clavatadines B (168) 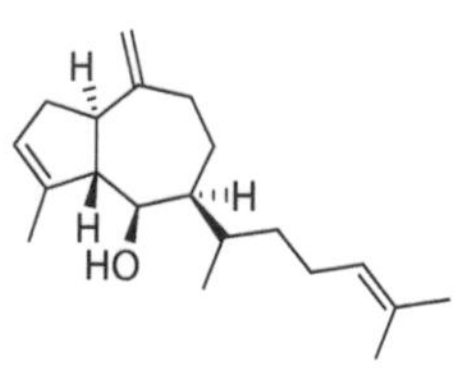   Pachydictyol A (169)

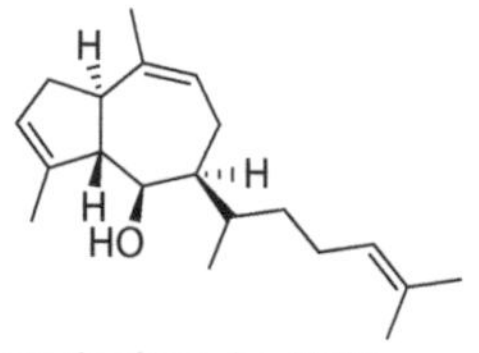

Isopachydictyol A (170) 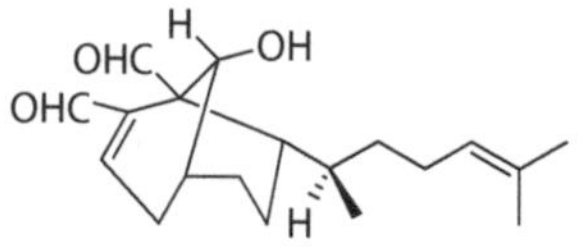   Dichotomanol (171)

FIGURE 10.13 (Continued)
Chemical structures of marine anticoagulant compounds.

10.7 Antidiabetic Compounds

SQA (161), also known as aleglitazar, reached phase III of clinical trials for the treatment of type-2 diabetes but was withdrawn due to its serious side effects such as bone fractures, heart failure, and gastrointestinal bleeding (Younk et al. 2011). SQA (161) and SHQA (160) increased adipocyte differentiation suggesting that these PPAR α/γ dual agonists may increase insulin sensitivity through adipogenesis regulation (Kim et al. 2008). A good deal of investigation has been performed in search of potential α-glucosidase and α-amylase inhibitors from natural sources. Mangrove endophytic fungus *Aspergillus sp.* 16-5B yielded aspergifuranone (172) which displayed significant inhibitory activity α-glucosidase with IC_{50} of 9.05 ± 0.60 μM (Liu et al. 2015). Two new lumazine-containing peptides terrelumamides A (173) and B (174) were purified from microbial culture of the marine-derived fungus *Aspergillus terreus*. These compounds improved insulin sensitivity in an adipogenesis model using human bone marrow mesenchymal stem cells (You et al. 2015). Biologically active polyphenol compounds exhibited multiple antidiabetic effects. Fucodiphloroethol G (175), dieckol (176), 6,6'-bieckol (177), 7-phloroeckol (178), and phlorofucofuroeckol A (179) were isolated from edible marine brown alga *Ecklonia cava*. Dieckol (176) showed significant inhibitory activities with IC_{50} of 10.8 μmol/L and 124.9 μmol/L with α-glucosidase and α-amylase, respectively (Lee et al. 2009; Lee and Jeon 2013). Bromophenols display their hyperglycemic effects by multiple mechanisms principally inhibiting the activities of proteins tyrosine phosphatase 1B and α-glucosidase (Lin and Liu 2012).

Two unsaturated fatty acids 7(Z)-octadecenoic acid (180) and 7(Z),10(Z)-octadecadienoic acid (181) were obtained from sea cucumber *Stichopus japonicus* with strong α-glucosidase inhibitory activity. Compound 180 showed α-glucosidase inhibitory activity with IC_{50} of 0.51 and 0.49 μg/mL, whereas compound 181 exhibited IC_{50} of 0.67 and 0.60 μg/mL against *Saccharomyces cerevisiae* α-glucosidase and *Bacillus stearothermophilus* α-glucosidase, respectively (Nguyen et al. 2011). Diphlorethohydroxycarmalol (182), previously isolated from the marine brown alga *Ishige okamurae*, a potent inhibitor of α-glucosidase as well as α-amylase enzymes (IC_{50} 0.16 and 0.53 nM, respectively) alleviated postprandial hyperglycemia in diabetic mice (Heo et al. 2009).

A sesquiterpene, dysidine (183), isolated from marine sponge *Dysidea villosa* inhibited human protein phosphatase 1B (IC_{50} 6.70 μM) as well as activated glucose uptake and glucose transporter-4 translocation (Li et al. 2009; Zhang et al. 2009). Albidopyrone (184), produced by *Streptomyces sp.* NTK 227, has moderate inhibitory activity against protein tyrosin phosphatase B (Hohmann et al. 2009). The O-Me nakafuran-8-lactone (185) isolated from marine sponge *Dysidea sp.* has good inhibitory activities for protein tyrosine phosphatase IB (Shao et al. 2006; Guo et al. 2007). Lukianol B (186), a pyrazine alkaloid, was isolated from tunicate showed aldose reductase inhibitory activity (Manzanaro et al. 2006; Figure 10.14).

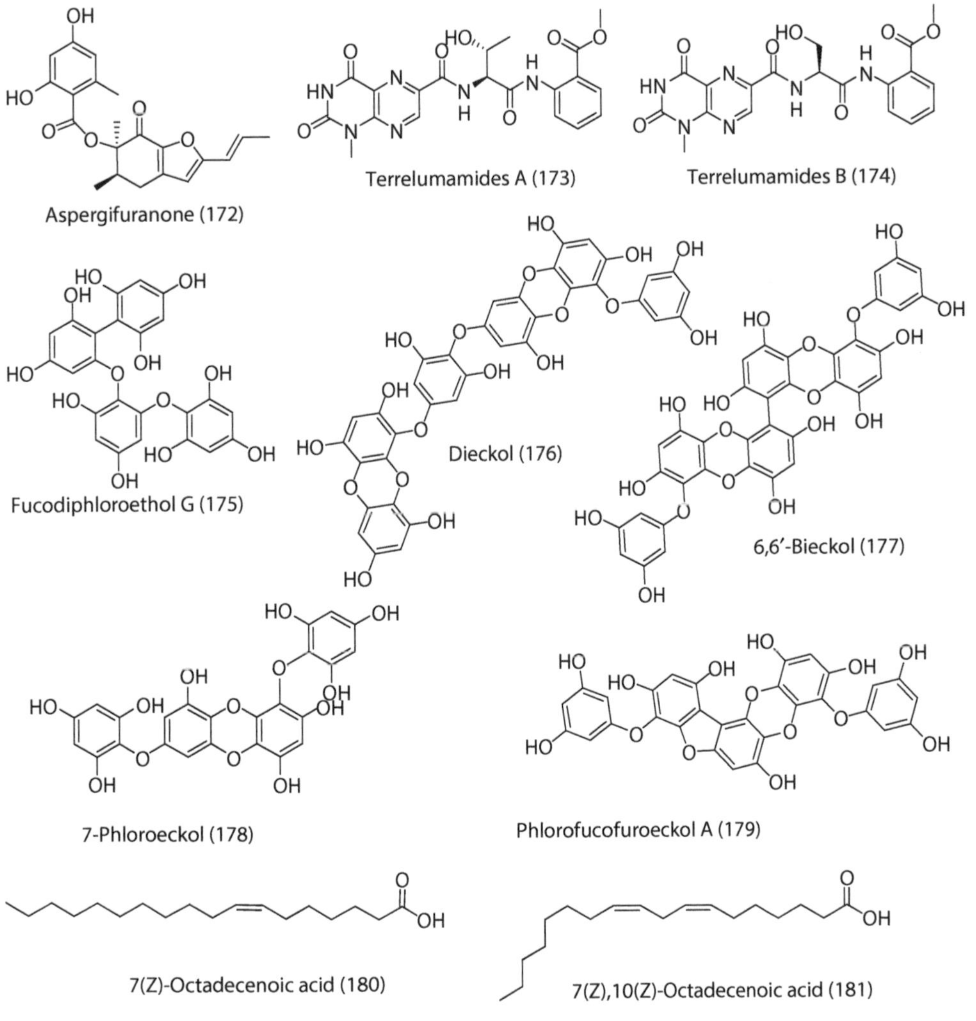

Aspergifuranone (172)

Terrelumamides A (173)

Terrelumamides B (174)

Fucodiphloroethol G (175)

Dieckol (176)

6,6′-Bieckol (177)

7-Phloroeckol (178)

Phlorofucofuroeckol A (179)

7(Z)-Octadecenoic acid (180)

7(Z),10(Z)-Octadecenoic acid (181)

FIGURE 10.14
Naturally occurring antidiabetic compounds from marine source.   *(Continued)*

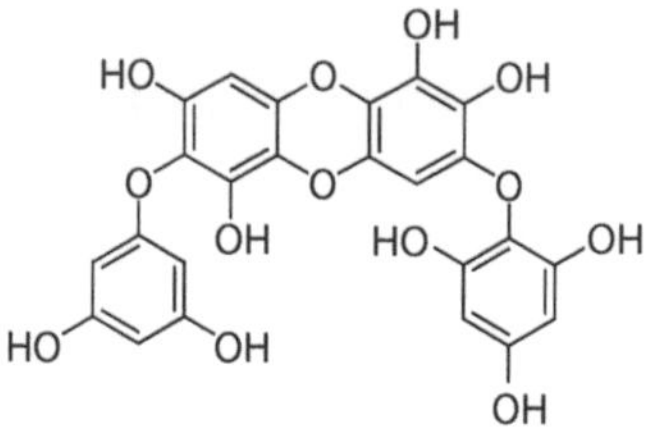

Diphlorethohydroxycarmalol (182)

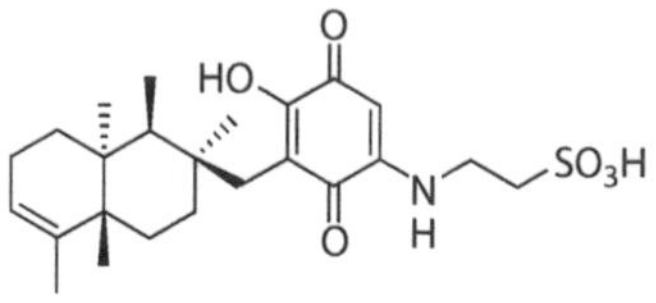

Dysidine (183)

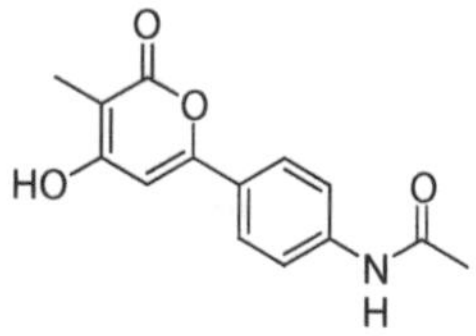

Albidopyrone (184)

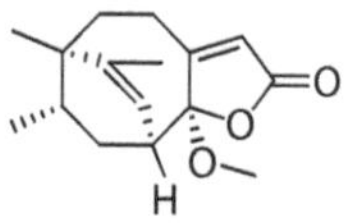

O-Me nakafuran-8-lactone (185)

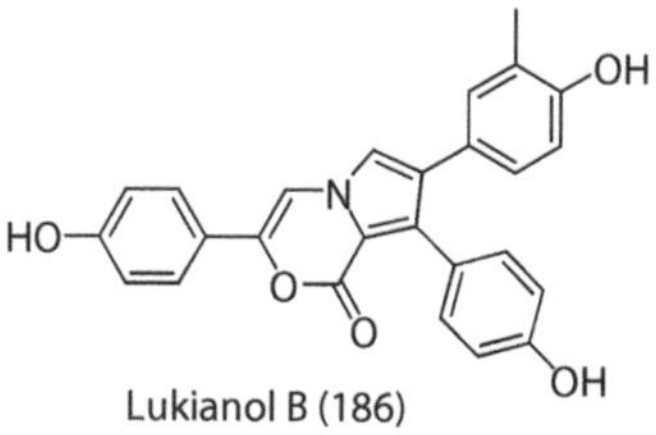

Lukianol B (186)

FIGURE 10.14 (Continued)
Naturally occurring antidiabetic compounds from marine source.

10.8 Autoimmune Disorder

Many MNP are active drugs for immunomodulating activities. They exert immune modulation by either activation or suppression of immune responses, thus representing invaluable leads in the drug discovery. Autoimmune diseases are basically treated with medication that suppresses the immune response. All immune response occurs as a consequence of some impaired immune response. Many autoimmune diseases have already been recognized, mainly type-1 diabetes, asthma, arthritis, and obesity. Moreover, immunosuppressant therapy has made organ transplantation, possible. Autoimmune disorders are managed by alleviation of inflammation by activation of anti-inflammatory genes. This section will cover compounds with promising IL suppression activity.

Splenocins displayed potent suppression of cytokine, exhibited minimal mammalian cell cytotoxicity, as well as inhibited the production of T Helper 2 (TH2) cytokines IL-5 and IL-13. These molecules showed significant antiasthmatic activity in an in vivo mouse model of allergen-induced TH2 splenocyte cytokine production characteristic of allergic asthma. Strangman et al, isolated splenocin B (187) from the marine bacterium *Streptomyces species* CNQ431. Splenocin B (187) inhibited antigen induced IL-5 by approximately 85% compared to the levels of OVA-stimulated control splenocytes (Strangman et al. 2009).

Thalassospiramide A (188) and B (189) are cyclic peptides isolated from marine α-proteobacterium *Thalassospira*. The compounds were screened for inhibition of cytokine IL-5 which plays an important role in TH2-mediated diseases like asthma. Thalassospiramides A (188) and B (189) showed IC_{50} of 10 and 5 μM, respectively, without any observable cytotoxicity at 10 μM (Oh et al. 2007). Verrucarin A (190), isolated from culture broth of *Myrothecium roridum*, inhibited PMA-stimulated IL-8 production in HL-60 cells (Oda et al. 2005; Figure 10.15).

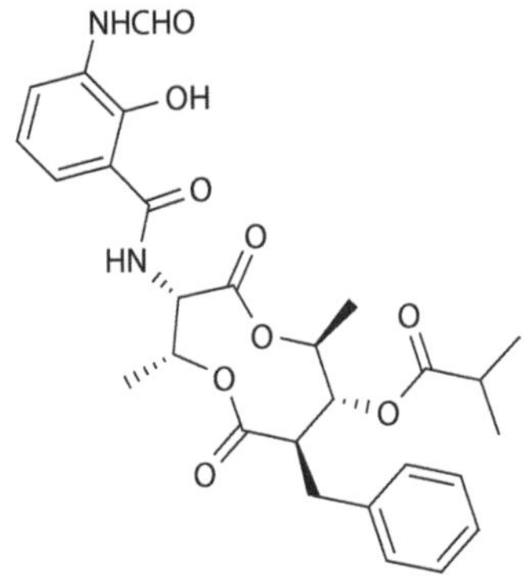

Splenocin B (187)

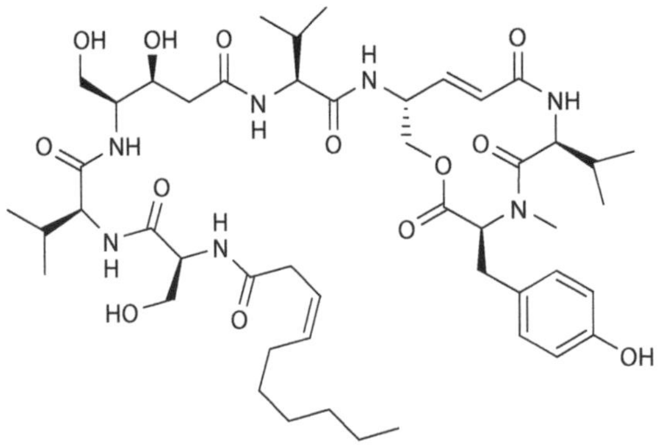

Thalassospiramide A (188)

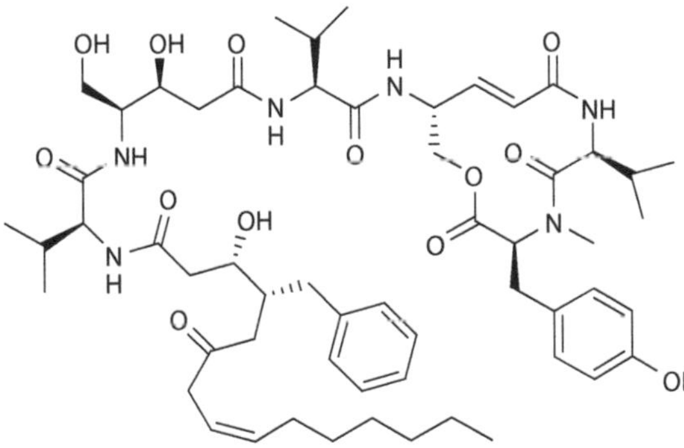

Thalassospiramide B (189)

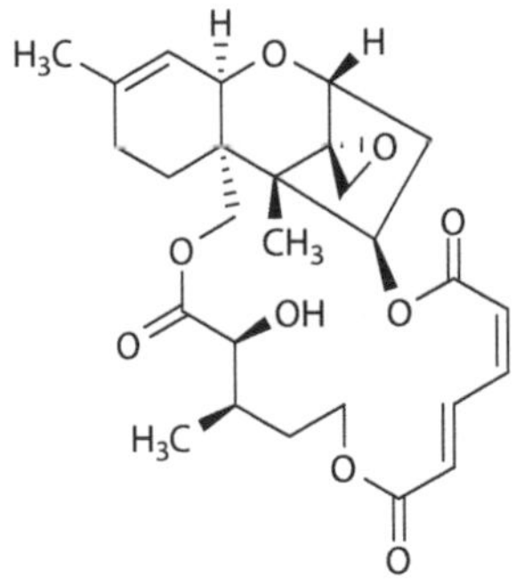

Verrucarin A (190)

FIGURE 10.15
Chemical structures of immunoprotective marine natural products.

10.9 Neuroprotective

Progression of neurodegenerative disease can be slowed by inhibition of glycogen synthase kinase-3 (GSK-3β), cyclin-dependent kinase-5 (CDK-5), and cdc2-like kinases (CLKs). Many of marine NPs are recognized to inhibit mammalian kinases. Several potent, selective, and structurally novel candidates having mammalian neurological activity have been isolated from marine sources.

Hymenialdisine (191), a sponge-derived brominated pyrrole isolated from marine sponges of the genera *Hymeniacidon, Acanthella, Axinella,* and *Pseudaxinyssa* (Nguyen and Tepe 2009), inhibited GSK-3, casein kinase 1 (CK1), and cyclin-dependent kinases (CDK 1 and 5) and consequently suppressed hyperphosphorylation of microtubule-associated protein (MAP)-1B and tau. Compound 191 exerted its effect by interacting with the ATP binding pocket of CDK2 (Meijer et al. 2000). A related compound leucettamine B (192) derived from the Palauan sponge *Leucetta microraphis* also inhibited cyclin-dependent kinases (Chan et al. 1993; Watanabe et al. 2000). An indole alkaloid, indirubin (193), has also been

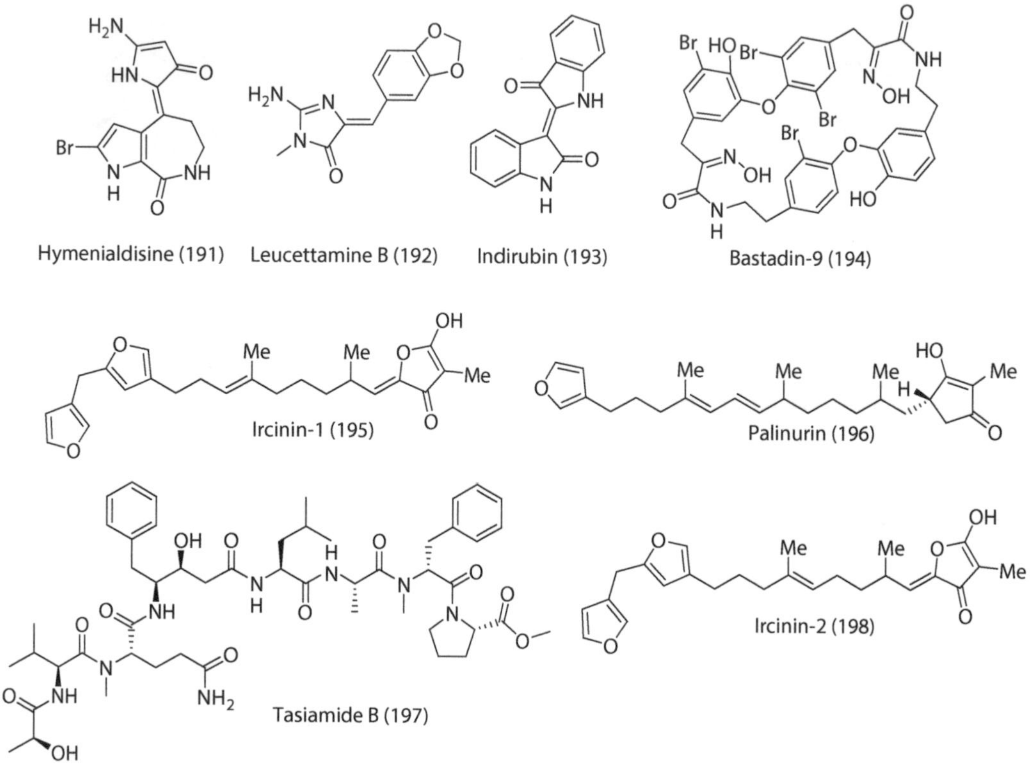

FIGURE 10.16
Chemical structures of neuroprotective marine natural products.

identified as an inhibitor of GSK-3, CDK5, and P25 that are involved in abnormal tau phosphorylation in Alzheimer's disease (Leclerc et al. 2001). Palinurin (194), a sponge-derived sesterterpene, inhibited both GSK-3α and GSK-3β, and related compounds ircinin-1 and ircinin-2 inhibited GSK-3β (Bidon-Chanal et al. 2013).

Another strategy for treating a neurodegenerative disease is the inhibition of β-site amyloid precursor peptide cleaving enzyme type 1 (BACE1) because this enzyme hydrolyzes amyloid precursor peptide to β-amyloid peptide, a causative agent of Alzheimer's disease. As a result of screen for BACE1 inhibitors from marine extracts, tasiamide B (195), a linear depsipeptide of cyanobacterial origin (Liu et al. 2012) and other compound bastadin-9 (196) have been identified (Wu et al. 2008; Williams et al. 2010; Figure 10.16).

10.10 Conclusions

The future of drug development from marine sources looks very optimistic as the number of new metabolite with interesting biological activities are continuously increasing along with several related designer molecules with improved pharmacological profile. Total synthesis, semisynthesis, and cultured production of marine compounds have given a new hope toward the development of marine NPs as drugs in the near future.

The clinical development of marine NPs is mainly focused on the treatment of cancer. Conjugation of a potential cytotoxic molecule with a tumor-specific antibody has given a new approach toward the development of a targeted therapy for the treatment of cancer. A number of ADCs have been tested in clinical trials for the better management of cancer patients.

This chapter presents an account of eight approved drugs, six compounds in clinical trial for different types of cancer, and one for Alzheimer's disease. Additionally, 183 compounds of marine origin have been discussed in this chapter, which hold new opportunities for the treatment of life-threatening diseases such as diabetes, blood disorders, inflammation, autoimmune disorders, neurological disorders, and some of the compounds exhibited promising antifungal, antituberculosis, anthelmintic, antiprotozoal, and antiviral activities.

In spite of the large number of marine NPs and related compounds in clinical and preclinical pipeline, a large portion of ocean is still unexplored. Despite many natural chemicals of potential economic importance, a large portion still awaits exploration of useful pharmaceutical activity from marine organisms. High throughput screening methods and mechanism-based biological assays have steadily improved the success rate of getting hit. The detailed knowledge of the mechanism of a disease has yielded several targets for target-based drug discovery. A metabolite identified in a marine organism may or may not be produced by the organism itself. Many of them have been found to be produced by symbiotic microorganism. Efforts are ongoing to identify the real microbial source of bioactive agents to develop suitable fermentation technology to ensure continuous supply of therapeutic agents. New approaches are needed to identify the cell type that produces the metabolite. Furthermore, the identification of gene clusters responsible for the biosynthesis of the molecule of interest is important to revolutionize the production of drugs from marine sources.

Ocean has supported early development of life and now they are serving a great reservoir of new pharmacological entities. The biodiversity of ocean is tremendous but not immortal. Marine resources should be used with great wisdom otherwise some organisms will be viewed only in pictures. Synthetic procedures must be followed for the production of marine-related compounds. Preserving the beauty and diversity of marine life, we must go further forward toward the development of drugs.

Acknowledgments

Sudhir Kumar is thankful to University Grant Commission, New Delhi, for providing him with Senior Research Fellowship. The CSIR-CDRI communication number for this manuscript is 9214.

References

Adamczeski, M. et al. 1988. Unusual anthelmintic oxazoles from a marine sponge. *J Am Chem Soc* 110 (5): 1598–1602.

Ahmadi, A. et al. 2015. Antiviral potential of algae polysaccharides isolated from marine sources: A review. *Biomed Res Int* 2015: 825203.

Aicher, T. D. et al. 1992. Total synthesis of halichondrin B and norhalichondrin B. *J Am Chem Soc* 114 (8): 3162–3164.

Albizati, K. F. et al. 1987. Luffariellolide, an anti-inflammatory sesterterpene from the marine sponge *Luffariella sp. Experientia* 43 (8): 949–950.

Anderson, H. J. et al. 1997. Cytotoxic peptides hemiasterlin, hemiasterlin A, and hemiasterlin B include mitotic arrest and abnormal spindle formation. *Cancer Chemother Pharmacol* 39 (3): 223–226.

Andjelic, C. D. et al. 2008. Characterizing the anti-HIV activity of papuamide A. *Mar Drugs* 6 (4): 528–549.

Andrianasolo, E. H. et al. 2005. Isolation of swinholide A and related glycosylated derivatives from two field collections of marine cyanobacteria. *Org Lett* 7 (7): 1375–1378.

Arefolov, A. and J. S. Panek 2002. Studies directed toward the total synthesis of discodermolide: Asymmetric synthesis of the C1-C14 fragment. *Org Lett* 4 (14): 2397–2400.

Armitage, J. O. et al. 2015. Managing risk in Hodgkin lymphoma. *Clin Adv Hematol Oncol* 13 (2 Suppl 1): 1–19.

Ayer, S. W. et al. 1984. Phidolopine, a new purine derivative from the bryozoan *Phidolopora pacifica*. *J Org Chem* 49 (20): 3869–3870.

Bai, R. et al. 1990. Dolastatin 10, a powerful cytostatic peptide derived from a marine animal: Inhibition of tubulin polymerization mediated through the vinca alkaloid binding domain. *Biochem Pharmacol* 39 (12): 1941–1949.

Bailey, H. et al. 2015. Panobinostat for the treatment of multiple myeloma: The evidence to date. *J Blood Med* 6: 269–276.

Bergmann, W. and R. J. Feeney 1950. The isolation of a new thymine pentoside from sponges. *J Am Chem Soc* 72 (6): 2809–2810.

Bidon-Chanal, A. et al. 2013. Evidence for a new binding mode to GSK-3: Allosteric regulation by the marine compound palinurin. *Eur J Med Chem* 60: 479–489.

Blunt, J. W. et al. 2005. Marine natural products. *Nat Prod Rep* 22 (1): 15–61.

Bubb, M. R. et al. 2002. Polylysine induces an antiparallel actin dimer that nucleates filament assembly: Crystal structure at 3.5-A resolution. *J Biol Chem* 277 (23): 20999–21006.

Buchanan, M. S. et al. 2008. Clavatadine A, a natural product with selective recognition and irreversible inhibition of factor XIa. *J Med Chem* 51 (12): 3583–3587.

Buckle, P. J. et al. 1980. The anti-inflammatory activity of marine natural products-6-n-tridecylsalicylic acid, flexibilide and dendalone 3-hydroxybutyrate. *Agents Actions* 10 (4): 361–367.

Cain, J. M. et al. 1992. Phase II trial of didemnin-B in advanced epithelial ovarian cancer. A southwest oncology group study. *Invest New Drugs* 10 (1): 23–24.

Cao, F. et al. 2014. Antiviral C-25 epimers of 26-acetoxy steroids from the South China Sea Gorgonian *Echinogorgia rebekka*. *J Nat Prod* 77 (6): 1488–1493.

Capon, R. J. et al. 1999. Geodin A magnesium salt: A novel nematocide from a Southern Australian marine sponge, *Geodia*. *J Nat Prod* 62 (9): 1256–1259.

Capon, R. J. et al. 2005a. (–)-Echinobetaine A: Isolation, structure elucidation, synthesis, and SAR studies on a new nematocide from a Southern Australian marine sponge, *Echinodictyum sp.* *J Nat Prod* 68 (2): 179–182.

Capon, R. J. et al. 2005b. (+)-Echinobetaine B: Isolation, structure elucidation, synthesis and preliminary SAR studies on a new nematocidal betaine from a Southern Australian marine sponge, *Echinodictyum sp. Org. Biomol. Chem.* 3 (1): 118–122.

Carroll, A. R. et al. 2002. Dysinosin A: A novel inhibitor of factor VIIa and thrombin from a new genus and species of Australian sponge of the family *Dysideidae*. *J Am Chem Soc* 124 (45): 13340–13341.

Chan, G. W. et al. 1993. New leukotriene B4 receptor antagonist: Leucettamine A and related imidazole alkaloids from the marine sponge *Leucetta microraphis*. *J Nat Prod* 56 (1): 116–121.

Chinworrungsee, M. et al. 2001. Antimalarial halorosellinic acid from the marine fungus *Halorosellinia oceanica*. *Bioorg Med Chem Lett* 11 (15): 1965–1969.

Christophersen, C. 1985. Secondary metabolites from marine bryozoans. A review. *Acta Chem Scand B* 39 (7): 517–529.

Cimino, G. et al. 1984. Antiviral agents from a gorgonian, *Eunicella cavolini*. *Experientia* 40 (4): 339–340.

Clarke, A. and S. Lidgard. 2000. Spatial patterns of diversity in the sea: Bryozoan species richness in the North Atlantic. *J Anim Ecol* 69 (5): 799–814.

Cohen, S. S. 1979. The mechanisms of inhibition of cellular and viral multiplication by aranucleosides and aranucleotides. Volume 26 of the series NATO Advanced Study Institutes Series. In R. T. Walker, E. De Clercq and F. Eckstein (eds.), *Nucleoside Analogues: Chemistry, Biology, and Medical Applications*, MA, Springer, pp. 225–245.

Corey, E. J. et al. 1996. Enantioselective total synthesis of ecteinascidin 743. *J Am Chem Soc* 118 (38): 9202–9203.

Corley, D. G. et al. 1988. Laulimalides. New potent cytotoxic macrolides from a marine sponge and a nudibranch predator. *J Org Chem* 53 (15): 3644–3646.

Cuadros, R. et al. 2000. The marine compound spisulosine, an inhibitor of cell proliferation, promotes the disassembly of actin stress fibers. *Cancer Lett* 152 (1): 23–29.

Cutignano, A. et al. 2000. Dragmacidin F: A new antiviral bromoindole alkaloid from the mediterranean sponge *Halicortex sp. Tetrahedron* 56: 3743–3748.

Dalisay, D. S. et al. 2011. Ptilomycalin A inhibits laccase and melanization in *Cryptococcus neoformans. Bioorgan med chem.* 19 (22): 6654–6657.

Davidson, S. K. et al. 2001. Evidence for the biosynthesis of bryostatins by the bacterial symbiont "Candidatus Endobugula sertula" of the bryozoan *Bugula neritina. Appl Environ Microbiol* 67 (10): 4531–4537.

Davis, R. A. et al. 2010. (+)-7-Bromotrypargine: an antimalarial β-carboline from the Australian marine sponge *Ancorina sp. Tetrahedron Lett.* 51 (4): 583–585.

Davyt, D. et al. 2006. Bisabolanes from the red alga *Laurencia scoparia. J Nat Prod* 69 (7): 1113–1116.

de Andrade Moura, L. et al. 2014. Antiplatelet and anticoagulant effects of diterpenes isolated from the marine alga, *Dictyota menstrualis. Mar Drugs* 12 (5): 2471–2484.

de, C. M. S. and R. S. Jacobs. 1991. Two-step inactivation of bee venom phospholipase A2 by scalaradial. *Biochem Pharmacol* 42 (8): 1621–1626.

de la Calle Gil, A. B. et al. 2015. Intrathecal ziconotide and morphine for pain relief: A case series of eight patients with refractory cancer pain, including five cases of neuropathic pain. *Neurol Ther* 4 (2): 159–168.

De Marino, S. et al. 2009. Coscinolactams A and B: New nitrogen-containing sesterterpenoids from the marine sponge *Coscinoderma matthewsi* exerting anti-inflammatory properties. *Tetrahedron* 65 (15): 2905–2909.

Digirolamo, J. A. et al. 2009. Reversal of fluconazole resistance by sulfated sterols from the marine sponge *Topsentia sp. J Nat Prod* 72 (8): 1524–1528.

Dyshlovoy, S. A. et al. 2014. Activity of aaptamine and two derivatives, demethyloxyaaptamine and isoaaptamine, in cisplatin-resistant germ cell cancer. *J Proteomics* 96: 223–239.

El Sayed, K. A. et al. 2000. Marine natural products as antituberculosis agents. *Tetrahedron* 56 (7): 949–953.

El Sayed, K. A. et al. 2002. Microbial and chemical transformation studies of the bioactive marine sesquiterpenes (s)-(+)-curcuphenol and -curcudiol isolated from a deep reef collection of the Jamaican sponge *Didiscus oxeata. J Nat Prod* 65 (11): 1547–1553.

Feling, R. H. et al. 2003. Salinosporamide A: A highly cytotoxic proteasome inhibitor from a novel microbial source, a marine bacterium of the new genus *Salinospora. Angew Chem Int Ed Engl* 42 (3): 355–357.

Francisco, J. A. et al. 2003. cAC10-vcMMAE, an anti-CD30-monomethyl auristatin E conjugate with potent and selective antitumor activity. *Blood* 102 (4): 1458–1465.

Furstner, A. et al. 2007. Total synthesis and evaluation of the actin-binding properties of microcarpalide and a focused library of analogues. *Chemistry* 13 (5): 1452–1462.

Gallimore, W. A. et al. 2005. Alkaloids from the sponge *Monanchora unguifera. J Nat Prod* 68 (9): 1420–1423.

Gerwick, W. H. et al. 1994. Structure of curacin A, a novel antimitotic, antiproliferative and brine shrimp toxic natural product from the marine cyanobacterium *Lyngbya majuscula. J Org Chem* 59 (6): 1243–1245.

Gerwick, W. H. and B. S. Moore. 2012. Lessons from the past and charting the future of marine natural products drug discovery and chemical biology. *Chem Biol (Oxford, U.K.)* 19 (1): 85–98.

Glaser, K. B. and A. M. S. Mayer. 2009. A renaissance in marine pharmacology: From preclinical curiosity to clinical reality. *Biochem Pharmacol* 78 (5): 440–448.

Glaser, K. B. and R. S. Jacobs. 1986. Molecular pharmacology of manoalide. Inactivation of bee venom phospholipase A2. *Biochem Pharmacol* 35 (3): 449–453.

Gochfeld, D. J. and M. T. Hamann. 2001. Isolation and biological evaluation of filiformin, plakortide F, and plakortone G from the Caribbean sponge *Plakortis sp. J Nat Prod* 64 (11): 1477–1479.

Guittat, L. et al. 2005. Ascididemin and meridine stabilise G-quadruplexes and inhibit telomerase in vitro. *Biochim Biophys Acta* 1724 (3): 375–384.

Gunasekera, S. P. et al. 1990. Discodermolide: A new bioactive polyhydroxylated lactone from the marine sponge *Discodermia dissoluta. J Org Chem* 55 (16): 4912–4915.

Guo, Y. et al. 2007. Extraction of O-methyl nakafuran-8-lactone from Dysidea and marine sponge and its use in treating diabetes mellitus, obesity and complication thereof, Shanghai Institute of Materia Medica, Chinese Academy of Sciences, People's Republic of China. p. 8

Hadaschik, B. A. et al. 2008. Targeting prostate cancer with HTI-286, a synthetic analog of the marine sponge product hemiasterlin. *Int J Cancer* 122 (10): 2368–2376.

Hamann, M. T. et al. 2007. Marine natural products: Key advances to the practical application of this resource in drug development. *Chimia* 61 (6): 313–321.

Hanessian, S. et al. 2004. The N-Acyloxyiminium ion aza-prins route to octahydroindoles: Total synthesis and structural confirmation of the antithrombotic marine natural product oscillarin. *J Am Chem Soc* 126 (19): 6064–6071.

Harris, G. H. et al. 1993. Isolation and structure determination of pycnidione, a novel bistropolone stromelysin inhibitor from a *Phoma sp. Tetrahedron* 49 (11): 2139–2144.

Hassan, W. et al. 2004. New imidazole alkaloids from the Indonesian sponge *Leucetta chagosensis. J Nat Prod* 67 (5): 817–822.

Heo, S.-J. et al. 2009. Diphlorethohydroxycarmalol isolated from Ishige okamurae, a brown algae, a potent α-glucosidase and α-amylase inhibitor, alleviates postprandial hyperglycemia in diabetic mice. *Eur J Pharmacol* 615 (1–3): 252–256

Hirata, K. et al. 2006. Structure basis for antitumor effect of aplyronine A. *J Mol Biol* 356 (4): 945–954.

Hirata, Y. and D. Uemura. 1986. Halichondrins-antitumor polyether macrolides from a marine sponge. *Pure Appl Chem* 58 (5): 701–710.

Hohmann, C. et al. 2009. Albidopyrone, a new α-pyrone-containing metabolite from marine-derived *Streptomyces sp.* NTK 227. *J Antibiot* 62 (2): 75–79.

Hood, K. A. et al. 2002. Peloruside A, a novel antimitotic agent with paclitaxel-like microtubule-stabilizing activity. *Cancer Res* 62 (12): 3356–3360.

Hsiao, C.-J. et al. 2012. Pycnidione, a fungus-derived agent, induces cell cycle arrest and apoptosis in A549 human lung cancer cells. *Chem-Biol Interact* 197 (1): 23–30.

Hu, Y. et al. 2015. Statistical research on the bioactivity of new marine natural products discovered during the 28 years from 1985 to 2012. *Marine Drugs* 13 (1): 202–221.

Ichiba, T. et al. 1991. Hennoxazoles, bioactive bisoxazoles from a marine sponge. *J Am Chem Soc* 113 (8): 3173–3174.

Ichiba, T. et al. 1994. 8-Hydroxymanzamine A, a β-carboline alkaloid from a sponge, *Pachypellina sp. J Nat Prod* 57 (1): 168–170.

Iguchi, K. et al. 1989. Novel 19-oxygenated sterols from the Okinawan soft coral *Litophyton viridis. Chem Pharm Bull* 37 (9): 2553–2554.

Inman, W. D. et al. 2010. A β-Carboline alkaloid from the Papua New Guinea marine sponge *Hyrtios reticulatus. J Nat Prod* 73 (2): 255–257.

Ivo de Medeiros, A. et al. 2012. 11-Oxoaerothionin isolated from the marine sponge *Aplysina fistularis* shows anti-inflammatory activity in LPS-stimulated macrophages. *Immunopharmacol Immunotoxicol* 34 (6): 919–924.

Jacob, M. R. et al. 2003. Reversal of fluconazole resistance in multidrug efflux-resistant fungi by the *Dysidea arenaria* sponge sterol 9alpha,11alpha-epoxycholest-7-ene-3beta, 5alpha, 6alpha, 19-tetrol 6-acetate. *J Nat Prod* 66 (12): 1618–1622.

Jefford, C. W. et al. 1996. Structures and absolute configurations of the marine toxins, latrunculin A and laulimalide. *Tetrahedron Lett* 37 (2): 159–162.

Jordan, M. A. et al. 1998. Suppression of microtubule dynamics by binding of cemadotin to tubulin: Possible mechanism for its antitumor action. *Biochemistry* 37 (50): 17571–17578.

Kanoh, K. et al. 1997. (−)-Phenylahistin: A new mammalian cell cycle inhibitor produced by *Aspergillus ustus*. *Bioorg Med Chem Lett* 7 (22): 2847–2852.

Kanoh, K. et al. 1999. Antitumor activity of phenylahistin in vitro and in vivo. *Biosci, Biotechnol, Biochem* 63 (6): 1130–1133.

Kashman, Y. et al. 1980. Latrunculin, a new 2-thiazolidinone macrolide from the marine sponge *Latrunculia magnifica*. *Tetrahedron Lett* 21: 3629–3632.

Kawauchi, K. et al. 2008. Cyprodigiosin hydrochloride activates the Ras-PI3K-Akt pathway and suppresses protein synthesis inhibition-induced apoptosis in PC12 cells. *Biosci, Biotechnol, Biochem* 72 (6): 1564–1570.

Kerr, R. G. et al. 2014. Synthesis and evaluation of anti-inflammatory activity of derivatives of the marine natural products fuscol and eunicol. *Bioorg Med Chem Lett* 24 (20): 4804–4806.

Kikuchi, H. et al. 1983. Marine natural products. XI. An anti-inflammatory scalarane-type bishomo-sesterterpene, foliaspongin, from the Okinawan marine sponge *Phyllospongia foliascens* (Pallas). *Chem Pharm Bull* 31 (2): 552–556.

Kim, S. N. et al. 2008. Sargaquinoic acid and sargahydroquinoic acid from *Sargassum yezoense* stimulate adipocyte differentiation through PPAR alpha/gamma activation in 3T3-L1 cells. *FEBS Lett* 582 (23–24): 3465–3472.

Kobayashi, M. et al. 1994. Arenastatin A, a potent cytotoxic depsipeptide from the Okinawan marine sponge *Dysidea arenaria*. *Tetrahedron Lett.* 35 (43): 7969–7972.

Kohmoto, S. et al. 1987. Puupehenone, a cytotoxic metabolite from a deep water marine sponge, *Stronglyophora hartmani*. *J Nat Prod* 50 (2): 336.

Konig, G. M. et al. 2000. Assessment of antimycobacterial activity of a series of mainly marine derived natural products. *Planta Med* 66: 337–342.

Kossuga, M. H. et al. 2004. (2S, 3R)-2-aminododecan-3-ol, a new antifungal agent from the ascidian *Clavelina oblonga*. *J Nat Prod* 67 (11): 1879–1881.

Kovtun, Y. V. and V. S. Goldmacher. 2007. Cell killing by antibody–drug conjugates. *Cancer Letts* 255 (2): 232–240.

Lamoth, F. and B. D. Alexander. 2015. Antifungal activities of SCY-078 (MK-3118) and standard antifungal agents against clinical non-Aspergillus mold isolates. *Antimicrob Agents Chemother* 59 (7): 4308–4311.

Lazaro, J. E. H. et al. 2002. Heptyl prodigiosin, a bacterial metabolite, is antimalarial in vivo and non-mutagenic in vitro. *J Nat Toxins* 11 (4): 367–377.

Leclerc, S. et al. 2001. Indirubins inhibit glycogen synthase kinase-3 beta and CDK5/p25, two protein kinases involved in abnormal tau phosphorylation in Alzheimer's disease. A property common to most cyclin-dependent kinase inhibitors? *J Biol Chem* 276 (1): 251–260.

Lee, D.-S. et al. 2013. Penicillinolide A: A new anti-inflammatory metabolite from the marine fungus *Penicillium sp.* sf-5292. *Mar Drugs* 11 (11): 4510–4526.

Lee, S.-H. and Y.-J. Jeon. 2013. *Antidiabetic Effect of Dieckol, a Marine Polyphenol, and Its Mechanisms of Blood Glucose Regulation*. Boca Raton, FL: CRC Press.

Lee, S.-H. et al. 2009. α-Glucosidase and α-amylase inhibitory activities of phloroglucinal derivatives from edible marine brown alga, *Ecklonia cava*. *J Sci Food Agric.* 89 (9): 1552–1558.

Li, Y. et al. 2009. A novel sesquiterpene quinone from Hainan sponge *Dysidea villosa*. *Bioorg Med Chem Lett* 19 (2): 390–392.

Li, S. and R. R. Breaker. 2012. Fluoride enhances the activity of fungicides that destabilize cell membranes. *Bioorg Med Chem Lett* 22 (9): 3317–3322.

Liu, Y. et al. 2012. Cyanobacterial peptides as a prototype for the design of potent beta-secretase inhibitors and the development of selective chemical probes for other aspartic proteases. *J Med Chem* 55 (23): 10749–10765.

Liu, Y. et al. 2015. Bioactive metabolites from mangrove endophytic fungus *Aspergillus sp.* 16-5B. *Mar Drugs* 13 (5): 3091–3102.

Lin, X. and M. Liu. 2012. Bromophenols from marine algae with potential anti-diabetic activities. *J Ocean Univ. China* 11 (4): 533–538.

Lo, M. C. et al. 2004. Probing the interaction of HTI-286 with tubulin using a stilbene analogue. *J Am Chem Soc* 126 (32): 9898–9899.

Lopez-Anton, N. et al. 2006. The marine product cephalostatin 1 activates an endoplasmic reticulum stress-specific and apoptosome-independent apoptotic signaling pathway. *J Biol Chem* 281 (44): 33078–33086.

Loya, S. et al. 1999. Polycitone A, a novel and potent general inhibitor of retroviral reverse transcriptases and cellular DNA polymerases. *Biochem J* 344 (Pt 1): 85–92.

Luesch, H. et al. 2001. Isolation of dolastatin 10 from the marine cyanobacterium symploca species vp642 and total stereochemistry and biological evaluation of its analogue symplostatin 1. *J Nat Prod* 64 (7): 907–910.

Madiraju, C. et al. 2005. Tubulin assembly, taxoid site binding, and cellular effects of the microtubule-stabilizing agent dictyostatin. *Biochemistry* 44 (45): 15053–15063.

Mandeau, A. et al. 2005. Isolation and absolute configuration of new bioactive marine steroids from *Euryspongia n. sp. Steroids* 70 (13): 873–878.

Mangalindan, G. C. et al. 2000. Agelasine F from a *Philippine Agelas sp.* sponge exhibits in vitro anti-tuberculosis activity. *Planta Med* 66 (4): 364–365.

Manzanaro, S. et al. 2006. Phenolic marine natural products as aldose reductase inhibitors. *J Nat Prod* 69 (10): 1485–1487.

Marrero, J. et al. 2004. Bielschowskysin, a gorgonian-derived biologically active diterpene with an unprecedented carbon skeleton. *Org Lett* 6 (10): 1661–1664.

Marshall, L. A. et al. 1994. Effects of scalaradial, a type II phospholipase A2 inhibitor, on human neutrophil arachidonic acid mobilization and lipid mediator formation. *J Pharmacol Exp Ther* 268 (2): 709–717.

Martin, M. J. et al. 2013. Isolation and first total synthesis of PM050489 and PM060184, two new marine anticancer compounds. *J Am Chem Soc* 135 (27): 10164–10171.

Martinez-Diez, M. et al. 2014. PM060184, a new tubulin binding agent with potent antitumor activity including P-glycoprotein over-expressing tumors. *Biochem Pharmacol* 88 (3): 291–302.

Matsui, Y. et al. 2009. Intravesical combination treatment with antisense oligonucleotides targeting heat shock protein-27 and HTI-286 as a novel strategy for high-grade bladder cancer. *Mol Cancer Ther* 8 (8): 2402–2411.

McCarthy, P. J. et al. 1992. Antifungal activity of meridine, a natural product from the marine sponge *Corticium sp. J Nat Prod* 55 (11): 1664–1668.

Meijer, L. et al. 2000. Inhibition of cyclin-dependent kinases, GSK-3beta and CK1 by hymenialdisine, a marine sponge constituent. *Chem Biol* 7 (1): 51–63.

Mickel, S. J. 2007. Total synthesis of the marine natural product (+)-discodermolide in multigram quantities. *Pure Appl Chem* 79 (4): 685–700.

Mickel, S. J. et al. 2005. A study of the paterson boron aldol reaction as used in the large-scale total synthesis of the anticancer marine natural product (+)-discodermolide. *Org Process Res Dev* 9 (1): 113–120.

Minale, L. et al. 1974. Avarol, a novel sesquiterpenoid hydroquinone with a rearranged drimane skeleton from the sponge *Disidea avara. Tetrahedron Lett* (38): 3401–3404.

Molinski, T. F. et al. 2009. Drug development from marine natural products. *Nat Rev Drug Discov* 8 (1): 69–85.

Monti, M. C. et al. 2007. Scalaradial, a dialdehyde-containing marine metabolite that causes an unexpected noncovalent PLA2 inactivation. *ChemBioChem* 8 (13): 1585–1591.

Monti, M. C. et al. 2011. The binding mode of cladocoran A to the human group IIA phospholipase A(2). *ChemBioChem* 12 (17): 2686–2691.

Mora, C. et al. 2011. How many species are there on Earth and in the ocean? *PLoS Biol* 9 (8): e1001127.

Morris, J. C. 2013. Marine natural products: Synthetic aspects. *Nat Prod Rep* 30 (6): 783–805.

Moya, C. E. and R. S. Jacobs. 2006. Pseudopterosin A inhibits phagocytosis and alters intracellular calcium turnover in a pertussis toxin sensitive site in *Tetrahymena thermophila*. *Comp Biochem Physiol, Part C: Toxicol Pharmacol* 143C (4): 436–443.

Müller, W. E. G. et al. 1985. Avarol, a cytostatically active compound from the marine sponge *Dysidea avara*. *Comp Biochem Phys Part C: Comp Pharmacol* 80 (1): 47–52.

Newman, D. J. and G. M. Cragg. 2004. Marine natural products and related compounds in clinical and advanced preclinical trials. *J Nat Prod* 67 (8): 1216–1238.

Nguyen, T. H. et al. 2011. Two unsaturated fatty acids with potent α-glucosidase inhibitory activity purified from the body wall of sea cucumber (*Stichopus japonicus*). *J Food Sci* 76 (9): H208–H214.

Nguyen, T. N. and J. J. Tepe. 2009. Preparation of hymenialdisine, analogues and their evaluation as kinase inhibitors. *Curr Med Chem* 16 (24): 3122–3143.

Oda, T. et al. 2005. Verrucarin A inhibition of MAP kinase activation in a PMA-stimulated promyelocytic leukemia cell line. *Mar Drugs* 3 (2): 64–73.

Oh, D.-C. et al. 2007. Thalassospiramides A and B, immunosuppressive peptides from the marine bacterium Thalassospira sp. *Org. Lett.* 9 (8): 1525–1528.

Oku, N. et al. 2004. Neamphamide A, a new HIV-inhibitory depsipeptide from the papua new guinea marine sponge *Neamphius huxleyi*. *J Nat Prod* 67 (8): 1407–1411.

Omura, S. et al. 2001. An anthelmintic compound, nafuredin, shows selective inhibition of complex I in helminth mitochondria. *Proc Natl Acad Sci USA* 98 (1): 60–62.

Ospina, C. A. et al. 2007. Bipinnapterolide B, a bioactive oxapolycyclic diterpene from the Colombian gorgonian coral *Pseudopterogorgia bipinnata*. *Tetrahedron Lett* 48 (42): 7520–7523.

Osterhage, C. et al. 2000. Ascosalipyrrolidinone A, an antimicrobial alkaloid, from the obligate marine fungus *Ascochyta salicorniae*. *J Org Chem* 65 (20): 6412–6417.

Ovenden, S. P. B. et al. 1999. Amphilactams A–D: Novel nematocides from Southern Australian marine sponges of the genus *Amphimedon*. *J Org Chem* 64 (4): 1140–1144.

Park, B.-G. et al. 2013. Anti-platelet aggregation and anti-thrombotic effects of marine natural products sargahydroquinoic acid and sargaquinoic acid. *Bull Korean Chem Soc* 34 (10): 3121–3124.

Periyasamy, N. et al. 2013. Isolation and characterization of anticoagulant compound from marine mollusc *Donax faba* (Gmelin, 1791) from Thazhanguda, Southeast coast of India. *Afr J Biotechnol* 12 (40): 5968–5974.

Pettit, G. R. et al. 1982. Isolation and structure of bryostatin 1. *J Am Chem Soc* 104 (24): 6846–6848.

Pettit, G. R. et al. 2004. Antineoplastic agents. 380. Isolation and X-ray crystal structure determination of isoaaptamine from the Republic of Singapore *Hymeniacidon sp.* and conversion to the phosphate prodrug hystatin 1. *J Nat Prod* 67 (3): 506–509.

Pfaller, M. A. et al. 2013. Activity of MK-3118, a new oral glucan synthase inhibitor, tested against *Candida spp.* by two international methods (CLSI and EUCAST). *J Antimicrob Chemother* 68 (4): 858–863.

Proteau, P. J. et al. 1993. The structure of scytonemin, an ultraviolet sunscreen pigment from the sheaths of cyanobacteria. *Experientia* 49 (9): 825–829.

Pruksakorn, P. et al. 2010. Trichoderins, novel aminolipopeptides from a marine sponge-derived *Trichoderma sp.*, are active against dormant mycobacteria. *Bioorg Med Chem Lett* 20 (12): 3658–3663.

Quang, T. H. et al. 2014. Tanzawaic acid derivatives from a marine isolate of *Penicillium sp.* (SF-6013) with anti-inflammatory and PTP1B inhibitory activities. *Bioorg Med Chem Lett* 24 (24): 5787–5791.

Randazzo, A. et al. 2001. Halipeptins A and B: Two novel potent anti-inflammatory cyclic depsipeptides from the Vanuatu marine sponge *Haliclona species*. *J Am Chem Soc* 123 (44): 10870–10876.

Rao, K. V. et al. 2006. Manzamine B and E and ircinal a related alkaloids from an Indonesian *Acanthostrongylophora* sponge and their activity against infectious, tropical parasitic, and Alzheimer's diseases. *J Nat Prod* 69 (7): 1034–1040.

Rao, M. R. et al. 1999. Two new 19-oxygenated polyhydroxy steroids from the soft coral *Nephthea chabroli*. *J Nat Prod* 62 (11): 1584–1585.

Rashid, M. A. et al. 2001. Microspinosamide, a new HIV-inhibitory cyclic depsipeptide from the marine sponge *Sidonops microspinosa*. *J Nat Prod* 64 (1): 117–121.

Ratnayake, A. S., et al. 2001. The structure of microcarpalide, a microfilament disrupting agent from an endophytic fungus. *Org Lett* 3 (22): 3479–3481.

Ravi, M. et al. 2005. Structure-based identification of the binding site for the hemiasterlin analogue HTI-286 on tubulin. *Biochemistry* 44 (48): 15871–15879.

Renner, M. K. et al. 1999. Cyclomarins A-C, new anti-inflammatory cyclic peptides produced by a marine bacterium (*Streptomyces sp.*). *J Am Chem Soc* 121 (49): 11273–11276.

Richter, R. K. et al. 2004. Differential modulation of the antifungal activity of amphotericin B by natural and ent-cholesterol. *Bioorg Med Chem Lett* 14 (1): 115–118.

Rifai, S. et al. 2004. Antimicrobial activity of untenospongin B, a metabolite from the marine sponge *Hippospongia communis* collected from the Atlantic Coast of Morocco. *Mar Drugs* 2 (3): 147–153.

Rinehart, K. L., Jr. et al. 1981. Didemnins: Antiviral and antitumor depsipeptides from a Caribbean tunicate. *Science* 212 (4497): 933–935.

Rizvi, S. A. et al. 2006. Structure of bistramide A-actin complex at a 1.35 angstroms resolution. *J Am Chem Soc* 128 (12): 3882–3883.

Rodriguez, A. D. and C. Ramirez. 2001. Serrulatane diterpenes with antimycobacterial activity isolated from the West Indian sea whip *Pseudopterogorgia elisabethae*. *J Nat Prod* 64 (1): 100–102.

Rodriguez, I. I. and A. D. Rodriguez. 2003. Homopseudopteroxazole, a new antimycobacterial diterpene alkaloid from *Pseudopterogorgia elisabethae*. *J Nat Prod* 66 (6): 855–857.

Rodríguez, I. I. et al. 2006. Ileabethoxazole: A novel benzoxazole alkaloid with antimycobacterial activity. *Tetrahedron Lett* 47 (19): 3229–3232.

Rowley, D. C. et al. 2002. Thalassiolins A-C: New marine-derived inhibitors of HIV cDNA integrase. *Bioorg Med Chem* 10 (11): 3619–3625.

Rowley, D. C. et al. 2003. Halovirs A-E, new antiviral agents from a marine-derived fungus of the genus *Scytalidium*. *Bioorg Med Chem* 11 (19): 4263–4274.

Rudi, A. et al. 2001. Clathsterol, a novel anti-HIV-1 RT sulfated sterol from the sponge *Clathria species*. *J Nat Prod* 64 (11): 1451–1453.

Sakai, R. et al. 1986. Manzamine A, a novel antitumor alkaloid from a sponge. *J Am Chem Soc* 108 (20): 6404–6405.

Sashidhara, K. V. et al. 2009. A selective account of effective paradigms and significant outcomes in the discovery of inspirational marine natural products. *J Nat Prod* 72 (3): 588–603.

Schmitz, F. J. et al. 1991. Cytotoxic aromatic alkaloids from the ascidian Amphicarpa meridiana and *Leptoclinides sp.*: Meridine and 11-hydroxyascididemin. *J Org Chem* 56 (2): 804–808.

Schwartsmann, G. et al. 2003. Marine-derived anticancer agents in clinical trials. *Expert Opin Invest Drugs* 12 (8): 1367–1383.

Senderowicz, A. M. et al. 1995. Jasplakinolide's inhibition of the growth of prostate carcinoma cells in vitro with disruption of the actin cytoskeleton. *J Natl Cancer Inst* 87 (1): 46–51.

Severin, I. et al. 2003. Toxic interaction between hydroxyurea and 1-beta-D-arabino-furanosylcytosine on the DNA of a human hepatoma cell line (HEPG2). *Toxicol Lett* 145 (3): 303–311.

Shao, Z. Y. et al. 2006. O-methyl nakafuran-8 lactone, a new sesquiterpenoid from a Hainan marine sponge *Dysidea sp*. *J Asian Nat Prod Res* 8 (3): 223–227.

Shenkar, N. and B. J. Swalla. 2011. Global diversity of *Ascidiacea*. *PLoS One* 6 (6): e20657.

Singh, I. P. et al. 1999. Tanikolide, a toxic and antifungal lactone from the marine cyanobacterium *Lyngbya majuscula*. *J Nat Prod* 62 (9): 1333–1335.

Singh, R. et al. 2013. Anticoagulant potential of marine polychaete (Nereis species). *J Biol Sci Opin* 1 (4): 337–340.

Statsuk, A. V. et al. 2005. Actin is the primary cellular receptor of bistramide A. *Nat Chem Biol* 1 (7): 383–388.

Stevenson, C. S. et al. 2002. Scytonemin—a marine natural product inhibitor of kinases key in hyperproliferative inflammatory diseases. *Inflamm Res* 51 (2): 112–114.

Strangman, W. K. et al. 2009. Potent inhibitors of pro-inflammatory cytokine production produced by a marine-derived bacterium. *J Med Chem* 52 (8): 2317–2327.

Supko, J. G. et al. 2000. A phase I clinical and pharmacokinetic study of the dolastatin analogue cemadotin administered as a 5-day continuous intravenous infusion. *Cancer Chemother Pharmacol* 46 (4): 319–328.

Suria, H. et al. 1999. Cytoskeletal disruption induces T cell apoptosis by a caspase-3 mediated mechanism. *Life Sci* 65: 2697–2707.

Talpir, R. et al. 1994. Hemiasterlin and geodiamolide TA: Two new cytotoxic peptides from the marine sponge *Hemiasterella minor* (Kirkpatrick). *Tetrahedron Lett* 35 (25): 4453–4456.

ter, H. E. et al. 1996. Discodermolide, a cytotoxic marine agent that stabilizes microtubules more potently than taxol. *Biochemistry* 35 (1): 243–250.

Terry, D. R. et al. 1997. Misakinolide A is a marine macrolide that caps but does not sever filamentous actin. *J Biol Chem* 272 (12): 7841–7845.

Toledo, T. R. et al. 2014. Potent anti-inflammatory activity of pyrenocine A isolated from the marine-derived fungus *Penicillium paxilli* Ma (G)K. *Mediat Inflamm*, pp. 1–11.

Ueoka, R. et al. 2009. Gracilioethers A-C, Antimalarial metabolites from the marine sponge *Agelas gracilis. J Org Chem* 74 (11): 4203–4207.

Ui, H. et al. 2001. Nafuredin, a novel inhibitor of NADH-fumarate reductase, produced by *Aspergillus niger* FT-0554. *J Antibiot (Tokyo)* 54 (3): 234–238.

Valoti, G. et al. 1998. Ecteinascidin-743, a new marine natural product with potent antitumor activity on human ovarian carcinoma xenografts. *Clin Cancer Res* 4 (8): 1977–1983.

Van Soest, R. W. M. et al. 2012. Global diversity of sponges (Porifera). *PLoS One* 7 (4): e35105.

Vashist, Y. K. et al. 2006. Inhibition of hepatic tumor cell proliferation in vitro and tumor growth in vivo by taltobulin, a synthetic analogue of the tripeptide hemiasterlin. *World J Gastroenterol* 12 (42): 6771–6778.

Vuong, D. et al. 2001. Onnamide F: A new nematocide from a Southern Australian marine sponge, *Trachycladus laevispirulifer. J Nat Prod* 64 (5): 640–642.

Wada, S.-I. et al. 1998. Actin-binding specificity of marine macrolide toxins, mycalolide B and kabiramide D. *J Biochem* 123 (5): 946–952.

Wagner, M. M. et al. 1999. In vitro pharmacology of cryptophycin 52 (LY355703) in human tumor cell lines. *Cancer Chemother Pharmacol* 43 (2): 115–125.

Wang, F. et al. 2016. Molecular basis of toxicity of N-type calcium channel inhibitor MVIIA. *Neuropharmacology* 101: 137–145.

Wang, H. et al. 2011. Anti-influenza virus polyketides from the acid-tolerant fungus *Penicillium purpurogenum* JS03-21. *J Nat Prod* 74 (9): 2014–2018.

Wang, W.-S. et al. 1997. High-dose cytarabine and mitoxantrone as salvage therapy for refractory non-Hodgkin's lymphoma. *Jpn J Clin Oncol* 27 (3): 154–157.

Wang, Y. et al. 1995. Cryptococcus neoformans melanin and virulence: Mechanism of action. *Infect Immun* 63 (8): 3131–3136.

Watanabe, K. et al. 2000. A new bioactive triene aldehyde from the marine sponge *Leucetta microraphis. J Nat Prod* 63 (2): 258–260.

Wei, X. et al. 2007. Novel ring B abeo-sterols as growth inhibitors of *Mycobacterium tuberculosis* isolated from a Caribbean Sea sponge, *Svenzea zeai. Tetrahedron Lett* 48 (50): 8851–8854.

Wei, X. et al. 2010. Neopetrosiamine A, biologically active bis-piperidine alkaloid from the Caribbean Sea sponge *Neopetrosia proxima. Bioorg Med Chem Lett* 20 (19): 5905–5908.

Weintraub, H. S. 2014. Overview of prescription omega-3 fatty acid products for hypertriglyceridemia. *Postgrad Med* 126 (7): 7–18.

Whitley, R. et al. 2012. Vidarabine: A preliminary review of its pharmacological properties and therapeutic use. *Drugs* 20 (4): 267–282.

Williams, P. et al. 2010. New methods to explore marine resources for Alzheimer's therapeutics. *Curr Alzheimer Res* 7 (3): 210–213.

Wipf, P. and W. Xu. 1996. Total synthesis of the antimitotic marine natural product (+)-curacin A. *J Org Chem* 61 (19): 6556–6562.

Wright, A. D. and N. Lang-Unnasch. 2005. Potential antimalarial lead structures from fungi of marine origin. *Planta Med* 71 (10): 964–966.

Wu, E. L. et al. 2009. Computational study for binding of oscillarin to human α-thrombin. *J Theor Comput Chem* 8 (4): 551–560.

Wu, Y.-J. et al. 2008. *Oxime-Containing Macrocyclic Acyl Guanidines as Beta-Secretase Inhibitors and Their Preparation*. New York: Bristol-Myers Squibb Company, p.20.

You, M. et al. 2015. Lumazine peptides from the marine-derived fungus *Aspergillus terreus*. *Mar Drugs* 13 (3): 1290–1303.

Younk, L. M. et al. 2011. Pharmacokinetics, efficacy and safety of aleglitazar for the treatment of type 2 diabetes with high cardiovascular risk. *Expert Opin Drug Metab Toxicol* 7 (6): 753–763.

Youssef, D. T. et al. 2014. Theonellamide G, a potent antifungal and cytotoxic bicyclic glycopeptide from the Red Sea marine sponge Theonella swinhoei. *Mar Drugs* 12 (4): 1911–1923.

Yu, M. J. et al. 2005. *Discovery of E7389, A Fully Synthetic Macrocyclic Ketone Analog of Halichondrin B*. Boca Raton, FL: CRC Press LLC.

Zhang, Y. et al. 2009. A sesquiterpene quinone, dysidine, from the sponge Dysidea villosa, activates the insulin pathway through inhibition of PTPases. *Acta Pharmacol Sin* 30 (3): 333–345.

Zheng, C. J. et al. 2012. Bioactive hydroanthraquinones and anthraquinone dimers from a soft coral-derived *Alternaria sp.* fungus. *J Nat Prod* 75 (2): 189–197.

11

Production of Life-Saving Drugs from Himalayan Herbs

Kriti Juneja, Shashank Sagar Saini, Mariam Gaid,
Ludger Beerhues, and Debabrata Sircar

CONTENTS

ABSTRACT Plant natural products constitute the most diversified source of unique chemical structures, providing the basis for identification of new lead molecules that serve as starting point for novel drug discovery. India has a rich heritage of using plant-derived or so-called herbal medicines since ancient times. Due to unique agro-climatic conditions and physio-geographical features, the Himalaya region harbors high-value medicinal plants, which serve as the source of many life-saving drugs. Despite extensive use in Indian traditional medicine, information on active constituents and scientific evidence for the disease-curing mechanisms of most of the Himalayan medicinal plants are still elusive. This chapter provides a brief introduction to plant-based medicine—present scenario and future prospects, information on the traditional Indian system of medicine, plant biodiversity of the Himalaya, important medicinal plants of this region and their medicinal uses, conservation procedure, marketing strategies, and various approaches to sustainable production of Himalayan medicinal plants.

11.1 Herbal Drug: Introduction

Plant natural products have a long history of use in the treatment of a wide variety of human ailments. Plant metabolites constitute the most diversified source of unique chemical structures, providing the basis for identification of new lead molecules that serve as starting point for novel drug discovery (Cragg and Newman, 2005; Koehn and Carter, 2005).

Several currently available and clinically approved drugs have been derived directly or indirectly from plant sources. For example, a wide number of commercially available anticancer drugs are derived from plant secondary metabolites (Shyur and Yang, 2008) They are either unmodified plant constituents or semisynthetic derivatives (DeVita et al. 2008; Mangal et al. 2013).

Initially, extracts from medicinal plants were only used for the treatment of diseases on an empirical basis, without knowing the bioactive principles and mechanisms of action of their pharmacologic activities. It was in the eighteenth century, when for the first time William Withering studied foxglove (*Digitalis purpurea*) for the treatment of edema, thereby setting the basis for the rational clinical investigation of medicinal plants (Sneader, 2005). In the nineteenth century, Friedrich Sertürner succeeded for the first time in isolating a plant alkaloid, which was probably the beginning of rational drug discovery from plants (Atanasov et al. 2015). Afterward, the German scientist H. E. Merck started large-scale isolation and marketing of morphine from plant sources (Kaiser, 2008). In subsequent years, many other bioactive natural products, such as quinine, colchicine, taxol, atropine, zingerol, resveratrol, artemisinin, cannabinol, and curcumin were isolated from plant sources. In recent years, total chemical synthesis of plant natural products and high-efficacy derivatives came into the picture in order to produce large quantities at less price (Atanasov et al. 2015).

However, in spite of the enormous ethno-medicinal potential, plant-based medicine has not gained enough popularity among the global scientific community (Harvey et al. 2015). One of the major reasons for this ignorance might be the absence of scientific validation of herbal medicines in terms of bioactive constituents and mode of action against molecular disease targets. Herbal extracts are usually complexes of various organic and inorganic components and some out of these ingredients may be toxic to various tissues in the body. As a result, these herbal medicines are not well defined and characterized, particularly with reference to the quality, safety, and mode of action. Naturally, it is of utmost importance at this point to screen potential medicinal plants and to perform proper elucidation of their phytochemical constituents and mode of action against molecular targets.

According to a recent estimation, only 36% of 1073 small molecules that had been clinically approved until 2010 were chemically synthesized, whereas the remaining portion was derived from natural sources including plants (Newman and Cragg, 2012; Atanasov et al. 2015). Some striking examples of plant-derived and clinically approved natural products are anticancer drugs, such as vincristine and vinblastine from *Catharanthus roseus*, paclitaxel from *Taxus brevifolia*, and camptothecin and its derivatives topotecan and irinotecan prepared from *Camptotheca acuminata* (Cragg and Newman, 2013). The drug galantamine used for the treatment of Alzheimer's disease was discovered in *Galanthus nivalis*. Furthermore, the famous antimalarial drug artemisinin was originally isolated from the Chinese medicinal herb *Artemisia annua*. Some commonly used plant-derived and clinically approved drugs are listed in Table 11.1.

TABLE 11.1
Clinically Approved Plant Metabolites Used in the Treatment of Various Ailments

Generic Name	Trade Name	Structure	Plant Source	Medicinal Use	Literature Reference
Paclitaxel	Taxol, Abraxane		*Taxus brevifolia*	Chemotherapy agent for treatment of cancer	Wani et al. 1971; Atanasov et al. 2015
Topotecan	Hycamtin		*Camptotheca acuminata* (semisynthetic derivative of natural camptothecin)	Chemotherapy agent for treatment of cancer	Garcia-Carbonero and Supko, 2002
Solamargine	Curaderm		*Solanum* species	Chemotherapy agent for treatment of cancer	Atanasov et al. 2015

(Continued)

TABLE 11.1 (Continued)

Clinically Approved Plant Metabolites Used in the Treatment of Various Ailments

Generic Name	Trade Name	Structure	Plant Source	Medicinal Use	Literature Reference
Vincristine sulfate	Oncovin		*Catharanthus roseus*	Treatment of leukemia	Lucas et al. 2010
Vinblastine sulfate	Velban		*Catharanthus roseus*	Treatment of lymphomas	Lucas et al. 2010
Etoposide	VePesid		*Podophyllum peltatum* (semisynthetic derivative of podophyllotoxin)	Treatment of lung cancer	Hande, 1998; Lucas et al. 2010
Teniposide	Vumon		*Podophyllum peltatum* (semisynthetic derivative of podophyllotoxin)	Treatment of cancer	Lucas et al. 2010

(Continued)

TABLE 11.1 (*Continued*)

Clinically Approved Plant Metabolites Used in the Treatment of Various Ailments

Generic Name	Trade Name	Structure	Plant Source	Medicinal Use	Literature Reference
Artemisinin	Artemisin		*Artemisia annua*	Treatment of malaria	Klayman et al. 1984; Atanasov et al. 2015
Capsaicin	Qutenza		*Capsicum annuum*	Treatment for postherpetic neuralgia	Jones et al. 2011
Galantamine	Razadyne		*Galanthus caucasicus*	Treatment of Alzheimer's disease	Scott and Goa, 2000; Atanasov et al. 2015
Colchicine	Colcrys		*Colchicum* species	Treatment of gout	Atanasov et al. 2015

11.2 Indian System of Traditional Medicine

India has a rich heritage of using plant-derived or so-called herbal medicines since ancient times. Ayurveda, Siddha, and Unani are the three main types of traditional medicine practiced in India for ages. Ayurveda remains the primary source of healthcare in present-day India and still more than 70% of the Indian population relies on Ayurvedic medicine for their primary healthcare (Vaidya and Devasagayam, 2007). The concept of Ayurvedic therapy is native to India, developed between 2500 and 500 BC (Pandey et al. 2013). Ayurvedic medicine mainly involves extracts from various aromatic and medicinal plants and believes in the concept of holistic treatment. Ayurvedic therapy mostly includes extracts from more than one herb/plant, which would probably reflect the modern concept of multitarget therapy. Ayurveda believes that a crude herbal formulation has a mixture of bioactive compounds, with few bioactive compounds being necessary for a particular pharmacological activity, whereas other compounds may either reduce the toxicity of associated constituents or increase the bioavailability of the bioactive compounds. For example, a mixture of turmeric and black pepper extracts always has better efficacy than the individual extracts. Now it is known that piperine, present in the black pepper extract, enhances the bioavailability of curcumin, the active principle of turmeric (Aggarwal et al. 2011).

India has a rich biodiversity of medicinal and aromatic plants, which are used in traditional medical treatments. Currently, there is a growing interest in herbal medicines mainly because of less or no side effects and affordable costs. Further, government agencies and different nongovernment organizations (NGOs) are giving acute attention to build up awareness among people to use health-friendly herbal medicines.

Presently, the number of registered medical practitioners of the Ayurvedic medicine in India is around 250,000, whereas there are more than 700,000 practitioners of the allopathic medicine. Indian traditional medicine including Ayurveda, Siddha, Unani, folk, and homeopathy is known to use more than 8000 herbs/plants for preparing various disease-curing formulations. Among them, Ayurveda alone utilizes more than 2000 plant extracts in preparing different Ayurvedic formulations. Currently, around 25,000 plant-based formulations and extracts are used in Indian traditional medicine. There are more than 7800 manufacturing units for preparing plant-based healthcare products and Ayurvedic formulations (Pandey et al. 2013). The annual requirement for medicinal herb raw material is around 2000 tons and a significantly high number of herbal products are marketed as dietary supplements (Patwardhan et al. 2005; Pandey et al. 2008).

In recent years, due to growing popularity of Ayurveda and other Indian traditional medication systems, there is a striking rise in the number of herbal-based drug industries, which have brought the magic remedy of Indian medicinal wealth into global focus. Recent statistics indicate that 80% of the global population cannot afford the drugs from Western pharmaceutical industries and have to rely upon the use of traditional medicines (Pant and Samant, 2010). According to the World Health Organization (WHO), out of 252 basic and essential drugs, 11% are exclusively derived from plants (Hasan et al. 2011).

Several research groups have embarked on research projects in the form of field surveys, wherein indigenous and traditional knowledge on ethno-botany has been documented in detail (Pant and Samant, 2010; Tiwari et al. 2010; Kala, 2015; Malik et al. 2015; Rathore et al. 2015; Sherpa et al. 2015). Such reports and documentations help the scientific and local communities to preserve the traditional ethno-medicinal knowledge as well as to provide raw material for advancement of biomedical research and development (Malik et al. 2015).

11.3 Himalayan Medicinal Plants

India is one of the 12 mega-biodiversity nations in the world, and it has two of the 18 plant biodiversity hotspots in the world (the Eastern Himalayas and the Western Ghats). India has an estimated number of 45,000 plant species, out of which 15,000 belong to medicinal plant species (Siwach et al. 2013). Among medicinal plants, approximately 7000 are used in Ayurveda, 700 in the Unani system of medicine, 600 in Siddha, 450 in homeopathy medicine, and around 30 plant species are routinely used in modern medicines.

Due to unique agro-climatic conditions and physio-geographical characteristics, the Himalaya region harbors high-value medicinal plants, which are the source of many life-saving drugs. The Himalayan medicinal plant diversity varies from the rain forests in the North-East to dry deciduous forests and alpine meadows in the North-West. Interestingly, about 30% of the endemic plant species found in India belong to the Himalaya region (Kaul, 2010), although the geographical area of the country that is occupied by the Himalaya is only 15% (Alam, 2004). The earliest record of medicinal use of Himalayan plants is found in the Rigveda, written between 4500 BC and 1600 BC (Rathore et al. 2015). It mentions 67 plants of therapeutic use (Rahul et al. 2010). The Himalayan Mountains extend over 3500 km and cover India, China, Nepal, Myanmar, Bhutan, Bangladesh, Pakistan, and Afghanistan. The Indian Himalayan region is home to over 18,440 species of plants including 8000 species of Angiosperms, 44 species of Gymnosperms, 600 species of Pteridophytes, 1736 species of Bryophytes, and 1159 species of Lichens (Singh et al. 2009). Out of 8000 species of Himalayan angiosperms, around 1750 species are used for medicinal purposes (Kumar et al. 2011; Malik et al. 2015).

A significantly large number of tribal and rural populations of the Himalayan region depend on locally available medicinal plants to meet their routine healthcare needs. Moreover, the collection and selling of these economically important plants provide an important source of income generation for local people inhabiting the Himalayan region (Alam, 2004; Pant and Samant, 2010; Kala, 2015). As of today, Himalayan plants are a major contributor to the herbal pharmaceutical industry, both in India and in other countries (Umadevi et al. 2013). A compilation of the Himalayan herbs used traditionally to treat various diseases is depicted in Table 11.2.

Unfortunately, the accelerating demand for herbal products, herbal cosmetics, and herbal medicine has resulted in very high market needs for raw parts of many medicinally important plants, causing tremendous pressure on their natural habitat. Further, overexploitation from natural sources, absence of novel technologies for sustainable harvesting, lack of serious efforts for commercial scale cultivation and micropropagation of important high-demand medicinal plants, cutting of medicinal trees for fuel, timber, and so on, changes in the climate and weather pattern, as well as lack of proper conservation measures, and so on, have badly affected the natural populations of many therapeutically important medicinal plants of the Himalayan region. As a result, the populations of certain medicinal plants are becoming endangered and rare and others are facing genetic erosion (Appasamy, 1993; Pandey et al. 2006). The Red Data Book of IUCN lists 215 threatened plant taxa native to India, out of which 121 species belong to the Himalayan region (Pandey et al. 2006).

Some therapeutically important and endangered high trade value (trade value >100 MT per year) medicinal plants of the Himalayan region are listed in Table 11.3.

TABLE 11.2

Medicinal Himalayan Herbs Traditionally Used against Life-Threatening Diseases

Serial No.	Ailment	Herbs Used	Family	Part Used
1	Diarrhoea (infants)	*Launaea asplenifolia*	Asteraceae	Root juice
		Paris polyphylla	Liliaceae	Root powder
		Melilotus indica	Fabaceaes	Whole plant
		Codonopsis viridis	Campanulaceae	Leaf
2	Asthma	*Euphorbia hirta*	Euphorbiaceae	Whole plant
		Boerhavia diffusa	Nyctaginaceae	Whole plant infusion
		Cassia fistula	Fabaceae	Fruit pulp
		Terminalia chebula	Combretaceae	Seeds
		Phyllanthus emblica	Euphorbiaceae	Fruit
		Selinium tenuifolium	Apiaceae	Root
		Fritillaria roylei	Liliaceae	Bulb
		Gerardinia diversifolia	Urticaceae	Root, leaf
		Viola canescens	Violaceae	Whole plant
		Stephania glabra	Menispermaceae	Root
		Artemisia vulgaris	Asteraceae	Leaf decoction
		Hedychium spicatum	Zingiberaceae	Rhizomes
		Zingiber officinale	Zingiberaceae	Rhizomes
		Ephedra gerardiana	Gnetaceae	Shoot
		Taxus baccata	Taxaceae	Leaf, bark
		Ranunculus arvensis	Ranunculaceae	Whole plant
		Saccharum spontaneum	Poaceae	Leaf
		Anisomeles indica	Lamiaceae	Leaf
3	Bronchitis	*Barleria cristata*	Acanthaceae	Root decoction
		Euphorbia hirta	Euphorbiaceae	Whole plant
		Origanum vulgare	Lamiaceae	Whole plant extract
		Boerhavia diffusa	Nyctaginaceae	Whole plant infusion
		Angelica glauca	Apiaceae	Root
		Acorus calamus	Acoraceae	Root
		Adiantum venustum	Pteridaceae	Leaf
		Bidens pilosa	Asteraceae	Whole plant
		Raphanus sativus	Brassicaceae	Whole plant
		Viola canescens	Violaceae	Whole plant
		Hedychium spicatum	Zingiberaceae	Rhizomes
4	Pneumonia	*Barleria cristata*	Acanthaceae	Root decoction
		Achyranthes aspera	Achyranthaceae	Whole plant
		Drymaria cordata	Caryophyllaceae	Whole plant
		Chenopodium ambrosioides	Chenopodiaceae	Leaf
		Trichosanthes tricuspidata	Cucurbitaceae	Roots, leaf, seed
		Terminalia chebula	Combretaceae	Fruit, bark
		Callicarpa arborea	Verbenaceae	Leaf, stem, bark
5	Snake bite	*Cyathula tomentosa*	Amaranthaceae	Leaf
		Arisaema jacquemontii	Araceae	Tuber
		Delphinium denudatum	Ranunculaceae	Root
		Parnassia nubicola	Saxifragaceae	Root
		Acorus calamus	Acoraceae	Root
		Paris polyphylla	Liliaceae	Root decoction
		Ageratum conyzoides	Asteraceae	Leaf
		Trichosanthes tricuspidata	Cucurbitaceae	Root, leaf, seed
		Euphorbia hirta	Euphorbiaceae	Whole plant
		Oxalis corniculata	Oxalidaceae	Whole plant
		Rubia manjith	Rubiaceae	Stem, root, leaf

(Continued)

TABLE 11.2 (*Continued*)

Medicinal Himalayan Herbs Traditionally Used against Life-Threatening Diseases

Serial No.	Ailment	Herbs Used	Family	Part Used
		Leucas cephalotes	Lamiaceae	Whole plant
		Piper longum	Piperaceae	Fruits, root
6	Malaria	*Achyranthes aspera*	Amaranthaceae	Root infusion, leaf
		Royale cinerea	Lamiaceae	Leaf decoction
		Viola canescens	Violaceae	Whole plant
		Picrorhiza kurroa	Scrophulariaceae	Rhizome powder
		Aconitum bisma	Ranunculaceae	Tuber, root
		Swertia angustifolia	Gentianaceae	Whole plant
		Swertia chirata	Gentianaceae	Plant extracts
		Satyrium nepalense	Orchidaceae	Root
		Berberis asiatica	Berberidaceae	Root, bark
		Berberis chitria	Berberidaceae	Whole plant, root, bark
		Justicia adhatoda	Acanthaceae	Root, flower, fruit, leaf, Wood
		Trachelospermum lucidum	Apocynaceae	Bark, leaf
		Bauhinia variegata	Caesalpiniaceae	Bark, flower, root, leaf
		Myrica esculenta	Myricaceae	Bark, fruit
7	Diabetes	*Syzygium cumini*	Myrtaceae	Bark, fruits
		Berberis asiatica	Berberidaceae	Root
		Coccinia grandis	Cucurbitaceae	Leaf, root, fruit
		Kickxia ramosissima	Scrophulariaceae	Whole plant
		Trigonella corniculata	Fabaceae	Young plants, fruit
		Stephania glabra	Menispermaceae	Root
		Aconitum bisma	Ranunculaceae	Tuber, root
		Melothria heterophylla	Cucurbitaceae	Root, leaf, fruit
		Osbeckia nepalensis	Melastomataceae	Root decoction
		Tupistra nutans	Liliaceae	Whole plants
		Panax pseudoginseng	Araliaceae	Root
		Ficus hookeriana	Moraceae	Fruits
		Terminalia chebula	Combretaceae	Fruit, bark
8	Tuberculosis	*Edychium spicatum*	Zingiberaceae	Root
		Stephania glabra	Menispermaceae	Root
9	Cancers and Tumors	*Stephania glabra*	Menispermaceae	Tuber, root
		Paris polyphylla	Liliaceae	Root, rhizome powder Leaf
		Ageratum conyzoides	Asteraceae	Rhizome oil
		Curcuma aromatic	Zingiberaceae	Leaf, bark extract
		Taxus baccata	Taxaceae	Bark, flower, root, leaf
		Bauhinia variegata	Caesalpiniaceae	Stem, flower, root
		Woodfordia fruticosa	Lythraceae	Bark, leaf, stem, seed
		Syzygium cumini	Myrtaceae	Leaf, root, stem
		Clematis buchananiana	Ranunculaceae	
10	Measles	*Artemisia vulgaris*	Asteraceae	Leaf decoction
		Cestrum fasciculatum	Solanaceae	Fruits
		Terminalia chebula	Combretaceae	Fruit, bark
11	Traumatic injury and Wounds	*Paris polyphylla*	Liliaceae	Roots, rhizome powder
		Rumex hastatus	Polygonaceae	Leaf
		Dactylorhiza hatagirea	Orchidaceae	Root
		Pteris biaurita	Pteridaceae	Stem
		Anaphalis contorta	Asteraceae	Flower
		Punica granatum	Punicaceae	Root, stem, fruit, leaf

(Continued)

TABLE 11.2 (*Continued*)

Medicinal Himalayan Herbs Traditionally Used against Life-Threatening Diseases

Serial No.	Ailment	Herbs Used	Family	Part Used
12	Cholera	*Rhus parviflora*	Anacardiaceae	Leaf infusion
		Spermadictyon suaveolens	Rubiaceae	Root
		Rhus javanica	Anacardiaceae	Bark, fruit
		Rhus parviflora	Anacardiaceae	Leaf
		Centella asiatica	Apiaceae	Whole plant
		Clerodendron serratum	Verbenaceae	Root, leaf, flower
		Saccharum spontaneum	Poaceae	Leaf
		Woodfordia fruticosa	Lythraceae	Stem, flower, root, inflorescence
		Ficus. religiosa	Moraceae	Leaf, fruit, bark
		Myrica esculenta	Myricaceae	Bark, fruit
		Zanthoxylum armatum	Rutaceae	Fruit, seed
13	Influenza	*Chenopodium ambrosioides*	Chenopodiaceae	Leaf
		Kalanchoe pennata	Crassulaceae	Leaf
		Origanum vulgare	Lamiaceae	Whole plant
		Heracleum wallichii	Apiaceae	Fruit, root
14	Typhoid	*Chenopodium ambrosioides*	Chenopodiaceae	Leaf
		Fagopyrum esculentum	Polygonaceae	Root, fruit, leaf
		Asplenium dalhousiae	Aspleniaceae	Frond
		Onychium contiguum	Cryptogrammaceae	Frond
		Aconitum bisma	Ranunculaceae	Tuber, root
15	Sepsis	*Cassia fistula*	Fabaceae	Fruit, bark
		Euphorbia royleana	Euphorbiaceae	Latex
		Melia azedarach	Meliaceae	Fruit, bark, leaf, shoot
		Nardostachyus jatamansi	Caprifoliaceae	Root
		Artemisia vulgaris	Asteraceae	Leaf decoction
		Ageratum conyzoides	Asteraceae	Leaf
		Toona ciliata	Meliaceae	Bark, fruit, leaf
		Urtica hyperborea	Urticaceae	Fruit, leaf

Source: Pant, S. and S.S. Samant, *Ethnobot Leaflets*, 14:193–217, 2010; Tiwari et al. *New York Sci. J.* 3(4):123–126, 2010; Kumar et al. *Indian J. Med. Plants Res.* 5(11):2252–2260, 2011; Shah et al. *J. Herb. Med. and Toxicol.* 6(1):27–33, 2012; Malik et al. *J. Ethnopharmacol.* 172:133–144, 2015; Kala, *Appl. Ecol. Env. Res.* 3(1):16–21, 2015; Rathore et al. *Global J. Res. Med. Plants & Indigen Med.* 4(4):65–78, 2015; Sherpa et al. *World J. Pharm. Pharm. Sci.* 4(2):161–184, 2015; Semwal and Semwal, *Nat. Pro. Res.* 29(5):396–410, 2015.

11.4 Trade of Himalayan Medicinal Plants

As of today, Himalayan medicinal plants are a major contributor to the herbal pharmaceutical industry both in India and other countries; however, the information on Himalayan medicinal plants, especially on the composition of bioactive constituents and the mode of action, still remains fragmentary. Due to special climatic conditions, Himalayan medicinal plants offer greater opportunities for discovering novel bioactive molecules (Dhawan, 1997). To date, the total export business of Indian medicinal plants is negligible despite having rich traditional knowledge and heritage. This is mainly because of the lack of standardization and quality control of herbal products. Approximately 25,000 pharmaceutical companies currently manufacture traditional medicines in India. The major traditional pharma sectors of India involved in producing herbal products are Himalaya, Zandu,

TABLE 11.3

List of High-Trade-Value and Endangered Medicinal Plants of the Himalayan Region

Plant Name	Medicinal Use
Aconitum heterophyllum	As anti-inflammatory, antipyretic, astringent
Aconitum ferox	Treatment of paralysis
Angelica glauca	Treatment of dyspepsia and constipation
Atropa acuminata	As diuretic
Cinnamomum tamala	As antispasmodic, antifungal, antibacterial agent
Nardostachys jatamansi	Rhizome used as stimulant tonic, antispasmodic
Picrorhiza kurroa	As antibacterial, agent, laxative
Rheum emodi	As antitumor, antiseptic agent
Rheum moorcroftiana	Treatment of diarrhea
Swertia chirata	As antipyretic, hypoglycemic, antimicrobial agent
Taxus wallichiana	As antifungal, antibacterial and antitumor agent
Zanthoxylum armatum	Treatment of tumors, headache, hepatitis, fever

Source: Siwach et al. *Int. J. Biodivers. Conserv.* 5(9):529–540, 2013.

Dabur, Patanjali, Baidyanath, Hamdard, among others. The approximate yearly turnover of the Indian herbal medicinal industry is about Rs. 2,300 crore, as compared to the pharmaceutical industry's annual turnover of Rs. 14,500 crore (Sharma et al. 2008).

The export of medicinal plants and herbs from India has been quite substantial in the last few years. The WHO has projected that the global herbal market will grow to US$ 5 trillion by 2050 from the current level of US$ 62 billion (Purohit and Vyas, 2004). The economic return potential for most of the Himalayan medicinal plants is quite high with end prices somewhere between Rs. 7150 and 55,000 per hectare (Nautiyal, 1995). The average yearly income is estimated Rs. 120,000 per hectare through mixed cultivation of high-altitude medicinal herbs (Rao and Saxena, 1994).

A list of important high-altitude Himalayan medicinal plants with established therapeutic claims as life-saving drugs is depicted in Table 11.4. The cultivation of medicinal plants plays an important role in the healthcare management of the mountain communities together with providing important income sources.

TABLE 11.4

Life-Saving Drugs from Himalayan Medicinal Plants

Plant Name	Bioactive Principles	Therapeutic Actions	Reference
Artemisia annua	Artemisinin	Treatment of malaria	Kaul, 2010
Atropa acuminata	Tropane alkaloid	Anticholinergic	Kaul, 2010
Dioscorea deltoidea	Diosgenin	Steroidal effect	Kaul, 2010
Hypericum perforatum	Hypercin, hyperforin	Antidepressant	Kaul, 2010
Podophyllum hexandrum	Podophyllotoxin, Etoposide	Treatment of cancer	Kaul, 2010
Swertia chirata	Secoiridoides	Antipyretic	Kaul, 2010
Taxus wallichiana	Paclitaxel	Treatment of cancer	Kaul, 2010
Valeriana officinalis	Sesquiterpene	Sedative	Kaul, 2010
Rhododendron anthopogon	Sesquiterpene, delta-cadinene	Antibacterial, anticancer	Innocenti et al. 2010
Picrorhiza kurroa	Picroside I and II	Liver disease, anti-inflamatory	Engels et al. 1992

11.5 Loss of Biodiversity of Himalayan Medicinal Plants

A large proportion of Himalayan medicinal plants are collected from the wild habitat. The continuously growing demand for medicinal plants by pharma industries and destructive harvesting is putting tremendous load on the existing natural resources in India. Almost every Himalayan medicinal plant, which is marketed to high extent, has come under extreme threat of depletion.

The price trends of most of the Himalayan medicinal plant species traded in the market, such as *Aconitum, Jatamansi, Picrorhiza*, have been continuously increasing, suggesting a huge demand and value for these species in world markets. The present scale of commercial cultivation and production of medicinal plants is strikingly low compared with the huge demand by the industries, resulting into mass-scale illegal harvesting from the wild sources to meet the market demands (Banerji and Basu, 2011). It is estimated that approximately 80%–90% of the raw material for pharma industries are available from wild populations of plants and only less than 20% of medicinal plants are commercially cultivated (Uniyal et al. 2002; Siwach et al. 2013).

More than 70% of the plant collections involve destructive harvesting mainly due to use of plant parts such as roots, bark, stem, and the whole plant in case of herbaceous plants. Since the majority of medicinal plant species of the Himalayan region are endemic to this region, they are more susceptible to extinction due to overexploitation. Multifold reasons (Alam, 2004; Banerji and Basu, 2011) for rapid depletion of Himalayan medicinal plants are as follows:

- Nonsustainable destructive harvesting by plant collectors to generate huge biomass and profit in short time.

- Gradual decline in the use of traditional knowledge and medicinal plants by local people, resulting in significantly reduced interest of these communities in conservation.

- Shift from local use of medicinal plants to commercial sale due to the growing popularity of herbal medicines, leading to the harvest of enormous amount of medicinal plant species.

- There is a huge gap in the economic return between medicinal plant growers and user industries. As the number of agents involved between the collection of plants and their final sale to the user industry/consumers is large, the prices paid to the collectors are relatively low and thus force them to overharvest the material to supplement their income.

- Ecological changes as a result of global warming and deforestation adversely affect the survival of several Himalayan medicinal plants. The impact of climate change is going to be more adverse for the highly sensitive subalpine and alpine species, such as *Saussurea* species.

- Lack of proper agro-technologies for propagation of plants as well as long gestation periods of medicinal plants compel farmers not to cultivate medicinal plants.

- Increasing human population in the Himalayan region has led to large-scale conversion of forests, wetlands, and grasslands for agriculture and settlements. Road construction has contributed to both fragmentation of habitats and facilitation of the spread of invasive species, resulting in great loss of genetic diversity.

Further, there is a lack in proper R&D activities and policy making to promote more and more cultivation of medicinal plants. The high-altitude belt of the Himalayan region seriously suffers from lack of proper information on the status of biodiversity, development, and its impact. There are only few high-altitude measurement centers to support scientific research on high-altitude medicinal plants. Furthermore, the high-altitude regions with the rarest species of medicinal plants gain almost no attention. Thus, there is an urgent need for conservation strategies to protect the medicinal wealth of the Himalayas. Commercial cultivation of high-demand species would ensure protection of the wild resources. However, most of these species suffer from lack of proven and successful propagation technologies followed by efficient harvesting techniques. Further, the herbal product should be scientifically validated in terms of bioactive constituents and mode of action. Hence, there is a vital need to promote research and development on Himalayan medicinal plants to foster their rapid propagation.

11.6 Conservation Strategies for Himalayan Medicinal Plants

Although the importance of the Himalayan herbal wealth and ecosystem and its protection has been realized by the Government of India, as well as by the world conservation community, the management of forests and medicinal plants has heavily suffered due to overexploitation and habitat degradation as well as deficiencies in proper conservation action. It is clear that the conservation of medicinal plants is vitally important for the maintenance of biodiversity and the preservation of indigenous knowledge.

The governments in the Himalayan states are seriously concerned about the accelerated depletion of the valuable medicinal plant species. Governments have now introduced polices both to conserve plant resources *in situ* and to promote cultivation by farmers. To protect unsustainable and illegal harvesting, the institutions and mechanisms involved in the collection of medicinal plants from the wild are being reformulated (Alam, 2004). Most of the Himalayan states have also established State Medicinal Plant Boards, which are responsible for promoting the development of the medicinal plant wealth.

In India, a number of ministries, government bodies, and NGOs are working in this direction to conserve Himalayan wealth. Some examples are the Ministry of AYUSH, National Medicinal Plants Board (NMPB), Ministry of Environment and Forests (MoEF), Jawahar Lal Nehru Tropical Botanic Garden and Research Institute (JNTBGRI), Indian Council of Forest Research and Education (ICFRE), Indian Council of Agricultural Research (IARI), Institute of Himalayan Bioresource Technology (IHBT), and Indian Institute of Integrative Medicine (IIIM).

The government is also making campaigns to involve the private sector and NGOs in the activities of promoting large-scale propagation of medicinal plants. Recently, banking sectors are also coming forward to provide financial aid to farmers for conservation programs. The "National Bank for Agriculture and Rural Development (NABARD)" supports the development of the medicinal plant sector (Siwach et al. 2013).

The government has now introduced a number of control measures to promote medicinal plant production.

- Providing subsidies and financial support for the purchase of planting material, land improvement, and cultivation of medicinal plants.
- Promotional activities to raise awareness among farmers to grow medicinal plants as cash crop.

- Proper training of farmers for micropropagation, harvesting and storage of medicinal plants.
- Establishing nurseries and parks to provide planting material to farmers.

Two broad types of conservation strategies—"*in situ* and *ex situ* conservations" are generally followed for conservation of medicinal plants. In *in situ* conservation, the plants are conserved in their natural habitat either as wild plant communities or as cultivated varieties. As *in situ* conservation measure, different Himalayan states have established biosphere zones, national parks, and wildlife sanctuaries for the conservation of wild flora and fauna of the respective area (Siwach et al. 2013).

One of the major problems associated with medicinal plant cultivation are the long gestation period and extreme dependency on natural pollinators, unpredictable seed germination patterns, and sometimes poor seed viability. Unfortunately, modern technology for large-scale cultivation and rapid propagation is available for only few of rare and endangered medicinal plants. Further, there is a lack of proper guidelines to make efficient harvesting from medicinal plants. In this direction, the "National Medicinal Plants Board" and other government and nongovernment research institutions are trying to develop cultivation manuals and monographs on agro-technologies for propagation of medicinal plants. Less economic returns and a complex procedure of getting permission for medicinal plant cultivation are another two reasons that demotivate farmers to cultivate medicinal plants.

Presently, at the national level, detailed statistics on the species-wise cultivation of medicinal plants, their medicinally useful raw material, bioactive constituents, and mode of action remain elusive. NMPB has prioritized 14 medicinal plants of the Himalayan region for *in situ* cultivation (Kala and Sajwan, 2007). In the present scenario, there is urgent need to make intensive efforts for training and guidance of farmers to promote medicinal plant cultivation.

In *ex situ* conservation, plants are protected outside the native habitat, such as seed storage, DNA storage, field gene banks, and botanical gardens. The gene banks [Jawaharlal Nehru Tropical Botanic Garden and Research Institute (JNTBGRI), National Bureau of Plant Genetic Resources (NBPGR), Central Institute of Medicinal and Aromatic Plants (CIMAP)] set up by the DBT and GOI are conserving almost thousand accessions of important plant species and germplasm. The "Indian Institute of Integrative Medicine (IIIM)" is involved in the conservation of threatened plants of the north-western Himalayas. Plants hard to seed or propagate by vegetative means are conserved in field gene banks.

11.7 Sustainable Cultivation of Himalayan Medicinal Plants

The long-term availability of medicinal plants is of critical importance, not only for the doctors but also for the preservation of indigenous traditional knowledge. For the conservation of rare medicinal plants, cultivation is often considered an alternative to wild collection, but the cultivation is not easy and impossible for some of the medicinal plants. However, if harvesting and trade are sustainable and controlled, then even natural collection may be beneficial for both local economy and habitat conservation. It is always advisable to promote the sustainable use of natural resources rather than to prohibit wild harvesting. A definite guideline should be framed for sustainable cultivation of medicinal plants in the Himalayan region.

i. *Species prioritization:* Medicinal plants of the Himalayan region are agro-climatically adapted and hence less susceptible to diseases and need little maintenance. Mostly, the high-value Himalayan species are endemic to the region, giving the mountain farmer a definite competitive advantage with these species. Prioritization of the medicinal plants becomes important in this context, and it should include analysis of the specific supply and demand side issues affecting a particular species. Other points to be considered are concerns of users (e.g., farmers, intermediate agents, and pharmaceutical industry), resource availability (i.e., quantity and quality), accessibility, cost-effectiveness of the product, and economic return. Species prioritization data will provide "Conservation Value" as well as "High Cultivation Prospect" of selected medicinal plants.

ii. *Documentation of ethno-botanical knowledge:* Ethno-botanical information should be collected on the following three aspects. (a) Interview with the informant while visiting the forest; (b) preparation of herbarium of collected plant along with getting all necessary information from local people; (c) interactive discussion with various stakeholders, such as traditional practitioners, farmers, traders, scientists and forest officers (Phondani et al. 2016).

iii. *Propagation techniques:* Development of scientifically validated propagation technologies, such as micropropagation techniques for rare and endangered medicinal plants. Data on propagation parameters using newly developed technology should be recorded.

iv. *Sustainable harvesting technologies:* To train farmers and medicinal plant growers for sustainable harvesting of medicinal plants without harming/killing the mother plants.

v. *Conservation and cultivation:* The priority species with established market demand should be placed in this group. A number of such species are also rated to possess high cultivation prospects, indicating an availability of cultivation techniques and hence good chances of lucrative returns to farmers.

vi. *Market enhancement:* To create market opportunities for the priority species. Research institutions and NGOs promoting plant conservation may act as a bridge between farmers and buyers, ensuring quality product selling and high economic return. For better marketing of selected medicinal plants, a memorandum of understanding should be signed between farmers and traders for a buy-back process. Promoting demand creation and proper product positioning will create adequate markets for such species.

vii. *Authentication:* Authentication and quality testing of medicinal plant material is very necessary for efficient and long-term marketing. Testing facilities also need to be established nearer to production hubs and with relatively less costs.

viii. *Community conservation of medicinal plant resources:* The main aim of this action is to promote local people to conserve medicinal plants. Community-based "Natural Heritage Conservation Councils" (NHCC) have been formed in all Himalayan valleys with the aim of conserving natural resources, involving local community activities.

ix. *Awareness and exposure visits:* Routine workshops and e-learning approaches should be undertaken to raise awareness among local people for sustainable use of medicinal plants. Discussions on cultivation technology, market demand, transportation process, conservation strategies, and so on, would be a priority. The selection of progressive farmers should be based on their willingness to cultivate medicinal plants.

11.8 Conclusions

Plant-based traditional systems of medicine continue to play an important role in the world healthcare system because of little or no side effects, easy availability, low costs, and longer shelf-life. Himalayan medicinal plants have been a rich source of many commercially available life-saving drugs and they still represent an important future stock for the identification of new drugs and pharmaceuticals.

However, unsustainable agricultural practices and destructive harvesting lead to rapid depletion of many medicinally important plants from Himalayan regions. Despite the existing conservation measures, there is an urgent need to develop novel technologies and to formulate new effective policies and programs for the conservation and sustainable cultivation of the medicinal wealth of the Himalayas. Different *in-situ* and *ex-situ* conservation strategies involving local communities need to be realized to combat the existing challenges of biodiversity loss. Authentication of medicinal plants and documentation of their properties are very important for ensuring high-quality herbal products. Large-scale cultivation of the prioritized medicinal plants for commercial purposes using micropropagation technologies holds excellent promises for rapid propagation of endangered medicinal plants.

Herbal medicines and formulations are mostly complexes of various organic and inorganic components and some may be toxic to various tissues in the body. As a result, these herbal medicines are not very well defined and characterized, particularly with reference to quality, safety points, bioactive constituents, and mode of action. In current scenario, importance should be given to scientifically validate the existing and debuting herbal products in terms of their phytochemical constituents, safety, associated toxicity, bioavailability, and mode of action against various disease targets. Further, proper training on nondestructive harvesting and conducting awareness campaigns on the importance and vulnerability of threatened medicinal plants for sustainable cultivation, along with high-economy generation through strategic marketing plans, may tempt people to incline toward medicinal plant cultivation.

References

Aggarwal, B.B., Prasad, S., Reuter, S. et al. 2011. Identification of novel anti-inflammatory agents from ayurvedic medicine for prevention of chronic diseases: "Reverse pharmacology" and "bedside to bench" approach. *Curr Drug Targets* 12(11):1595–1653.

Alam, G. 2004. *Database on Medicinal Plants*. CUTS Centre for International Trade, Economics and Environment, Calcutta.

Appasamy, P.P. 1993. Role of NTFPs in subsistence economy: The case of joint forestry project in India. *Econ Bot* 47(3):258–267.

Atanasov, A.G., Waltenberger, B., and Pferschy-Wenzig, E.M. 2015. Discovery and resupply of pharmacologically active plant-derived natural products: A review. *Biotech Adv* 33:1582–1614.

Banerji, G., and Basu, S. 2011. S Management of the Herbal Wealth of the Himalayas: prioritising biodiversity for conservation and development. Pre-Congress Workshop of 1st Indian Forest Congress, HFRI. http://www.pragya.com.

Chulet, R., Pradhan, P., Sharma, K.S., and Jhajharia, K.M. 2010. Phytochemical screening and antimicrobial activity of *Albizzia lebbeck*. *J Chem Pharm Res* 2(5):476–484.

Cragg, G.M., and Newman, D.J. 2005. Biodiversity: A continuing source of novel drug leads. *Pure Appl Chem* 77:7–24.

Cragg, G.M., and Newman, D.J. 2013. Natural products: A continuing source of novel drug leads. *Biochim Biophys Acta* 1830:3670–3695.

DeVita, V.T., Lawrence, T.S., and Rosenberg, S.A. 2008. *Cancer: Principles and Practice of Oncology.* 8th ed. LWilliams & Wilkins, Philadelphia, PA.

Dhawan, B.N. 1997. Biodiversity–A valuable resource for new molecules. In *Himalayan Biodiversity: Action Plans*, ed. U. Dhar. Gyanodaya Prakashan, National government publication.

Engels, F., Renirie, B.F., Hart, B.A., Labadie, R.P., and Nijkamp, F.P. 1992. Effects of apocynin, a drug isolated from the roots of *Picrorhiza kurroa*, on arachidonic acid metabolism. *FEBS Lett* 305(3):254–256.

Garcia-Carbonero, R., and Supko, J.G. 2002. Current perspectives on the clinical experience, pharmacology, and continued development of the camptothecins. *Clin Cancer Res* 8:641–661.

Hande, K.R. 1998. Etoposide: Four decades of development of a topoisomerase II inhibitor. *Eur J Cancer* 34:1514–1521.

Harvey, A.L., Edrada-Ebel, R., and Quinn, R.J. 2015. The re-emergence of natural products for drug discovery in the genomics era. *Nat Rev Drug Discov* 14:111–129.

Hasan, T., Das, B.K., Qibria, T., Morshed, M.A., and Uddin, M.A. 2011. Phytochemical screening and evaluation of analgesic activity of *Xanthium strumarium* L. *Asian J of Biochem and Pharm Res* 1(3):455–463.

Innocenti, G., Dall'Acqua, S., Scialino, G., Banfi, E., Sosa, S., Gurung, K., Barbera, M., Carrara, M., 2010. Chemical composition and biological properties of *Rhododendron anthopogon* essential oil. *Molecules* 15(4):2326–2338.

Jones, V.M., Moore, K.A., and Peterson, D.M. 2011. Capsaicin 8% topical patch (Qutenza)–A review of the evidence. *J Pain Palliat Care Pharmacother* 25(1):32–41.

Kaiser, H. 2008. From the plant to the chemical-The early history of "rheumatism". *Z Rheumatol* 67:252–262.

Kala, C.P. 2015. Medicinal and aromatic plants of tons watershed in Uttarakhand Himalaya. *Appl Ecol Env Res* 3(1):16–21.

Kala, C.P., and Sajwan, B.S. 2007. Revitalizing Indian systems of herbal medicine by the national medicinal plants board through institutional networking and capacity building. *Curr Sci* 93:797–806.

Kaul, M.K. 2010. High altitude botanicals in integrative medicine–Case studies from Northwest Himalaya. *Ind J Trad Knowledge* 9(1):18–25.

Klayman, D.L., Lin, A.J., Acton, N., Scovill, J.P., Hoch, J.M., and Milhous, W.K. 1984. Isolation of artemisinin (qinghaosu) from Artemisia annua growing in the United States. *J Nat Prod* (47): 715–717.

Koehn, F.E., and Carter, G.T. 2005. The evolving role of natural products in drug discovery. *Nat Rev Drug Discov* (4):206–220.

Kumar, M., Bussmann, R.W., Mukesh, J., and Kumar, P. 2011. Ethnomedicinal uses of plants close to rural habitation in Garhwal Himalaya. *Indian J Med Plants Res* 5(11):2252–2260.

Lucas, D.M., Still, P.C., Pérez, L.B., Grever, M.R., and Kinghorn, A.D. 2010. Potential of plant-derived natural products in the treatment of leukemia and lymphoma. *Curr Drug Targets* 11(7):812–822.

Malik, Z.A., Bhat, J.A., Ballabha, R., Bussmann, R.W., and Bhatta, A.B. 2015. Ethnomedicinal plants traditionally used in health care practices by inhabitants of Western Himalaya. *J Ethnopharmacol* 172:133–144.

Mangal, M., Sagar, P., Singh, H., Raghava, G.P.S., and Agarwal, S.M. 2013. NPACT: Naturally Occurring Plant-based Anti-cancer Compound-Activity-Target database Nucleic Acids Research (NAR 2013, 41, D1124-D1129).

Nautiyal, M.C. 1995. Agro-technique of some high altitude medicinal herbs. In *Cultivation of Medicinal Plants and Orchids in Sikkim Himalaya*, ed. R.C. Sundriyal and E. Sharma, 53–64. Bishen Singh Mahendra Pal Singh, Dehra Dun, India.

Newman, D.J., and Cragg, G.M. 2012. Natural products as sources of new drugs over the 30 years from 1981 to 2010. *J Nat Prod* 75: 311–335.

Pandey, H.K., Deendayal, and Das, S.C. 2006. Threatened medicinal plant biodiversity of Western Himalaya and its conservation. In: S. John William (ed.) *Biodiversity: Life to Our Mother Earth*, ed. S. John William, 281–294. SECNARM. Loyola College, Chennai, Tamil Nadu, India.

Pandey, M.M., Rastogi, S., and Rawat, A.K.S. 2008. Indian herbal drug for general healthcare: An overview. *Internet J Altern Med* 6(1):1–12.

Pandey, M.M., Rastogi, S., and Rawat, A.K.S. 2013. Indian traditional ayurvedic system of medicine and nutritional supplementation. *J Evid Based Complementary Altern Med* 1–12.

Pant, S., and Samant, S.S. 2010. Ethnobotanical observations in the Mornaula reserve forest of Kumoun, west Himalaya India. *Ethnobot Leaflets* 14:193–217.

Patwardhan, B., Warude, D., Pushpangadan, P., and Bhatt, N. 2005. Ayurveda and traditional Chinese medicine: A comparative overview. *J Evid Based Complementary Altern Med* 2(4):465–473.

Phondani, P.C., Bhatt, I.D., Negi, V.S., Kothyari, B.P., Bhatt, A., and Maikhuri, R.K. 2016. Promoting medicinal plants cultivation as a tool for biodiversity conservation and livelihood enhancement in Indian Himalaya. *J Asia Pac Biodivers* 1–8.

Purohit, S.S., and Vyas, S.P. 2004. Marketing of medicinal and aromatic plants in Rajasthan. Paper presented at the National Consultative Workshop on Medicinal and Aromatic Plants, GBPUAT, Pantnagar.

Rao, K.S., and Saxena, K.G. 1994. Sustainable development and rehabilitation of degraded village lands in Himalaya. HIMAVIKAS Publication-G.B. Pant Institute of Himalayan Environment and Development (India) 8:301.

Rathore, S., Tiwari, J.K., and Malik, Z.A. 2015. Ethnomedicinal survey of herbaceous flora traditionally used in health care practices by inhabitants of Dhundsir Gad watershed of Garhwal Himalaya, India. *Global J Res Med Plants & Indigen Med* 4(4):65–78.

Scott, L.J., and Goa, K.L. 2000. Galantamine: A review of its use in Alzheimer's disease. *Drugs* 60(5):1095–122.

Semwal, D.K., and Semwal, R.B. 2015. Efficacy and safety of *Stephania glabra*: An alkaloid-rich traditional medicinal plant. *Nat Prod Res* 29(5):396–410.

Shah, S.A., Mazumder, P.B., and Choudhury, M.D. 2012. Medicinal properties of *Paris polyphylla* Smith: A review. *J Herb Med and Toxicol* 6(1):27–33.

Sharma, A., Shanker, C., Tyagi, L.K., Singh, M., and Rao, V. 2008. Herbal medicine for market potential in India: An overview. *Acad J Plant Sci* 1(2):26–36.

Sherpa, M.T., Mathur, A., and Das, S. 2015. Medicinal plants and traditional medicine system of Sikkim: A review. *World J Pharm Pharm Sci* 4(2):161–184.

Shyur, L.F., and Yang, N.S. 2008. Metabolomics for phytomedicine research and drug development. *Cur Opin Chem Biol* 12:66–71.

Singh, P., Singh, B.K., Joshi, G.C., and Tewari, L.M. 2009. Veterinary ethno-medicinal plants in Uttarakhand Himalayan Region. *Nat Sci* 7(8):44–52.

Siwach, M., Siwach, P., Solanki, P., and A.R. Gill 2013. Biodiversity conservation of Himalayan medicinal plants in India: A retrospective analysis for a better vision: Review. *Int J Biodivers Conserv* 5(9):529–540.

Sneader, W. 2005. *Drug Discovery: A History*. Wiley, Chichester.

Tiwari, J.K., Ballabha, R., and Tiwari, P. 2010. Ethnopaediatrics in Garhwal Himalaya, Uttarakhand, India (Psychomedicine and Medicine). *New York Sci J* 3(4):123–126.

Umadevi, M., Kumar, K.P.S., Bhowmik, D., and Duraivel, S. 2013. Traditionally used anticancer herbs in India. *J Med Plants Stud* 1:56–74.

Uniyal, S.K., Awasthi, A., and Rawat, G.S. 2002. Current status and distribution of commercially exploited medicinal and aromatic plants in Upper Gori Valley, Kumaon Himalaya, Uttaranchal. *Curr Sci* 82:1246–1252.

Vaidya, A.D.B., and Devasagayam, T.P.A. 2007. Current status of herbal drugs in India: An overview. *J Clin Biochem Nutr* 41(1):1–11.

Wani, M.C., Taylor, H.L., Wall, M.E., Coggon, P., and McPhail, A.T. 1971. Plant antitumor agents. VI. The isolation and structure of taxol, a novel antileukemic and antitumor agent from *Taxus brevifolia*. *J Am Chem Soc* 93:2325–2327.

12

Energy Production through Microbial Fuel Cells

Ravi Shankar, Niren Pathak, Amit Kumar Chaurasia,
Prasenjit Mondal, and Shri Chand

CONTENTS

ABSTRACT In an era of climate change, alternate energy sources including living microorganisms and nonliving renewable resources are gaining interest. Microbial fuel cells (MFCs) have the potential to simultaneously treat wastewater for reuse and to generate electricity using living microorganisms. Microbial electrolysis cells (MECs) have the potential to generate biohydrogen along with water treatment. Different types of wastewater including domestic and industrial types can be treated through these routes. Although the MFC and MEC technologies have generated interest in the simultaneous treatment of wastewater generation of electricity and hydrogen, respectively, understanding of these technologies is still limited and many fundamental and technical problems remain to be solved. This chapter includes a comprehensive and critical review on several aspects of MFCs and MECs such as electron transfer mechanisms, functioning of mediators, configuration of MFCs, reactor design, components of MFCs, and role of microorganisms in the performance of MFCs. Research and development work being carried out around the world on these technologies as well as future prospects is also reported.

12.1 Introduction

Utilization of nonconventional energy resources is increasing day by day around the world to produce sustainable energy. Renewable bioenergy is considered as one of the ways to reduce the future energy crisis due to its potential for producing less greenhouse gases than fossil fuels. Both cellulosic biomass, abundantly available in the form of agriculture residue and woody biomass, and living microorganisms can be utilized for the production of energy. The nonliving biomass can be used to produce different fuels such as methane, butanol, ethanol, and hydrogen through thermos-chemical and biological routes (Khanal et al. 2010, Oh et al. 2010), whereas living microorganisms can be used to produce electricity, hydrogen, and so on thorough a biological route using other biomass or wastes as substrates (Lovley 2006a, 2006b, Davis and Higson 2007). Microbial fuel cells (MFCs) and microbial electrochemical cells are two emerging technologies that produce electricity and hydrogen respectively from organic substrate or wastewater containing organics. Large volumes of wastewaters are produced each year from domestic, industrial, and agricultural activities (Khanal et al. 2010). Recently, energy recovery from wastewater has been investigated by many researchers employing microorganisms.

MFC is an efficient device that can convert the chemical energy of a wide range of organic compounds present in pure solution or wastewater into electrical energy using the catalytic activity of microorganisms (Rabaey and Verstraete 2005, Zhang et al. 2016). Generally, MFC contains two parts: anodic and cathodic chambers (Figure 12.1).

In the anodic chamber, the organic component of wastewater is degraded by bacteria under the anaerobic condition to form H+ ions and free electrons as per Equation 12.1.

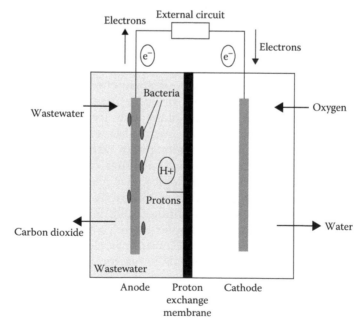

FIGURE 12.1

Schematic diagram of a double-chamber MFC. (From Mathuriya, A.S., *Critical Reviews in Environmental Science and Technology*, 44, 97–153, 2014.)

$$C_xH_yO_z + H_2O \rightarrow mCO_2 + nH^+ + ne^- + \text{Heat} \tag{12.1}$$

In both anode and cathode compartments, respective electrodes (anode and cathodes) are immersed and connected externally by highly conductive lead wires. This circuit allows free electron transport from the anodic to cathodic chamber as shown in Figure 12.1. However, either the hydrogen transfer membrane or the salt bridge (in the dual-chamber MFC) is employed for H^+ movement from the anodic to cathodic chamber. In the cathodic chamber, the electrons and protons combine in the presence of oxygen from cathodic solution/air to produce a water molecule as per Equation 12.2.

$$4H^+ + O_2 + 4e^- \rightarrow 2H_2O \tag{12.2}$$

In the absence of oxygen electricity production is not spontaneous. Nevertheless, by inducing a little voltage in the order of 0.5 V between electrodes, proton reduction takes place at the cathode and hydrogen production commences. This modified MFC is called the bioelectrochemically assisted microbial reactor or the microbial electrolytic cell. Hydrogen is produced in this cell instead of electricity through electrohydrogenesis.

Membrane-less MFCs do not require the proton exchange membrane (PEM). The capital, operational, and maintenance costs of these MFCs are also less than MFCs with membranes. These can be used in continuous mode and may be suitable for industrial applications. However, the efficiency of membrane-less MFCs is less. Hence, efforts are on for improving the performance of membrane-less MFCs so that these can be utilized in industrial applications (Sleutels et al. 2009, Sevda and Sreekrishnan 2012). Both batch (Min et al. 2005) and column (continuous) studies have been described in the literature on membrane-less MFCs (Jang et al. 2004, Lee and Nirmalakhandan 2011). At present, the efficiency of continuous MFCs is low; however, there is ample scope to enhance the efficiency by exploring better electrode material and suitable cell configuration. The efficiencies of both MFC, and MECs are affected by several parameters such as biochemical oxygen demand (BOD) and chemical oxygen demand (COD) value of the wastewater/solution, availability of mixed liquor suspended solid in the anodic chamber, temperature, pH, anode and cathode materials, nature of electrode, catalyst, and mediators, surface area for anode and cathode, space between electrodes, and so on (Timmins and Davies 1998, Angenent and Sung 2001, Park and Zeikus 2003, Rabaey et al. 2003, Logan et al. 2005, Franks and Nevin 2010). With the help of mediators, a rapid movement of electrons from bulk of the anodic solution to the anode surface becomes feasible. Suitable electrode spacing is essential to get rid of transport losses. To increase electron transfer efficiency it is necessary to use appropriate catalyst, mediator, or electrolyte solution that may help enhancing the coefficient of electron transfer. It is to be noted that added mediator or chemicals must be inert toward the working solution and materials of the MFC reactor; otherwise it may impart color to the solution or may chemically react with material of construction of the reactor.

This chapter discusses electricity generation and hydrogen production employing MFCs and microbial electrolytic cells, respectively. The mechanisms of electron transport for different microbial systems including working principles of different mediators are reviewed in detail. It also describes the possible ways to reduce different types of losses to improve the power generation in MFCs, summarizes important reports available in the literature on these technologies, and describes the design of different reactor types, electrodes as well as improvement in electrode materials, microorganisms, and so on, for the overall improvement in the performance of MFCs and MECs.

12.2 Mechanism of a Microbial Fuel Cell

In an MFC, microorganisms degrade (oxidize) organic matter and produce free electrons and H+ ions in the anodic chamber as described above. The H+ ions are transported to the cathode through the hydrogen transfer membrane or through the filter media (in case of membrane-less MFCs). However, the electrons are transported to the anode through various routes depending on the nature of the microorganisms used (Huang et al. 2011).

Two major routes, that is, direct and indirect routes, for the electron transfer from the bulk phase to the anode surface in an MFC have been reported, as shown in Figure 12.2. In the direct route, microorganisms themselves transfer electrons to the anode through a physical contact between the bacterial cell membrane and the electrode surface via cytochrome or putative nanowire structures (Sund et al. 2007, Huang et al. 2011). However, a soluble redox active compound (mediator) is responsible for electron transfer in the indirect route (Chaudhuri and Lovley 2003, Turick et al. 2003, Bond and Lovley 2005, Reguera et al. 2006). The working principle of the redox mediator is shown in Figure 12.2.

Some microorganisms are capable of releasing redox-active compounds that are responsible for indirect electron transfer to electrodes (Huang et al. 2011). Mediators are added to promote energy production in an MFC when bacteria cannot transfer electrons alone (Bullen et al. 2006). The reduced mediator is readily oxidized at the anode, and the oxidized form of the mediator shows favorable kinetics for reduction by the microbes (Sund et al. 2007). The addition of mediators may influence the metabolic properties of organisms (Peguin and Soucaille 1995, Park and Zeikus 2000).

In a mediator MFC, commonly used electron shuttles or exogenous mediators are various dyes such as neutral red, methylene blue, thionine, azure A, and anthraquinone-2,6-disulfonate (Vega and Fernandez 1987, Park and Zeikus 2000, Liu et al. 2004, Min and Logan 2004, Ieropoulos et al. 2005, Kiran and Gaur 2013). Mediators are generally expensive and toxic compounds having short lifetime. Potassium ferricyanide and humic acids have also been used as a mediator. High consumption rate and the risk associated with these compounds make their use practically not viable (Liu et al. 2004, Min and Logan 2004).

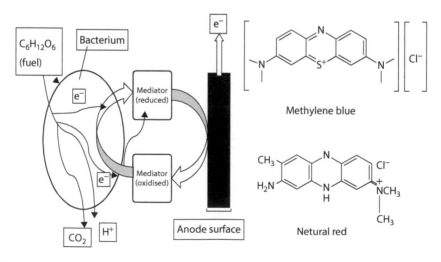

FIGURE 12.2
Working principle of redox mediators in an MFC. (From Sevda, S., and T.R. Sreekrishnan, *J. Environ. Sci. Health A,* 47, 878–886, 2012. With permission.)

TABLE 12.1

Some Microorganisms with Different Metabolic Paths and Electron Transport Systems

Metabolic Type	Transfer Type	Examples of Organisms
Oxidative metabolism	Membrane-driven	*Rhodoferax ferrireducens*
		Geobacter sulfurreducens
		Aeromonas hydrophila
	Mediator-driven	*Escherichia coli*
		Shewanella putrefaciens
		Pseudomonas aeruginosa
		Erwinia dissolvens
		Desulfovibrio desulfuricans
Fermentative metabolism	Membrane-driven	*Clostridium butyricum*
	Mediator-driven	*Enterococcus faecium*

Source: Rabaey, K., and W. Verstraete, *TRENDS Biotechnol.*, 23, 291–298, 2005.

In the direct route, the electrons travel through a series of respiratory enzymes in the cell and make energy for the cell in the form of adenosine triphosphate. Eventually, electron acceptors accept the electrons and thus get reduced. Many terminal electron acceptors that are most easily reduced such as oxygen, nitrate, and sulfate can readily diffuse into the cell. These compounds readily accept electrons inside the cell forming products that can diffuse out of the cell (Logan 2008).

Both oxidative and fermentative metabolisms of microorganisms have been observed in the MFC. Some microbes having different types of metabolic paths and electron transport systems are reported in Table 12.1.

Though the electricity generation mechanism is very much similar to chemical fuel cells, microbes work as a catalyst in the MFC anode. However, microbes obtain energy from decomposition of organic molecules for their growth.

12.3 Performance of a Microbial Fuel Cell

The performance of an MFC can be measured in terms of coulombic efficiency, which is a measure of the number of coulombs harnessed as electrical current compared to the theoretical maximum number of coulombs that can be recovered from the organic feed present in the MFC. The coulombic efficiency depends on the microorganisms that are responsible for the oxidation of organic compound as well as the capability of the organic compounds for generating electrons. This can be attributed to the fact that different microorganisms adopt different metabolic pathways and different anodic electron transfer mechanisms. Table 12.2 summarizes different types of substrate solution and source of inoculums and current density generation. Due to incomplete oxidation of organic compounds the capability of the cell is not fully exploited (Zeng et al. 2010). Both synthetic solution containing pure organic compounds and real wastewater have been used in an MFC.

The performance of an MFC also depends on mediator concentration, types of hydrogen transfer membrane/system, electron acceptors, operating conditions such as pH and temperature, substrate concentration and its nature, external resistance, and electrode spacing (Ghangrekar and Shinde 2007, Sun et al. 2016). The type of reactor also influences the

TABLE 12.2

Substrate Used, Source of Microorganisms, and Current Generation in an MFC

Type of Substrate	Source Inoculums	Current Density (mA/cm^2) at Maximum Power Density	Reference
Acetate	Pre-acclimated bacteria from MFC	0.8	Logan et al. 2007
Azo dye with glucose	Mixture of aerobic and anaerobic sludge	0.09	Sun et al. 2009a
Corn stover biomass	Domestic wastewater	0.15	Zuo et al. 2006
Cysteine	Sediment sample from 30 cm depth	0.0186	Logan et al. 2005
Ethanol	Anaerobic sludge from wastewater plant	0.025	Kim et al. 2007b
Farm manure	Self build-up of anaerobic environment	0.004	Scott and Murano 2007
Landfill leachate	Leachate and sludge	0.0004	Greenman et al. 2009
Brewery wastewater	Full strength brewery wastewater	0.2	Feng et al. 2008
Brewery wastewater	Anaerobic mixed consortia	0.18	Wen et al. 2009
Meat processing wastewater	Domestic wastewater	0.115	Heilmann and Logan 2006
Paper recycling wastewater	Diluted paper recycling wastewater	0.25	Huang and Logan 2008
Glucose	Mixed bacterial culture maintained on sodium acetate for one year (*Rhodococcus* and *Paracoccus*)	0.70	Catal et al. 2008

performance. It has been reported that an MFC performs better in the batch scale reactor than the continuous reactor (Min and Logan 2005).

12.4 Factors Limiting Microbial Fuel Cell Performance

To understand the performance of an MFC it is necessary to consider sources of energy loss occurring in an MFC. These losses comprise the resistance to electron transport in terms of charge-transfer resistance (activation loss), ohmic resistance, and diffusion resistance (mass transfer). Among these losses, ohmic resistance loss plays a major role as far as the performance of an MFC is concerned (You et al. 2008). When designing an MFC, each type of polarization loss should be minimized to decrease overall potential losses and thus increase the power generation.

12.4.1 Activation Losses

Activation losses occur due to the activation energy needed to overcome the initial energy barriers for an oxidation/reduction reaction during the electron transfer to or from a compound reacting at the electrode. This compound can be a soluble mediator, a mediator present at the surface of bacteria, or an electron acceptor at the cathode (Logan et al. 2006). It is essential to break energy barrier to initiate electron transfer from exoelectrogens to electrode or to transfer electrons to an oxidizing agent that works as the final electron acceptor. This results in voltage loss or activation over potential shown by a steep fall of MFC voltage especially in the early phase of electricity production. It has been suggested that attachment of firm thick biofilm to the electrode, an increase in the electrode surface area, catalyst modification for electrodes, and a slight increase in temperature may result in reducing activation losses (Nwokocha et al. 2012).

Ramasamy et al. (2008) used electrochemical impedance spectroscopy and found that the impedance associated with the anode is the dominant loss in an MFC, even with an established biofilm. They also reported that the charge transfer resistance dropped from 2.6 to 1.5 kΩ cm^2 after five days of closed circuit operation and decreased further to 0.48 kΩ cm^2 after three weeks of operation. This indicates that as the biofilm is established, biocatalytic oxidation of the substrate is enhanced and thus activation losses are reduced.

12.4.2 Ohmic Losses

One of the significant and problematic parameter that limits the electricity generation in an MFC is high internal resistance of the system. The electrons and ions have to overcome minimum ohmic resistance to flow through the electrochemical system. This internal resistance is rooted in the MFC internal structure. Indeed, higher ohmic resistance and current density lead to higher ohmic loss in the MFC system. The electrolyte ohmic loss and electrode ohmic loss are two major ohmic losses relevant for MFCs. The electrolyte ohmic loss is caused by the movement of electrons through electrodes, while the electrolyte voltage loss is caused by the movement of ions through the electrolyte (Rozendal et al. 2008). Many real wastewaters have very low conductivity causing considerable electrolyte ohmic losses. However, MFC performance improvement by reducing internal losses can be achieved either by engineering advancements or by conducting an in-depth microbiological study to better understand bacterial morphology and classification (Nam et al. 2010a, b). Furthermore, it is convenient to achieve high conductivity by keeping high salt concentration or with the help of employing buffer solution during the laboratory study, but in case of actual sewage and industrial effluent electrical conductivities are too low typically of the order of 1 ms/cm (Rozendal et al. 2008). To reduce the internal resistance of an MFC, one of the options is to reduce the distance between the anode and cathode. Another way is to increase ionic strength, which increases the solution conductivity of the MFC system (Liu et al. 2005a). According to Kim et al. (2007a), high ohmic resistance and energy losses are related to the restricted proton movement to and from the electrodes in an MFC. Although ionic species controls the charge balance, protons generated at the anode and consumed at the cathode must be in balance (Kim et al. 2007a). Another approach to reduce ohmic losses is up-flow anode design in which the organic substrate is allowed to flow through biofilm continuously, which results in higher power production and less internal resistance (You et al. 2008). Several researchers have focused on methods to reduce the MFC internal resistance as this is the major impediment to harness more power.

12.4.3 Mass Transfer

A limited mass transfer of substrate or electron acceptors toward the electrodes leads to concentration or mass transfer losses, which results in a sharp drop in cell voltage. This phenomenon is more predominant at maximum current densities during polarization. Concentration polarization also occurs when there are different anodic and cathodic mass transfer rates. This means anodic oxidation occurs at a faster rate than transport of species toward the cathode. As discussed before, this phenomenon observed near maximum current density (due to nonconductive thick biofilm formed at the anode) leads to a very slow rate of diffusion. In addition, the competition between organisms that can consume the substrate before an anodophilic organism is encountered also plays a role in concentration polarization to occur (Rozendal et al. 2008, Nwokocha et al. 2012).

It has been observed that mass transfer limitations in the bulk electrolyte solution also hinder substrate transport, which is another kind of concentration loss. With the help of polarization curves, the starting point of concentration losses can be determined (Logan et al. 2006).

12.4.4 Bacterial Metabolic Losses and Electron Quenching Reactions

Bacterial metabolic losses are caused by the energy used by bacteria, for example, growth and maintenance, instead of the energy captured by an MFC. The higher the difference in redox potential between the substrate and anode, the higher the energy gained by bacteria, but the lower the energy captured by the MFC or the lower the MFC voltage. The anode potential should be low enough to minimize metabolic losses but high enough to prevent bacteria switching to fermentative metabolism instead of using the anode as the electron acceptor (Logan et al. 2006). Besides this, other side reactions such as fermentation, methanogenesis, and respiration (in the presence of oxygen) lead to electron loss and partial substrate conversion into anodophilic mass. Another possibility is substrate leakage into the cathode compartment. This can contribute to electron loss in terms of load, and this is a kind of short circuiting since actual path (circuit) is bypassed (Mohan et al. 2007, Nwokocha et al. 2012).

12.5 Batch and Continuous Microbial Fuel Cells

MFCs can be operated in either batch mode or continuous mode. In case of batch mode operation, definite quantity of feed is charged to the chamber and thus the fuel cell is operated under unsteady-state conditions. In case of continuous MFCs, the feed flows in and out continuously and the fuel cell is operated under steady-state conditions. Treated effluent quality as well as electricity generation rate generally remains uniform in continuous mode of operation. Further, batch mode operation also suffers from disadvantages such as a lag time for start-up and downtime for filling, substrate depletion with respect to limited nutrients, and toxicity of byproducts (Khanal et al. 2010, Rahimnejad et al. 2011). Nevertheless, batch operation is used in laboratory and pilot scale studies due to its simplicity of operation and reduced risk of contamination (Khanal et al. 2010). In comparison to batch mode operation the continuous mode of operation is less labor intensive. The continuous mode operation can produce reproducible results and reliable data. It can also provide high productivity per unit volume with the possibility to accurately determine the growth kinetics and kinetic constants (Rahimnejad et al. 2011). A review of the important investigations on batch and continuous MFCs by different researchers is provided below.

12.5.1 Batch Microbial Fuel Cells

Investigation has been carried out by many researchers on a batch MFC in the laboratory scale. The performance of the batch MFC has been evaluated with solution containing noble organic compounds such as glucose and acetic acid (Du 2007), as well as real wastewater containing organics (expressed in terms of BOD and COD) from domestic and industrial sectors. Removal of toxic pollutants such as phenols and chromium has also been investigated along with COD removal. MFC has also been tested for the desalination of wastewater. The effects of process parameters such as pH, temperature, dissolved oxygen, mediators, conductivity of anodic solution, and catholytes have been investigated on electricity generation as well as COD removal. Different types of microorganisms have been used to explore the possibility of simultaneous treatment of wastewater and power generation. Modifications of reactor configuration, electrodes, hydrogen transfer membranes, and so on, and their effects on the performance of an MFC, have also been investigated by some researchers. Efforts have also been made to increase the electricity production using the number of MFCs in series and or parallel connections. A summary of some important investigations is provided below.

Li et al. (2016) evaluated the performance of a batch-operated MFC used for food waste disposal and electricity generation. A maximum power density of 5.6 W/m^3 and a voltage of 0.51 V were obtained. It has been observed that during the MFC operation aromatic compounds in the hydrophilic fraction are more preferentially removed than nonaromatic compounds.

Ye et al. (2016) used a series of biological buffers with positive, negative, and neutral charges to improve the performance by stabilizing the electrode pH and increasing the electrolyte conductivity in an air-cathode MFC. It has been reported that at lower pH, buffers with positive charge increases cathodic current by as much as 95% within certain ranges (potential windows) of cathode potentials. The reason for this increase with the net positive charge buffers is likely a more stable electrode pH produced by the electric field driving the positively charged ions toward the cathode.

An et al. (2016) studied a dual-chamber MFC with biocathode arrangements using ethanolamine and nitrate as substrates. It has been demonstrated that removal of ethanolamine, electricity generation, and denitrification take place simultaneously. In this MFC ethanolamine oxidation at the anode released electrons, which were transported to biocathode. Then biofilm at the cathode performed complete denitrification accepting electrons. The biocathode MFC produced current, cell voltage, and maximum power density of 2.9 mA, 170 mV, and 8.41 W/m^3 respectively.

Cheng et al. (2015) investigated power production using recalcitrant organics substrates in a dual-chamber MFC. By exposing bioanode to air, aniline removal efficiency of more than 91.2% was achieved in batch mode operation. It was found that oxygen plays a significant role in biodegradation of recalcitrant organics substrates.

Zhang et al. (2015) compared substrate removal rates in MFCs having readily biodegradable compound (acetate) and domestic wastewater as a feed solution and fed-batch mode operation. As electricity production increased COD removal rate also accelerated for both substrate solutions. It is established that MFCs can be used for energy recovery from wastewater, but a combination with another post-treatment process is required to reduce CODs to meet discharge standards.

Zheng et al. (2015) developed a novel binder-free carbon black/stainless steel mesh (CB/SSM) composite anode using the dipping/drying method. This 3D-CB/SSM electrode produced a current density of 10 mA/cm^2, which is higher than conventional carbon felt electrodes in batch-mode operation. Furthermore, this electrode material is robust, low-cost, and suitable for up-scaling.

Shankar et al. (2015a) studied the performance of a membrane-less MFC using glucose and glutamic acid (1:1 mole/mole) as a substrate. Simultaneous removal of COD and generation of electricity has been studied along with the removal of heavy metals (Zn^{2+}, Cr^{6+}, and Fe^{3+}). The difference in anodic and cathodic pH during the experimentation for seven days has been noted. Initial COD $\leq$ 1250 mg/L was reduced below the permissible limit of COD for surface water.

Mansoorian et al. (2013) tested the performance of a two-chamber MFC using protein and food industry wastewater (initial COD 1900 mg/L) and activated sludge from the municipal wastewater treatment plant as a bacterial source. Maximum current density and power density were observed as 230 mW/m^2 and 527 mA/m^2, respectively.

Karra et al. (2013) used different anode materials such as activated carbon nanofibers (ACNF), granular activated carbon (GAC), and carbon cloths (CC) for studying the performance of an MFC using synthetic wastewater (initial COD 1500–3000 mg/L). The maximum volumetric power density of 3.50 W/m^3 was obtained with ACNF. *Pseudomonas aeruginosa* (ATCC-97) and *Shewanella oneidensis* (ATCC-BAA1096) were used as biocatalysts for this study.

Wang et al. (2012) investigated an MFC-membrane bioreactor (MFC-MBR) system for the production of electricity and removal of COD of wastewater with an initial COD value of 234–400 mg/L. Anaerobic and activated sludge from a laboratory bioreactor was used as a microbial source. An average COD removal of ~89% was obtained with the maximum volumetric power density of 6.0 W/m^3.

Antonopoulou et al. (2010) examined a dual-chamber membrane-less MFC for cheese whey wastewater treatment. Under experimental conditions they achieved power density, current density, voltage, and coulombic efficiency as 18.4 mW/m^2, 80 mA/m^2, 0.23 V, and 1.9%, respectively. A similar power density value was obtained when glucose and lactose substrates were tested.

Mohanakrishna et al. (2010) investigated the multiple functions of an MFC as a fuel generator and as an integrated wastewater treatment unit. The MFC was operated under acidophilic pH range and at three different loading rates under a fed-batch mode cycle. MFC operation reduced COD, color, and total dissolved solids to the extent 72.84%, 31.67%, and 23.96%, respectively.

An et al. (2009) treated organic-contaminated water in a floating-type MFC (FT-MFC). They fed the anaerobic digester fluid obtained from brewery wastewater (initial COD, 1700 mg/l) to four FT-MFCs and achieved an open circuit voltage and final COD of around 0.4–0.5 V and 380 mg/L, respectively, over a 2-day period.

Daniel et al. (2009) connected five MFCs in series and tested for both batch-fed and continuous modes of operation, using methylene blue and neutral red as mediators. Methylene blue was found to be a more effective mediator than neutral red. A higher voltage generation was achieved when anode was retained in batch-fed mode and cathode was operated in continuous mode. An open-circuit voltage of 2.2 V, a power density of 979 μW/m^2, a current density of 1.15 mA/m^2, and COD removal of 10% were obtained.

Jadhav and Ghangrekar (2009) investigated the performance of an MFC under various operating conditions of pH, feed concentration, external load, and temperature when treating synthetic wastewater. The coulombic efficiency and COD removal rate were measured during experiments. They analyzed the MFC in two temperature ranges such as 20°C–35°C and 8°C–22°C and a COD range 100–600 mg/L. Operation in a higher temperature range favored higher COD removal efficiency up to 90% and low current generation (0.7 mA) as well as columbic efficiency (1.5%). However, in a lower temperature range, higher current generation and columbic efficiency of 1.4 mA and 5%, respectively, were achieved although lower COD removal rate (~59%) was noted. The highest current was generated at a pH of 6.5 in the anodic chamber with a coulombic efficiency of 4%.

Jiang et al. (2009) produced a stable electrical power in a dual-chamber MFC employing a potassium ferricyanide electron acceptor. The average voltage and maximum power density during 250h operation time were 0.687 V and 8.5 W/m^3, respectively. Further, 46.4% total COD removal was achieved when anodic pH was 7.2 at high initial COD input.

Lefebvre et al. (2009) developed a cathode that uses a low-cost catalyst (Co) at a reduced load using the sputtering technology. In an MFC, best performance was achieved when the catalyst was applied on the air side of the cathode over a backing layer made of nafion and further protected by a gas diffusion layer made of polytetrafluoroethylene.

Li et al. (2009) developed an overflow-type wetted-wall MFC to produce a stable voltage using air cathode and acetate as a substrate. The maximum power density generated in the MFC from 1000 mg/L acetate was 18.21 W/m^3 and the internal resistance was 400 Ω.

Luo et al. (2009) used phenol as an organic substrate in a two-chamber MFC. Phenol having initial 400 mg/L concentration in an aqueous air-cathode MFC achieved almost 15% higher degradation rate as compared to the open circuit controls. Under such operating conditions, a maximum volumetric power density of 9.1 W/m^3 was achieved.

Lu et al. (2009) studied an air-cathode MFC operated for 140 days over four batch cycles when treating starch processing wastewater as a substrate containing 4852 mg/L of COD. Maximum voltage and power density were achieved in the third cycle due to 120 Ω minimum internal resistance. The maximum coulombic efficiency of 8.0% was obtained.

Oh et al. (2009) studied the effects of applied voltages and dissolved oxygen on energy production in a dual-chamber MFC. Their results indicate that oxygen seepage into the anaerobic anode compartment is the major concern that seriously affects power generation rather than charge reversal for sustained performance of MFCs.

Thygesen et al. (2009) explained the effect of humic acid (mediator) on the performance of a double-chamber MFC with different substrates such as acetate, glucose, and xylose. The highest voltage and maximum power density of 570 mV with 1000 Ω and 123 mW/m^2, respectively, were reported with the acetate substrate followed by xylose and glucose, respectively. However, the addition of humic acid at 2 g/L concentration improved power density to 84% for glucose and 30% for xylose.

Trinh et al. (2009) investigated the energy production in an MFC with a sodium acetate substrate and *Geobacter sulfurreducens* as the biocatalyst on the anodic electrode. At 30°C–32°C optimum temperature and at an external resistance of 1000 Ω, 0.20–0.24 mA stable current was obtained under batch mode operation with a maximum power density of 418–470 mW/m^2.

Cao et al. (2009) demonstrated a new water desalination method and simultaneous electricity generation. In this work, two membranes were mounted between anode and cathode thereby the middle chamber was created for water desalination purpose. The anion and cation exchange membranes were positioned adjacent to the anode and cathode electrode, respectively. When electrons were produced by microbes on the anodic chamber, ions from saline water from the middle chamber were transferred into the anodic and cathodic compartments from the middle chamber. The experiments were performed using sodium acetate as a feed for the bacteria and saline water at 5, 20, and 35 g/L salt concentrations for desalination.

Zhang et al. (2009) used wheat straw hydrolysate as a substrate in a dual-chamber MFC study. They achieved 123 mW/m^2 power density and 37.1% maximum columbic efficiency with 1000 mg/L initial COD concentration. They also classified bacterial species in the reactor and reported that *Bacillus* (22.2%) was found as predominant species in suspended consortia.

Zhao et al. (2009) developed an MFC for removal of sulfur-based pollutants as a target compound for simultaneous wastewater treatment and energy production. A carbon-based composite anode material, air cathode, and the sulfate-reducing species *Desulfovibrio desulfuricans* were used in a dual-chamber MFC. The MFC removed 1.16 g/L sulfite and 0.97 g/L thiosulfate from the wastewater at 22°C.

Zhu and Ni (2009) demonstrated that an MFC not only can remove biodegradable organics and generate electricity, but also can remove biorefractory pollutants like *p*-nitrophenol. They reported that in an MFC, *p*-nitrophenol was completely degraded after 12 h, and about 85% TOC was removed after 96 h. Simultaneously, a maximum power output of 143 mW/m^2 was generated.

Feng et al. (2008) examined the efficiency of a full strength brewery wastewater using a air-cathode MFC. The performance of the MFC was measured for maximum power densities, coulombic efficiencies, and COD removal. The operating variables were temperature (20°C and 30°C) and influent concentration. As the temperature reduced to 20°C, the power density decreased to 170 mW/m^2 from 205 mW/m^2. However, no significant effect of a temperature change was observed on columbic efficiency and COD removals. Besides this, the addition of a 50 mM and 200 mM phosphate buffer increased the power output by 136% and 158%, respectively.

Li et al. (2008) employed a dual-chamber MFC to treat real wastewater from electroplating operation with graphite paper cathode. They accomplished more promising results with graphite paper cathode in terms of both chromium removal and electricity generation than with previously employed carbon-based cathodes such as carbon paper and carbon felt cathode material.

Shimoyama et al. (2008) developed a novel MFC design comprising a series of cassette electrodes to improve power generation capacity during treating wastewater with organic contaminants such as fish extract, peptone, and starch. Each cassette electrode (CE) chamber was made up of a box-shaped flat cathode fitted in between two proton-exchange membranes and two graphite felt anodes. This 12-chamber CE-MFC showed the possibility for scaling up and it was capable of treating wastewater and simultaneously electricity generation.

Huang and Logan (2008) examined the performance of MFCs in simultaneously treating paper recycling plant effluent and energy production. The power density and coulombic efficiency of the order of 501 mW/m^2 and 16%, respectively, were achieved on the addition of 50 mM phosphate buffer having 5.9 mS/cm conductivity to the paper recycling effluent. Over a 500h batch cycle, reasonably good total COD (76%) and soluble COD (73%) removal efficiency were noted. Further, increasing the conductivity to 100 mM PBS, the power density was increased to 672 ± 27 mW/m^2.

Mohan et al. (2008b) evaluated the performance of a single-chamber MFC at two different substrate loading rates (COD) enriched by acidogenic bacterial culture. They used plain graphite electrodes and PEM made of glass wool. It has been observed that higher current density is produced under acidophilic conditions, while effective substrate degradation was observed at neutral pH.

Zhang et al. (2008) studied the effect of aerobic bacterial culture as a catalyst in a mediatorless MFC and using glucose as a substrate. The MFC startup took nine day lag time when anaerobic sludge and aerobic sludge were used in the anode and cathode, respectively. At 300 mL/min aeration rate, the voltage, power density, and current density were achieved as 0.324 V, 24.7 W/m^3 and 117.2 A/m^2, respectively.

12.6 Continuous Microbial Fuel Cells

Continuous MFCs are relatively less reported in comparison to batch MFCs. Both membrane and membrane-less MFCs have been used in continuous mode. Single- and double-chamber MFC configuration has also been used. The effects of retention time, mediator, electrode area, electrode distance have been investigated. Investigation has also been carried out to understand different types of losses in continuous MFCs as well as optimization of power generation. Removal of toxic chemicals has also been studied. Some studies are also carried out to improve the performance by connecting different MFCs in series. A comparison of batch and continuous mode operation, using the same substrate and microorganisms, has been made by some researchers. A summary of some important investigations is provided below.

Kubota et al. (2016) treated molasses wastewater in two single-chamber MFCs connected in series and operated in a continuous mode. Two serially connected MFCs were capable to reach COD removal and columbic efficiency of 79.8% and 11.6%, respectively, at 26h hydraulic retention time (HRT). The volumetric power densities for individual MFCs, MFCs in series, and MFCs in parallel modes were 0.54, 0.34, and 0.40 W/m^3, respectively.

Pushkar and Mungray (2015) treated domestic wastewater and actual textile wastewater in a novel cross-linked MFC reactor operated continuously. In this newly designed cross-linked MFC, sequential aerobic and anaerobic treatments were achieved in respective biocathodes and bioanodes when two H-shaped MFCs were stacked hydrodynamically. Under the control operating conditions, cross-linked MFCs achieved 337 W/m^3 volumetric power density, 82% COD removal, and 60% color removal from combined effluent.

Shankar et al. (2015b) studied a membrane-less MFC with an internal resistance of 4.9 $M\Omega$ in continuous mode for simultaneous generation of electricity and removal of organics from 500 mg/L glucose and glutamic acid (1:1 mole/mole) solution with graphite sheets as electrodes. The effects of operational parameters such as retention time, electrode surface area, inter electrode distance and mediator concentration on the voltage generation of MFC were studied.

Yu et al. (2012) studied the performance of a submerged-type MFC consisting of six readily exchangeable air cathodes for treating low strength domestic wastewater and production of electricity. With an HRT of 6–8 h, the maximum power density of 191 mW/m^2 was achieved.

Huang et al. (2009) demonstrated the use of MFCs using carbon brush anode to reduce the COD of a paper-plant wastewater at short hydraulic retention time and simultaneously generated electricity in continuous-flow conditions. In an MFC, around 26% COD in wastewater could be removed at a 6h HRT when the initial COD was 506 mg/L and organic loading rate was 2.0 kg COD/m^3 d. Under these conditions around 6 W/m^3 volumetric power density and 210 mW/m^2 power density could be achieved.

Rahimnejad et al. (2011) used pure glucose solution (30 g/L) in batch and continuous mode dual-chamber MFC for the evaluation of the performance of the MFC. With an HRT of 6.7 h, the maximum power density of 283 mW/m^2 was achieved.

Luo et al. (2010) used recalcitrant indole as a substrate in the continuous mode (C-MFC) and batch-fed MFC (B-MFC) for electricity production. With an initial indole concentration of 250 mg/L, the maximum volumetric power density achieved was 2.1–3.3 W/m^3 in the C-MFC, whereas it was 3.3 W/m^3 in the B-MFC. Further, 250 mg/L indole was removed within 6h time in the B-MFC, while it took about 30 h for the complete removal of indole in the continuous mode MFC.

Wen et al. (2010) studied a single-chamber MFC treating brewery wastewater in continuous mode. In this MFC design, carbon fiber anode and air cathode were employed. The MFC demonstrated a maximum power density of 669 mW/m^2 when 23.3 Ω internal resistance was connected and the initial feed COD concentration was 1501 mg/L.

Chung and Okabe (2009) operated a mediator-less dual-chamber MFC continuously in concurrent mode for more than 1.5 years with simulated influent containing glucose at 10 mM concentration. When the HRT was set as 4.5 h for the anodic compartment and phosphate buffer was used as the cathode electrolyte, the volumetric power density and columbic efficiency of 28 W/m^3 and 46% were accomplished.

Greenman et al. (2009) investigated the electricity generation from landfill leachate by using MFCs. They employed four experimental columns. The first column (C_1) was examined as a biological aerated filter and columns C_2 and C_3 were connected with load whereas column C_4 was operated in open circuit mode. All four columns were examined at different loading rates, and it has been deduced that higher electricity was generated at higher flow rates. However, more BOD_5 removal was obviously achieved at lower flow rates.

Wen et al. (2009) demonstrated the feasibility of power generation and modeling of an MFC treating brewery wastewater in continuous mode. A single air-cathode MFC with carbon fiber anode was constructed to quantify various losses occurring in the MFC. When treating brewery wastewater having 626.58 mg/L COD the MFC displayed a maximum power density and open-circuit voltage of 264 mW/m^2 and 0.578 V, respectively. To quantify losses, a polarization curve-based model was developed. At a current density of 1.79 A/m^2, 0.046 V ohmic loss and 0.248 V mass transport loss were calculated. This study shows that the reaction kinetic loss and mass transport loss are the most significant factors that affect MFC performance.

Woodward et al. (2009) demonstrated real-time maximization of power in a stack of two continuous flow MFCs. A multiunit optimization algorithm was used to control external resistances of two air-cathode membrane-less MFCs for maximum power output.

12.7 Design of Microbial Fuel Cells

Generally, an MFC comprises an anodic and cathodic chamber, which are separated by a PEM as shown in Figure 12.1. In the case of a single-chamber MFC, one chamber (cathodic) of the MFC can be eliminated by exposing the cathode directly to the air. In the case of a membrane-less MFC, use of a specific membrane can be eliminated. Table 12.3 shows the MFC components and the materials used to construct these. The photographs of some commonly used anode materials are shown in Figure 12.3.

Different types of MFCs such as a single-chamber MFC, double-chamber MFC, and stacked MFC, with different designs and configurations, have been developed in advancement of the technology to improve the power density (Vineetha and Shibu 2013, Cheng and Logan 2011). Some common designs of MFCs are described below.

The simplest form of an MFC is an H-type cell (Figure 12.4). It usually consists of two glass bottles or cylinder-shaped chambers interconnected by a membrane such as nafion. H-type MFCs operated in batch mode are suitable for basic research in laboratories such as for investigating power generation using novel materials, types of bacterial colonies, and so on (Logan et al. 2006, Asodariya and Patel 2011). In an H-type MFC, large electrode

TABLE 12.3

Basic Components of Microbial Fuel Cells

Items	Materials	Remarks
Anode	Graphite, graphite felt, carbon paper, carbon cloth, Pt, Pt black, reticulated vitreous carbon (RVC)	Necessary
Cathode	Graphite, graphite felt, carbon paper, carbon cloth, Pt, Pt black, RVC	Necessary
Anodic chamber	Glass, polycarbonate, Plexiglas	Necessary
Cathodic chamber	Glass, polycarbonate, Plexiglas	Optional
Proton exchange system	Proton exchange membrane: nafion, ultrex, salt bridge, porcelain septum, or solely electrolyte polyethylene, poly(styrene-co-divinylbenzene)	Necessary
Electrode catalyst	Pt, Pt black, MnO_2, Fe^{3+}, polyaniline, electron mediator immobilized on anode	Optional

Source: Du, Z. et al., *Biotechnol. Adv.*, 25, 464–482, 2007.

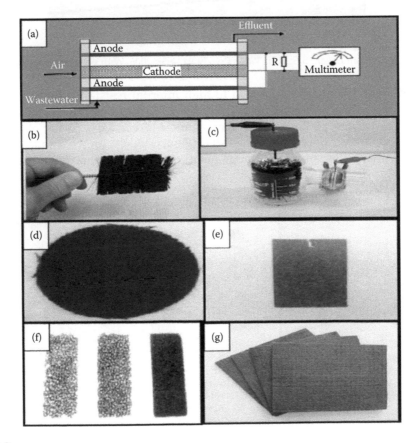

FIGURE 12.3
Anode materials used in MFCs: (a) graphite rod anode in a single-chamber MFC, (b) graphite brush, (c) graphite brush anode in a single-chamber MFC, (d) carbon cloth, (e) carbon paper anode in a dual-chamber MFC, (f) reticulated vitreous carbon (RVC), and (g) carbon felt. (From Zhou, M. et al., *J. Power Sources*, 196, 4427–4435, 2011. With permission.)

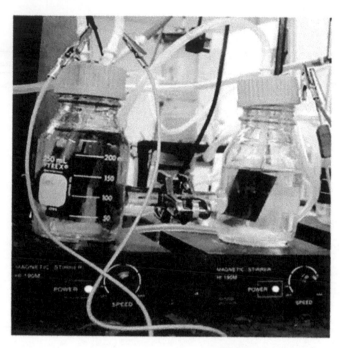

FIGURE 12.4
H-cell MFC design. (From Logan, B.E. et al., *Water Res.*, 39, 942–952, 2005. With permission.)

spacing and small membrane area lead to high internal resistance that reduces power produced in this design (Krishnaraj and Yu 2015). To improve power densities of two-chamber MFCs various configurations such as rectangular MFC or flat plate MFC have been developed. A basic double-chamber MFC (Figure 12.5) comprises anode and cathode chambers partitioned by an ion exchange membrane and coupled to an external circuit

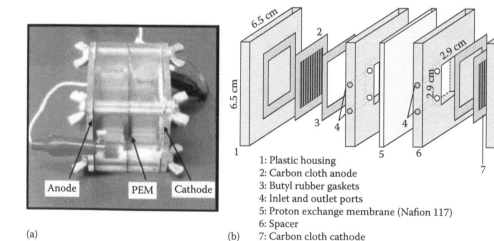

1: Plastic housing
2: Carbon cloth anode
3: Butyl rubber gaskets
4: Inlet and outlet ports
5: Proton exchange membrane (Nafion 117)
6: Spacer
7: Carbon cloth cathode

(a)   (b)

FIGURE 12.5
(a) Two-chamber MFC with aqueous cathode (From Kim, J.R. et al., *Environ. Sci. Technol.*, 41, 1004–1009, 2007a), (b) typical schematic for a two-chamber MFC. (From Nevin, K.P. et al., *Environ. Microbiol.*, 10, 2505–2514, 2008. With permission.)

(Rinaldi et al. 2008). MFCs can be operated under steady-state conditions either with continuous or batch feed of substrate for electricity production (Xu et al. 2016). Min and Logan (2004) designed a flat plate MFC to operate in a plug flow manner without mixing by using a combined electrode/PEM design. For such plug flow reactor architecture a single channel was formed between two nonconductive plates (polycarbonate) that were parted into two halves by the electrode/PEM assembly. The PEM (nafion 117) was hot pressed to the cathode and placed on top of the anode, forming the PEM/electrode assembly sandwiched between the two polycarbonate plates (Min and Logan 2004). Rabaey et al. (2003) examined a rectangular MFC comprising four cells, which were joined into a single block, thus giving rise to four graphite electrodes (Rabaey et al. 2003). This dual-chamber assembly would be difficult to apply to larger systems for continuous treatment of wastewater (Liu et al. 2004). Although stirring and bubbling in MFC operation are helpful, their energy demands are usually greater than the power yields fromthe MFC (Du et al. 2007).

Among the different MFC configurations and designs that have been technologically advanced, the air-cathode MFC (Figure 12.6) is the most prospective configuration to be scaled up for wastewater treatment due to its simple structure, relatively low cost, and high power production. Energy production is greatly affected by solution conditions such as organic substrate concentration (Liu et al. 2004), temperature, pH, and conductivity (Liu et al. 2005b, Cheng and Logan 2011). To increase energy output and reduce the cost of MFCs, Liu and Logan (2004) examined electricity production in an air-cathode MFC in the presence and absence of a polymeric PEM. Power density was found to be much higher than usually reported for aqueous-cathode MFCs. Removing the PEM increased the power density; however, without PEM microbes in the anode chamber could attack organic feed due to aerobic oxidation and hence substrate loss was observed (Liu and Logan 2004). Therefore, power generation from an air-cathode MFC still remains less than that from dual-chamber MFCs with aqueous cathode (You et al. 2006, 2008). To overcome major bottleneck in terms of the limited efficiency of the open-air cathodes, novel catalysts are required to be developed that directly reduce oxygen without unacceptable activation losses (Pham et al. 2006).

Another approach to improve power generation is by adopting up-flow anode configuration. In this novel anode design, substrate flows over the attached biofilm. Thus during continuous mode of operation, increased energy production and less internal resistance could be realized (Rabaey et al. 2005a, He et al. 2006). Recently, a tubular air-cathode MFC has been developed by Kim et al. (2009) (Figure 12.7). The tubular up-flow MFC consists of a cylindrical anode tube with equispaced holes at regular intervals and the region with holes is wrapped with the membrane. The cathode remains in contact with the membrane;

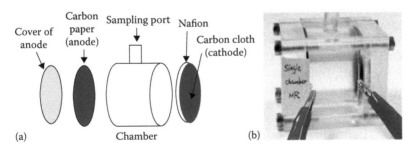

FIGURE 12.6
(a) Schematic and (b) laboratory-scale design of a single chamber. (From Liu, H., and B.E. Logan, *Environ. Sci. Technol.*, 38, 4040–4046, 2004. With permission.)

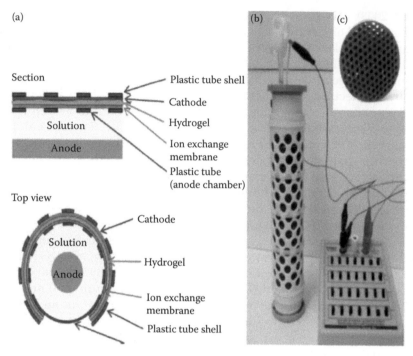

FIGURE 12.7
Tubular up-flow air-cathode MFC. (a) Section and top views, (b) system view, and (c) porous monolithic carbon anode. (From Kim, J.R. et al., *J. Power Sources*, 187, 393–399, 2009. With permission.)

at the same time the cathode is exposed to the air at outer periphery. A shell is formed by another longitudinal tube that has holes of the same diameter and space opening as of the inner tube. This shell is fastened around the cathode to hold it onto the membrane, thus providing mechanical strength to the entire structure (Kim et al. 2009). Rabaey et al. (2005a) tested a tubular, single-chamber up-flow MFC using a granular graphite matrix as the anode material and ferricyanide liquid as the cathode material. In another study, Kangping et al. (2010) constructed a single-chamber tubular MFC without PEM using a sponge-like stainless steel wire as anode material reducing the construction costs to a very low level and reducing the operation costs by no aeration during run-time. Its performance is superior to an ordinary dual-chamber MFC due to low cost, simple structure, and high energy efficiency (Kangping et al. 2010). Tubular designs are very popular (at least in MFCs) because it is believed that cylindrical reactors enable near optimal cross-sectional dimensions and thus result in minimal dead space (Janicek et al. 2014, Escapa et al. 2016). The issues to focus on with tubular MFCs are the accumulation of nonsoluble organics in the reactor and the presence of an alternative electron acceptor that attracts the electrons to divert toward them. The current operating cost is too high compared with the cost for other sources of renewable electricity (Rabaey et al. 2005a).

He et al. (2005) developed a continuously fed up-flow microbial fuel cell (UMFC), by combining the advantages of the UASB system with the requirements for a dual-chamber MFC. It is important to note that the batch-fed MFC is most studied configuration, but it is not easy to scale up; nevertheless the UMFC can be easily scaled up especially for continuous mode operation. However, the power production of this UMFC is less due to the significant internal resistance of the up-flow that minimizes advantage of microbial electricity production in such membrane-less reactors (Rabaey et al. 2005b). In addition, due to recirculation of the fluid, the energy consumption for pumping fluid through the system is much more than net power gain from the system. Therefore, the primary function of this system is wastewater purification rather than energy harnessing (Du et al. 2007).

The use of series or parallel stacked MFCs can achieve enhanced voltages and currents produced by MFCs (Pham et al. 2006). Aelterman et al. (2006) connected six individual MFC units as a stacked configuration in series or parallel. With continuous advances in MFC configuration, electricity production can be improved and at the same time production and operating cost can be reduced (Aelterman et al. 2006). Still scaling up of an MFC for the field study is in a budding stage (Vineetha and Shibu 2013).

12.8 Hydrogen Production

It has been discovered that hydrogen gas can be produced using electrolysis principles in a MFC in the presence of bacteria. In normal MFC operation, organic compounds are attacked by microbes to release carbon dioxide, protons, and electrons. Electrons transport from anode to cathode via the external electric circuit. At the cathode oxygen accepts electrons and spontaneous electricity generation in the MFC takes place with the formation of H_2O as per Equation 12.2, but if oxygen is absent in the cathodic chamber, then current generation is not spontaneous. However, in the absence of oxygen, by applying little voltage of the order of ~0.2 V it is possible to practically produce hydrogen gas at the cathode through the proton reduction process (Equation 12.3)

$$2H^+ + 2e^- = H_2 \qquad (12.3)$$

The process is known as electrohydrogenesis or microbial electrolysis, the bacteria are exoelectrogens, and the reactors are named as microbial electrolysis cells (MECs).

In an MFC, due to air cathodes, oxygen can leak toward the anode compartment. Hence, the presence of oxygen in the anode chamber negatively affects the growth of obligate anaerobic bacteria. On the other hand, the MEC does not have any air cathode or oxygen seepage possibility. So, perfect anaerobic conditions are maintained in the MEC that promote the comfortable growth of exoelectrogenic microbes like *Geobacter spp.* species. Other methanogenic bacteria and nonexoelectrogenic fermentative microbes can also be sustained.

Like an MFC, both pure compounds and real wastewater samples have been used in an MEC in batch as well as continuous mode operation. Industrial wastewater, sludge, and so on have been used and different microbial sources have been investigated. Different types of anode and cathode materials, and reactor configurations have been tested. Few pilot scale studies are also available. A summary of some literature on hydrogen production as well as treatment of wastewater is described below.

Montpart et al. (2015) evaluated the performance of a single-chamber MEC for hydrogen generation using simulated wastewaters containing different carbon sources. The attained current intensities were 62.5 A/m^3 for milk, 50 A/m^3 for glycerol, and 9 A/m^3 for starch as a carbon source in the synthetic solution. At a voltage of 0.8 V, hydrogen production, current intensity, and cathodic gas recovery for milk were 0.94 $m^3/m^{-3}/d$, 150 A/m^3, and 91%, respectively.

Escapa et al. (2015) assessed the capability of two twin membrane-less MEC units to treat domestic wastewater in batch and continuous mode. During the first 48h batch experiments, more than 80% H_2 production and 92% COD removal efficiency were achieved. However, the same MEC could not produce satisfactory results in terms of COD removal and hydrogen generation when operated in continuous mode. The possible shortfalls could be poor mixing in the MEC compartment, poor design of the anodic chamber, and the lack of effective hydrogen management on the cathode side and sewage quality itself.

Gupta and Parkhey (2015) built a 2 L capacity single-chamber MEC employing a rice straw as an organic substrate. The anode and cathode were made of acrylic + graphite sheets and a stainless-steel mesh, respectively. During oxidation of the rice straw in the MEC reactor, the maximum H_2 yield per gram of COD, cathodic H_2 recovery, total H_2 recovery, and coulombic efficiency achieved were 801 mL H_2, 58%, 51%, and 88%, respectively when *Shewanella putrefaciens* was used as exoelectrogen. The total energy efficiency of the MEC was 74%.

Feng et al. (2014) successfully achieved significant amount of hydrogen production and high energy recovery at higher applied voltage (>0.5 V) in a sleeve-shape MEC. In this MEC construction, graphite felt anode, titanium coated platinum cathode, and acetate were used as a substrate. In this novel MFC, the anode was surrounded by the cathode when operated in batch mode experiments. Within 24 h the voltage was increased from 0.2 to 1 V and both hydrogen production and recovery rate were increased in the range 0.14–2.36 L/L/d and 40.15% to 86.13%, respectively. The better performance could be attributed to the bigger system capacity of the MEC, little gap between electrodes, higher surface area of the cathode, and improved configuration.

Brown et al. (2014) operated an MEC continuously for anodic treatment of actual wastewater. The hydraulic retention time and organic COD loading were set as 1.23 days and 0.5 g $O_2/d/L_{Reactor}$. They achieved 67% COD removal efficiency and 40% total nitrogen removal efficiency. The coulombic efficiency and current density were reported as 11% and 72 µA/cm², respectively.

Heidrich et al. (2014) treated raw sewage in an MEC from very low (1°C) to room temperature (22°C) condition for a 1-year period. They continuously produced 0.6 L/day of hydrogen from a 100 L capacity MEC reactor. Although gas collection reduced with time, coulombic efficiency and electrical energy input was recovered at 41.2% and 48.7%, respectively. Still, uneven COD removal was observed and COD removal achieved was below the prescribed standards. This could be attributed to the poor pumping system as well as huge overpotential accounted in the MEC system.

Hou et al. (2014) showed that hydrogen production in an MEC can be improved by using UV irradiation. In an MEC with UV irradiation, hydrogen concentration (>91%) was obtained with 1.5 times energy efficiency. In contrast, methane gas concentration reached up to 94% in an MEC without irradiation.

Heidrich et al. (2013) operated a pilot plant microbial electrolysis cell (MEC) for a period of three months using sewage as a carbon source in Northern England. They kept the volumetric organic loading rate of the order of 0.14 kg COD/m³/d lower than typically used for the conventional process. The estimated costing for energy consumption was 2.3 kJ/g COD for 120 L plant capacity. From this pilot plant process, 100% pure 0.015 L H_2/L/d equivalent hydrogen was produced. The reported value of coulombic efficiency was 55%. The authors opined that with the modified design of the reactor and better mechanism for capturing hydrogen more COD removal efficiency and electrical energy recovery could be achieved.

Manuel et al. (2010) compared the performance of an MEC using novel Ni alloy and multicomponent cathode in a continuous MEC. The applied loading of Ni catalyst of

0.6 mg/cm^2 on Ni-based alloy produced 4.1 L/L$_{Reactor}$/d hydrogen gas. The multicomponent cathode made of Ni, Mo, Cr, and Fe with overall catalyst loading of 1 mg cm^2 achieved 2.8–3.7 L/L$_{Reactor}$/d hydrogen gas generation.

Escapa et al. (2009) examined hydrogen production in a MEC using glycerol feed. In this MEC, an architecture graphite-felt anode and a gas-phase cathode were separated by very thin highly porous synthetic fabric.

Tartakovsky et al. (2009) demonstrated hydrogen production without the PEM in a MEC operated in continuous mode. In this MEC, configuration anode and cathode materials were carbon-felt and Pt (0.5 mg cm^2) loaded gas-phase cathode, respectively. J-cloth was used for partitioning electrodes. A very low internal resistance (19 Ω) was reported due to membrane-less design and small gap (0.3 mm) between electrodes. This MEC was capable of producing 6.3 LSTP/L$_{Reactor}$/d hydrogen.

Sun et al. (2009b) configured a hybrid MFC and MEC for hydrogen production. The attractive feature of this configuration is utilization of electricity produced by an energy-assisted MFC for powering an MEC. No additional energy is required for hydrogen production. The desired power input to the MEC can be controlled by connecting different resistances in series. The volumetric hydrogen production rate was varied from 2.9 to 0.2 mL/L/d and the circuit current was changed from 78 to 9 mA/m^2, respectively, when load was altered from 10 Ω to 10 kΩ. Also, it has been observed that by increasing load the coulombic efficiency and H$_2$ yield are reduced. To overcome this limitation, more numbers of MFCs were connected in series; as a result a significant rise in hydrogen production was achieved. In comparison to series connection, parallel connection resulted in less hydrogen production.

Wagner et al. (2009) reported that during swine wastewater treatment by using an MEC the generation of hydrogen and methane gas could help to lower the treatment costs. They reported that hydrogen generation rate and COD removal efficiency were in the range of 0.9–1.0 m^3/m^3 d H$_2$ and 8% to 29% respectively at 20h residence time. At 184h longer retention time, 69%–75% COD removal was reported with a graphite-fiber brush as an anodic material.

Mohan et al. (2008a) studied biohydrogen (H$_2$) production in an anaerobic sequencing batch biofilm reactor treating distillery wastewater. The reactor was tested at two different pH and 24h retention time in the batch study. The test results revealed that acidophilic pH is a favorable condition for hydrogen production.

Hu et al. (2008) designed single-chamber membrane-free MECs and used them to examine hydrogen production by both a mixed culture and a pure culture (*S. oneidensis MR-1*).

For the mixed culture MFC study, at fixed applied voltage (0.6 V) and under neutral pH conditions, 0.53 m^3/day/m^3 of hydrogen gas was produced and 9.3 A/m^2 current density was achieved. When pH was set at 5.8 at the same voltage, higher hydrogen production rate of the order of 0.69 m^3/day/m^3 and 14 A/m^2 current density were measured. Also, it has been reported that *S. oneidensis* from lactic acid has the capacity to produce stable hydrogen gas.

12.9 Conclusion and Future Prospects

MFCs and MECs have potential to convert the chemical energy of organic compounds present in solution or real wastewater to electricity and hydrogen, respectively, at suitable operating conditions. Significant development has taken place for the understanding of the mechanism of MFCs and MECs as well as the effects of process parameters on their performance. However, these technologies are still in the early stage of development and not matured. The MFCs are currently limited to laboratory studies due to their poor power output.

A significant number of reports are available for the improvement of the performance of MFCs and MECs through modification in anode and cathode materials, reactor design, mediator, and microorganisms. Extensive research is going on around the world on the application of catalyst, increment of active surface area of anode and cathode, use of innovative bacterial species with favorable electricity harnessing potential as well as development of low-cost PEM. For continuous generation of electricity, different circuit designs using capacitor are being investigated. Different modes of operation such as batch and continuous mode are being tested. Although batch mode operation is found to have more power generation efficiency, continuous mode operation seems promising for handling large volume of wastewater in industrial sector. Membrane-less MFCs and MECs are also gaining interest due to their relatively low cost and ease of operation. The MFC and MEC technologies show great potential in treating wastewater and generating power and hydrogen gas due to their great flexibility, but much more needs to be done to commercialize these technologies.

References

Aelterman, P., K. Rabaey, H.T. Pham, N. Boon, and W. Verstraete. 2006. Continuous Electricity Generation at High Voltages and Currents using Stacked Microbial Fuel Cells. *Environmental Science & Technology* 40(10):3388–3394.

An, B.M., Y.H. Song, J.W. Shin, and J.Y. Park. 2016. Two-Chamber Microbial Fuel Cell to Simultaneously Remove Ethanolamine and Nitrate. *Desalination and Water Treatment* 57(17):7866–7873.

An, J., D. Kim, Y. Chun, S.J. Lee, H.Y. Ng, and I.S. Chang. 2009. Floating-type Microbial Fuel Cell (FT-MFC) for Treating Organic-contaminated Water. *Environmental Science & Technology* 43(5):1642–1647.

Angenent, L.T., and S. Sung. 2001. Development of Anaerobic Migrating Blanket Reactor (AMBR), A Novel Anaerobic Treatment System. *Water Research* 35(7):1739–1747.

Antonopoulou, G., K. Stamatelatou, S. Bebelis, and G. Lyberatos. 2010. Electricity Generation from Synthetic Substrates and Cheese Whey Using a Two Chamber Microbial Fuel Cell. *Biochemical Engineering Journal* 50(1):10–15.

Asodariya, H., and P. Patel. 2011. Evolution of Microbial Fuel Cell as a Promising Technology for Waste Water Treatment. International Conference on Current Trends in Technology, 'NUiCONE—2011'. http://nuicone.org/site/common/proceedings/Chemical/poster/CH_31.pdf.

Bond, D.R., and D.R. Lovley. 2005. Evidence for Involvement of an Electron Shuttle in Electricity Generation by Geothrix Fermentans. *Applied and Environmental Microbiology* 71(4):2186–2189.

Brown, R.K., F. Harnisch, S. Wirth, H. Wahlandt, T. Dockhorn, N. Dichtl, and U. Schröder. 2014. Evaluating the Effects of Scaling up on the Performance of Bioelectrochemical Systems Using a Technical Scale Microbial Electrolysis Cell. *Bioresource Technology* 163:206–213.

Bullen, R.A., T.C. Arnot, J.B. Lakeman, and F.C. Walsh. 2006. Biofuel Cells and Their Development. *Biosensors and Bioelectronics* 21(11):2015–2045.

Cao, X., X. Huang, P. Liang, K. Xiao, Y. Zhou, X. Zhang, and B.E. Logan. 2009. A New Method for Water Desalination Using Microbial Desalination Cells. *Environmental Science & Technology* 43(18):7148–7152.

Catal, T., K. Li, H. Bermek, and H. Liu. 2008. Electricity Production from Twelve Monosaccharides Using Microbial Fuel Cells. *Journal of Power Sources* 175(1):196–200.

Chaudhuri, S.K., and D.R. Lovley. 2003. Electricity Generation by Direct Oxidation of Glucose in Mediatorless Microbial Fuel Cells. *Nature Biotechnology* 21(10):1229–1232.

Cheng, H.Y., B. Liang, Y. Mu, M.H. Cui, K. Li, W.M. Wu, and A.J. Wang. 2015. Stimulation of Oxygen to Bioanode for Energy Recovery from Recalcitrant Organic Matter Aniline in Microbial Fuel Cells (MFCs). *Water Research* 81:72–83.

Cheng, S., and B.E. Logan. 2011. Increasing Power Generation for Scaling up Single-Chamber Air Cathode Microbial Fuel Cells. *Bioresource Technology* 102(6):4468–4473.

Chung, K., and S. Okabe. 2009. Continuous Power Generation and Microbial Community Structure of the Anode Biofilms in a Three-stage Microbial Fuel Cell System. *Applied Microbiology and Biotechnology* 83(5):965–977.

Daniel, D.K., B.D. Mankidy, K. Ambarish, and R. Manogari. 2009. Construction and Operation of a Microbial Fuel Cell for Electricity Generation from Wastewater. *International Journal of Hydrogen Energy* 34(17):7555–7560.

Davis, F., and S.P. Higson. 2007. Biofuel Cells—Recent Advances and Applications. *Biosensors and Bioelectronics* 22(7):1224–1235.

Du, Z., H. Li, and T. Gu. 2007. A State of the Art Review on Microbial Fuel Cells: A Promising Technology for Wastewater Treatment and Bioenergy. *Biotechnology Advances* 25(5):464–482.

Escapa, A., M.F. Manuel, A. Morán, X. Gómez, S.R. Guiot, and B. Tartakovsky. 2009. Hydrogen Production from Glycerol in a Membraneless Microbial Electrolysis Cell. *Energy & Fuels* 23(9):4612–4618.

Escapa, A., R. Mateos, E.J. Martínez, and J. Blanes. 2016. Microbial Electrolysis Cells: An Emerging Technology for Wastewater Treatment and Energy Recovery. From Laboratory to Pilot Plant and Beyond. *Renewable and Sustainable Energy Reviews* 55:942–956.

Escapa, A., M.I. San-Martín, R. Mateos, and A. Morán. 2015. Scaling-up of Membraneless Microbial Electrolysis Cells (MECs) for Domestic Wastewater Treatment: Bottlenecks and Limitations. *Bioresource Technology* 180:72–78.

Feng, Y., Y. Cheng, Y. Du, Q. Teng, and H. Li. 2014. Hydrogen Production from Acetate in a Sleeve Shape Microbial Electrolysis Cell with a Mipor Cathode. *International Journal of Electrochemical Science* 9:6993–7002.

Feng, Y., X. Wang, B.E. Logan, and H. Lee. 2008. Brewery Wastewater Treatment Using Air-cathode Microbial Fuel Cells. *Applied Microbiology and Biotechnology* 78(5):873–880.

Franks, A.E., and K.P. Nevin. 2010. Microbial Fuel Cells, A Current Review. *Energies* 3(5):899–919.

Ghangrekar, M.M., and V.B. Shinde. 2007. Performance of Membrane-less Microbial Fuel Cell Treating Wastewater and Effect of Electrode Distance and Area on Electricity Production. *Bioresource Technology* 98(15):2879–2885.

Greenman, J., A. Gálvez, L. Giusti, and I. Ieropoulos. 2009. Electricity from Landfill Leachate Using Microbial Fuel Cells: Comparison with a Biological Aerated Filter. *Enzyme and Microbial Technology* 44(2):112–119.

Gupta, P., and P. Parkhey. 2015. Design of a Single Chambered Microbial Electrolytic Cell Reactor for Production of Biohydrogen from Rice Straw Hydrolysate. *Biotechnology Letters* 37(6):1213–1219.

He, Z., S.D. Minteer, and L.T. Angenent. 2005. Electricity Generation from Artificial Wastewater Using an Upflow Microbial Fuel Cell. *Environmental Science & Technology* 39(14):5262–5267.

He, Z., N. Wagner, S.D. Minteer, and L.T. Angenent. 2006. An Upflow Microbial Fuel Cell with An Interior Cathode: Assessment of the Internal Resistance by Impedance Spectroscopy. *Environmental Science & Technology* 40(17):5212–5217.

Heidrich, E.S., J. Dolfing, K. Scott, S.R. Edwards, C. Jones, and T.P. Curtis. 2013. Production of Hydrogen from Domestic Wastewater in a Pilot-scale Microbial Electrolysis Cell. *Applied Microbiology and Biotechnology* 97(15):6979–6989.

Heidrich, E.S., S.R. Edwards, J. Dolfing, S.E. Cotterill, and T.P. Curtis. 2014. Performance of a Pilot Scale Microbial Electrolysis Cell Fed on Domestic Wastewater at Ambient Temperatures for a 12 month Period. *Bioresource Technology* 173:87–95.

Heilmann, J., and B.E. Logan. 2006. Production of Electricity from Proteins Using a Microbial Fuel Cell. *Water Environment Research* 78(5):531–537.

Hou, Y., H. Luo, G. Liu, R. Zhang, J. Li, and S. Fu. 2014. Improved Hydrogen Production in the Microbial Electrolysis Cell by Inhibiting Methanogenesis Using Ultraviolet Irradiation. *Environmental Science & Technology* 48(17):10482–10488.

Hu, H., Y. Fan, and H. Liu. 2008. Hydrogen Production Using Single-chamber Membrane-free Microbial Electrolysis Cells. *Water Research* 42(15):4172–4178.

Huang, L., S. Cheng, F. Rezaei, and B.E. Logan. 2009. Reducing Organic Loads in Wastewater Effluents from Paper Recycling Plants Using Microbial Fuel Cells. *Environmental Technology* 30(5):499–504.

Huang, L., and B.E. Logan. 2008. Electricity Generation and Treatment of Paper Recycling Wastewater Using a Microbial Fuel Cell. *Applied Microbiology and Biotechnology* 80(2):349–355.

Huang, L., J.M. Regan, and X. Quan. 2011. Electron Transfer Mechanisms, New Applications, and Performance of Biocathode Microbial Fuel Cells. *Bioresource Technology* 102(1):316–323.

Ieropoulos, I., J. Greenman, C. Melhuish, and J. Hart. 2005. Energy Accumulation and Improved Performance in Microbial Fuel Cells. *Journal of Power Sources* 145(2):253–256.

Jadhav, G.S., and M.M. Ghangrekar. 2009. Performance of Microbial Fuel Cell Subjected to Variation in pH, Temperature, External Load and Substrate Concentration. *Bioresource Technology* 100(2):717–723.

Jang, J.K., T.H. Pham, I.S. Chang, K.H. Kang, H. Moon, K.S. Cho, and B.H. Kim. 2004. Construction and Operation of a Novel Mediator-and Membrane-less Microbial Fuel Cell. *Process Biochemistry* 39(8):1007–1012.

Janicek, A., Y. Fan, and H. Liu. 2014. Design of Microbial Fuel Cells for Practical Application: A Review and Analysis of Scale-up Studies. *Biofuels* 5(1):79–92.

Jiang, J., Q. Zhao, J. Zhang, G. Zhang, and D.J. Lee. 2009. Electricity Generation from Bio-treatment of Sewage Sludge with Microbial Fuel Cell. *Bioresource Technology* 100(23):5808–5812.

Kangping, C., W. Ye, and S. Shiqun. 2010. Electricity Generation and Wastewater Treatment Using an Air-Cathode Single Chamber Microbial Fuel Cell. *Power and Energy Engineering Conference (APPEEC), Asia-Pacific,* pp. 1–6. IEEE.

Karra, U., S.S. Manickam, J.R. McCutcheon, N. Patel, and B. Li. 2013. Power Generation and Organics Removal from Wastewater Using Activated Carbon Nanofiber (ACNF) Microbial Fuel Cells (MFCs). *International Journal of Hydrogen Energy* 38(3):1588–1597.

Khanal, S.K., R.Y. Surampalli, T.C. Zhang, B.P. Lamsal, R.D. Tyagi, and C.M. Kao. 2010. Bioenergy and Biofuel from Biowastes and Biomass. *American Society of Civil Engineers (ASCE)* 23(117):276–296.

Kim, J.R., S. Cheng, S.E. Oh, and B.E. Logan. 2007a. Power Generation Using Different Cation, Anion, and Ultrafiltration Membranes in Microbial Fuel Cells. *Environmental Science & Technology* 41(3):1004–1009.

Kim, J.R., S.H. Jung, J.M. Regan, and B.E. Logan. 2007b. Electricity Generation and Microbial Community Analysis of Alcohol Powered Microbial Fuel Cells. *Bioresource Technology* 98(13):2568–2577.

Kim, J.R., G.C. Premier, F.R. Hawkes, R.M. Dinsdale, and A.J. Guwy. 2009. Development of a Tubular Microbial Fuel Cell (MFC) Employing a Membrane Electrode Assembly Cathode. *Journal of Power Sources* 187(2):393–399.

Kiran, V., and B. Gaur. 2013. Microbial Fuel Cell: Technology for Harvesting Energy from Biomass. *Reviews in Chemical Engineering* 29(4):189–203.

Krishnaraj, R.N., and J.-S. Yu. 2015. *Bioenergy: Opportunities and Challenges*. Apple Academic Press, Oakville, Canada.

Kubota, K., T. Watanabe, T. Yamaguchi, and K. Syutsubo. 2016. Characterization of Wastewater Treatment by two Microbial Fuel Cells in Continuous Flow Operation. *Environmental Technology* 37(1):114–120.

Lee, Y., and N. Nirmalakhandan. 2011. Electricity Production in Membrane-less Microbial Fuel Cell Fed with Livestock Organic Solid Waste. *Bioresource Technology* 102(10):5831–5835.

Lefebvre, O., W.K. Ooi, Z. Tang, M. Abdullah-Al-Mamun, D.H. Chua, and H.Y. Ng. 2009. Optimization of a Pt-free Cathode Suitable for Practical Applications of Microbial Fuel Cells. *Bioresource Technology* 100(20):4907–4910.

Li, H., Y. Tian, W. Zuo, J. Zhang, X. Pan, L. Li, and X. Su. 2016. Electricity Generation from Food Wastes and Characteristics of Organic Matters in Microbial Fuel Cell. *Bioresource Technology* 205:104–110.

Li, Z., X. Zhang, and L. Lei. 2008. Electricity Production During the Treatment of Real Electroplating Wastewater Containing Cr 6+ Using Microbial Fuel Cell. *Process Biochemistry* 43(12):1352–1358.

Li, Z., X. Zhang, Y. Zeng, and L. Lei. 2009. Electricity Production by An Overflow-type Wetted-wall Microbial Fuel Cell. *Bioresource Technology* 100(9):2551–2555.

Liu, H., S. Cheng, and B.E. Logan. 2005a. Power Generation in Fed-batch Microbial Fuel Cells as a Function of Ionic Strength, Temperature, and Reactor Configuration. *Environmental Science & Technology* 39(14):5488–5493.

Liu, H., S. Cheng, and B.E. Logan. 2005b. Production of Electricity from Acetate or Butyrate Using a Single-chamber Microbial Fuel Cell. *Environmental Science & Technology* 39(2):658–662.

Liu, H., and B.E. Logan. 2004. Electricity Generation Using An Air-cathode Single Chamber Microbial Fuel Cell in the Presence and Absence of a Proton Exchange Membrane. *Environmental Science & Technology* 38(14):4040–4046.

Liu, H., R. Ramnarayanan, and B.E. Logan. 2004. Production of Electricity During Wastewater Treatment Using a Single Chamber Microbial Fuel Cell. *Environmental Science & Technology* 38(7):2281–2285.

Logan, B., S. Cheng, V. Watson, and G. Estadt. 2007. Graphite Fiber Brush Anodes for Increased Power Production in Air-cathode Microbial Fuel Cells. *Environmental Science & Technology* 41(9):3341–3346.

Logan, B.E. 2008. *Microbial Fuel Cells*. John Wiley & Sons, Wiley Publications, New Jersey, pp. 4–5.

Logan, B.E., B. Hamelers, R. Rozendal, U. Schröder, J. Keller, S. Freguia, P. Aelterman, W. Verstraete, and K. Rabaey. 2006. Microbial Fuel Cells: Methodology and Technology. *Environmental Science & Technology* 40(17):5181–5192.

Logan, B.E., C. Murano, K. Scott, N.D. Gray, and I.M. Head. 2005. Electricity Generation from Cysteine in a Microbial Fuel Cell. *Water Research* 39(5):942–952.

Lovley, D.R. 2006a. Bug Juice: Harvesting Electricity with Microorganisms. *Nature Reviews Microbiology* 4(7):497–508.

Lovley, D.R. 2006b. Microbial Fuel Cells: Novel Microbial Physiologies and Engineering Approaches. *Current Opinion in Biotechnology* 17(3):327–332.

Lu, N., S.G. Zhou, L. Zhuang, J.T. Zhang, and J.R. Ni. 2009. Electricity Generation from Starch Processing Wastewater Using Microbial Fuel Cell Technology. *Biochemical Engineering Journal* 43(3):246–251.

Luo, H., G. Liu, R. Zhang, and S. Jin. 2009. Phenol Degradation in Microbial Fuel Cells. *Chemical Engineering Journal* 147(2):259–264.

Luo, Y., R. Zhang, G. Liu, J. Li, M. Li, and C. Zhang. 2010. Electricity Generation from Indole and Microbial Community Analysis in the Microbial Fuel Cell. *Journal of Hazardous Materials* 176(1):759–764.

Mansoorian, H.J., A.H. Mahvi, A.J. Jafari, M.M. Amin, A. Rajabizadeh, and N. Khanjani. 2013. Bioelectricity Generation Using Two Chamber Microbial Fuel Cell Treating Wastewater from Food Processing. *Enzyme and Microbial Technology* 52(6):352–357.

Manuel, M.F., V. Neburchilov, H. Wang, S.R. Guiot, and B. Tartakovsky. 2010. Hydrogen Production in a Microbial Electrolysis Cell with Nickel-based Gas Diffusion Cathodes. *Journal of Power Sources* 195(17):5514–5519.

Mathuriya, A.S. 2014. Eco-affectionate face of Microbial Fuel Cells. *Critical Reviews in Environmental Science and Technology* 44(2):97–153.

Min, B., S. Cheng, and B.E. Logan. 2005. Electricity Generation Using Membrane and Salt Bridge Microbial Fuel Cells. *Water Research* 39(9):1675–1686.

Min, B., and B.E. Logan. 2004. Continuous Electricity Generation from Domestic Wastewater and Organic Substrates in a Flat Plate Microbial Fuel Cell. *Environmental Science & Technology* 38(21):5809–5814.

Mohan, S.V., G. Mohanakrishna, S.V. Ramanaiah, and P.N. Sarma. 2008a. Simultaneous Biohydrogen Production and Wastewater Treatment in Biofilm Configured Anaerobic Periodic Discontinuous Batch Reactor Using Distillery Wastewater. *International Journal of Hydrogen Energy* 33(2):550–558.

Mohan, S.V., S.V. Raghavulu, and P.N. Sarma. 2008b. Biochemical Evaluation of Bioelectricity Production Process from Anaerobic Wastewater Treatment in a Single Chambered Microbial Fuel Cell (MFC) Employing Glass Wool Membrane. *Biosensors and Bioelectronics* 23(9):1326–1332.

Mohan, S.V., S.V. Raghavulu, S. Srikanth, and P.N. Sarma. 2007. Bioelectricity Production by Mediatorless Microbial Fuel Cell Under Acidophilic Condition Using Wastewater as Substrate: Influence of Substrate Loading Rate. *Current Science* 92(12):1726.

Mohanakrishna, G., S.V. Mohan, and P.N. Sarma. 2010. Bio-electrochemical Treatment of Distillery Wastewater in Microbial Fuel Cell Facilitating Decolorization and Desalination Along with Power Generation. *Journal of Hazardous Materials* 177(1):487–494.

Montpart, N., L. Rago, J.A. Baeza, and A. Guisasola. 2015. Hydrogen Production in Single Chamber Microbial Electrolysis Cells with Different Complex Substrates. *Water Research* 68:601–615.

Nam, J.Y., H.W. Kim, K.H. Lim, and H.S. Shin. 2010a. Effects of Organic Loading Rates on the Continuous Electricity Generation from Fermented Wastewater Using a Single-Chamber Microbial Fuel Cell. *Bioresource Technology* 101(1):33–37.

Nam, J.Y., H.W. Kim, K.H. Lim, H.S. Shin, and B.E. Logan. 2010b. Variation of Power Generation at Different Buffer Types and Conductivities in Single Chamber Microbial Fuel Cells. *Biosensors and Bioelectronics* 25(5):1155–1159.

Nevin, K.P., H. Richter, S.F. Covalla, J.P. Johnson, T.L. Woodard, A.L. Orloff, H. Jia, M. Zhang, and D.R. Lovley. 2008. Power Output and Columbic Efficiencies from Biofilms of *Geobacter sulfurreducens* Comparable to Mixed Community Microbial Fuel Cells. *Environmental Microbiology* 10(10):2505–2514.

Nwokocha, J.V., J.N. Nwaulari, and A.N. Lebe. 2012. The Microbial Fuel Cell: The Solution to the Global Energy and Environmental Crises. *International Journal of Academic Research in Progressive Education and Development* 1:2226–6348.

Oh, S.E., J.R. Kim, J.H. Joo, and B.E. Logan. 2009. Effects of Applied Voltages and Dissolved Oxygen on Sustained Power Generation by Microbial Fuel Cells. *Water Science and Technology* 60(5):1311.

Oh, S.T., J.R. Kim, G.C. Premier, T.H. Lee, C. Kim, and W.T. Sloan. 2010. Sustainable Wastewater Treatment: How Might Microbial Fuel Cells Contribute. *Biotechnology Advances* 28:871–881.

Park, D.H., and J.G. Zeikus. 2000. Electricity Generation in Microbial Fuel Cells Using Neutral Red as an Electronophore. *Applied and Environmental Microbiology* 66(4):1292–1297.

Park, D.H., and J.G. Zeikus. 2003. Improved Fuel Cell and Electrode Designs for Producing Electricity from Microbial Degradation. *Biotechnology and Bioengineering* 81(3):348–355.

Peguin, S., and P. Soucaille. 1995. Modulation of Carbon and Electron Flow in Clostridium Acetobutylicum by Iron limitation and Methyl Viologen Addition. *Applied and Environmental Microbiology* 61(1):403–405.

Pham, T.H., K. Rabaey, P. Aelterman, P. Clauwaert, L. De Schamphelaire, N. Boon, and W. Verstraete. 2006. Microbial Fuel Cells in Relation to Conventional Anaerobic Digestion Technology. *Engineering in Life Sciences* 6(3):285–292.

Pushkar, P., and A.K. Mungray. 2015. Real Textile and Domestic Wastewater Treatment by Novel Cross-linked Microbial Fuel Cell (CMFC) Reactor. *Desalination and Water Treatment* 57(15):6747–6760.

Rabaey, K., P. Clauwaert, P. Aelterman, and W. Verstraete. 2005a. Tubular Microbial Fuel Cells for Efficient Electricity Generation. *Environmental Science & Technology* 39(20):8077–8082.

Rabaey, K., G. Lissens, S.D. Siciliano, and W. Verstraete. 2003. A Microbial Fuel Cell Capable of Converting Glucose to Electricity at High Rate and Efficiency. *Biotechnology Letters* 25(18):1531–1535.

Rabaey, K., W. Ossieur, M. Verhaege, and W. Verstraete. 2005b. Continuous Microbial Fuel Cells Convert Carbohydrates to Electricity. *Water Science & Technology* 52(1):515–523.

Rabaey, K., and W. Verstraete. 2005. Microbial Fuel Cells: Novel Biotechnology for Energy Generation. *TRENDS in Biotechnology* 23(6):291–298.

Rahimnejad, M., A.A. Ghoreyshi, G. Najafpour, and T. Jafary. 2011. Power Generation from Organic Substrate in Batch and Continuous Flow Microbial Fuel Cell Operations. *Applied Energy* 88(11):3999–4004.

Ramasamy, R.P., Z. Ren, M.M. Mench, and J.M. Regan. 2008. Impact of Initial Biofilm Growth on the Anode Impedance of Microbial Fuel Cells. *Biotechnology and Bioengineering* 101(1):101–108.

Reguera, G., K.P. Nevin, J.S. Nicoll, S.F. Covalla, T.L. Woodard, and D.R. Lovley. 2006. Biofilm and Nanowire Production Leads to Increased Current in *Geobacter sulfurreducens* Fuel Cells. *Applied and Environmental Microbiology* 72(11):7345–7348.

Rinaldi, A., B. Mecheri, V. Garavaglia, S. Licoccia, P. Di Nardo, and E. Traversa. 2008. Engineering Materials and Biology to Boost Performance of Microbial Fuel Cells: A Critical Review. *Energy & Environmental Science* 1(4):417–429.

Rozendal, R.A., H.V. Hamelers, K. Rabaey, J. Keller, and C.J. Buisman. 2008. Towards Practical Implementation of Bioelectrochemical Wastewater Treatment. *Trends in Biotechnology* 26(8):450–459.

Scott, K., and C. Murano. 2007. A Study of a Microbial Fuel Cell Battery Using Manure Sludge Waste. *Journal of Chemical Technology and Biotechnology* 82(9):809–817.

Sevda, S., and T.R. Sreekrishnan. 2012. Effect of Salt Concentration and Mediators in Salt Bridge Microbial Fuel Cell for Electricity Generation from Synthetic Wastewater. *Journal of Environmental Science and Health, Part A* 47(6):878–886.

Shankar, R., P. Mondal, and S. Chand. 2015a. Simultaneous Generation of Electricity and Removal of Organic Load from Synthetic Wastewater in a Membrane Less Microbial Fuel Cell: Parametric Evaluation. *Environmental Progress & Sustainable Energy* 34(1):255–264.

Shankar, R., P. Mondal, R. Singh, and S. Chand. 2015b. Simultaneous Electricity Production and Removal of Organics from Synthetic Wastewater in a Continuous Membrane Less MFC: Effects of Process Parameters. *Environmental Progress & Sustainable Energy* 34(5):1404–1413.

Shimoyama, T., S. Komukai, A. Yamazawa, Y. Ueno, B.E. Logan, and K. Watanabe. 2008. Electricity Generation from Model Organic Wastewater in a Cassette-electrode Microbial Fuel Cell. *Applied Microbiology and Biotechnology* 80(2):325–330.

Sleutels, T.H., H.V. Hamelers, R.A. Rozendal, and C.J. Buisman. 2009. Ion Transport Resistance in Microbial Electrolysis Cells with Anion and Cation Exchange Membranes. *International Journal of Hydrogen Energy* 34(9):3612–3620.

Sun, H., S. Xu, G. Zhuang, and X. Zhuang. 2016. Performance and Recent Improvement in Microbial Fuel Cells for Simultaneous Carbon and Nitrogen Removal: A Review. *Journal of Environmental Sciences* 39:242–248.

Sun, J., Y.Y. Hu, Z. Bi, and Y.Q. Cao. 2009a. Simultaneous Decolorization of Azo Dye and Bioelectricity Generation Using a Microfiltration Membrane Air-cathode Single-Chamber Microbial Fuel Cell. *Bioresource Technology* 100(13):3185–3192.

Sun, M., G.P. Sheng, Z.X. Mu, X.W. Liu, Y.Z. Chen, H.L. Wang, and H.Q. Yu. 2009b. Manipulating the Hydrogen Production from Acetate in A Microbial Electrolysis Cell–microbial Fuel Cell-coupled System. *Journal of Power Sources* 191(2):338–343.

Sund, C.J., S. McMasters, S.R. Crittenden, L.E. Harrell, and J.J. Sumner. 2007. Effect of Electron Mediators on Current Generation and Fermentation in a Microbial Fuel Cell. *Applied Microbiology and Biotechnology* 76(3):561–568.

Tartakovsky, B., M.F. Manuel, H. Wang, and S.R. Guiot. 2009. High Rate Membrane-less Microbial Electrolysis Cell for Continuous Hydrogen Production. *International Journal of Hydrogen Energy* 34(2):672–677.

Thygesen, A., F.W. Poulsen, B. Min, I. Angelidaki, and A.B. Thomsen. 2009. The Effect of Different Substrates and Humic Acid on Power Generation in Microbial Fuel Cell Operation. *Bioresource Technology* 100(3):1186–1191.

Timmins, G.S., and M.J. Davies. 1998. *Biological Free Radicals. Electron Paramagnetic Resonance.* The Royal Society of Chemistry, Cambridge, UK.

Trinh, N.T., J.H. Park, and B.W. Kim. 2009. Increased Generation of Electricity in a Microbial Fuel Cell Using *Geobacter sulfurreducens*. *Korean Journal of Chemical Engineering* 26(3):748–753.

Turick, C.E., F. Caccavo, and L.S. Tisa. 2003. Electron Transfer from Shewanella Algae BrY to Hydrous Ferric Oxide is Mediated by Cell-associated Melanin. *FEMS Microbiology Letters* 220(1):99–104.

Vega, C.A., and I. Fernández. 1987. Mediating Effect of Ferric Chelate Compounds in Microbial Fuel Cells with Lactobacillus Plantarum, Streptococcus Lactis, and Erwinia Dissolvens. *Bioelectrochemistry and Bioenergetics* 17(2):217–222.

Vineetha, V., and K. Shibu. 2013. Electricity Production Coupled with Wastewater Treatment Using Microbial Fuel Cell. In *Energy Efficient Technologies for Sustainability (ICEETS) International Conference on*. 821–826. IEEE.

Wagner, R.C., J.M. Regan, S.E. Oh, Y. Zuo, and B.E. Logan. 2009. Hydrogen and Methane Production from Swine Wastewater Using Microbial Electrolysis Cells. *Water Research* 43(5):1480–1488.

Wang, Y.P., X.W. Liu, W.W. Li, F. Li, Y.K. Wang, G.P. Sheng, R.J. Zeng, and H.Q. Yu. 2012. A Microbial Fuel Cell–membrane Bioreactor Integrated System for Cost-effective Wastewater Treatment. *Applied Energy* 98:230–235.

Wen, Q., Y. Wu, D. Cao, L. Zhao, and Q. Sun. 2009. Electricity Generation and Modeling of Microbial Fuel Cell From Continuous Beer Brewery Wastewater. *Bioresource Technology* 100(18):4171–4175.

Wen, Q., Y. Wu, L. Zhao, and Q. Sun. 2010. Production of Electricity From the Treatment of Continuous Brewery Wastewater Using a Microbial Fuel Cell. *Fuel* 89(7):1381–1385.

Woodward, L., B. Tartakovsky, M. Perrier, and B. Srinivasan. 2009. Maximizing Power Production in A Stack of Microbial Fuel Cells Using Multiunit Optimization Method. *Biotechnology Progress* 25(3):676–682.

Xu, L., Y. Zhao, L. Doherty, Y. Hu, and X. Hao. 2016. The Integrated Processes for Wastewater Treatment Based on the Principle of Microbial Fuel Cells: A Review. *Critical Reviews in Environmental Science and Technology* 46(1):60–91.

Ye, Y., X. Zhu, and B.E. Logan. 2016. Effect of Buffer Charge on Performance of Air-Cathodes Used in Microbial Fuel Cells. *Electrochimica Acta* 194:441–447.

You, S., Q. Zhao, J. Zhang, H. Liu, J. Jiang, and S. Zhao. 2008. Increased Sustainable Electricity Generation in Up-flow Air-cathode Microbial Fuel Cells. *Biosensors and Bioelectronics* 23(7):1157–1160.

You, S.J., Q.L. Zhao, J.Q. Jiang, and J.N. Zhang. 2006. Treatment of Domestic Wastewater with Simultaneous Electricity Generation in Microbial Fuel Cell under Continuous Operation. *Chemical and Biochemical Engineering Quarterly* 20(4):407–412.

Yu, J., J. Seon, Y. Park, S. Cho, and T. Lee. 2012. Electricity Generation and Microbial Community in a Submerged-Exchangeable Microbial Fuel Cell System for Low-Strength Domestic Wastewater Treatment. *Bioresource Technology* 117:172–179.

Zeng, Y., Y.F. Choo, B.H. Kim, and P. Wu. 2010. Modelling and Simulation of Two-Chamber Microbial Fuel Cell. *Journal of Power Sources* 195(1):79–89.

Zhang, J.N., Q.L. Zhao, P. Aelterman, S.J. You, and J.Q. Jiang. 2008. Electricity Generation in A Microbial Fuel Cell with a Microbially Catalyzed Cathode. *Biotechnology Letters* 30(10):1771–1776.

Zhang, Q., J. Hu, and D.J. Lee. 2016. Microbial Fuel Cells as Pollutant Treatment Units: Research Updates. *Bioresource Technology* 217:121–128.

Zhang, X., W. He, L. Ren, J. Stager, P.J. Evans, and B.E. Logan. 2015. COD Removal Characteristics in Air-cathode Microbial Fuel Cells. *Bioresource Technology* 176:23–31.

Zhang, Y., B. Min, L. Huang, and I. Angelidaki. 2009. Generation of Electricity and Analysis of Microbial Communities in Wheat Straw Biomass-powered Microbial Fuel Cells. *Applied and Environmental Microbiology* 75(11):3389–3395.

Zhao, F., N. Rahunen, J.R. Varcoe, A.J. Roberts, C. Avignone-Rossa, A.E. Thumser, and R.C. Slade. 2009. Factors Affecting the Performance of Microbial Fuel Cells for Sulfur Pollutants Removal. *Biosensors and Bioelectronics* 24(7):1931–1936.

Zheng, S., F. Yang, S. Chen, L. Liu, Q. Xiong, T. Yu, F. Zhao, U. Schröder, and H. Hou. 2015. Binder-free Carbon Black/Stainless Steel Mesh Composite Electrode for High-Performance Anode in Microbial Fuel Cells. *Journal of Power Sources* 284:252–257.

Zhou, M., M. Chi, J. Luo, H. He, and T. Jin. 2011. An Overview of Electrode Materials in Microbial Fuel Cells. *Journal of Power Sources* 196(10):4427–4435.

Zhu, X., and J. Ni. 2009. Simultaneous Processes of Electricity Generation and *p*-nitrophenol Degradation in a Microbial Fuel Cell. *Electrochemistry Communications* 11(2):274–277.

Zuo, Y., P.C. Maness, and B.E. Logan. 2006. Electricity Production from Steam-Exploded Corn Stover Biomass. *Energy & Fuels* 20(4):1716–1721.

13

Harvesting Solar Energy: Fundamentals and Applications

Syed Shaheer Uddin Ahmed, Sayedus Salehin,
Md. Mustafizur Rahman, and A. K. M. Sadrul Islam

CONTENTS

ABSTRACT Energy is an indispensable need for the mankind. Most of the methods used for energy conversion utilize fossil fuel as the primary energy source. This phenomenon associates with greenhouse gas emission responsible for ozone layer depletion and global warming. Clean energy technologies offer an alternative solution, among which solar energy has the potential to play a pivotal role for the sustainable development. The heat and light received from the sun are referred to as solar energy. In this chapter, a comprehensive analysis of the utilization of solar energy as a natural resource, along with the fundamental principles, is done. Solar energy is converted into usable energy through conversion processes, that is, solar photovoltaic (PV) and solar thermal energy conversion. Solar energy can be converted into electricity by using photovoltaic effect in solar cells. Different types of solar cells are currently present, which are used on variety of applications. Owing to its modularity, solar PV energy can be used in various applications, including electricity generation in power plants; buildings; transport sector; rural electrification; heating, ventilation, and air conditioning (HVAC); refrigeration; water pumping for irrigation; telecommunications; and space industry. Solar thermal energy is collected using different types of collector available, for example, flat-plate collector and concentrating collector, and can be used for numerous applications. These include electricity generation in power plants, HVAC, refrigeration, water heating, room heating, passive heating, and cooling industrial process heat, drying, and cooking. Solar energy paved ways of rural development in many underdeveloped and developing countries, where social entrepreneur organizations utilized solar energy to improve the livelihood of people living in rural areas. Solar home systems, solar mini grid, and solar PV-powered water irrigation pumps are the most common projects having great socioeconomic impact.

13.1 Introduction

The fact that the fossil fuels are rapidly decreasing, along with the increasing pollution they cause when used, is no open secret today. For a permanent solution to this problem, a prompt, clean, reliable, and robust solution is a dire need. What better solution can be conceived than to utilize the energy that is all around us?

Sun or, more specifically, solar energy (the light and heat energies from the sun) reaching the earth is solely responsible for almost all of the sources that we use to harness energy for our day-to-day work. Traditionally, before the industrial revolution, sun has been the constant indirect source of energy for almost all activities or our lives. Whether it be the wood that we use in the kilns for production of merchandize or for warming our houses, the food that we eat to sustain ourselves, or the wind or water flow, the presence of sun is mandatory. Postindustrial times, we saw the rapid use of fossil fuels, which again were the indirect gift that the sun stored in the earth for us from the prehistoric times! However, the problem arises when we seek to convert the solar energy into the energy that we need to displace the fossil fuels. While trees produce fruits, they lack the capabilities to produce electricity, the most useful form of secondary energy source, directly. The potential for solar energy is massive and is a route toward energy independence. Also, there is a dire need to reduce our carbon footprint per energy production and usage. For that kind of photoconversion (converting light into useful form of energy), we need a more technological solution. The most popular one being "photovoltaic (PV) cells." The technical potential of PV depends on the land area available and the solar irradiation potential. Its potential is significantly large, as per unit of area of the output of solar PV is relatively high, as compared with that of other renewable energy sources. Another popular way of harnessing the solar energy is to use the heat from solar energy in myriads of application. The potential of solar energy for heating purposes is virtually limitless. The mostly used application is passive use in the built environment, the use of solar energy for drying agricultural products, and the use of solar water heating.

Both of these technologies will be covered in this chapter, along with some case studies of the scope of solar energy in Bangladesh, a developing country with high solar potential and emerging with proper use of solar power. However, before that, for a deeper understanding, initially, the solar radiation will be discussed. This is followed by a discussion regarding the PV cells. After that, the concept of solar thermal energy will be discussed, and the chapter will end with a case study and the general challenges to solar energy harvesting.

13.2 Solar Radiation

13.2.1 Sun and Earth

The sun is the never-tiring lamp for us, a continuous chain of nuclear fusion reaction is what fuels it. The great ball of fire in the center of solar system incessantly sends us heat and light, which ensure our survival. Table 13.1 gives a comparison of the sun and the earth.

TABLE 13.1

Comparison of the Sun and the Earth

	Sun	Earth	Ratio
Diameter (km)	1,392,520	12,756	1:109
Circumference (km)	4,373,097	40,075	1:109
Surface (km²)	6.0874×10^{12}	5.101×10^{8}	1:11,934
Volume (km³)	1.4123×10^{18}	1.0833×10^{12}	1:1,303,670
Mass (kg)	1.9891×10^{30}	5.9742×10^{24}	1:332946
Average density (g/cm³)	1.409	5.516	1:0.26
Gravity (surface) (m/s²)	274.0	9.81	1:28
Surface temperature (K)	5777	288	1:367
Center temperature (K)	15,000,000	6700	1:2200

Thus, it can be reasonably concluded, given the size difference between the earth and the sun that the earth is only privy to a tiny fraction of the vast amount of radiation that the sun emits. While the solar radiation incident on the earth's outer atmosphere is fairly constant, the radiation that is reaching the surface varies. This is due to the given factors (Hu and White, 1983):

- Atmospheric effects, including absorption and scattering
- Local variations in the atmosphere, such as water vapor and clouds pollution
- Latitude of the location
- The season of the year and the time of the day

The scattering occurs due to the suspended particles in the atmosphere. The most common scattering is "Rayleigh scattering," which is caused by the dust specks or nitrogen or oxygen molecules with diameter less than the wavelength of the light. Blue light is scattered more strongly than red, and it is due to this scattering that the sky appears blue when the sun is up. Another common scattering is called "Mie scattering"; this is caused by the dust aerosols and molecules with the diameter greater than the wavelength of light, and this scattering results in the white light from mist or cloud.

The absorption is due to the gasses in the atmosphere, such as ozone layer in the stratosphere filters out radiations with a wavelength less than 300 nm, thus absorbing ultraviolet (UV) rays. Also, trace gasses (H_2O, O_2, CO_2, CH_4, and so on) absorb the infrared rays. The water droplets or the ice crystals in the clouds reflect the solar radiation. The reflected or scattered rays, which reach the surface, are called diffuse solar radiation. The global solar radiation is the sum of the direct and the diffuse solar radiations (Kalogirou, 2009). Figure 13.1 gives a schematic model of the journey of the solar radiation from the sun to the earth.

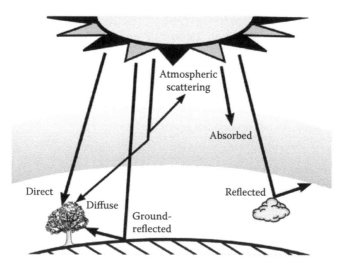

FIGURE 13.1

Solar radiation from the sun to the earth's surface. (From Newport. [2015, November 23]. *Introduction to Solar Radiation.* http://www.newport.com/Introduction-to-Solar-Radiation/411919/1033/content.aspx.)

13.2.2 Measures of Solar Radiation

The radiation from the sun to the earth is studied in various details, and thus, multiple standards exist in quantifying the radiation. For convenience, the most important terms that can describe the radiation are given below:

Radiant Energy: The energy of the radiation reaching any place. The unit of measurement is Joule (J); the conventional expression is "Q."

Radiant Flux: It is the radiant energy per unit time (dQ/dt), also known as "radiant power." The unit of measurement is Watt (W) or Joule per second (J/s). The conventional expression is given as "φ."

Irradiance: Radiant power incident on surface with a given area "A" (dφ/dA), also called "radiant flux density." The unit of measurement is Watt per square meter (W/m^2). The conventional expression is "E."

Insolation: This is the total amount of solar radiation energy received on a given surface area during a given time (in hours), ([dt.dφ]/dA). The unit of measurement is Watt-hour per square meter (Wh/m^2). A popular expression is "H." It is sometimes also referred to as "irradiation" (Figure 13.2).

From the global annual irradiation, we can judge, on a general scale, where solar energy harvesting will be profitable. As expected, the zones near the equator have higher irradiation rating in comparison with the zones away from it (northern and southern extremes). Also, known arid areas such as Middle East, Central Africa, west coast of South America show extreme irradiation ratings. However, an annual study is merely a weighted average and to decide on solar panel type and rating for a specific country or place and for more detailed study, a more transient study of the irradiation throughout the year for that place is needed. Also, a measure of the direct and diffuse radiations is also needed to be considered. Direct irradiation is much more profitable, and thus, areas with high direct irradiation will have greater potential than the ones with lower potential. Table 13.2 gives the average daily diffuse and direct irradiation in a few cities.

FIGURE 13.2
Global annual irradiation. (From Meteonorm. [2015, November 23]. *Meteonorm Software*. http://meteonorm.com/.)

TABLE 13.2

Annual Average Daily Direct and Diffuse Irradiation in kWh/m² day

City	Bergen	Berlin	London	Rome	LA	Cairo	Bombay	Upington	Sydney
Direct	0.86	1.2	0.99	2.41	3.03	3.39	2.75	4.70	2.42
Diffuse	1.29	1.61	1.47	1.78	2.07	1.95	2.39	1.47	2.13

Source: Quaschning, V. (2005). *Understanding Renewable Energy Systems* Vol 1. Earthscan: Sterling, VA.

With the concepts of solar radiation in mind, we can augment the simple PV panel to ensure better solar radiation reception to obtain better power output. Two of such attempts are discussed in the following section to understand how the knowledge of solar radiation can be used to develop ideas to increase efficiency of PV.

13.2.3 Solar Radiation Estimation

Design of a solar energy conversion system requires precise knowledge regarding the estimate of global solar radiation at the location of interest. Based on the estimation, the viability or solar energy conversion system is ruled. There are various methods at hand to estimate the solar radiation; one popular way is to calculate the "extraterrestrial radiation." The solar radiation reaching the earth is then calculated using the eccentricity correction factor of the earth's orbit, the time of the day, and the angle it makes with the sun. The extraterrestrial radiation is rather easy to know, since all it requires are latitude and longitude of the position in concern for a specific day of the year when it's being calculated.

However, when solar radiation data are unavailable, it is possible to get decent radiation estimates by using different models. A widely used method is based on empirical relations between solar radiation and commonly measured meteorological variables such as precipitation, dew point temperature, and relative humidity. The solar radiation itself can be estimated via measuring maximum and minimum air temperatures. It is done by assuming that clear sky will increase maximum temperature due to higher short wave radiation, and thus, the minimum temperature will decrease.

Thus, it can be seen that there are multiple means by which solar radiation can be estimated, with which one can proceed to take decision accurately regarding the solar energy conversion system.

13.2.4 Solar Tracking

The sun's trajectory is a function of the geographical location, time of day, and season. Thus, for a PV to have optimal input from the sun throughout the day is difficult, since best output is achieved when the direct radiation is incident to the normal of the solar panel. Thus, to overcome this issue, the concept of solar tracking is developed. The solar tracker, in contrast to a static solar system, follows the sun's trajectory and ensures that the solar panels are positioned for maximum exposure to sunlight. The tracker can be operated only in one axis, or it can have two-axis rotational freedom. Two-axis tracking systems move a surface always into an ideal position; however, two-axis tracking systems are relatively complicated, and thus, one-axis tracking systems are preferred in some instances. It has been estimated that static solar system would use about 40% more modules than a two-axis tracking system for equal annual energy (Gay et al. 1982). More detailed studies show that during days with high direct irradiation, tracking can achieve energy gains over horizontal orientation in the order of 50% in summer and up to 300% in winter, depending on the latitude of the location.

The two-axis system, without any doubt, gives higher power output; however, it introduces many complexities, since it depends on hydraulic or electrical motors to maneuver the panel, in installing and maintaining the system. Also, it is possible that if the system fails, the panel is stuck in a rather-inefficient angle, which would result in bad power output. Not to mention, the cost of the tracking system is an extra investment. Thus, often times, one-axis system, which is designed to follow annual path of the sun, is preferred, since all it needs is a weekly or monthly attention. Alternatively, the whole tracking is dropped all together if regular maintenance is an issue.

13.2.5 Fresnel Lens

Optical systems, usually Fresnel lens, can be used to concentrate the solar radiation in multiple PV cells stacked one above the other. The technology can be used to concentrate radiation to the factor of 500! However, the major drawback in this system is that only direct normal irradiance can be used. Thus, it cannot elevate the solar power output in places with low direct irradiation (Table 13.2).

Also, as stated, the radiations that are normal to the PV array can be concentrated. Thus, it would require a two-axis tracker system to ensure that the radiations from the sun are always normal to the solar panel plane. Given the issues with the two-axis solar tracker and the issue of the need of direct irradiation, it is thus seldom used, despite its promising outputs.

Thus, from this section, we learned a bit about the sun in comparison with the earth, to better understand it. We learned what factors affect the solar radiation while it travels to the surface of the earth. Also, we learned the different measures that are standard in measuring the solar radiation. We then discussed two ways of using this knowledge in the optimization of solar output and their respective limitations. From here on, we are ready to start studying the "solar panel" and how it actually performs the photoconversion.

13.2.6 Luminescent Solar Concentrators

Problem with PV cells is that the absorbers can only accept photons of certain wavelength. So, no matter how much the sunlight is tracked and concentrated, there are losses of energy, since sunlight is a motley of photons with different wavelength. Research to tackle this problem has given rise to a device called a luminescent solar concentrator (LSC). The operation of an LSC is based on the idea of light-pipe trapping of molecular or ionic luminescence. This trapped light can be coupled out of the LSC into photovoltaic cells in such a way that the LSC provides a concentrated flux that is spectrally matched to the PV. This can reduce the PV heating and massively increase the output. An LSC does not need to track the sun and in fact can produce highly concentrated light output under either diffuse or direct insolation (Batchelder et al. 1979, Tosun, 2015).

13.3 Solar Photovoltaic

With the passing of time, solar panel is becoming more and more central to the grid production, especially in the developed world. Its greatest strength by far is modularity. Truly from delivering as a plant, which supplies powers in megawatt range to small calculators or in miliwatt range to watches, solar panel can be relied upon. With the development of the ideas like micro-grid, solar panel demand is set to more than ever, especially in the developing world. In order to understand how it works, it is imperative to have a basic idea of the material that is used to construct the panel. Hence, the first subsection of this section will deal with the material itself.

13.3.1 Semiconductor for Photovoltaic

The solar panel is essentially, construction-wise, a diode. Instead of applying a potential difference externally, it is created using photon from light, and connecting the two ends (p and n junctions) facilitates a current flow. A more detailed explanation would make the internal process clearer.

Semiconductors come in different kinds. Some are elemental, such as silicon (Si) and germanium (Ge). They can also be alloys from elements of group III and group V or, respectively, from group II and group VI (Figure 13.3). Since they are made from different elements or are different elements themselves, their properties vary. Silicon is the most commonly used semiconductor material, as it forms the basis for integrated circuit (IC) industry. It has been heavily researched, and thus, its devices represent a more mature technology. Most solar cells are also Si-based; hence, from here on, Si will be used as a model material when talking about semiconductor, unless specified otherwise.

Silicon, as seen in Figure 13.3, is a group 4 element and thus has the capability of forming bonds with four other silicon elements, which it does. In its unadulterated form, single silicon atom forms covalent bond with four other. There is no surplus of any charge, and thus, overall, it is totally neutral. Depending on the fabrication technique, silicon can get different material structure, which affects properties greatly (Wolfe et al. 1988). The three common structures are given below (also in Figure 13.4):

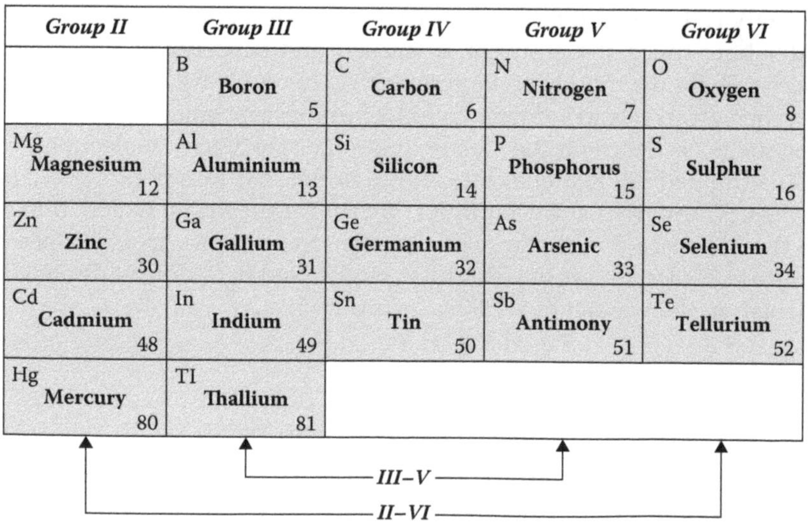

Group II	*Group III*	*Group IV*	*Group V*	*Group VI*
	B Boron 5	C Carbon 6	N Nitrogen 7	O Oxygen 8
Mg Magnesium 12	Al Aluminium 13	Si Silicon 14	P Phosphorus 15	S Sulphur 16
Zn Zinc 30	Ga Gallium 31	Ge Germanium 32	As Arsenic 33	Se Selenium 34
Cd Cadmium 48	In Indium 49	Sn Tin 50	Sb Antimony 51	Te Tellurium 52
Hg Mercury 80	Tl Thallium 81			

III–V
II–VI

FIGURE 13.3
Part of periodic table showing elements involved in forming semiconductor. (From Palankovski, V. [2015, November 11]. Simulation of Heterojunction Bipolar Transistors. http://www.iue.tuwien.ac.at/phd/palankovski/node18.html.)

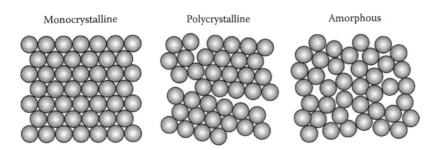

FIGURE 13.4
Schematic of (a) c-Si, (b) mc-Si, and (c) a-Si.

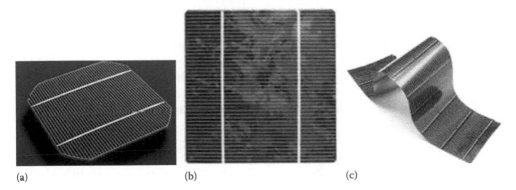

(a)　　　　　　　　　　(b)　　　　　　　　　　(c)

FIGURE 13.5
PV module of (a) c-Si, (b) mc-Si, and (c) a-Si.

Single crystal (c-Si): atoms are periodically arranged in a lattice.

Poly- or multicrystal (mc-Si): there are regions of crystallinity but no long range order exists.

Amorphous silicon (a-Si): only short range order exists and no periodicity.

These different kinds of silicon are used to make different kinds of silicon-based solar panels. More recently, the a-Si type is taking center of attention for research, as theoretically, the optical absorption is greatly enhanced with this type of silicon. The resulting PV module from the a-Si is very thin and ribbon-like. Figure 13.5 shows the three kinds of solar panel from the three kinds of silicon. To note, at present, the c-Si- and mc-Si-based PV are much more common for their matured technology and higher efficiency than a-Si based PV (13%–18% in crystalline against 12%–13% [at best] in case of amorphous silicon). However, owing to its better optical absorption, amorphous silicon has a potential in low-sunlight applications.

The bonded electron in the silicon structure is not free to participate in any electricity transfer under normal condition; however, in elevated temperature, the electrons gain enough energy to escape the bond (leaving behind an empty space called hole and is deemed to be positively charged) and thus can be considered relatively free and in a high-energy state. This higher-energy state is a discrete energy level, and thus, there is no such thing as an intermediate level of energized electrons. The minimum energy required for the electron to reach this high energy state or conduction band is called the "band gap." In case of solar panel, the solar radiation provides this energy.

The most important parameters of a semiconductor for solar cell operation are:

- The band gap (Eg)
- The number of free carriers available for conduction
- The generation and recombination of free carriers (electron or holes) in response to light shining on the material

The concentration of the free carriers (electron and holes) depends mainly on the Eg and on the temperature at which the semiconductor is operating. The greater the Eg is for a particular material, the harder it is to excite the electrons to form the free carrier and thus less available free carriers. Also, at elevated temperature, it is more likely that the electrons would rise to the conduction band, and thus, higher temperature gives greater number of free carriers.

For a pure silicon, the number of hole and free electrons is always same. However, if impurities are added by doping it with elements from different groups, then an imbalance can be created. If doped with group V, the resulting product contains greater number of free electron than holes and is thus called "n-type." If doped with group III, which has one less valence electron, the number of holes in the structure is higher than the free electrons and is thus called the "p-type." Thus, in "p-type," holes are the "majority carrier" and free electron the "minority carrier." The statistics are exactly opposite for the "n-type" material.

Recent research is trying to ascertain the potential of using the natural process in tandem with the artificial means to generate energy. The resulting PV is called biohybrid PV. It is constructed using a combination of organic and inorganic matter. The basic concept behind the photosynthesis is used to generate energy in the process, increasing efficiency and generating greater energy. Multiple layers of photosystem gather photonic energy, convert it into chemical energy, and create a current that goes through the cell (Gizzie et al. 2015).

13.3.2 Working Principle of Solar Photovoltaic

A solar panel is designed like a diode. The p-type- and n-type-based silicon (or other suitable semiconductor) are seamed together with a common contact (Figure 13.6). Unlike

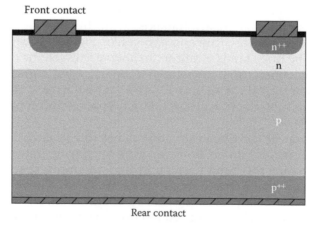

FIGURE 13.6
Schematic of the cross section of PV system with n- and p-type zones.

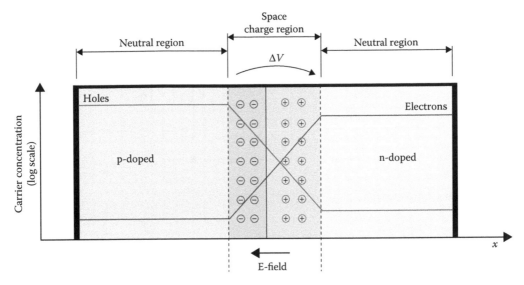

FIGURE 13.7
Space charge region formation at the p–n junction, owing to diffusion of holes and electrons from p and n zones. (From Solar Cell Central. [2016, January]. *Solar Cell Central.* http://solarcellcentral.com/images/pn_junction_w_diagrams.jpg.)

diode, there is no applied voltage; however, the light is shone, and thus, electron and holes are created. In diode, the terminals are used to provide potential difference for current flow, while in solar panel, the terminals are to allow the produced current to flow and complete the circuit.

For both diode and solar panel, however, at the junction of the p- and n-type, there lies what is called a space charge region. Initially, owing to the attraction of the opposite charge, the holes from the p-type zone and the electron from the n-type zone diffuse into each other. Thus, there is a localized positive zone and a negative zone at the interface in the n-type and p-type silicon, respectively. This sets up a potential at the p-n junction, which counteracts the potential, owing to the charge imbalance due to charge carrier density in the n and p zones overall (Figure 13.7). Hence, a direct flow of electron from the n-type to the p-type is hampered and doesn't occur, unless there is an external charge applied, which overrides the localized field set up at the interface. This occurs in the diode. Alternatively, current can also flow by connecting the p- and n-types via a closed circuit gives a way out. Thus, the electrons from the n-type travel through the circuit to the p-type material, thereby creating a direct current (DC)! This, in brief, is how solar panel works (Quaschning, 2005).

The standard test conditions (STC) for the solar panel are

- Irradiance of 1000 W/m^2
- Cell temperature of 25°C

In order for the solar panel to utilize the available solar radiation, the reception of solar radiation needs to be optimized. For better optical reception, certain measures are to be taken during the construction of solar panel, and those are listed below:

- Minimize the front contact coverage
- Anti-reflection coatings (ARC) on the front surface

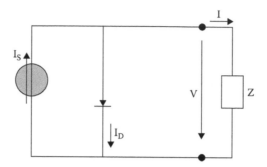

FIGURE 13.8
Three Parameter 1 diode model.

- Reflection reduced via surface texturing
- Make solar cell thicker to increase absorption
- Increase optical path length in solar cell via light trapping

This concludes the basics of the solar panel construction. Now, it is time to start discussing, in brief, the equivalent circuit of the solar panel.

13.3.3 Three-Parameter One-Diode Model

The simplest equivalent circuit that can encapsulate the PV system is the three-parameter one-diode model (Figure 13.8), where the PV acts as the DC source, with the source current "I_S" and the output current across the load "I." The current across the diode is "I_D."

The output power from the solar panel, "P_{DC}," can be obtained from the product of the current "I" and the voltage "V," which is across the load provided. The efficiency of the solar panel can be calculated by the ratio of the output and input. The input is the radiant power for the area of solar panel, "A." Thus, the formula of the efficiency at STC is given below.

$$\eta_{STC} = \frac{P_{DC}}{AE_{STC}} \qquad (13.1)$$

The characteristic curve resulting from the setup is given below (Figure 13.9).

13.3.4 Applications

The first solar cell built by Bell laboratories in 1954 had the efficiency around 5%. Initially, the use of solar panel was limited to space application, and thus, its cost, though very high, was not an issue. However, with passing days and with the rapid development of the semiconductor technology, its efficiency increased modestly and the cost decreased radically. The efficiency of the common solar panel ranges around 20% at the present time. Also, the cost has decreased to the point that it is affordable for a wristwatch!

13.3.4.1 Photovoltaic-Powered Systems

The strongest points of solar panel are its modularity, robustness, and noiseless nature. These make it suitable for application where public safety should be taken into account,

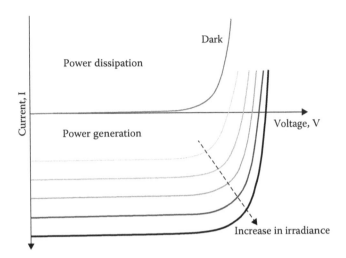

FIGURE 13.9
I–V curve of a PV, with varying irradiance.

such as household rooftop application. A whole PV module consists of multitudes of PV cells in array. The common PV modules have a rated power output of 50–180 W and deliver about 12 V of DC. Thus, to integrate into the grid or be used in house with alternate current (AC) machine, inverters are needed. It is a common practice in many countries these days to have PV roofs, which are used to provide for the home and feed the surplus to the grid. In such integration, the grid acts as a sort of an energy storage.

However, for local use, it is possible for a solar panel to appliance connection, which eliminates the need of an inverter, nor does it feed the grid any power. Such type of application is common in systems such as pumping for irrigation. Such system seldom needs any form of storage or grid integration, and thus, a direct panel to appliance connection is preferred. Stand-alone systems are used in remote places, where grid connection is difficult. Thus, using solar panel and other decentralized power sources such as wind power or modular diesel engines, micro-grids are set up. This kind of system has solar panel in array for high output to meet the daily load and is equipped with batteries for storage during the time when solar radiation is very low. Inverters are used, when necessary, to convert the DC power into AC. A schematic of the stand-alone system is given in Figure 13.10.

To mitigate intermittency, hybridization of renewable-energy-producing system is preferred. Especially since battery storage is highly expensive and can work only for medium power range, hybridization with small diesel or biofuel generator is done sometimes. A particular favorite concept is to couple wind and solar systems together. This provides a good trade-off in some sites, since one source supplements the other, keeping the issue of intermittency at the bay.

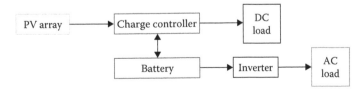

FIGURE 13.10
Schematic of stand-alone system.

13.3.4.2 Types of Applications

With the knowledge of different kinds of PV system, we can proceed to the actual applications where PVs are used.

Off-grid application: Solar panels can be used to set up micro-grid to set up decentralized grid in remote areas, where electrification is an issue. These set-ups are for long-term use and the loads include small-appliance communication equipment and water pumps. PV provides a unique benefit of not requiring any fuel or trained operating personnel, or it does not contribute to any environmental pollution. The load demand does not vary much, since the population itself is usually numbered. Thus, a PV array with some form of hybridization or even with a battery bank can suffice. Hence, the use of PV in such areas is highly appreciated.

Different kinds of communication towers in remote places, such as radio relay stations, emergency call units, and military communication facilities, are sometimes powered by PV. These systems are stand-alone and have medium to low power demand; thus, a PV array with battery bank can relive such system from the grid.

Low power demand application: Certain small applications also enjoy solar power. Many wristwatches and calculators come with a very small PV cell array to power these systems. Since the load demand is very low and hovers around microwatt, such small-sized PV can provide the system with what is adequate. A slightly larger application is to power city lights or larger applications such as providing a small charging current to the larger storage battery, for example, car battery. This kind of application improves battery life and prevents the battery running out of charge due to self-discharging over time. Also, PVs are directly connected pumps, which are used in outdoor pool heating; sometimes, it is also used to power parking meters and streetlights.

PV-powered water pumping: Solar PV-powered (SPP) pumping for crop irrigation and supplying drinking water are two of the many diverse applications of solar energy. Photovoltaic pumps are considered an environmentally sound and resource-conserving technology. By using SPP pumping, contamination of soil and ground-water resources can be avoided (Hahn, 2000). In locations where electricity is not available from the main grid, SPP pumping is a viable option for crop irrigation and for supplying drinking water (Abu-Aligah, 2011). The feasibility of SPP pumping for crop irrigation depends on many factors such as location, crop type, water depth, energy cost, and government incentive. In some areas, irrigation may be required only during dry seasons, whereas in many areas, year-round irrigation may be required (Figures 13.11 and 13.12).

The SPP irrigation system is considered technically feasible when sufficient land is present to mount the solar PV array. This types of system is considered economically feasible when the life cycle cost of SPP pumping system is lower than that of the alternative pumping systems, for example, diesel-powered pumping system and electric grid-based system (Kelley et al. 2010). Considering the operating cost, diesel-run pumps are profitable for the period of less than five years, whereas SPP pumps are more profitable for five or more

FIGURE 13.11
Solar PV-powered irrigation system in Bihar, India. (From GIZ, German Technical Cooperation, Solar water pumping for Irrigation: Opportunities in Bihar, India, Technical report, 2013.)

FIGURE 13.12
Solar Irrigation Project in Panchagar, Bangladesh. (From SREDA, Sustainable and Renewable Energy Authority, Government of Peoples' Republic of Bangladesh, www.sreda.gov.bd, 2015.)

years for the same capacity. SPP pumps also help reduce the greenhouse gas (GHG) emission significantly (Hossain et al. 2015).

Building Integrated Photovoltaics (BIPV): One of the most common applications of grid-integrated system is the building integrated photovoltaics (BIPV). The PV is installed on the roof or the building façade. This, however, can increase the thermal load of the building, and thus the PV panels, placing needs to be optimized to allow proper ventilation. A concept of zero-energy house is getting popular, especially in Europe, where the building is set to produce the energy that it needs, thereby reliving the grid. Since wind requires an open space, decent height, and is very noisy, it is not used for such application.

Solar parks: A solar park is a large-scale generating station designed to supply bulk power to the electricity grid. It is also called solar farm. The installation of solar park requires installing multitudes of PV modules on multiple acres of land to provide a large overall power output. Solar park typically produces electricity on the scale of 10–50 MW, which is enough to power 3000–14000 homes.

FIGURE 13.13
Waldpolenz Solar Park. (From JUWI. [2015, December]. JUWI energy is here. Waldpolenz Solar Park: http://www.juwi.com/solar_energy/references/details/waldpolenz_solar_park/-/-/5.html.)

The Waldpolenz Solar Park in Germany is an example in this power range. Completed in 2008, it is designed to deliver a power of 52 MW. It consists of thin-film CdTe modules spreading across 220 hectares. The cost of the whole system is recorded to be 130 million euros. Figure 13.13 shows the solar park.

At present, the largest solar park is in Neuhardenberg, Germany, with a total of 145 MWp. It can provide green power to about 48,000 households.

13.3.5 Cost and Economic Analysis

The cost of a PV system varies with country and capacity and scope of use. Hence, it is sensible to represent the cost of the system with USD/W for different countries, given different scope/application. Table 13.3 gives a summarized representation of the cost of PV, given all the mentioned factors.

Levelized cost of electricity (LCOE) is a USD/MWh value representing the total life cycle costs of producing an MWh of power, using a specific technology. Therefore, for a project to breakeven, it must earn per unit of output the LCOE amount. It takes into account not just the cost of generating electricity but also the initial capital and development expense, the cost of equity and debt finance, and operating and maintenance fees (Bloomberg New Energy Finance, 2015). Thus, LCOE is a standard means of calculating and comparing the cost of renewable energy system. Figure 13.14 and Table 13.4 will give a summary of the LCOE of PV systems.

TABLE 13.3

Typical PV System Prices in 2013 in Selected Countries (USD)

USD/W	Australia	China	France	Germany	Italy	Japan	United Kingdom	United States
Residential	1.8	1.5	4.1	2.4	2.8	4.2	2.8	4.9
Commercial	1.7	1.4	2.7	1.8	1.9	3.6	2.4	4.5
Utility scale	2.0	1.4	2.2	1.4	1.5	2.9	1.9	3.3

Source: International Energy Agency. (2014). Technology Roadmap Solar Photovoltaic Energy. Paris: International Energy Agency.

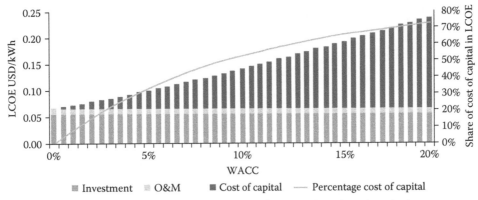

Note: This example is based on output of 1 360 kWh/kW/y, investment costs of USD 1 500/W, annual operations and maintenance (O&M) of 1% of investment, project lifetime of 20 years, and residual value of 0.

FIGURE 13.14
Share of the costs of capital in the LCOE of PV systems. (From International Energy Agency. [2014]. Technology Roadmap Solar Photovoltaic Energy. Paris: International Energy Agency.)

TABLE 13.4

Projections for LCOE for New Built Utility Scale PV Plants in Different Years Till 2050 in hi-Ren Scenario

USD/MWh	2013	2020	2025	2030	2035	2040	2045	2050
Minimum	119	96	71	56	48	45	42	40
Average	177	133	96	81	72	68	59	56
Maximum	318	220	180	139	119	109	104	97

Source: International Energy Agency. (2014). Technology Roadmap Solar Photovoltaic Energy. Paris: International Energy Agency.

13.4 Solar Thermal

Till now, our concentration has been the use of light (specifically photon) to generate electricity or, to be more general, useful power. However, if we recall, solar radiation is not just light; it is both light and heat. Thus, it is imperative for us to know how the thermal part of the solar radiation is used for useful power generation.

Solar thermal is a very old form of technology, and it encompasses the thermal use of the solar radiation to generate useful power. Owing to our mature knowledge of heating process, the solar thermal enjoys wide range of application, from simple heating to using the heat to generate electricity or even as energy storage. Solar thermal can be used in a versatile manner, and thus, for each application, the finer detail will vary. Thus, different kind of popular collectors will be described, followed by the different applications of solar thermal.

13.4.1 Collectors

The solar energy collectors are a type of heat exchangers that convert solar radiation to internal energy. These structures are designed to absorb the incoming solar radiation and then heat up a fluid (oil, water, and so on), and it is circulated as the primary fluid or as a heat-exchanger fluid. The solar radiation collectors can be classified into two major types:

- Stationary
- Tracking

The tracker that is not designed to follow the sun's movement during the collection of solar irradiance is a stationary tracker. These trackers generally have the same area for receiving and absorbing the solar radiation. The two-axis tracking system follows the solar pathway on both daily and seasonal bases to optimize the absorption of solar radiation. The concentrating technology is usually used in collectors that are mounted with a two-axis solar tracking system, though there are some stationary and single-axis tracking collectors that use the concentration technology. These types use reflective surface to concentrate the intercepting solar radiation to the receiving area, thus increasing the overall reception of the solar irradiance.

The most common forms of stationary collectors are

- Flat-plate collectors
- Stationary compound parabolic collectors
- Evacuated tube collectors

The general construction of these collectors is that they have a glass covering, which allows the solar radiation to enter into the collector cavity and prevent heat loss via convection. The solar radiation then heats up the fluid that usually flows from one end of the collector to other via some kind of conduit. The glass covering also serves to enact a sort of greenhouse effect, trapping the outgoing radiations.

The more effective collectors are, of course, the two-axis trackers with concentration technology augmentation. There are some distinct advantages of such system over plain non-concentrating stationary types. Since the solar radiation is received by a large arc and is then concentrated to a relatively smaller one, the area across which energy loss occurs is thus smaller. Hence, the overall efficiency is very high. The working fluid can get heated more quickly and to a higher temperature than the stationary non-concentrating types. However, the concentration technology doesn't work very well with diffuse or scattered radiation. Moreover, a solar tracker introduces its own limitations, as discussed previously, and periodic cleaning and maintenance are a must.

The popular solar collectors of these type are

- Parabolic trough collector
- Linear Fresnel reflector
- Parabolic dish
- Central receiver

Figure 13.15 shows the "flat-plate" and "parabolic trough" collectors.

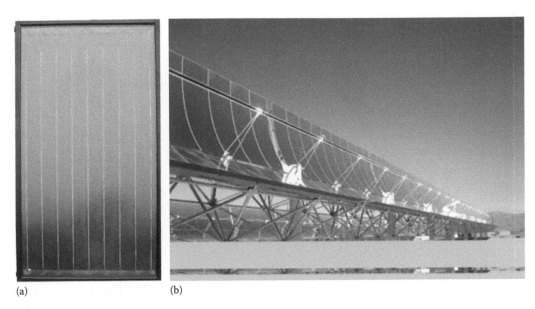

(a)   (b)

FIGURE 13.15
(a) Flat-plate collector, including a top transparent layer and pipe for fluid flow. (From Rain Harvest. [2016, January]. *Rain Harvest*. http://www.rainharvest.co.za/wp-content/uploads/2010/08/Flat-Plate-Solar-Collector. jpg.) (b) Parabolic trough collector showing the parabolic reflector and a pipe at the focal point carrying the heat absorption fluid. (Business Wire, 2015.)

13.4.2 Applications

13.4.2.1 Domestic Water Heating

While swimming pool heating requires a rise of 2°C–3°C, domestic water heating requires a lot more, especially during the winter time. Hence, if solar thermal systems are installed for domestic heating, the collectors shouldn't be the simple pipes but more sophisticated ones, as mentioned in the collector section of the chapter. The collectors should be able to resist heat loss via convection and radiation from the heated water in the absorber. However, it should be noted that a very simplistic system for water heating is possible. This is done by simply having a black-colored reservoir, with water just above the water outlet, and this should be ensured that the water reservoir is intercepting solar radiation. Such construction is done usually in outdoors, for example, camps. Such systems lack any means of self-filling usually and are also very inefficient. Thus, for a regular domestic use, more sophisticated solution, as alluded above, is needed.

13.4.2.1.1 Thermosiphon System

Thermosiphon system is a self-circulating system. It is designed such that the heated water from the collector would cause the circulation, thereby eliminating the need of an external pump in the system. The hot water is lighter than cold water and thus seeks to rise up. The reservoir is constructed above the collector, and thus, the hot water naturally moves up to the reservoir. Also, the water in the reservoir being relatively colder

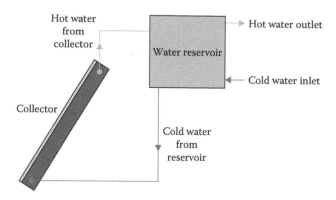

FIGURE 13.16
Thermosiphon system.

and thus heavier moves downward to the collector. Thus, by using the natural convection, the system displaces the need of a pump. As long as the collectors receive the solar radiation, the system functions spontaneously. The collectors used in this system are generally either "flat-plate" or "evacuated tube" collector. To minimize the effect of friction, the tubes used should have large diameter. A simple circuit structure of the whole system is shown in Figure 13.16.

   The advantage of such system is the fact that it does not rely on any actuators, such as motor, or any kind of controllers. This makes the setup relatively simple, and the operation cost is minimized. However, this system needs to be comparatively very tall, and thus, aesthetically, it is not very desired. Also, since these systems are open systems, extreme acidic or basic water might cause structural erosion or scale formation, thus limiting the life span. In countries where temperature goes below the freezing point, it needs to be designed with freeze-protection devices, such as external heat source, to warm up the water at the entrance of the collector, where the cold water goes.

13.4.2.1.2 *Forced-Circulation Heaters*

Forced-circulation system uses pump to force the hot water into the reservoir to heat the reservoir water. A popular construct for such system is double-circulation system, where both the solar collector using thermosiphon and another auxiliary system using pumped circulation are used to heat the reservoir water. Fossil fuel or alternative forms of heater may be employed in the forced-circulation circuit to ramp the performance.

13.4.2.2 *Solar Thermal Engines*

The solar thermal concept can be used to power plants to produce electricity. For higher efficiency, solar radiation is concentrated using specially constructed collectors/reflectors. This kind of concept is popularly named as "concentrated/concentrating solar power" (CSP).

   One such concept is the power tower, where multiple mirrors are used to concentrate the solar radiation to a small area. The water or system fluid is vaporized using

FIGURE 13.17
Solar Two power tower. (From Eeremultimedia, https://www.eeremultimedia.energy.gov/solar//sites/default/files/photo_csp_tower_development_solartwo_barstow_2000_high.jpg, 2015.)

the heat from the concentrated solar radiation, and the vapor is then used to run a turbine, thereby producing electricity via a coupled generator. A pilot plant of 10 MW was constructed using this technology in California, called "Solar One" (which was later renamed as "Solar Two" after some developments). The project consist of 1818 of 7 × 7 meter mirrors in a circular arrangement, designed to track the sun. It concentrates the solar radiation to a central receiver 91 meters above the ground. The receiver is designed to produce 50,900 kg/h of steam at 565°C, with the absorber operating at 620°C. The system uses molten salt as heat-transfer fluid to store the heat for nighttime operation (Figure 13.17).

Another way in which solar thermal is used to produce electricity is to use parabolic trough collectors in series. Generally, it is also supplemented with fossil fuel to heat the working fluid during the nighttime to ensure a continuous operation. The solar Energy Generating Systems (SEGS), a series of plants in California, is world's first commercial parabolic trough plant. It was commissioned during 1984 to 1990 and is still running today with a capacity of 354 MW (Figure 13.18).

Another popular method for generating power is to use disc-shaped parabolic mirrors to concentrate the solar power to a receiver. These are generally stand-alone system consisting of the reflector receiver and the engine. The engine, which consist of a turbine, then converts the heat into mechanical and then electrical energy. Dish engine systems use a dual-axis tracking system to follow the sun. Concentration ratios usually range from 600 to 2000, and they can achieve temperatures in excess of 1500°C. A good example of this system in practice is EURODISH, developed by a European consortium, with partners from industry and academia. It is a cost-effective 10 kW dish that produces electricity by using Stirling engine.

FIGURE 13.18
Areal view of the solar power station SEGS III-IV, KJC, Kramer Junction, California, with installed capacity of 30 MW each. (From L.P. Daniel Engineers and Contractor Inc. [2016, November 19]. http://www.lpdaniel.com/SEGS.html.)

13.4.2.3 Solar Chimney

Solar chimney is a device that uses solar energy to heat the air by convection heat transfer, which can be used to improve the natural ventilation of buildings. It can also be used to generate electricity. Owing to its potential benefits in terms of low operational cost, energy uses resulting GHG emissions. Solar chimney systems are being increasingly proposed in Asia, Middle East, Africa, and Europe for ventilation and power generation. The sun is a huge source of energy supply. This gigantic source of energy can be used in home ventilation as well as in power generation by using solar thermal chimney. It is becoming popular in the areas where the intensity of solar radiation is very high.

Solar chimney utilizes the solar insolation, which enhances the buoyancy effect to get sufficient air flow inside the room. One of the walls of the solar chimney is made of transparent glass and is exposed to the sun. Solar radiation enters the chimney through the transparent wall and heats up the absorber wall. The absorber wall is made of concrete or aluminum sheet. The surface of the absorber wall is coated with black color to increase the heat gain from the sun. Solar radiation increases the temperature inside the chimney, and as a result, the air is heated up and starts to flow through the chimney. The solar chimney sucks the air from the inside of the room, which is exhausted through the opening at the top. The fresh air enters the room through the inlet openings and pushes out the room air to the outlet openings. Solar chimney can also be used during winter season by doing the reverse, using dampers in the chimney (Figure 13.19).

Manitoba Hydro Place, headquarters building of Manitoba Hydro, the electric power and natural gas utility of Manitoba, a province in Canada, is using solar chimney to provide ventilation in the building. Dimensions of the chimney: 115 m height, 15.5 m length, and 2.85 m width (Gontikaki et al. 2010). It is utilized as a natural ventilation element in summer and a heat-recovery device in winter. Around 2000 people are getting 100% fresh outdoor air for the whole year. The building consumes only 88 kWh/m² energy yearly and is 66% more efficient than model national energy code for buildings (Manitoba Hydro, 2015).

FIGURE 13.19
Manitoba Hydro Place, where the concept of solar chimney is utilized. (From Manitoba Hydro. Available at http://manitobahydroplace.com/. [2015, November 23].)

A solar chimney can also be used to generate electricity. Collector is a part of solar chimney that increases the temperature of air by using greenhouse effect. The collector is made of plastics or glass, but the latter one is more efficient. The electricity production depends on the surface area of the collector. The air inside the collector is heated up and starts to flow upward within the chimney. Turbines are located in the chimney, which extracts the power from the hot air. To get maximum power, turbine blades should cover all the area of the chimney. Power production also depends on the chimney height. The efficiency of a solar chimney is directly proportional to the ratio between the height of the chimney and the outside temperature (Schlaich, 1995). Figure 13.20 shows the working principle of solar chimney for power generation.

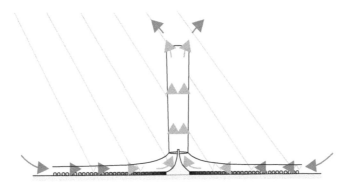

FIGURE 13.20
Working principle of solar chimney power generation, showing various parts. (From Schlaich Bergermann. The Solar Chimney. http://www.math.purdue.edu/~lucier/The_Solar_Chimney.pdf.)

FIGURE 13.21
Prototype of solar chimney at Manzanares. (From Schlaich Bergermann. The Solar Chimney. http://www.math.purdue.edu/~lucier/The_Solar_Chimney.pdf.)

An experimental setup was built at Manzanares, about 150 km south of Madrid, Spain, with a peak output of 50 kW in 1981–1982. The aim of this project was to verify through field experiments the performance that has been based on theoretical calculations and to examine the factors affecting the plant output and its efficiency. Dimensions of the plant: chimney height, 195 m; chimney diameter, 10 m; and collector diameter, 240 m.

This experimental plant operated for 15,000 hours from 1982 onward, thus establishing that solar chimney can be built, run, and maintained in the long term. The reliability of the plant was about 95%. Five percent non-operational time was because of Spain's grid breakdown (Schlaich, 2015) (Figure 13.21).

13.4.2.4 Thermal Storage

Swimming pool heating or air heating using solar thermal technology can accommodate the issue of intermittent nature of solar radiation; however, for rest of the solar-powered application, it is not as desired. Considering that on average, half of the year, if not more, is basically without sun, the solar-powered application faces an obvious handicap. It is common knowledge that one cannot store light, and thus, thermal storage of the solar radiation is one of the key applications of solar thermal concept. By introducing this thermal capacitance, the efficiency of the system improves markedly.

The concept of thermal storage is relatively unsophisticated. The heat from the sun is used to heat mediums with heat capacity decent enough to retain a significant amount of heat for any planned application. When needed, the medium is used either as some kind of heat exchanger or as the system fluid itself. The solar energy can be stored in liquids, solids, or phase-change materials (PCM). Water is used most of the time, owing to its availability, non-toxic nature, and high heat capacity. In addition, being a liquid, it can be transported with relative ease. For maximum efficiency, the water entering the solar thermal collectors should be as cold as possible.

The water needs to be stored in tanks that are highly insulated, thus minimizing the heat loss. The storage system can be pressurized (usually for small-service water heating system) or unpressurized (usually for system dealing with volumes greater than 30 m³). The heat exchanger used by the system can be external, which consumes electricity, since water is needed to pump out of the storage tank to the heat exchanger; however, it provides the system with more flexibility and ease of maintenance. Thus, for small system, internal heat exchanger is preferred, while for larger system, external heat exchangers are a luxury.

System that uses air as heat-transfer fluid generally uses gravels as thermal capacitor. Alternatively, water, PCM, or even the whole building itself is used for storage. In its most basic construction, gravels are loaded into a container made out of concrete or wood and hot air from the collector is introduced from one end and relatively colder air exits from the other. The flow inside can be designed to be vertical or horizontal. During the flow, it increases the temperature of the gravel. The gravels can be used to heat air, when needed.

13.4.2.5 Swimming Pool Heating

Water, as we know, has a very high specific heat capacity. Maintaining a bearable temperature thoughout the winter for a huge water mass, such as a swimming pool, requires a decent expenditure of fossil fuel. In countries like Germany, where the average ambient temperature, even in summer, is around 20°C, there is a huge market for solar thermal application. This is because the average temperature of the pool in summer is around 16°C–19°C and a few degree rise in temperature would suffice, and with the available solar irradiation, such rise is possible. A simple and economical solar energy system is good enough for the pool heating, thus moving from fossil fuel to solar thermal provides an economical advantage.

The system can be a very basic one, with roof top collector being simply black-colored plastic pipes, with water flowing through it. A simple pump can be used to force the water from the pool into the collector circuit. The water that exits the collector will have elevated temperature and will be circulated back into the pool. Since the absorber circuit in the collector will be exposed to solar radiation constantly, materials that can degrade due to solar radiation should be avoided. Some suitable materials are polyethylene (PE), polypropylene (PP), and ethylene propylene diene monomer (EPDM).

This kind of heating will not be very effective in the bad weather condition, owing to lower level of solar irradiance. However, outdoor pools are not used very often in bad weather; hence, this is not a very big caveat to this technology. From estimate, the solar absorber surface should be 50%–80% of the pool surface; however, this will depend mainly on the climate. Since this system won't be of use when solar irradiance is poor, a small solar panel can be used to power the motor that drives the water into the heating circuit, thus making the whole system autonomous from the grid and renewable energy-based. Figure 13.22 gives a simple illustration of the basic system.

13.4.2.6 Nanofluid-Based Solar Thermal Collectors

Nanofluid is a colloidal mixture of nanoparticles, usually metals or metal oxides, in a base fluid, with enhanced thermo-physical property, as compared with the base fluid. This type of fluid can be used in solar thermal energy applications for its enhanced heat-transfer property, along with the rheological properties. The experimental results presented in Arthur et al. show enhancement in thermo-physical and rheological properties, for example, specific heat, thermal conductivity, and viscosity of different molten salt nanofluids that are used in solar thermal energy systems (Arthur and Karim, 2016). The effect of gold nanoparticles in nanofluids has been investigated by Chen et al. for enhancing photothermal conversion in direct solar absorption solar collector. Localized surface plasmon

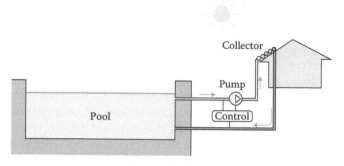

FIGURE 13.22
Outdoor pool heating using solar thermal technology.

resonance effect enhances the solar light absorption in gold nanoparticles as compared with the base fluids (Chen et al. 2016).

The performance of an alumina-water nanofluid-based direct-absorption solar collector was investigated by Tyagi et al. The results show that the collector's efficiency increases with the increase in nanoparticle size. Ten percent higher collector efficiency was obtained as compared with collectors using water as the working fluid, under the same operating conditions (Tyagi et al. 2009). Khullar et al. have assessed the performance of amorphous carbon nanoparticles' dispersion in ethylene glycol and multiwalled carbon nanotubes' (MWCNTs) dispersion in distilled water in comparison with commercial material as selective solar absorber. The study shows that the volumetric solar collector with nanoparticle dispersion shows the best performance at an optimum volume fraction (Khullar et al. 2014). The potential use of different nanoparticles (aluminum, copper, graphite, and silver nanoparticles) dispersed in Therminol VP-1 heat-transfer fluid to be used in power tower solar collectors has been presented by Taylor et al. The experimental data show an increase of 10% efficiency as compared with surface-based collectors for solar concentration ratio range of 100–1000 (Taylor et al. 2011).

13.4.2.7 Photon-Enhanced Thermionic Emission

Since PV and thermal means of solar harvesting introduce limits of their own (discrete wavelength absorption for PV and Carnot limit for thermal engines), combining the two concept can ramp up the overall output. Photon-enhanced thermionic emission (PETE) is one such attempt, where both are combined into a single physical process to take advantage of both the high per-quanta energy of photons and the available thermal energy due to thermalization and absorption losses. PETE occurs in a three-step process: first, electrons in the PETE cathode (constructed by p-type semiconductor) are excited by solar radiation into the conduction band. Second, they rapidly thermalize within the conduction band to the equilibrium thermal distribution according to the material's temperature and diffuse throughout the cathode.

Finally, electrons that encounter the surface with energies greater than the electron affinity can emit directly into vacuum and are collected at the anode, generating current. Each emitted electron thus harvests photon energy to overcome the material band gap and also thermal energy to overcome the material's electron affinity. The total voltage produced can thus be higher than for a PV of the same band gap, owing to this "thermal boost," thus more completely using the solar spectrum (Schwede et al. 2010).

13.5 Solar Heating and Air Conditioning

One of the most practical applications that directly effects the end user is the scope of solar heating and air conditioning. It includes solar thermal concept to maintain the building HVAC and electricity from PV module to power the system, which contributes in the HVAC. The space heating or cooling system of a building requires determination of the thermal resistance of the building. The heat from the sun essentially is transferred to the building through all three means of conduction, convection, and radiation. Thus, a very careful study of the whole building to determine the thermal resistance due to material is necessary to optimize the solar heating and cooling system.

The solar heating and cooling system can be active or passive. Passive system is integral to the building itself. The building is designed to optimize its heat intake for the required amount of heating and/or cooling. For this kind of system there is no need for installing an auxiliary system. Generally, a combination of both types of heating system is employed to obtain optimal heating. The combination of both in buildings gives rise to the concept known as solar architecture.

13.5.1 Passive Space System

Passive heating is the most spontaneous form of warming a house, and it is nigh impossible to construct a building that will never be heated passively. The sun warms all buildings, as it shines on them during the day. Passive solar heating, cooling, and lighting are a function of the building orientation, surrounding conditions, insulation, thermal storage mass, window placement, design, natural ventilation system, and so on.

The building insulation plays a very crucial role in both passive and active heating. The building must be well insulated to reduce the thermal loads. The roof of the building needs special care in this regard, especially during summer, when it receives considerable amount of radiation. Also, if it is ill-insulated, then the heat inside the building cannot be contained, thereby causing a big energy loss in the winter time. Heat can also be stored in the structural materials of the building. The storage material is referred to as thermal mass. In the winter, during the morning and noon, the solar radiation could be used to heat up the thermal mass, and during late afternoon and evening, the thermal storage would release the heat into the surrounding. To take advantage of this, a south-facing heavy masonry wall could be glazed, and thus, it would absorb the heat from the solar radiation in the day. As the night approaches, the wall can radiate heat to maintain a feasible temperature. Glazing the wall reduces heat loss back to the atmosphere and increases the efficiency. Figure 13.23 illustrates this method.

Possibly, the most important aspect in passive heating or cooling is the building's shape and orientation. It needs to be designed in such a way that it can provide both heating and cooling. For better heating, the building must be designed to be more receptive to solar radiation and for longest period possible in the appropriate surfaces. The building should be such that it can receive the radiation during the solar peak periods. For passive cooling, the inlet of breeze and shading must be taken into consideration. Proper ventilation and air flow through the building are needed. The ratio of the building's volume to the exposed surface area is very important in passive heating and cooling. For example, in buildings with high volume-to-surface ratio, the passive heating is slower, owing to smaller available heating area.

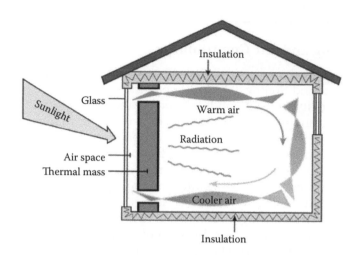

FIGURE 13.23

Passive solar heating using thermal mass. (From Iklim. [2015, December]. *Passive Solar Heating.* iklimnet: http://www.iklimnet.com/save/passive_solar_heating.html.)

Windows are simple way of boosting passive heating or cooling. During daytime, sunlight can enter into the house, heating it up via the window. With a transparent glass to cover it, the whole house can act as a green house. Alternatively, windows provide a means for breeze to flow in and cool the interior and thus, during summer, provide a very important source of both heating and cooling. In hotter countries, windows usually have a sun shade, an overhanging structure that blocks direct solar radiation from entering the windows, thus reducing the cooling load.

13.5.2 Active Solar Space Heating System

Active solar space system uses collectors to heat a fluid and a storage unit to store solar energy, until needed. The controlled distribution system provides solar energy to the heated space in a controlled manner. The complete system requires pump or fans, which are usually powered through the grid and hence, most of the time, from non-renewable sources.

During the day, collectors are used to absorb the solar radiation into the working fluid, and it is stored in a suitable medium, which acts like a thermal capacitor. As the building requires heat, it is simply retrieved from the thermal capacitor. Temperature can be easily controlled using temperature sensors, coupled with a proportional–integral–derivative controller. A heat pump can be used to regulate the internal temperature with the heat input gained from the thermal capacitor. The solar heating system follows the following modes, depending on the situation it is in (Kalogirou, 2009);

- When solar energy is available and heat is not required in the building, solar energy is added to storage.
- When solar energy is available and heat is required in the building, solar energy is used to supply the building load demand.
- When solar energy is not available, heat is required in the building, and storage unit has stored energy, the stored energy is used to supply the building load demand.

- When solar energy is not available, heat is required in the building, and the storage unit has been depleted, auxiliary energy is used to supply the building load demand.
- When storage unit is fully heated, there are no loads to meet, and the collector is absorbing heat, solar energy is discarded.

In order to heat up the house, when needed, collectors are used to heat up working fluid, which might be air or water. The heat is then stored in materials with high heat capacity, such as gravel/pebble and water. Air to be heated is flown over the pebbles or heat exchanger, and the hot air is then fed into the air duct, which, in turn, warms the house (Figure 13.24). Usually, auxiliary heating systems are integrated into the system to offset the problem that occurs when the heat in thermal storage is low but the demand for room heating is high.

The heat from the collector can be used to run a heat pump as well. Heat pumps are setups/machines that transfer heat from low temperature to high temperature. The high-temperature source can be the thermal storage and the low-temperature end is the air in the air duct, which is pumped into the space to be heated. The electricity requirements can be obtained from PV modules or can be taken from the grid.

13.5.3 Active Solar Space Cooling System

Although this may come as a surprise, solar radiation can be used for air conditioning. The most basic and obvious way is to use PV to power the space cooling systems, such as air conditioner and heat pumps in hot countries. There are other popular forms of active cooling which involve the sun-like evaporative cooling and desiccant cooling.

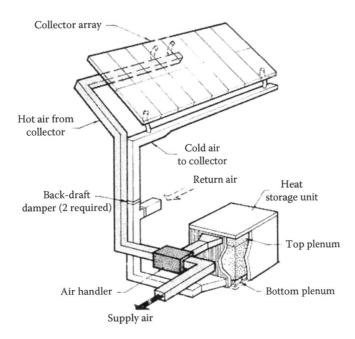

FIGURE 13.24
Active solar air heating. (From Warmair. [2015, December]. *Active Solar Heating*. http://www.warmair.com/html/active_solar_heating.htm.)

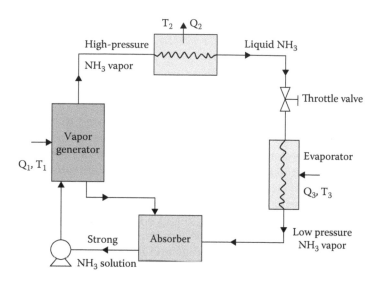

FIGURE 13.25

Absorption cycle schematic, where Q_1 is the heat from the sun via the collector/storage. (From Ustudy. [2015, December]. Vapour absorption refrigeration system. http://www.ustudy.in/node/3197.)

In intermediate seasons in hot countries, good ventilation and, at best, evaporative cooling suffice. Evaporative cooling takes the concept of water taking heat from the environment to evaporate, thereby reducing the temperature of the surrounding. However, in hot climate, such processes are not enough to maintain a comfortable temperature in the interior of the house. Since solar energy is already available in these countries at this time, it is possible and energy-wise more economical to use solar energy to power absorption cycles to cool the interior.

Absorption cycle use desiccants, materials that can attract other gasses or liquids that have affinity for water, to attract and hold moisture. In general, the absorbent absorbs an evaporating refrigerant. The most usual combinations of fluid include lithium bromide-water (LiBr-H$_2$O), where water vapor is the refrigerant, and ammonia-water (NH$_3$-H$_2$O), where ammonia is the refrigerant. The high-pressure refrigerant vapor is formed using the heat from the solar collector/thermal storage in the vapor generator. It is then flown to the condenser, where the refrigerant is condensed to liquid, rejecting the heat to outside. It then gets throttled to a low pressure and goes to evaporator. The evaporator takes in the heat from the surrounding, cooling it in the process. The absorber unit then absorbs the evaporated refrigerant and is then pumped to the generator, completing the cycle. The basic principle of the absorption air conditioning system is illustrated in Figure 13.25.

Solar cooling and heating provide a wide scope, and if used properly, they can bring us one step closer to energy-zero buildings. However, much research is needed for proper integration of all of the concepts to ensure optimal utilization.

13.6 Solar to Chemical Energy Conversion (Artificial Photosynthesis)

Perhaps, the most archaic harvesting of solar energy is the solar to chemical energy conversion, and the ones who did it are none other than plants! Plants perform this conversion through natural photosynthesis (NPS), in which oxygen and carbohydrates are produced from water and carbon dioxide by using sunlight, thus closing the carbon cycle and giving us oxygen to breathe. It might come as a surprise to some that the efficiency of NPS is very low. For agricultural crop, over the life cycle, the efficiency in harvesting energy is lower than 1%! In the most optimum possible condition, the efficiency of NPS may range around 7%. The artificial photosynthesis (APS) is a manmade attempt to harvest solar energy by converting it into chemical energy of any usable kind. It is much more controllable in terms of size and output; that is, it can be designed to produce chemicals that can be directly used as fuels. An example of that would be to produce hydrogen by splitting water, which can be used as fuel cells, instead of carbohydrate. For NPS, that kind of control is never possible. However, the current technology has not evolved enough to have APS device capable of producing molecular fuels, such as hydrogen, at a scale and cost that can compete with fossil fuels. The efficiency of the process is a major hindering factor.

13.6.1 Efficiency of Artificial Photosynthesis

Like Betz limit for wind turbine and Carnot efficiency for thermal engines, the APS has a theoretically imposed upper limit. This limit is associated with band cap excitation and can vary from solar collector to collector. Owing to the fact that the band gaps are discrete and not continuous, solar photons that have greater wavelength than the band gaps wavelength cannot be absorbed, and thus, they will be totally lost. On the other hand, the ones with lower wavelength will be absorbed, but the extra energy that they had will be lost. Only the photons with wavelength equal to the wavelength of the band gap will be absorbed, with no loss in photon energy. Thus, the choice of material for absorber limits the amount of energy it could potentially absorb (Bolton, 1996).

13.6.2 Light Absorption and Fuel Production

To mirror the NPS, the APS uses antenna/reaction center complex to harvest sunlight and generate electrochemical potential, which is used for further reaction. The oxidation of water or other electron source and reduction of precursors to hydrogen or reduced carbon are done by catalyst. The antenna is the main hub for accepting the solar radiation. It increases the overall efficiency of collection of solar radiation by absorbing light over a relatively large area and delivering the excitation energy to a single reaction center. In NPS, chlorophylls are used to absorb light and generate electrons at high-energy state. In APS, there are two different kinds of material structures that act as absorbers. The first kind comprises a single light excitation site, which absorbs the light attached to an electron donor on one side and electron acceptor on other. A dye molecule or visible light-absorbing semiconductor is usually employed as the excitation site. In the second kind, the whole process of donating electron and accepting the exited electron occurs in two steps. The single-step kind has the advantage of the structure being simple. While the two-step process makes the whole thing more complicated, but since in both the steps, light is absorbed, a greater fraction of the incident light can be utilized in it.

There are a number of ways in which APS can be useful; however, none has seen the light of being commercialized fully yet. One unique way to use it is to upgrade the available biofuels for later use. Some photochemical cells can be used to generate hydrogen gas from biofuels such as glucose, ethanol, and methanol. Alternatively, hydrogen can be produced from water, where the active electron produced from the APS system is used to create hydrogen from the proton, which results from the splitting of water (Gust et al. 2009).

13.7 Case Study: Solar Home Systems in Bangladesh

The installation of solar home systems in Bangladesh is so far a successful story and has attracted the attention of researchers and policy makers from around the world. The Renewable Energy Policy of the Bangladesh Government envisions that 10% of total energy production will have to be achieved by 2020. Infrastructure Development Company Limited (IDCOL) started the Solar Home Systems (SHS) program in 2003 to ensure access to clean electricity for the energy-starved off-grid rural areas of Bangladesh. The initiative has been taken as "Rural Electrification and Renewable Energy Development Project" (REREDP) in 2003. Till now, 3.6 million solar home systems (total 150 MW) have been installed around the country (SREDA, 2015). More than 86% of SHS installations in Bangladesh are carried out under the auspices of the publicly owned IDCOL of Bangladesh, with grants and soft loans from donor agencies such as the International Development Association (IDA), Global Environment Facility (GEF), Asian Development Bank (ADB), Islamic Development Bank (IDB), Gesellschaft für Technische Zusammenarbeit (GTZ), and Kreditanstalt für Wiederaufbau (KfW) (Baten et al. 2009) (Chowdhury et al. 2011). About 13 million beneficiaries are getting solar electricity, which is around 9% of the total population of Bangladesh (IDCOL). IDCOL has a target to finance 6 million SHS by 2017, with an estimated generation capacity of 220 MW of electricity (IDCOL). More than 65,000 SHSs are now being installed every month under the program, with average year-to-year installation growth of 58% (SREDA, 2015). Solar home systems typically use 10–135 W_p solar photovoltaic panels. The program replaces 180,000 tons of kerosene, which has an estimated value of USD 225 million per year. Around 70,000 people are directly or indirectly involved with the program (IDCOL, 2015) (Figures 13.26 and 13.27).

A typical Solar Home Systems (SHS) in Bangladesh generally consists of solar PV panels, battery, charge controller, and DC appliances (Chakrabarty and Islam, 2011). The common DC appliances that are used by SHS are lamps, small fans, and televisions (Ahammed and

FIGURE 13.26

Solar home systems in a village in Bangladesh. (From Deutsche Welle, http://www.dw.com/en/modern-energy-technology-can-help-combat-poverty/a-5533266.)

FIGURE 13.27
A drug store uses solar PV for lighting at Adorsho Char Bazaar, Feni, Bangladesh. (From Salehin, S. et al. (2014, May). Optimized model of a solar PV-biogas-diesel hybrid energy system for Adorsho Char Island, Bangladesh. In *Developments in Renewable Energy Technology (ICDRET), 2014 3rd International Conference on the IEEE*, pp. 1–6.)

Taufiq, 2008). Some SHSs also use mobile phone chargers. The electrification by SHS offers tangible benefits to the people living in rural areas, including better quality of life, reduced indoor pollution, and better education to children (Mondal et al. 2011). The SHSs are economically sustainable and attractive for small rural businesses and household lighting, coupled with entertainment (TV). However, the SHSs that are used only for household lighting are not economically and financially viable without considering the associated social benefits (Mondal and Hossain, 2010).

13.8 The Challenge

While the chapter showed multitudes of application that solar energy can harness and its potential and asserted advantage of using solar power, the question that remains now is, Why didn't we go solar fully yet? That is because any technology comes with its limitation, and the concept of solar power is no exception to this rule. It is always better to know the challenges and limitations of a particular technology to better fit it in the whole spectrum. Also, it opens up the scope for research and development of the particular technology. Thus, no study of any system is complete unless one gathers knowledge of the limitation and challenges. Hence, this section of the chapter will deal with the commonly known challenges that this technology faces.

The biggest issue regarding solar energy is its intermittency, or irregular nature. This fact has been hinted to multiple times in this chapter. It is obvious that something that depends integrally on solar radiation will cease to give output in the absence of it. Also, given the fact that late afternoons and early mornings suffer from poor direct radiation, the time window for proper production of power from solar radiation becomes very limited. This limitation is magnified especially for the time of the year when rain is abundant or during winters, when days are substantially smaller.

Intermittency can be tackled if there is a possibility of a proper storage and if the power produced via the solar energy system is high enough to be surplus after quenching the need of the time. Both of these issues are met with problem. The energy storage is an issue, which still needs a lot of work to be reliable for long periods and cheap. Different forms of energy storage, as discussed, and also the others that are in use are still not up to the mark. Thus, solar power, though available all around us for free, provides a big caveat of being too costly to harvest at times.

The issue of production is met with inefficiency. The standard state-of-the-art solar panel of today has the efficiency around 20%–30% at best. Thus, to produce a power of substantial amount, solar parks or CSP are the only solution. These solutions, given the cost of production and low efficiency of single producing panel, are very costly. Adding to the fact that present-day infrastructures are built with fossil fuel in mind makes overnight change difficult. Most of our energy-harnessing plants are built for fossil fuel. The cars are run by petrol, not electricity. Homes are fitted with natural gas outlets, thus making electric stove a luxury at times. Overhauling all of it would require immense financial support and thus rendering the prospect of rapid transition very bleak.

However, challenges are meant to be surpassed. Every change came with challenge, but if history has shown anything, it is that persistence and diligence have shown the light in the end of the tunnel. Solar power or renewable energy resources are no exception.

References

Abu-Aligah, M. 2011. Design of photovoltaic water pumping system and compare it with diesel powered pump. *Jordan Journal of Mechanical Industrial Engineering* 5(3): 273–280.

Ahammed, F., D. A. Taufiq. 2008. Applications of solar PV on rural development in Bangladesh. *Journal of Rural and Community Development* 3(1): 93–103.

Arthur, O., and M. A. Karim. 2016. An investigation into the thermophysical and rheological properties of nanofluids for solar thermal applications. *Renewable and Sustainable Energy Reviews* 55: 739–755.

Batchelder, J. S., A. H. Zewai, and T. Cole. 1979. Luminescent solar concentrators. 1: Theory of operation and techniques for performance evaluation. *Applied Optics* 18(18): 3090–3110.

Baten, M. Z., E. M. Amin, A. Sharin, R. Islam, and S. A. Chowdhury. 2009. Renewable energy scenario of bangladesh: Physical perspective, in: Proceedings of the 1st International Conference on the Developments in Renewable Energy Technology, United International University, Dhaka, Bangladesh, 225–229.

Bloomberg New Energy Finance. (2015, December). *Wind and Solar Boost Cost-Competitiveness Versus Fossil Fuels*. Retrieved from Bloomberg New Energy Finance: http://about.bnef.com/press-releases/wind-solar-boost-cost-competitiveness-versus-fossil-fuels.

Bolton, J. R. 1996. Solar photo production of hydrogen: A review. *Solar energy* 57(1): 37–50.

Business Wire. (2015, December). *Business Wire*. Retrieved from http://mms.businesswire.com/bwapps/mediaserver/ViewMedia?mgid=321682&vid=4.

Chakrabarty, S., and T. Islam. 2011. Financial viability and eco-efficiency of the solar home systems (SHS) in Bangladesh. *Energy* 36(8): 4821–4827.

Chen, M., Y. He, J. Zhu, and D. R. Kim. 2016. Enhancement of photo-thermal conversion using gold nanofluids with different particle sizes. *Energy Conversion and Management*, 112: 21–30.

Chowdhury, S. A., S. M. Monjurourshed, R. Kabir, M. Islam, T. Morshed, M. Rezwan Khan, and M. N. Patwary. 2011. Technical appraisal of solar home systems in Bangladesh: A field investigation. *Renewable Energy* 36(2): 772–778.

Daniel Engineers and Contractor Inc., L.P. (2016, November 19). Retrieved from http://www.lpdaniel.com/SEGS.html.

Deutsche Welle http://www.dw.com/en/modern-energy-technology-can-help-combat-poverty/a-5533266, accessed January 3, 2016.

Eeremultimedia, https://www.eeremultimedia.energy.gov/solar//sites/default/files/photo_csp_tower_development_solartwo_barstow_2000_high.jpg, 2015.

Gay, C., J. Wilson, and J. Yerkes. 1982. Performance advantages of two-axis tracking for large flat-plate photovoltaic energy systems. *IEEE Photovoltaic Spec. Conf.* Arco Solar Industries, Woodland Hills, CA: IEEE.

GIZ. 2013. German Technical Cooperation, Solar water pumping for Irrigation: Opportunities in Bihar, India, Technical report.

Gizzie, E. A., G. LeBlanc, G. K. Jennings, and D. E. Cliffel. 2015. Electrochemical Preparation of Photosystem I–Polyaniline Composite Films for Biohybrid Solar Energy Conversion. *ACS Applied Materials & Interfaces*, 7(18): 9328–9335.

Gontikaki, M., H. L. Jan, and H. Pieter-Jan. 2010. Optimization of a solar chimney to enhance natural ventilation and heat harvesting in a multi-storey office building. *Buildings, Physics and Services*.

Gust, D., T. A. Moore, and A. L. Moore. 2009. Solar fuels via artificial photosynthesis. *Accounts of Chemical Research* 42(12): 1890–1898.

Hahn, A., R. Schmidt, A. Torres, and A. Torres. (2000 May). Resource-conserving irrigation with photovoltaic pumping systems. In s16th European Photovoltaic Solar Energy Conference, Glasgow, 1–5.

Hossain, M. A., M. S. Hassan, M. A. Mottalib, and M. Hossain. 2015. Feasibility of solar pump for sustainable irrigation in Bangladesh. *International Journal of Energy and Environmental Engineering* 6(2): 147–155.

Hu, C., and R. White. 1983. Solar cells: from basic to advanced systems McGraw-Hill Book Company, pp. 20–23.

IDCOL, Infrastructure Development Company Limited, www.idcol.org, 2015.

Iklim. (2015, December). *Passive Solar Heating*. Retrieved from iklimnet: http://www.iklimnet.com/save/passive_solar_heating.html. (Accessed December 16, 2015).

International Energy agency. (2014). Technology Roadmap Solar Photovoltaic Energy. Paris: International Energy Agency.

JUWI. (2015, December). JUWI energy is here. Retrieved from Waldpolenz Solar Park: http://www.juwi.com/solar_energy/references/details/waldpolenz_solar_park/-/-/5.html.

Kalogirou, S. A. (2009). *Solar Energy Engineering Process and System*. Elsevier: Burlington, MA.

Kelley L. C., E. Gilbertson, A. Sheikh, S. D. Eppinger, and S. Dubowsky. 2010. On the feasibility of solar-powered irrigation. *Renewable and Sustainable Energy Reviews* 14(9): 2669–2682.

Khullar, V., H. Tyagi, N. Hordy, T. P. Otanicar, Y. Hewakuruppu, P. Modi, and R. A. Taylor. 2014. Harvesting solar thermal energy through nanofluid-based volumetric absorption systems. *International Journal of Heat and Mass Transfer*, 77: 377–384.

Manitoba Hydro. Available at http://manitobahydroplace.com/. (Accessed December 30, 2015).

Meteonorm. (2015, November 23). *Meteonorm Software*. Retrieved from http://meteonorm.com/.

Mondal, A. H., and D. Klein. 2011. Impacts of solar home systems on social development in rural Bangladesh. *Energy for Sustainable Development* 15(1): 17–20.

Mondal, Md., A. H. 2010. Economic viability of solar home systems: Case study of Bangladesh. *Renewable Energy* 35(6): 1125–1129.

Newport. (2015, November 23). Introduction to Solar Radiation. Retrieved from Newport: http://www.newport.com/Introduction-to-Solar-Radiation/411919/1033/content.aspx.

Palankovski, V. (2015, November 11). Simulation of Heterojunction Bipolar Transistors. Retrieved from http://www.iue.tuwien.ac.at/phd/palankovski/node18.html.

Quaschning, V. 2005. *Understanding Renewable Energy Systems* Vol 1. Earthscan: Sterling, VA.

Rain Harvest. (2016, January). *Rain Harvest.* Retrieved from http://www.rainharvest.co.za/wp-content/uploads/2010/08/Flat-Plate-Solar-Collector.jpg.

Salehin, S., A. S. Islam, R. Hoque, M. Rahman, A. Hoque and E. Manna. (2014, May). Optimized model of a solar PV-biogas-diesel hybrid energy system for Adorsho Char Island, Bangladesh. In *Developments in Renewable Energy Technology (ICDRET), 2014 3rd International Conference on the IEEE,* pp. 1–6.

Schlaich Bergermann. The Solar Chimney. Available at http://www.math.purdue.edu/~lucier/The_Solar_Chimney.pdf. (Accessed December 30, 2015).

Schlaich, J. 1995. *The Solar Chimney.* Edition Axel Menges.

Schwede, J. W., I. Bargatin, D. C. Riley, B. E. Hardin, S. J. Rosenthal, Y. Sun, F. Schmitt et al. 2010. Photon-enhanced thermionic emission for solar concentrator systems. *Nature materials* 9(9): 762–767.

Solar Cell Central. (2016, January). *Solar Cell Central.* Retrieved from http://solarcellcentral.com/images/pn_junction_w_diagrams.jpg.

SREDA, Sustainable and Renewable Energy Authority, Government of Peoples' Republic of Bangladesh, www.sreda.gov.bd, 2015.

Taylor, R. A., P. E. Phelan, T. P. Otanicar, C. A. Walker, M. Nguyen, S. Trimble, and R. Prasher. 2011. Applicability of nanofluids in high flux solar collectors. *Journal of Renewable and Sustainable Energy,* 3(2): 023104.

Tosun, T. 2015. Luminescent solar concentrator. Technical University of Delft *SPOOL,* 2(1): 21–23.

Tyagi, H., P. Phelan, and R. Prasher. 2009. Predicted efficiency of a low-temperature nanofluid-based direct absorption solar collector. *Journal of Solar Energy Engineering,* 131(4): 041004.

U.S. Department of Energy's Office of Energy Efficiency and Renewable Energy (EERE). (2016, November 19). Energy.gov. Retrieved from http://energy.gov/eere/sunshot/downloads/solartwo-tower-system.

Ustudy. (2015, December). Vapour absorption refrigeration system. Retrieved from ustudy: http://www.ustudy.in/node/3197. (Accessed December 15, 2016).

Warmair. (2015, December). *Active Solar Heating.* Retrieved from WARMAIR: http://www.warmair.com/html/active_solar_heating.htm.

Wolfe, C., J., Holonyak, and G. Stillman. (1988). *Physical Properties of Semiconductors.* Prentice-Hall, Inc: Englewood Cliffs, NJ.

14

Utilization of CO_2 for Fuels and Chemicals

Fatin Hasan and Paul A. Webley

CONTENTS

ABSTRACT Since the industrial revolution, carbon dioxide (CO_2) concentration in the atmosphere has increased significantly owing to human activities, causing climate change and global warming. Furthermore, the depletion of fossil fuel reserves owing to ever-growing consumption is another critical problem that threatens our civilization. Thus, innovative and global-scale solutions are required. One potential solution involves capturing CO_2 from concentrated sources or the atmosphere and then converting it into high-value chemicals and liquid fuels. Urea, salicylic acid, carbonates, and so on, are examples of some commercial industries where CO_2 is used as a feedstock since the nineteenth century. Carbon dioxide is an attractive feedstock, because it has advantageous characteristics such as being non-toxic, non-flammable, abundant, and inexpensive. Also, it exhibits unique supercritical properties that make it suitable for applications such as supercritical fluid extraction (SFE) and CO_2-enhanced oil recovery (CO_2-EOR). Carbon dioxide can also be converted into energy-rich products such as methanol, dimethyl ether (DME), formic acid, and syngas, which can be used as alternatives to fossil fuels. However, as CO_2 is a stable molecule, energy input is required to upgrade it to a higher-value material. It is crucial that the input energy is renewable in order to be a sustainable process. This chapter presents an overview of different routes of CO_2 utilization as a feedstock for fuels and chemicals production.

14.1 Introduction

The biological sequestration of atmospheric CO_2 over hundreds of millions of years produced fossil fuels equivalent to 4000–6000 gigatons of carbon (Gt-C) on our planet (Yu et al. 2008, Rackley 2009). Since the middle of the eighteenth century, about 5% of those resources have been combusted, producing 280 Gt-C in the form of CO_2 released to the atmosphere. The natural flux of CO_2 among three inventories, namely, land, ocean, and the atmosphere, is about 60 Gt-C per year. However, after the industrial revolution, the carbon cycle has been disrupted owing to the change of land use (urbanization and agricultural practices) and the increase of CO_2 concentration in the ocean. The ability of the land and ocean to take up more CO_2 has declined, resulting in the accumulation of CO_2 from both the natural fluxes and the anthropogenic emissions in the atmosphere. Consequently, the concentration of atmospheric CO_2 has increased, from an average of 100 ppm before the industrial era to 384 ppm in 2004 and then to 388 ppm in 2010 (Yu et al. 2008). The rise of the concentration of CO_2 in the atmosphere, in addition to other greenhouse gases such as water vapor, has led to an enhanced greenhouse effect, resulting in global warming (Yu et al. 2008, IPCC 2007). According to instrumental records, the rate of temperature rise on the surface of the earth has increased over the last 50 years as almost double that of the last 100 years, that is, 0.13°C versus 0.07°C per decade (Yu et al. 2008, IPCC 2007). In addition to the rise of surface temperature, other climate change effects resulting from higher temperatures have also been observed, including higher sea levels, higher atmospheric water vapor concentration, stronger winds, and longer droughts over wider areas (IPCC 2007). Despite these dire consequences of continual emissions of CO_2 to the atmosphere, forecast studies predict that between 2010 and 2050, the energy consumption of the world will double and fossil fuels will remain the dominant source for energy. This growing demand for fossil-fuel-based energy translates to an increase of global CO_2 emissions (Rackley 2009). Therefore, enormous efforts have been devoted to finding potential solutions to mitigate the carbon content in the atmosphere, some of which are long-term solutions that emphasize a no-carbon economy by switching to renewable energy sources and others are short-term solutions such as CO_2 capture and storage (CCS) and utilization of CO_2 as feedstock for fuels and chemicals production. The main challenge is how to build an emissions-free, sustainable world, without a substantial impact on economic growth and standards of living.

14.2 CO_2 Capture and Storage

There are several technologies used for CO_2 capture from flue gas streams of power plants; some of them are mature and already demonstrated at a million ton per annum scale and others are recent and still suffer limitations and uncertainties (Aresta and Dibenedetto 2007, Yang et al. 2008). Power generation with carbon capture and its storage fall into three categories: postcombustion capture, precombustion capture, and oxy-fuel capture. Within each strategy, a variety of carbon-capture technologies have been examined. The four most promising CO$_2$-capture technologies are absorption (Øi 2007, Rochelle 2009, Peters et al. 2011), adsorption (Li et al. 2008, Xiao et al. 2008, Webley 2014), cryogenic separation (Valencia and Victory 1990, Chiesa et al. 2011), and membrane technology (Hägg and Lindbråthen 2005, Ho et al. 2008, Merkel et al. 2010). Each of those technologies is restricted to particular conditions and scale; for example, adsorption is typically applicable when CO$_2$ concentration is low, whereas cryogenic separation is practical for streams with high CO$_2$ concentration (Riemer and Ormerod 1995). Regardless of capture technology, the captured CO$_2$ is compressed (to a supercritical phase) and then sequestered into geological reservoirs, such as saline aquifers and depleted gas or oil wells. All of these processes are very expensive; therefore, CCS has not yet been used at a very large scale (Yu et al. 2008), except where some revenue can be earned, for example, enhanced oil recovery (EOR). Ideally, the storage should be as close as possible to the CO$_2$ collection site to reduce the transportation cost. In the absence of a global price on carbon, or a clear economic driver (such as sale of CO$_2$ for EOR), there is little incentive for any business to engage in deployment of a CCS facility. A recent economic assessment indicates that the integration of CCS to power plants like integrated gasification combined cycle (IGCC), natural gas-fired combined cycle (NGCC), oxyfuel combustion, and pulverised coal- supercritical (PC supercritical) results in an increase in the capital cost up to 37%, 40%, 53%–65%, and 61%–76%, respectively (Parsons 2011).

In addition to the additional cost of carbon capture, storing the captured CO$_2$ underground may present environmental risks such as the possible leakage of the injected CO$_2$ back to the atmosphere if there is an earthquake or any other type of earth movement. These challenges suggest the necessity for more work/research to find solutions and alternatives to enhance the validity of this technology. One way to avoid storage of large amounts of CO$_2$ is to seek marketable products that require CO$_2$ as a feedstock.

14.3 Utilization of CO$_2$ in Chemicals and Fuels Industries

Since the nineteenth century, CO$_2$ has been utilized as a feedstock in various commercial industries to produce chemicals such as urea, salicylic acid, pigments, and carbonates. Also, it has been used directly as a fluid for food processing, welding, refrigeration, and so on. More recently, because of global warming, expanding the utilization of CO$_2$ in processes at industrial scale has attracted great attention. Furthermore, the depletion of fossil fuels has added an additional incentive to use CO$_2$ as a carbon source to produce high-energy products such as methanol, dimethyl carbamate (DMC), and dimethyl ether (DME).

The utilization of CO$_2$ to produce high-value chemicals/materials has the potential to reduce carbon emissions and also to generate revenue that can help offset the cost of CO$_2$ capture. This can be achieved through efficiently designed process, in which CO$_2$ feedstock is taken from the atmosphere or the emissions of industrial/power plants, using renewable energy sources, and the waste products are kept to a minimum and CO$_2$-free (Aresta et al. 2005). Figure 14.1 illustrates a conceptual anthropogenic chemical carbon cycle (Olah et al. 2011b).

In addition to environment sustainability, other factors that make CO$_2$ an attractive feedstock are: (1) CO$_2$ is abundant, inexpensive, non-flammable, and non-toxic; (2) it can be a good carbon source for existing applications, replacing toxic precursors and intermediates such as phosgene in the production of polycarbonates; (3) it can be used as a building block to construct new materials and polymers; (4) it exhibits unique supercritical properties,

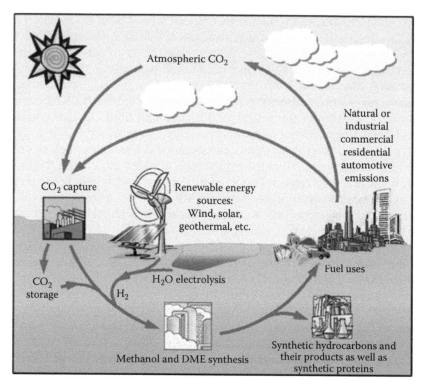

FIGURE 14.1
Conceptual anthropogenic chemical carbon cycle. (From Olah, G.A. et al. *J. Am. Chem. Soc* 133, 12881–12898, 2011b. With permission.)

which make it a good substitute for organic solvents in recovery or separation applications. In summary, CO_2 can be directly used as a fluid or can be converted into other products.

14.3.1 Direct Utilization of CO_2 as a Fluid

In this type of application, CO_2 is used as it is, without undergoing any chemical conversion. There is a wide range of applications where CO_2 is used as a fluid: refrigeration, packaging, fire extinguishers, cooling, propellant, additives to food and beverages, and so on (Song 2002, Aresta and Dibenedetto 2007).

14.3.1.1 CO_2 Utilization in Supercritical Fluid Extraction

Fluids, such as CO_2, at temperatures and pressures above their critical points are called supercritical fluids. They show unique physicochemical properties between those of a liquid and a gas. Supercritical fluids have gas-like diffusivity and viscosity and liquid-like density (Jan 2003). They exhibit a dissolving capacity similar to liquid solvents but at a rapid rate, owing to the gas-like diffusivity (Aresta 2010). The solute can be separated easily by a small change in the temperature and/or the pressure to lower than the solubility; hence, the solute is no longer soluble in the supercritical fluid. Because of these unique characteristics, supercritical fluids have found wide application in extraction and separation processes in a technology referred to as supercritical fluid extraction (SFE) (Rozzi and Singh 2002).

Supercritical CO$_2$ has been widely used in SFE, especially in the food industry, for example, for coffee decaffeination to extract caffeine from green coffee (Kazlas et al. 1994), fats and oils extraction (Fujimoto et al. 1989), extraction of lipids and cholesterol from beef (Chao et al. 1991), food flavor production (Yonei et al. 1995), extraction of essential oils (Reverchon et al. 1995), and fractionation of fats and oils (Arul et al. 1987). Supercritical CO$_2$ has replaced the traditional solvents in industry, because it is non-toxic, non-flammable, inexpensive, readily available, and can be recovered completely, without any traces of residue. Supercritical CO$_2$ is also attractive for distillation/separation of volatile and heat-sensitive materials owing to its moderate critical conditions ($P_c = 74.98$ bar, $T_c = 31.1°C$) (Raventós et al. 2002).

Supercritical CO$_2$ has also been useful for different chemical industries, such as the extraction of heavy metals from environmental contamination, remediating heavy-metal-contaminated soil (Lin et al. 2014), production of micro- and nanoparticles of pharmaceutical ingredients (Esfandiari 2015), and as a medium for chemical reactions (van den Broeke et al. 2001) and polymerization reactions (Kendall et al. 1999).

CO$_2$ has also found application in processing microalgae for biofuels. The extraction of lipids is the key factor in the process of generating biofuels from algae. This efficient extraction method results in a higher yield of lipids and hence more biofuel. However, lipid extraction is challenging, especially with matured microalgae, because lipid molecules are enclosed within strong cell walls; this hinders the mass transfer of the lipid (Mendes et al. 2003). New extraction techniques such as accelerated solvent extraction and SFE have been developed and tested using fluids like methanol, ethanol, and CO$_2$ (Mouahid et al. 2013). Carbon dioxide offers advantages of low toxicity and environmental footprint and also low viscosity, which allows the CO$_2$ molecules to readily penetrate the walls of microalgae cells, thus extracting more lipids (Mendes et al. 2003, Santana et al. 2012). In another study, the efficiencies of three different lipid extraction methods were evaluated. The first method was conventional Soxhlet extraction, which used hexane as a solvent; the second method was a mixed organic solvent extraction (ethyl acetate and methanol, toluene and methanol, and hexane and methanol); and the third method was SFE, which used supercritical CO$_2$ at 306 bar and 60°C. The type of microalgae tested was *Pavlova sp.* The results indicate that the highest fatty acid methyl ester (FAME) extraction yield of 98.7% was obtained with supercritical CO$_2$, whereas the extraction yield obtained with the Soxhlet and mixed organic solvent extractions were 58.5% and 98.1%, respectively (Cheng et al. 2011).

14.3.1.2 CO$_2$-Enhanced Oil Recovery

CO$_2$-EOR is a term used to describe the process of injecting large amounts of CO$_2$ into an existing well for additional oil recovery (Cooney et al. 2015). CO$_2$-EOR is the tertiary production technique that recovers the oil left behind by the conventional primary and secondary techniques. In the primary stage, about 6%–15% of the original oil in place (OOIP) flows out of the field driven by the force of the natural pressure inside the well. As the pressure decreases, water is injected in the well, recovering about 6%–30% of OOIP; this is the secondary stage. Then, the tertiary stage is applied, recovering about 8%–20% of OOIP (Meyer 2007, Cooney et al. 2015).

In the second stage, water is used for oil recovery because it is cheap and available; however, the immiscibility of water and oil produces capillary forces that trap a significant amount of oil inside the small pores of the rocks (Orr Jr et al. 1982). In the tertiary stage, CO$_2$ is used, because under particular conditions, it acts as a completely soluble solvent in oil. Thus, it is essential to inject CO$_2$ in the well under the right conditions; otherwise, it forms two phases with oil (Orr Jr et al. 1982). Up to 40% of the remaining oil can be recovered if CO_2-EOR is operated under the optimal conditions (Blunt et al. 1993).

The number of projects that use CO_2-EOR for oil recovery is increasing steadily since 1972, when it was first introduced commercially in Texas (Meyer 2007). The CO_2 injected in those projects was sourced from natural geologic CO_2 deposits and from captured CO_2 from fossil fuel power plants. For example, in 2012, in the United States, about 60 Mt CO_2 per year was purchased to recover 352,000 bbl/d of oil in 120 projects (Middleton et al. 2015). Three quarters of the CO_2 used in those projects was from natural CO_2 deposits and one quarter came from industrial plants. Some of the injected CO_2 is produced with the oil; however, most of it is regenerated and reused multiple times before it ends up being stored in the well, when the well is abandoned (Middleton et al. 2015).

Some studies and analysis suggested that *via* CO_2-EOR technology, substantial amounts of CO_2 emissions could be productively utilized and then sequestrated inside the depleted wells. CO_2-EOR provides a secure and permanent storage for CO_2 emissions, in addition to the revenue from selling the CO_2 to the EOR operators; this can offset the cost of CO_2 capture (Kuuskraa et al. 2013). Other analysts argue that oil fields would provide only a small volume capacity compared to the amount of CO_2 emitted, and also, the carbon-rich fuel produced will ultimately be combusted, emitting CO_2 to the atmosphere (Stevens et al. 2001). In their study, Jaramillo et al. assessed the overall life cycle of emissions associated with CO_2-EOR in five different existing fields (Jaramillo et al. 2009). They proposed that about 3.7–4.7 Mt of CO_2 is emitted for every 1 Mt CO_2 injected and 0.2 Mt CO_2 is sequestrated per 1 bbl oil produced. In another words, to reach zero CO_2 emissions, about 0.62 Mt CO_2 is required to be sequestrated per every 1 bbl of oil produced (Jaramillo et al. 2009).

14.3.2 Conversion of CO_2 into Synthetic Chemicals

In this type of application, CO_2 is converted into other products *via* chemical reactions. There are two classes of reactions: (1) reactions where CO_2 molecules are fixed on organic substrates such as amines and unsaturated hydrocarbons and (2) reactions where CO_2 molecules are converted into "energy" molecules such as C_1 and C_n. Figure 14.2 presents the prospective reactions for CO_2 transformations into organic materials; the industrialized reactions are signed with $ (Sakakura et al. 2007).

Unfortunately, CO_2 is the most oxidized state of carbon; thus, it is chemically and thermodynamically a stable molecule. Hence, owing to the low energy level and non-reactivity, a large energy input is required to convert CO_2 into other chemicals with higher Gibbs free energy (Sakakura et al. 2007). The energy input could be *via* providing physical energy such as heat (thermal processes), electricity (electrochemical processes), light (photochemical processes), or by using highly reactive starting materials such as hydrogen (Aresta et al. 1992). To enhance CO_2 transformation into other chemicals, a catalyst is usually required in the reaction. Both homogeneous and heterogeneous catalyst systems are used. The former exhibits higher activity compared with the latter, while the latter shows higher stability, can be recovered and reused, and is also more convenient for reactor design (Ma et al. 2009). Table 14.1 presents the production of chemicals and fuels in the world utilizing CO_2 (Omae 2012).

14.3.2.1 Production of Urea from CO_2

The annual world production of urea is 175 (megaton) (Heffer and Prud'homme 2009). About 90% of this urea production is used as fertilizer owing to high nitrogen content (46%), and the remainder is used as an intermediate to produce polymers such as polyurethane

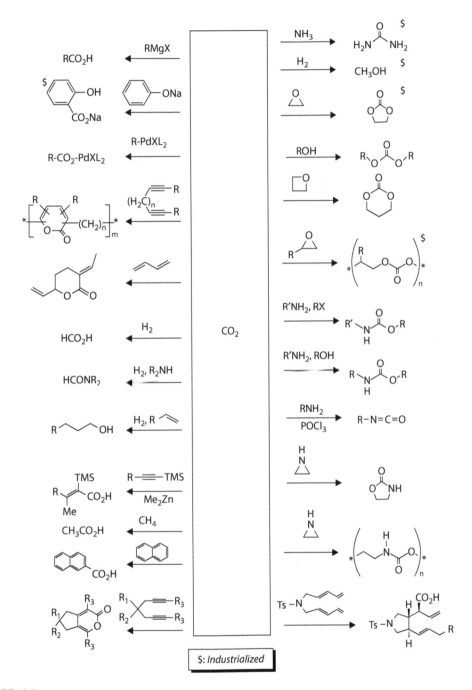

FIGURE 14.2
Summary of the prospective organic chemicals that can be synthesized utilizing CO_2 as a feedstock. Four important organics are produced at industrial scale (signed with $). (From Sakakura, T. et al. *Chem. Rev.* 107, 2365–2387, 2007. With permission.)

TABLE 14.1

Worldwide Production Rate of Organic Chemicals, Starting
from CO_2 (Units are Modified to the SI System)

Product	Production Rate (kg/s)
Cyclic carbonates	2.30 (80,000 ton/y)
Polypropylene carbonate	2.18 (76,000 ton/y)
Polycarbonate (Asahi Kasei process)	17.39 (605,000 ton/y)
Urea	4513.36 (175,000,000 ton/y)
Acetylsalicylic acid	0.46 (16,000 ton/y)
Salicylic acid	2.59 (90,000 ton/y)
Methanol	0.11 (4000 ton/y)

Source: Omae, I. *Coord. Chem. Rev.* 256 (13–14):1384–1405, 2012.

(Chih-Hung and Chung-Sung 2014). There are about 330 urea plants worldwide; 54% of these plants are located in South East Asia (Ricci 2003). Urea is an organic compound with a chemical structure consisting of two NH_2 groups connected *via* carbonyl group [C=O]. Urea is formed *via* a two-step reaction between two moles of ammonia and one mole of CO_2 at 185°C–190°C and 152–223 bar. In the first step, ammonia and CO_2 react, forming ammonium carbamates [$H_2NCOO^- NH_4^+$], which dehydrate, thus producing one mole of urea [H_2NCONH_2] (Ricci 2003).

It might be inferred that the utilization of CO_2 to produce millions of tons of urea every year would have a significant contribution to carbon reduction; however, this is not the case, because the lifetime of CO_2 in urea is short. Urea is converted back into CO_2 within 2–3 months when it is used as a fertilizer (Aresta and Tommasi 1997). On the other hand, if renewable energy is used for urea production, then the process will not affect the CO_2 global balance, because the CO_2 released from the fertilizer will be fixed as a reactant.

14.3.2.2 Conversion of CO_2 into Salicylic Acid and Derivatives

Salicylic acid is produced at industrial scale *via* the Kolb-Schmitt reaction, where dry sodium phenoxide is carboxylated by CO_2 at 150°C–160°C and 5 bar in a closed system (Lindsey and Jeskey 1957). In this system, CO_2 is incorporated as it is into the organic substrate, forming a compound with carboxylate moiety (COO); thus, this system can be considered as a sustainable process (Aresta and Dibenedetto 2007). Worldwide, yearly production of salicylic acid is about 90,000 tons; most of it is used as a pharmaceutical ingredient and the rest is used in food, dyes, and phenolic resins industries (Omae 2006).

One of the well-known salicylic acid derivatives is acetylsalicylic acid (Aspirin), which is produced at a rate of 16,000 ton/y worldwide. In 1895, Felix Hoffmann prepared Aspirin for the first time by acetylating the hydroxyl group on the benzene ring of salicylic acid (Vane and Botting 2003).

14.3.2.3 Conversion of CO_2 into Carbonates and Polycarbonates

The addition of CO_2 to epoxides to produce cyclic carbonates and polycarbonates is a process that has been gaining interest since it was discovered by Inoue et al. 50 years ago (Inoue et al. 1969). Before this discovery, phosgene, a very toxic compound, was the main feedstock for organic carbonates production. Thus, the utilization of non-toxic CO_2 as a feedstock offers a safer and greener copolymerization process.

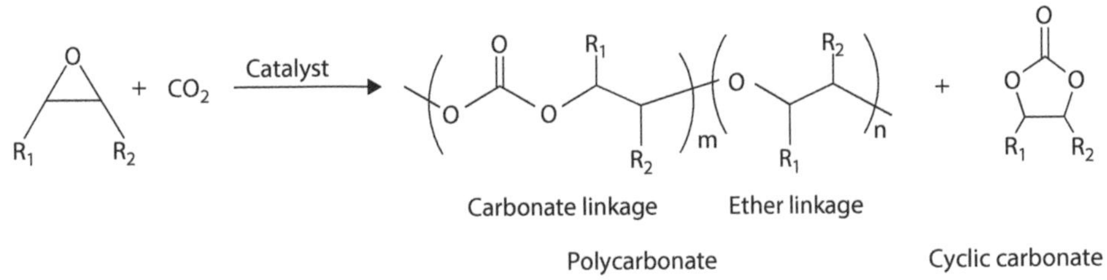

Carbonate linkage   Ether linkage

Polycarbonate   Cyclic carbonate

SCHEME 14.1
The coupling reaction of CO$_2$ and epoxides. (From Lu, X.B. and D.J. Darensbourg. *Chem. Soc. Rev.* 41, 1462–1484, 2012. With permission.).

In CO$_2$/epoxide copolymerization, the product could be cyclic carbonates, polycarbonates, or cyclic carbonates plus polycarbonates, depending on the type of the catalyst and the reaction conditions (Scheme 14.1).

Various catalyst systems have been investigated for CO$_2$/epoxides copolymerization, starting from simple alkali salts to advanced organometallic complexes (Darensbourg and Holtcamp 1996). The copolymerization process starts with the activation of epoxide through the interaction of the oxygen atom with Lewis acid sites (metal center), followed by nucleophilic attack, which results in opening the epoxide ring. Carbon dioxide as a non-polar molecule is activated through an electrophilic (carbon atom) and a nucleophilic (oxygen atom) attack. At this stage of the reaction, carbonate intermediate precursors are formed, which then either react with other CO$_2$ and epoxide molecules, forming polycarbonates, or close the ring by a backbiting reaction, forming cyclic carbonates (Pescarmona and Taherimehr 2012).

Inoue et al. produced poly (propylene carbonates) by copolymerizing CO$_2$ with propylene oxide (PO) in a system containing organometallic catalyst, co-catalyst, and solvent (Inoue et al. 1969). Since then, enormous efforts were devoted to finding a catalyst that is suitable for industrial production of polycarbonates, starting from CO$_2$. However, the process always suffered from low reaction rates, low conversion, high catalyst loadings, and also the need for a co-solvent and for high CO$_2$ pressure (Lu and Darensbourg 2012). In 2004, Lu and Wang used [(salen)CO^{III}X] as a catalyst and quaternary ammonium salt as a co-catalyst for CO$_2$/PO copolymerization system (Lu and Wang 2004). They produced poly (propylene carbonate) with up to 99% selectivity at mild temperatures and pressures; however, the yield was always less than 50%. This is probably due to high conversion, which leads to an increase in the backbiting of poly (propylene carbonate) into cyclic propylene carbonate. Significant progress was achieved by Noh et al. when they coupled (salen)CO complex and quaternary ammonium salt unit in one molecule of catalyst (Noh et al. 2007). They achieved >90% selectivity of polycarbonates over cyclic carbonates and also achieved the highest turnover frequency and turnover number records at 90°C of 3500 h^{-1} and 14,500, respectively.

Ethyl carbonate (EC), propylene carbonate (PC), dimethyl carbonate (DMC), and diphenyl carbonate (DPC) are the most important organic carbonates in industry. They are used as intermediates to produce polycarbonates, aprotic solvents, and electrolytes in lithium-ion batteries. DMC has attracted special attention, because it exhibits excellent properties as a fuel additive. (Sakakura et al. 2007). Polycarbonates are used as packaging materials and in the synthesis of resins and thermoplastics (Clements 2003, Lu and Darensbourg 2012). PPC is one of the most widely used copolymers in industry, because it is mechanically stable and biodegradable (Qin and Wang 2010). The biodegradable polymers are attractive, because they are: (1) decomposed into non-toxic and naturally metabolized substances; (2) very useful in biomedical applications, such as controlled drug delivery, tissue engineering, and adhesion barriers; and (3) are good substitutes for the conventional thermoplastics, which are responsible for an ever-increasing plastic waste problem (Yang et al. 2004, Albertsson and Srivastava 2008). Other epoxides such as styrene oxide, epichlorohydrin, and indene oxide are also copolymerized with CO_2, producing different types of polycarbonates (Darensbourg and Wilson 2012).

More than 2 Mt polycarbonates are produced every year worldwide; however, about 80%–90% of this production is synthesized through the phosgene process (Sakakura et al. 2007). Table 14.1 presents the worldwide production of polycarbonates utilizing CO_2. As with urea manufacture, the amounts of CO_2 used to produce those carbonates and polycarbonates represent only a small fraction of yearly CO_2 emissions; however, this route of CO_2 utilization is still considered advantageous, because it reduces the amount of phosgene used in industry.

14.3.3 Conversion of CO_2 into Energy Products

The worldwide energy crisis of fossil fuels depletion raised an urgent need to develop new and renewable sources of energy. Carbon dioxide is an excellent candidate that can be converted into energy-rich products, such as methanol, CO, C_n hydrocarbons, and their derivatives. This pathway allows mitigating the CO_2 level into the atmosphere and also provides inexhaustible energy source. However, it is essential to point out that since CO_2 is usually the product of combustion (a "spent fuel"), any attempt to upgrade it to a higher-value material will face a positive change in Gibbs free energy, requiring energy input. There is little point in providing this energy input by combusting fossil fuel (with associated release of CO_2). If renewable energy is used to upgrade the CO_2 to higher-value molecule (especially a fuel), the process needs to be analyzed to determine the feasibility of this process compared with direct utilization of the renewable energy.

14.3.3.1 Conversion of CO_2 into Methanol

Carbon dioxide hydrogenation to methanol is a promising approach toward solving the global warming and the fossil fuels depletion crises. Carbon dioxide provides a virtually inexhaustible source of energy, because it is abundant and renewable. Hydrogen, the other component of the reaction, is also abundant on the earth in the form of water. Water electrolysis using regenerative power source such as solar, wind, and geothermal energy is a well-known technology for H_2 production. Hydrogen can also be generated from fossil fuels through the CO_2 emission free "Carnol process" technology (Steinberg 1996, 1997, 1998).

Sun, wind, waves, hydrothermal, and geothermal energies have been proposed as alternative energy sources to fossil fuels. Even though those sources have good potential and are increasingly used, their supply is still very small compared to the amount of energy that people need (Olah et al. 2011a). Hydrogen is another promising alternative for clean fuels; however, it requires special storing conditions, because it is a very light gas. Therefore, Olah has proposed the concept of a "Methanol economy," which is based on replacing fossil fuels, with methanol as means of energy, and is also based on producing that methanol through chemical recycling of CO_2 as an energy carrier (Olah et al. 2011a). Methanol is a safe and convenient liquid fuel with a very good potential to substitute fossil fuels.

The world production capacity of methanol is about 100 Mt/y, which is mainly used as raw material for a wide range of synthetic chemicals, such as formaldehyde, ethylene, propylene, methyl tert-butyl ether, and acetic acid. Methanol exhibits excellent combustion characteristics, which make it a better and safer fuel than gasoline for internal combustion engines (ICE). Therefore, it has been used as a fuel for race cars since the 1960s (Olah et al. 2009). Furthermore, methanol is used in a Direct Methanol Fuel Cell (DMFC) as a fuel to generate electricity through direct oxidation with air to CO$_2$ and water (McGrath et al. 2004).

Conventionally, methanol is produced *via* gas-phase reaction of syngas over a solid catalyst Cu/ZnO/Al$_2$O$_3$ at ~250°C and 50–100 bar (CO + 2H$_2$ = CH$_3$OH) (Hansen and Nielsen 2008). In the early 1960s, it was discovered that adjusting the H$_2$:CO ratio in the syngas feedstock to 2:1 by adding a small amount of CO$_2$ (3%–10%) results in enhancing the methanol production significantly (Joó et al. 2000).

Carbon dioxide can be converted into methanol *via* one of two possible routes: (1) direct hydrogenation of CO$_2$ to methanol (CO$_2$ + 3H$_2$ = CH$_3$OH + H$_2$O) or (2) CO$_2$ is first reduced to CO *via* Reverse Water-Gas Shift (RWGS) reaction (CO$_2$ + H$_2$ = CO + H$_2$O) and subsequently CO is hydrogenated to methanol (CO + 2H$_2$ = CH$_3$OH). The catalyst in the CO$_2$-to-methanol system faces the challenge of water formation, which is adsorbed on the active sites of the catalyst, suppressing its activity. Various additives such as ZrO$_2$, Ga$_2$O$_3$, and SiO$_2$ were added to Cu/ZnO/Al$_2$O$_3$ to improve its ability to handle the presence of water and also the high oxidation power of CO$_2$ (Joó et al. 2000, Ampelli et al. 2015). The catalyst stability was improved; however, the productivity of methanol from hydrogenation of CO$_2$ is 3 to 10 times lower than that from hydrogenation of CO (Ganesh 2014). The low productivity makes the use of CO$_2$ as feedstock less economically feasible compared with syngas system. Numerous studies have been devoted to finding an effective catalyst for industrial production of methanol from CO$_2$, including those of Süd-Chemie, who developed a new catalyst, which was thoroughly tested by Lurgi in the 1990s and showed an economical service life similar to the commercial catalyst used for the CO-to-methanol system (Lorenzo et al. 2014).

The route of CO$_2$ to methanol is not a recent discovery, as it was developed in the United States in 1920–1930s by using CO$_2$ generated as a by-product in fermentation processes (Olah et al. 2009). A demonstration plant with a capacity of 50 kg/day methanol was built in Japan. It was operated at 250°C and 50 bar, with Cu/ZnO catalyst modified with SiO$_2$. It achieved a space-time yield of methanol up to 600 gL^{-1} h^{-1}, with 99.9% selectivity (Aresta and Dibenedetto 2007). Another plant with a capacity of 100 ton/y has been developed by Mitsui chemicals in Osaka, Japan, in 2009. The CO$_2$ required for this plant is extracted from the waste gases from nearby ethylene-oxide-manufacturing unit (Lorenzo et al. 2014). Furthermore, a plant with a commercial production scale (2 million litters of methanol per year) has been built in Iceland by Carbon Recycling International (CRI). The plant was named "George A. Olah Renewable Methanol Plant" after George A. Olah, the Noble Prize Laureate and founder of the "Methanol Economy" concept. (Wernicke et al. 2014). This plant is green because geologic CO$_2$ is used as feedstock and hydrothermal heat is used to provide energy for water electrolysis and for CO$_2$ hydrogenation (Lorenzo et al. 2014).

14.3.3.2 Conversion of CO$_2$ into Dimethyl Ether

Dimethyl ether (DME) is another synthetic chemical that has the potential to be a clean substitute for conventional fossil fuels. Unlike methanol, DME has a very high cetane number (55–60), which is a measure of the ignitibility of fuel in compression engines; therefore, it is considered an excellent replacement for diesel (cetane number, 40–50).

DME is already produced at industrial scale in countries such as Japan, Korea, and China, with a total production rate of 40 Mt/y. There are two distinct methods to synthesize

DME: (1) indirect or two-step DME synthesis, which includes the conversion of syngas into methanol over $Cu/ZnO/Al_2O_3$ in one reactor, and then, the produced methanol is dehydrated over an acidic catalyst, forming DME and water in a second reactor and (2) direct DME synthesis includes the conversion of syngas into methanol and hence into DME in one reactor over a bifunctional catalyst (Lange 2001, Ogawa et al. 2003, Akarmazyan et al. 2014). In both routes, DME is formed *via* methanol intermediate (Zhang et al. 2010, Naik et al. 2011, Chen et al. 2012, Azizi et al. 2014).

DME can be produced from CO_2 indirectly by using methanol produced from CO_2. One-step CO_2-to-DME process is also possible by using a bifunctional catalyst (Naik et al. 2011).

Mitsubishi Heavy Industries is currently building a CO_2-to-DME plant in Iceland, with production capacity of 500 metric ton/day (Thorbjornsson et al. 2012). The feedstock CO_2 is captured from the exhaust gas of the ELKEM plant, which is the largest ferrosilicon producer in the world, with a production capacity of 120,000 ton/y. The required H_2 gas is generated from water by electrolysis. The catalyst to be used is developed by Mitsubishi Chemical (MGC). The process consists of two stages: CO_2 to methanol and methanol to DME.

14.3.3.3 Conversion of CO₂ to Formic Acid and Derivatives

Every year, about 300,000 tons of formic acid are produced in the homogeneously catalyzed reaction of CO and methanol. The catalyst used is sodium methoxide ($NaOCH_3$) (Baiker 2000). Carbon dioxide can also be converted into formic acid or its derivatives (such as formamides) by using catalysts containing metal compounds such as Ti, Pd, Ru, and Rh through the hydrogenation reaction ($CO_2 + H_2 = HCOOH$). The reaction is carried out in organic solvents, water, ionic liquids, or supercritical CO_2. Further, reaction promoters such as amines and KOH are also added to the system (Omae 2006).

Formic acid is a very important chemical that can be used in wide range of applications, such as in leather and textile industry, coagulation of rubber latex, pharmaceuticals, crop protection, and additive for cleaning agent (Aresta 2010). More importantly, formic acid can directly be used in formic acid fuel cells and also as carrier for H_2 (Centi et al. 2013).

BP Chemicals has built a process in which formic acid is produced from CO_2 hydrogenation over a Rh complex catalyst in the presence of triethylamine (Beevor et al. 1990). The produced ammonium formate reacts with a high-boiling-point base, yielding formate, which can thermally decompose producing formic acid. Significant progress has been achieved when a very high reaction rate, up to 1400 mol formic acid per mol catalyst per h, was demonstrated by the Jessop group. They used supercritical CO_2 as a solvent, with Ru(II) phosphine complex as catalyst. They attributed the high reaction rate to the high solubility of H_2 in $SCCO_2$ (Jessop et al. 1994). However, the high toxicity of formic acid (maximum formic acid level allowed in the work environment is 5 ppm) limits the possibility of production at large scale such as that required for CO_2 chemical recycling (Ampelli et al. 2015).

14.3.3.4 Utilization of CO₂ for Syngas Production

The worldwide shift from coal to natural gas is an approach that leads to lower CO_2 emissions. However, methane, the major component of natural gas, is another powerful greenhouse gas. Thus, a small leakage of methane has a strong global warming potential (Howarth et al. 2011). In recent years, significant research has been dedicated to incorporate those two greenhouse gases in reactions that generate useful products. Dry reforming, for example, is a reaction in which methane reacts with CO_2, producing syngas, a very important intermediate for various chemicals and fuels ($CH_4 + CO_2 = 2CO + 2H_2$). In other words, dry reforming is another approach of chemical conversion of CO_2 into higher-value chemicals.

The commercialization of dry reforming faces a challenge of severe catalyst deactivation due to the deposition of carbon generated from methane decomposition ($CH_4 = C + 2H_2$) and the Boudouard reaction ($2CO = C + CO_2$). Extensive studies have been conducted to minimize the carbon formation and to improve the catalyst activity and selectivity of desired product (Fan et al. 2010, Bosko et al. 2010). The low H_2:CO ratio (<1) of syngas produced from dry reforming makes it suitable for applications that require pure or concentrated CO feedstock. Calcor is an example of a commercial process of CO_2 reforming of natural gas, developed by the German company called Caloric. The downstream gas of the reforming step in the Calcor process shows H_2:CO ratio of 0.42:1 and 0.0005 vol% methane (Teuner et al. 2001). The integration of low-temperature purification with the CO_2 reforming in the Calcor process results in producing CO with purity of 99.98%. Such a pure CO feedstock is useful for applications such as carbonic acid production.

Conventionally, conversion of syngas to high-value hydrocarbon processes needs feedstocks with high H_2:CO ratio. For example, methanol and Fischer-Tropsch synthesis require syngas with H_2:CO ratio of 2:1 and 1.7–2.2:1, respectively (Noureldin et al. 2015). It is not beneficial to use the dry-reforming-generated syngas for methanol synthesis, without adjusting the H_2:CO ratio to the desired value. Steam reforming, the traditional method of natural gas reforming, produces syngas with H_2:CO ratio between 3 and 6, which is higher than the required value. Thus, the ratio adjustment can be achieved by combining the steam reforming with the dry reforming in one process (Noureldin et al. 2015). This process is named the bi-reforming process ($CH_4 + H_2O + CO_2 = 2CO + 4H_2$). The H_2:CO of the syngas generated by bi-reforming is variable, depending on the ratio of CH_4, steam, and CO_2. For example, syngas with H_2:CO ratio 1:2 can be obtained when the CH_4:steam:CO_2 ratio is 3:2:1. Such a gas is named metgas and is suitable, particularly, for methanol synthesis (Olah et al. 2013).

Another approach for CO_2 utilization for syngas synthesis is tri-reforming. The advantage of this approach lies in the possibility of using CO_2 in flue gases from power plants directly, without separation. Tri-reforming is a combination of three catalytic reforming reactions in one vessel, including steam reforming, dry reforming, and partial oxidation of methane. After the pollutants' removal, the flue gas containing mainly CO_2, H_2O, and O_2 is fed to the reformer, along with natural gas and O_2 or O_2-enriched air (Song and Pan 2004).

14.3.4 Fixation of CO_2 in Biofuels (Microalgae Cultivation)

One of the CO_2 mitigation options put forward to help mitigate the global warming problem is displacing the petroleum-derived fuels with biofuels, which are generated from crops such as oil seeds and sugarcanes. Despite cultivating large amounts of those crops, the produced fuel is still much less than the current or projected market demand. Therefore, alternatives such as microalgae have been proposed and preferred owing to their high oil content (about 80% of the dry weight in some cases), relatively easy cultivation conditions, high growth rate, and no need for fertile land (Chisti 2007, 2008). Microalgae are microscopic unicellular or simple multicellular organisms that grow rapidly even under harsh conditions. There are about 50,000 different species of microalgae on the earth, some of which are terrestrial and others are aquatic (Mata et al. 2010). Microalgae use CO_2, water, and sunlight to produce algal biomass *via* the photosynthesis process. They grow rapidly by using CO_2 as building blocks, doubling their biomass volume in less than 24 hours (Chih-Hung and Chung-Sung 2014). Microalgae have attracted a great deal of attention since the middle of the last century, because they were proposed as a means for CO_2 fixation (Oswald and Golueke 1960). Every 1 ton of microalgae require about 1.8 ton CO_2 (Acién Fernández et al. 2012). However, the annual worldwide production of microalgae is only 5 kt, which consumes about 9 kt CO_2 (Pulz and Gross 2004).

Microalgae can capture CO_2 from the atmosphere or flue gases and can be cultivated in open ponds and bioreactors and in fresh and saline waters. They can survive a wide range of environmental conditions; however, every species has specific conditions under which their growth rate is accelerated considerably (Aslan and Kapdan 2006). Several factors, such as CO_2 concentration, quality and quantity of light, temperature, salinity, nutrient concentration, pH, pollutants, and continuous mixing, affect the growth of biomass inside the cultured microalgae.

Traditionally, microalgae cultures (in open and closed systems) are aerated to adsorb CO_2 from the atmosphere. However, CO_2 supplied by the atmosphere is very small (0.03%–0.06%), resulting in slow cell growth. Therefore, higher-CO_2-concentration sources, such as flue gases (15% CO_2), coming directly from their source (power plants or industrial processes), and even CO_2 that was already captured and sequestrated underground (Wang et al. 2008, Yoo et al. 2010), are used. Flue gases, in general, contain components such as CO, NO_x, SO_x, N_2, H_2O, and O_2, in addition to CO_2. These components, particularly, SO_x, prohibits the growth of some microalgal species; it has a toxic effect, even at low concentration (Chih-Hung and Chung-Sung 2014). Also, the presence of NO and NO_x at the early stage of microalgae growth causes reduction in the pH of the media, restraining the growth (Negoro et al. 1991). However, at a later stage, both components represent effective nutrients for microalgae (Travieso et al. 1993, Yu et al. 2008). The adsorption of CO_2 by microalgae is a mass-transfer-controlled process, suggesting that higher CO_2 concentration gas injected to the system would improve the driving force and result in more microalgae biomass being produced. Owing to the high cost of complete capture of CO_2 for biosequestration, a new technique was developed by González López et al. 2009, in which they used water enriched with bicarbonate/carbonate as a buffer to adsorb CO_2 from flue gas in a bubble column (González-López et al. 2009). Then, the produced liquid was used as media to culture *Anabaena sp.* in a photobioreactor. About 80% of the dissolved CO_2 was regenerated by microalgae (González-López et al. 2012). Other techniques, such as modified reactor design and flue gas injection methods, have been tested as functions for optimal CO_2 uptake. A flue gas stream was bubbled continuously into an outdoor vertical flat-plate photobioreactor to cultivate cyanobacterium, *Synechocystis aquatilis* SI-2 sp., using the sun as the source of light (Zhang et al. 2001). The results showed that the maximum CO_2 usage efficiency obtained was 8.1%, because the culture medium became too acidic. In another study, an on-demand injection mode was applied on an outdoor open thin-layer photobioreactor used to cultivate *Chlorella sp.* (Kurano et al. 1995, Doucha et al. 2005). Carbon dioxide use efficiency at about 50% was achieved when flue gas containing 6%–8% CO_2 was injected to the system.

The cultivation of microalgae in open ponds is economically favorable compared with that in bioreactors; however, the fluctuation of the sunlight strength and the temperature during the seasonal changes affects the growth and quality of the product (Benemann 1997, Chisti 2007, Acién Fernández et al. 2012). Furthermore, there is a possibility that the CO_2 injected into the pond might be released back to the atmosphere. Bioreactors, on the other hand, offer advantages such as better control over the cultivation conditions, larger surface-area-to-volume ratios, ability to prevent the contamination, higher CO_2 use efficiency, and ability to produce higher biomass density (Rosello-Sastre et al. 2007). However, this technology is still very expensive for a practical large-scale application, and it requires more technical solutions to reduce the cost (Dibenedetto and Tommasi 2003).

For a more efficient cultivation process, continuous mixing is needed to guarantee that light and CO_2 reach all of the cultivated cells. Similarly, temperature and salinity, if they are adjusted corresponding to the need of the cultivated species, can provide very high growth rate. For example, *Anabaena sp.* was grown at a rate of 0.1 h^{-1} when cultivated at 45°C and 6–10 pH (Moreno et al. 2003).

The microalgal biomass is converted into biofuel by one of the following methods: direct combustion, extraction with supercritical CO_2 ($SCCO_2$) or organic solvents, pyrolysis, gasification, liquefaction, and anaerobic fermentation (Aresta et al. 2005). The most preferable method consists of extracting the lipids from microalgae and then converting those lipids into biofuel by transesterification (Section 14.3.1.1).

Using microalgal biomass as biofuel source suggests zero increase in atmospheric CO_2, because the CO_2 generated by burning that fuel is previously adsorbed by the microalgae *via* photosynthesis (Acién Fernández et al. 2012). Theoretically, zero CO_2 emission process is achievable if the flue gases resulting from the biofuel combustion are recycled and injected in the cultivation media of microalgae.

14.3.5 Photo- and Electrocatalytic Conversion of CO_2

In this approach, the energy required for the reduction of CO_2 is in the form of electrochemical energy, which can be generated from sources such as sun or wind. Both two- or single-cell processes have been investigated. In the former, the electrons are generated in a separate cell (photovoltaic cell or wind, and so on) and then used to reduce CO_2 in another cell. Otherwise, both steps are carried out in a single cell; hence, it is called photoelectrocatalytic process (Centi and Perathoner 2009). The protons and electrons supplied to the reduction cell generate a potential between the electrodes and subsequently allow the CO_2 to transform into valuable chemicals such as CO, methane, formic acid, methanol, formaldehyde, and even long-chain hydrocarbons (DuBois 2007, Spinner et al. 2012, Albo et al. 2015).

Table 14.2 summarizes the reaction pathways of electrochemical conversion of CO_2 and also presents the prospective products (Augustynski et al. 2003).

TABLE 14.2

Summary of Electrochemical Conversion of CO_2 with Associated Reduction Potentials

Reaction	Product Name	E^0 (V)[a]	E_{eq} (V)[b]
$CO_2 + 2H^+ + 2e^- = HCOOH$	Formic acid	−0.199	−0.61
$CO_2 + 2H^+ + 2e^- = CO + H_2O$	Carbon monoxide	−0.103	−0.52
$CO_2 + 4H^+ + 4e^- = CH_2O + H_2O$	Formaldehyde	−0.028	−0.44
$CO_2 + 8H^+ + 6e^- = CH_3OH$	Methanol	+0.031	−0.38
$CO_2 + 8H^+ + 8e^- = CH_4 + 2H_2O$	Methane	+0.169	−0.25
$2CO_2 + 2H^+ + 2e^- = H_2C_2O_4$	Oxalic acid	−0.49	−0.90

Source: Augustynski, J. et al. Electrochemical conversion of carbon dioxide. In *Carbon Dioxide Recovery and Utilization*, (eds.) M. Aresta, 279–292. Netherlands, NL: Springer.

[a] Standard reduction potentials.
[b] Reduction potentials measured at neutral solution (pH 7).

Recently, many studies have investigated the factors that determine the reaction pathways and hence the selectivity toward the products such as the reaction media (gaseous and aqueous and non-aqueous electrolytes), catalyst, electrodes, pH, cell configuration, temperature, and pressure (Hori et al. 1989, Hori et al. 2005, Das and Daud 2014).

In aqueous electrolytes, low CO_2 conversion was observed. The reason is that water is reduced to H_2 simultaneously with CO_2 reduction, consuming the electrons provided to the system. In fact, H_2 formation is thermodynamically preferable over CO_2 reduction, which decreases the Faradaic efficiency of CO_2 reduction (Kaneco et al. 2002). The low solubility of CO_2 in water at standard conditions is another reason that suppresses the CO_2 electro-reduction. High pressure was applied to increase the dissolved species; however, a quick deactivation of the electrodes was observed (Frese and Leach 1985). Non-aqueous reagents, such as organic solvents and supercritical CO_2, in which CO_2 is highly soluble, were tested. However, a rapid deactivation of the electrodes was detected due to the high current density required for CO_2 solubility. Some recent studies claimed that CO_2 can be reduced electrocatalytically without a reducing agent, as the protons required for the reduction reactions are transferred to the cell through a membrane integrated to the unit (Centi et al. 2003, Ampelli et al. 2016).

The electrode material also plays a major role in determining the products of the CO_2 electro-reduction process. For example, the use of metals such as Ni, Pt, Co, and Rh, which exhibit low hydrogen overpotentials, results in dominating the formation of hydrogen over hydrocarbons. On the other hand, metals such as Au, Ag, Pd, Ga, Cu, and Zn promote the formation of CO. Copper (Cu) is well known as a metal that promotes the reaction of CO, with H_2 forming hydrocarbons and oxygenated hydrocarbons (Hori et al. 1989). Carbon dioxide reduction in an aqueous solution of $KHCO_3$ and pre-purified Cu electrode produced CH_4 and C_2H_4, with Faradic efficiency of 70%; however, the electrodes were degraded after a short time (Hori et al. 2005). Hydrocarbons of up to C_6 in length were synthesized with non-electro-polished Cu electrodes (Shibata et al. 2008). Gas-phase electro-reduction of CO_2 was performed in an electrochemical cell catalyzed with nanocarbon and integrated with a membrane. Hydrocarbons up to C_{10} were produced in addition to isopropyl alcohol (the main product), ethanol, methanol, acetone, acetaldehyde and traces of alkanes and aromatics (Centi et al. 2007).

14.4 Conclusion

It is well known that climate change, the defining problem of our age, was brought about through the growing global demand for energy. Many solutions have been proposed, some of which are based on switching to environmentally benign energy sources such as solar, wind, and nuclear energy, and others are based on controlling the CO_2 emissions through: (1) carbon capture and storage; (2) utilization of the captured CO_2 chemically to produce high-value chemicals and energy products. It is likely that all of these options will contribute to the overall solution, given the constraints of continued economic growth. No one option alone will be possible. While current market for CO_2-based products is small and not on a sufficient scale to offset CO_2 emissions to the atmosphere, CO_2 utilization does offer a business case for initial deployment of carbon capture technologies, until a carbon price is established. The time has come to regard CO_2 not as a waste material but rather as a non-toxic, inexpensive, non-flammable, abundant chemical that can be converted into a wide range of synthetic chemicals and fuels. Only a few processes based on chemical utilization of CO_2 are currently industrialized, including urea and salicylic acid production. There are many other promising approaches; however, they are non-feasible economically because of the high cost of the renewable energy input required for the chemical conversion of the non-reactive CO_2, in addition to the cost of the CO_2 capture, separation, purification, and transportation to the user site. Further research and development should be aimed at reducing these costs and creating new markets for CO_2-based products.

Acknowledgment

The authors thank the Australia-India Strategic Research Fund (AIRSF) for the financial support for this work, which forms part of the Mini-DME project. The aim of Mini-DME project is the conversion of natural gas to the clean and low-carbon-emission fuel, dimethyl ether (DME).

References

Acién Fernández, F. G., C. V. González-López, J. M. Fernández Sevilla, and E. Molina Grima. 2012. Conversion of CO_2 into biomass by microalgae: How realistic a contribution may it be to significant CO_2 removal? *Applied Microbiology and Biotechnology* 96 (3):577–586.

Akarmazyan, S. S., P. Panagiotopoulou, A. Kambolis, C. Papadopoulou, and D. I. Kondarides. 2014. Methanol dehydration to dimethylether over Al_2O_3 catalysts. *Applied Catalysis B: Environmental* 145:136–148.

Albertsson, A. C., and R. K. Srivastava. 2008. Recent developments in enzyme-catalyzed ring-opening polymerization. *Advanced Drug Delivery Reviews* 60 (9):1077–1093.

Albo, J., M. Alvarez-Guerra, P. Castano, and A. Irabien. 2015. Towards the electrochemical conversion of carbon dioxide into methanol. *Green Chemistry* 17 (4):2304–2324.

Ampelli, C., G. Centi, R. Passalacqua, and S. Perathoner. 2016. Electrolyte-less design of PEC cells for solar fuels: Prospects and open issues in the development of cells and related catalytic electrodes. *Catalysis Today* 259, Part 2:246–258.

Ampelli, C., S. Perathoner, and G. Centi. 2015. CO_2 utilization: An enabling element to move to a resource-and energy-efficient chemical and fuel production. *Philosophical Transactions of the Royal Society of London A: Mathematical, Physical and Engineering Sciences* 373 (2037):2014.0177.

Aresta, M. 2010. *Carbon dioxide Recovery and Utilization*. Boston, MA: Kluwer Academic Publishers.

Aresta, M., and A. Dibenedetto. 2007. Utilisation of CO_2 as a chemical feedstock: Opportunities and challenges. *Dalton Transactions* (28):2975–2992.

Aresta, M., A. Dibenedetto, M. Carone, T. Colonna, and C. Fragale. 2005. Production of biodiesel from macroalgae by supercritical CO_2 extraction and thermochemical liquefaction. *Environmental Chemistry Letters* 3 (3):136–139.

Aresta, M., E. Quaranta, and I. Tommasi. 1992. Proceedings of the first international conference on carbon dioxide removal prospects for the utilization of carbon dioxide. *Energy Conversion and Management* 33 (5):495–504.

Aresta, M., and I. Tommasi. 1997. Carbon dioxide utilisation in the chemical industry. *Energy Conversion and Management* 38:373–378.

Arul, J., A. Boudreau, J. Makhlouf, R. Tardif, and M. R. Sahasrabudhe. 1987. Fractionation of anhydrous milk fat by superficial carbon dioxide. *Journal of Food Science* 52 (5):1231–1236.

Aslan, S., and I. K. Kapdan. 2006. Batch kinetics of nitrogen and phosphorus removal from synthetic wastewater by algae. *Ecological Engineering* 28 (1):64–70.

Augustynski, J., C. J. Sartoretti, and P. Kedzierzawski. 2003. Electrochemical conversion of carbon dioxide. In *Carbon Dioxide Recovery and Utilization*, edited by M. Aresta, 279–292. Netherlands, NL: Springer.

Azizi, Z., M. Rezaeimanesh, T. Tohidian, and M. R. Rahimpour. 2014. Dimethyl ether: A review of technologies and production challenges. *Chemical Engineering and Processing: Process Intensification* 82:150–172.

Baiker, A. 2000. Utilization of carbon dioxide in heterogeneous catalytic synthesis. *Applied Organometallic Chemistry* 14 (12):751–762.

Beevor, R. G., D. J. Gulliver, M. Kitson, and R. M. Sorrell. 1990. The production of formate salts of nitrogenous bases. EP0357243 A3.

Benemann, J. R. 1997. CO_2 mitigation with microalgae systems. *Energy Conversion and Management* 38, Supplement:S475–S479.

Blunt, M., F. J. Fayers, and F. M. Orr. 1993. Proceedings of the international energy agency carbon dioxide disposal symposium carbon dioxide in enhanced oil recovery. *Energy Conversion and Management* 34 (9):1197–1204.

Bosko, M. L., J. F. Múnera, E. A. Lombardo, and L. M. Cornaglia. 2010. Dry reforming of methane in membrane reactors using Pd and Pd–Ag composite membranes on a NaA zeolite modified porous stainless steel support. *Journal of Membrane Science* 364 (1):17–26.

Centi, G., and S. Perathoner. 2009. Opportunities and prospects in the chemical recycling of carbon dioxide to fuels. *Catalysis Today* 148 (3):191–205.

Centi, G., S. Perathoner, and G. Iaquanillo. 2013. Realizing resource and energy efficiency in chemical industry by using CO. In *CO_2: A Valuable Source of Carbon*, edited by M. De Falco, G. Iaquaniello and G. Centi, 37. London: Springer.

Centi, G., S. Perathoner, and Z. S. Rak. 2003. Reduction of greenhouse gas emissions by catalytic processes. *Applied Catalysis B: Environmental* 41 (1–2):143–155.

Centi, G., S. Perathoner, G. Winè, and M. Gangeri. 2007. Electrocatalytic conversion of CO_2 to long carbon-chain hydrocarbons. *Green Chemistry* 9 (6):671–678.

Chao, R. R., S. J. Mulvaney, M. E. Bailey, and L. N. Fernando. 1991. Supercritical CO_2 conditions affecting extraction of lipid and cholesterol from ground beef. *Journal of Food Science* 56 (1):183–187.

Chen, W. H., B. J. Lin, H. M. Lee, and M. H. Huang. 2012. One-step synthesis of dimethyl ether from the gas mixture containing CO_2 with high space velocity. *Applied Energy* 98:92–101.

Cheng, C.-H., T.-B. Du, H.-C. Pi, S.-M. Jang, Y.-H. Lin, and H.-T. Lee. 2011. Comparative study of lipid extraction from microalgae by organic solvent and supercritical CO_2. *Bioresource Technology* 102 (21):10151–10153.

Chiesa, P., S. Campanari, and G. Manzolini. 2011. CO_2 cryogenic separation from combined cycles integrated with molten carbonate fuel cells. *International Journal of Hydrogen Energy* 36 (16):10355–10365.

Chih-Hung, H., and T. Chung-Sung. 2014. A review: CO_2 utilization. *Aerosol and Air Quality Research* 14:480–499.

Chisti, Y. 2007. Biodiesel from microalgae. *Biotechnology Advances* 25 (3):294–306.

Chisti, Y. 2008. Biodiesel from microalgae beats bioethanol. *Trends in Biotechnology* 26 (3):126–131.

Clements, J. H. 2003. Reactive applications of cyclic alkylene carbonates. *Industrial & Engineering Chemistry Research* 42 (4):663–674.

Cooney, G., J. Littlefield, J. Marriott, and T. J. Skone. 2015. Evaluating the climate benefits of CO_2-enhanced oil recovery using life cycle analysis. *Environmental Science & Technology* 49 (12):7491–7500.

Darensbourg, D. J., and M. W. Holtcamp. 1996. Catalysts for the reactions of epoxides and carbon dioxide. *Coordination Chemistry Reviews* 153:155–174.

Darensbourg, D. J., and S. J. Wilson. 2012. What's new with CO_2? Recent advances in its copolymerization with oxiranes. *Green Chemistry* 14 (10):2665–2671.

Das, S., and W. W. Daud. 2014. Photocatalytic CO_2 transformation into fuel: A review on advances in photocatalyst and photoreactor. *Renewable and Sustainable Energy Reviews* 39:765–805.

Dibenedetto, A., and I. Tommasi. 2003. Biological utilization of carbon dioxide: The marine biomass option. In *Carbon Dioxide Recovery and Utilization*, edited by M. Aresta, 315–324. Dordrecht, EC: Springer Science & Business Media.

Doucha, J., F. Straka, and K. Lívanský. 2005. Utilization of flue gas for cultivation of microalgae (Chlorella sp.) in an outdoor open thin-layer photobioreactor. *Journal of Applied Phycology* 17 (5):403–412.

DuBois, D. L. 2007. Electrochemical reactions of carbon dioxide. In *Encyclopedia of Electrochemistry*. Weinheim, DE: Wiley-VCH Verlag GmbH & Co. KGaA.

Esfandiari, N. 2015. Production of micro and nano particles of pharmaceutical by supercritical carbon dioxide. *The Journal of Supercritical Fluids* 100:129–141.

Fan, M.-S., A. Z. Abdullah, and S. Bhatia. 2010. Utilization of greenhouse gases through carbon dioxide reforming of methane over Ni–Co/MgO–ZrO₂: Preparation, characterization and activity studies. *Applied Catalysis B: Environmental* 100 (1):365–377.

Frese, K. W., and S. Leach. 1985. Electrochemical reduction of carbon dioxide to methane, methanol, and CO on Ru electrodes. *Journal of the Electrochemical Society* 132 (1):259–260.

Fujimoto, K., Y. Endo, S.-Y. Cho, R. Watabe, Y. Suzuki, M. Konno, K. Shoji, K. Arai, and S. Saito. 1989. Chemical characterization of sardine meat powder produced by dehydration with high osmotic pressure resin and defatting with high pressure carbon dioxide. *Journal of Food Science* 54 (2):265–268.

Ganesh, I. 2014. Conversion of carbon dioxide into methanol–A potential liquid fuel: Fundamental challenges and opportunities (a review). *Renewable and Sustainable Energy Reviews* 31:221–257.

González-López, C. V., F. G. Acién-Fernández, J. M. Fernández-Sevilla, J. F. Sánchez-Fernández, M. C. Cerón-García, and E. Molina-Grima. 2009. Utilization of the cyanobacteria anabaena sp. ATCC 33047 in CO_2 removal processes. *Bioresource Technology* 100 (23):5904–5910.

González-López, C. V., F. G. Acién-Fernández, J. M. Fernández-Sevilla, J. F. Sánchez-Fernández, and E. Molina-Grima. 2012. Development of a process for efficient use of CO_2 from flue gases in the production of photosynthetic microorganisms. *Biotechnology and Bioengineering* 109 (7): 1637–1650.

Hägg, M.-B., and A. Lindbråthen. 2005. CO_2 capture from natural gas fired power plants by using membrane technology. *Industrial & Engineering Chemistry Research* 44 (20):7668–7675.

Hansen, J. B., and P. E. H. Nielsen. 2008. Methanol Synthesis. In *Handbook of Heterogeneous Catalysis*, edited by G. Ertl, H. Knozinger, F. Schuith and J. Weitkamp, 2920–2949. Weinheim, DE: Wiley-VCH Verlag GmbH & Co. KGaA.

Heffer, P., and M. Prud'homme. 2009. Fertilizer Outlook 2009–2013. 77th IFA Annual Conference, Shanghai (China P. R.).

Ho, M. T., G. W. Allinson, and D. E. Wiley. 2008. Reducing the cost of CO_2 capture from flue gases using membrane technology. *Industrial & Engineering Chemistry Research* 47 (5):1562–1568.

Hori, Y., H. Konishi, T. Futamura, A. Murata, O. Koga, H. Sakurai, and K. Oguma. 2005. "Deactivation of copper electrode" in electrochemical reduction of CO₂. *Electrochimica Acta* 50 (27):5354–5369.

Hori, Y., A. Murata, and R. Takahashi. 1989. Formation of hydrocarbons in the electrochemical reduction of carbon dioxide at a copper electrode in aqueous solution. *Journal of the Chemical Society, Faraday Transactions 1: Physical Chemistry in Condensed Phases* 85 (8):2309–2326.

Howarth, R. W., R. Santoro, and A. Ingraffea. 2011. Methane and the greenhouse-gas footprint of natural gas from shale formations. *Climatic Change* 106 (4):679–690.

Inoue, S., H. Koinuma, and T. Tsuruta. 1969. Copolymerization of carbon dioxide and epoxide. *Journal of Polymer Science Part B: Polymer Letters* 7 (4):287–292.

IPCC. 2007. Climate Change 2007: The Physical Science Basis. In *IPCC Report*, edited by S. Solomon, D. Qin, M. Manning, M. Marquis, K. Averyt, M. M. B. Tignor, H. L. Miller and Z. Chen. Cambridge University Press, Cambridge, UK. pp. 237–241.

Jan, V. 2003. Carbon dioxide emission and merchant market in the European union. In *Carbon Dioxide Recovery and Utilization*, edited by M. Aresta. Boston, MA: Kluwer Academic Publishers.

Jaramillo, P., W. M. Griffin, and S. T. McCoy. 2009. Life cycle inventory of CO_2 in an enhanced oil recovery system. *Environmental Science & Technology* 43 (21):8027–8032.

Jessop, P. G., T. Ikariya, and R. Noyori. 1994. Homogeneous catalytic hydrogenation of supercritical carbon dioxide. *Nature* 368 (6468):231–233.

Joó, F., G. Laurenczy, P. Kárády, J. Elek, L. Nádasdi, and R. Roulet. 2000. Homogeneous hydrogenation of aqueous hydrogen carbonate to formate under mild conditions with water soluble rhodium(I)– and ruthenium(II)–phosphine catalysts. *Applied Organometallic Chemistry* 14 (12):857–859.

Kaneco, S., K. Iiba, M. Yabuuchi, N. Nishio, H. Ohnishi, H. Katsumata, T. Suzuki, and K. Ohta. 2002. High efficiency electrochemical CO_2-to-methane conversion method using methanol with lithium supporting electrolytes. *Industrial & Engineering Chemistry Research* 41 (21):5165–5170.

Kazlas, P. T., R. D. Novak, and R. J. Robey. 1994. Supercritical carbon dioxide decaffeination of acidified coffee. U.S. 5,288,511 A.

Kendall, J. L., D. A. Canelas, J. L. Young, and J. M. DeSimone. 1999. Polymerizations in supercritical carbon dioxide. *Chemical Reviews* 99 (2):543–564.

Kurano, N., H. Ikemoto, H. Miyashita, T. Hasegawa, H. Hata, and S. Miyachi. 1995. Fixation and utilization of carbon dioxide by microalgal photosynthesis. *Energy Conversion and Management* 36 (6–9):689–692.

Kuuskraa, V. A., M. L. Godec, and P. Dipietro. 2013. CO_2 utilization from "next generation" CO_2 enhanced oil recovery technology. *Energy Procedia* 37:6854–6866.

Lange, J.-P. 2001. Methanol synthesis: A short review of technology improvements. *Catalysis Today* 64 (1):3–8.

Li, G., P. Xiao, P. Webley, J. Zhang, R. Singh, and M. Marshall. 2008. Capture of CO_2 from high humidity flue gas by vacuum swing adsorption with zeolite 13X. *Adsorption* 14 (2–3):415–422.

Lin, F., D. Liu, S. M. Das, N. Prempeh, Y. Hua, and J. Lu. 2014. Recent progress in heavy metal extraction by supercritical CO_2 fluids. *Industrial & Engineering Chemistry Research* 53 (5):1866–1877.

Lindsey, A. S., and H. Jeskey. 1957. The Kolbe-Schmitt reaction. *Chemical Reviews* 57 (4):583–620.

Lorenzo, M., M. Bertau, F. Schmidt, and L. Plass. 2014. Methanol generation. In *Methanol: The Basic Chemical and Energy Feedstock of the Future*, edited by M. Bertau, H. Offermanns, L. Plass, F. Schmidt and H.-J. Wernicke, 269. Heidelberg, MS: Springer.

Lu, X. B., and D. J. Darensbourg. 2012. Cobalt catalysts for the coupling of CO_2 and epoxides to provide polycarbonates and cyclic carbonates. *Chemical Society Reviews* 41 (4):1462–1484.

Lu, X.-B., and Y. Wang. 2004. Highly active, binary catalyst systems for the alternating copolymerization of CO_2 and epoxides under mild conditions. *Angewandte Chemie International Edition* 43 (27):3574–3577.

Ma, J., N. Sun, X. Zhang, N. Zhao, F. Xiao, W. Wei, and Y. Sun. 2009. A short review of catalysis for CO_2 conversion. *Catalysis Today* 148 (3):221–231.

Mata, T. M., A. A. Martins, and N. S. Caetano. 2010. Microalgae for biodiesel production and other applications: A review. *Renewable and Sustainable Energy Reviews* 14 (1):217–232.

McGrath, K. M., G. K. S. Prakash, and G. A. Olah. 2004. Direct methanol fuel cells. *Journal of Industrial and Engineering Chemistry* 10 (7):1063–1080.

Mendes, R. L., B. P. Nobre, M. T. Cardoso, A. P. Pereira, and A. F. Palavra. 2003. Supercritical carbon dioxide extraction of compounds with pharmaceutical importance from microalgae. *Inorganica Chimica Acta* 356:328–334.

Merkel, T. C, H. Lin, X. Wei, and R. Baker. 2010. Power plant post-combustion carbon dioxide capture: An opportunity for membranes. *Journal of Membrane Science* 359 (1):126–139.

Meyer, J. P. 2007. Summary of carbon dioxide enhanced oil recovery (CO_2EOR) injection well technology. *American Petroleum Institute*. pp. 1–2.

Middleton, R. S., J. S. Levine, J. M. Bielicki, H. S. Viswanathan, J. W. Carey, and P. H. Stauffer. 2015. Jumpstarting commercial-scale CO_2 capture and storage with ethylene production and enhanced oil recovery in the US Gulf. *Greenhouse Gases: Science and Technology* 5 (3):241–253.

Moreno, J., M. Á. Vargas, H. Rodrıguez, J. Rivas, and M. G. Guerrero. 2003. Outdoor cultivation of a nitrogen-fixing marine cyanobacterium, Anabaena sp. ATCC 33047. *Biomolecular Engineering* 20 (4–6):191–197.

Mouahid, A., C. Crampon, S.-A. A. Toudji, and E. Badens. 2013. Supercritical CO_2 extraction of neutral lipids from microalgae: Experiments and modelling. *The Journal of Supercritical Fluids* 77:7–16.

Naik, S. P., T. Ryu, V. Bui, J. D. Miller, N. B. Drinnan, and W. Zmierczak. 2011. Synthesis of DME from CO_2/H_2 gas mixture. *Chemical Engineering Journal* 167 (1):362–368.

Negoro, M., N. Shioji, K. Miyamoto, and Y. Micira. 1991. Growth of microalgae in high CO_2 gas and effects of SO_X and NO_X. *Applied Biochemistry and Biotechnology* 28–29 (1):877–886.

Noh, E. K., S. J. Na, S. Sujith, S.-W. Kim, and B. Y. Lee. 2007. Two components in a molecule: Highly efficient and thermally robust catalytic system for CO_2/epoxide copolymerization. *Journal of the American Chemical Society* 129 (26):8082–8083.

Noureldin, M. M. B., N. O. Elbashir, K. J. Gabriel, and M. M. El-Halwagi. 2015. A process integration approach to the assessment of CO_2 fixation through dry reforming. *ACS Sustainable Chemistry & Engineering* 3 (4):625–636.

Ogawa, T., N. Inoue, T. Shikada, and Y. Ohno. 2003. Direct dimethyl ether synthesis. *Journal of Natural Gas Chemistry* 12 (4):219–227.

Øi, L. E. 2007. Aspen HYSYS simulation of CO_2 removal by amine absorption from a gas based power plant. The 48th Scandinavian Conference on Simulation and Modeling (SIMS 2007), 30–31 October, 2007, Göteborg (Särö).

Olah, G. A., A. Goeppert, and G. K. S. Prakash. 2009. Chemical recycling of carbon dioxide to methanol and dimethyl ether: From greenhouse gas to renewable, environmentally carbon neutral fuels and synthetic hydrocarbons. *The Journal of Organic Chemistry* 74 (2):487–498.

Olah, G. A., A. Goeppert, and G. K. S. Prakash. 2011a. *Beyond Oil and Gas: The Methanol Economy*. Weinheim, DE: John Wiley & Sons.

Olah, G. A., A. Goeppert, M. Czaun, and G. K. S. Prakash. 2013. Bi-reforming of methane from any source with steam and carbon dioxide exclusively to metgas ($CO–2H_2$) for methanol and hydrocarbon synthesis. *Journal of the American Chemical Society* 135 (2):648–650.

Olah, G. A., G. K. S. Prakash, and A. Goeppert. 2011b. Anthropogenic chemical carbon cycle for a sustainable future. *Journal of the American Chemical Society* 133 (33):12881–12898.

Omae, I. 2006. Aspects of carbon dioxide utilization. *Catalysis Today* 115 (1–4):33–52.

Omae, I. 2012. Recent developments in carbon dioxide utilization for the production of organic chemicals. *Coordination Chemistry Reviews* 256 (13–14):1384–1405.

Orr Jr, F. M., J. P. Heller, and J. J. Taber. 1982. Carbon dioxide flooding for enhanced oil recovery: Promise and problems. *Journal of the American Oil Chemists Society* 59 (10):810A–817A.

Oswald, W. J., and C. G. Golueke. 1960. Biological transformation of solar energy. *Advances in Applied Microbiology* 2:223–262.

Parsons, W. 2011. Economic assessment of carbon capture and storage technologies: 2011 update. In *Report for the Global CCS Institute*. Canberra, Australia: Global CCS Institute.

Pescarmona, P. P., and M. Taherimehr. 2012. Challenges in the catalytic synthesis of cyclic and polymeric carbonates from epoxides and CO_2. *Catalysis Science and Technology* 2 (11):2169–2187.

Peters, L., A. Hussain, M. Follmann, T. Melin, and M. B. Hägg. 2011. CO_2 removal from natural gas by employing amine absorption and membrane technology—A technical and economical analysis. *Chemical Engineering Journal* 172 (2–3):952–960.

Pulz, O., and W. Gross. 2004. Valuable products from biotechnology of microalgae. *Applied Microbiology and Biotechnology* 65 (6):635–648.

Qin, Y., and X. Wang. 2010. Carbon dioxide-based copolymers: Environmental benefits of PPC, an industrially viable catalyst. *Biotechnology Journal* 5 (11):1164–1180.

Rackley, S. 2009. *Carbon capture and storage*. Boston, MA: Elsevier Inc.

Raventós, M., S. Duarte, and R. Alarcón. 2002. Application and possibilities of supercritical CO_2 extraction in food processing industry: An overview. *Food Science and Technology International* 8 (5):269–284.

Reverchon, E., G. Della Porta, and F. Senatore. 1995. Supercritical CO_2 extraction and fractionation of lavender essential oil and waxes. *Journal of Agricultural and Food Chemistry* 43 (6):1654–1658.

Ricci, M. 2003. Carbon dioxide as a building block for organic intermediates: An industrial perspective. In *Carbon Dioxide Recovery and Utilization*, edited by M. Aresta, 395–402. Netherlands, NL: Springer.

Riemer, P. W. F., and W. G. Ormerod. 1995. International perspectives and the results of carbon dioxide capture disposal and utilisation studies. *Energy Conversion and Management* 36 (6–9):813–818.

Rochelle, G. T. 2009. Amine scrubbing for CO_2 capture. *Science* 325 (5948):1652–1654.

Rosello-Sastre, R., Z. Csögör, I. Perner-Nochta, P. Fleck-Schneider, and C. Posten. 2007. Scale-down of microalgae cultivations in tubular photo-bioreactors—A conceptual approach. *Journal of Biotechnology* 132 (2):127–133.

Rozzi, N. L., and R. K. Singh. 2002. Supercritical fluids and the food industry. *Comprehensive Reviews in Food Science and Food Safety* 1 (1):33–44.

Sakakura, T., J. C. Choi, and H. Yasuda. 2007. Transformation of carbon dioxide. *Chemical Reviews* 107 (6):2365–2387.

Santana, A., S. Jesus, M. A. Larrayoz, and R. M. Filho. 2012. Supercritical carbon dioxide extraction of algal lipids for the biodiesel production. *Procedia Engineering* 42:1755–1761.

Shibata, H., J. A. Moulijn, and G. Mul. 2008. Enabling electrocatalytic Fischer–Tropsch synthesis from carbon dioxide over copper-based electrodes. *Catalysis Letters* 123 (3–4):186–192.

Song, C. 2002. CO_2 conversion and utilization: An overview. *ACS Symposium Series*, 2–30, Washington, DC.

Song, C., and W. Pan. 2004. Tri-reforming of methane: A novel concept for catalytic production of industrially useful synthesis gas with desired H_2/CO ratios. *Catalysis Today* 98 (4):463–484.

Spinner, N. S., J. A. Vega, and W. E. Mustain. 2012. Recent progress in the electrochemical conversion and utilization of CO_2. *Catalysis Science & Technology* 2 (1):19–28.

Steinberg, M. 1996. The Carnol process for CO_2 mitigation from power plants and the transportation sector. *Energy Conversion and Management* 37 (6–8):843–848.

Steinberg, M. 1997. The Carnol process system for CO_2 mitigation and methanol production. *Energy* 22 (2–3):143–149.

Steinberg, M. 1998. Production of hydrogen and methanol from natural gas with reduced CO_2 emission. *International Journal of Hydrogen Energy* 23 (6):419–425.

Stevens, S. H., J. M. Pearce, and A. A. J. Rigg. 2001. Natural analogs for geologic storage of CO_2: An integrated global research program. *Proceedings of the First National Conference on Carbon Sequestration, National Energy Technology Laboratory*, Washington, DC.

Teuner, St. C., P. Neumann, and F. Von Linde. 2001. The calcor standard and calcor economy processes. *Oil Gas European Magazine* 117:44–46.

Thorbjornsson, I., G. H. Oskarsdottir, T. I. Sigfusson, G. Gudmundsson, G. Gunnarsson, and M. Gudmundsson. 2012. Future prospects of renewable energy production in Iceland. World Renewable Energy Forum (WREF) Denver, Colorado.

Travieso, L., E. P. Sànchez, F. Benitez, and J. L. Conde. 1993. Arthospira sp. intensive cultures for food and biogas purification. *Biotechnology Letters* 15 (10):1091–1094.

Valencia, J. A., and D. J. Victory. 1990. *Method and Apparatus for Cryogenic Separation of Carbon dioxide and Other Acid Gases from Methane*. USA: Exxon Production Research Company.

van den Broeke, L. J. P., E. L. V. Goetheer, A. W. Verkerk, E. de Wolf, B.-J. Deelman, G. van Koten, and J. T. F. Keurentjes. 2001. Homogeneous reactions in supercritical carbon dioxide using a catalyst immobilized by a microporous silica membrane. *Angewandte Chemie International Edition* 40 (23):4473–4474.

Vane, J. R., and R. M. Botting. 2003. The mechanism of action of aspirin. *Thrombosis Research* 110 (5–6):255–258.

Wang, B., Y. Li, N. Wu, and C. Q. Lan. 2008. CO_2 bio-mitigation using microalgae. *Applied Microbiology and Biotechnology* 79 (5):707–718.

Webley, P. A. 2014. Adsorption technology for CO_2 separation and capture: A perspective. *Adsorption* 20 (2–3):225–231.

Wernicke, H.-J., L. Plass, and F. Schmidt. 2014. "Methanol Generation." In *Methanol: The Basic Chemical and Energy Feedstock of the Future*, edited by M. Bertau, H. Offermanns, L. Plass, F. Schmidt and H.-J. Wernicke, 51–301. Heidelberg, MS: Springer.

Xiao, P., J. Zhang, P. Webley, G. Li, R. Singh, and R. Todd. 2008. Capture of CO_2 from flue gas streams with zeolite 13X by vacuum-pressure swing adsorption. *Adsorption* 14 (4–5):575–582.

Yang, H., Z. Xu, M. Fan, R. Gupta, R. B. Slimane, A. E. Bland, and I. Wright. 2008. Progress in carbon dioxide separation and capture: A review. *Journal of Environmental Sciences* 20 (1):14–27.

Yang, J., Q. Hao, X. Liu, C. Ba, and A. Cao. 2004. Novel biodegradable aliphatic poly(butylene succinate-co-cyclic carbonates)s with functionalizable carbonate building blocks. 1. Chemical synthesis and their structural and physical characterization. *Biomacromolecules* 5 (1):209–218.

Yonei, Y., H. Ōhinata, R.i Yoshida, Y. Shimizu, and C. Yokoyama. 1995. Extraction of ginger flavor with liquid or supercritical carbon dioxide. *The Journal of Supercritical Fluids* 8 (2):156–161.

Yoo, C., S.-Y. Jun, J.-Y. Lee, C.-Y. Ahn, and H.-M. Oh. 2010. Selection of microalgae for lipid production under high levels carbon dioxide. *Bioresource Technology* 101 1, Supplement:S71–S74.

Yu, K. M. K., I. Curcic, J. Gabriel, and S. C. E. Tsang. 2008. Recent advances in CO_2 capture and utilization. *ChemSusChem* 1 (11):893–899.

Zhang, K., S. Miyachi, and N. Kurano. 2001. Photosynthetic performance of a cyanobacterium in a vertical flat-plate photobioreactor for outdoor microalgal production and fixation of CO_2. *Biotechnology Letters* 23 (1):21–26.

Zhang, L., W. Junhua, W. Pei, H. Zhaoyin, F. Jinhua, and X. Zheng. 2010. Synthesis of dimethyl ether via methanol dehydration over combined Al_2O_3-HZSM-5 solid acids. *Chinese Journal of Catalysis* 31 (8):987–992.

15

Hydrogen from Water

Pramod H. Borse

CONTENTS

ABSTRACT   Water is seen as a natural source of hydrogen, which is an environment-friendly alternative to the non-renewable energy sources. Hydrogen is an energy carrier that shows tremendous potential to replace the fossil fuel dependence of mankind; thus, there are several efforts with respect to existing hydrogen production methodologies to refine the hydrogen generation efficiency and its economy. The basic fact relies with the splitting of water molecule by providing thermodynamic energy of 1.23 eV to obtain hydrogen. The evolution of this technology is hindered due to inefficient ways of generation as high cost is involve in the hydrogen generation. The present chapter deals with the discussion of existing status of diffident methodologies being used for hydrogen generation in industries. Apart from this, it provides an exhaustive status of electrocatalytic hydrogen generation. The evolving fields of solar thermal as well as photocatalytic hydrogen production are discussed to present recent status on the solar-based water splitting.

15.1 Introduction: Background and Driving Forces

World is envisaging tremendous energy crisis, which is predicted to be experienced in next few decades due to vanishing fossil fuels on the earth. The ever-increasing population on the earth and improving standard of living have led to a steep increase in energy demand with respect to power requirements and transportation needs. In order to have an energy security in future, it is necessary to explore alternate energy-producing technologies from the renewable natural sources, which will not act inversely on the ecological balance of the planet earth. There are several natural resources, such as sunlight, wind, and water, that have the capability to replace fossil fuels (Muradov and Veziroğlu 2005; Holladay et al. 2009; Konieczny et al. 2008). Also, hydrogen is one of the most desirable fuels and is capable of replacing vanishing hydrocarbons, provided it is generated from water. Hydrogen is a constituent of abundantly existing water on the earth. The reaction $2H_2O \rightarrow 2H_2 + O_2$ indicates that, in principle, if a water molecule is subjected to an energy exceeding 1.2 V, it can yield two moles of hydrogen molecules. This concept is very important with respect to the importance of hydrogen as energy carrier and in view of existing abundance of water on the planet earth. The concept is well known as *water splitting (WS)*. Water splitting can be facilitated by making use of various forms of energies *viz.*, light, thermal energy, wind, electric energy, bioenergy (Bridgewater et al. 2003, 2007; FNR 2009), and chemical energy (Dincer and Acar 2015). Depending on the type of energy used, there are specific mechanisms involved to facilitate WS. These can be photosplitting, thermosplitting, steam reforming, electrolysis, photoelectrolysis, bioelectrolysis and so on, as displayed in Figure 15.1. Each of hydrogen production methodologies has its own advantages with respect to efficiency, economy, and *eco-friendliness*. The demand of hydrogen in industries is huge. It is expected that in future, the transport sector and "residential energy" would also use hydrogen to fulfill the energy demands in the form of *"chemical energy."* Presently, the hydrogen utilized in industries is prepared by *steam reforming technique*, as it is most cost-effective technique. However, the coal/natural gas used to produce hydrogen is hampering the ecology. Consequently, the worldwide efforts are underway to identify efficient *eco-friendly* hydrogen-producing methodologies, which can be specifically obtained from *renewable resources*. This chapter deals with the present and future energy solutions to generate hydrogen from water, thereby giving summarized updates on status and progress of the methodologies with respect to their industrial usability (Borse and Das 2013).

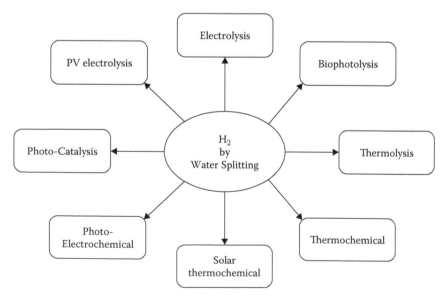

FIGURE 15.1
Schematic diagram showing different ways to generate hydrogen from water splitting.

15.2 Methods to Generate Hydrogen from Water

15.2.1 Non-Renewable Sources

It is known that about 50% of global demand of hydrogen is fulfilled through *steam reforming* of natural gas, whereas around 30% is fulfilled through *naphtha reforming* from chemical industry off-gases, 18% from coal gasification (Elliott et al. 1981), 3.9% from electrolysis, and 0.1% by other methods of hydrogen production (Muradov and Veziroğlu 2005). The details of other methods are described in the following sections.

15.2.2 Renewable Sources

Figure 15.1 shows different possible methodologies that produce hydrogen through water. These can be categorized based on the energy used for *water splitting* to produce hydrogen. It is noteworthy to mention that at present, most of these methods are at the stage of infancy but are showing tremendous potential as future technologies. Nonetheless, electrolysis and PV electrolysis are being exploited at various levels for industrial applications (Muradov and Veziroğlu 2005). In the following sections, each of the techniques has been described in detail, so as to portray their present status as well as industrial credibility to date. The techniques that are at the infancy of their development are briefly described with respect to their possible technological development.

15.3 Electrolysis

15.3.1 Description and Status

Electrolysis is a process of *water splitting* that occurs by the utilization of electrical energy. In this process, generally an anode and a cathode are immersed in an aqueous solution of KOH to form an electrolyzer, as shown in Figure 15.2. Electric energy facilitates the dissociation of water molecule and migration of the positively charged H^+ ions to the negatively charged cathode, leading to reduction reaction, forming gaseous hydrogen (H_2) at that end (Santos et al. 2013; Turner 2008). Similarly, gaseous oxygen (O_2) is formed at anode in the cell. This mechanism yields stoichiometric amounts of H_2 and O_2 in a ratio of 2:1. Another important constituent of the electrolytic cell is a separation membrane that allows the passage of positive or negative ions across it, so as to facilitate ion transport between electrodes. The membrane enables to keep the gaseous H_2 and O_2 in respective regions of the cell (Ursua et al. 2012). There are several well-documented reports with respect to electrolytic hydrogen generation that described various criticalities of electrodes, membrane, reactor, and so on.

Currently, several electrolyzers (Ursua et al. 2012) exist, which work on different mechanisms, *viz.* (1) alkaline electrolyzer, (2) proton exchange membrane electrolyzer, and (3) steam electrolyzer. As hydrogen generation by electrolytic water splitting is a matured technology, there exists a number of electrolyzers from various manufacturers, as listed in Table 15.1. *Siemens* and *Linde* (Germany) are the oldest manufacturers; they produced electrolyzers during 1840s to manufacture alkaline polar electrolyzers. There are several new manufacturers who fabricated various electrolyzers per developments

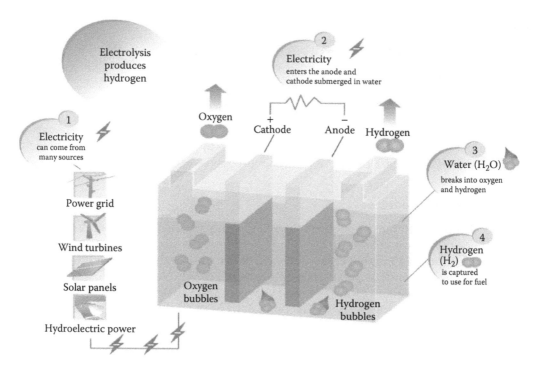

FIGURE 15.2
Schematic of electrolysis cell, describing the principle of hydrogen generation by electrolysis.

in electrode materials, separator membrane materials, gas pressure, gas purity, and economy requirements. Recently, *Hydrogenics* (Canada) and *HOGEN* (United States) came up as new companies that launched new products of the alkaline and polymer electrolyte membrane (PEM) electrolyzers. Among all the listed companies, ELT (Germany) provides alkaline bipolar electrolyzer, which is capable of producing highest hydrogen production rate of ~1400 Nm³/h (Turner 2008). It is important to note that there have been several efforts toward electrolysis of water by making use of solar photovoltaic (PV) instead of utilization of grid electricity (Khaselev et al. 2001). However, at present times, PV economy proves to be expensive, which, in turn, inhibits the commercial viability of the PV-based electrolysis.

Therefore, the PV-based electrolysis has not been elaborated in this chapter. There are several reports of interest that may be referred (Ghosh et al. 2003; Kelly and Gibson 2008; Gazey et al 2006; Varkaraki et al 2008; Ulleberg et al 2010; Harrison et al. 2009; Perez et al. 2015).

It is worthwhile to mention that other than PV-based electrolysis, concentrated PV (CPV) and wind electrolysis offer hydrogen at potentially lower cost, specifically the CPV electrolysis technology, which offers hydrogen in the same range as gasoline-based hydrogen. However, it needs a thorough cost analysis and technology demonstration.

TABLE 15.1

List of Commercial Electrolyzer Manufacturers with Their Specifications (1880–2015)

S. No.	Manufacturer/ Supplier (Year)	Country	Technology (Configuration)	Rated Production (Nm³/h)[a]	Rated Power (KW)[b]	Specific Energy Consumption (KWh/Nm³)[c]	Efficiency (%)[d]	Maximum Pressure (bar)	Hydrogen Purity (vol%)
1	AccaGen (2003–2007)	Switzerland	Alkaline (bipolar)	1–100	6.7–487	6.7–4.87[g]	52.8–72.7	10 (optional 30 & 200[f])	99.9 (99.999[k])
2	Acta (2004–2015)	Italy	Alkaline	1	n.a.	53.2[l]	n.a.	29	99.94
3	AREVA (2000–2015)	France	PEM	5–120	n.a.	55.6[f]	n.a.	35	99.9995
4	Avalence (2004)	United States	Alkaline (monopolar)	0.4–4.6 (139[f])	2–25 (750[f])	5.43–5[h]	65.2–70.8	448	99.7
5	CETH2 (2011–2015)	France	PEM	240	n.a.	54.5[l]	n.a.	14	99.9
6	Claind (2010–2012)	Italy	Alkaline (bipolar)	0.5–30	n.a.	n.a.	n.a.	15	99.7 (99.999[k])
7	ELT (1995–2015)	Germany	Alkaline (bipolar)	2–480	13.8–1518	4.6–4.3[h]	76.9–82.3	Atmospheric	99.8–99.9
8	ELT (1995–2015)	Germany	Alkaline (bipolar)	200–1400	465–3534	4.65–4.3[h]	76.1–82.3	30	99.8–99.9
9	Erredue (2000–2015)	Italy	Alkaline (bipolar)	0.6–21.3	3.6–108	6–5.1[g]	59–69.8	2.5–4	99.3–99.8 (99.999[k])
10	FuelGen (2005–2012)	United States	PEM	13 kg/day	n.a.	n.a.	n.a.	n.a.	99.999
11	Giner (1973–2015)	United States	PEM (bipolar)	3.7	20	5.4[i]	65.5	85	n.a.
12	HOGEN (2011–2015)	United States	Alkaline	1.05–30	n.a.	n.a.	n.a.	n.a.	99.995
13	Hydrogen Technologies, division of Statoil (2015)	Norway	Alkaline (bipolar)	100–500	43–2150	4.3[h]	82.3	Atmospheric	99.9 (99.999[k])
14	H-TECH SYSTEMS (1997–2015)	Germany	PEM	3.6	n.a.	55.6[l]	n.a.	29	n.a.

(Continued)

TABLE 15.1 (Continued)

List of Commercial Electrolyzer Manufacturers with Their Specifications (1880–2015)

S. No.	Manufacturer/ Supplier (Year)	Country	Technology (Configuration)	Rated Production (Nm³/h)[a]	Rated Power (KW)[b]	Specific Energy Consumption (KWh/Nm³)[c]	Efficiency (%)[d]	Maximum Pressure (bar)	Hydrogen Purity (vol%)
15	Hydrogenics (2014)	Canada	Alkaline (bipolar)	10–60	54–312	5.4–5.2[g]	65.5–68.1	10 (optional 25)	99.9 (99.998[k])
16	Hydrogenics (2014)	Canada	PEM (bipolar)	1	7.2	7.2[g]	49.2	7.9	99.99
17	H2 Logic (2003–2015)	Denmark	Alkaline (bipolar)	0.66–42.62	3.6–213	5.45–5[g]	64.9–70.8	4 (optional 12)	99.3–99.8 (99.999[k])
18	H2 Nitidor	Italy	Alkaline	200	n.a.	52.3[l]	n.a.	30	99.9
19	Idroenergy (2008)	Italy	Alkaline (bipolar)	0.4–80	3–377	7.5–4.71[h]	47.2–75.2	1.8–8	99.5
20	Industrie Haute Technologie (1950–2010)	Switzerland	Alkaline (bipolar)	110–760	511.5–3534	4.65–4.3[h]	76.1–82.3	32	99.8–99.9
21	ITM POWER (2001–2015)	United Kingdom	PEM (Alkaline in dev.)	2.4	n.a.	53.4[l]	n.a.	15	99.9
22	Linde (1880–2015)	Germany	Alkaline (bipolar)	5–250	n.a.	n.a.	n.a.	25	99.9 (99.998[k])
23	McPhy (2000–2015)	Germany	Alkaline	60 100–400	n.a.	57.8[l]	n.a.	10 30	>99.3 >99.999
24	Nel Hydrogen (1927–2015)	Norway	Alkaline	50–485	n.a.	4.4 ± 0.1	n.a.	Atmospheric	99.9 ± 0.1
25	PIEL, division of ILT Technology (2008–2015)	Italy	Alkaline (bipolar)	0.4–400	2.8–80	7–5[g]	50.6–70.8	1.8–18	99.5
26	Proton Onsite (2005–2015)	United States	PEM (bipolar)	0.265–30	1.8–174	7.3–5[g]	48.5–61	13.8–15 (optional 30)	99.999
27	Sagim (1922–2015)	France	Alkaline (monopolar)	1–5	5–25	5[h]	70.8	10[j]	99.9
28	Siemens (1847–2015)	Germany	PEM	~250	n.a.	~60[l]	n.a.	n.a.	n.a.

(Continued)

TABLE 15.1 (*Continued*)

List of Commercial Electrolyzer Manufacturers with Their Specifications (1880–2015)

S. No.	Manufacturer/ Supplier (Year)	Country	Technology (Configuration)	Rated Production (Nm³/h)[a]	Rated Power (KW)[b]	Specific Energy Consumption (KWh/Nm³)[c]	Efficiency (%)[d]	Maximum Pressure (bar)	Hydrogen Purity (vol%)
29	Teledyne Energy Systems (1968–2014)	United States	Alkaline (bipolar)	2.8–80	n.a.	n.a.	n.a.	10	99,999
30	Treadwell Corporation (1896–2015)	United States	PEM (bipolar)	1.2–10.2	n.a.	n.a.	n.a.	75.7	n.a.
31	Wasserelektrolyse Hydrotechnik (1956–2015)	Germany	Alkaline	225	n.a.	58.7[l]	n.a.	Atmospheric	99.9

Source: (Ursua, A. et al., *Proc. IEEE*, 100(2), 410–425, 2012.)

n.a., not available; PEM, polymer electrolyte membrane.

[a] Gas production rates are commonly given in normal cubic meters per hour (Nm³/h) and kilograms per hour (kg/h): 1 Nm³/h = 0.0899 kg/h.

[b] Rated power has been obtained either directly from the manufacturers, when the data are available, or as the product of third and fifth columns (rated production and specific energy consumption, respectively).

[c] Specific energy consumption has been obtained either directly from the manufacturers, when the data are available, or as the ratio between fourth and third columns (rated production and specific energy consumption, respectively).

[d] Efficiency has been calculated as the ratio between the HHV of hydrogen (3.54 kWh/Nm³) and the specific energy consumption (fifth column).

[e] Module configuration by series-connected monopolar cells.

[f] In development.

[g] The manufacturer indicates that the specific energy consumption refers to the global hydrogen production system.

[h] The manufacturer does not indicate if the specific energy consumption is exclusively of the electrolysis process (module) or if it refers to the global hydrogen production system.

[i] The manufacturer indicates that the specific energy consumption refers only to the electrolysis process (module).

[j] Pressurization with internal compressor, not by isothermal process inside the electrolysis module.

[k] With an additional purification system (to remove oxygen and water vapor from the produced hydrogen gas).

[l] Electricity consumption in kWh/kg.

15.4 Solar Thermochemical Method

15.4.1 Description and Status

Solar thermochemical (STC), as the name suggests, is based on the utilization of thermal energy obtained from solar radiation to split the water molecule. Ultimately, the heat energy drives an endothermic chemical transformation, that is, thermolysis of water. Generally, direct thermal water splitting requires temperatures of more than 2500 K (Steinfeld 2005; Funk 2001; Steinfeld 2014). This basically involves spontaneous generation of hydrogen *via* following reactions, with respective dissociation constant (K) under corresponding pressure (Kogan 1998):

$$H_2O \leftrightarrow HO + H \quad K_1 \tag{15.1}$$

$$HO \leftrightarrow H + O \quad K_2 \tag{15.2}$$

$$2H \leftrightarrow H_2 \quad K_3 \tag{15.3}$$

$$2O \leftrightarrow O_2 \quad K_4 \tag{15.4}$$

The thermochemical route can be categorized into five types, as displayed in Figure 15.3, based on the solar hydrogen production from *water splitting* or *decarbonization (cracking/*

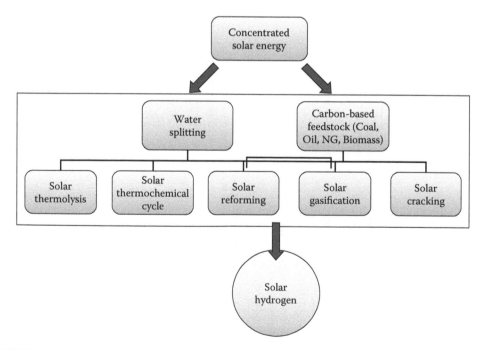

FIGURE 15.3

Schematic diagram showing the thermochemical routes for solar hydrogen production, using concentrated solar radiation obtained from sunlight.

reforming/gasification) of fossil fuels. The primary component of this technique is the generation of high temperature by concentration of solar energy. By the utilization of such energy, WS can be facilitated by certain mechanisms such as thermolysis and thermochemical cycle, while hydrogen can be generated by decarbonization *via* reforming/gasification/cracking of carbon-based fuel stock (Steinfeld 2014). The present section focuses only on the thermolytic WS that is facilitated by using renewable solar energy and abundant water. Thermally assisted electrolysis under solar light is mainly dependent on the light-harvesting capability; components of thermal (*sub-band gap*) and electronic (*band gap*) radiations, which facilitate *"temperature and pressure changes"*; as well as *"photoelectrochemical charge transfer"* for splitting of the water molecule. It can be understood that thermodynamically, the water dissociation is favored at elevated temperature. Practically, this is challenging task with respect to reactor. The solar reactor (material) would be subjected to the exposure of large amount of heat energy from more than 10,000 suns (solar objects). This becomes a stringent condition with respect to *the selection of reactor material* and *reactor design*. It is important to note that the common stainless steels are unable to sustain temperatures above few hundreds of degree centigrade. Thus, there is a necessity of using those materials that can work in the range of 1800°C–3000°C. These may include Al_2O_3 and Mullite/fused SiO_2 (Kogan 1998). Therefore, there is a great requirement for exploring *well-suited material* for reactor design. A list of prospective materials is given in Table 15.2.

TABLE 15.2

Assessment and Criteria of Material Selection by Melting Points of Various Refractory Materials

Material Type	Material System	Melting Point (in °C)
Oxide	SiO_2 (silica)	1720
Oxide	Quartz	1610
Oxide	TiO_2 (titania)	1840
Oxide	Cr_2O_3	1990–2200
Oxide	Al_2O_3 (alumina)	2050
Oxide	UO_2	2280
Oxide	Y_2O_3 (yittria)	2410
Oxide	BeO	2550
Oxide	CeO_2 (ceria)	2660–2800
Oxide	ZrO_2 (zirconia)	2715
Oxide	MgO	2800
Oxide	HfO_2	2810
Oxide	ThO_2	3050
Carbide	SiC	2200 (decomposition)
Carbide	B_4C	2450
Carbide	WC	2600 (decomposition)
Carbide	TiC	3400–3500
Carbide	Electrolyte-grade graphite	3650 (sublimation)
Carbide	HfC	4160
Nitride	Si_3N_4	1900
Nitride	*h*-BN	3000 (decomposition)

Source: Kogan, A., *Int. J. Hydrogen. Energ.* 23(2), 89–98, 1998.

Another challenge in this technology is related to the separation of hydrogen gas from the mixture of the WS *by-products*, which may form explosive combinations. A porous ceramic membrane $ZnO\text{-}TiO_2\text{-}Y_2O_3$ oxide has been reported in the past. However, it demands a *Knudsen flow region-type* reactor design (Figure 15.4), so as to have a mean free path of H_2 molecule greater than the average pore diameter. These challenges have limited the studies in this field. The figure shows solar reactor design for direct thermochemical WS (Kogan 1998).

In an attempt to develop this technology, a multi-step chemical cycle has been suggested and reported, using ZnO/Zn (i.e., M_xO_y/M) system, as

$$1^{st}\text{step} \qquad M_xO_y \rightarrow xM + y/2(O_2) \quad \text{-- -- -- Solar} \qquad (15.5)$$

$$2^{nd}\text{step} \qquad xM + yH_2O \rightarrow M_xO_y + H_2 \quad \text{-- -- -- Non-solar} \qquad (15.6)$$

Till recently, ZnO/Zn has been reported as the best metal-oxide redox pair for a two-step cycle (Steinfeld 2005).

The technology interfaces solar and thermal material aspects for hybrid option for hydrogen generation, which is decided by *sub-band gap* and *super-band gap* material capability to yield solar electrolysis at elevated temperatures in an efficient manner. High-temperature electrolysis for hydrogen generation will be ultimately attainable and will be commercially available.

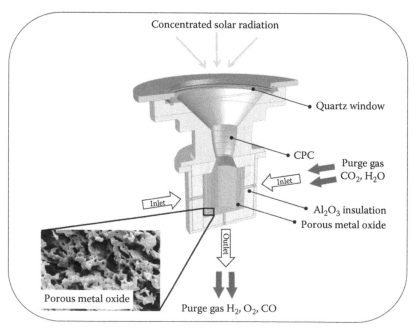

FIGURE 15.4
Solar reactor for direct thermochemical water splitting and solar hydrogen generation. (From Steinfeld, A., *CRC Handbook of Hydrogen Energy—Solar Thermochemical Production of Hydrogen*, CRC Press, pp. 421–444, 2014.)

15.5 Photocatalysis and Photoelectrochemical Method

15.5.1 Description and Status

Photo-induced WS has been described as an ideal artificial photosynthesis, like that is found in plants. Photosynthesis is a metabolic process occurring in plants. In this process, energy (glucose) is produced by *chlorophyll* (in leaf) by series of bioprocesses, using solar energy, water, and carbon dioxide. This is an exciting example of generation of useful energy by making use of available natural resources. Unfortunately, till date, there does not exist any artificial replica of leaf that could generate useful energy. Consequently, it is thought that if water molecule is used to generate hydrogen in the presence of sunlight over some light-absorbing system (*e.g., leaf*). one can imitate artificial photosynthesis-type mechanism to generate hydrogen energy. Such mechanism, where solar energy photons can be used for splitting water molecule, is termed *photocatalytic WS* or *photoelectrochemical WS*. The present section deals with photocatalysis/photoelectrochemical WS over various photoactive systems. This method of hydrogen generation was well known in context of UV-induced WS over TiO_2 electrode (E_g ~3.2 eV) (Fujishimaand Honda 1972). However, in order to utilize solar energy, it is necessary to identify best-suited low-band-gap material that is capable of splitting water molecule under low-energy visible-light photons or sunlight (Acar and Dincer 2016; Mettenborger et al. 2016). Photoelectrochemical (PEC) cell is the simplest design. It consists of a conducting counter electrode (Pt) and a semiconducting working electrode (e.g., TiO_2, Si, WO_3, and CdS) immersed in an electrolyte, which on light illumination, facilitates a redox chemical reaction of WS and yields H_2 and O_2 (Kang et al. 2015). Figure 15.5 shows the schematic band representation of such PEC cell.

The electrode material is the most important constituent of PEC cell, and its physicochemical properties determine its usability for WS electrode (Rajeshwar 2007). It can be easily observed in Figure 15.5 that valence band edge (E_V), conduction band edge (E_C), and Fermi energy level (E_F) of photoanode and its alignment with electrolyte-related level (chemical potential), as well as the Fermi level of the counter-electrode, altogether play a crucial role in driving the electron and hole generated due to photoexcitation of

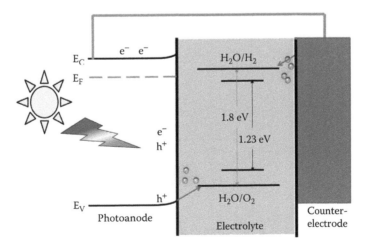

FIGURE 15.5
Schematic diagram showing the principle of photo electrochemical cell that is used for solar hydrogen production.

semiconducting photoanode, thus facilitating the reduction of water (hydrogen evolution) at counter-electrode and oxygen generation at photoanode. Effectively, as shown in Figure 15.6, when the band edge positions of semiconductors are straddle, the *redox* levels of water are said to be capable of WS (Chen and Wang 2012). As mentioned earlier, TiO_2 seems to be the most suited candidate for WS; however, its limited absorption capability in UV range makes it unsuited for visible-light hydrogen generation application. Despite the fact that there are plenty of low-band-gap semiconductors available, there is no ideal leaf-like system that has been found to be capable of efficiently and stably evolving hydrogen under solar light. Fe_2O_3 is best suited with respect to its band gap, cost-effectiveness, and eco-friendly nature, but it suffers by poor hole mobility, thus hampering the solar-to-hydrogen conversion efficiency. Similarly, on other hand, although the low-band-gap chalcogenides such as CdS, CdSe, and CdTe exhibit low band gap and energy, they severely suffer from the problem of photocorrosion. Similar severity of photocorrosion is also found in case of Si, GaAs, and GaP. Tremendous research work (Pareek et al. 2013) is in progress around the world to identify best photoelectrode material via the technique of nanostructuring, composite formation, and band gap tuning; however, there has not been any success in obtaining material with solar-to-hydrogen (STH) of >15% and a sustainable hydrogen evolution of around >1000 h.

There are several reviews in the literature; however, in the present section, we will discuss a few reviews that are assumed to be contributory achievements with respect to photocatalytic or PEC WS and have proven a step closer to the technological advancement.

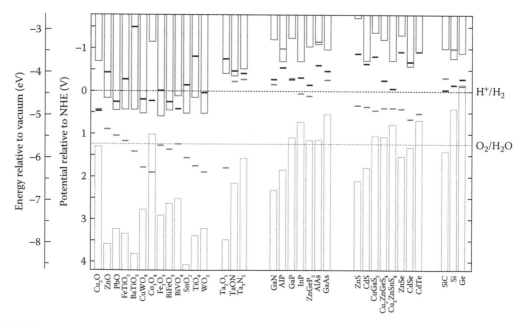

FIGURE 15.6
The calculated oxidation potential ϕ_{ox} (lower dashes) and reduction potential ϕ_{re} (upper dashes) relative to the NHE and vacuum level for a series of semiconductors in solution at pH = 0, the ambient temperature 298.15 K, and pressure 1 bar. The water redox potentials $\phi(O_2/H_2O)$ and $\phi(H + /H_2)$ (dashed lines) and the valence (lower vertical columns) and conduction (upper vertical columns) band edge positions at pH = 0 are also plotted. The alignment for these potentials at different pH values. (a simplified Pourbaix diagram) can be derived according to the relations. (Adapted from Chen, S., and Wang, L.W., *Chem. Mater.*, 24(18), 3659–3666, 2012. With permission.)

15.5.1.1 CASE I

The work by (Reece et al. 2011) has shown that by using amorphous Si-based system, constituted by *triple junction* made in conjunction with oxygen- and hydrogen-evolving catalyst, one is able to obtain WS efficiencies of 4.7% for a wired configuration and 2.5% for a wireless configuration. As shown in Figure 15.7a, the wired photoanode assembly of steel/Si/ITO/Co oxygen evolving catalyst (OEC), when coupled with Ni/NiMoZn electrode, yielded solar–to–fuel efficiency (**SFE**) of more than 4.7% in potassium borate electrolyte (black curve-flat curve). The efficiency of conversion of light and water to hydrogen is given as

$$\mathbf{SFE}(\%) = j \cdot \frac{\Delta E}{S} \cdot 100\% \tag{15.7}$$

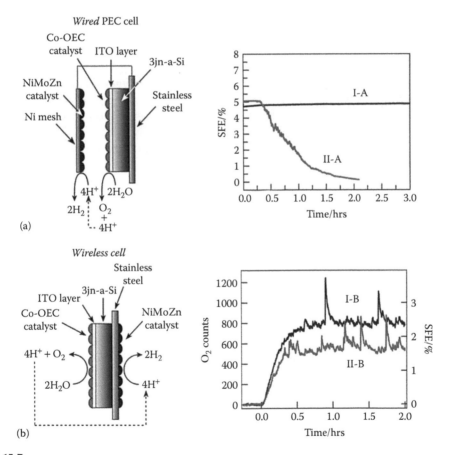

FIGURE 15.7

(a) Plot of the efficiency versus time for Co-OEC | 3jn-a-Si | NiMoZn PEC cell (left) in 1 M potassium borate (pH 9.2, Curve I-A) and in 0.1 M KOH (pH 13, Curve II-A) under AM 1.5 illumination. The traces are for solar cells of 7.7% PV efficiency. The cells were operated in a two-electrode cell configuration. (b) MS signal and SFE values for a wireless Co-OEC | 3jn-a-Si | NiMoZn cell under AM 1.5 illumination in 1 M KBi (Curve II-B) and in 0.5 M KBi and 1.5 M KNO₃ (Curve I-B). The cell was illuminated over the 2 hours of the experiment; MS signal corresponds to the concentration of O₂ in the carrier gas of the cell. The spikes in the data originate from sudden release of gas bubbles that were adhered to the cells, resulting in a temporary increase in the O₂ concentration in the headspace. SFE values were calculated as described in the SOM. (From Reece, et al., *Sci.*, 334, 645, 2011. With permission.)

where:

j is current density obtained at photoanode,

ΔE is the stored energy of water-splitting reaction

S is the total solar power incident electrode while illumination by 100 mW/cm² solar simulator with AM 1.5 filter.

The SFE (Curve II-A) showed a decrease after 0.5 h, when 0.1 M KOH was used as electrolyte. KOH facilitates pitting corrosion of ITO layer. The SFE has been expressed as direct product of efficiency of solar conversion by PV cell—φ (PV) and fuel generation efficiency by water-splitting—φ (WS); this includes the losing arising from catalyst overpotential and ohmic resistances,

$$\text{SFE}(\%) = \varphi(\text{PV}) \cdot \varphi(\text{WS}) \tag{15.8}$$

Effectively, this discussion indicates that with the improvement in efficiency of PV material, the SFE can be improved further.

Figure 15.7b shows the wireless photoanode configuration based on the steel coated on either side by steel/Si/ITO/Co-OEC and by NiMoZn coatings. The illumination of this assembly yielded oxygen evolution from Co-OEC, that is, oxygen-evolving catalyst (front side) and hydrogen evolution from NiMoZn, that is, *back side* of the wireless cell, when immersed in 1 M potassium borate electrolyte. The oxygen evolution was found to be more efficient for potassium borate electrolyte than KOH electrolyte.

Based on this work of artificial leaf, Daniel Nocera launched a company called "*Sun Catalytix*"; however, the plan to utilize this research has been put on hold, indicating that commercial viability of such *artificial leaf* has to go a long way.

15.5.1.2 CASE II

A work by (Maeda et al. 2006) demonstrated a well-fabricated photocatalytic water splitter material with $(Ga_{1-x}Zn_x)(N_{1-x}O_x)$-based system mixed with rhodium and chromium oxide. This yellow powder, as shown in Figure 15.8, yielded a quantum efficiency of >4.0% with a long stability of photocatalytic water during year of 2006. Although this novel photocatalyst made by solid solution of GaN and ZnO showed complete water splitting in 2006 and was found as main contribution, its complicated preparation methodology, and thus, industrial applicability, was doubtful. Figure 15.8 shows the actual yellow-colored photocatalyst formed by two wide-band-gap material systems. The preparation involved nitridation of Ga_2O_3 and ZnO at 1123 K for around 15 h under ammonia atmosphere.

This photocatalyst showed a wonderful response toward photons of long wavelength range, that is, 300 nm to 500 nm, and yielded maximum quantum yield of greater than 4% for complete water splitting, as shown in Figure 15.9a. Its long duration is demonstrated in Figure 15.9b, which shows time course of overall WS under visible light photons. It shows that novel photocatalyst has the capability to generate stoichiometric hydrogen and oxygen for more than 35 h of photocatalytic reaction in a slurry type of reactor.

Despite several attempts in PEC hydrogen generation, there is still no discovery that is expected to yield sustainable hydrogen production. However, there is a hybrid concept that implements the solar energy absorption for electrolysis of water by making use of titania-based electrode (Nanoptek USA year 2008). The product called "solar hydrogen generator" was launched in 2002; however, the cost of hydrogen generation and longevity still remains a point in case of PEC cell.

FIGURE 15.8
Photograph of the $(Ga_{1-x}Zn_x)(N_{1-x}O_x)$ solid solution, showing GaN and ZnO. (From Maeda, K. et al., *Nature.*, 440(16), 295, 2006.)

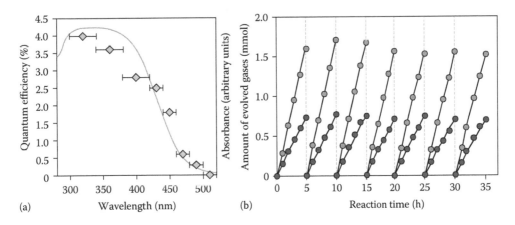

FIGURE 15.9
(a) Quantum efficiency of the photocatalyst, plotted as a function of wavelength of the incident light. Quantum efficiency was estimated by using several cut-off filters; each plot is in the middle of two cut-off wavelengths. and (b) time course of overall WS under visible light (of wavelength longer than 400 nm) on the $(Ga_{1-x}Zn_x)$ $(N_{1-x}O_x)$ photocatalyst. The reaction was continued for 35 h, with evacuation every 5 h (dotted line). Upper curve, hydrogen; lower curve, oxygen. (Adapted from Maeda, K. et al., *Nature.*, 440(16), 295, 2006. With permission.)

References

Acar C., and I. Dincer, 2016, A review and evaluation of photoelectrode coating materials and methods for Photoelectrochemical hydrogen production, *International Journal of Hydrogen Energy* 41 (19), 7950–7959.

Borse P.H., and D. Das, 2013, Advance workshop report on evaluation of hydrogen producing technologies for industry relevant application ARCI, Hyderabad, India 8–9 February 2013, *International Journal of Hydrogen Energy* 38 (26), 11470–11471 & references therein.

Bridgewater A. V. 2007, Applications for utilisation of liquids produced by fast pyrolysis of *biomass. Biomass and Bioenergy* 31 (8), 1–7.

Bridgewater A. V., ed., 2003, *Pyrolysis and Gasification of Biomass and Waste*. Newbury, UK: CPL Scientific Press.

Chen S., and L.W. Wang, 2012, Thermodynamic oxidation and reduction potentials of photocatalytic semiconductors in aqueous solution, *Chemistry of Materials* 24 (18), 3659–3666.

Dincer I., and C. Acar, 2015, Review and evaluation of hydrogen production methods for better sustainability, *International Journal of Hydrogen Energy* 40 (34), 11094–11111.

Elliott M. A., ed., 1981, Chemistry of Coal Utilization, 2nd Suppl. Volume. New York, John Wiley.

FNR, 2009, Bioenergy—Plants, Raw Materials, Products. Report Fachagentur Nachwachsende Rohstoffe, 2nd edn, www.fnr.de.

Fujishima. A., and K. Honda, 1972, Electrochemical photolysis of water at a semiconductor electrode, *Nature*, 238 (5358), 37–38.

Funk J. E., 2001, Thermochemical hydrogen production: Past and present, *International Journal of Hydrogen Energy* 26, 185–190.

Gazey R., S. K. Salman, and D. D. Aklil-D'Halluin, 2006, A field application experience of integrating hydrogen technology with wind power in a remote island location, *Journal of Power Sources* 157, 841–847.

Ghosh P. C., B. Emonts, H. Janben, J. Mergel, and D. Stolten, 2003, Ten years of operational experience with a hydrogen-based renewable energy supply system, *Solar Energy* 75, 469–478.

Harrison K. W., G. D. Martin, T. G. Ramsden, and W. E. Kramer, 2009, The wind-to-hydrogen project: Operational experience, performance testing, and systems integration, NREL Report, NREL/TP-550-44082.

Holladay D., J. Hu, D. L. King, and Y. Wang, 2009, An overview of hydrogen production technologies, *Catalysis Today*, 139 (4), 244–260.

Kang D. et al., 2015, Electrochemical synthesis of photoelectrode and catalysts for use in water splitting, *Chemical Reviews* 115, 12839–12887.

Kelly N. A., and T. L. Gibson, 2008, A solar-powered, high-efficiency hydrogen fueling system using high-pressure electrolysis of water: Design and initial results, *International Journal of Hydrogen Energy* 33, 2747–2764.

Khaselev O, A. Bansal, and J. A Turner, 2001. High-efficiency integrated multijunction photovoltaic/electrolysis systems for hydrogen production, *International Journal of Hydrogen Energy* 26 (2), 127–132.

Kogan A., 1998, Direct solar thermal splitting of water and on site separation of the products-II. Experimental feasibility study, *International Journal of Hydrogen Energy* 23 (2), 89–98.

Konieczny A., K. Mondal, T. Wiltowski, and P. Dydo, 2008, Catalyst development for thermocatalytic decomposition of methane to hydrogen, *International Journal of Hydrogen Energy* 33(1), 264–272.

Maeda K, et al., 2006, Photocatalyst releasing hydrogen from water, *Nature* 440 (16), 295.

Mettenborger et al., 2016, Inetrfacial insight in multi-junction metal oxide photoanodes for water splitting applications, *Nano Energy* 19, 415–427.

Muradov N. Z., and T. N. Veziroğlu, 2005, From hydrocarbon to hydrogen-carbon to hydrogen economy, *International Journal of Hydrogen Energy* 30 (3), 225–237.

Pareek A., N. Y. Hebalkar, and P. H. Borse, 2013, Fabrication of a highly efficient and stable nano-modified photoanode for solar H_2 generation, *RSC Advances*. 3 (43), 19905–19908.

Perez G. I., A.J. Calderon Godoy, M. Calderon Godoy, and J.L. Herrero Agustin, 2015, Monitoring of electric power systems: Application to self-sufficient hybrid renewable energy systems, Compatibility and Power Electronics (CPE), 9th International Conference on, 560–565.

Rajeshwar K., 2007, Fundamentals of semiconductor electrochemistry and photoelectrochemistry, in *Encyclopedia of Electrochemistry*, A. J. Bard, M. Stratmann, and S. Licht, eds., 6, 1–53, Weinheim, Germany: Wiley-VCH Verlag GmbH & Co. KGaA.

Reece et al., 2011, Wireless Solar Water Splitting Using Silicon-Based Semiconductorsand Earth-Abundant Catalysts, *Science* 334, 645.

Santos D., C. Sequeira, and J. L. Figueiredo, 2013. Hydrogen production by alkaline water electrolysis, Quim, *Nova* 36 (8), 1176–1193.

Steinfeld A., 2005, Solar thermochemical production of hydrogen—a review, *Solar Energy* 78 (5), 603–615.

Steinfeld A., 2014, *CRC Handbook of Hydrogen Energy—Solar Thermochemical Production of Hydrogen*, 421–444, CRC Press, Taylor Francis Group, Boca Raton, FL.

Turner J., 2008, Renewable hydrogen energy, *International Journal of Energy Research* 32, 379–407.

Ulleberg O., T. Nakken, and A. Eté, 2010, The wind/hydrogen demonstration system at Utsira in Norway: Evaluation of system performance using operational data and updated hydrogen energy system modelling tools, *International Journal of Hydrogen Energy* 35, 1841–1852.

Ursua A., M. Luis Gandı́a, and Pablo Sanchis, 2012, Hydrogen Production from water electrolysis: Current status and future trends, *Proccedings. of IEEE* 100 (2), 410–425.

Varkaraki E., G. Tzamalis, and E. Zoulias, 2008, Operational experience from the RES2H2 wind-hydrogen plant in Greece, Proccedings 17th World Hydrogen Energy Conf. 218–221.

16

Integrated Water Resources Management

Narayan C. Ghosh

CONTENTS

ABSTRACT   The article highlights the concept of integrated water resources management; its basic functions; and linkages among functions, challenges, and implementation strategies in river basins, as well as also a way forward toward its implementation. A numerical conjunctive use model for management of surface and groundwater on line to the managed aquifer recharge, together with aquifer storage recovery, was also presented. Performance of the model was also demonstrated by an illustrated example. The derived model can be integrated into the Integrated Water Resources Management strategy as a tool.

16.1 Introduction: Background and Driving Forces

Water, on and beneath the land surfaces, available as surface and groundwater, originates from the same source, precipitation. Its distribution and availability in the surface and groundwater domains depend on the meteorological, hydrological, and hydrogeological characteristics and variability. Water cycle of different continents of the world is different.

It is primarily driven by the hydroclimatic conditions, land uses and land covers, water practices pattern, water footprints, and so on. Despite having good annual precipitation in many countries, spatiotemporal distribution of surface and groundwater is many a time such that either one domain's water is not sufficient to meet the expected demands. Sometimes, the areas supported by surface water alone may be inadequate to meet increasing demands; alternatively, the areas supported by groundwater alone by its own may be inadequate to meet expected demands or both may be limiting in many cases. Demands of water in various sectoral uses are continuously rising because of population growth and their allied demands, together with prospective socioeconomic development. The scenarios of water demands in fast-developing countries, such as India, are changing so rapidly that there is little chance to reduce the gap between the supply side and demand side, unless a judicious management of water resources is perceived.

Water is a key driver of economic and social development. It also has a basic function in maintaining the integrity of the natural environment. Water managers, whether in the government or private sectors, have to make difficult decisions on water allocation by apportioning balance between supplies and ever-increasing demands. Drivers such as demographic and climatic changes further increase the stress on water resources.

Most developed countries have, in large measure, artificially overcome natural variability by supply-side infrastructure to assure reliable supply and reduce risks, albeit at high cost and often with negative impacts on the environment and, sometimes, on human health and livelihoods. Many developing countries, and some developed countries, are now finding that supply-side solutions alone are not adequate to address the ever-increasing demands from demographic, economic, and climatic pressures. Wastewater treatment, water recycling, and demand management measures are being introduced to counter the challenges of inadequate supply. In addition to problems of water quantity, there are also problems of water quality. Pollution of water by the source originates from anthropogenic and geogenic sources poses a major threat to the water users as well as for maintaining natural ecosystems. In many regions, the availability of water, both quantity and quality, is being severely affected by climate variability and climate change, with more or less precipitation in different regions and more extreme weather events. In many regions, too, demand is increasing as a result of population growth and other demographic changes (in particular urbanization) and agricultural and industrial expansion, following changes in consumption and production patterns. As a result, some regions are now in a perpetual state of demand, outstripping supply, and in many more regions, this is the case at critical times of the year or in years of low water availability.

The traditional fragmented approach is no longer viable, and a more holistic approach to water management is essential. This is the rationale for the integrated water resources management (IWRM) approach that has now been accepted internationally as the way forward for efficient, equitable, and sustainable development and management of the world's limited water resources and for coping with conflicting demands.

16.2 Defining Integrated Water Resources Management

Integrated water resources management (IWRM), as defined by the global water partnership (GWP), is "a process which promotes the coordinated development and management of water, land and related resources, in order to maximize the resultant economic and social welfare in an equitable manner without compromising the sustainability of vital ecosystems" (GWP 2009; UNESCO 2009).

IWRM is thus an empirical concept (Biswas 2004), which was built up from the on-the-ground experience of practitioners. Although many parts of the concept were around for several decades, it was made more inclusive after the first global water conference in Mar del Plata in 1977 and the World Summit on Sustainable Development in 1992 in Rio. The GWP's definition of IWRM is now widely accepted.

IWRM basically represents three principles: social equity, economic efficiency, and environmental sustainability. The meaning of these principles is as follows:

> *Social equity* means ensuring equal access for all users to an adequate quantity and quality of water necessary to sustain human well-being. The right of all users to the benefits gained from the use of water also needs to be considered when making water allocations. Benefits may include enjoyment of resources through recreational use or the financial benefits generated from the use of water for economic purposes.
>
> *Economic efficiency* means bringing the maximum benefit to the highest number of users with the available financial and water resources. The economic value is not only about price— it should also consider current and future social and environmental costs and benefits.
>
> *Ecological sustainability* requires that aquatic ecosystems are acknowledged as users and that adequate allocation is made to sustain their natural functioning.

16.2.1 How to Address the Challenges?

Many of the challenges facing water managers are not new. However, because the nature and size of the problems differ from one region to another and from one basin to another, the responses vary widely. There cannot be a single blueprint solution to the problems. However, addressing these challenges usually needs three pillars of IWRM getting right (Figure 16.1). Three key areas to address these three pillars are (i) structural issues, which include data acquisition, assessment, and operations and maintenance; and (ii) institutional issues, which include central-local working arrangement, scale of planning, and public and private participation; and (iii) the others are policies and pricing, legislation for implementation of IWRM. All three kinds of responses are important and interrelated, putting in place the institutional framework through which the policies, strategies, and legislation can be implemented. Whereas the management instruments help doing the jobs by these institutions. Largely, IWRM addresses the following:

- Improved water governance and management
- Potential competition and conflicts among different stakeholders from all sectors and among individuals, communities, and governments
- Environmental degradation
- Gender and social disparities in terms of equitable access to and control over resources, benefits, costs, and decision making
- Need for sustainable water resources development as a key to poverty eradication

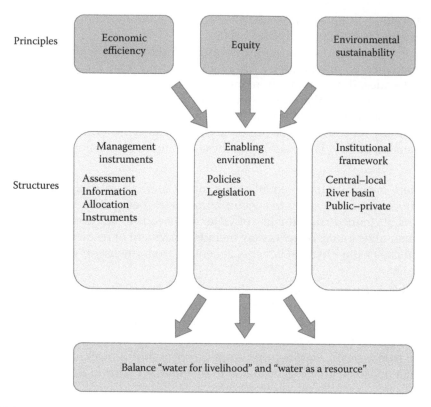

FIGURE 16.1
IWRM components and key areas for addressing challenging issues. (From GWP [Global Water Partnership] and the International Network of Basin Organizations [INBO], *A Handbook for Integrated Water Resources Management in Basins*, ISBN: 978-91-85321-72-8, 103, 2009. https://www.gwpforum.org.)

16.2.2 Integrated Water Resources Management in River Basins

The concept of IWRM has been accompanied by promotion of the river basin as the logical geographical unit for its practical realization. The river basin offers many advantages for strategic planning, particularly at higher levels of government. Groundwater aquifers frequently cross catchment boundaries, and more problematically, river basins rarely conform to existing administrative entities or structures. A basin-level perspective enables integration of downstream and upstream issues, quantity and quality, surface water and groundwater, and land use and water resources in a practical manner. On the river basin scale, there are many actors that have roles and responsibilities in the management of the environment and society, which are all linked to the status of the water resources. For successful implementation of IWRM all, these actors have to be involved. It is therefore logical that IWRM on the river basin scale should be focused on a set of basic water resources management functions. The processes involved in the implementation of IWRM in a river basin follow a cyclic framework of activities (Figure 16.2).

16.2.3 Basic Functions for Water Resources Management

The basic functions for water resources management in a river basin (Gunawardena 2010) are presented in Figure 16.3, and Table 16.1 gives a definition of these functions.

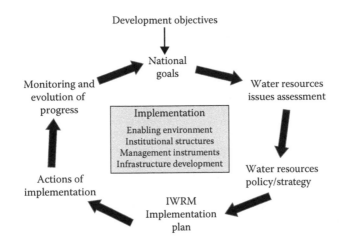

FIGURE 16.2
Management cycle of planning and implementation of IWRM in a river basin. (From GWP [Global Water Partnership] and the International Network of Basin Organizations [INBO], *A Handbook for Integrated Water Resources Management in Basins*, ISBN: 978-91-85321-72-8, 103, 2009. https://www.gwpforum.org.)

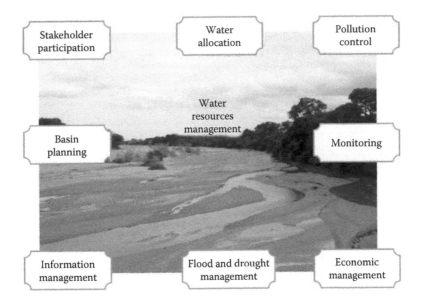

FIGURE 16.3
Basic functions for water resources management. (From Gunawardena, E. R. N., Operationalizing IWRM Through River Basin Planning and Management, Lecture Note Prepared for Training of Trainers in Integrated Water Resources Management, Held at Kandy, Sri Lanka, 16–25 September, 2010. https://www.saciwaters.org/CB/iwrm/Lecture%20Notes/8.3.%20Restraining%20conflicst%20through%20institutional%20interventions_Gunawardena%20ERN.pdf.)

16.2.4 Linkage to Other Functions

Water allocation, together with pollution management, supports the other functions. Basin planning provides the setting for water allocation. It provides the naturally available surface and groundwater resources and the environmental flow requirements. It also gives present and projected future socioeconomic conditions, water demand, and

TABLE 16.1

Functions of Water Resources Management in a River Basin

Functions	Examples of Activities
Stakeholder participation—implementing stakeholder participation as a basis for decision making that takes into account the best interests of society and the environment in the development and use of water resources in the basin.	• Develop and maintain an active stakeholder participation process through regular consultation activities. • Provide specialist advice and technical assistance to local authorities and other stakeholders in IWRM.
Water allocation—allocating water to major water users and uses and maintaining minimum levels for social and environmental use while addressing equity and development needs of society.	• License of water uses, including enforcement of these.
Pollution control—managing pollution by using polluter pays principles and appropriate incentives to reduce most important pollution problems and minimize environmental and social impact.	• Identify major pollution problems. • License and manage polluters.
Monitoring of water resources, water use, and pollution—implementing effective monitoring systems that provide essential management information and identifying and responding to infringements of laws, regulations, and permits.	• Carry out hydrological, geographical, and socioeconomic surveys for the purposes of planning and development of water resources. • Develop, update, and maintain a hydrometric database required for controlling compliance of water use allocation.
Information management—providing essential data necessary to make informed and transparent decisions for the development and sustainable management of water resources in the basin.	• Define the information outputs that are required by the water managers and different stakeholder groups in a river basin. • Organize, coordinate, and manage the information management activities, so that the water managers and stakeholders get the information that they require.
Economic and financial management—applying economic and financial tools for investment, cost recovery, and behavior change to support the goals of equitable access and sustainable benefits to society from water use.	• Set fees and charges for water use and pollution.
River basin planning—preparing and regularly updating the basin plan, incorporating stakeholder views on development and management priorities for the basin.	• Conduct situation analysis with stakeholders. • Assess future developments in the basin.

infrastructural development. All this information is the basis for how much water is needed and how much can be allocated in the river basin. Financial management gives the tools to encourage and, if necessary, force efficient use of water. Stakeholder participation and information management give transparency and ownership to the decided allocation. Through participatory activities, coordination between different water uses is also made possible. Monitoring of water use and water resources is necessary to enforce the water allocation. Figure 16.4 illustrates various functions linked to the water allocation and pollution control.

16.2.4.1 Pollution Management

Water resources management involves two closely related elements, that is, the maintenance and development of adequate quantities of water of appropriate quality. Thus, water quality plays an important role in the management of water resources. Managing

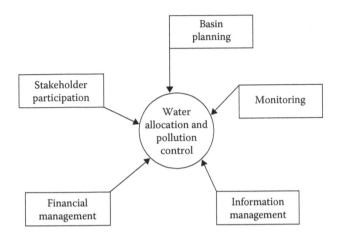

FIGURE 16.4
The water allocation and pollution control functions are dependent on input from the other functions. (From Gunawardena, E. R. N., Operationalizing IWRM through River Basin Planning and Management, Lecture Note Prepared for Training of Trainers in Integrated Water Resources Management, Held at Kandy, Sri Lanka, September 16–25, 2010. https://www.saciwaters.org/CB/iwrm/Lecture%20Notes/8.3.%20Restraining%20conflicst%20through%20institutional%20interventions_Gunawardena%20ERN.pdf.)

water pollution is one of the most critical challenges to sustainable management of water resources. Institutional and legislative systems to address the problem of water pollution control and management are essentially required to attain sustainability in water resources management.

16.3 Ways Forward to Integrated Water Resources Management

GWP (2009) brought out *A Handbook for Integrated Water Resources Management in Basins*, which spelt out the roles, functions, and implementation strategies to be followed for IWRM in river basin level. In gist, an IWRM plan involves the following:

- A realistic IWRM plan requires the design of functions, structures, and procedures to take into account the financial and human resource constraints, the existing institutional structures, the management capacity, and the capacity for change.
- Successful IWRM plans need to be aligned with the high-priority national development processes and broad cross-sectoral and stakeholder support, even if these are outside the water sector.
- Multi-stakeholder involvement in the decision-making processes is essential for the acceptability of the outcome.
- Economic arguments for financing water resources management must be developed and well communicated.

Climate change (CC) has emerged as a global issue (IPCC 2007; 2014). Its realistic impact assessment on water resources is a concern to the policy makers, researchers, water managers, and so on. Adaptation measures to fight in response to the projected CC impact, while become a gigantic task to the governments, on the other hand, for suggesting about adaptation measures one needs to know the magnitude and variability of different CC parameters on the hydrological processes. Frequent torrential rains of high magnitude and drought-like situation thereafter are commonly observed in different parts of the world. It is also projected that, in many parts of the world, although there are chances of temporal variability and uncertainty of rainfall in local/regional scale, the total rainfall in a year may remain same (IPCC 2007; 2014). Question then arises: when there could be a number of high-frequency/magnitude rainfalls and drought-like situations, with total rainfall in a year, by and large, remaining same, how would water resources from available rainfall be conserved and managed for expected demands in a year?

South Asian part of the world has similar climatological and meteorological variability. This region largely receives 75%–80% of rainfall, mainly during monsoon months (May/June through September/October); the remaining period of the year receives about 20%–25% of annual rainfall (Ranade et al. 2007; Sivakumar and Stefanski 2011). One of the potential strategies familiar in this region is the managed aquifer recharge (MAR), by conserving excess monsoon surface runoff and recharge to aquifer for subsequent recovery and uses. MAR as a potential solution tool can be integrated to IWRM framework, together with its other beneficial functions. MAR, as one of the IWRM strategies, is described in detail, together with a mathematical model for the management of surface and groundwater.

16.3.1 A Conjunctive Use Model for the Management of Surface and Groundwater

India has a long tradition of practicing rainwater harvesting (RWH) and artificial recharge (AR) by employing indigenously developed techniques and methods to fulfill requirements of agricultural and drinking water supply, particularly in rural areas. For the last one and a half decade, RWH and AR are promoted as the government-supported national program for augmentation of groundwater resources in water-stressed and groundwater-problematic areas. Many countries are turning to MAR for conjunctive management of surface water, groundwater, and recycled water resources to stretch water supplies. An integrated approach of surface and groundwater management by conserving excess surface runoff from monsoon rainfall and recharge to aquifer with appropriate care to quality of recharge water in basin scale can be regarded as one of the strategies for IWRM, if not in its full scale, but it could satisfy a major part of it. MAR, together with aquifer storage treatment and recovery (ASTR), nearly covers the main "quantity and quality" issues of IWRM, except its operational, management, economic, and stakeholders parts.

MAR refers to intentional storage and treatment of water in aquifers for subsequent recovery or environmental benefits (Dillon 2005; Dillon et al. 2009; EPA Victoria 2009). It has similar connotation to "artificial recharge (AR)" commonly used in India, except the component of water quality treatment. AR to aquifers is promoted on a large scale in India (CGWB 1996; 2002; 2007; 2013; SAPH PANI 2012) by conserving monsoon surface runoffs for augmentation of groundwater resources in areas facing depleted water table, scarcity of water during lean flow months, deteriorating groundwater quality, saline water ingress threat, and so on. MAR (Figure 16.5) as a tool can be used for replenishing and re-pressurizing depleted aquifers, controlling saline intrusion or land subsidence, and improving water quality through filtration and chemical and biological processes. Depending on the field requirements, MAR can be used as tool to IWRM.

Recognizing the potential prospect of MAR, a generalized conjunctive use model for the management of surface and ground water is presented here. The model is derived on the basis of basic water balance of a groundwater recharge basin and hence can successfully be used as one of the computational tools for IWRM.

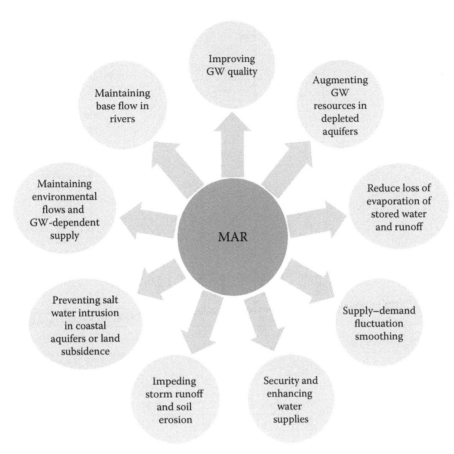

FIGURE 16.5
Some important groundwater conservation and management issues that can be addressed by MAR as a tool to IWMR.

16.3.2 Statement of the Problem

A schematic diagram showing different hydrological components associated with a recharge basin is given in Figure 16.6a. Let $Q_i(t)$, $Q_0(t)$, $R(t)$, and $E(t)$ be the hydrological components representing runoff from the catchment (L^3T^{-1}), outflow from the basin (L^3T^{-1}), rainfall over the basin (LT^{-1}), and water surface evaporation from the basin (LT^{-1}), respectively. The recharge basin has porous bed material and hydraulic passage between the basin and the aquifer, which allow infiltrated water to recharge to the underneath aquifer. The aquifer is unconfined, having characteristics of homogeneous and isotropic, with vertical hydraulic conductivity of K_v and storage coefficient of S. Let $Q_{gw}(t)$ be the groundwater recharge from the basin to aquifer due to potential head difference between the basin and groundwater table in the underneath aquifer (L^3T^{-1}). Let $P(1)$, $P(2)$, $P(3)$, and $P(4)$ be the four pumping wells with coordinates of (a_1, b_1), (a_2, b_2), (a_3, b_3), and (a_4, b_4), respectively, to be operated as recovery of aquifer storage, as shown in Figure 16.6b. Let H be the height of initial groundwater table measured upward from the impervious stratum. Let the recharge basin be rectangular in shape, having dimension of $2a$ (length) and $2b$ (breadth). The origin $(0,0)$ of the axes is located at the center of the basin, and the coordinates of the four corners are as shown in Figure 16.6b. With reference to the center of axes, the coordinates of the four

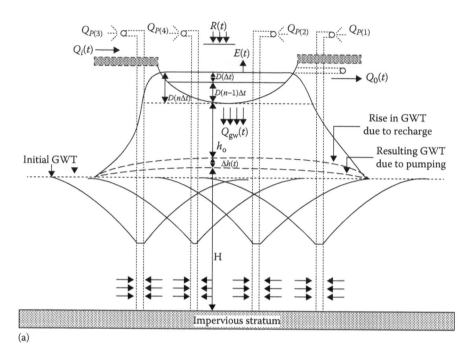

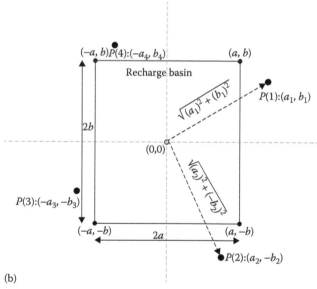

FIGURE 16.6
Groundwater recharge basin showing hydrological components associated with it, including arrangement of pumping for aquifer storage recovery: (a) Sectional view and (b) plan view, along with coordinates of the pumping wells with reference to the basin origin.

pumping wells are $P(1):(a_1, b_1)$; $P(2):(a_2, -b_2)$; $P(3):(-a_3, -b_3)$, and $P(4):(-a_4, b_4)$. It is assumed that initially, the groundwater table below the recharge basin is in dynamic equilibrium and horizontal and the recharge to the underneath groundwater table is reckoned since time $t = 0$. Further, it is assumed that water to be pumped from the groundwater will be used to meet some sectoral demands, as envisaged in IWRM strategy.

The recharge rate, $Q_{gw}(t)$, due to the interaction of different hydrological components in the recharge basin and also the enhanced recharge rate due to the pumping of wells, when all the hydrological components are time-varying, required to be determined.

16.3.2.1 Model Development

The basic concept followed in estimating the unsteady groundwater recharge consequent from variable inflows and outflows is the water balance of the recharge basin (Figure 16.6a).

In mathematical terms, the water balance equation, derived from the difference of all inflows and all outflows to/from the basin at a given time is equal to the change in storage at that time; using notations of Figure 16.6, the equation can be written as (Ghosh et al. 2015):

$$\left[Q_i(t) + AR(t)\right]t - \left[Q_0(t) + A_i(t)E(t) + Q_{gw}(t)\right]t = \Delta V(t) \tag{16.1}$$

where:

$Q_i(t)$ is the inflow rate from its catchment to the recharge basin at time t (L^3T^{-1})
A is the gross surface area of the recharge basin (L^2)
$R(t)$ is the rainfall rate at time t (LT^{-1})
$Q_0(t)$ is the outflow rate from the basin at time t (L^3T^{-1})
$A_i(t)$ is the water surface area of the basin at time t(L^2)
$E(t)$ is the water surface evaporation rate at time t (LT^{-1})
$Q_{gw}(t)$ is the groundwater recharge rate from the basin at time t (L^3T^{-1})
$\Delta V(t)$ is the change in storage between time t and $t + \Delta t$ (L^3)
Δt is the time step size (T)

For non-monsoon period, when there is no rainfall and surface runoff, the components, $Q_i(t)$, $R(t)$, and $Q_0(t)$ in Equation 16.1 will be zero, and the equation in such case will turn to:

$$-\left[A_i(t)E(t) + Q_{gw}(t)\right]t = \Delta V(t) \tag{16.2}$$

where, (−) sign indicates that the storage will decrease with time.

In Equation 16.1, all components are time-varying and those, except $Q_{gw}(t)$, if known externally, then $Q_{gw}(t)$ can be computed. The responses of $Q_{gw}(t)$ are governed by the potential difference of hydraulic heads between the basin and in the aquifer underneath the basin bed, which are influenced by the inflows and outflows to/from the basin and its geometry, including the subsurface hydraulic parameters; therefore, the computation of $Q_{gw}(t)$ from Equation 16.1 appears to be a time-varying implicit function of hydraulic head.

For obtaining expression of $Q_{gw}(t)$ as a function of hydraulic head, Hantush's (1967) approximate analytical equation—for the rise of water table in a homogeneous and isotropic unconfined aquifer of infinite areal extent due to uniform percolation of water from a spreading basin in the absence of a pumping well—together with well pumping equation given by Theis (1935), are integrated into Equation 16.1 to obtain expression for resulting rise/fall of head beneath the recharge basin.

The Hantush's expression for rise of water table in the absence of a pumping well is given by

$$h(x, y, t) = \left(\frac{w\,t}{4\,S}\right) f(x, y, t) \tag{16.3}$$

where:

$h(x,y,t)$ is the rise in groundwater table below the rectangular spreading basin at location x, y (L)

w is the constant rate of percolation per unit area (LT^{-1})

S is the storage coefficient of the aquifer (dimensionless)

t is the time since the percolated water joins the water table (T)

$f(x, y, t)$ is the analytical expression derived by Hantush (1967), which is given by

$$f(x,\ y,\ t) = \left\{ F\left[\frac{(a+x)}{2\sqrt{(Tt/S)}}, \frac{(b+y)}{2\sqrt{(Tt/S)}}\right] + F\left[\frac{(a+x)}{2\sqrt{(Tt/S)}}, \frac{(b-y)}{2\sqrt{(Tt/S)}}\right] \right.$$
$$\left. + F\left[\frac{(a-x)}{2\sqrt{(Tt/S)}}, \frac{(b+y)}{2\sqrt{(Tt/S)}}\right] + F\left[\frac{(a-x)}{2\sqrt{(Tt/S)}}, \frac{(b-y)}{2\sqrt{(Tt/S)}}\right] \right\}$$

(16.4)

where:

T is the transmissivity = KH

H is the weighted mean of the depth of saturation during the period of flow

K is the coefficient of permeability (i.e., hydraulic conductivity) of the aquifer material

a is the half of the length, and b is the half of the width of the rectangular basin

x and y are the coordinates at which response is to be determined $F(p,q) = \int_0^1 erf(p/\sqrt{z})$. $erf(q/\sqrt{z})\,dz$ and $erf(x) = 2/\sqrt{\pi}\int_0^X e^{-u^2}\,du$

Theis (1935) well response equation for drawdown in an observation well located at a distance r from the pumping well with a uniform thickness and infinite in areal extent is given by

$$s = \frac{Q_p}{4\pi T}W(u)$$

(16.5)

where:

s is the drawdown (L)

Q_p is the constant pumping rate (L^3T^{-1})

T is the transmissivity (L^2T^{-1})

$W(u)$ is the well function defined by an infinite series

the dimensionless parameter, u, which depends on time t, is defined as

$$u = \frac{r^2 S}{4Tt}$$

(16.6)

where, r is the distance between the observation point and the well.

Let the time domain be discretized into uniform time steps, each of size Δt, such that $t = n\Delta t$. Let the recharge from the bed of the recharge basin be approximately equal to a train of pulses. Let Q_{gw} $(\gamma\Delta t)$ be the total recharge from the basin during the duration from $(\gamma-1)\Delta t$ to $(\gamma\Delta t)$. The average rise/fall in water table height, $\Delta h(t)$, consequent to a train of pulse recharge and pumping can be derived using Duhamel's principle, as

$$\Delta h(n\Delta t) = \sum_{\gamma=1}^{n} \frac{Q_{gw}(\gamma\Delta t)}{4\,a\,b} \delta_H(n - \gamma + 1, \Delta t) - \sum_{\xi=1}^{\xi_{max}}\sum_{\gamma=1}^{n} Q_p(\xi,\gamma)\ \delta_p(\xi,(n - \gamma + 1)\Delta t)$$

(16.7)

where:
  a is the half of the length
  b is the half of the width of the rectangular basin
  $\delta_H(*)$ is the Hantush's discrete kernel coefficients
  ξ_{max} is the number of pumping wells
  Q_p is the pumping rate of each well
  δp is the discrete kernel coefficients of the pumping well

6.3.2.1.1 Deriving Discrete Kernel Coefficients

6.3.2.1.1.1 Hantush's Kernel Coefficients The unit step response function, $U_h(t)$, for rise in water table height due to unit recharge per unit area per unit time, that is, $w = 1$ ($L^3\,L^{-2}\,T^{-1}$), from Equation 16.3, is given by

$$U_H(t) = \left(\frac{t}{4\,S}\right) f(x,y,t) \tag{16.8}$$

The unit pulse response function coefficients, $\delta_H(n\Delta t)$, in discrete time steps of size, Δt ($t = n\Delta t$), which takes place during the first time period, Δt, and no recharge afterward, is given by

$$\delta_H(n\Delta t) = \frac{U_H(n\Delta t) - \{U_H(n-1)\Delta t\}}{\Delta t} \tag{16.9}$$

6.3.2.1.1.2 Theis's Pumping Kernel Coefficients The unit step response function for drawdown, $U_p(t)$, due to unit pumping rate, that is, $Q_p = 1$ ($L^3\,T^{-1}$), from Equation 16.5, is given by

$$U_p(t) = \left(\frac{1}{4\,\pi\,T}\right) W\left(\frac{r^2 S}{4\,T\,t}\right) \tag{16.10}$$

The unit pulse response function coefficients, $\delta_p(n\Delta t)$, in discrete time steps of size, Δt ($t = n\Delta t$), is given by

$$\delta_p(n\Delta t) = \frac{U_p(n\Delta t) - \{U_p(n-1)\Delta t\}}{\Delta t} \tag{16.11}$$

Assuming a linear relationship between the influent seepage and potential difference, seepage during nth unit time step can be expressed by using Darcy's equation (Figure 16.6a), as

$$Q_{gw}(n\text{A}t) = \frac{4\,a\,b\,K_v\left[H + h_0 + \overline{D}(n\Delta t) - (H + \Delta h(n\Delta t))\right]}{h_0} \tag{16.12}$$

Putting $\Delta h(n\Delta t)$ from Equation 16.7 into Equation 16.12, we obtain:

$$Q_{gw}(n\text{A}t) = \frac{4\,a\,b\,K_v\left[h_0 + \overline{D}(n\Delta t) - \left\{\begin{array}{l}\sum\limits_{\gamma=1}^{n}\dfrac{Q_{gw}(\gamma)}{4\,ab}\delta_H(x,y,(n-\gamma+1)\Delta t)\\[2ex] -\sum\limits_{\xi=1}^{\xi_{max}}\sum\limits_{\gamma=1}^{n}Q_p(\xi,\lambda)\delta_p(\xi,(n-\gamma+1)\Delta t)\end{array}\right\}\right]}{h_0} \tag{16.13}$$

where:

K_v is the vertical hydraulic conductivity (LT^{-1})

$Q_{gw}(n\Delta t)$ is the total recharge from the basin during nth period (L^3T^{-1})

H is the height of the initial groundwater table, measured upward from the impervious stratum (L)

h_0 is the height of the porous material between the basin bed and the initial groundwater table (L)

$\bar{D}(n\Delta t)$ is the average water level in the basin, measured from its bed (L) = $(D(n\Delta t)+D(n-1)\Delta t)/2$

By separating the nth term and rearranging, Equation 16.13 can be rewritten as

$$Q_{gw}(n\Delta t) = \frac{4\,a\,b\,K_v \left[h_0 + \bar{D}(n\Delta t) - \left\{ \begin{array}{l} \left(\sum\limits_{\gamma=1}^{n-1} \dfrac{Q_{gw}(\gamma\Delta t)}{4\,a\,b} \delta_H(x,y,(n-\gamma+1)\Delta t) \right) \\ - \sum\limits_{\xi=1}^{\xi_{max}} \sum\limits_{\gamma=1}^{n} Q_p(\xi,\gamma)\delta_p(\xi,(n-\gamma+1)\Delta t) \end{array} \right\} \right]}{\left\{ h_0 + K_v\,\Delta t\,\delta_s(x,y,\Delta t) \right\}} \tag{16.14}$$

From Equation 16.13, $Q_{gw}(n\Delta t)$ can be obtained by solving it in succession of time steps, starting from time step 1.

The equation for computing the time-varying depth of water in the recharge basin, $D(n\Delta t)$, is obtained by substituting the recharge component from Equation 16.13 into Equation 16.1. The equation is given by

$$D(n\Delta t) = D(n\Delta - \Delta t)\frac{A_{ws}(n\Delta t - \Delta t)}{A_{ws}(n\Delta t)}$$

$$+ \frac{1}{A_{ws}(n\Delta t)}\left[Q_i(n\Delta t) + R(n\Delta t)A_s - E(n\Delta t)\overline{A_{ws}} - Q_0(n\Delta t) \right]\Delta t$$

$$- \frac{\overline{A_{rs}}}{A_{ws}(n\Delta t)} \left[K_v \frac{\left\{ h_0 + \bar{D}(n\Delta t) - \left| \begin{array}{l} \left(\sum\limits_{\gamma=1}^{n-1} \dfrac{Q_{gw}(\gamma\Delta t)}{4\,a\,b} \delta_H(x,y,(n-\gamma+1)\Delta t) \right) \\ - \sum\limits_{\xi=1}^{\xi_{max}} \sum\limits_{\gamma=1}^{n} Q_p(\xi,\gamma)\delta_p(\xi,(n-\gamma+1)\Delta t) \end{array} \right| \right\}}{\left\{ h_0 + K_v\delta_H(x,y,\Delta t) \right\}} \right] \tag{16.15}$$

where, $\Delta V(n\Delta t)=\left[A_{ws}(n\Delta t)D(n\Delta t) - A_{ws}(n\Delta t-\Delta t)D(n\Delta t-\Delta t)\right]$; $D(n\Delta t)$ is the depth of water in the recharge basin at time $t = n\Delta t$; $D(n\Delta t - \Delta t)$ is the depth of water in the basin at the proceeding time step, that is $(n - 1)\Delta t$; $A_{ws}(n\Delta t - \Delta t)$ is the water surface area of the basin at the proceeding time $(n-1)\Delta t$; $A_{ws}(n\Delta t)$ is the water surface area of the basin at time step $n\Delta t$; A_s is the gross top surface area of the recharge basin; $\overline{A}_{ws} = (A_{ws}(n\Delta t) + A_{ws}(n-1)\Delta t)/2$; and $\overline{A}_{rs} = 4ab$.

Equation 16.15 has a single unknown $D(n\Delta t)$- and several other $D(n\Delta t)$-dependent variables, when all other variables are known externally. The unknown $D(n\Delta t)$ and its dependent variables, $A_{ws}(n\Delta t)$ and $A_{rs}(n\Delta t)$, are in different mathematical forms at the right-hand side (RHS) of Equation 16.15. $D(n\Delta t)$ in such case can be solved by iteration procedures in succession of time steps. Substituting estimated $D(n\Delta t)$ from Equation 16.15 into Equation 16.14, $Q_{gw}(n\Delta t)$ corresponding to the value of $D(n\Delta t)$ can be computed. The time-varying $D(n\Delta t)$ and $Q_{gw}(n\Delta t)$ can thus be calculated in succession of time for $n = 1, 2, 3, \dots$

For computing $D(n\Delta t)$ and $Q_{gw}(n\Delta t)$ corresponding to unsteady inflows and outflows, using Equations 16.15 and 16.14, respectively, the variables $Q_i(n\Delta t)$, $R(n\Delta t)$, $Q_0(n\Delta t)$, $E(n\Delta t)$, $A_{ws}(n\Delta t)$, A_s, and $D(n\Delta t - \Delta t)$ and the parameters H, h_0, T, S, and K_v are to be known.

16.3.2.1.2 Recharge from Recharge Basin of Variable-Spread Area

The water-spread area may change with time owing to the fluctuation of water depth in the basin by the variable inflows and outflows. This will be more pronounced when the basin sides will have low slopes. The algorithm presented in Equations 16.3 and 16.14 can be extended for variable recharge area. Let $(2a[1], 2b[1])$, $(2a[2], 2b[2])$ … $(2a[n], 2b[n])$ be the length and the width of the water spread area during 1st, 2nd, and nth unit time period, respectively. Let $\delta_H(a[1], b[1], m)$, $\delta_H(a[2], b[2], m)$ … $\delta_H(a[n], b[n], m)$, where $m = 1, 2, \dots, n$, be the unit pulse response coefficients, which can be generated by using Equation 16.9. The water table rise at any point below the lake bed can be expressed as:

$$\Delta h(n\Delta t) = \sum_{\gamma=1}^{n} \frac{Q_{gw}(\gamma\Delta t)}{4\, a(\gamma)\, b(\gamma)}\, \delta_s(a(\gamma), b(\gamma), n-\gamma+1, \Delta t)$$

$$- \sum_{\xi=1}^{\xi_{max}} \sum_{\gamma=1}^{n} Q_p(\xi,\gamma)\, \delta_p(\xi, (n-\gamma+1)\Delta t) \tag{16.16}$$

Substituting $\Delta h(n, \Delta t)$ from Equation 16.16 into Equation 16.12 and rearranging, as done for Equations 16.13 and 16.14, the expression for estimate of recharge can be obtained.

6.3.2.1.3 Time-Varying Depth of Water and Recharge Rate

When there are inflows to the recharge basin from external sources, viz. from basin catchment, direct rainfall over the basin, and outflows from the basin, viz. losses due to evaporation and groundwater recharge, the depth of water in the recharge basin will vary, which will lead to variable recharge rate and evaporation losses. Further, pumping of groundwater will increase the groundwater table depletion, which will result in more potential hydraulic head difference, thereby enhancing the recharge rate. The recharge rate and evaporation losses in such cases will depend on the water spread area and the corresponding depth of water present in the basin at that time.

Substituting the recharge component, $Q_{gw}(n\Delta t)$, derived using Equation 16.16, the expression for time-varying depth of water in the basin, $D(n\Delta t)$, can be obtained.

16.4 An Illustrated Example

To demonstrate the use of the derived models, a hypothetical case is illustrated with the databases in Box 1. Let us consider a rectangular groundwater recharge basin in a depleted groundwater area. The recharge basin is meant for augmentation of groundwater resources. The basin has a catchment area, from which it receives surface inflows through runoffs from rainfall. To recover part of the aquifer storage created by recharge, groundwater withdrawal by four pumping wells located around the recharge basin, as in Figure 16.6, is taken into consideration. It is assumed that the pumped water shall be used for some beneficial purposes, viz. drinking and irrigation, and the scheme is a part of an IWRM strategy. It is intended to find the rate of groundwater recharge consequent to the hydrological interventions in the basin, the enhanced recharge due to pumping, and also the rise/fall of groundwater table beneath the basin due to the resulting effect of recharge and pumping.

A 243-day daily data of observed rainfall and the corresponding runoffs calculated using SCS-CN model as well as the observed evaporation rate data (Table 16.2) are used as inputs. The method for calculating runoffs by SCS-CN model is given in Annexure I. One can also use observed runoff. In the absence of the observed water surface evaporation rate, one can calculate evaporation rate based on the meteorological data, using an appropriate method (preferably, Priestley-Taylor method) given in Annexure II.

Making use of $T = 500$ m^2/day, $S = 0.01$, $a = 100$ m, and $b = 100$ m in Equation 16.4 and using Equations 16.8 and 16.9, the Hantush's discrete kernel coefficients are calculated by taking average of two locations: the one at the center and the other one at the extreme corner of the basin, as follows.

$$U_H(t) = \frac{t}{8\,S} \left[f(0,0,t) + f(a,b,t) \right] \tag{16.17}$$

Theis's pumping discrete kernel coefficients are calculated using Equations 16.10 and 16.11. For the calculation of the well function, $W(u)$, using Equation 16.6 corresponding to each well, the distance between the observation point and the well, r, is required to be

---

BOX 16.1   INPUT DATA

Dimension of the recharge basin = 200 m × 200 m, that is, $2a = 200$ m and $2b = 200$ m; maximum depth of the basin, $D_{max} = 3.5$ m; free board = 0.5 m; outlet at 0.5 m below the basin top; initial groundwater table above impervious stratum, $H = 50$ m; height of porous material below the basin, $h_0 = 8$ m; coordinates of the four pumping wells with reference to the origin at center of the basin = $P(1)$: (500, 70); $P(2)$: (70, –500); $P(3)$: (–500, –70); and $P(4)$: (–70, 500); pumping wells operate for 8 hours in a day, with pumping rate, $Q_p = 25$ m^3/h and pumping starts from 31 days; initial depth of water in the basin, $D(0) = 1$ m; vertical hydraulic conductivity, $K_v = 0.1$ m/day; transmissivity, $T = 100$ m^2/day; storage coefficient, $S = 0.1$; and time step size, $\Delta t = 1$ day.

The inflow rate, $Q_i(t)$, is considered variable, and the value calculated using SCS-CN model of a particular location corresponding to the rainfall of that location is used. The evaporation rate, $E(t)$, is also considered variable, and observed data of that specific location are used (Table 16.2).

TABLE 16.2

The Rainfall, $R_i(t)$; Runoff, $Q_i(t)$; and Evaporation Rate, $E(t)$ Data

Day	$R_i(t)$ (m)	$Q_i(t)$ (m³/day)	$E(t)$ (m/day)	Day	$R_i(t)$ (m)	$Q_i(t)$ (m³/day)	$E(t)$ (m/day)	Day	$R_i(t)$ (m)	$Q_i(t)$ (m³/day)	$E(t)$ (m/day)
1	0	0	0.0134	82	0.0038	239.4	0.0013	163	0	0	0.003
2	0	0	0.0146	83	0.0025	34.2	0.0019	164	0	0	0.0028
3	0	0	0.0108	84	0.0025	34.2	0.001	165	0	0	0.0033
4	0.0065	0	0.0108	85	0.0004	0	0.0019	166	0	0	0.0033
5	0	0	0.0071	86	0	0	0.0032	167	0.0036	0	0.0035
6	0.0016	0	0.0093	87	0	0	0.0047	168	0	0	0.0037
7	0	0	0.0109	88	0.0004	0	0.0023	169	0	0	0.0039
8	0.0186	4685.4	0.01	89	0.0846	77850.6	0.0015	170	0.0027	0	0.004
9	0.028	17886.6	0.0073	90	0.0302	20007	0.0038	171	0.0021	0	0.0031
10	0.0006	0	0.006	91	0	0	0.0013	172	0.0012	0	0.0029
11	0.0008	0	0.0039	92	0	0	0.0049	173	0	0	0.0033
12	0	0	0.0053	93	0.0056	775.2	0.0023	174	0	0	0.0036
13	0.021	11434.2	0.006	94	0	0	0.0037	175	0	0	0.0036
14	0.0766	69038.4	0.0048	95	0.0224	6919.8	0.0063	176	0	0	0.0033
15	0.0056	775.2	0.0035	96	0.0066	216.6	0.0016	177	0	0	0.0031
16	0.0012	0	0.0024	97	0.017	8048.4	0.003	178	0	0	0.0032
17	0	0	0.0052	98	0	0	0.0032	179	0	0	0.0033
18	0	0	0.0054	99	0	0	0.004	180	0	0	0.0032
19	0	0	0.0064	100	0	0	0.006	181	0	0	0.0034
20	0	0	0.0093	101	0.0088	649.8	0.0036	182	0	0	0.0029
21	0	0	0.0063	102	0	0	0.0039	183	0	0	0.0027
22	0.0428	32718	0.008	103	0	0	0.0043	184	0	0	0.003
23	0	0	0.0048	104	0	0	0.0044	185	0	0	0.0032
24	0.0036	193.8	0.0031	105	0	0	0.0037	186	0	0	0.0029
25	0.0144	6007.8	0.0022	106	0	0	0.0036	187	0	0	0.003
26	0	0	0.0029	107	0	0	0.0033	188	0	0	0.0031
27	0.0084	558.6	0.0028	108	0.0116	0	0.0022	189	0	0	0.0032
28	0	0	0.0027	109	0.042	31885.8	0.0027	190	0	0	0.0032
29	0.0008	0	0.0034	110	0	0	0.0024	191	0	0	0.0028
30	0	0	0.0052	111	0.011	3613.8	0.0035	192	0.0005	0	0.0021
31	0.0174	4047	0.0078	112	0.0352	24954.6	0.0039	193	0.0064	0	0.0022
32	0.0528	43228.8	0.009	113	0.0034	159.6	0.0026	194	0.0003	0	0.0028
33	0.013	4981.8	0.0019	114	0.0006	0	0.0028	195	0	0	0.0027
34	0	0	0.004	115	0.0004	0	0.0019	196	0	0	0.0028
35	0.0077	1698.6	0.0047	116	0	0	0.0039	197	0	0	0.0028
36	0	0	0.0032	117	0	0	0.0034	198	0	0	0.0029
37	0	0	0.0048	118	0	0	0.0046	199	0	0	0.0029
38	0	0	0.0053	119	0	0	0.0036	200	0.0004	0	0.003
39	0.0008	0	0.0062	120	0	0	0.0037	201	0	0	0.0035
40	0.037	17556	0.0065	121	0.0016	0	0.0033	202	0.0002	0	0.003
41	0.0014	0	0.0063	122	0.027	9963.6	0.0053	203	0.0018	0	0.0026
42	0.0798	72549.6	0.0026	123	0.0068	239.4	0.0035	204	0	0	0.0029
43	0.0044	387.6	0.0042	124	0.013	4981.8	0.0018	205	0	0	0.003
44	0.017	8048.4	0.0029	125	0.0082	1960.8	0.0016	206	0	0	0.003

(Continued)

TABLE 16.2 (*Continued*)

The Rainfall, $R_i(t)$; Runoff, $Q_i(t)$; and Evaporation Rate, $E(t)$ Data

Day	$R_i(t)$ (m)	$Q_i(t)$ (m³/day)	$E(t)$ (m/day)	Day	$R_i(t)$ (m)	$Q_i(t)$ (m³/day)	$E(t)$ (m/day)	Day	$R_i(t)$ (m)	$Q_i(t)$ (m³/day)	$E(t)$ (m/day)
45	0.004	285	0.0015	126	0.0092	2508	0.0047	207	0	0	0.0028
46	0.0008	0	0.0014	127	0.008	1858.2	0.003	208	0	0	0.0029
47	0.0028	0	0.0021	128	0	0	0.0032	209	0	0	0.0032
48	0.0196	10225.8	0.0032	129	0	0	0.0031	210	0	0	0.0029
49	0	0	0.0038	130	0.002	0	0.0042	211	0	0	0.0028
50	0.0074	342	0.0053	131	0.002	0	0.0023	212	0.0013	0	0.0031
51	0.0066	216.6	0.002	132	0	0	0.0032	213	0	0	0.0028
52	0.0074	1550.4	0.0027	133	0.0004	0	0.0029	214	0	0	0.0025
53	0.055	45577.2	0.0024	134	0	0	0.0035	215	0	0	0.0031
54	0.0062	1014.6	0.003	135	0.004	0	0.0052	216	0.0004	0	0.0031
55	0.0012	0	0.0028	136	0.0002	0	0.0028	217	0.0008	0	0.003
56	0.011	3613.8	0.0022	137	0	0	0.0036	218	0	0	0.003
57	0.013	4981.8	0.0004	138	0	0	0.0031	219	0	0	0.003
58	0.0002	0	0.0016	139	0	0	0.0033	220	0	0	0.003
59	0.0016	0	0.0026	140	0	0	0.0039	221	0	0	0.0026
60	0.0562	46865.4	0.0021	141	0.006	0	0.0035	222	0	0	0.0028
61	0.1224	119973.6	0.0016	142	0.0003	0	0.0041	223	0	0	0.0029
62	0.0138	5563.2	0.0022	143	0.0001	0	0.0039	224	0	0	0.003
63	0.0294	19231.8	0.0016	144	0	0	0.0037	225	0	0	0.0029
64	0.0014	0	0.003	145	0.0006	0	0.0038	226	0	0	0.0028
65	0	0	0.0035	146	0.0034	0	0.0037	227	0	0	0.003
66	0.0326	22355.4	0.0048	147	0.0037	0	0.0042	228	0	0	0.0026
67	0.0038	239.4	0.0041	148	0.0008	0	0.0042	229	0	0	0.0029
68	0.0638	55062	0.0038	149	0	0	0.0042	230	0	0	0.0031
69	0	0	0.0035	150	0	0	0.0042	231	0	0	0.0033
70	0.01	2986.8	0.0031	151	0	0	0.003	232	0	0	0.0036
71	0.0082	1960.8	0.0055	152	0	0	0.0038	233	0	0	0.0035
72	0	0	0.003	153	0.001	0	0.0032	234	0	0	0.0035
73	0.0016	0	0.0022	154	0	0	0.0032	235	0	0	0.0035
74	0.0208	11263.2	0.0016	155	0.0005	0	0.0032	236	0	0	0.0032
75	0	0	0.0045	156	0.001	0	0.0024	237	0	0	0.0034
76	0	0	0.0024	157	0.0001	0	0.0026	238	0	0	0.0041
77	0.0092	752.4	0.0043	158	0	0	0.0033	239	0	0	0.0035
78	0.0398	29617.2	0.0048	159	0	0	0.0035	240	0	0	0.0031
79	0.0698	61582.8	0.0023	160	0	0	0.0034	241	0	0	0.0033
80	0.0374	27166.2	0.0024	161	0	0	0.0034	241	0	0	0.0034
81	0.0374	27166.2	0.001	162	0	0	0.0033	243	0	0	0.0037

calculated. Here, the aim is to calculate the average resulting drawdown of water levels consequent to the pumping of four wells with respect to the origin and four corners of the basin. The distances corresponding to the observation points in respect of the four pumping wells are calculated, as given in Table 16.3.

Having calculated the respective distances by expressions given in Table 16.3, using coordinates of the pumping wells as given in Figure 16.6b (i.e., in accordance with the sign of coordinate system), the well functions corresponding to the respective observation points

TABLE 16.3

Distances of the Pumping Wells with Respect to the Observation Points

For Origin (0,0)	For Corner (a, b)	For Corner (a, −b)	For Corner (−a, −b)	For Corner (−a, b)
$r_{O1} = \sqrt{(a_1)^2 + (b_1)^2}$	$r_{C1} = \sqrt{(a_1 - a)^2 + (b - b_1)^2}$	$r_{C1} = \sqrt{(a_1 - a)^2 + (b + b_1)^2}$	$r_{C1} = \sqrt{(a + a_1)^2 + (b + b_1)^2}$	$r_{C1} = \sqrt{(a + a_1)^2 + (b - b_1)^2}$
$r_{O2} = \sqrt{(a_2)^2 + (b_2)^2}$	$r_{C2} = \sqrt{(a - a_2)^2 + (b - b_2)^2}$	$r_{C2} = \sqrt{(a - a_2)^2 + (-b_2 - b)^2}$	$r_{C2} = \sqrt{(a + a_2)^2 + (-b_2 - b)^2}$	$r_{C2} = \sqrt{(a + a_2)^2 + (b - b_2)^2}$
$r_{O3} = \sqrt{(a_3)^2 + (b_3)^2}$	$r_{C3} = \sqrt{(-a_3 + a)^2 + (b - b_3)^2}$	$r_{C3} = \sqrt{(-a_3 + a)^2 + (b + b_3)^2}$	$r_{C3} = \sqrt{(-a_3 - a)^2 + (b + b_3)^2}$	$r_{C3} = \sqrt{(-a_3 - a)^2 + (b - b_3)^2}$
$r_{O4} = \sqrt{(a_4)^2 + (b_4)^2}$	$r_{C4} = \sqrt{(a - a_4)^2 + (b_4 - b)^2}$	$r_{C4} = \sqrt{(-a_4 + a)^2 + (b_4 + b)^2}$	$r_{C4} = \sqrt{(a + a_4)^2 + (b + b_4)^2}$	$r_{C4} = \sqrt{(a + a_4)^2 + (b_4 - b)^2}$

(r_{O1} represents with respect to origin for pumping well, $P(1)$, r_{C1} represents with respect to corner for pumping well, $P(1)$, and so on; coordinates of the pumping well should be as shown in Figure 16.6b)

are calculated. The pumping discrete kernel coefficients corresponding to the well functions of five observation points (center and four corners) are generated by using Equations 16.11 and 16.12. The resulting discrete kernel coefficients are obtained by taking average of discrete kernel coefficients of five locations for all the wells, as follows:

$$U_p\left(\xi, n\Delta t\right) = \sum_{\gamma=1}^{n} \sum_{\xi=1}^{\xi_{max}} \frac{1}{5} \sum_{j=1}^{5} U_p\left(j, \xi, n\Delta t\right) \tag{16.18}$$

where:
  j is the location number
  ξ is the well number

16.4.1 Results and Discussion

The dimensionless plot of variation of groundwater recharge $\left(Q_{gw}(\gamma\Delta t)/(4\,ab\,K_v)\right)$ versus time $\left((\gamma\Delta t)/t\right)$, with and without pumping wells, is shown in Figure 16.7. The dimensionless plot of variation of depth of water in the recharge basin $\left(D(\gamma\Delta t)/D_{max}\right)$ versus time $\left((\gamma\Delta t)/t\right)$, with and without pumping, is shown in Figure 16.8. The cumulative inflows, outflows, rainfall, total evaporation, and groundwater recharge are shown in Figure 16.9. The recharge component (Figure 16.7) continued till there was water in the basin (Figure 16.8) and it ceased after the basin got emptied. Figure 16.7 also clearly showed that because of pumping, the groundwater recharge component increased more than that of without pumping and the recharge with pumping ceased much before than without pumping because of the nonavailability of water in the basin (Figure 16.8). The variation of depth of water in the basin also confirmed the fast reduction of water depth due to pumping. Figure 16.9 can be regarded as the account of the water balance components of the basin. The performance of the model showed that it can be used as a tool for conjunctive use

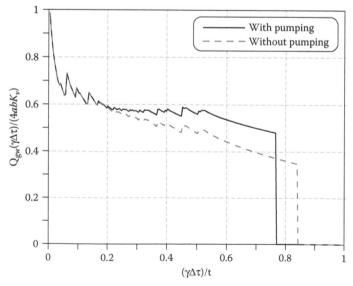

FIGURE 16.7
Dimensionless plot of variation of groundwater recharge with and without pumping.

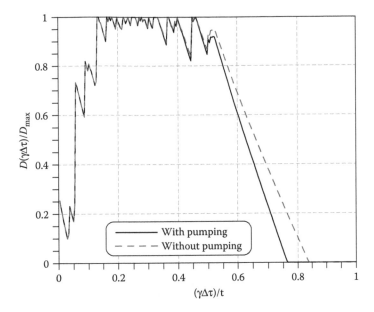

FIGURE 16.8
Dimensionless plot of variation of depth of water in the recharge basin with and without pumping.

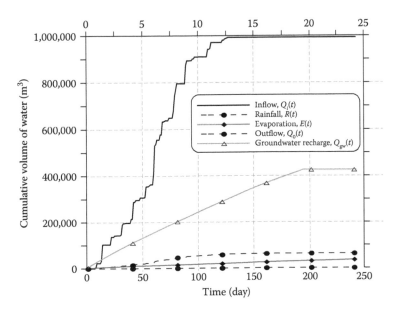

FIGURE 16.9
The cumulative variation of hydrological components; inflows to, rainfall over, outflows and evaporation from, and groundwater recharge from the basin.

management of surface and groundwater. The position selection of pumping well(s) is an important criterion. Locating pumping wells near to the recharge basin can help enhance the groundwater recharge rate, but there can be a chance of associating risk of less treatment of recharged water because of inadequate travel time.

16.5 Concluding Remarks

Integrated water resources management (IWRM) is an empirical concept. Its successful implementation would require involvement of a number of stakeholders, right from policy makers to direct beneficiaries. Many countries are yet to decide the methodology for quantification of water required for environmental sustainability. The social equity on water is another area of concern, on which many countries will have to evolve and adopt a rational approach. Climate change impact on water resources is an added complexity to the problem solving. A holistic river basin approach can help resolve many of the water-related issues.

As a scientific tool for conjunctive management of surface and groundwater, including achieving improved groundwater quality on similar line to the managed aquifer recharge (MAR) and aquifer storage recovery (ASR), a numerical model has been presented. The model can be integrated into the prospect of IWRM.

Notations

The notations used in the article have been explained wherever they appear first; however, for the benefits of the reader, these are elaborated here:

A	gross surface area of the recharge basin
$A_i(t)$	water surface area of the recharge basin at time t
a	half of the length of the recharge basin
b	half of the width of the recharge basin
$D(t)$	depth of water in the basin, measured from its bed
$E(t)$	water surface evaporation rate at time t
$f(x, y, t)$	analytical function derived by Hantush for rectangular recharge basin
H	height of the initial groundwater table measured upward from the impervious stratum
h_0	height of the porous material between the basin bed and the initial groundwater table
$h(x, y, t)$	rise in groundwater table below the rectangular spreading basin at location x and y and time, t
K	coefficient of permeability (i.e., hydraulic conductivity) of the aquifer material
K_v	vertical hydraulic conductivity of the subsurface formation
$P(*)$	pumping well number
$Q_i(t)$	inflow rate from the catchment to the recharge basin at time t
$Q_0(t)$	outflow rate from the basin at time t
$Q_{gw}(t)$	groundwater recharge rate from the basin at time t
Q_p	constant pumping rate by the well

$R(t)$ rainfall rate at time t

r distance between the observation point and the well

S storage coefficient of the aquifer

s drawdown of the water table

T aquifer transmissivity

t time

$U.(*)$ unit step response functions

w constant rate of percolation per unit area

$W(u)$ Theis well function, defined by an infinite series

$\Delta h(t)$ rise/fall of the groundwater table height

$\Delta V(t)$ change in storage between time t and $t + \Delta t$

Δt time step size

$\delta.(*)$ discrete kernel coefficients or unit pulse response functions

γ an integer number

ξ an integer number

Appendix-I: SCS-CN Model

The runoff yield, Q by the SCS-CN model (SCS 1993; Mishra and Singh 1999) is given by

$$Q = \frac{(P - \lambda S)^2}{P + (1 - \lambda)S} \quad \text{for } P > \lambda S \tag{A.1}$$

$$Q = 0 \quad \text{for } P \leq \lambda S$$

$$S = \frac{25400}{CN} - 254 \tag{A.2}$$

where:

 S is the maximum potential retention (mm)

 λ is the initial abstraction weight as a fraction of S, normally $0 \leq \lambda \leq 0.3$, but conventionally 0.2; 25400 and 254 in Equation AI:2 are arbitrary constants in unit of S

 CN is the Curve Number, dimensionless

Theoretically, S varies between 0 to ∞ for the CN ranges from 100 to 0. The $CN = 100$ represents $S = 0$ an impermeable watershed. Conversely, the $CN = 0$ represents $S = \infty$ an infinitely abstracting watershed.

Substituting S and $\lambda = 0.2$ in Equation A.1 yields to:

$$Q = \frac{25.4\left[\dfrac{P}{25.4} - \dfrac{200}{CN} + 2\right]^2}{\left[\dfrac{P}{25.4} + \dfrac{800}{CN} - 8\right]}; \quad \text{valid for } P \geq 0.2S \tag{A.3}$$

The watershed specific-CNs relating to the antecedent moisture condition (AMC) (SCS 1985; Lewis, Singer and Take 2000) is calculated as

$$CN_I = \frac{4.2\,CN_{II}}{10 - 0.058\,CN_{II}} \tag{A.4}$$

$$CN_{III} = \frac{23\,CN_{II}}{10 - 0.13\,CN_{II}} \tag{A.5}$$

where, subscripts indicate the AMC, I for dry, II normal, and III wet condition.

Appendix-II: Evaporation Estimate Models

Pan Evaporation (PE) Method

Annual or seasonal evaporation rate can be estimated using standard pan evaporation (Abtew 2001) data, as

$$E = K_p E_{pan} \tag{A.1}$$

where:
  E is the evaporation rate (mm/day)
  K_p is the pan coefficient, defined as the ratio of the theoretically free surface evaporation
       to the pan evaporation
  E_{pan} is the pan evaporation (mm/day)

Mass Transfer (MT) Method

The mass transfer method (Harbeck 1962) is expressed as

$$E = \mu\,U_2\left(e_s - e_a\right) \tag{A.2}$$

where:
  E is the evaporation rate (mm/day)
  μ is the mass transfer coefficient (mm s m^{-1} kPa^{-1})
  U_2 is the wind speed at 2 m above the water surface (ms^{-1})
  e_s is the saturated vapor pressure at water temperature (kPa)
  e_a is the actual vapor pressure at air temperature (kPa)

$$\mu = 2.909 A^{-0.05} \tag{A.3}$$

where A is the surface area of water body (km^2).

The daily wind speed measured at z m above ground surface is converted at 2 m height by using the equation:

$$U_2 = U_Z \left[\frac{4.87}{\ln(67.8z - 5.42)} \right] \quad \text{(A.4)}$$

where:
  U_z is the wind speed at z m above the ground surface (ms^{-1})
  z is the height of measurement of wind speed above the ground surface

The saturated vapor pressure (e_s) can be estimated by

$$e_s = 0.6108 \exp \left[\frac{17.269 T_w}{T_w + 237.3} \right] \quad \text{(A.5)}$$

The actual vapor pressure (e_a) can be estimated by using the relative humidity and air temperature, as follows:

$$e_a = \frac{e^0(T_{min}) \dfrac{RH_{max}}{100} + e^0(T_{max}) \dfrac{RH_{min}}{100}}{2} \quad \text{(A.6)}$$

where:
  $e^0(T_{max})$ is the saturated vapor pressure at maximum air temperature (kPa)
  $e^0(T_{min})$ is the saturated vapor pressure at minimum air temperature (kPa)
  RH_{max} is the maximum relative humidity (%)
  RH_{min} is the minimum relative humidity (%)

Priestley-Taylor (PT) Method

Priestley-Taylor method (1972) is a simplified form of the combined equations of energy balance and aerodynamic, where the aerodynamic component is left out, but a coefficient, α, value of 1.26 is included as a multiplier. The Priestley-Taylor method (Ries and Dias 1998; Abtew 2001) is given by

$$E = \frac{1}{\lambda} \alpha \left[\frac{\Delta}{\Delta + \gamma} \right] (R_n - G) \quad \text{(A.7)}$$

where:
  E is the evaporation rate estimated by the Priestley-Taylor method (mm/day)
  α is the dimensionless proportionality parameter or Priestley-Taylor's coefficient, equal to 1.26
  Δ is the slope of saturated vapor pressure at water temperature versus temperature curve (kPa °C^{-1})
  γ is the psychometric constant (kPa °C^{-1})
  λ is the latent head of evaporation of water, generally taken as 2.45 MJ/kg
  R_n is the net radiation on the water surface (MJ/m^2/day)
  G is the head gained or lost by the upper layer of water body (MJ/m^2/day)

The parameter α is calculated as

$$\alpha = \frac{\Delta + \lambda}{\Delta\,(1 + \beta)} \tag{A.8}$$

The parameter Δ can be obtained by

$$\Delta = \frac{4098 \left[0.6108 \exp\left(\dfrac{17.27 T_w}{T_w + 237.3} \right) \right]}{\left(T_w + 237.30 \right)^2} \tag{A.9}$$

The psychometric constant (γ) can be computed by

$$\gamma = \frac{C_{pa} P}{0.622 \times 1000 \times \lambda} \tag{A.10}$$

where:
C_{pa} is the specific heat capacity (= 1.013 MJ/kg/°C)
0.622 represents the ratio of water to dry air molecular weights (= 18.016/28.996)
P is the atmospheric pressure (kPa)

The atmospheric pressure at elevation z can be calculated as

$$P = 101.3 \left(\frac{293 - 0.0065\,z}{293} \right)^{5.26} \tag{A.11}$$

z is the elevation in meter.
The Bowen ratio, β, can be calculated by

$$\beta = 0.665 \times 10^{-3} P \left(\frac{T_w - T_a}{e_s - e_a} \right) \tag{A.12}$$

where:
P is the atmospheric pressure, (kPa)
T_w is the water temperature at the water surface (°C)
T_a is the air temperature above the water surface (°C)
e_s is the saturated vapor pressure at water temperature (kPa)
and e_a is the actual vapor pressure at air temperature (kPa)

References

Abtew W. 2001. Evaporation estimation for Lake Okeechobee in south Florida. *Journal of Irrigation and Drainage Engineering*, American Society Civil Engineer. 127(3):140–146.
Biswas Asit K. 2004. Integrated water resources management: A reassessment: A water forum contribution, *Water International*. 29(2):248–256. https://www.adb.org/Documents/Books/AWDO/2007/dp05.pdf, accessed April, 2016.
Central Ground Water Board (CGWB). 1996. *National Perspective Plan for Recharge to Groundwater by Utilizing Surplus Monsoon Runoff*, Ministry of Water Resources, Govt. of India, New Delhi.

Central Ground Water Board (CGWB). 2002. *Master lan for Artificial Recharge to Ground Water in India*, Ministry of Water Resources, Govt. of India, New Delhi. p. 117.

Central Ground Water Board (CGWB). 2007. *Mannual on Artificial Recharge to Ground Water*, Ministry of Water Resources, Govt. of India, New Delhi. p. 185. http://cgwb.gov.in/documents/Manual%20on%20Artificial%20Recharge%20of%20Ground%20Water.pdf, accessed May, 2016.

Central Ground Water Board (CGWB). 2007. Groundwater scenarios in major cities of India, https://www.cgwb.gov.in/documents/GW-Senarioin%20cities-May2011.pdf.

Central Ground Water Board (CGWB). 2013. *Master Plan for Artificial Recharge to Ground Water in India*, Ministry of Water Resources, Govt. of India, New Delhi. p. 208.

Dillon P. J. 2005. Future management of aquifer recharge, *Hydrogeology Journal*. 13(1):313–316.

Dillon P. J., Pavelic P., Page D., Beringen H., and Ward J. 2009. *Managed Aquifer Recharge: An Introduction*, Waterlins Report Series No. 13, Canberra, Australia: National Water Commission. p. 37.

EPA Victoria. 2009. *Guidelines for Managed Aquifer Recharge (MAR)-Health and Environmental Risk Management*, Publication 1290, p. 21. http://www.epa.vic.gov.au/our-work/publications/publication/2009/july/1290, accessed April, 2016.

Ghosh N. C., Kumar S., Grützmacher G., Ahmed S., Singh S., Sprenger C., Singh R. P., Das B., and Arora T. 2015. Semi-analytical model for estimation of unsteady seepage from a large water body influenced by variable flows, *Journal Water Resources Management, Springer*, 29(9):3111–3129.

Gunawardena E. R. N. 2010. *Operationalizing IWRM through River Basin Planning and Management*, Lecture note prepared for Training of Trainers in Integrated Water Resources Management held at Kandy, Sri Lanka from 16–25 September 2010. https://www.saciwaters.org/CB/iwrm/Lecture%20Notes/8.3.%20Restraining%20conflicst%20through%20institutional%20interventions_Gunawardena%20ERN.pdf accessed March, 2016.

GWP (Global Water Partnership) and the International Network of Basin Organizations (INBO). 2009. A handbook for integrated water resources management in basins, ISBN: 978-91-85321-72-8, p. 103. https://www.gwpforum.org, accessed March, 2016.

Hantush M. S. 1967. Growth and decay of groundwater mounds in response to uniform percolation. *Water Resources Research*, 3:227–234.

Harbeck G. E. 1962. *A Practical Field Technique for Measuring Reservoir Evaporation Utilizing Mass-transfer Theory*. U. S. Geological Survey Professional Paper, Washington, DC.

IPCC (Intergovernmental Panel on Climate Change). 2007. Climate Change-2007: Synthesis Report. Working Group I, II &III contribution to fourth Assessment Report of the IPCC. p. 73. https://www.ipcc.ch/pdf/assessment-report/ar4/syr/ar4_syr.pdf.

IPCC (Intergovernmental Panel on Climate Change). 2014. Climate Change-2014: Impacts. Adaptation, and Vulnerability: Summary for Policy Makers. Working Group II contribution to Fifth Assessment Report of the IPCC. p. 32. https://www.ipcc-wg2.gov/AR5/images/uploads/WG2AR5_SPM_FINAL.pdf, accessed April, 2016.

Mishra S. K. and Singh V. P. 1999. Another look at SCS-CN method, *Journal Hydrological Engineering ASCE*, 4(3), 257–264.

Priestley C. H. B. and R. J. Taylor. 1972. On the assessment of surface heat flux and evaporation using large scale parameters. *Monthly Weather Review*, 100: 81–92.

Ranade Ashwani A., Singh Nityananda, Singh H. N. and Sontakke N. A. 2007. Characteristics of hydrological wet season over different river basins of India. IITM, Pune, ISSN 0252-1075. Research Report No RR-119, p. 155.

Reis R. J. D. and Dias N. L. 1998. Multi-season lake evaporation: energy budget estimates and CRLE model assessment with limited meteorological observations. *Journal of Hydrology*, 208: 135–147.

SAPH PANI. 2012. Report on existing MAR practice and experience in India, especially in Chennai, Maheshwaram, Raipur. Project supported by the European Commission within the Seventh Framework Programme Grant agreement No. 282911. p. 86.

SCS. 1993. *Hydrology-National Engineering Handbook, Supplement A, Section 4, Chapter 10, Soil Conservation Service*. United State Department of Agriculture (USDA), Washington, DC.

Sivakumar M. V. K. and Stefanski R. 2011. Climate Change in South Asia, R. Lal et al., eds., *Climate Change and Food Security in South Asia*, Springer. DOI: 10.1007/978-90-481-9516-9_2.

Theis C. 1935. The relation between the lowering of the piezometric surface and the rate and duration of discharge of a well using groundwater storage. *Transactions of the American Geophysical Union*, 16:519–524.

The United Nations World Water Assessment Programme. Introduction to the IWRM guidelines at River Basin level, Dialogue paper, UN Educational, Scientific and Cultural Organization. http://unesdoc.unesco.org/images/0018/001850/185074e.pdf, accessed March, 2016.

UNESCO. 2009. *Integrated Water Resources Management in Action: Dialogue Paper*, The United Nations Educational, Scientific and Cultural Organization Publications, ISBN:978-92-3-104114.3, p. 18. http://www.gwp.org/Global/ToolBox/About/IWRM/IWRM%20in%20Action%20(UNESCO, UNEP-DHI,%202009).pdf, accessed March, 2016.

17

Water Optimization in Process Industries

Elvis Ahmetović, Ignacio E. Grossmann, Zdravko Kravanja, and Nidret Ibrić

CONTENTS

ABSTRACT Global consumption of natural resources has been significantly increased over recent decades. Consequently, the research regarding sustainable utilization of natural resources, including water and energy, has received considerable attention throughout academia and industry. The main goals have been to find promising solutions, with reduced water and energy consumption within different sectors (i.e., domestic, agricultural, and industrial). Those solutions are also beneficial from the aspects of wastewater and emission minimization and protection of the environment. The focus of this chapter is on optimization of water consumption within the industrial sector, including process industries (i.e., chemical, food, petrochemical, pulp, and paper). This chapter first briefly presents the global water consumption and water usages within the process industries. Then, a concept of process water networks involving wastewater reuse, wastewater regeneration, and reuse/recycling is explained, followed by a brief description of systematic methods, based on water-pinch analysis and mathematical programming. An illustrative large-scale case study of the total water network, including multiple contaminants, is used to demonstrate a superstructure-based optimization approach. The results of the optimal water network show that significant savings of freshwater consumption and wastewater generation can be obtained when compared with a conventional water network design.

17.1 Introduction

Global water consumption. Water is a valuable natural resource, which is used for different purposes in daily life and various sectors, namely, domestic, agricultural, and industrial. Average global water consumption within those sectors varies, and it depends on the development level of countries. In developed and industrialized countries (i.e., United

States, Germany, France, and Canada), the average industrial water usage varies between 46% and 80%, while in developing countries (i.e., India and Brazil), it is between 4% and 17% (Davé 2004). The predictions show that the populations' needs for freshwater as well as for different products obtained within industrial and agricultural sectors will increase. According to the predictions of global water consumption, the important issues that can be addressed in the future are rational utilizations of natural resources and sustainable water management within the above-mentioned sectors.

Water use within the process industries. In a typical industrial process, water is used for different purposes (i.e., washing, extraction, absorption, cooling, and steam production). After using water within a process, wastewater is generated and discharged into the environment. Generally speaking, large amounts of freshwater are consumed, and, consequently, large amounts of wastewater are generated within industrial processes. Accordingly, freshwater usage and wastewater generation should be minimized for achieving sustainable and more efficient processes. Some of the works focused on freshwater and wastewater minimization considering separate networks (Wang and Smith 1994a,b; Wang and Smith 1995; Kuo and Smith 1997; Galan and Grossmann 1998; Castro et al. 2009), namely, process water-using network and/or wastewater treatment network. Also, an overall network consisting of process water-using and wastewater treatment networks (Takama et al. 1980; Huang et al. 1999; Gunaratnam et al. 2005; Karuppiah and Grossmann 2006; Ahmetović and Grossmann 2011; Faria and Bagajewicz 2011) was the focus of the research, in order to explore additional freshwater and wastewater minimization opportunities. In other words, in the process, its overall water system was studied for identifying process units, which consume water and/or generate wastewater. The main research challenges were devoted to systematically exploring all water integration opportunities within the process in order to achieve solutions with a reduced water usage and wastewater generation. It is important to mention that the synthesis problem of an overall or total water network is more complex, and this problem has been addressed in fewer papers compared with the synthesis problem of separate networks (water-using network or wastewater treatment network).

In order to address the synthesis problems of separate and total water networks, different tools and targeting techniques based on water-pinch analysis, namely, the limiting composite curve (Wang and Smith 1994a,b), the water surplus diagram (Hallale 2002), and the water cascade analysis (Manan et al. 2004), and methods based on mathematical programming (Takama et al. 1980).

(Takama et al. 1980; Doyle and Smith 1997; Alva-Argáez et al. 1998; Yang and Grossmann 2013) have been proposed. Over recent decades, it has been shown that water-pinch technology (Foo 2009) and mathematical programming (Bagajewicz 2000; Jeżowski 2010) are very useful approaches for solving industrial water reuse and wastewater minimization problems. Those approaches are used for determining the minimum freshwater usage and wastewater generation before designing the network (targeting phase) as well as for synthesizing the water network (design phase). It has been demonstrated that by applying systematic approaches, very promising solutions can be obtained, with the increased water reuse (18.6%–37%) within different processes and very short payback times (0–10 months) (Mann and Liu 1999). Accordingly, it is worth pointing out that a significant progress has been made within this field. The reader is referred to several review papers (Bagajewicz 2000; Yoo et al. 2007; Foo 2009; Jeżowski 2010; Klemeš 2012; Grossmann et al. 2014; Khor et al. 2014; Ahmetović et al. 2015), and books (Mann and Liu 1999; Smith 2005; Klemeš et al. 2010; El-Halwagi 2012; Klemeš 2013; Foo 2013) for more details about water network synthesis and recent progress within this field.

17.2 Concepts of Water Use and Water Networks within an Industrial Process

Figure 17.1 shows typical water users and water treatment within an industrial process (Mann and Liu 1999). Raw water, usually taken from lakes, rivers, or wells, is first treated in raw-water treatment units in order to be purified and used within the process in water-using units, steam boiler, and cooling tower (deaeration, demineralization, deionization, dealkalization, pH control, and so on). After using freshwater within the process in water-using units, cooling tower, and steam boiler, wastewater is generated and directed to a final wastewater treatment. The main role of the wastewater treatment is to remove contaminants from wastewater and to satisfy environmental constraints on the effluent discharged into the environment. Also, note that water from wastewater treatment can be reused or recycled within the process (not shown in Figure 17.1) and that in this way, freshwater consumption and wastewater generation can be significantly reduced.

Figure 17.2 shows those concepts of water networks. A water network problem is a special case of a mass exchanger network (MEN) problem (El-Halwagi and Manousiouthakis 1989). Figure 17.2a presents a process water network consisting of water-using units. Within this network, mass loads of contaminants are transferred from process streams to water streams and wastewater is generated within a process water network. Mass loads of contaminants transferred to the water streams are usually too small when compared with the water flow rate. In such cases, it can be assumed that the inlet and the outlet water flows of water-using units are the same. Also, it should be mentioned that wastewater can be generated by utility systems, a boiler, and a cooling tower (Figure 17.1). In those systems, parts of wastewater, namely, boiler blow-down and cooling tower blow-down, must be periodically discharged (Allegra et al. 2014) and make-up water introduced, in order to keep the water quality at certain acceptable levels. There is loss of water within the cooling tower owing to water evaporation (Figure 17.1), which cannot be reused again within a process. In the cooling tower, there is no direct contact between process and water streams. The contaminants' concentrations within this unit

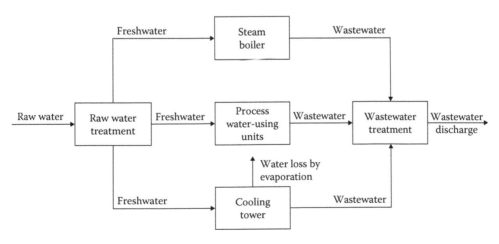

FIGURE 17.1
Water use and wastewater treatment within an industrial process.

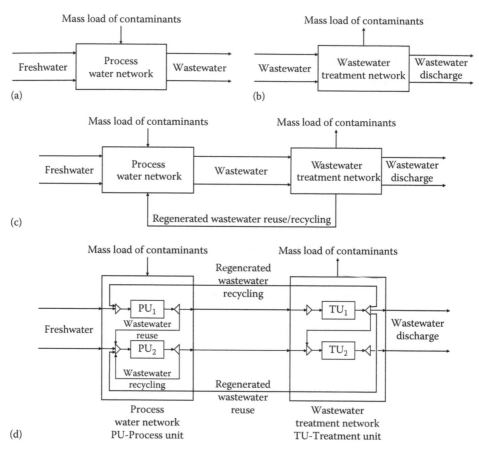

FIGURE 17.2
Concepts of water networks: (a) Process water network, (b) wastewater treatment network, (c) combined process water network and wastewater treatment network, and (d) different water integration options within the combined network.

are increased owing to water evaporation. Within the process, water network, freshwater consumption, and wastewater generation can be minimized by water reuse and the local recycling of water. Figure 17.2b shows a wastewater treatment network. It consists of wastewater treatment units that remove contaminants from wastewater. The removed mass loads of contaminants from wastewater are usually too small when compared with the wastewater flow rates, and in such cases, it can be assumed that the inlet and the outlet wastewater flows of wastewater treatment units are the same. A wastewater treatment network can be a centralized or a distributed system (Galan and Grossmann 1998; Zamora and Grossmann 1998). In the centralized system, consisting of more wastewater treatment units, all wastewater streams from different processes are mixed and directed to wastewater treatment. In this case, the total flow of wastewater streams goes through each wastewater treatment unit. However, in the distributed system, consisting of more wastewater treatment units, all flows of wastewater streams are not treated within each wastewater treatment unit. Consequently, the total cost of the distributed wastewater treatment network can be decreased compared with the centralized wastewater treatment network.

It should be mentioned that different technologies can be used for removing contaminants, and in most cases, wastewater treatment models are based on the fixed removal ratios of contaminants. However, also note that more realistic models (Yang et al. 2014) have been recently proposed for wastewater treatment units.

A process water network (Figure 17.2a) and a wastewater treatment network (Figure 17.2b) can be integrated, thus producing an overall or total network (Figure 17.2c). The following water integration opportunities are enabled within this network (see Figure 17.2d):

- Wastewater reuse (direct wastewater reuse from process unit 1 [PU_1] to process unit 2 [PU_2])
- Wastewater recycling (wastewater recycling within PU_2)
- Wastewater regeneration and recycling (wastewater from PU_1 is regenerated within wastewater treatment unit 1 [TU_1] and recycled to the same unit [PU_1], where it has previously been used)
- Wastewater regeneration and reuse (wastewater from PU_1 is regenerated within TU_1 and reused in PU_2)

Note that in the case of wastewater regeneration and reuse, wastewater from TU_1 does not enter the same unit (PU_1), where it has previously been used. By solving the overall network (Figure 17.2c, d), an improved solution can be obtained simultaneously when compared with the overall network obtained by sequential solutions of networks shown in Figure 17.2a, b. This will be shown later in this chapter, as demonstrated on an illustrative case study.

17.3 Systematic Methods for Water Network Design

This section presents a brief description of systematic methods based on water-pinch analysis and mathematical programming, which are used for water network design. Water-pinch technology/analysis (Wang and Smith 1994a,b; Wang and Smith 1995) is a graphical method, which represents an extension of pinch analysis for heat integration (Linnhoff and Hindmarsh 1983). It consists of two phases, namely, targeting and design. Assuming a single contaminant, the main goal of the targeting phase is to determine the minimum freshwater consumption (maximum water reuse) (Doyle and Smith 1997) before a water network design, while within the design phase, a water network is constructed, satisfying the minimum freshwater consumption. Also, water-targeting models for multiple contaminants have been proposed, based on mathematical programming, in order to perform simultaneous flow-sheet optimization (Yang and Grossmann 2013).

The mathematical programming approach is based on a water network superstructure optimization (Takama et al. 1980). A superstructure includes all feasible alternatives, from which the best is selected. The main steps within the mathematical programming approach (Biegler et al. 1997) are to develop a superstructure of alternative designs, develop a superstructure optimization model, solve the model in order to extract the optimum design from the superstructure, and analyze the obtained results. This approach can easily deal with multiple contaminant problems, different constraints (i.e., forbidden connections), and trade-offs between investment and operational costs, when compared with water-pinch analysis, which can have difficulties when addressing those issues, especially for large-scale problems. In some cases, it can be very useful to combine both approaches in order to solve the overall water network synthesis problem.

17.4 Steps of Mathematical Programming Approach for Water Network Design

As will be shown later, the mathematical programming approach was applied for the synthesis of water networks consisting of process water-using and wastewater treatment units. Accordingly, this section describes the main steps of the mathematical programming approach for the synthesis of water networks, namely, problem formulation, superstructure development, optimization model formulation, and solution strategies' development. The reader is referred to the paper (Ahmetović and Grossmann 2011) for more details regarding the superstructure, model, and solution strategy described within this chapter. Here, only a brief description is given.

Problem formulation. The first step in the application of the mathematical programming approach for the synthesis of water networks is a problem formulation. The problem formulation of the water network synthesis problem can be stated as follows. Given is a set of water sources, a set of water-using units, and a set of wastewater treatment operations. For the set of water sources, concentrations of contaminants within water sources and the cost of water are specified. Water-using units can operate with fixed or variable flow rates. In the case of the fixed flow rates through water-using units, for the set of process water-using units, the fixed flow rates, maximum concentrations of contaminants within inlet streams at process units, and mass loads of contaminants within the process units are given. However, in the case that the flow rate of water is not fixed, the maximum concentrations of contaminants within inlet and outlet streams of process units and mass loads of contaminants within process units are given. For a set of wastewater treatment units, percentage removals (removal ratios) of each contaminant within wastewater treatment units are specified and corresponding cost relationships between investment and operation cost, which depend on the flow rates of the wastewater treated within treatment units. Wastewater discharged from the network has to satisfy environmental constraints specified by regulations. Accordingly, the maximum acceptable level of concentrations of contaminants within effluent is given. The problem formulation can be extended to include a set of water demand units and a set of water source units (non-mass transfer operations), as presented in Ahmetović and Grossmann (2011). In that case, flow rates and contaminant concentrations of those units are usually specified.

The main goal of the water network synthesis problem is to determine the minimum freshwater usage and wastewater generation, the interconnections, flow rates, and contaminants' concentrations within each stream of the water network. The objective can be formulated as the minimization of the total annualized cost of the water network, which consists of the cost of freshwater usage, the cost of wastewater treatment, the cost of piping, and the cost of pumping water through pipes. It is assumed that the water network operates continuously under isothermal and isobaric conditions. However, those assumptions can be easily relaxed and the model extended (Ahmetović et al. 2014).

Superstructure development. On the basis of problem formulation, a superstructure should be synthesized, including all feasible network alternatives. Figure 17.3 shows a superstructure consisting of freshwater sources, process water-using units, wastewater treatment units, internal water source units, and water demand units (Ahmetović and Grossmann 2011). This superstructure incorporates both the mass-transfer and non-mass-transfer operations. In the mass-transfer process operations (process units [PU_p]), there is a direct contact between a contaminant-rich process stream and a contaminant-lean water stream. In that case, during the mass-transfer processes, the contaminants mass load $LPU_{p,j}$ (pollutant) is transferred from the process streams to the water. The contaminant concentration within the process stream is reduced, while the contaminant concentration within the water stream increases. The main purpose of water use in mass-transfer operations (i.e., absorption and washing) is to remove contaminants. However, in non-mass-transfer operations (i.e., cooling tower, boiler, and reactor), the main purpose of water use is not to remove contaminants. In those cases, water can be required, for example, in a cooling tower, owing to a water loss by evaporation, or in a reactor, owing to a water demand by chemical reaction. In the case that there is a loss of water in a unit, water cannot be reused from that unit in other water-using operations. That unit is represented by a water demand unit (DU_d) in Figure 17.3. Water can be produced in a unit

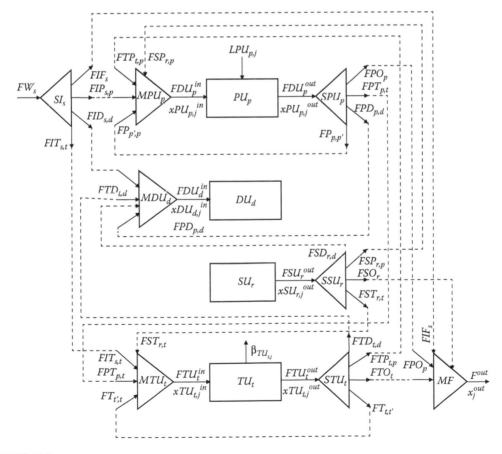

FIGURE 17.3
The superstructure of an integrated water network. (From Ahmetović, E., and Grossmann, I.E., *AIChE J.*, 57(2), 434–457, 2011. doi: http://dx.doi.org/10.1002/aic.12276. With permission.)

as by-product (i.e., in reactor) and be available for reusing within other operations. That unit is represented in Figure 17.3 by an internal water source unit (SU_r), which is a typical non-mass-transfer operation. There is also wastewater treatment unit (TU_t) within the superstructure in order to remove contaminants (percentage removal $[RR_{t,j}]$ is specified) from wastewater stream (FTU_t^{in}).

The superstructure considers all feasible connections between units, including options for water integration (water reuse, regeneration reuse, and regeneration recycling). The freshwater (FWs) from the freshwater splitter (SI_s) can be directed toward process water-using units (stream $FIP_{s,p}$); wastewater treatment units (stream $FIT_{s,t}$) in order to enable freshwater pretreatment, if required; final wastewater mixer (stream FIF_s); and demand units (stream $FID_{s,d}$). Streams from other process units (stream $FP_{p',p}$) and source units $(FSP_{s,p})$ are directed to process unit mixer (MPU_p), enabling water reuse options. Also, water regeneration reuse and recycling are enabled by the existence of stream $FTP_{t,p}$, connecting treatment unit t with process unit p. The wastewater stream $FP_{p,p'}$ leaving process unit p (from splitter SPU_p) is directed to another process unit p', demand unit d (stream $FPD_{p,d}$), wastewater treatment unit t (stream $FPT_{p,t}$) for wastewater regeneration and final mixer (stream FPO_p). The inlet streams to demanding unit mixer (MDU_d) are those directed from the external freshwater source and internal source ($FID_{s,d}$ and $FSD_{r,d}$), respectively, as well as streams leaving process and treatment units ($FPD_{p,d}$ and $FTD_{t,d}$). From the internal water source, r, water can be sent toward process unit p ($FSP_{r,p}$), demand unit d ($FSD_{r,d}$), or treatment unit t ($FST_{r,t}$), and final mixer (FSO_r). Streams from all the splitters within the network are directed to the treatment unit mixer (MTU_t), enabling freshwater pretreatment, if required ($FIT_{s,t}$), and wastewater regeneration ($FPT_{p,t}$, $FST_{r,t}$). Also, streams from other treatment units ($FT_{t',t}$) are directed to mixer MTU_t. Wastewater stream F^{out} from final mixer (MF) is discarded into the environment.

Optimization model formulation. For a given superstructure, an optimization model is formulated in order to perform optimization and extract the optimal solution embedded within the superstructure. The optimization model of water network synthesis problem consists of an objective function and constraints. The objective function can be formulated using various economic criteria (Pintarič et al. 2014). The objective functions used in this chapter represent the minimization of total annualized cost of the networks (Ahmetović and Grossmann 2011) and consist of the following:

a. The freshwater cost, the investment and operating costs of wastewater treatment units, the cost of piping, and the operational cost of water pumping through pipes (integrated water network involving process water-using and wastewater treatment units)

b. The freshwater cost, the cost of piping, and the operational cost of water pumping through pipes (water network involving only process water-using units)

c. The investment and operating costs of wastewater treatment units, the cost of piping, and the operational cost of water pumping through pipes (water network involving only wastewater treatment units)

Note that in the first case (a), the objective function is used for the integrated water network, while in other cases, it is used for its separate networks, namely, water-using network (b) and wastewater treatment network (c).

In addition to the economic nature of the objective function of the water network problem, it can also be formulated as the minimization of the total freshwater consumption or the minimization of the total flow rate of the freshwater consumption and wastewater treated within treatment units, because the cost of the wastewater treatment units depends on wastewater flow rates treated within wastewater treatment units.

The model constraints can be formulated as equalities and inequalities. The equalities are, for example, the overall mass and contaminant mass balance equations, while inequalities can represent constraints on variables, for example, flow rates, contaminant concentrations, and design constraints corresponding to the existence of pipes or wastewater treatment technologies. The variables within an optimization model of a water network synthesis problem can be continuous and binary. Continuous variables are, for example, flow rates and contaminant concentrations, while binary variables can only have values 0 or 1 and they are used for selecting, for example, a wastewater treatment technology or a piping connection. If a binary variable has the value 1, in that case, the wastewater treatment technology or the piping connection is selected, and vice versa. The optimization problem can be formulated as linear programming (LP), nonlinear programming (NLP), mixed-integer linear programming (MILP), or mixed-integer nonlinear programming problem (MINLP). The latter one is the most difficult to solve, owing to the nonlinear nature of the problem and many design alternatives within the network. For the purposes of solving the case study presented in this chapter, an MINLP optimization model of water network proposed by Ahmetović and Grossmann (2011) was used. This model was implemented within the General Algebraic Modeling System (GAMS) (Rosenthal 2014). It is worth mentioning that this model is generic and independent of initial data, and it can be used for solving different types of water network synthesis problems. For example, process water networks or wastewater treatment networks can be optimized separately (sequentially) or as an integrated network (simultaneously). Also, the proposed model can be used for the synthesis problems with fixed or variable flow rates through process water-using units and for establishing trade-offs between the network cost and network complexity. The generic nature of the model enables easy manipulation within the proposed model, for example, excluding some units from the superstructure can be done only by specifying an empty set for the given units. Figure 17.4 shows different options of the water network model, starting from a data input to the optimal network design solution. The following data input is required, for example, in the case that an overall network consisting of a process water network (fixed flow rates through process water-using units) and a wastewater treatment network should be synthesized:

- A number of freshwater sources, a number of contaminants and corresponding concentrations within the freshwater source stream, and the cost of freshwater
- A number of process water-using units, mass flow rates of process units, mass loads of contaminants transferred to water streams within the process units, and maximum inlet concentrations of contaminants at the inlet streams to the process units
- A number of wastewater treatment units, removal percentages of contaminants within the wastewater treatment units, operation and investment cost coefficients, cost function exponents, and an annualized factor for investment on treatment units
- Cost coefficients corresponding to existence of pipes, investment cost coefficients, cost function exponent for the investment cost of pipes, and an annualized factor for investment on pipes
- Operating cost coefficient for pumping water through pipes
- Hours of network plant operation per annum
- Limiting the concentrations of contaminants within the wastewater stream discharged into the environment

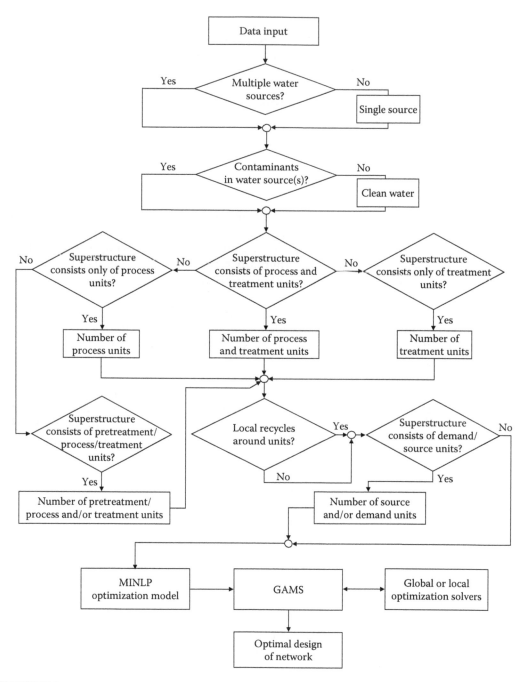

FIGURE 17.4
Different options of the water network model proposed by Ahmetović and Grossmann (2011).

Accordingly, only the number of units has to be specified next to data input, in order to solve the model using MINLP optimization solvers, for example, BARON (Tawarmalani and Sahinidis 2005), in order to obtain the global optimum solution of the network design. The presented model can be also applied for multiple pinch-water network problems (Hallale 2002; Manan, Tan, and Foo 2004; Parand 2014).

Solution strategies development. The overall synthesis problem consisting of process water-using and wastewater treatment networks can be solved sequentially and simultaneously. In a sequential approach, a process water-using network is first solved in order to determine the minimum freshwater usage and the water network design, followed by solving a wastewater treatment network in order to determine a wastewater network design (see Figure 17.2a, b, as well as Figure 17.4). On the basis of the obtained network design solutions, an overall network consisting of a water-using network and a wastewater treatment network can be constructed. In a simultaneous approach, process water-using and wastewater treatment networks are solved together as an overall synthesis problem (see Figures 17.2c and 17.4) in order to extract the optimal design of integrated network from the superstructure. Water-pinch technology and mathematical programming can be used for sequentially solving the overall water network synthesis problem. However, only the mathematical programming approach can address the overall water network synthesis problems simultaneously. The obtained solution by the sequential strategy is suboptimal, while the simultaneous strategy enables obtaining locally optimal as well as globally optimal solutions, depending on the type of solver that is used. In the cases of using local optimization solvers, a good initialization point should be provided as well as good bounds for optimization variables, while an initialization point does not need to be provided for the global optimization solver, for example, BARON. The proposed MINLP optimization model can be solved by global optimization solver BARON directly or by using a two-stage solution strategy. The reader is referred to works (Ahmetović and Grossmann 2010, 2011) for more details about the MINLP model and those strategies. For solving of case studies in this chapter, MINLPs were solved directly with BARON, and global optimal solutions of all case studies were found within the specified optimality tolerance (1%) and reasonable computational times. It is worth pointing out that the special redundant constraint for the overall contaminant mass balance (Karuppiah and Grossmann 2006) and good variable bounds were incorporated within the model, and this had a very big impact in improving the strength of the lower bound for the global optimum, as well as in reducing the CPU time for BARON.

17.5 Case Study

The case study considered in this section represented a large-scale example involving five process water-using units, three treatment units, and three contaminants (A, B, and C). The contaminants' concentrations within the wastewater stream discharged into the environment were limited to 10 ppm. Data for the process units (flow rates, discharge load/contaminant mass load, and maximum inlet concentrations of contaminants) were taken from the literature (Karuppiah and Grossmann 2006) and are given in Table 17.1. Table 17.2 shows data for the process units modified to address a case with variable flow rate through water-using units. Table 17.3 shows data for wastewater treatment units (percentage removal of contaminants, operating and investment cost coefficients, and cost function exponent).

The freshwater cost is assumed to be $1/t, the annualized factor for investment 0.1 (1/y), and the total time for the network plant operation in a year 8000 h. The fixed cost coefficient pertaining to the pipe is assumed to be $6, while the investment cost coefficient for each individual pipe is taken to be $100/(t·h⁻¹). The operating cost coefficient for pumping

TABLE 17.1

Data for Process Units (Fixed Flow Rate Problem)

Process Unit	Flow Rate (t/h)	Discharge Load (kg/h)			Maximum Inlet Concentration (ppm)		
		A	B	C	A	B	C
PU1	40	1	1.5	1	0	0	0
PU2	50	1	1	1	50	50	50
PU3	60	1	1	1	50	50	50
PU4	70	2	2	2	50	50	50
PU5	80	1	1	0	25	25	25

Source: Karuppiah, R., and Grossmann, I.E., *Comput. Chem. Eng.*, 30(4), 650–673, 2006, doi:10.1016/j.compchemeng.2005.11.005.

TABLE 17.2

Data for Process Units (Variable Flow Rates through the Process Units)

Process Unit	Discharge Load (kg/h)			Maximum Inlet Concentration (ppm)			Maximum Outlet Concentration (ppm)			Limiting Water Flow Rate (t/h)
	A	B	C	A	B	C	A	B	C	
PU_1	1	1.5	1	0	0	0	25	37.5	25	40
PU_2	1	1	1	50	50	50	70	70	70	50
PU_3	1	1	1	50	50	50	66.67	66.67	66.67	60
PU_4	2	2	2	50	50	50	78.57	78.57	78.57	70
PU_5	1	1	0	25	25	25	37.5	37.5	25	80

Source: Ahmetović, E., and Grossmann, I.E., *AIChE J.*, 57(2), 434–457, 2011. doi:http://dx.doi.org/10.1002/aic.12276.

TABLE 17.3

Data for Treatment Units

Treatment Units	% Removal of Contaminant			Investment Cost Coefficient	Operating Cost Coefficient	Cost Function Exponent
	A	B	C			
TU_1	95	0	0	16,800	1	0.7
TU_2	0	0	95	9,500	0.04	0.7
TU_3	0	95	0	12,600	0.0067	0.7

Source: Karuppiah, R., and Grossmann, I.E., *Comput. Chem. Eng.*, 30(4), 650–673, 2006, doi: 10.1016/j.compchemeng.2005.11.005.

water through pipe is taken to be $0.006/t. Data for piping and water pumping costs were taken from the literature (Karuppiah and Grossmann 2008).

Several cases are presented in this section for the same case study, addressing the issues of a conventional water network, centralized and distributed wastewater treatment systems, water reuse and recycling, sequential and simultaneous synthesis of process water and wastewater treatment networks, fixed and variable flow rates through process water-using units. For all cases, optimality tolerances were set at 0.01, and MINLPs of all cases were directly solved by BARON.

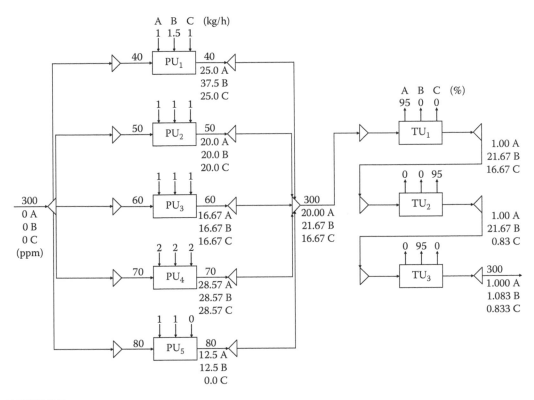

FIGURE 17.5
Base case of process water-using and wastewater treatment network (cases $BC_1 + BC_2$).

First, a conventional water network without water reuse and with centralized wastewater treatment was considered as a base-case 1 (BC_1). In this case, the freshwater was used in all process units (see Figure 17.5). The freshwater consumption for this case was 300 t/h, and the concentrations of contaminants (A, B, and C) within the mixed wastewater stream were 20, 21.67, and 16.67 ppm. The total annual cost (TAC) of the network, including freshwater, water pumping, and piping cost, is 2,429,964.5 $/y (see BC_1 in Table 17.4). The concentrations of the contaminants in the wastewater stream should not exceed the limiting concentrations of 10 ppm. In order to satisfy this constraint, in the next base case 2 (BC_2), the wastewater stream was first treated within the centralized wastewater treatment system, with the same wastewater flow rate going through all treatment units subsequently (see Figure 17.5).

The centralized wastewater treatment system is characterized by very high operating and investment costs of treatment units (see BC_2 in Table 17.4), owing to maximum wastewater flow rate through all three treatment units. The TAC of the centralized wastewater treatment for the BC_2 is 2,781,739.0 $/y. As can be seen from Figure 17.5, the network solutions of the base cases BC_1, and BC_2 were merged, representing an overall network ($BC_1 + BC_2$) (Figure 17.5). The TAC of the merged base case networks shown in Figure 17.5 was 5,211,703.5 $/y.

In the previous base case (BC_1), water reuse options were unconsidered within the network, and consequently, the network required the maximum amount of freshwater (300 t/h). In order to consider water reuse options, on the basis of the general superstructure (Figure 17.3), the superstructure was constructed for five process units (Figure 17.6). This superstructure included all possible connections between a freshwater source splitter, process units, and a final wastewater mixer. The objective in this

TABLE 17.4

Base Case Results of Conventional Networks

	Base Case Networks		
	BC_1	BC_2	$BC_1 + BC_2$
FW (t/h)	300	0.0	300
$Cost_{FW}$ ($/y)	2,400,000.0	0.0	2,400,000.0
IC_{TU} ($/y)	0.0	210,831.0	210,831.0
OC_{TU} ($/y)	0.0	2,512,080.0	2,512,080.0
IC_{Pipes} ($/y)	1,164.5	1,228.0	2,392.5
$OC_{Pumping}$ ($/y)	28,800.0	57,600.0	86,400.0
TAC ($/y)	2,429,964.5	2,781,739.0	5,211,703.5
CPU (s)	0.1	0.1	0.2

BC_1—base case process water network, which uses freshwater in all process units; BC_2—base case centralized wastewater treatment network.

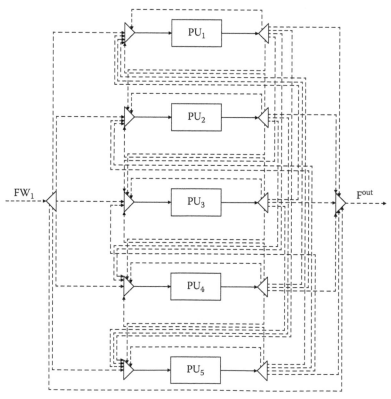

FIGURE 17.6
Superstructure of process water network (case C_1).

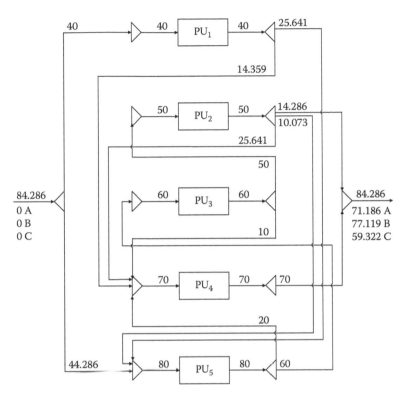

FIGURE 17.7
Optimal design of process water network (case C_1).

case (C_1) was to find the optimal network design by minimizing TAC of the network. From the initial data (Table 17.1) regarding the maximum inlet concentrations of contaminants (0 ppm for A, B, and C) within the water stream directed to the process unit 1 (PU_1), it was noted that water reuse options from all process units to PU_1 were infeasible, because the PU_1 required only clean freshwater without contaminants (0 ppm). Accordingly, these connections could be removed from the superstructure and a simplified network could be obtained (five connections could be removed). The optimal network design obtained by using the MINLP model (Ahmetović and Grossmann 2011) is given in Figure 17.7. The network design (case C_1) exhibited the minimum freshwater consumption of 84.286 t/h. The freshwater consumption was reduced by approximately 71.9% (84.286 vs. 300 t/h) compared with the base case BC_1. Note that the investment cost for piping (see solution C_1 in Table 17.5) was also significantly reduced compared with the base case design, even though the base case design exhibited a lower number of connections. The reason for this was that piping costs were directly proportional to the water flow rate, which was at the maximum through all connections in the base case. The TAC of the network, when comparing BC_1 and C_1 designs, was reduced from 2,429,964.5 \$/y to 693,655.5 \$/y.

Two wastewater streams leaving the process units PU_2 and PU_4 were collected and discharged as a single wastewater stream (84.286 t/h) and given contaminants' concentrations as shown in Figure 17.7. As can be seen, contaminants' concentrations within the

TABLE 17.5

Sequential Solution Results of Integrated Water Network

	Sequential Solution 1			Sequential Solution 2	
	C_1	C_2	$C_1 + C_2$	C_{2a}	$C_1 + C_{2a}$
FW (t/h)	84.286	0	84.286	0	84.286
Cost$_{FW}$ ($/y)	674,285.7	0.0	674,285.7	0.0	674,285.7
IC$_{TU}$ ($/y)	0.0	80,602.3	80,602.3	80,041.9	80,041.9
OC$_{TU}$ ($/y)	0.0	637,813.5	637,813.5	632,454.8	632,454.8
IC$_{Pipes}$ ($/y)	924.0	618.3	1,542.4	621.8	1,545.8
OC$_{Pumping}$ ($/y)	18,445.7	14,953.4	33,399.1	14,839.7	33,285.4
TAC ($/y)	693,655.5	733,987.5	1,427,642.9	727,958.1	1,421,613.6
CPU (s)	0.28	0.12	0.40	0.16	0.44

C_1—optimal solution of water-reuse network; C_2—optimal solution of wastewater treatment network (wastewater streams from process water units are mixed and directed to wastewater treatment network).

C_{2a}—optimal solution of wastewater treatment network (wastewater streams from process water units are distributed to wastewater treatment network).

wastewater stream were above their maximum allowable concentrations of 10 ppm. Accordingly, the wastewater needed to be treated before it was discharged into the environment. Two cases were studied: the first one (case C_2) when the wastewater stream was directed as a single stream to a wastewater treatment network and the second one (case C_{2a}) when two separate wastewater streams leaving the process units PU_2 and PU_4 were directed to a wastewater treatment network. The superstructure representation of the wastewater treatment network for these two cases is given in Figure 17.8. The superstructure presented includes options of distributed wastewater treatment, where wastewater streams can be distributed among different treatment units in order to reduce wastewater treatment cost that depends on wastewater flow rate through the treatment units. The objective was to find the optimal wastewater treatment design by minimizing the TAC. Figure 17.9 shows the optimal network designs obtained for two cases: single wastewater stream (Figure 17.9a) and two separate (Figure 17.9b) wastewater streams. As can be seen, wastewater streams were distributed within three different treatment units compared with base case design (BC$_2$), where a centralized wastewater treatment was used. The TACs of the wastewater treatment networks for cases C_2 and C_{2a} were 733,987.5 and 727,958.1 $/y, respectively (see Table 17.5). An additional reduction in wastewater treatment cost for case C_{2a} was possible owing to the existence of more options for distributive wastewater treatment, caused by an increased number of wastewater connections. Note that not merging the wastewater streams enabled more than a half of the first wastewater stream to be directed to the effluent, thereby reducing the loads of the treatment units. The wastewater treatment operating costs for cases C_2 and C_{2a} were 637,813.5 and 632,454.8 $/y, respectively. The centralized wastewater system (BC$_2$), which treated wastewater stream generated within the process units, had a substantial increase in operating treatment cost (2,512,080.0 $/y) compared with both cases of distributive wastewater treatments (C_2 and C_{2a}). The solutions presented for cases C_1 and C_2/C_{2a} were sequential solutions, where a process water network and a wastewater treatment network were synthesized separately and sequentially. This approach could not produce the best solutions, because all interactions were not considered between the process water network and the wastewater treatment network. Table 17.5 shows the operating, investment, and total annual costs of the

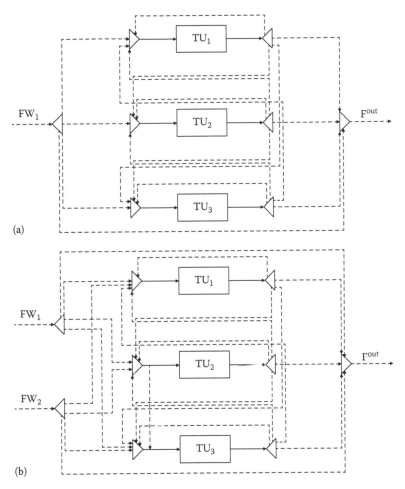

(a)

(b)

FIGURE 17.8
Superstructures of wastewater treatment networks (a) C_2 and (b) C_{2a}.

individual networks cases (cases C_1, C_2, and C_{2a}), as well as merged individual networks ($C_1 + C_2$ and $C_1 + C_{2a}$), in order to produce overall network designs. Note that this approach could not produce best solutions, because only one-way interactions were considered—those from the process water network to the wastewater treatment network. Nevertheless, the TACs of both total network designs ($C_1 + C_2$ and $C_1 + C_{2a}$) were decreased almost to one quarter, when compared with the one of the conventional networks ($BC_1 + BC_2$).

In the next case (C_3), a superstructure of integrated water and wastewater treatment networks (Figure 17.10) was constructed in order to explore all interactions between the two different networks. Additional interactions, those directed from the wastewater treatment network to the process water network, enabled additional options of wastewater reuse and recycling within the process and, hence, possible reductions in freshwater consumption and the total annualized cost of the network. Similarly, as explained earlier in this chapter, owing to the specified contaminants' concentrations (0 ppm; A, B, and C) within the inlet water stream to PU_1, all connections from all process units (PU_1–PU_5) and all treatment units (TU_1–TU_3) to PU_1 could be excluded from the superstructure, because they were infeasible in this case study. Accordingly, the superstructure given in Figure 17.10 can be simplified by eight connections.

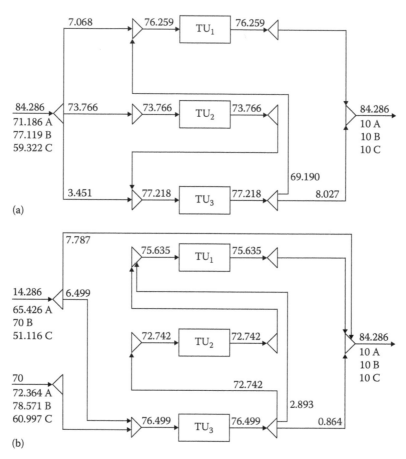

FIGURE 17.9
Optimal designs of wastewater treatment networks (a) C_2 and (b) C_{2a}.

The optimal network design (Figure 17.11) exhibited a minimum freshwater consumption of 40 t/h, which was reduced by 52.5% when compared with the solution obtained with the sequential approach. An additional reduction of freshwater consumption was enabled by the existence of treatment units, in which the contaminants were removed from the wastewater streams, thus increasing the potential for water reuse.

The dotted lines in Figure 17.11 represent the regenerated wastewater streams reused in the process units. The freshwater consumption of 40 t/h was the theoretically minimum consumption determined by the existence of process unit PU_1, requiring only freshwater without contaminants (see Table 17.1). The optimal network design obtained by the simultaneous optimization approach, using the integrated superstructure, exhibited lower operating and investment costs in treatment units and piping installations (see solution C_3 in Figure 17.11 and Table 17.6). The TAC was reduced by approximately 25%, when compared with the sequential solution (1,062,675.6 vs. 1,427,642.9 \$/y). This clearly showed that the simultaneous approach integrating process water network with wastewater treatment networks was the better approach to synthesizing water networks. Note that the wastewater stream (0.042 t/h), leaving the wastewater treatment unit TU_3 and directed to the final wastewater mixer, was too small and could be impractical. Accordingly, the same model (C_3) was solved again when a wastewater

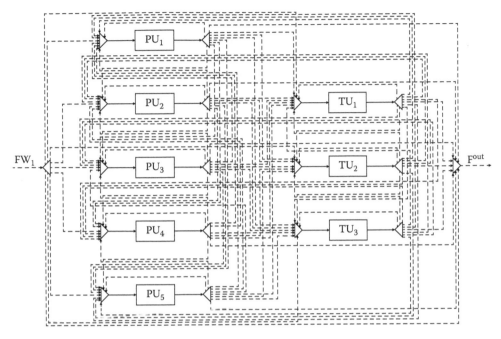

FIGURE 17.10
Superstructure of integrated process water and wastewater treatment networks (cases C_3–C_6).

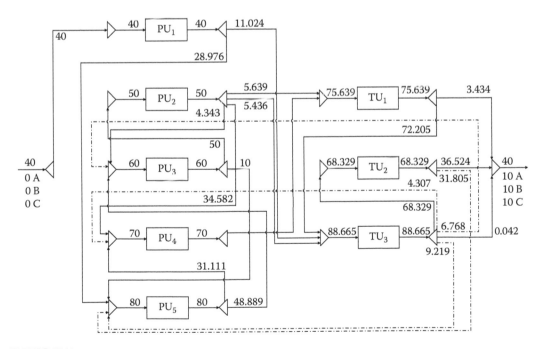

FIGURE 17.11
Optimal solution of integrated process water and wastewater treatment networks for fixed flow rates through water-using units (case C_3).

TABLE 17.6

Simultaneous Solutions of Integrated Process Water and Wastewater Treatment Network

	Simultaneous Solutions			
	C_3	C_4	C_5	C_6
FW (t/h)	40	40	40	40
$Cost_{FW}$ ($/y)	320,000.0	320,000.0	320,000.0	320,000.0
IC_{TU} ($/y)	82,080.5	82,088.4	79,963.0	79,963.0
OC_{TU} ($/y)	631,730.4	631,743.9	635,257.6	635,257.6
IC_{Pipes} ($/y)	1,378.3	1,388.4	1,169.0	1,162.6
$OC_{Pumping}$ ($/y)	27,486.4	27,488.4	22,439.2	21,890.5
TAC ($/y)	1,062,675.6	1,062,709.2	1,058,828.8	1,058,273.8
CPU (s)	38.5	41.1	75.5	39.0

C_3, C_4—optimal network solutions with fixed flow rates through process units.
C_5, C_6—optimal network solutions with variable flow rates through the process units.

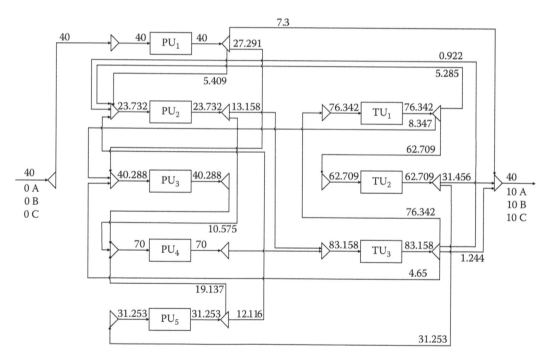

FIGURE 17.12
Optimal solution of integrated process water and wastewater treatment networks for variable flow rates through water-using units (case C_5).

stream of 0.042 t/h was fixed at zero. As can be seen from Table 17.6 in this case (C_4), an insignificant increase in TAC (1,062,709.2 vs. 1,062,675.6 $/y) was obtained by excluding the impractical flow rate. In addition, the integrated model of a process water network and a wastewater treatment network was solved for the case of variable flow rates throughout the process units (fixed mass load problem). Figure 17.12 shows the optimal network design.

As can be seen, water flow rates through the process units PU_2, PU_3, and PU_5 were significantly reduced when compared with the case of the fixed flow rates (see Figures 17.11 and 17.12). However, this did not affect the minimum freshwater consumption of 40 t/h. On the other hand, treatment units' operating and investment costs, as well as pumping and piping costs, were reduced because of the decreased water flow rates within the network (see comparison between cases C_3 and C_5 in Table 17.6). The TAC was reduced only by about 0.4% in this case. In the previous case studies considered, local recycles around the process and treatment units were disallowed. Figure 17.13 shows the optimal network design for the integrated water network (variable flow rates through process units) when local recycles were allowed (case C_6) within the superstructure. However, local recycle options were not selected by optimization, and most of the network connections were the same as in the previous case (C_5).

The dotted lines in Figure 17.13 represent connections not existing within the network design given in Figure 17.12. The TAC of the network for the case C_6 was only very slightly reduced compared with network design C_5 (1,058,273.8 vs. 1,058,828.8 $/y). Note that both solutions were within the selected optimality tolerance of 1% when solving the model by using global optimization solver BARON. As can be seen from Table 17.6, all simultaneous solutions were obtained within reasonable computational times.

Table 17.7 shows the summarized results of the various presented cases for the fixed flow rate problem. Those results clearly show significant savings in freshwater consumption (FW) and TAC obtained by the simultaneous optimization of the integrated network.

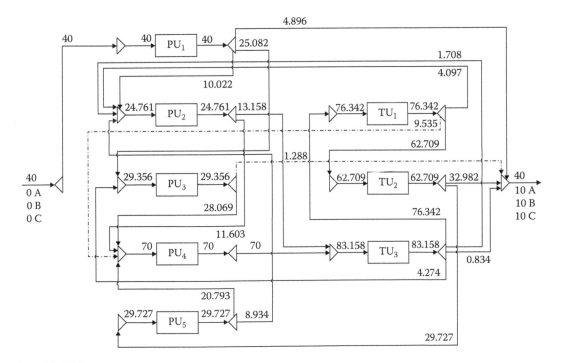

FIGURE 17.13
Optimal solution of integrated process water and wastewater treatment networks for variable flow rates through water-using units (local recycling allowed but not selected) (case C_6).

TABLE 17.7

The Summarized Results of Freshwater Consumption and Total Annual Cost

	No Optimization $(BC_1 + BC_2)$	Sequential Optimization $(C_1 + C_{2a})$	Simultaneous Optimization (C_3)
FW (t/h)	300	84.286	40
FW reduction (%)	0	71.9	86.7
TAC ($/y)	5,211,703.5	1,421,613.6	1,062,675.6
TAC reduction (%)	0	72.7	79.6

Reductions of FW and TAC are given when compared with the base case (no optimization).

17.6 Conclusions

This chapter has presented water optimization in process industries as a challenging problem that should be addressed to achieve sustainable solutions, with the minimum freshwater consumption and wastewater generation. It has been shown that using systematic methods and considering water integration opportunities (wastewater reuse and wastewater regeneration reuse/recycling), a significant reduction in freshwater consumption and wastewater generation can be achieved. Different cases of a large-scale example, involving total water network and multiple contaminants, have been solved in order to present the development of increasingly better network solutions obtained by applying sequential and simultaneous strategies, as well as solutions of a fixed flow rate and a fixed mass load problem. The results showed that by using the simultaneous approach, more than 50% of savings in water usage and wastewater generation can be obtained when compared with the design obtained by the sequential approach and more than 86% can be obtained when compared with the conventional network design. The obtained results in all cases correspond to global optimal solutions. The applied model is general, data-independent, and can be used for solving different water network problems to global optimality.

Acknowledgments

The authors are grateful to the Swiss National Science Foundation (SNSF) and the Swiss Agency for Development and Cooperation (SDC) for providing financial support within the SCOPES 2013–2016 (Scientific Co-operation between Eastern Europe and Switzerland) joint research project (CAPE–EWWR: IZ73Z0_152622/1), as well as support from the Slovenian Research Agency (Program No. P2–0032) and Bilateral Project (No. 05–39–116–14/14) between Bosnia and Herzegovina and Slovenia. Also, the supports from the Center for Advanced Process Decision-making (CAPD) at Carnegie Mellon University and different mobility programmes (Fulbright, EM JoinEU SEE Penta, EM STEM, DAAD, and CEEPUS) are appreciated.

Nomenclature

Abbreviations

BARON	Branch and Reduce Optimization Navigator
CPU	central processing unit time
FW	freshwater
GAMS	General Algebraic Modeling System
IC_{Pipes}	investment cost for piping
IC_{TU}	investment cost for treatment units
LP	linear programming
MEN	mass exchange network
MILP	mixed-integer linear programming
MINLP	mixed-integer nonlinear programming
NLP	nonlinear programming
OC_{TU}	operating cost for treatment units
$OC_{Pumping}$	pumping operating cost
PU	process unit
TU	treatment unit
TAC	total annual cost

Sets and indices

j	contaminant
DU	set of demand units
d	demand unit
PU	set of process units
p	process unit
SU	set of source units
r	source unit
SW	set of freshwater sources
s	freshwater source
TU	set of treatment units
t	treatment unit

Parameters

FDU_d^{in}	mass flow rate of inlet water stream in demand unit d
FPU_p^{in}	mass flow rate of inlet water stream in process unit p
FSU_r^{out}	mass flow rate of outlet water stream from source unit r
$LPU_{p,j}$	mass load of contaminant j in process unit p
$RR_{t,j}$	percentage removal of contaminant j in treatment unit t
$x_j^{out,max}$	maximum concentration of contaminant j in stream discharged to the environment
$xDU_{d,j}^{in,max}$	maximum concentration of contaminant j in inlet stream into demand unit d
$xPU_{p,j}^{in,max}$	maximum concentration of contaminant j in inlet stream into process unit p
$xSU_{r,j}^{out}$	concentration of contaminant j in outlet stream from source unit r
$xW_{s,j}^{in}$	concentration of contaminant j in freshwater source s

Continuous variables

F^{out}	mass flow rate of outlet wastewater stream from final mixer
$FID_{s,d}$	mass flow rate of water stream from freshwater source s to demand unit d
FIF_s	mass flow rate of water stream from freshwater source s to final mixer
$FIP_{s,p}$	mass flow rate of water stream from freshwater source s to process unit p
$FIT_{s,t}$	mass flow rate of water stream from freshwater source s to treatment unit t
$FP_{p',p}$	mass flow rate of water stream from other process unit p' to process unit p
$FPD_{p,d}$	mass flow rate of water stream from process unit p to demand unit d
FPO_p	mass flow rate of water stream from process unit p to final mixer
$FPT_{p,t}$	mass flow rate of water stream from process unit p to treatment unit t
FPU_p^{out}	mass flow rate of outlet water stream from process unit p
$FSD_{r,d}$	mass flow rate of water stream from source unit r to demand unit d
FSO_r	mass flow rate of water stream from source unit r to final mixer
$FSP_{r,p}$	mass flow rate of water stream from source unit r to process unit p
$FST_{r,t}$	mass flow rate of water stream from source unit r to treatment unit t
$FT_{t',t}$	mass flow rate of water stream from other treatment unit t' to treatment unit t
$FTD_{t,d}$	mass flow rate of water stream from treatment unit t to demand unit d
FTO_t	mass flow rate of water stream from treatment unit t to final mixer
$FTP_{t,p}$	mass flow rate of water stream from treatment unit t to process unit p
FTU_t^{in}	mass flow rate of inlet water stream in treatment unit t
FTU_t^{out}	mass flow rate of outlet water stream from treatment unit t
FW_s	mass flow rate of water for freshwater source s
x_j^{out}	concentration of contaminant j in discharge stream to the environment
$xDU_{d,j}^{in}$	concentration of contaminant j in inlet stream into demand unit d
$xPU_{p,j}^{in}$	concentration of contaminant j in inlet stream into process unit p
$xPU_{p,j}^{out}$	concentration of contaminant j in outlet stream from process unit p
$xSPU_{t,j}^{out}$	concentration of contaminant j in outlet stream from splitter process unit p
$xSTU_{t,j}^{out}$	concentration of contaminant j in outlet stream from splitter treatment unit t
$xTU_{t,j}^{in}$	concentration of contaminant j in inlet stream into treatment unit t
$xTU_{t,j}^{out}$	concentration of contaminant j in outlet stream from treatment unit t

Subscripts/superscripts

in	inlet stream
out	outlet stream

References

Ahmetović, E., and I.E. Grossmann. 2010. Strategies for the Global Optimization of Integrated Process Water Networks. In *Computer Aided Chemical Engineering*, ed. S. Pierucci and G. Buzzi Ferraris, 901–906. Elsevier, Oxford, UK.

Ahmetović, E., and I.E. Grossmann. 2011. Global superstructure optimization for the design of integrated process water networks. *AIChE Journal* no. 57 (2):434–457. doi:http://dx.doi.org/10.1002/aic.12276.

Ahmetović, E., N. Ibrić, and Z. Kravanja. 2014. Optimal design for heat-integrated water-using and wastewater treatment networks. *Applied Energy* no. 135:791–808. doi: http://dx.doi.org/10.1016/j.apenergy.2014.04.063.

Ahmetović, E., N. Ibrić, Z. Kravanja, and I.E. Grossmann. 2015. Water and energy integration: A comprehensive literature review of non-isothermal water network synthesis. *Computers and Chemical Engineering* no. 82:144–171. doi: 10.1016/j.compchemeng.2015.06.011.

Allegra, K., P.E. da Silva, and P.E. Al Goodman. 2014. Reduce water consumption through recycling. *Chemical Engineering Progress* no. 110 (4):29–35.

Alva-Argáez, A., A.C. Kokossis, and R. Smith. 1998. Wastewater minimisation of industrial systems using an integrated approach. *Computers and Chemical Engineering* no. 22 (0):S741–S744. doi: 10.1016/s0098-1354(98)00138-0.

Bagajewicz, M. 2000. A review of recent design procedures for water networks in refineries and process plants. *Computers and Chemical Engineering* no. 24 (9–10):2093–2113. doi: 10.1016/s0098-1354(00)00579-2.

Biegler, L.T., I.E. Grossmann, and A.W. Westerberg. 1997. *Systematic methods of chemical process design*. New Jersey: Prentice-Hall.

Castro, P., J. Teles, and A. Novais. 2009. Linear program-based algorithm for the optimal design of wastewater treatment systems. *Clean Technologies and Environmental Policy* no. 11 (1):83–93. doi: 10.1007/s10098-008-0172-5.

Davé, B. 2004. Water and Sustainable Development: Opportunities for the Chemical Sciences: A Workshop Report to the Chemical Sciences Roundtable. Washington (DC): National Academies Press (US). 2004. *www.ncbi.nlm.nih.gov/books/NBK83724/*. Accessed August 29, 2013.

Doyle, S.J., and R. Smith. 1997. Targeting water reuse with multiple contaminants. *Process Safety and Environmental Protection* no. 75 (3):181–189. doi: 10.1205/095758297529020.

El-Halwagi, M.M. 2012. *Sustainable Design Through Process Integration, Fundamentals and Applications to Industrial Pollution Prevention, Resource Conservation, and Profitability Enhancement*. Amsterdam, The Netherlands: Butterworth-Heinemann/Elsevier.

El-Halwagi, M.M., and V. Manousiouthakis. 1989. Synthesis of mass exchange networks. *AIChE Journal* no. 35 (8):1233–1244. doi: 10.1002/aic.690350802.

Faria, D.C., and M.J. Bagajewicz. 2011. Global Optimization of Water Management Problems Using Linear Relaxation and Bound Contraction Methods. *Industrial and Engineering Chemistry Research* no. 50 (7):3738–3753. doi: 10.1021/ie101206c.

Foo, D.C.Y. 2009. State-of-the-Art review of pinch analysis techniques for water network synthesis. *Industrial and Engineering Chemistry Research* no. 48 (11):5125–5159. doi: 10.1021/ie801264c.

Foo, D.C.Y. 2013. *Process Integration for Resource Conservation*. Boca Raton, Florida, USA: CRC Press.

Galan, B., and I.E. Grossmann. 1998. Optimal design of distributed wastewater treatment networks. *Industrial and Engineering Chemistry Research* no. 37 (10):4036–4048. doi: 10.1021/ie980133h.

Grossmann, I.E., M. Martín, and L. Yang. 2014. Review of optimization models for integrated process water networks and their application to biofuel processes. *Current Opinion in Chemical Engineering* no. 5 (0):101–109. doi: http://dx.doi.org/10.1016/j.coche.2014.07.003.

Gunaratnam, M., A. Alva-Argáez, A. Kokossis, J.K. Kim, and R. Smith. 2005. Automated Design of Total Water Systems. *Industrial and Engineering Chemistry Research* no. 44 (3):588–599. doi: 10.1021/ie040092r.

Hallale, N. 2002. A new graphical targeting method for water minimisation. *Advances in Environmental Research* no. 6 (3):377–390. doi: http://dx.doi.org/10.1016/S1093-0191(01)00116-2.

Huang, C.H., C.T. Chang, H.C. Ling, and C.C. Chang. 1999. A mathematical programming model for water usage and treatment network design. *Industrial and Engineering Chemistry Research* no. 38 (7):2666–2679.

Jeżowski, J. 2010. Review of water network design methods with literature annotations. *Industrial and Engineering Chemistry Research* no. 49 (10):4475–4516. doi: 10.1021/ie901632w.

Karuppiah, R., and I.E. Grossmann. 2006. Global optimization for the synthesis of integrated water systems in chemical processes. *Computers and Chemical Engineering* no. 30 (4):650–673. doi: 10.1016/j.compchemeng.2005.11.005.

Karuppiah, R., and I.E. Grossmann. 2008. Global optimization of multiscenario mixed integer non-linear programming models arising in the synthesis of integrated water networks under uncertainty. *Computers and Chemical Engineering* no. 32 (1–2):145–160. doi: http://dx.doi.org/10.1016/j.compchemeng.2007.03.007.

Khor, C.S., B. Chachuat, and N. Shah. 2014. Optimization of water network synthesis for Single-Site and Continuous Processes: Milestones, Challenges, and Future Directions. *Industrial & Engineering Chemistry Research* no. 53 (25):10257–10275. doi: 10.1021/ie4039482.

Klemeš, J.J. 2012. Industrial water recycle/reuse. *Current Opinion in Chemical Engineering* no. 1 (3):238–245. doi: 10.1016/j.coche.2012.03.010.

Klemeš, J.J., ed. 2013. *Handbook of Process Integration (PI): Minimisation of energy and water use, waste and emissions.* Cambridge: Woodhead Publishing Limited.

Klemeš, J., F. Friedler, I. Bulatov, and P. Varbanov. 2010. *Sustainability in the Process Industry: Integration and Optimization.* New York, USA: McGraw-Hill.

Kuo, W.-C.J., and R. Smith. 1997. Effluent treatment system design. *Chemical Engineering Science* no. 52 (23):4273–4290. doi: http://dx.doi.org/10.1016/S0009-2509 (97)00186-3.

Linnhoff, B., and E. Hindmarsh. 1983. The pinch design method for heat exchanger networks. *Chemical Engineering Science* no. 38 (5):745–763. doi: http://dx.doi.org/10.1016/0009-2509(83)80185-7.

Manan, Z.A., Y.L. Tan, and D.C.Y. Foo. 2004. Targeting the minimum water flow rate using water cascade analysis technique. *AIChE Journal* no. 50 (12):3169–3183. doi: 10.1002/aic.10235.

Mann, J.G., and Y.A. Liu. 1999. *Industrial water reuse and wastewater minimization.* New York: McGraw Hill.

Parand, R. 2014. Water and wastewater optimization through process integration for industrial processes. http://espace.library.curtin.edu.au/R/?func=dbin-jump-full&object_id=203501&local_base=GEN01-ERA02. Accessed May 1, 2016.

Pintarič, Z.N., N. Ibrić, E. Ahmetović, I.E. Grossmann, and Z. Kravanja. 2014. Designing optimal water networks for the appropriate economic criteria. *Chemical Engineering Transactions* no. 39:1021–1026. doi: 10.3303/CET1439171.

Rosenthal, R.E. 2014. *GAMS—A User's Guide.* Washington, DC, USA: GAMS Development Corporation.

Smith, R. 2005. *Chemical process design and integration.* Chichester, England: John Wiley & Sons Ltd.

Takama, N., T. Kuriyama, K. Shiroko, and T. Umeda. 1980. Optimal water allocation in a petroleum refinery. *Computers & Chemical Engineering* no. 4 (4):251–258. doi: http://dx.doi.org/10.1016/0098-1354(80)85005-8.

Tawarmalani, M., and N.V. Sahinidis. 2005. A polyhedral branch-and-cut approach to global optimization. *Mathematical Programming* no. 103 (2):225–249. doi: 10.1007/s10107-005-0581-8.

Wang, Y.P., and R. Smith. 1994a. Design of distributed effluent treatment systems. *Chemical Engineering Science* no. 49 (18):3127–3145. doi: http://dx.doi.org/10.1016/0009-2509(94)E0126-B.

Wang, Y.P., and R. Smith. 1994b. Wastewater minimisation. *Chemical Engineering Science* no. 49 (7):981–1006. doi: http://dx.doi.org/10.1016/0009-2509(94)80006-5.

Wang, Y.P., and R. Smith. 1995. Wastewater minimization with flowrate constraints. *Trans. IChemE* no. 73 (Part A):889–904.

Yang, L., and I.E. Grossmann. 2013. Water Targeting Models for Simultaneous Flowsheet Optimization. *Industrial and Engineering Chemistry Research* no. 52 (9):3209–3224. doi: 10.1021/ie301112r.

Yang, L., R. Salcedo-Diaz, and I.E. Grossmann. 2014. Water Network Optimization with Wastewater Regeneration Models. *Industrial and Engineering Chemistry Research* no. 53 (45):17680–17695. doi: 10.1021/ie500978h.

Yoo, C., T. Lee, J. Kim, I. Moon, J. Jung, C. Han, J.-M. Oh, and I.-B. Lee. 2007. Integrated water resource management through water reuse network design for clean production technology: State of the art. *Korean Journal of Chemical Engineering* no. 24 (4):567–576. doi: 10.1007/s11814-007-0004-z.

Zamora, J.M., and I.E. Grossmann. 1998. Continuous global optimization of structured process systems models. *Computers and Chemical Engineering* no. 22 (12):1749–1770. doi: http://dx.doi.org/10.1016/S0098-1354(98)00244-0.

18

Energy Integration in Process Plants

Pratik N. Sheth, Suresh Gupta, and Sushil Kumar

CONTENTS

ABSTRACT The availability of useful energy is less and resources are fewer as compared with its demand, which is increasing immensely day by day. Owing to the situation of high electricity demand, increasing amount of use of imported coal and low contribution of renewable energy sector, it is very much necessary to minimize the energy requirement. The minimization of energy requirement requires careful examination of process plant energy utilities and energy losses and wastages. This chapter includes the energy audit, which is required to identify the available energy sources, energy wastage, and utilization pattern in the process plants. It also discusses the energy integration among various unit operations to minimize the need. The energy optimization is discussed at length, including case studies containing multi-objective optimization with constraints.

18.1 Introduction

The energy demand is increasing immensely day by day. However, the availability of useful energy is less and resources are fewer as compared with its demand. Hence, for any process plant, proper and efficient technique for effective utilization of available energy should be in place. Despite industry having invested in energy-efficiency technologies for more than 35 years, a considerable amount of energy is still wasted in gas, liquid, or solid form, and there is still scope for significant efficiency improvement [1]. The energy can be classified into various forms such as kinetic, potential, chemical, mechanical, electrical, nuclear, and thermal. As far as process plants are concerned, electrical and thermal energies are of prime importance.

Central Electricity Authority (CEA), Government of India, anticipated a base load energy deficit of 5.1% for the fiscal year 2014–2015 in their Load Generation Balance report [2]. Apart from developing domestic energy sources to satisfy the growing demand, increasing amounts of fossil fuels are imported; this is exacerbating the trade deficit and can also be harmful to the environment. Coal imports hit a record high during the last fiscal year and will likely rise further over the next five years, since India aims to expand its power-generation capacity by 44% (Ministry of coal). According to data reported by Ministry of Coal, Government of India, the total import of coal and product, that is, coke, for the year 2013–2014 is 154.55 million tons [3]. There is a rise of 6.4% in the coal import from the previous year. It is observed that there is a hike in power tariff rates continually in both categories: domestic and non-domestic. Power generation from renewable sources is on the rise in India, with the share of renewable energy in the country's total energy mix rising from 7.8% in 2008 to 12.3% in 2013 [3]. To reduce the dependence on the fossil fuels and to replace them with the alternative fuels, an extensive research work is performed across the globe [4]. Having low sulfur content compared with petroleum fuels, CO_2-neutral, and being easy to transport, biomass-based liquid fuels would emerge as a prominent alternative to fossil fuels in the near future [5]. Biomass gasification is considered a prominent technology for decentralized power generation [6]. Considering the prevailing situation of high electricity demand, increasing amount of use of imported coal, and low contribution of renewable energy sector, it is very much necessary to minimize the energy requirement.

The minimization of energy requirement requires careful examination of process plant energy utilities and energy losses and wastages. In process industries, thermal energy loss is an important issue owing to the requirement of high temperatures and involvement of multiple heat-intensive processes. Generally, high-grade thermal energy is utilized within processes and low-grade heat is mostly rejected to the environment. Fetching and utilization of low-grade thermal energy depend on the properties of heat in the waste streams. The most important parameter is the temperature of the low-grade heat stream. The effectiveness and efficiency of recovery of residual heat from the low-grade heat resources depend on the temperature difference between the source and a suitable sink, for example, another process or space heating/cooling. In most cases, the temperatures of these waste heat sources are too low to produce electricity and are difficult to find potential stream for direct heat utilization.

Nowadays, for any process plant, apart from turnover and capacity, an important parameter is energy usage per ton of production. This chapter includes the energy audit, which is required to identify the available energy sources, energy wastage, and utilization pattern in the process plants. It also discusses the energy integration among various unit operations to minimize the need. The energy optimization is discussed at length, including case studies containing multi-objective optimization with constraints.

18.2 Energy Audit

Energy audit is must for decision making in the area of energy management. A continuous energy audit of industrial machines and of a process plant as a whole is essential to reduce energy consumptions for sustainable and energy-efficient manufacturing. It is a systematic approach that comprehensively evaluates the actual performance of a plant.

The energy using systems and equipment are compared against the designed performance level or the industry best practices. The difference between observed performance and best practice is the potential for energy and cost savings. The energy input is an essential part of any manufacturing process and often forms a significant part of expenditure of the plant. Any saving in energy directly adds to the profit of the company. Energy audit helps identify the actions for improving the energy performance, prioritizing projects, and setting targets and track progress.

As per the Energy Conservation Act, 2001, energy audit is defined as "the verification, monitoring and analysis of use of energy including submission of technical report containing recommendations for improving energy efficiency with cost benefit analysis and an action plan to reduce energy consumption." Depending on the function and type of industry, depth to which final audit is needed, and the potential and magnitude of cost reduction desired, energy audit can be divided into two types: (1) preliminary audit and (2) detailed audit.

Preliminary audit is a quick exercise to achieve following goals by using easily available data.

- Understanding of energy consumption (data and pattern) in the organization.
- Assessment of the scope for saving.
- Identification of the easiest areas for attention and areas that need immediate attention.
- Setting up a "reference point" based on past experience or on literature or market trends.
- Identification of areas or systems for more detailed assessment.

A detailed energy audit provides a comprehensive energy project implementation plan for a whole plant or part of a plant, since it evaluates all major energy-using systems. The most accurate estimate of energy savings and cost are to be incorporated in this analysis. The interactive effects of all projects, accounts for the energy use of all major equipment, and detailed energy cost-saving calculations and project cost should be included in this exercise. The detailed energy balance could be an important tool to carry out this exercise. It is based on an inventory of energy-using systems, assumptions of current operating conditions, and calculations of energy use.

Grinbergs and Gusta [7] have given a five-step methodology to carry out the industrial energy audit. The five steps are data collection, data processing, analysis of results, recommendations for improvement, and economic foundations. These five steps are elaborated in Figure 18.1.

18.2.1 Case Study

Engine and Ari [8] have presented a case study of an energy auditing of dry-type rotary kiln of a cement-manufacturing process. The salient features of this case study are included in this chapter.

18.2.1.1 System Description

Rotary kilns are tubes lined with refractory bricks, having diameter up to 6 m. Generally, kilns are placed inclined at an angle of 3°–3.5° from the horizontal axis and their speed

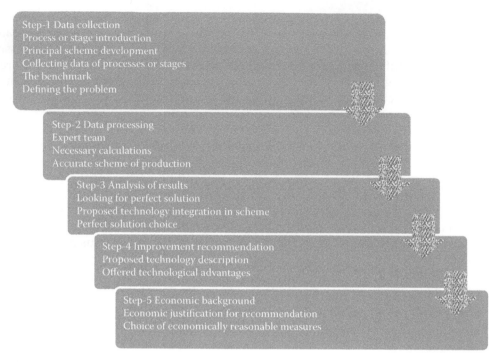

FIGURE 18.1
Methodology of industrial energy audit. (From Grinbergs, K., Energy Audit Method for Industrial Plants. In *Civil Engineering. International Scientific Conference. Proceedings (Latvia)*, 2013.)

of rotation lies within 1 to 2 rpm. To preheat the raw material before it enters the kiln, cyclone-type preheaters are used. Hence, precalcination starts in the preheaters, and approximately one-third of the raw material would be calcined before entering the kiln. The precalcined raw material (at 850°C) passes through the rotary kiln toward the flame. The initial mixing of alumina, ferric oxide, and silica with lime takes place in the calcination zone (700°C–900°C). The clinker component, $2CaO \cdot SiO_2$, forms between 900°C and 1200°C. After that, the other component, $3CaO \cdot SiO_2$, forms in a subsequent zone (temperature of 1250°C). The molten phase, $3CaO \cdot Al_2O_3$, forms during cooling stage. Alite component may dissolve back into the liquid phase, thus forming secondary belite if rate of cooling is slow. Fast cooling of the product (clinker) improves the product quality and also enables heat recovery from the clinker [8].

The plant uses a dry process with a series of cyclone-type preheaters and an inclined kiln. The diameter of the kiln is 3.60 m and the length is 50 m. The specific energy consumption is 3.68 GJ/ton-clinker and plant clinker capacity is 600 tons/day. The rotary kiln system considered for the energy audit is schematically shown in Figure 18.2. The control volume for the system includes the preheaters group, rotary kiln, and cooler. The streams to and from the control volume and all measurements are indicated in the same figure [8].

18.2.1.2 Mass and Energy Balance

The mass balance is shown in Figure 18.3. Table 18.1 shows the complete energy balance, with all details of parameters, assuming steady state condition and negligible air leakage.

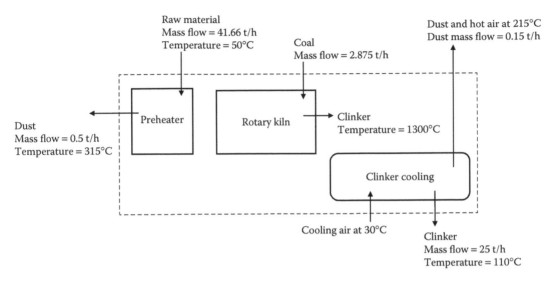

FIGURE 18.2
System description: A rotary kiln. (From Engin, T. and V. Ari., *Energ. Convers. Manage.* 46, 551–562, 2005.)

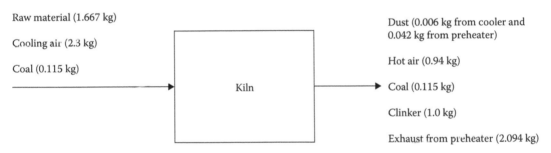

FIGURE 18.3
Mass balance of the system. (From Engin, T. and V. Ari., *Energ. Convers. Manage.*, 46, 551–562, 2005.)

It can be seen from the complete energy balance that only ~7% of energy losses are unaccounted; this is very obvious with assumptions taken here. The overall efficiency can be calculated as the ratio of enthalpy of clinker to the total energy input, that is, 48%. The kiln exhaust gases account up to 19% of the total energy input. The second highest heat output is heat loss due to radiation, which accounts up to 10%. Hence, it is very clear by this analysis that to improve the efficiency, radiation heat losses are required to be minimized and heat recovery from the kiln exhaust gases has to be performed. Hence, we can conclude that action plan to be made to get maximum recovery from these two heat outputs (approximately 30% combined of the energy losses). Many other case studies [9–11] are reported in the literature, which can be referred further.

TABLE 18.1

Complete Energy Balance

Item	Equations Applied	Value of Variables	Value of Heat (kJ/kg-clinker)
Energy inputs			
Heat content of coal (Q_1)	$Q_1 = m_c H_c$ Where H_c is enthalpy of coal	$m_c = 0.115$ kg/kg-clinker, $H_c = 30{,}600$ kJ/kg	3519 (95.47%)
Sensible heat of coal (Q_2)	$Q_2 = m_c h_c$, $h_c = CT$ Where C is specific heat of coal	$m_c = 0.115$ kg/kg-clinker, $C = 1.15$ kJ/kg °C, $T = 50$°C	7 (0.19%)
Heat of raw material (Q_3)	$Q_3 = m_{rm} h_{rm}$, $h_{rm} = CT$	$m_{rm} = 1.667$ kg/kg-clinker, $C = 0.86$ kJ/kg °C, $T = 50$°C	72 (1.95%)
Heat of organics in the kiln feed (Q_4)	$Q_4 = FKh_{os}$ Where h_{os} is enthalpy of organics	$F = 0.10$, $K = 0.9\%$, $h_{os} = 21{,}036$ kJ/kg	19 (0.52%)
Heat of cooling air (Q_5)	$Q_5 = m_{ca} h_{ca}$	$m_{ca} = 2.300$ kg/kg-clinker, $h_{ca} = 30$ kJ/kg, ($T = 30$°C)	69 (1.87%)
Total energy input			3686 (100%)
Energy outputs			
Heat of formation of clinker (Q_6)	$Q_6 = 17.196(\mathrm{Al_2O_3}) + 27.112(\mathrm{MgO})$ $+ 32(\mathrm{CaO}) - 21.405(\mathrm{SiO_2}) - 2.468(\mathrm{Fe_2O_3})$		1795 (48.70%)
Heat of kiln exhaust gas (Q_7)	$Q_7 = m_{eg} C_{p-eg} T_{eg}$	$m_{eg} = 2.094$ kg/kg-clinker, $C_{p-eg} = 1.071$ kJ/kg °C, $T = 315$°C	706 (19.15%)
Heat of moisture in raw material and coal (Q_8)	$Q_8 = m_{H2O}(h_{fg(50°C)} + h_{315°C} - h_{50°C})$	$m_{H2O} = 0.008835$ kg/kg-clinker (in coal+ raw material), $h_{fg(50°C)} = 2384$ J/kg, $h_{50°C} = 2591$ J/kg, $h_{315°C} = 3104$ J/kg	26 (0.71%)
Heat of hot air from cooler (Q_9)	$Q_9 = m_{air\text{-}co} h_{air\text{-}co}$	$m_{air\text{-}co} = 0.940$ kg/kg-clinker, $h_{air\text{-}co} = 220$ kJ/kg ($T = 215$°C)	207 (5.61%)
Heat loss by dust (Q_{10})	$Q_{10} = (m_{dust\text{-}preheater} + m_{dust\text{-}air\ cooler}) h_{dust\text{-}ave}$	$m_{dust\text{-}preheater} = 0.042$ kg/kg-clinker, $m_{dust\text{-}air\ cooler} = 0.006$ kg/kg-clinker, $h_{dust\text{-}ave} = 275$ kJ/kg	11 (0.30%)
			(Continued)

TABLE 18.1 (*Continued*)

Complete Energy Balance

Item	Equations Applied	Value of Variables	Value of Heat (kJ/kg-clinker)
Heat of clinker discharge (Q_{11})	$Q_{11} = m_{cli} h_{cli}, T = 110°C$	$m_{cli} = 1$ kg/kg-clinker, $h_{cli} = 86$ kJ/kg	86 (2.33%)
Heat by radiation from kiln surface (Q_{12})	$Q_{12} = \sigma\varepsilon A_{kiln}((T_s)^4 - (T_\infty)^4)/1000(\text{mclinker})$ where: σ is Stefan Boltzmann's constant ε is emissivity, and A_{kiln} is surface area of the kiln	$\sigma = 5.67 * 10^{-8}$ W/m²K⁴ $\varepsilon = 0.78, A_{kiln} = 565.48$ m², $T_s = 581$ K	386 (10.47%)
Heat by convection from kiln	$Q_{13} = h_{con} A_{kiln}(T_s - T_\infty), h_{con} = Nu * k_{air}/D_{kiln}$ where: R_e is Reynolds number Nu is Nusselt number h_{con} is convective heat transfer coefficient	$R_e = 373,110$ ($V_{air} = 3$ m/s), $Nu = 733$, at $T_f = 161°C$ (film temp.)	171 (4.64%)
Heat by radiation from pre-heater surface	$Q_{14} = \sigma\varepsilon A_{ph}((T_s)^4 - (T_\infty)^4)/1000(\text{mclinker})$	$\sigma = 5.67 * 10^{-8}$ W/m²K⁴, $\varepsilon = 0.78$, $A_{kiln} = 150.8$ m², $T_s = 348$ K	7 (0.19%)
Heat by natural convection from pre-heater surface	$Q_{15} = h_{ncon} A_{ph}(T_s - T_\infty), h_{ncon}$ $= Nu * k_{air}/L_{ph}, Nu$ $= 0.1(Ra)^{1/3}$	Pre-heater is considered as a vertical cylinder With $D = 4$ m, $L_{ph} = 12$ m, $Ra = 0.649 * 10^3$, $Nu = 1865$, at $T_f = 45°C$ (film temp.)	5 (0.14%)
Heat by radiation from cooler surface	$Q_{16} = \sigma\varepsilon A_{ac}((T_s)^4 - (T_\infty)^4)/1000(\text{mclinker})$	$\sigma = 5.67 * 10^8$ W/m²K⁴, $\varepsilon = 0.78$, $A_c = 64$ m², $T_s = 35°C$	4 (0.11%)
Heat by natural convection from cooler surface	$Q_{17} = h_{ncon} A_c(T_s - T_\infty), h_{ncon} = Nu * k_{air}$ $Lc, Nu = 0.1(Ra)^{1/3}$ Where Ra is Rayleigh number	Cooler surface is considered as a vertical plate $8 m*8$ m, $Ra = 0.7277 * 10^{12}, Nu = 899$, at $T_f = 70°C$ (film temp.)	9 (0.24%)
Heat losses (unaccounted)			273 (7.41%)
Total energy output	3686 (100%)		

Source: Engin. T. and V. Ari., *Energ. Convers. Manage.*, 46, 551–562, 2005.

18.3 Energy Integration

In the past three decades, extensive efforts have been made in the fields of energy integration and energy recovery technologies owing to gradual increase in energy cost and shortage of energy resources [12]. A heat recovery system consisting of a set of heat exchangers can be treated as a heat exchanger network (HEN), which is widely used in process industries such as gas processing and petrochemical industries. In HEN, heat energy is exchanged among several process streams with different supply and target temperatures. The aim of the heat exchangers network synthesis (HENS) is to design an optimum HEN for the available process streams, which is corresponding to minimum total annualized cost. HENS has been considered one of the most important research subjects in process engineering owing to its significant benefits in saving energy and equipment costs. Optimum HEN may be produced by optimizing the area of heat exchangers and utility (hot and cold) requirements.

Pinch analysis is a methodology for minimizing energy consumption of chemical processes by calculating thermodynamically feasible energy targets (or minimum energy consumption) [13]. The feasible energy targets can be achieved by optimizing heat recovery systems, energy supply methods, and process operating conditions. It is also known as process integration, heat integration, energy integration, or pinch technology. The process data are represented as a set of energy flows or streams, as a function of heat load (kW) against temperature (°C). These data are combined for all the available streams in the process plant and are represented in terms of composite curves. Hot composite curve for all hot streams (releasing heat) and cold composite curve for all cold streams (requiring heat) can be made. The point of closest approach between the hot and cold composite curves is the pinch temperature (pinch point), at which the design is most constrained. Hence, identification of pinch point is important. The design of HEN starts from pinch, and the energy targets can be achieved using heat exchangers to recover heat between hot and cold streams.

HEN analysis requires the three pieces of information for each process stream: flow rate and heat capacity, the source temperature in °C or °F, and the target temperature in °C or °F. The enthalpies of all streams are calculated and combined in order to generate plots of cumulative enthalpy against temperature, which are referred as composite curves. There are four types of composite curves required for the analysis are (i) hot composite curve, (ii) cold composite curve, (iii) combined plot of hot and cold composite curves, and (iv) grand composite curve. In the combined composite curves, the point at which the distance between the hot and cold curves is at a minimum is referred to as pinch point. Identification of pinch is an important constraint for the design of heat exchanger network. This is helpful in the design of heat exchanger network that can meet our ideal minimum energy requirements. The optimum utility requirement and total area of heat exchanger network could be calculated by adjusting the minimum approach distance, known as ΔT_{min}. Pinch analysis was successfully applied to carry out the HENS for various case studies reported in the literature [14–16].

18.3.1 Case Study

Considering minimum approach temperature difference, $\Delta T_{min} = 10°C$, carry out the energy integration analysis by using pinch technology by determining the following:

    a. Net amount of heat available in the streams, based on the first law of thermodynamics.

b. Shifted-temperature scales diagram with net heat in respective intervals.

c. Construction of cascade diagram.

d. Minimum hot utility and cold utility requirements.

e. Pinch temperature.

f. Temperature enthalpy diagram.

g. Total area requirement, *in priori*, the network proposal.

h. Number of heat exchangers based on the first law analysis.

i. Number of heat exchangers based on the second law analysis.

j. Hot end design.

k. Cold end design.

l. Heat exchanger network for the maximum energy recovery (MER).

m. Number of loops crossing the pinch.

n. Identification of the loops.

o. Final heat exchanger network after breaking all the loops and restoring ΔT_{min} as and when there is a violation.

p. Total area requirement for the final network.

q. Comparison of the results obtained in (g) and (p).

r. Estimation of energy saving.

Assume that the supply temperatures of hot utility and cold utility are 330°C and 15°C, respectively. Heat transfer coefficient of hot utility and cold utility is 0.5 kW/m²°C (Table 18.2).

Solution 1:

a. First law of thermodynamics:
   Heat available in any process stream

$$(Q) = FC_p (T_1 - T_2)$$

where:
    T_1 and T_2 are the inlet and outlet temperatures, respectively
    F is the mass flow rate
    C_p is the specific heat

TABLE 18.2

Case Study Data

Stream No.	Condition	FC_p (kW/°C)	h (kW/m²°C)	Source Temperature (°C)	Target Temperature (°C)
1	Hot	10	0.4	175	45
2	Hot	40	0.55	125	65
3	Cold	20	0.75	20	155
4	Cold	15	0.65	40	112

TABLE 18.3

First Law Calculation

Stream No.	Condition	FC_p (kW/°C)	Source Temperature (°C)	Target Temperature (°C)	Q available (kW)
1	Hot	10	175	45	1300
2	Hot	40	125	65	2400
3	Cold	20	20	155	−2700
4	Cold	15	40	112	−1080
Net heat available					−80

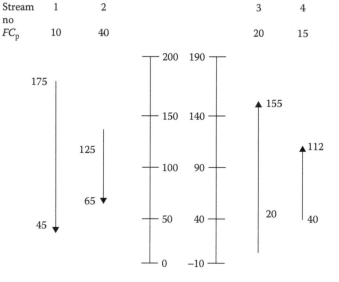

FIGURE 18.4
Shifted temperature scales diagram.

First law calculation is given in Table 18.3:

Net heat available in hot streams = 3700 kW

Net heat required for cold streams = 3780 kW

Net amount of heat to be supplied = 80 kW

b. Shifted-temperature scales diagram:

Shifted-temperature scales diagram is shown in Figures 18.4 and 18.5.

Net heat available in any temperature interval = $Q_i = \left[\sum (FC_p)_{\text{hot},i} - \sum (FC_p)_{\text{cold},i} \right] \Delta T_i$

$Q_1 = 10(175 - 165) = 100 \, \text{kW}$

$Q_2 = (10 - 20)(165 - 125) = -400 \, \text{kW}$

$Q_3 = (10 + 40 - 20)(125 - 122) = 90 \, \text{kW}$

$Q_4 = (10 + 40 - 20 - 15)(122 - 65) = 855 \, \text{kW}$

$Q_5 = (10 - 20 - 15)(65 - 50) = -375 \, \text{kW}$

$Q_6 = (10 - 20)(50 - 45) = -50 \, \text{kW}$

$Q_7 = (-20)(45 - 30) = -300 \, \text{kW}$

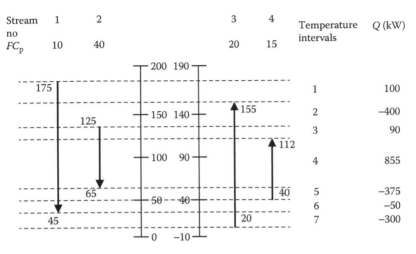

FIGURE 18.5
Shifted temperature scales diagram with heat loads in each temperature interval.

Net heat available based on the second law of thermodynamics (Q):
$$Q = 100 - 400 + 90 + 855 - 375 - 50 - 300$$
$$= -80 \text{ kW}$$

Net heat available based on the first law of thermodynamics (Q) = –80 kw.

Hence, as expected, the net heat available obtained from both the first and second laws of thermodynamics is the same.

c. Cascade diagram.

Cascade diagram for the above shifted-temperature scales diagram is drawn and shown in Figure 18.6.

The temperature where there is no energy transfer between second and third temperature intervals is called pinch temperature.

d. Minimum hot and cold utilities requirements:

Minimum hot utility requirement = 300 kW

Minimum cold utility requirement = 220 kW

e. Pinch temperature = 125°C (based on hot temperature scale)

= 115°C (based on cold temperature scale)

= 120°C (based on an average)

f. Temperature enthalpy diagram.

Define the enthalpy corresponding to the lowest temperature of any hot stream as base condition, that is, $H = 0$, to calculate the enthalpy in each temperature interval by using the cumulative approach. Similarly, for cold streams, calculate the cumulative enthalpy by taking the minimum cold utility requirement (220 kW in the present case) as the enthalpy corresponding to the lowest temperature of any cold stream.

Estimation of cumulative enthalpy for hot and cold ends in each temperature interval is given in Tables 18.4 and 18.5.

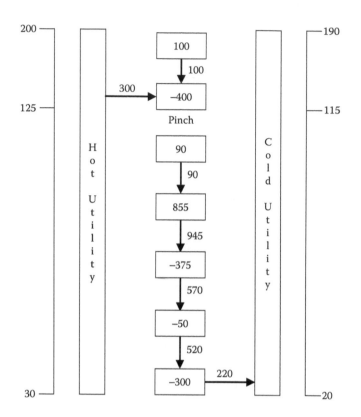

FIGURE 18.6
Cascade diagram.

TABLE 18.4

Cumulative Enthalpy for Hot Ends

Hot Streams, °C	Enthalpy, H	Cumulative H, kW
$T = 45$	$H_0 = 0$	0
$T = 50$	$H_1 = 10\,(50 - 45) = 50$	50
$T = 65$	$H_2 = 10\,(65 - 50) = 150$	200
$T = 125$	$H_3 = (10 + 40)(125 - 65) = 3000$	3200
$T = 175$	$H_4 = 10\,(175 - 125) = 500$	3700

TABLE 18.5

Cumulative Enthalpy for Cold Ends

Cold Streams, °C	Enthalpy, H	Cumulative H, kW
$T = 20$	$H_0 = 220$	220
$T = 40$	$H_1 = 20\,(40 - 20) = 400$	620
$T = 112$	$H_2 = (20 + 15)(112 - 40) = 2520$	3140
$T = 155$	$H_3 = 20\,(155 - 112) = 860$	4000

The temperature-enthalpy diagram is plotted by using hot and cold ends data and is shown in Figure 18.7.

The point of closest approach (pinch) between the curve, the temperature difference, is 10°C.

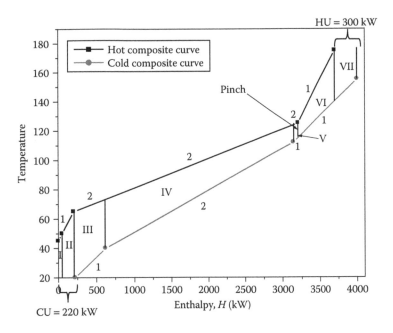

FIGURE 18.7
Temperature enthalpy diagram.

Hot utility requirement = 300 kW

Cold utility requirement = 220 kW

g. The area for each interval can be estimated by Equation 18.1.

$$A = \frac{Q}{\Delta T_{LM}} \left(\sum_i^{hot} \frac{1}{h_i} + \sum_j^{cold} \frac{1}{h_j} \right) \tag{18.1}$$

where:

Q is the heat load

ΔT_{LM} is the logarithmic mean temperature different $= \Delta T_1 - \Delta T_2 \Big/ \ln \dfrac{\Delta T_1}{\Delta T_2}$

h is the heat transfer coefficient

The total area is estimated by adding the areas obtained from each of the intervals.

First interval:

Area for the first interval can be estimated by three different approaches.

(i) Driving force for heat transfer at both ends can be assumed as $\Delta T_{min} = 10°C$

$$\Delta T_1 \quad = 45 - 15 = 30°C$$
$$\Delta T_2 \quad = 65 - 20 = 45°C$$
$$\Delta T_{LM} = 36.99°C$$
$$Q \quad = 200 \text{ kW}$$
$$h_{CU} \quad = 0.5 \text{ kW/m°C}$$
$$h_{H1} \quad = 0.4 \text{ kW/m°C}$$
$$A_1 \quad = 24.33 \text{ m}^2$$

Second interval:

$$\Delta T_1 = 65 - 20 = 45°C$$
$$\Delta T_2 = 65.5 - 20 = 45.5°C$$
$$\Delta T_{LM} = 45.25°C$$
$$Q = 20 \text{ kW}$$
$$h_{H1} = 0.4 \text{ kW/m°C}$$
$$h_{H2} = 0.55 \text{ kW/m°C}$$
$$h_{C1} = 0.75 \text{ kW/m°C}$$
$$A_2 = 2.498 \text{ m}^2$$

Third interval:

$$\Delta T_1 = 65.5 - 20 = 45.5°C$$
$$\Delta T_2 = 73 - 40 = 33°C$$
$$\Delta T_{LM} = 38.92°C$$
$$Q = 400 \text{ kW}$$
$$h_{H1} = 0.4 \text{ kW/m°C}$$
$$h_{H2} = 0.55 \text{ kW/m°C}$$
$$h_{C1} = 0.75 \text{ kW/m°C}$$
$$A_3 = 58.104 \text{ m}^2$$

Fourth interval:

$$\Delta T_1 = 73 - 40 = 33°C$$
$$\Delta T_2 = 24 - 112 = 12°C$$
$$\Delta T_{LM} = 20.76°C$$
$$Q = 2520 \text{ kW}$$
$$h_{H1} = 0.4 \text{ kW/m°C}$$
$$h_{H2} = 0.55 \text{ kW/m°C}$$
$$h_{C1} = 0.75 \text{ kW/m°C}$$
$$h_{C2} = 0.65 \text{ kW/m°C}$$
$$A_4 = 872.77 \text{ m}^2$$

Fifth interval:

$$\Delta T_1 = 124 - 112 = 12°C$$
$$\Delta T_2 = 125 - 115 = 10°C$$
$$\Delta T_{LM} = 10.97°C$$
$$Q = 60 \text{ kW}$$
$$h_{H1} = 0.4 \text{ kW/m°C}$$
$$h_{H2} = 0.55 \text{ kW/m°C}$$
$$h_{C2} = 0.65 \text{ kW/m°C}$$
$$A_5 = 32.033 \text{ m}^2$$

Sixth interval:

$$\Delta T_1 = 125 - 115 = 10°C$$
$$\Delta T_2 = 175 - 140 = 35°C$$
$$\Delta T_{LM} = 19.96°C$$
$$Q = 500 \text{ kW}$$
$$h_{H1} = 0.4 \text{ kW/m}°C$$
$$h_{C2} = 0.65 \text{ kW/m}°C$$
$$A_6 = 101.164 \text{ m}^2$$

Seventh interval:

$$\Delta T_1 = 175 - 140 = 35°C$$
$$\Delta T_2 = 330 - 155 = 175°C \text{ (if Tout of HU is not given)}$$
$$\Delta T_{LM} = 86.98°C$$
$$Q = 300 \text{ kW}$$
$$h_{HU} = 0.5 \text{ kW/m}°C$$
$$h_{C2} = 0.65 \text{ kW/m}°C$$
$$A_7 = 12.20 \text{ m}^2$$

Total area requirement for heat transfer before network proposal = 1103.10 m²

h. Estimation of number of heat exchangers based on the first law analysis:

Consider the heating and cooling loads for each of the process streams as well as the minimum utility requirements that correspond to the second law. Ignore the minimum approach temperature and consider the number of possible paths (heat exchangers) to transfer the heat from the sources to the sinks.

First way of representing the heat transfer from sources to sink is given in Figure 18.8.

Number of heat exchangers (I law) = Number of streams (N_S)
+ Number of utilities (N_U)
+ Number of loops (N_L)
− Number of independent problems (N_I)
= 4 + 2 + 0 − 1
= 5

Another way of representing the heat transfer from sources to sink is given in Figure 18.9.

In the first representation, five heat exchangers are required to transfer the heat from sources to sinks. However, in the second way of representation, four heat exchangers are needed.

Number of heat exchangers (I law) = Number of streams
+ Number of utilities
+ Number of loops
− Number of independent problems
= 4 + 2 + 0 − 2
= 4

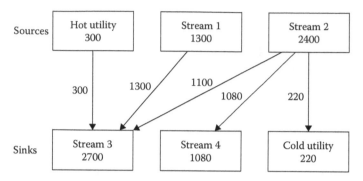

FIGURE 18.8
Minimum number of exchangers using the first law of thermodynamics.

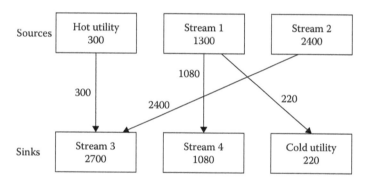

FIGURE 18.9
Minimum number of exchangers using the first law of thermodynamics by incorporating the independent problems.

i. Estimation of number of heat exchangers based on the second law analysis:

In calculating minimum heating and cooling requirements, we found that there was a pinch temperature that decomposed the problem into two distinct points: (1) Above the pinch: only supply heat from a utility (hot). (2) Below the pinch: only remove heat and give to a utility (cold). For incorporating the second law into our analysis, we need to bring in the pinch, which means decomposing the problem into two parts (above the pinch and below the pinch), as shown in Figure 18.10. Hence, we need to calculate the number of exchangers separately for above and below the pinch.

Assuming that there are no loops and one independent problem for both above and below the pinch, we can readily calculate total number of heat exchangers (N_E) by using Equation 18.2:

$$N_E = N_S + N_U - 1 \tag{18.2}$$

$$\text{Above pinch: } N_E = 2 + 1 - 1$$
$$= 2$$
$$\text{Below pinch: } N_E = 4 + 1 - 1$$
$$= 4$$

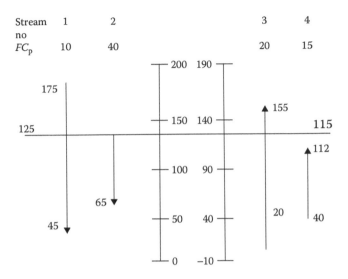

FIGURE 18.10
Decomposing of the problem into two parts (above pinch and below pinch).

Total number of heat exchangers using the second law analysis $= 2 + 4 = 6$

To satisfy the minimum heating and cooling requirements, we need a total of six heat exchangers (second law).

However, to satisfy the first law, we requires only five heat exchangers.

$$N_E = N_S + N_U - N_I = 4 + 2 - 1 = 5$$

Thus, network for the minimum energy requirements has one loop that crosses the pinch.

j. Hot end design:

Estimation of heat loads for each stream:

For stream 1: $Q = FC_p(T_{in} - T_{out}) = 30(300 - 150) = 4500$ KW

Similarly, the heat loads (Q) for other streams are also calculated.

The heuristics for feasibility of matches above the pinch (near pinch matches) are:

1. No. of hot streams (N_H) ≤ No. of cold streams (N_C)

2. $(FC_p)_{Hot\ stream}$ ≤ $(FC_p)_{Cold\ stream}$

In the present case, first condition is satisfied ($N_H = N_C$).

FC_p of stream 1 (hot) is less than FC_p of stream 3, which means that stream 1 can be matched with stream 3.

Based on this analysis, heat exchangers are placed in different streams (see Figure 18.11).

k. Cold end design:

The heuristics for feasibility of matches below the pinch (near pinch matches) are:

1. No. of hot streams (N_H) ≥ No. of cold streams (N_C)

2. $(FC_p)_{Hot\ stream}$ ≥ $(FC_p)_{Cold\ stream}$

In the present case, first condition is satisfied ($N_H = N_C$).

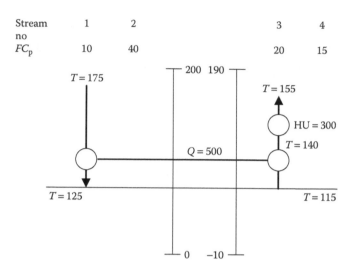

FIGURE 18.11
Hot end design.

Stream 2 can be matched either with stream 3 or with stream 4, but stream 1 cannot be matched with either stream 3 or stream 4.

$$[(FC_P)_1 < (FC_P)_3 \text{ and } (FC_P)_1 < (FC_P)_4]$$

So, we have to split stream 3; then, $N_H < N_C$.

In order to satisfy this criterion, we also have to split stream 2.

Then, the matches are shown in Figure 18.12.

1. After combining hot end and cold end designs, heat exchanger network for the maximum energy recovery (MER) is given in Figure 18.13.

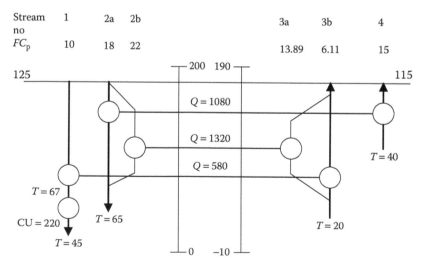

FIGURE 18.12
Cold end design.

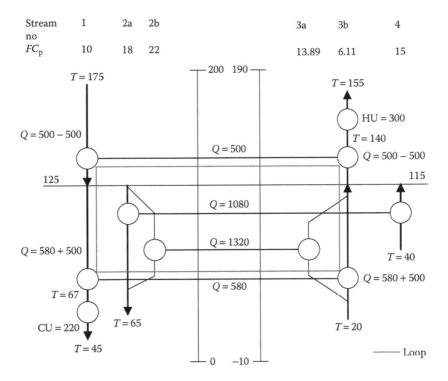

Stream no 1, 2a, 2b, 3a, 3b, 4 with FC_p = 10, 18, 22, 13.89, 6.11, 15

FIGURE 18.13
Heat exchanger network for maximum energy recovery and identification of loop.

m. Identification of loop.

There is only one loop that crosses the pinch.

n. Breaking the loop and restoring ΔT_{min}, if any:

According to the heuristics, remove the lowest load in the loop first. Smallest load in the loop corresponds to heat exchanger with $Q = 500$ kW. When the smallest load ($Q = 500$ kW) from the loop is removed by breaking it, the resultant network is shown in Figure 18.14.

After breaking the loop, the temperatures of various streams at entrance and exit of each heat exchanger are calculated and shown. As can be seen from Figure 18.11, there is a ΔT_{min} violation for heat exchanger with load $Q = 1080$ kW (ΔT_{min} violation is encircled in the figure below).

$$
\begin{array}{c}
175 \longrightarrow \quad \boxed{Q = 1080} \quad \longrightarrow 67 \\
197 \longleftarrow \quad \qquad \quad \longleftarrow 20
\end{array}
$$

Outlet cold stream temperature has to be made to 165°C.

ΔT_{min} violation has to be restored by adjusting the load along the path.

Calculation for Q_E:

$$1080 - Q_E = 6.11(165 - 20)$$
$$Q_E = 194.05 \approx 194 \text{ kW}$$

A heat load of $Q_E = 194$ kW has to be shifted along the path to restore $\Delta T_{min} = 10$°C.

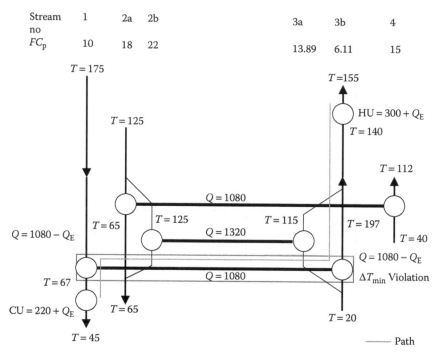

FIGURE 18.14
Heat exchanger network after breaking loop.

o. The resultant network is shown in Figure 18.15.

The above network corresponds to minimum heat exchanger area.

Total number of heat exchangers = 5

Hot utility requirement = 494 kW

Cold utility requirement = 414 kW

p. Total area required can be calculated by adding the area of each heat exchanger in the network.

The area for heat exchanger can be estimated by Equation 18.3.

$$A = \frac{Q}{\Delta T_{LM}} \left(\frac{1}{h_{hot}} + \frac{1}{h_{cold}} \right)$$

(18.3)

where:

Q is the heat load

ΔT_{LM} is the logarithmic mean temperature difference = $\Delta T_1 - \Delta T_2 \Big/ \ln \dfrac{\Delta T_1}{\Delta T_2}$

h is the heat transfer coefficient

Heat exchanger of $Q = 1080$ kW

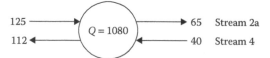

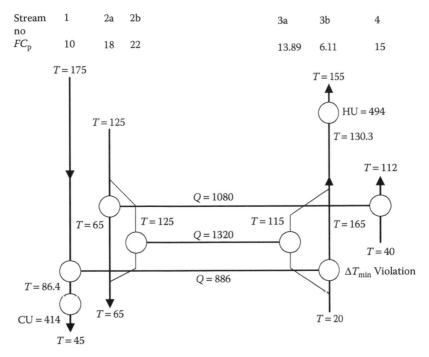

FIGURE 18.15
Final heat exchanger network.

$$\Delta T_1 = 125 - 112 = 13°C$$
$$\Delta T_2 = 65 - 40 = 25°C$$
$$\Delta T_{LM} = 18.35°C$$
$$Q = 1080 \text{ kW}$$
$$h_{H2a} = 0.2475 \text{ kW/m°C}$$
$$h_{C2} = 0.65 \text{ kW/m°C}$$
$$A_1 = 328.35 \text{ m}^2$$

Heat exchanger of $Q = 1320$ kW

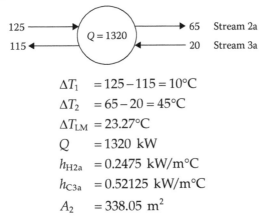

$$\Delta T_1 = 125 - 115 = 10°C$$
$$\Delta T_2 = 65 - 20 = 45°C$$
$$\Delta T_{LM} = 23.27°C$$
$$Q = 1320 \text{ kW}$$
$$h_{H2a} = 0.2475 \text{ kW/m°C}$$
$$h_{C3a} = 0.52125 \text{ kW/m°C}$$
$$A_2 = 338.05 \text{ m}^2$$

Heat exchanger of $Q = 884.5$ kW

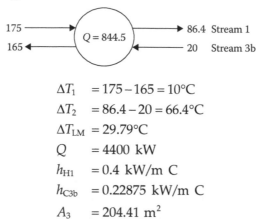

$$\Delta T_1 \quad = 175 - 165 = 10°C$$
$$\Delta T_2 \quad = 86.4 - 20 = 66.4°C$$
$$\Delta T_{LM} = 29.79°C$$
$$Q \quad = 4400 \ kW$$
$$h_{H1} \quad = 0.4 \ kW/m \ C$$
$$h_{C3b} \quad = 0.22875 \ kW/m \ C$$
$$A_3 \quad = 204.41 \ m^2$$

Heat exchanger of HU = 494 kW

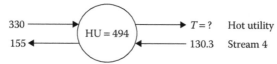

The target temperatures of hot and cold utilities are to be supplied, as they are based on the design criteria of boiler and cooling tower, respectively, in the actual practice. However, in the absence of these data, as a first approximation, the target temperatures of hot and cold utilities could be estimated by using the heat loads prevailing.

So, the estimation of outlet temperature of hot utility is

$$494 \quad = 20(330 - T_{HUout})$$
$$T_{HUout} = 305.3°C$$
$$\Delta T_1 \quad = 330 - 155 = 175°C$$
$$\Delta T_2 \quad = 305.3 - 130.3 = 175°C$$
$$\Delta T_{avg} = 175°C$$
$$Q \quad = 494 \ kW$$
$$h_{HU} \quad = 0.5 \ kW/m°C$$
$$h_{C1} \quad = 0.75 \ kW/m°C$$
$$A_4 \quad = 9.41 \ m^2$$

Heat exchanger of CU = 414 kW

Similarly, as we estimated the target temperature of hot utility, the estimation of outlet temperature of cold utility is

$$414 \quad = 10(15 - T_{CUout})$$
$$T_{CUout} = 56.4°C$$
$$\Delta T_1 \quad = 86.4 - 56.4 = 30°C$$
$$\Delta T_2 \quad = 45 - 15 = 30°C$$
$$\Delta T_{avg} = 30°C$$
$$Q \quad = 414 \text{ kW}$$
$$h_{CU} \quad = 0.5 \text{ kW/m°C}$$
$$h_{H1} \quad = 0.4 \text{ kW/m°C}$$
$$A_5 \quad = 62.11 \text{ m}^2$$

Total area of heat exchangers for the final network $= A_1 + A_2 + A_3 + A_4 + A_5 = 942.33 \text{ m}^2$

q. Comparison of results obtained in (g) and (p):

Before design, the area required $= 1103.10 \text{ m}^2$

Area required with area targeting (network) $= 942.33 \text{ m}^2$

r. Estimation of energy saving:

Net heat available in hot streams $= 3700 \text{ kW}$

Net heat available in cold streams $= 3780 \text{ kW}$

Heat required from cold utility $= 3700 \text{ kW}$

Total heat required from hot utility $= 3780 \text{ kW}$

After energy integration analysis (figure mentioned above)

Total heat required from cold utility $= 414 \text{ kW}$

Total heat required from hot utility $= 494 \text{ kW}$

Energy saving estimation

Energy saved for cold utility $= 3700 - 414 = 3286 \text{ kW}$

% Energy saving in cold utility $= (3286/3700)100 = 88.8\%$

Energy saved for hot utility $= 3780 - 494 = 3286 \text{ kW}$

Similarly, % Energy saving in hot utility $= (3286/3780)100 = 86.9\%$

It may be noted that these savings are possible at the expense of additional heat exchangers, the capital cost of which has to be accounted for the overall cost savings.

The above case study indicates the importance of heat exchanger network synthesis (HENS). The hot and cold utility requirements are 3780 kW and 3700 kW, respectively, before proposing the HEN. In the shifted-temperature scale diagram, transfer all the heat available at the highest interval to the next lower-temperature interval and repeat for all intervals. Since the heat is transferred to a lower-temperature interval, the second law is satisfied. The total minimum hot utility requirement and minimum cold utility requirement are 300 kW and 220 kW, respectively. The difference between the two values, 80 kW, is still consistent with the first law requirement. However, the minimum hot and cold utility requirements have now been fixed to satisfy the second law. After proposing the final heat exchanger network, this requirement is further increased to 494 kW and 414 kW for hot and cold utility requirement, respectively. However, the overall energy savings in terms of hot and cold utility requirements are 86.9% and 88.8%, respectively. The total area for heat exchange and the number of heat exchanger required for this case study are estimated to be 942.33 m^2 and 5, respectively. This study indicates that with proper heat exchanger network design, lot of energy can be saved in the process industry.

18.4 Summary

In process design, energy conservation has always been an important activity. Energy auditing and HENS are found to be key tools for energy savings in the process plants. Energy auditing is considered an efficient tool to identify the available energy sources, energy wastage, and utilization pattern in the process plants. Pinch analysis is now widely used method to perform HENS. This method can be applied efficiently to almost any process plant (large or small scale). The case study demonstrated that approximately 87% of the energy can be saved with the application of appropriate heat exchanger network design.

References

1. Ammar, Y. et al., Low grade thermal energy sources and uses from the process industry in the UK. *Applied Energy*, 2012. 89(1): pp. 3–20.
2. Mathur, N., *Load Generation Balance Report 2014–15*. 2014, Central Electricity Authority: New Delhi, India.
3. Patra, T.K. and P.N. Sheth, *Biomass gasification models for downdraft gasifier: A state-of-the-art review*. Renewable and Sustainable Energy Reviews, 2015. 50: pp. 583–593.
4. Sharma, S. and P.N. Sheth, Air–steam biomass gasification: Experiments, modeling and simulation. *Energy Conversion and Management*, 2016. 110: pp. 307–318.
5. Sharma, R., P.N. Sheth, and A.M. Gujrathi, Kinetic modeling and simulation: Pyrolysis of Jatropha residue de-oiled cake. *Renewable Energy*, 2016. 86: pp. 554–562.
6. Sheth, P. and B. Babu, Effect of moisture content on composition profiles of producer gas in downdraft biomass gasifier. *Proceedings of International Congress Chemistry and Environment (ICCE)*, 2005: pp. 356–360.
7. Grinbergs, K., Energy Audit Method for Industrial Plants. In *Civil Engineering. International Scientific Conference. Proceedings (Latvia)*. 2013: Latvia University of Agriculture, Jelgava.
8. Engin, T. and V. Ari, Energy auditing and recovery for dry type cement rotary kiln systems—A case study. *Energy Conversion and Management*, 2005. 46(4): pp. 551–562.
9. Zhu, Y., Applying computer-based simulation to energy auditing: A case study. *Energy and Buildings*, 2006. 38(5): pp. 421–428.
10. Klugman, S., M. Karlsson, and B. Moshfegh, A Scandinavian chemical wood pulp mill. Part 1. Energy audit aiming at efficiency measures. *Applied Energy*, 2007. 84(3): pp. 326–339.
11. Chaudhary, V.P. et al., Energy auditing of diversified rice–wheat cropping systems in Indo-gangetic plains. *Energy*, 2009. 34(9): pp. 1091–1096.
12. Linnhoff, B. and E. Hindmarsh, The pinch design method for heat exchanger networks. *Chemical Engineering Science*, 1983. 38(5): pp. 745–763.
13. Douglas, J.M., *Conceptual Design of Chemical Engineering*. 1988, New York: McGraw Hill.

14. Agarwal, A. and S.K. Gupta, Multiobjective optimal design of heat exchanger networks using new adaptations of the elitist nondominated sorting genetic algorithm, NSGA-II. *Industrial & Engineering Chemistry Research*, 2008. 47(10): pp. 3489–3501.

15. Chang, H. and J.-J. Guo, Heat exchange network design for an ethylene process using dual temperature approach. *Tamkang Journal of Science and Engineering*, 2005. 8(4): p. 283.

16. Lotfi, R. and R. Bozorgmehry, Development of a New Approach for Synthesizing Heat Exchanger Networks to Save Energy in Chemical Plants. In *WSEAS International Conference on Energy Planning, Energy Saving, Environmental Education*, 2007. Arcachon, France.

19

Green Building

Shahriar Shams and M. Motiar Rahman

CONTENTS

ABSTRACT   The process of designing, constructing, and inhabiting the buildings has a profound influence on a community's economy, environment, and quality of life. Building contributes almost 40% of CO_2 emissions and is a major contributor to the greenhouse gas (GHG) emissions. It is therefore necessary to harmonize the relationships between the buildings, the environment, and the inhabiting communities through sustainable or green buildings. This chapter discusses history and concepts of green buildings, impacts of buildings on the environment with embodied energy, and CO_2 emission from common construction materials. Major focus is given on green or sustainable design practices for buildings. It is proposed that both architectural and structural designers should first analyze the building scenario based on available "green" design strategies and then collate their outcomes with various design dimensions, with the help of Building Information Modeling, to finally decide on a building-specific model. The practice of green building is growing around the world. A number of examples of green buildings from both developed and developing countries, with their key features, are discussed. The rating criteria for assessing the green buildings, along with certified body to assess the energy performance, are discussed in details. It is observed that green buildings offer a wide range of benefits, including reduced use of materials and energy, as well as low carbon dioxide emission. On the other hand, main barriers seem to be the initial increased investment. It is expected that the readers will gain valuable knowledge on green building, involving architectural, structural, and passive designs for reduction of energy use and waste, leading to an affordable life in the future.

19.1 Introduction

Buildings are one of the vital components of the built environment; they provide the setting for human activity in this modern world. They include the place at which we work, live, and do other things. Internal air quality is therefore important for human health and well-being, which depend on the materials used in construction and on the manner (i.e., the process through which) they are used. Buildings result in loss of farmland and consume naturally occurring construction materials, a part of which becomes waste. They are major energy-consumption sector of many countries and are the cause of huge carbon dioxide emissions. According to the U.S. Environmental Protection Agency (EPA), buildings in the United States generate 160 million tons of construction and demolition (C&D) waste every year and consume 13% of water and 40% of total energy [1]. The latter is attributed as the principal reason of climate change, since consumption of energy is the major contributor to carbon dioxide emissions [2]. Moreover, unsustainable building practices such as urban sprawl, brownfields' redevelopment, access to healthy food, loss of open space, and health impacts due to decreased physical activity can lead to unintended social and economic consequences [1].

Therefore, the process of designing, constructing, and inhabiting the buildings has a profound influence on a community's economy, environment, and quality of life. It is therefore necessary to harmonize the relationships between the buildings, the environment, and the inhabiting communities through sustainable or green buildings. Many construction firms (51%) are shifting their business toward green building [3]. Nevertheless, many attempt to achieve this by ensuring the design and construction requirements of different established sustainability assessment methodologies. Our attempt is to achieve the same through the practice of sustainable design practices. The aim is to integrate the buildings with the local ecology, while simultaneously attempting to lessen their impacts on natural resources, minimize energy consumption and carbon dioxide emissions, use environmentally friendly products, and improve operation and maintenance practices [1].

The previous chapters have highlighted the integrated management of water with optimization of water and energy use in processing industries. This chapter focuses on the sustainable design principles for green buildings. However, we first discuss the concept of green building, highlight some impacts of buildings on environment, and summarize benefits of green buildings. Sustainable design is then discussed at length, and some examples of successful green buildings from around the world are provided. The chapter then concludes with the assessment and economics of green buildings.

19.2 History and Concept of Green Building

"Green building" is also known as "sustainable building," "sustainable construction," or "green construction." While there are many definitions for it, green building refers to both a structure (i.e., the constructed facility) and the processes responsible for constructing that structure, to be environmentally accountable and resource-efficient, throughout a building's life cycle. Therefore, relevant activities span from conceptualizing, planning, designing, and constructing through to operation, maintenance, renovation, and demolition. These are carried out with a focused consideration of a few core issues of energy use (and thereby carbon dioxide emissions), building's effect on the adjacent ecology, selection of construction materials, and indoor environmental quality [2].

The concept of green building appears to have emerged in 1960s, when many cities saw further, and rapid, urbanization as a result of constructing high-rise-tower buildings, for example, as happened in Chicago and New York for providing our dwelling homes, work places, and other public buildings. Such physical infrastructures reflect the economic strength or growth of a nation and also validate the success and progress of human society [4]. However, they (i.e., the buildings) not only consumed high volume of concrete (thereby creating demand on naturally occurring construction materials, e.g., boulders for aggregates, limestones for cement, and woods) but also led to the clearance of forest cover and replacement of communities. Moreover, they created huge demand of large amounts of energy for running and maintaining the ventilation and other services of those buildings, heating during the winter, and air conditioning in the hot summer. As a result, architects, civil engineers, and property developers began to rethink about their crucial roles as planners, designers, and developers of the buildings, the most cherished environment of where we live and work.

Green building started to emerge as reality from its research and development phase to overcome the energy crisis of the 1970s. Developers and designers focused on reducing the energy uses derived from fossil fuel to reduce reliance on fossil fuels, additional expenses from relevant higher costs, and carbon dioxide emissions. Solar panels were introduced as an alternate, and renewable energy was used to lessen the dependence on conventional fossil fuels and to make more environmentally friendly homes. However, the uses of solar panels were very limited owing to their high initial costs. Therefore, emphases were given on manufacturing solar panels that were more efficient and less expensive, making solar energy more of a reality. Designers, developers, individual users, and companies also focused on reducing energy bills through various types of design, such as "passive solar" buildings made of non-toxic and low-impact materials. These are designed with good insulation, adequate ventilation, wider use of natural light through integrated design, greening of surrounding areas, and optimum use of water through various water-saving devices [2].

The growing "green movement," particularly during 1990s, responded with sustainable and wise use of natural resources for the reduction of greenhouse gas (GHG) emission, which is responsible for global warming and climate change [5]. The green building then received increased popularity. The American Institute of Architects (AIA) formed the Committee on the Environment in 1989, and it led to one of the pioneering steps toward green movement in the United States. Also, the Environmental Protection Agency (EPA) and the Department of Energy launched the Energy Star program in 1992. Austin, Texas, was the first U.S. city to introduce a local green-housing program. The following year, that is, 1993, saw many more advancements in the green revolution, and the U.S. Green Building Council (USGBC) was founded. A large number of people were motivated by USGBC initiatives, and they made changes in their homes by having appliances with energy star ratings. In 1998, the USGBC launched their Leadership in Energy and Environmental Design (LEED) program and promoted the practice and benefits of having green building [2].

As mentioned above, the history of green-building movement evolved owing to demand for more energy-competent and environmentally less-damaging construction practices. As it is practiced now, the approach includes an array of design strategies that focus on the selection of a suitable location and also of construction materials. Buildings consume almost 40% of energy in most countries [6]. This may be more in many mega-cities such as Beijing, Shanghai, Mumbai, Jakarta, Dhaka, and Mexico City, where construction is also booming, implying that those will consume more energy. The concept of sustainable construction (or green building) takes into consideration the environmental, socioeconomic and cultural issues. It also addresses the issues such as design, construction and management of buildings, performance of buildings, and energy and resource consumption. This is now regarded as the guideline for the building industry to move toward achieving sustainable development [7–8].

Green building or sustainable building is the process of having a constructed facility that encompasses ultimate energy efficiency, forward-thinking resources management, and general sustainable construction [9]. Larsson [10] explained how integrated design process (IDP) assists in applying green-building practices and ensures the prospect of introducing environmentally friendly and resource-efficient buildings. According to Sorrel [11], the barriers to and complexity of developing green building arising from building services, structural and architectural features, and related costs can be overcome through an integrated team having a wide range of different specialists. While Section 5 of this chapter explains five examples (or case study with visuals) of green buildings from around the world, we summarize below the salient features of green buildings:

- Resource conservation, including energy efficiency, renewable energy, and water-saving features
- Consideration of environmental impacts and waste minimization
- Reduction of operation and maintenance costs and use of environment friendly construction materials through life cycle assessment during the planning and development process [12]
- Creation of a healthy and comfortable environment, leading to overall quality of life [12]

19.3 Impacts of Building on the Environment

Climate of a region significantly influences the nature of a building and its energy demands. For instance, the demand for heating energy is the highest in colder climates. Intertek Report [13] observed that 68.4% of all energy consumption in dwellings is attributed to space heating. On the other hand, hotter regions require more attention to cooling. On the whole, in either case, more use of energy means more emission of carbon dioxide, which is the leading cause of climate change [2]. Thus, the nature of the buildings creates counter-influence onto the environment.

The higher demand of energy plays a significant role in the design of building. For example, buildings in colder climates give more focus on better air tightness and insulation. Humidity, rainfall, and temperature are crucial factors that should be taken into consideration during the building design. However, it is impossible to come with a single technical solution to energy efficiency that will work in all cities and countries when all those meteorological parameters need to be considered. As such, American Society of Heating, Refrigeration, and Air-Conditioning Engineers (ASHRAE) assesses energy-related design conditions through a combination of factors, including the numbers of heating and cooling days.

Moreover, despite the remarkable benefits of green buildings, the activities for constructing them can be a major source of environmental damage through depletion of the natural resource base, degradation of fragile eco zones, chemical pollution, and the use of building materials harmful to human health [14].

As seen above, the design, construction, and maintenance of buildings have a remarkable impact on the environment and natural resources. Impacts of buildings can be analyzed from multiple perspectives, such as the type of buildings (private and commercial, office, dwellings, civic centers, schools, and hospitals), life cycle phases (design, construction, operation, and refurbishment), and type of environmental impacts (soil, energy, CO_2 emissions, material, water, and waste). Nevertheless, residential and commercial buildings consume approximately 60% of the world's electricity [15]. On the other hand, Augenbroe et al. [16] summarize that, globally, buildings account for:

- One-sixth (17%) of the world's freshwater withdrawals
- One-quarter (25%) of world's wood harvest

- Two-fifths (40%) of world's material and energy flows
- Nearly one-quarter (25%) of all ozone-depleting chlorofluorocarbons (CFCs) emissions generated from building air conditioners and the processes used to manufacture building materials

Indiscriminate use of building material, without giving a thought to the environment, possesses a challenge to the effort of reducing GHG contributed from buildings. A report from the Worldwatch Institute stated that building construction uses 55% of non-fuel wood [17], 40% of global resources, and 40% of global energy [15]. Moreover, resources, such as ground cover, water, and tillable lands, are depleted to accommodate space required for buildings. The energy consumption in both residential and commercial buildings is likely to remarkably increase in the next 30 years, particularly in the developing countries. On the other hand, energy consumption of developed countries will increase very steadily, in general. This steady increase in developed countries indicates their use of sustainable technologies and various energy-efficient construction materials in the built environment. This is frequently measured in terms of "embodied energy," a term to express "the total energy required for the extraction, processing, manufacture and delivery of building materials to the building site." It is imperative to use materials with less embodied energy in buildings, in an attempt to reduce overall GHG emission. As such, the embodied energy and CO_2 emission of common construction materials [18,19] are shown in Table 19.1, as ready reference.

TABLE 19.1

Embodied Energy and CO_2 Emission of Common Construction Materials

Materials	Embodied Energy (MJ/kg)	CO_2 Emission (kg/kg)	
Gravel and sand	0.5	0.0018	
Timber	3	0.003	
Concrete	1.3	0.1311	
Aggregate	0.1	0.16	
Brick	2.5	0.189	
Ceramics	5	0.349	
Glass	30.3	0.748	Lower to Higher
Cement	5.8	0.9638	Emission
Lime	6	1.352	
Plastics	61	2.2	
Paint	93.3	2.42	
PVC	70	2.6904	
Steel	32	2.95	
Plywood	10.4	4.2	
Aluminum	227	9.964	

Source: Kibert, C.J., *Sustainable Construction, Green Building Design and Delivery*, John Wiley & Sons, Hoboken, NJ, 2013; Mulavdic, E., *Therm. Sci.*, 9, 39–52, 2005; Franchetti, M.J., and Apul, D., *Carbon Footprint Analysis: Concepts, Methods, Implementation, and Case Studies*, CRC Press, Boca Raton, FL, 2013.

19.4 Benefits, Drawbacks, and User Satisfaction of Green Building

Benefits of green buildings are manifold. Lallanilla [20] summarized eight interdependent benefits of green buildings: (1) lower costs of green buildings in terms of return from initial additional investment in sustainable design; (2) healthy occupants in green buildings owing to the use of non-toxic construction materials and improved ventilation; (3) improved quality of life of the occupants, perceived to be the outcome of healthy lifestyle; (4) improved productivity of occupants; (5) improved sales from and reduced electricity costs of the retail stores constructed by applying green-building principles; (6) reduced utility bills; (7) green buildings retain higher resale market value compared with conventional buildings; and (8) tax benefits for green buildings. Many authors/organizations also reported the benefits in greater detail and grouped them as economic, environmental, and social or health and community benefits, for example, USGBC [21], Lennox [22], and The City of Bloomington [23]. A cross section of those benefits is shown in Figure 19.1. It can be argued that some specific items of benefits can be arranged under a different category, for example, waste reduction, since it saves money. However, our primary aim is to show the divergence of the benefits, while also loosely categorizing them.

Benefits of green buildings		
Environmental benefits:	Economic benefits:	Social benefits:
• Emissions reduction	• Energy and water savings	• Improved health
• Water conservation	• Increased property values and profits	• Improved schools
• Stormwater management	• Decreased infrastructure strain	• Healthier lifestyles and recreation
• Temperature moderation	• Improved employee attendance	• Improved employee satisfaction
• Waste reduction	• Increased employee productivity	• Improve air, thermal, and acoustic environments
• Improved air and water quality	• Sales improvements	• Enhance occupant comfort and health
• Reduced solid waste	• Development of local talent pool	• Possibly limiting growth of mold and other airborne contaminants that can affect workers' productivity and/or health
• Conserve natural resources	• Reduced operating costs	
• Enhance and protect ecosystems and biodiversity	• Optimize life cycle performance	
	• Qualifying for various tax rebates, zoning allowances, and other incentives in many cities	

FIGURE 19.1
A cross section of benefits of green buildings. (Adopted from USGBC, What is Green building? United States Green Building Council (USGBC), 2014, Available at: http://www.usgbc.org/articles/what-green-building, accessed April 5, 2016; Kibert, C.J., *Sustainable Construction, Green Building Design and Delivery*, Hoboken, NJ, John Wiley & Sons, 2013; Ding, G.K.C., *J. Environ. Manage.*, 86, 451–464, 2008; Robichaud, L.B. and Anantatmula, V.S., *J. Manage. Eng.*, 27, 48–57, 2010; USGBC, Benefits of Green Buildings. United States Green Building Council (USGBC), 2015. Available at: http://www.usgbc.org/articles/benefits-green-homebuilding, accessed April 27, 2016; The City of Bloomington, Green Building Benefits, 2016. Available at: https://bloomington.in.gov/green-building-benefits, accessed April 27, 2016; Zuo, J. and Zhao, Z.Y., *Renew. Sust. Energ. Rev.*, 30, 271–281, 2014; Shams, S. et al., *Int. J. Eng. Technol.*, 3, 2011.)

From environmental perspective, green buildings help improve the urban biodiversity and protect the ecosystem through sustainable use of land [24–25]. Green buildings generally provide higher performance than conventional buildings owing to efficient use of energy and water and higher reduction in carbon dioxide emission owing to the use of materials with low embodied energy [26–27]. Furthermore, Turner and Frankel [28] observed that the LEED-certified buildings could achieve more than 28% of energy savings compared with the national average level.

Contrary to the benefits summarized above, green buildings have a number of drawbacks or disadvantages too, including high initial costs, unavailability of sustainable construction materials, problems in getting loan, and lack of suitable builders of green buildings [29–32]. However, the core drawback seems to be the cost—that sustainable buildings cost more than conventional buildings. In fact, sustainable buildings cost 2%–7% more, in general, if compared with conventional buildings [29], although the homes are peaceful and quiet and dwellers/occupants are more comfortable inside, irrespective of the outside weather. Moreover, decision makers rarely take into account the savings from reduced operating costs and utility bills of the sustainable buildings, as their estimates are hardly based on whole-life costs [33]. In this regard, many authors reported that the initial higher expenses are recovered from the incredible savings in utility bills. Al-Yami and Price [29] urged decision makers to focus on value, instead of cost, and on long-term perspectives, instead of short-term perspectives. However, Lallanilla [20] argues, "… green buildings save money, starting (from) the very first day of construction. This is true for green homes as well as sustainable office buildings, factories, churches, schools and other structures." He gave an example citing the 2003 case study conducted by the California Sustainable Building Task Force that: "… an initial green design investment of just two percent … (which may be) … $40,000 … in a $2 million project will be repaid in just two years. Over 20 years, the savings will amount to $400,000." In other words the savings will be 10 times higher over 20 years.

For any new concept or product, the satisfaction level or acceptability of the users plays a considerable role. As such, opinion of the users/occupants will greatly impact the future of green buildings. Foregoing paragraphs show that benefits of green buildings grossly outweigh the relevant drawbacks. Along the same line, studies have shown that occupants are more comfortable in green buildings, in general, compared with conventional buildings [34–35]. Moreover, occupants of green buildings are usually energy-conscious, so they highly appreciate natural lighting, cooling/heating, and ventilation systems, as well as the relevant building control system [36]. Such occupants are ready to change their lifestyle and sacrifice their comfort levels to conserve energy. Nevertheless, the building control system could be significantly different for non-energy-conscious occupants. Therefore, the designers have to appreciate the energy-conservation behavior of different types of occupants.

Santin [37] observed that less-energy-conscious occupants conserve more energy with systems that do not require active involvement, and on the other hand, energy-conscious occupants conserve more energy with systems that require active involvement. Azizi [36] observed that occupants in green buildings are less likely to adjust temperature and use personal heaters/fans when feeling hot or cold, as compared with conventional-building occupants, even though they have higher access to building control systems. Whether hot and cold, they tend to cope more with discomfort in comparison with occupants in conventional buildings, because they consider that the green buildings are designed to be energy-efficient, and therefore, the building operation should support energy-saving features.

19.5 Green or Sustainable Design

Achieving zero carbon or green building is a concept inspired to bring minimum impact on the natural environment on which development was brought. The concept mainly covers two broad intentions: (i) increasing the efficiency in the way buildings use energy, materials, and water; and (ii) reducing the impact of building on both the environment and human health, thereby ensuring comfort of occupants. A great deal of these are arguably achieved through design. Moreover, initiatives targeting sustainability improvement are more effective if they are considered in the design. As such, it is now becoming almost a truism that design must consider sustainable construction principles crafted in design strategies by all concerned professionals, that is, sustainable or green design. Sustainable design is mostly concerned with energy conservation features, use of renewable energy, water conservation features, use of low-GHG-emitting and recycled materials, reduced construction waste, and less environmentally destructive site development. We present here sustainable design strategies in three groups: (i) as in architectural practice; (ii) for structural engineers; and (iii) those that follow natural phenomenon, that is, passive design. Some argue that the latter should not be considered a separate stream rather all "designers" should consider it along with any practice-specific (i.e., architectural or structural) design strategy. However, we then provide a brief overview of how Building Information Modeling (BIM) can help use the above design strategies in building design, in general, and in green-building design, in particular, and lead to the development of structure-specific green-building model. This model can then be used for the entire life of the green building, namely in construction, operation, maintenance, renovation, and demolition.

19.5.1 Sustainable Design in Architecture

As shown in Figure 19.2, the University of Michigan proposed a framework for sustainable design in architecture, with different levels of principles, strategies, and methods [38]. The three principles are aligned with those of sustainability in architecture, which are economy of resources, life cycle design, and humane design. Economy of resources is concerned with the reduction, reuse, and recycling of the natural resources that are used in buildings, focusing on the conservation of energy water and building materials. Life cycle design (LCD), the second principle, offers a methodology for analyzing the building process and relevant impact of buildings on the environment. Humane design, the third principle, focuses on the interactions between the natural world and humans. These principles are applied in design by using a number of methods, which are summarized below.

19.5.1.1 Strategies and Methods under Principle One—Economy of Resources

There are three strategies under this principle, in order to focus on the conservation of energy, water, and building materials. Relevant methods to focus on energy conservation, which is strategy one, are energy-conscious urban and site planning, alternative sources of energy, avoidance of heat gain/heat loss, use of materials with low-embodied energy, use of energy-efficient appliances with timing devices, and passive heating and cooling.

Strategy two focuses on water conservation through the methods of reduction and reuse. Reduction deals with indigenous landscaping, vacuum-assist toilets or smaller toilet tanks, and low-flow showerheads. On the other hand, reuse method deals with

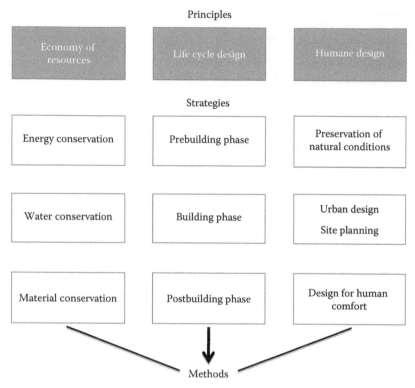

Principles

FIGURE 19.2
Framework for sustainable design in architecture. (From Kim, J.J. and Rigdon, B., *Introduction to Sustainable Design*, National Pollution Prevention Center for Higher Education, The University of Michigan, Ann Arbor, MI, 1998.)

collection and use of rainwater and gray water from domestic uses. Strategy three deals with the conservation of building materials, which are channeled through the methods of material-conserving design and construction, rehabilitation of existing structures, use of reclaimed/recycled materials and components, proper sizing of building systems, and use of nonconventional building materials.

19.5.1.2 *Strategies and Methods under Principle 2—Life Cycle Design*

The three strategies under this principle are aligned with the stages of the building, namely preconstruction stage, construction stage, and postconstruction stage. Preconstruction stage is concerned with the methods of minimizing energy needed to distribute/supply the materials and using the materials that are recycled, recyclable, made of renewable resources, long-lasting, harvested/extracted without ecological damage, and require low maintenance. The methods applied during construction stage include scheduling of construction activities to minimize site impact, provision for waste-separation facilities, specification of regular maintenance with nontoxic cleaners, and use of nontoxic materials to protect construction workers and end users. Strategy three of postconstruction stage is approached through the methods of adapting existing structures to new users and programs, recycling building materials and components, reusing building materials and components, and reusing the existing infrastructure and land.

19.5.1.3 Strategies and Methods under Principle Three—Humane Design

Among the three strategies under this principle, strategy one deals with the preservation of natural conditions; it is addressed through the methods of understanding the impact of design on nature, preserving the existing flora and fauna, respecting topographical contours, and not disturbing the water table. Strategy two is concerned with urban design site planning, which is appreciated through the methods of avoiding pollution contribution, providing for human-powered transportation, promoting mixed-use development, integrating design with public transportation, and creating pedestrian pockets. Strategy three is design for human comfort. Relevant methods are to provide clean and fresh air; thermal, visual, and acoustic comfort; operable windows; and visual connection to exterior, as well as accommodate persons with differing physical abilities.

19.5.2 Sustainable Structural Design

One of the goals for structural engineers in sustainable design is to minimize the impact of the project/building on the environment and natural resources, irrespective of whether designing high/low-rise buildings, long/short-span bridges or any structure in between. Danatzko and Sezen [39] compiled five methods that structural engineers can apply when designing a system or any structural element, as discussed below. Also, Figure 19.3 shows how the outcomes from the analysis of the five methods, jointly with the other factors

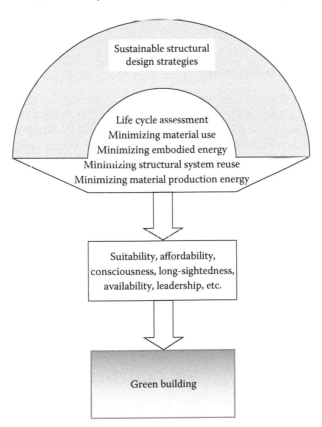

FIGURE 19.3
Green-building design process following sustainable design strategies.

affecting the decision makers' choice, lead to the eventual structure-specific design of the green building.

19.5.2.1 Minimizing Material Use

The objective of this method is reduction of the amount of materials in the use of the building/structure, aiming to reduce project's impact on environment. This is achieved either by optimization of a structural model, employing a single material type, or by combinations of various material types to form more efficient structural members and systems. Positive attributes of this method are to reduce requirement of raw materials and lower the impact on natural environment, which can lead to innovative designs and practices. Negative attributes are longer design and analysis time, possible greater structural system complexity, requirement of more drawings and details, and possible longer approvals process.

19.5.2.2 Minimizing Material Production Energy

The objective of this method is to reduce the production energy cost through reducing the amount of natural resources and energy required for the production of building materials. Positive attributes of this method are reduction of by-product and conservation of natural resources, which can lead to simultaneous assessment of strength and sustainability properties of materials and allow innovative designs. Negative attributes are lack of input from building industry, limitations to sustainability from material choice, and that the eventual design may not be "most" sustainable.

19.5.2.3 Minimizing Embodied Energy

The objective is to minimizing embodied energy by assessing the energy cost of construction and operational phase of the building. The core is to design long-life, durable, and adaptable buildings to reduce the impact of embodied energy. Positive attributes are considerations of both sustainable form and function, focus on operating energy use, and attention to service core. Negative attributes are that this method is highly sensitive to location or region, it can result in less-efficient structural system, its surrounding built environment can limit methodology, and its design is limited to most effective use of ambient energy (i.e., naturally occurring).

19.5.2.4 Life Cycle Analysis/Inventory/Assessment

The objective of this method is to justify/qualify the economic impact or net-cost-to-benefit ratio of a design decision. Positive attributes are greater inclusion of representative project parties, consideration on sustainability over project life, and that it encourages cross-discipline interaction. Negative attributes are risk and uncertainty included in analyses, potential model accuracy, and other sustainable issues that can detract from most sustainable structural design.

19.5.2.5 Maximizing Structural System Reuse

The objective of this method is to generate designs and layouts that generate minimum solid waste at end of life or to allow for the greatest amount of whole or partial system and/or structural component reuse. Positive attributes are financial incentives, extended service life, design to suit surrounding built environment, and potential innovation in standardized designs. Negative attributes are adverse effects from minimal design changes, possibility for reduced primary-use functionality, and the requirement of inspection for structural element reuse.

19.5.3 Passive Design

Buildings are highly sensitive to climate, and therefore, adequate attention should be given during their design to allow maximum natural ventilation and day lightings to minimize the cost of energy use. A study carried out by Azizi [37] found that people staying in green buildings are more used to personal and psychological adjustments and less concerned with environmental adjustments, when compared with people living in conventional buildings. It is therefore necessary for designers to understand the level on interaction of the occupants with the building control system, in order to design buildings that encourage occupants to practice energy-saving behavior. These are addressed through a number of "passive design" strategies, which considers two major aspects: (i) the use of the building's location and site to reduce the building's energy profile; and (ii) the design of the building itself, that is, its aspect ratio, orientation, ventilation paths, massing, fenestration, and other measures. The target is to optimize the design to remarkably reduce the energy costs of heating, cooling, ventilation, and lighting. Its success is highly dependent on factors such as local climate, site conditions, building aspect ratio, and building orientation. These are briefly discussed in the following subsections.

19.5.3.1 Incorporation of Nature

There should be shading in terms of vegetation and trees. The plants directly absorb the sunlight and carbon dioxide as well as produce oxygen, which contributes to natural ventilation.

19.5.3.2 Orientation

Building should be located on the long side on a true east-west axis to minimize solar loads on the east and west surfaces.

19.5.3.3 Aspect Ratio

The aspect ratio is the ratio of a building's length to its width, which is an indicator of the general shape of a building. For hot and humid climate, it is recommended to increase a building's aspect ratio, with the building becoming longer and narrower. This minimizes the east and west surfaces that experience the greatest sun load. It is further minimized by installing smaller windows [5]. However, under cold and dry climate, buildings require a decrease in their aspect ratio, with the building becoming shorter and wider and requiring installment of large windows.

19.5.3.4 Roofing

The roof has the highest heat gain within the building structure, with approximately 70% of total heat gain [40]. Thickness and reflective texture, along with types of color used for slanting roof used with tiles, can be effective in reducing heat transmission. Apart from the thickness, texture, and color of roof, green building focuses on development of eco-roof or green roof, which can reduce indoor temperature significantly. An eco-roof is made of plants, root barrier, soil mix, filter fabric, water-retention layer, drainage layer, waterproof layer, or insulation layer [41–42], allowing it to blend with surrounding environment to reduce heat island effect and assist in climate stabilization; it is particularly beneficial in wet climate [43].

19.5.3.5 Thermal Comfort

Thermal comfort comprising complex dynamics of temperature and humidity [44–46] is one of the key features that building users look for. Green buildings, which are mostly mechanically and naturally ventilated buildings, adopt the ASHRAE 552010 [47] and ISO7730 [48] guidelines, which suggest a thermal comfort temperature between 20°C and 24°C.

19.5.3.6 External Painting and Coloring

Various color surfaces can reflect invisible solar ray; for example, light-colored surface reflects 80%, dark-colored surface reflects 40%, and a normal dark-colored surface reflects 20% of incoming sunlight [42].

19.5.3.7 Flooring

Wood is rather preferred as flooring within the house, because it is a good conductor of heat in the context of buildings in cold climate. It is also convenient and cheaper to maintain in the long haul than carpet application. However, ceramic tiles/mosaics are more preferred for floors in hot climate to keep the interior cool.

19.5.3.8 Passive Cooling

Building should have minimal mass for storing energy and should generally be lightweight and well-insulated. Moujalled et al. [49] discovered that occupants prefer more naturally ventilated buildings as compared with air-conditioned buildings.

19.5.3.9 Daylighting

Based on scientific research, it was shown that using natural lighting provides great physical and psychological benefits to the building occupants. This, in return, generate an increase in productivity. The use of shading devices is vital for facades with large, glazed portions for providing adequate daylight in buildings. Highly glazed facades are widely used in new buildings to provide natural light and external view. There are different shading device types used to improve building energy performance, such as overhangs [50], external roller shades [51], venetian blinds [52], and internal shading. The field measurements and simulations showed that movable solar shading devices had crucial effect on energy performance and indoor thermal and visual comfort [53].

19.5.4 BIM and Green-Building Design

Autodesk, [54] one of the leading software developers and suppliers to construction industry, defines Building Information Modeling (BIM) as "an intelligent 3D model-based process that equips architecture, engineering, and construction professionals with the insight and tools to more efficiently plan, design, construct, and manage buildings and infrastructure." Although the BIM process is largely generic, even building information model is project- or structure-specific. A brief overview of the development of BIM and relevant conceptualization is given in the following paragraph.

The invention of AutoCAD allowed designers to visualize their designs (or projects) in 3D form (i.e., x, y, and z axes). However, 3D in construction goes beyond the design's geometric dimensions and replicates visual attributes such as color and texture. Time sequencing of such 3D geometric model in visual environments is commonly referred to as 4D modeling [55]. 4D modeling allows demonstration of the building construction process before any real construction activity takes place. It helps identify any possible mistakes and conflicts at early project stages and enables prediction of construction methods and schedule [56]. This concept can be extended to further integrate "n" number of design dimensions in a holistic model to portray and visually project the building design over whole project life cycle [57]. This is based on the concept that a computer model database will contain all the information of building design, which will also contain information about the building's construction, management, operations, and maintenance [57–58] and both graphical (i.e., drawings) and nongraphical (e.g., specifications and schedule) data, in a logical structure, forming a highly coordinated repository. Changes in each item are made at only one place, and all project participants see the same information.

BIM can significantly help in implementing sustainable construction, in general, and in delivering green buildings, in particular. As shown in Figure 19.4, BIM can consider all sustainable design strategies and various design dimensions relating to client and project, as well as those relating to sustainability and environment, and can lead to the development of a structure-specific green-building model. This model can be used through all the phases of the building, namely construction, operation, maintenance, and demolition. A few specific examples of the help or support that BIM can offer are as follows:

- Full-scale analysis of passive design issues, for example, relating to geography, climate, place, surrounding systems, and resources [5], [59]
- To efficiently consider various aspects of energy use, emission of GHG or CO_2, waste generation, life cycle cost, and facilities management, thereby improving sustainability rating [5]
- To reduce planning time and insurance cost and improve project quality, risk management, and life cycle performance [60]
- To consider innovative ideas and adopt best practices in construction, for example, lean construction and prefabrication [61]

19.6 Case Study from Developed and Developing Countries

19.6.1 The San Francisco Federal Building, California, United States

The 18-story building having an area of 600,000 square feet and located on a 3-acre site was completed in 2007. It takes into account the benefits of natural ventilation by utilizing the breeze coming from the sea through adjustable front windows fixed with glass. Concrete is one of the major construction materials to integrate modernization with

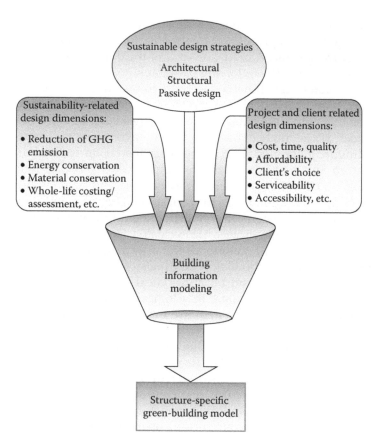

FIGURE 19.4
Integrated green building design process using BIM.

tradition. The front windows are operated using remote controls in order to admit cooler air by expelling warmer air and to create a comfortable environment by reducing temperature inside the building. The building comprises a shear wall built from concrete, with a dimension of 60 ft, to overcome the effect of potential shocks resulting from earthquakes, as shown in Figure 19.5. Besides, cast-in-place fluted surfaces and floor system provide natural air circulation and thus save energy used for cooling. The floor also increases reflectivity through the mixture of Portland cement with slag cement and reduces the cost of indoor lightings. USGBC awarded LEED® Silver Certification to the San Francisco Federal Building, and it was awarded The International Outstanding Building of the Year (TOBY) owing to its various green features such as sustainability and materials and energy conservation.

19.6.1.1 Key Features

- Incorporates energy-efficient technologies through roof-mounted 96-panel photovoltaic solar panels to reduce energy use by 30% annually.
- Energy-efficient office lighting equipped with daylight/occupancy sensors and electronic dimming ballasts provides adjustable levels of light to match the prevailing conditions; this reduces electric lighting loads by more than 50%.

FIGURE 19.5
Federal Office Building, San Francisco, California. (From Federal Office Building, San Francisco, CA, https://en.wikipedia.org/wiki/San_Francisco_Federal_Building. With permission.)

- Green roof design decreases storm-water runoff by 75%, reduces the urban heat island effect, and provides a safe haven for bird, butterfly, and insect populations.
- Water-efficient plumbing fixtures and a drip irrigation system provide for 30% indoor water reduction and 50% outdoor water reduction.
- 98% of the existing walls, floors, and roof, as well as 52% of the building's interior, nonstructural elements, were reused.
- Building's unique "D" shape with central courtyard ensures that workspaces are within 6 m of daylight.
- Mixed-mode ventilation system relies on operable windows for thermal conditioning and ventilation for perimeter spaces, and 100% outside air units for interior spaces.

19.6.1.2 Awards

- 2008 Design Award from the AIA San Francisco
- 2012 EPA Designed to Earn Energy Star Challenge
- 2014 GSA Public Buildings Service Commissioner Award
- 2014 GSA Design Excellence Award

19.6.2 K2 Housing Project, Melbourne, Australia

Design Inc Melbourne designed the K2 buildings for housing; they have 96 units with balconies and communal garden [63]. K2 housing project is Australia's first ecologically and environmentally sustainable public housing, as shown in Figure 19.6. The housing complex made partially of recycled timber generates and consumes only renewable energy on site through solar water heating and photovoltaic panels and uses half water of the average water used by other buildings by practicing rainwater harvesting and gray water reuse.

FIGURE 19.6
K2 Housing Project, Melbourne. (From K2 Housing Project, Melbourne, Australia. Available at: http://aquacell. com.au/case-studies/k2-sustainable-housing-project/.)

19.6.2.1 Key Features

According to the Victorian State Government's Office of Housing, it has following key features:

- The building has a life span of 200 years.
- Annual reduction in energy consumption by 55% or 716 tons of CO_2 in comparison with a standard building.
- Saving of 46% due to the solar hot-water system.
- Water-efficient fittings, rainwater harvesting, and recycling of wastewater contributed to 54% saving of water.
- Special consideration given with selection of materials such as biodegradable linoleum used for kitchen floor and recycled polymer used for carpet tiles.

19.6.3 Clinton Presidential Library, Little Rock, Arkansas, United States

Clinton Presidential Library integrates various key features of green building, which include floors made of recycled rubber tire, rooftop garden for reducing emission and rainwater runoff, regulation of temperature with solar panels as part of its renewable energy use, and supply of power-charging station for electric cars. As a result, USGBC awarded LEED Silver Certification in 2004 and Platinum Certification in 2007 [64], as shown in Figure 19.7.

19.6.3.1 Key Features

- The library uses 24% less energy, because of 306 roof-mounted solar panels than comparable code-compliant buildings [64].

FIGURE 19.7
Clinton Presidential Library. (From Leslie Fisher, Clinton Presidential Library. Available at: http://www.datravelmagazine.com/the-clinton-presidential-library/. With permission.)

- Recycles 100% of goods.
- Reduces water use by 90% for maintenance of natural green space in the park attached to the library.

19.6.3.2 Awards

- Applied Arts Awards Annual, Environmental Design
- Industrial Designers Society of America
- Gold Industrial Design Excellence Award

19.6.4 Khoo Teck Puat Hospital, Yishun Central, Singapore

Singapore has become a pioneer in evolving architectural, structural, and passive design in Asia, with particular focus on energy and material conservation, waste minimization, and maximum utilization of space. It is a small country with highly urbanized area, where utmost focus is given to improve energy efficiency for various types of buildings, which are assessed by Singapore Green Building Council. Khoo Teck Puat Hospital in Singapore [66] is an example of a green building that incorporates green vegetation to reduce surface runoff and indoor temperature and to maintain a comfortable and healing environment for its patients.

19.6.4.1 Key Features

- The hospital is 27% more energy-efficient due to solar heating system.
- Hot water required for the hospital is obtained by solar thermal system.

19.6.5 Turbo Energy Limited, Chennai, India

Turbo Energy Systems [67] in Chennai utilizes solar power for air conditioning. The building uses solar thermal energy for water heating and solar panel for generation of electricity.

19.6.5.1 Key Features

- Reduction of electrical load of 117 kW due to the 90-TR hot-water-fired vapour absorption machines system.
- Working turbine of 5 kW capacity.

19.7 Assessment of Green Buildings

Building sector has higher potential for cost-effective energy savings. For example, it can reduce 11% of total energy consumption in EU in 2020. As such, the European Commission has launched the Energy Performance of Buildings Directive (EPBD) (2002/91/EC and recast 2010/31/EU). This compilation applies to both EU residential and nonresidential buildings and aims to substantially augment investments in energy-efficiency measures taken for buildings. EPBD suggests all EU member states to undertake a methodology for energy performance of newly constructed buildings, focusing on, at least, thermal and characteristics, space and hot water heating, ventilation and air conditioning (HVAC) systems, lighting installations, orientation of the building, and indoor climatic conditions. Meanwhile, the United Kingdom consulted the relevant groups and introduced the Code for Sustainable (or Greener) Homes in 2007. The code is voluntary to adopt. It focuses on improving the overall sustainability of new homes through nine criteria: energy and CO_2 emissions, pollution, water, health and well-being, materials, management, surface water runoff, ecology, and waste [67]. The code is set in a framework and is intended for the home building industry to design and construct homes of higher environmental standards. It also offers a tool for developers to differentiate themselves within the market. If used, the code gives new homebuyers better information about the environmental impact of their new home, along with potential running costs.

However, different countries use different rating systems, as shown in Table 19.2, and each country's rating system follows different level of certification based on scoring range. LEED uses four levels of certification, such as Certified for 40–49 points, Silver for 50–59 points, Gold for 60–79 points, and Platinum for 80+ points.

19.7.1 Coding and Rating System

The Code for Greener Homes (CGH) of the United Kingdom assesses environmental performance of new homes by using objective criteria and verification in two stages: design and postconstruction phase. The outcomes of the evaluation are documented on a certificate assigned to the dwelling. The code especially focuses on energy, water, materials, waste, ecology, and health and well-being. Each of these criteria is given a weight, based on the relative importance of criteria over one another, as shown in Table 19.3. The weighting factors, in turn, are derived from extensive studies involving a wide range of stakeholders (e.g., design engineers, architects, city planners, policy

TABLE 19.2

Rating System of Various Countries of the World

Countries	Green Rating System	Description
United Kingdom	BREEAM: Building Research Establishment Environmental Assessment Method	Launched in 1990 and is a widely used to review and improve the environmental performance of buildings.
United States	LEED: Leadership in Energy and Environmental Design	Inspires and instigates global adoption of sustainable green-building practices relative to performance, including sustainable site development, energy and atmosphere, water efficiency, indoor environmental quality, materials and resources, and innovation in design.
Australia	NABERS: National Australian Built Environment Rating System Also known as Green Star	National rating system that measures the energy efficiency, water usage, waste management, and indoor-environment quality of buildings, tenancies, and homes. It uses a scale of 1–6 stars for certification, where a 6-star rating demonstrates market-leading performance, while a 1-star rating means that the building or tenancy has considerable scope for improvement. For example, utility bills are converted into an easy-to-understand star rating scale from 1 to 6 stars.
Japan	CASBRE: Comprehensive Assessment System for Built Environment Efficiency	Introduced in 2004; the implementation is voluntary for both government and private sectors. The system is maintained by Japan Green Building Council.
China	CEPAS: Compressive Environmental Performance Assessment Scheme	The implementation is voluntary for both government and private sectors. It is maintained by building department.
Malaysia	GBI: Green Building Index	Launched in 2009 and comprises six key criteria for rating the green building in Malaysia. It is maintained by Malaysia Green Building Confederation.
Singapore	Green Mark	The implementation is enforced at minimum-level incentives for both government and private sectors. It is managed by the Building and Construction Authority of Singapore.
India	IGBC: Indian Green Building Council Also known as GRIHA	The Indian Green Building Council (IGBC), part of the Confederation of Indian Industry (CII), was formed in the year 2001 and offers a wide array of services, including development of new green-building rating programmes, certification services, and green-building training programmes. GRIHA assesses a building out of a total of 34 criteria and awards points on a scale of 100; a project must achieve at least 50 points to qualify for GRIHA certification.

(Continued)

TABLE 19.2 (*Continued*)

Rating System of Various Countries of the World

Countries	Green Rating System	Description
Indonesia	Greenship	Launched in January 2010 and maintained by Green Building Council of Indonesia. The implementation is voluntary for both government and private sectors.
Thailand	TREES: Thai Rating of Energy and Environmental Sustainability	Launched in December 2012 and developed by Thai Green Building Institute, which was modified from USGBC's LEED to fit Thailand's environment. The system is categorized in seven basic areas.

Source: USGBC, What is Green building? United States Green Building Council (USGBC), 2014, Available at: http://www.usgbc.org/articles/what-green-building, accessed April 5, 2016; Kibert, C.J., *Sustainable Construction, Green Building Design and Delivery*, John Wiley & Sons, Hoboken, NJ, 2013; USGBC, Benefits of Green Buildings. United States Green Building Council (USGBC), 2015. Available at: http://www.usgbc.org/articles/benefits-green-homebuilding, accessed April 27, 2016; Zuo, J. and Zhao, Z.Y., *Renew. Sust. Energ. Rev.*, 30, 271–281, 2014; Shafii, F., Status of Sustainable Building in South-East Asia: Report prepared for Sustainable Building Conference (SB08), Melbourne, Australia, Institute Sultan Iskandar of Urban Habitat & Highrise, Universiti Teknologi Malaysia, Malaysia, 2008. Available at: http://www.iisbe.org/sb08_regional_reports, accessed April 27, 2016; Esa, M.R. et al., *Aust. J. Basic Appl. Sci.*, 5, 1806–1812, 2011; Shams, S. et al., *Int. J. Eng. Technol.*, 3, 2011.

TABLE 19.3

Relative Importance of Each Criterion in the Coding Assessment

Criteria	Weight	Justification
Energy	0.3	Energy is one of the most essential criteria, as prudent use of energy can reduce emissions. It is measured as how much conventional energy can be saved by alternative renewable sources, for example, solar, thermal, wind, and biomass energy.
Water	0.2	Clean and potable water is the prerequisite for a healthy environment. Good water quality as well as quantity can ensure health and well-being and protect ecosystem surrounding the built environment.
Waste	0.2	Proper and efficient solid waste management can reduce health hazards and promote well-being. Properly managed waste can be regarded as resources, for example, composting and recycling.
Materials	0.1	Materials that emit less embodied energy and CO_2 during their life cycle should be selected. Use of locally available materials is encouraged, and emphasis has been given to the use of materials that can be recycled.
Ecology	0.1	Buildings often possess a threat to ecology as natural area is converted into built-up area, often causing serious damages to ecosystem. Greening of roof by using vegetation can substantially cool down (4°C–5°C) indoor air temperature and protect from heat flux [69–71].
Health and well-being	0.1	Health and well-being are directly correlated with water, waste disposal, and ecology. The building should ensure adequate natural sunlight and ventilation, as the both of them can improve mental health and work productivity [70].

Source: Shams, S. et al., *Int. J. Eng. Technol.*, 3, 480, 2011; Hien, W.N., *Build. Environ.*, 42, 25–54, 2007; Younger, M. et al., *Am. J. Prev. Med.*, 35, 517–526, 2008; Rumana, R. and Ahmed, M.H.B., *J. Architect. Built Environ.*, 37, 41–50, 2009.

makers, environmental specialist, doctors, and dwellers). They were asked to rank a range of environmental impacts in order of importance. The feedback from stakeholders was obtained through a questionnaire and recompiled to assess the relative importance of each criterion over one another. Therefore, the weighting factors may change based on the socioeconomic and cultural perspective of each country.

19.8 Economics of Green Building

It is estimated that the building sector contributes to 10% of global GDP (USD 7.5 trillion) and employs 111 million people [15]. As with all business decisions, the "build or not to build" decision depends on a cost-benefit analysis, that is, if perceived benefits outweigh costs, the green buildings will be constructed. However, there are important additional considerations that influence cost-benefit analysis of green buildings, that is, the price of going green and the value that it will impart. However, there is growing recognition that "green" should not be considered a discreet "add-on" feature grafted on to an otherwise-normal project and should be evaluated independently to assess its financial burdens and benefits. Rather, it is becoming ever clearer that sustainable building requires changes of both paradigm and process that, when embraced and applied to the entire building process, can make green building an attractive option, without being an expensive one.

As mentioned earlier, constructing a green building naturally costs more than a conventional building, since green buildings use premium materials, high-efficiency equipment, and additional layers of process workflow. The mindset that paying extra is beginning to give way to more holistic designs and a life cycle view of costs and relevant benefits. Recent studies show that green buildings cost a little or no more than conventional buildings, if undertaken through an integrated approach. A study conducted in 2003 found that 25 office buildings and 8 schools seeking certification through LEED had a small cost premium associated with completing the steps necessary for basic levels of certification and a single building seeking platinum status had a more substantial cost premium associated with completing the necessary steps of certification [72].

Although the clients pay higher premiums to designers and contractors for their efforts toward potential savings in energy, it is the occupants who actually enjoy the benefits from energy savings. However, recent studies show that green buildings are sold at higher price, indicating that clients are getting their extra-spent money back. McGraw Hill measured the price premium for the sale of Energy Star®-labeled buildings to be 12% [2]. Another study estimated the premium on LEED-certified buildings at 31% [73].

However, many of the international systems do not involve the economic evaluation of green buildings. For example, LEED, BREEAM, Evaluation Standard for Green Building, and Comprehensive Assessment System for Building Environment Efficiency do not contain such economic evaluation. Although the GBTool system proposes to evaluate cost benefits, it does not provide specific evaluation contents and methods. Moreover, current awareness of many people about green buildings is not sufficiently comprehensive and accurate. They do not develop or purchase green buildings, owing to their incomprehensive perceptions that green buildings require high investment and high cost. This has been the case that is hindering development of green buildings in China. Hence, it is very necessary to construct the theoretical method system of green building's cost–benefit analysis from a technical and economic point of view, which has important theoretical value and practical significance for the healthy development of green buildings.

19.9 Summary and Concluding Observations

Construction industry is frequently blamed for degrading, or at least for not uplifting, sustainable development, mainly due to its vast use of natural resources and consumption of energy, thereby emitting greenhouse gas or carbon dioxide, which is the main reason for climate change. As the most important and most frequent product of construction industry, as well as the largest consumer of natural resources and energy, buildings are grossly held responsible for this blame. As such, efforts from around the globe are being made for the efficient use of naturally occurring resources (e.g., land, water, and construction materials) and fossil-fuel-based energy in buildings and, at the same time, optimal use of sunlight (both for light and heat) and renewable energy to reduce the demand of fossil fuel. The solution appears to be, what nowadays is widely known as, "green building." The concept of green building refers to a constructed facility, as well as the processes involved in constructing that facility, to be environmentally responsible and resource-efficient, throughout a building's life cycle, that is, in all the phases of its construction, operation, maintenance, renovation, and demolition. This is achieved through an IDP and at conceptual and design stage of a building, by crafting all suitable "green" features into the building. As such, this chapter focused on the practice of green or sustainable design practices for buildings. It is proposed that both architectural and structural designers should first analyze the building scenario on the basis of available "green" design strategies and then collate their outcomes with various design dimensions (i.e., both client- and project-related and environment-related design dimensions), with the help of BIM, to finally decide on a building-specific model. This model will be used throughout the entire life cycle of the building, that is, during construction, operations, maintenance, renovation, and the eventual demolition.

The practice of green building is growing around the world. A number of examples of green buildings from both developed and developing countries with their key features are discussed. The potential benefits leading to sustainable constructions resulting from green buildings are adequately addressed. It is observed that green buildings offer a wide range of benefits, including reduced use of materials, energy, and carbon dioxide emission. On the other hand, main barriers seem to be the initial increased investment. On the whole, the benefits seem to grossly outweigh the drawbacks. Current research also appears to delve on issues to justify the wider acceptability of green buildings. For example, the research agenda of USGBC spans from the delivery process of green buildings through to individual aspects of integrated building systems, building interactions with surrounding ecology, as well as indoor air quality and occupants' satisfaction level. Other leading researches include effective utilization of IDP in delivering green buildings [10], occupants' behavior to thermal discomfort in New Zealand [36], application of BIM and teamwork, suitability of building materials, and relevant cost analysis under various climatic conditions. All these grossly support wider adoption of green buildings; however, they have to be suited to local ecology and climatic condition. It appears that the practice of green building will continue to grow day by day.

References

1. EPA, *Sustainable Design and Green Building Toolkit for Local Governments*, Washington, DC: United States Environmental Protection Agency (EPA), 2010.
2. USGBC, What is Green building? United Stages Green Building Council (USGBC), 2014, Available at: http://www.usgbc.org/articles/what-green-building, accessed April 5, 2016.
3. Global Green Building SmartMarket Report, 2008. http://www.worldgbc.org/files/8613/6295/6420/World_Green_Building_Trends_SmartMarket_Report_2013.pdf., accessed 27 March, 2016
4. R.W. Hudson, R. Haas and W. Uddin, *Infrastructure Management: Design, Construction, Maintenance, Rehabilitation, Renovation*, New York: McGraw-Hill, 1997.
5. C.J. Kibert, *Sustainable Construction, Green Building Design and Delivery*, Hoboken, NJ: John Wiley & Sons, 2013.
6. WBCSD, *Energy Efficiency in Buildings, Business Realities and Opportunities*, Geneva, Switzerland: The World Business Council for Sustainable Development, 2007.
7. Y. Miyatake, Technology development and sustainable construction, *Journal of Management in Engineering*, 12(4), 23–27, 1996.
8. G.K.C. Ding, Sustainable construction—The role of environmental assessment tools, *Journal of Environmental Management*, 86(3), 451–464, 2008.
9. G. Rajgor, Sustainable Building: Firing on target—US building designers aim for green, *Refocus*, 5(2), 6–63, 2004.
10. N. Larsson, The Integrated Design Process: History and Analysis. International Initiative for a Sustainable Built Environment (iiSBE), Canada Government, Ottawa, Canada, 2009. Available at: http://iisbe.org/node/88, accessed April 27, 2016.
11. S. Sorrell, Making the link: Climate policy and the reform of the UK construction industry, *Energy Policy*, 31(9), 865–878, 2003.
12. L.B. Robichaud and V.S. Anantatmula, Greening project management practices for sustainable construction, *Journal of Management Engineering*, 27(1), 48–57, 2010.
13. Intertek Report R66141 Household Electricity Survey A study of domestic electrical product usage. Available at: http://www.google.com.sg/url?sa=t&rct=j&q=&esrc=s&source=web&cd=2&cad=rja&uact=8&ved=0ahUKEwiG6rSht7XMAhVBUo4KHXs3D-QQFgggMAF.&url=http%3A%2F%2Frandd.defra.gov.uk%2FDocument.aspx%3FDocument%3D10043_R66141HouseholdElectricitySurveyFinalReportissue4.pdf&usg=AFQjCNFyTkKNDJsVedJMVmHP_0-D7A3EIg&bvm=bv.121070826,d.c2E, accessed February 13, 2016.
14. Earthwatch, Promoting sustainable human settlement Development, Agenda 21 Chapter 7.67, UNEP. 2009. Available at: http://earthwatch.unep.net/agenda21/7.php, accessed June 19, 2015.
15. UNEP, Why Buildings. United Nations Environment Programme. Available at: http://www.unep.org/sbci/AboutSBCI/Background.asp, accessed April 27, 2016.
16. G. Augenbroe, A.R. Pearce and C.J. Kibert, Sustainable Construction in the United States of America: A perspective to the year 2010, CIB-W82 Report, 1998, Available at: http://www.p2pays.org/ref/14/13358.htm, accessed December 7, 2015.
17. D.M. Roodman and N. Lenssen, A Building Revolution: How Ecology and Health Concerns are Transforming Construction, Worldwatch Paper 124, Worldwatch Institute, Washington, DC, 1995.
18. E. Mulavdic, Multi-criteria optimization of construction technology of residential building upon the principles of sustainable development, *Thermal Science*, 9(3), 39–52, 2005.
19. M.J. Franchetti and D. Apul, *Carbon Footprint Analysis: Concepts, Methods, Implementation, and Case Studies*, Boca Raton, FL: CRC Press, 2013.

20. M. Lallanilla, 8 Benefits of Green Buildings, 2014. Available at: http://greenliving.about.com/od/architecturedesign/tp/green_building_advantages.htm, accessed April 27, 2016.
21. USGBC, Benefits of Green Buildings. United States Green Building Council (USGBC), 2015. Available at: http://www.usgbc.org/articles/benefits-green-homebuilding, accessed April 27, 2016.
22. Lennox, Long Lasting Impact, 2015. Available at: http://www.lennoxcommercial.com/green-building/benefits-of-green-building.asp, accessed April 27, 2016.
23. The City of Bloomington, Green Building Benefits, 2016. Available at: https://bloomington.in.gov/green-building-benefits, accessed April 27, 2016.
24. A. Henry and N. Frascaria-Lacoste, Comparing green structures using lifecycle assessment: A potential risk for urban biodiversity homogenization, *International Journal of Life Cycle Assessment*, 17(8), 949–950, 2012.
25. F. Bianchini and K. Hewage, How green are the green roofs? Life cycle analysis of green roof materials, *Built Environment*, 48, 57–65, 2012.
26. J. Zuo and Z.Y. Zhao, Green building research–current status and future agenda: A review, *Renewable and Sustainable Energy Reviews*, 30, 271–281, 2014.
27. J.H. Jo, J.S. Golden, and S.W. Shin, Incorporating built environment factors into climate change mitigation strategies for Seoul, South Korea: A sustainable urban systems framework, *Habitat International*, 33(3), 267–275, 2009.
28. Turner, C. and M. Frankel, *Energy Performance of LEED for New Construction Buildings*, Vancouver, WA: New Buildings Institute, 2008.
29. A.M. Al-Yami and A.D.F. Price, A framework for implementing sustainable construction in building briefing project. In: Boyd, D. (Ed.), *Proceedings 22nd Annual ARCOM Conference*, September 4–6, 2006, Birmingham, UK, Association of Researchers in Construction Management, pp. 327–337.
30. F. Shafii, Status of Sustainable Building in South-East Asia: Report prepared for Sustainable Building Conference (SB08), Melbourne, Australia, Institute Sultan Iskandar of Urban Habitat & Highrise, Universiti Teknologi Malaysia, Malaysia, 2008. Available at: http://www.iisbe.org/sb08_regional_reports, accessed April 27, 2016.
31. RoofingPost, What are the Disadvantages of Green Buildings, 2016. Available at: http://roofingpost.com/zinc-roofs/what-are-the-disadvantages-of-green-buildings/, accessed April 27, 2016.
32. Everything Log Homes, Disadvantages of Green Building, 2016. Available at: http://www.everything-log-homes.com/sustainable-house/disadvantages-of-green-building/, accessed April 27, 2016.
33. R. Castillo and N. Chung, Last update, the value of sustainability, 2005. Available at: http://www.stanford.edu/group/CIFE/online.publications/WP091.pdf, accessed January 14, 2015.
34. A. Leaman and B. Bordass, Are users more tolerant of "green" buildings? *Building Research and Information*, 35, 662–673, 2007, doi:10.1080/09613210701529518.
35. M.R. Esa, M.A. Marhani, R. Yaman, A.A. Hassan, N.H.N. Rashid and H. Adnan, Obstacles in implementing Green Building Projects in Malaysia, *Australian Journal of Basic Applied Science*, 5, 1806–1812, 2011.
36. N.S.M. Azizi, S. Wilkinson and E. Fassman, Analysis of occupants response to thermal discomfort in green and conventional buildings in New Zealand, *Energy and Buildings*, 104, 191–198, 2015.
37. O.G. Santin, Behavioural patterns and user profiles related to energy consumption for heating, *Energy and Buildings*, 43, 2662–2672, 2011.
38. J.J. Kim and B. Rigdon, *Introduction to Sustainable Design*, Ann Arbor, MI: National Pollution Prevention Center for Higher Education, The University of Michigan, 1998.
39. J.M. Danatzko and H. Sezen, Sustainable structural design methodologies, *ASCE Journal of Practice Periodical on Structural Design and Construction*, 16(4), 186–190, 2011.
40. B. Urban and K. Roth, *Guidelines for Selecting Cool Roofs*, U.S. Department of Energy, Washington DC, 2010.
41. M. Gibler, *Comprehensive Benefits of Green Roofs*, World Environmental and Water Resources Congress 2015, American Society of Civil Engineers (ASCE), pp. 2244–2251, 2015.

42. Living System Design Group. Available at: http://www.livingsystemsdesign.net/, accessed October 21, 2015.
43. D.J. Sailor, T.B. Elley and M. Gibson, Exploring the building energy impacts of green roof design decisions—A modeling study of buildings in four distinct climate, *Journal of Building Physics*, 35(4), 372–391, 2012.
44. Y. Zhang and H. Altan, A comparison of the occupant comfort in a conventional high-rise office block and a contemporary environmentally-concerned building, *Built Environment*, 46(2), 535–545, 2011.
45. S. Mekhilef, A. Safari, A., W.E.S. Mustaffa, R. Saidur, R. Omar and M.A.A. Younis, Solar energy in Malaysia: Current state and prospects, *Renewable & Sustainable Energy Reviews*, 16(1), 386–96, 2012.
46. T.S. Bisoniya, A. Kumar and P. Baredar, Experimental and analytical studies of earth-air heat exchanger (EAHE) systems in India: A review, *Renewable & Sustainable Energy Reviews*, 19, 238–246, 2013.
47. ASHRAE, Thermal environmental conditions for human occupancy, in ASHRAE Standard 55, American Society of Heating, Refrigerating and Air-Conditioning Engineers, Atlanta, 2010.
48. ISO, Ergonomics of the thermal environment: Analytical determination and interpretation of thermal comfort using calculation of the PMV and PPD indices and local thermal comfort criteria, in:ISO7730, ISO, 2005.
49. B. Moujalled, R. Cantin and G. Guarracino, Comparison of thermal comfort algorithms in naturally ventilated office buildings, *Energy and Buildings*, 40, 2215–2223, 2008, doi:10.1016/j.enbuild.2008.06.014.
50. E.S. Lee and A. Tavil, Energy and visual comfort performance of electrochromic windows with overhangs, pp. 1–13, 2005.
51. A. Tzempelikos and A.K. Athienitis, The impact of shading design and control on building cooling and lighting demand, *Solar Energy*, 81(3), 369–382, 2007.
52. S. Hans and B. Binder, Experimental and numerical determination of the total solar energy transmittance of glazing with venetian blind shading, *Building and Environment*, 43, 197–204, 2008.
53. J. Yao, An investigation into the impact of movable solar shades on energy, indoor thermal and visual comfort improvements, *Built Environment*, 71, 24–32, 2014.
54. Autodesk, What is BIM? Available at: http://www.autodesk.com/solutions/bim/overview, accessed April 27, 2016.
55. L. Rischmoller, M. Fisher, R. Fox and L. Alarcon, 4D Planning & Scheduling: Grounding Construction IT Research in Industry Practice. *Proceedings of CIB W78 Conference*, Construction Information Technology: Iceland, 2000.
56. A. Lee, A.J. Marshall-Ponting, G. Aouad, S. Wu, I. Koh, C. Fu, R. Cooper, M. Betts, M. Kagioglou and M. Fischer, *Developing a vision of nD-enabled construction*, Construct IT Report, 2003.
57. Lee, S. Wu, A. Marshall-Ponting, A., G. Aouad, J. Tah, R. Cooper and C. Fu. 2005. *nD Modelling: A driver or enabler for construction improvement?* RICS research paper series, 5(6), 2005.
58. Graphisoft, The Graphisoft Virtual Building: Bridging the Building Information Model from Concept into Reality. Graphisoft Whitepaper, 2013.
59. B. Hardin, *BIM and Construction Management: Proven Tool, Methods and Workflows*, Indianapolis, IN, Wiley, 2009.
60. M.K., Hossain, A. Munns and M.M. Rahman. Enhancing team integration in building information modellings (BIM) projects, ARCOM Doctoral Workshop on BIM: Management and Interoperability, edited by Prof David Boyd, June 20, Birmingham City University, Birmingham, UK, pp. 78–92, 2013.
61. M.M. Rahman, Barriers of implementing modern methods of construction, *Journal of Management in Engineering*, 30(1), 69–77, 2014.
62. Federal Office Building, San Francisco, CA, https://en.wikipedia.org/wiki/San_Francisco_Federal_Building, accessed January 14, 2016.

63. K2 Housing Project, Melbourne, Australia. Available at: http://aquacell.com.au/case-studies/k2-sustainable-housing-project/, accessed January 2, 2016.
64. *Clinton Presidential Library Earns a LEED for Existing Buildings Platinum Rating. Green Progress.* November 14, 2007. Available at: http://www.greenprogress.com/green_building_article.php?id=1392, accessed January 2, 2016.
65. Clinton Presidential Library. Available at: http://www.datravelmagazine.com/the-clinton-presidential-library/, accessed January 2, 2016.
66. Khoo Teck Puat Hospital. Available at: http://www.greenroofs.com/projects/pview.php?id=1622, accessed January 2, 2016.
67. Turbo Energy Limited, Chennai. Available at: http://vvarchitect.com/GreenBuildingsTurbo Energy.html, accessed January 2, 2016.
68. S. Shams, K. Mahmud and M. Al-Amin. Building Greener Homes Based on Coding and Rating System, *International Journal of Engineering and Technology*, 3(5), 480, 2011.
69. W.N. Hien, Study of thermal performance of extensive roof top greenery system in the tropical climate, *Building and Environment*, 42(1), 25–54, 2007.
70. M. Younger, H.R. Morrow-Almeida, S.M. Vindigni and A.L. Dannenberg. The Built Environment, Climate Change, and Health Opportunities for Co-Benefits, *American Journal of Preventive Medicine*, 35(5), 517–526, 2008.
71. R. Rumana and M.H.B. Ahmed, Thermal performance of rooftop greenery system at the tropical climate of Malaysia, a case study of a 10 storied building R.C.C flat rooftop, *Journal of Architecture and Built Environment*, 37(1), 41–50, 2009.
72. A Report to California's Sustainable Building Task Force–October 2003. http://www.calrecycle.ca.gov/greenbuilding/Design/CostBenefit/Report.pdf, accessed June 11, 2016.
73. F. Franz and P. McAllister, New Evidence on the Green Building Rent and Price Premium, Paper presented at the *Annual Meeting of the American Real Estate Society*, Monterey, CA, April 3, 2009.

20

Life Cycle Analysis as the Sustainability Assessment Multicriteria Decision Tool for Road Transport Biofuels

Vineet Kumar Rathore, Lokendra Singh Thakur, and Prasenjit Mondal

CONTENTS

ABSTRACT Considering the progressive depletion of fossil resources and the rising concern of global warming, renewable energy resources like biofuels are becoming more attractive in recent years. Extensive research is going on around the world to improve the quality and yield of biofuels produced from different biomass feedstocks. Although, apparently, it seems that the biofuels are attractive over fossil fuels, the strategies for sustainability have to be thoroughly assessed at several levels, such as relevance of biofuels with respect to sustainable development, sustainable transport and its strategies, and alternative fuel strategy, to assess their suitability as transport fuels. Sustainability assessment is performed through different methodologies for analyzing the social, economic, and environmental impacts. Life cycle analysis (LCA, also known as life cycle assessment) determines the effects of any process/product throughout its life period. Multicriteria decision analysis method considers different alternative scenarios and ranks them to find out the most suitable one to achieve the targeted goal. It is very useful when the problem is complex in nature. Since biofuels can be generated from different feedstocks by using various routes, as well as there is large variation in vehicles types, traveling mode, and so on, the sustainability assessment for road transport biofuels deals with a complex problems. To assess the sustainability, a large number of scenarios have to be considered, and hence multicriteria decision analysis methods coupled with LCA practices seems to be more suitable. Many researchers have also applied integrated approach for sustainability analysis of transport systems and biofuel production chain in recent years.

20.1 Introduction

In road transport system, the conventional vehicles normally use liquid fuels such as diesel and petrol. Gaseous fuels and electricity are also being used in some vehicles in recent years. The liquid fuels emit different types of pollutants, including CO and soot, which are produced by the incomplete combustion of these fuels at lower air-to-fuel ratio. These fuels also create more air pollution, including CO_2 emissions, than gaseous fuels and electricity on road. However, electricity and gaseous fuels production also emit large amount of emissions in the production units. It is reported that ~0.973 kg CO_2 is emitted in the production of 1 kwh electricity from coal, whereas 1 kg natural gas production emits 0.9538 kg CO_2 into the atmosphere (Wang and Mu 2014; Yue et al. 2014). Therefore, total emissions generated through the whole life cycle of these fuel systems may be comparable. Further, world reserve of petroleum crude is estimated to be 1,184–1,342 billion barrels, whereas the world consumes over 86 million barrels of liquid and other petroleum fuels/day (EIA 2010; Curran 2013). In addition, the demand of energy is also increasing gradually; it is forecasted that in the years 2020 and 2035, the world consumption of liquid and other petroleum fuel will be 103.9 and 110.6 million barrel/day, respectively (EIA 2010). Climate change due to greenhouse gas emissions, increasing demand for energy, volatile oil prices, and energy poverty have led to a search for energy options that can be socially equitable, economically efficient, and environmentally sound.

Biofuels such as biodiesel and bioethanol possess indigenous oxygen content, and hence, these biofuels produce lesser soot and CO during combustion with the same air-to-fuel ratio than petrodiesel. Therefore, it seems that these biofuels may be attractive alternatives of fossil fuels to achieve better environmental quality. However, these also have some drawbacks, such as more nitrogen content, high viscosity, less heating value, and less stability. Further, the technologies for the production of third-and fourth-generation biofuels are not well proven (Biofuel digest 2010). Although the use of biofuels in road transport is apparently seen as an important pathway to achieve sustainable transport, the strategies for sustainability have to be developed and assessed at several levels (Holden and Gilpin 2013).

Assessment of sustainability of any process/product is normally done by considering its environmental, economic, and social impacts. A number of scenarios are considered on the basis of identified goals. Different indicators related to environmental, economic, and social factors, under these scenarios, are measured, and the most suitable scenario is identified to achieve the goal. When the goal is based on complex problems, decision making is normally based on multicriteria decision analysis (MCDA) method (Pohekar and Ramachandran 2004).

The MCDA process includes different alternatives of a system (scenarios), which are associated with various criteria. Each criterion is weighted to make a grouped decision matrix and evaluated for final decision (Halog and Bortsie-Aryee 2013). It is applicable for a system with high uncertainty, multiple interests, different forms of information, and biophysical and socioeconomic problems (Pór et al. 1990; Munda 2003). In MCDA method, criteria are based on economic, technical, social, and environmental aspects of the system. Similarly, weighing methods include objective weighing, subjecting weighing, and combined method (Halog and Bortsie-Aryee 2013). An example of objective weighing is entropy method, whereas analytical hierarchy process (AHP) is an example of subjective weighing. Different MCDA approaches are applied to evaluate the most critical criteria, matrix, and indicators that most efficiently meet the interest of stakeholders (Halog and Bortsie-Aryee 2013).

Life cycle assessment (LCA) assists decision makers in evaluating the comparative potential cradle-to-grave, multimedia environmental impacts of different processes to prevent unintended consequences such as global warming, human toxicity, eutrophication, and ozone layer depletion. The role of LCA is crucial in determining the values of the various matrix and emissions along the entire chain of biofuel production and, as such, must be applied to different processing techniques available now and those that might become available after research, development, and demonstration (RD&D). An effective life cycle approach can identify where potential trade-offs may occur across different media and across the life cycle stages (Fava et al. 1993). Integrated approach, combining MCDA process with different LCA methods, is more suitable for evaluating sustainable development of very complex systems (Pineda-Henson et al. 2002).

Before adopting the biofuel for road transport fuel, it has become a necessity to assess the policies very carefully in terms of their compliance with key characteristics and criteria, as stated below (Curran 2013).

1. The three main dimensions (economic, environmental, and social) for sustainable development must be satisfied if biofuel is used.
2. Gains from biofuel strategies must be competitive to the gains from other sustainable transport strategies, such as reducing transport volume and altering transport patterns.
3. Gains from using one generation of biofuels (e.g., first generation) must compare favorably to gains from using others (e.g., second through fourth generations).
4. Benefits from the use of biofuel-powered vehicles must be competitive to those from the use of other alternative-fueled vehicles.

In addition, development of robust and scientific sustainability criteria demands a reliable theoretical perspective and a well-established methodological base. In this chapter, relevance of biofuels in road transport applications, sustainable transport and its strategies, sustainability assessment methods and LCA, and integrated approach for the sustainability assessment in road transport biofuels, along with some real examples from literature on this area, are described.

20.2 Relevance of Biofuels in Road Transport Applications

Biofuels are produced from biomass. Depending on the nature of the biomass, these are classified into first to fourth generations. The first generation includes the biofuels generated from edible food grains (e.g., bioethanol from sugar and corn); the second-generation biofuels are produced from nonedible and lignocellulosic biomass (e.g., biodiesel from jatropha seeds and bioethanol from rice straw); and the third-generation biofuels are produced from microalgae (biodiesel from algae) (Biofuel digest 2010; Sims et al. 2010). The fourth-generation biofuels are photobiological solar fuels and electrofuels, which can be produced by a combination of photovoltaics or inorganic water-splitting catalysts with metabolically engineered microbial fuel production pathways (electrobiofuels) (Aro 2016). Bio-oil from genetically modified microalgae is also included under the fourth-generation biofuels (Lu et al. 2011; Vogelpohl 2011). Production of first-generation biofuels is not recommended, owing to food crisis issue around the world (Sims et al. 2010). The second- and third-generation biofuels seem very attractive and promising as alternatives to fossil fuels. The second-generation biofuels (e.g., cellulosic ethanol) are produced commercially, but availability of nonedible oil seeds is an issue (Sims et al. 2010). The third-generation biofuels are gradually moving toward commercialization (Biofuel digest 2010). The fourth-generation biofuels are in early stage (Lu et al. 2011). Hence, it seems that second- and third-generation biofuels are more relevant in present days for application in road transport. The biofuels (biodiesels) are produced from nonedible oil seeds by their oil extraction and subsequent upgradation to biodiesel through transesterification. However, there are a number of ways to produce biofuels (such as bioethanol and biodiesel) from lignocellulosic biomass and microalgae. Figure 20.1 shows some important paths for the production of biofuels for transport from lignocellulosic biomass and microalgae.

From Figure 20.1, it seems that road transport biofuels can be produced through different routes, such as fermentation (microbial processing), Fischer-Tropsch synthesis (FTS), hydrotreating, and liquid-phase processing. Pretreatment is required for second-generation biofuels, but for third-generation biofuel production, it may not be essential. Thus, complexities of these routes are different. The utilization of biofuels as a blend with petroleum-based fuels helps reduce soot and CO in the exhaust and produce less greenhouse gas emissions. Since the petroleum reserves are localized and nonrenewable, use of biofuels in transport sector is becoming attractive around the world. In many developed countries, ethanol-blended gasoline and biodiesel-blended diesel are being used in road transport.

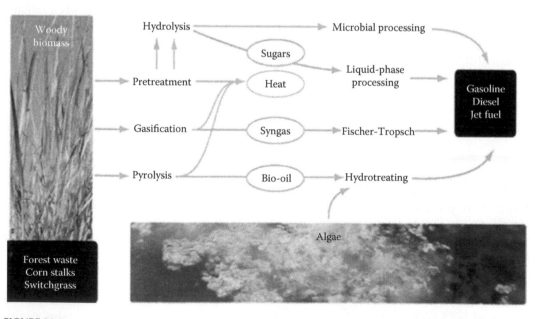

FIGURE 20.1
Important pathways to biomass hydrocarbons. (From Regalbuto, J. R., *Science*, 325, 822–824, 2009. With permission.)

20.3 Sustainable Transport and Its Strategies

Sustainable transport meets the present transport/mobility needs in an efficient way and also ensures the future generation's mobility needs. Different types (simple and comprehensive) of indicators have been developed to evaluate the sustainability of transport system as well as to achieve those goals that society ultimately wants. Simple indicators are based on relatively easily available data such as emission of pollutants from vehicles, injuries, death due to accident, mileage of vehicle, and aesthetic conditions on road; however, comprehensive indicators include complex issues related to economical (accessibility, diversity of vehicles, affordability, cost for facilities, and so on), environmental (climate change, impact on land use, water pollution, other air pollutants, noise pollution, and so on), and social (safety, equity, health and fitness, and so on) aspects (Litman and Burwell 2006).

For achieving sustainable transport, two approaches, that is, "reductionist" and "comprehensive," are reported in transportation planning. In the first approach, individual problems from a narrow set are ranked and decision is taken. Per this approach, growing travel volumes can continue if vehicle design addresses the most critical sustainability objectives. The latter approach considers a set of integrated problems, in which solutions to one problem may intensify other problems. Per this approach, travel volumes reduction is required for sustainability. A combination of these two approaches is actually employed for most of the sustainable transport plans. Some important criteria for sustainable transport are improved travel choices, pricing and road design incentives, land use patterns to reduce travel distances, technical improvements to the motor vehicles, alternative fuels, super-efficient vehicles, technological innovation, and so on (Dudson 1998; Newman and Kenworthy 1998). Biofuels are emerging as alternative fuels and are expected to play a significant role in sustainable transport. Different energy sources such as gasoline, diesel, methanol, ethanol, dimethyl ether, hydrogen, and electricity can be used in transport sector through different vehicle types such as internal combustion engine, electric, and hybrid electric vehicles. From the above discussion, it seems that to understand the sustainability issue, a large number of scenarios have to be considered and sustainability indicators of these scenarios need to be compared.

20.4 Sustainability Assessment Methods and Life Cycle Analysis

Sustainability assessment of a system is performed through different steps such as identification of goal and scope on the basis of problems; creation of different scenarios on the basis of available data; and estimation of environmental, economic, and social impacts on the society under each scenario, as well as to identify the scenario for optimum impacts. Different phases applied for sustainability analysis of a system are shown in Figure 20.2.

A system boundary is considered, for which sustainability assessment is required, which can be decided on the basis of the problem (Sala et al. 2015). Different sustainability indicators are defined, including functional indicators such as adaptability, durability, robustness, reliability, maintenance, and effluent quality; economic indicators such as cost on investment, operation, and maintenance; environmental indicators such as optimal utilization of water, land, and energy; and social indicators such as acceptance and

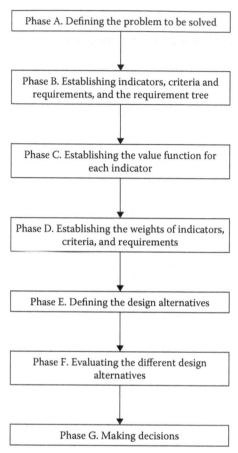

FIGURE 20.2
Phases for sustainability assessment. (From Barros, J. J. C. et al., *Energy*, 89, 473–489, 2015. With permission.)

institutional requirement. Goal and objectives are fixed on the basis of problem types. The recognition of problems from field data and formulation of policy are shown in Figure 20.3.

Evaluation of the scenarios is done through different methods, which include both quantitative and qualitative approaches. Different models, including those for uncertainty management, are used for analyzing the criteria, and different indicators are also used to assess the sustainability. Some important methods for sustainability analysis are multicriteria analysis (MCA), cost-benefit analysis (CBA), environmental impact assessment (EIA), strategic environmental assessment (SEA), life cycle assessment (LCA), life cycle costing (LCC), social LCA (sLCA), life cycle sustainability assessment (LCSA), European social impact assessment (EuSIA), and so on (Sala et al. 2015). For complex systems, integrated methods are used, as shown in Figure 20.4.

For benchmarking and ranking of problems, data envelopment analysis (DEA) and multicriteria decision-making (MCDM) approaches are extensively applied (Iribarren et al. 2014; Barros et al. 2015). DEA, a linear programming-based model, applicable for performance assessment area of homogeneous set of decision-making units (DMUs), has been used jointly with LCA to calculate eco-efficiency in transportation (Dyckhoff and Allen 2001; Iribarren et al. 2010; Egilmez and Park 2014). MCDM accounts for the impact of multiple criteria in complex decision-making problems (Achillas et al. 2013). It uses rigorous

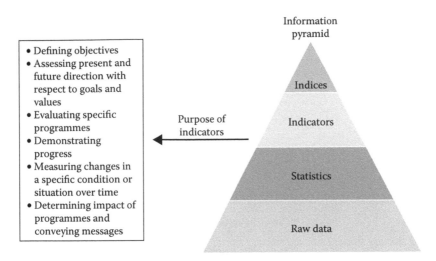

FIGURE 20.3
Recognition of problems and formulation of policy. (From Waas, T. et al., *Sustainability*, 6, 5512–5534, 2014.)

mathematical formulation and follows iterative decision-making process (Egilmez et al. 2015). It can deal with multiple and confronting criteria such as job creation versus emission of air pollutants, multi-objectives (economic development, improved environment, equity in society, and so on), and multi-attributes (jobs, climate change, accidents, economy, and so on). It considers different scenarios and ranks them to provide alternatives and select the most suitable one (Achillas et al. 2013). In this process, different alternatives of a system (scenarios) are considered with various criteria, each of which is weighted to make a grouped decision matrix and evaluated for final decision (Halog and Bortsie-Aryee 2013). Values of weight are decided on the basis of historical information or by expert groups, including different stakeholders (von Doderer and Kleynhans 2014).

To perform the assessment, inventory analysis is done, which includes mass and energy balance, cost-benefit analysis, risk analysis, and factor analysis. Different scenarios are explored to get optimum outputs/results. Optimization is done to minimize cost, energy and water requirement, land requirement, and waste production, as well as to maximize score of quantitative sustainability indicators (Gumus et al. 2016). Indicators have also been classified as descriptive indicators, used to characterize environmental situation (no damage or financial consequences, slight damage with systems physical boundaries, and so on); performance indicators, used to measure achievement degree of established targets (they may use the same variables as descriptive indicators but are connected with target values, e.g., no damage or financial consequences is indicated by level 0, slight damage with system's physical boundaries is indicated by level 1, and localized effect is indicated by level 3); efficiency indicators, used to measure efficiency degree of products and processes (such as fuel performance); policy-effectiveness indicators, used to relate detected environmental changes with political efforts (local regional impact on society due to minor environmental damage and national-level impact due to short-term damage); and total welfare indicators (Narayanan et al. 2007). Indicators based on exergy analysis, economic analysis, life cycle assessment, and general system analysis have also been used in some sustainability studies (Lundin and Morrison 2002; Sala et al. 2015).

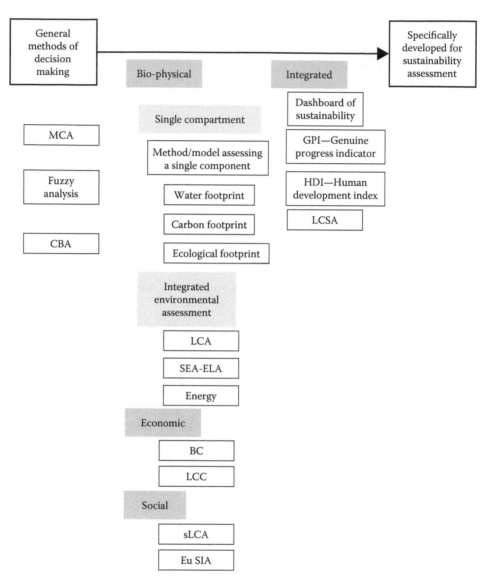

FIGURE 20.4
Categorization scheme for the integratedness of sustainability assessment methodologies. (From Sala, S. et al., *Ecol. Econ.*, 119, 314–325, 2015. With permission.)

LCA process includes goal and scope identification, inventory analysis, and impact assessment and interpretation. Process-based LCA (P-LCA) captures the environmental impact of all life cycle processes, with specific focus on production. However, economic input-output-based LCA (EIO-LCA) critically increases the scope of analysis in LCA research, considering both process- and supply-chain-level impacts (Suh et al. 2004; Tukker and Jansen 2006; Onat et al. 2014).

Some important indicators for P-LCA, along with the methodologies to determine these indicators, are shown in Figure 20.5. Different methodologies mentioned in

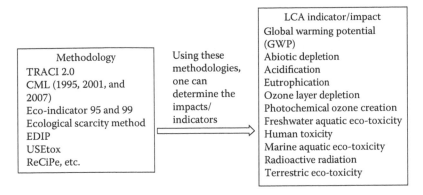

FIGURE 20.5
Methodologies for determination of different LCA and indicators.

Figure 20.5 have been adapted and applied for life cycle impact assessment (LCIA) by different researchers/organizations. Some methodologies consider midpoint impacts such as CML developed by Center of Environmental Science of Leiden University in 2001, whereas others consider end point (final) impact such as eco-indicator, which analyzes three different types of damage: human health, ecosystem quality, and resources (Chevalier et al. 2011). Some methodologies consider both midpoint and end point impacts such as ReCiPe, which provides a method that combines Eco-Indicator 99 and CML, in an updated version. Ecological scarcity method calculates environmental impacts such as pollutant emissions and resource consumption by applying "eco-factors" (Acero et al. 2014). The Environmental Development of Industrial Products (EDIP) is a method developed by the Institute for Product Development (IPU) at the Technical University of Denmark for LCA. TRACI 2.0 (Tool for the Reduction and Assessment of Chemical and other environmental Impacts 2.0) is used to quantify stressors that have potential effects, including ozone depletion, global warming, acidification, eutrophication, tropospheric ozone (smog) formation, human-health-criteria-related effects, human health cancer, human health noncancer, ecotoxicity, and fossil-fuel-depletion effects. USEtox is endorsed by UNEP/SETAC life cycle initiative and is used to calculate environmental impact based on scientific consensus to identify and obtain human and eco-toxicological impact values of chemicals in LCIA (GreenDelta 2014). Some important software used to find out the indicator(s), using any of the methodology, are GaBi, SimaPro, openLCA, and so on.

The above environmental impact indicators make the decision making complex in sustainability problems, and hence, an overall sustainability performance score called "eco-efficiency" is introduced, which integrates multiple environmental and economic impacts. Eco-efficiency has been considered a robust benchmarking and sustainability performance instrument in various environmental impact assessment and sustainable policy-making studies (Tatari and Kucukvar 2011; Egilmez et al. 2013; Egilmez and Park 2014; Gumus et al. 2016).

From the above discussion, it seems that integrated approach of MCDM and LCA for eco-efficiency assessment can be used for the analysis of sustainability of biofuels in road transport, which is also supported recently by the proposed scheme on sustainable development (SD), sustainable transport (ST), and biofuels (Holden and Gilpin 2013).

20.5 Integrated Approach for the Sustainability Assessment in Road Transport Biofuels

From the above discussion, it is clear that sustainability of road transport biofuels depends on many factors, which are associated with different levels important for sustainable development, as shown in Figure 20.6, as well as on the life cycle phases of the biofuels. From Figure 20.6, it is clear that for the analysis of sustainability of road transport system, we need to focus on the various options at different levels (L_1–L_7).

Owing to the variations in the options in each level (L_1–L_7), a large number of scenarios could be available, which need to be assessed and ranked per their superiority in terms of environmental, economic, and social impacts quantified by some cumulative indicator/index values. Depending on the complexity of the scenario, different assessment tools (L_7) can be used to quantify the impacts. Owing to large variation in options and complex nature of most of the problems in each level, it seems that the scenarios in the present case will be complex in nature. Further, since it involves different types of fuels, LCA is also essential to assess the impacts throughout the life cycle of the fuels. Thus, MCDA, coupled with LCA, will be the most suitable assessment option.

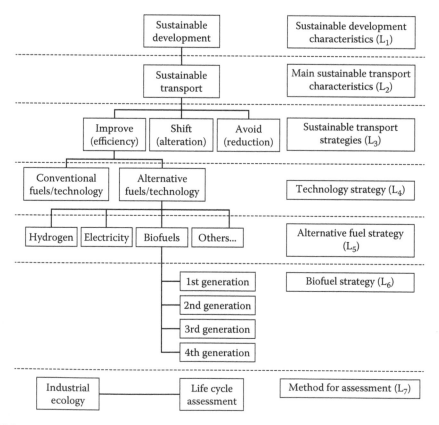

FIGURE 20.6

Different levels important for sustainability analysis of road transport biofuels. (From Holden, E., and G. Gilpin, *Sustainability*, 5, 3129–3149, 2013.)

From Figure 20.6, it is evident that like biofuel strategy, the sustainable transport strategy (including reduced vehicles, improved engine efficiency, and shifting vehicle types) and alternative technology and fuel strategy play significant roles in sustainable transport. It has also been reported that environmental and economic impacts can be improved by improving fuel efficiency of an engine by its modification as well as by using alternate fuels such as compressed natural gas and reformulated gasoline (Maclean and Lave 2000). It seems that achieving sustainable transport will most probably require a full portfolio of strategies. No single strategy such as improving public transport; reducing traffic volumes; and increasing the use of biofuels, plug-in hybrids, and long-range-battery electric vehicles will achieve it. Some real examples on the application of integrated approach for the assessment of sustainability of biofuels and road transports are provided below.

Recently, MCDA, coupled with LCA, has been used to assess the sustainability of transport in Portugal (Domingues et al. 2015). In this study, alternative technology and fuel strategy were considered. Gasoline and diesel internal combustion engine vehicles, hybrid electric vehicle (HEV), plug-in hybrid electric vehicles (10- and 40-mile battery range), and battery electric vehicle (BEV) were included. Six environmental indicators, global warming, abiotic depletion, acidification, eutrophication, ozone layer depletion, and photochemical oxidation, were considered and determined through LCIA. Four indicators, fuel consumption, CO, NO_x, and particulate matter emissions, were also considered for vehicle operation. Six alternative scenarios were considered along with the reference scenario. A multiple-criteria sorting method named ELECTRE TRI was used to rank the alternative scenarios on the basis of cumulative indicator values (Mousseau et al. 2000). ELECTRE TRI is an improved version of ELECTRE (Elimination and Choice Expressing Reality) method originated in Europe in the mid-1960s for MCDA. The alternative scenarios were ranked as much worse, worse, similar, similar but slightly worse, similar but slightly better, and better than the reference scenario. Based on MCDA, the performance of BEV was identified as the best one. It was also observed that the use of MCDA brings new perspectives into traditional LCA by increasing its comprehensiveness and providing new ways of communicating results for impact analysis and environmental decision making, instead of presenting a list of performance indicators for the considered alternatives.

Sustainability analysis of biofuel production chain (i.e., from biomass production to biofuel end use) has been reported recently (Perimenis et al. 2011). In this study, the basic framework of a decision support tool was proposed to consolidate all relevant aspects, namely social, economic, technical, and environmental. Both MCA and LCIA were used to determine the sustainability of the process. Greenhouse gas emission (combined gases CO_2, CH_4, and N_2O) and primary energy demand were determined through LCIA and included in MCDA method, along with other socioeconomic indicators such as energy efficiency, conversion ratio, complexity, development status, production cost, and employment indicators. Pair-wise comparison method was used for weighing the above criteria; 3, 2, and 1 points were assigned to the criteria that were more important, equally important, and less important with respect to the reference criteria, respectively. Further, since some criteria were qualitative and some were quantitative in nature; these were graded per their performance as poor performance (point 1) to good performance (point 5). A score was computed for a pathway (scenario) by multiplying the weight of the criteria with their respective grades and summing up the results. In this study, the authors recommended a pathway for biodiesel production from rapeseed in Germany with a score of 3.28.

20.6 Conclusion

Biofuels are becoming attractive fuel option for their renewable and ecofriendly nature. Biodiesel and bioethanol are being used as blends with diesel and gasoline, respectively, in road transport. Biogas is also being developed, which can be used as a fuel for road transport. There are a large number of potential feedstocks for biofuels production, which can be processed through various routes. However, the level of maturity of the technology options for biofuel production differs for different feedstocks. Each route has its own advantages and disadvantages. Therefore, sustainability assessment is necessary for the application of biofuels in road transport. Further, the environmental and economic impact can also be improved by increasing fuel efficiency of an engine by its modification and by application of alternate fuels such as compressed natural gas and reformulated gasoline. All these factors, along with social factors such as equity, choice of transport, and road conditions, need to be considered during the assessment of sustainability. Assessment of sustainability of this system becomes very complex owing to a large number of criteria, and hence, MCDA method coupled with LCA will be the most suitable option for sustainability analysis of the road transport biofuels. Process-based LCA (P-LCA) and economic input-output-based LCA (EIO-LCA) jointly supplement MCDA method.

References

Acero, A. P., C. Rodríguez, and A. Ciroth. 2014. LCIA methods–Impact assessment methods in Life Cycle Assessment and their impact categories. GreenDelta GmbH, Berlin, Germany. http://www.openlca.org/wp-content/uploads/2016/08/LCIA-METHODS-v.1.5.5.pdf, Accessed May 25, 2016.

Achillas, C., N. Moussiopoulos., A. Karagiannidis., G. Banias., and G. Perkoulidis. 2013. The Use of Multi-Criteria Decision Analysis to Tackle Waste Management Problems: A Literature Review. *Waste Management & Research* 31(2):115–129.

Aro, E. M. 2016. From First Generation Biofuels to Advanced Solar Biofuels. *Ambio* 45(Suppl. 1): 24–31, doi:10.1007/s13280-015-0730-0.

Barros, J. J. C., M. L. Coira., M. P. de la Cruz López., and A. del CañoGochi. 2015. Assessing the Global Sustainability of Different Electricity Generation Systems. *Energy* 89:473–489.

Biofuel digest. 2010. What are–and who's making–2G, 3G and 4G biofuels? http://www.biofuelsdigest.com/bdigest/2010/05/18/3g-4g-a-taxonomy-for-far-out-%E2%80%94-but-not-far-away-%E2%80%94-biofuels/, Accessed May 25, 2016.

Chevalier, B., T. Reyes-Carrillo., and B. Laratte. 2011. Methodology for Choosing Life Cycle Impact Assessment Sector-specific Indicators. In *DS 68-5: Proceedings of the 18th International Conference on Engineering Design (ICED 11), Impacting Society through Engineering Design, Vol. 5: Design for X/Design to X*, Lyngby/Copenhagen, Denmark, August 15–19.

Curran, M. 2013. A Review of Life-Cycle Based Tools Used to Assess the Environmental Sustainability of Biofuels in the United States. U.S. Environmental Protection Agency, Washington, DC, EPA/600/R/12/709. https://cfpub.epa.gov/si/si_public_record_report.cfm?dirEntryId=262400, Accessed May 15, 2016.

Domingues, A. R., P. Marques, R. Garcia, F. Freire., and L. C. Dias. 2015. Applying Multi-Criteria Decision Analysis to the Life-Cycle Assessment of Vehicles. *Journal of Cleaner Production* 107:749–759.

Dudson, B. 1998. When Cars are Clean and Clever: A Forward-Looking View of Sustainable and Intelligent Automobile Technologies. *Transportation Quarterly* 52(3):103–120.

Dyckhoff, H., and K. Allen. 2001. Measuring Ecological Efficiency with Data Envelopment Analysis (DEA). *European Journal of Operational Research* 132(2):312–325.

Egilmez, G., S. Gumus., and M. Kucukvar. 2015. Environmental Sustainability Benchmarking of the US and Canada Metropoles: An Expert Judgment-Based Multi-Criteria Decision Making Approach. *Cities* 42:31–41.

Egilmez, G., M. Kucukvar., and O. Tatari. 2013. Sustainability Assessment of US Manufacturing Sectors: An Economic Input Output-Based Frontier Approach. *Journal of Cleaner Production* 53:91–102.

Egilmez, G., and Y. S. Park. 2014. Transportation Related Carbon, Energy and Water Footprint Analysis of US Manufacturing: An Eco-efficiency Assessment. *Transportation Research Part D: Transport and Environment* 32:143–159.

Fava, J. A., F. Consoli., R. Denison., K. Dickson., T. Mohin., and B. Vigon. 1993. A Conceptual Framework for Life-Cycle Impact Assessment, Workshop Report, Society of Environmental Toxicology and Chemistry (SETAC), SETAC Foundation for Environmental Education, Pensacola, FL.

Gumus, S., G. Egilmez., M. Kucukvar., and Y. S. Park. 2016. Integrating Expert Weighting and Multi-criteria Decision Making into Eco-efficiency Analysis: The Case of US Manufacturing. *Journal of the Operational Research Society* 67:616–628.

Halog, A., and N. A. Bortsie-Aryee. 2013. The Need for Integrated Life Cycle Sustainability Analysis of Biofuel Supply Chains. In Z. Fang (ed.), *Biofuels—Economy, Environment and Sustainability*, INTECH Open Access Publisher, Rijeka, Croatia, pp. 197–216.

Holden, E., and G. Gilpin. 2013. Biofuels and Sustainable Transport: A Conceptual Discussion. *Sustainability* 5(7):3129–3149.

International Energy Outlook. 2010. DOE/EIA-0484. http://large.stanford.edu/courses/2010/ph240/riley2/docs/EIA-0484-2010.pdf, Accessed May 25, 2016.

Iribarren, D., I. Vázquez-Rowe., M. T. Moreira., and G. Feijoo. 2010. Further Potentials in the Joint Implementation of Life Cycle Assessment and Data Envelopment Analysis. *Science of the Total Environment* 408(22): 5265–5272.

Iribarren, D., I. Vázquez-Rowe., B. Rugani., and E. Benetto. 2014. On the Feasibility of Using Emergy Analysis as a Source of Benchmarking Criteria Through Data Envelopment Analysis: A Case Study for Wind Energy. *Energy* 67:527–537.

Litman, T., and D. Burwell. 2006. Issues in Sustainable Transportation. *International Journal of Global Environmental Issues* 6(4):331–347.

Lu, J., C. Sheahan., and P. Fu. 2011. Metabolic Engineering of Algae for Fourth Generation Biofuels Production. *Energy and Environmental Science* 4:2451–2466.

Lundin, M., and G. M. Morrison. 2002. A Life Cycle Assessment Based Procedure for Development of Environmental Sustainability Indicators for Urban Water Systems. *Urban Water* 4(2):145–152.

Maclean, H. L., and L. B. Lave. 2000. Environmental Implications of Alternative-Fueled Automobiles: AirQuality and Greenhouse GasTradeoffs. *Environmental Science & Technology* 34(2):225–231.

Mousseau, V., R. Slowinski., and P. Zielniewicz. 2000. A User-Oriented Implementation of the ELECTRE-TRI Method Integrating Preference Elicitation Support. *Computers & Operations Research* 27:757–777.

Munda, G. 2003. Social Multi-Criteria Evaluation: Methodological Foundations and Operational Consequences. *European Journal of Operations Research* 3:662–677.

Narayanan, D., Y. Zhang., and M. S. Mannan. 2007. Engineering for Sustainable Development (Esd) In Bio-Diesel Production. *Process Safety and Environmental Protection* 85(5):349–359.

Newman, P., and J. Kenworthy. 1998. *Sustainability and Cities: Overcoming Automobile Dependency*. Island Press, Covelo, CA, www.islandpress.org.

Onat, N. C., M. Kucukvar., and O. Tatari. 2014. Scope-Based Carbon Footprint Analysis of US Residential and Commercial Buildings: An Input–Output Hybrid Life Cycle Assessment Approach. *Building and Environment* 72:53–62.

Perimenis, A., H. Walimwipi, S. Zinoviev, F. Muller-Langer., and S. Miertus. 2011. Development of a Decision Support Tool for the Assessment of Biofuels. *Energy Policy* 39:1782–1793.

Pineda-Henson, R., A. B. Culaba., and G. A. Mendoza. 2002. Evaluating Environmental Performance of Pulp and Paper Manufacturing Using the Analytic Hierarchy Process and Life-Cycle Assessment. *Journal of Industrial Ecology* 6(1):15–28.

Pohekar, S. D., and M. Ramachandran. 2004. Application of Multi-Criteria Decision Making to Sustainable Energy Planning—A Review. *Renewable and Sustainable Energy Reviews* 8(4):365–381.

Pór, A., J. Stahl., and J. Temesi. 1990. Decision Support System for Production Control: Multiple Criteria Decision Making in Practice. *Engineering Costs and Production Economics* 20:213–218.

Regalbuto, J. R. 2009. Cellulosic Biofuels-Got Gasoline. *Science* 325(5942):822–824.

Sala, S., B. Ciuffo., and P. Nijkamp. 2015. A Systemic Framework for Sustainability Assessment. *Ecological Economics* 119:314–325.

Sims, R. E. H., W. Mabee, J. N. Saddler., and M. Taylor. 2010. An Overview of Second Generation Biofuel Technologies. *Bioresource Technology* 101:1570–1580.

Suh, S., M. Lenzen, G. J. Treloar, H. Hondo, A. Horvath, G. Huppes, O. Jolliet et al. 2004. System Boundary Selection in Life-Cycle Inventories Using Hybrid Approaches. *Environmental Science & Technology* 38(3):657–664.

Tatari, O., and M. Kucukvar. 2011. Eco-Efficiency of Construction Materials: Data Envelopment Analysis. *Journal of Construction Engineering and Management* 138(6):733–741.

Tukker, A., and B. Jansen. 2006. Environmental Impacts of Products: A Detailed Review of Studies. *Journal of Industrial Ecology* 10(3):159–182.

Vogelpohl, T. 2011. First and Next Generation Giofuels. Workshop on "Exploring New Technologies," Berlin, September 28. http://www.ioew.de/uploads/tx_ukioewdb/Thomas_Vogelpohl_28.09.2011_CONTEC.pdf, Accessed June 5, 2016.

von Doderer, C.C.C., and T.E. Kleynhans. 2014. Determining the Most Sustainable Lignocellulosic Bioenergy System Following a Case Study Approach. *Biomass and Bioenergy* 70:273–286.

Waas, T., J. Hugé., T. Block., T. Wright., F. Benitez-Capistros., and A. Verbruggen. 2014. Sustainability Assessment and Indicators: Tools in a Decision-Making Strategy for Sustainable Development. *Sustainability* 6(9):5512–5534.

Wang, C., and D. Mu. 2014. An LCA Study of an Electricity Coal Supply Chain. *Journal of Industrial Engineering and Management* 7(1):311–335.

Yue, W., Y. Cai, Q. Rong, L. Cao., and X. Wang. 2014. A Hybrid MCDA-LCA Approach for Assessing Carbon Foot-prints and Environmental Impacts of China's Paper Producing Industry and Printing Services. *Environmental Systems Research* 3(1):1–9.

21

Optimization of Life Cycle Net Energy of Algal Biofuel Production

Varun Sanwal, Raja Chowdhury, and Jhuma Sadhukhan

CONTENTS

ABSTRACT This study investigates the net energy balance of algal bioenergy production, using nutrients present in the dairy manure. Algal bioenergy production system includes dairy waste treatment through anaerobic digestion, uses of solubilized nutrients for algal growth, uses of algal biomass for biodiesel production, and then recovery of nutrients and energy from the residual biomass through anaerobic digestion. The useful energy outputs from the algal bioenergy production include methane and biodiesel. An attempt has been made to minimize the energy burdens of algal bioenergy production by using constraint optimization, such that highest net energy gain was possible. For optimization purpose, an energy input-output function was developed for the production of bioenergy through different processes, as described before. The net energy of a process was defined as the difference between the sum of total energy required for the process and the useful energy output from the process. The simulation was carried out for one year, taking residence time of completing one cycle of bioenergy production as 45 days. Net energy ratio (NER) was varied from 1.91 to 2.39 for different scenarios considered in our study. For the highest NER scenario, net energy value for one year of algal bioenergy production was estimated at 3184.861 GJ (Giga-Joule), where the total input was 3260.541 GJ and the output was 6445.402 GJ. Drying of algal biomass was the most energy-intensive process in the algal bioenergy production, contributing about 40% to 60% of the total input, depending on the scenario.

21.1 Introduction

Concerns over rapidly depleting fossil energy sources and escalating greenhouse gas emissions have resulted in an increased interest in alternative energy sources. Biofuels, as a result of bioenergy systems, is a renewable alternative to fossil fuels and is compatible with the existing automobile and fuel infrastructure. Bioenergy systems are also considered inherently carbon-neutral, since carbon fixation during plant growth largely offsets the carbon emissions generated during biomass combustion (Cruz et al. 2009).

In recent years, interest in bioenergy systems has grown rapidly in response to the rising costs of fossil fuels and increasing public awareness and concerns about environmental issues such as climate change. In the developing world, it offers energy independence for nations that have been heavily dependent on fossil fuel imports. In addition to more traditional uses of solid biofuels (mainly fuel wood and charcoal) in rural areas, many countries have initiated biofuel programs for the production of biodiesel, ethanol, industrial heat, and electricity (Castanheira et al. 2015, Wan et al. 2015).

One of the major criticisms directed at bioenergy system is that bioenergy production diverts valuable agricultural resources from food production to the energy crops. The thermodynamic efficiency at which bioenergy systems convert sunlight into useful energy is very low. Also, the technological know-how for fully converting this energy into useful biofuel is lagging. This implies that large tracts of land may be required to produce commercially significant quantities of biofuel. It will increase land use change, which itself generates substantial amount of greenhouse gases (Malca et al. 2014). The majority of the current commercially available biofuels are bioethanol derived from sugarcane or corn starch and biodiesel derived from oil crops, including soybean and oilseed rape. The above criticism was true for first- and second-generation biofuels, but the arrival of third-generation biofuels, that is microalgae-based bioenergy systems, cancelled the above criticism.

Owing to its higher growth rates and higher accumulation of lipids (from 20% to 80% of dry weight), microalgae has evolved as a potential candidate for the production of renewable biofuel. Microalgae exceeds in oil yield per hectare than conventional crops, such as palm, rapeseed, and sunflower, owing to lower lipid accumulation of conventional crops (less than 5% dry weight). The biochemical composition of microalgae (lipids or protein) can be altered by altering the cultivation conditions, for example, higher lipid accumulation due to nitrogen starvation. In addition, microalgae can be grown in seawater as well as on nonarable lands with the help of photobioreactors (Park et al. 2011, Schlagermann et al. 2012, Sutherland et al. 2015). Therefore, there is no competition of microalgae with conventional agriculture for resources. Microalgae can also take wastewater as a nutrient source for its growth and thus can be integrated in wastewater management schemes.

In this study, algae-based bioenergy system, which produces biodiesel through the process of transesterification, was evaluated. In this system, the dairy waste was the source of nutrients for the algae growth. Dairy waste, taken from dairy industries, was processed in the anaerobic digester, where the nutrients were solubilized. The processed supernatant from anaerobic digester was then transferred to algal pond, where cultivation of algae took place. Algae was then dewatered and dried. The concentrated algal biomass was then sent for the process of transesterification for producing biodiesel in the form of methyl ester.

The objective of the present study was to carry out the net energy input-output analysis of the aforementioned algae bioenergy system. The study was primarily focused on obtaining the optimum value of the parameters, in the above system, such that the net energy function (input–output) could be minimized over different time scales. Optimized net energy obtained from the exercise would help estimate the optimum values of the processes for sustainable production of bioenergy.

21.2 Methodology

The present study focuses on the "Life Cycle Analysis" of bioenergy systems, which in the present study is an integrated microalgal biodiesel production system that facilitates energy and nutrient recovery through anaerobic digestion (Figure 21.1). To evaluate the environmental sustainability of algal bioenergy system, process simulation was carried out because of the unavailability of performance data of large-scale algal bioenergy production system. Simulation results help in the analysis of various production and design scenarios and thus assist in improving the life cycle performance of algal bioenergy production systems.

During early phase of process design, as in the present problem, model-based approach is utilized for simulation of the process performance. For model creation, available knowledge obtained from laboratory and pilot experiments is used to define various process parameters (Boxtel et al. 2015). Most of the life cycle analysis study has used laboratory data for different processes to simulate the environmental profiles of the produced bioenergy. Several authors tried to include various processes, including anaerobic digestion, algal biodiesel production, pyrolysis, enzymatic hydrolysis, and hydrothermal liquefaction (Chowdhury and Freire, 2015, Hende et al. 2015, Mehrabadi et al. 2015, Caporgno et al. 2016, Chowdhury and Franchetti, 2016). However, these studies did not use any optimization routine to understand the optimal parameters that can help minimize the environmental burdens of the produced bioenergy. For this purpose, a mathematical model was developed,

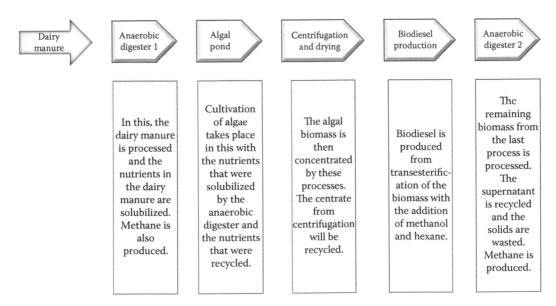

FIGURE 21.1
Schematic representation of various processes used in this study.

which includes (i) the available biological, physical, and chemical knowledge obtained from the laboratory and pilot experiments; (ii) creating a net energy input-output function from the above model; (iii) optimizing the net energy function with a suitable routine and in the presence of various constraints; and (iv) estimating the energy demand and production of different processes in the model at the optimum values of various parameters.

The system boundary of the present study includes the following: (i) Nitrogen and phosphorous nutrients, present in the dairy waste, were used for the growth of algae. (ii) Open raceway pond-based algal growth system, where the CO_2 was injected (energy demand for sparging was included; however, construction- and maintenance-related activities were not included). (iii) Algae water slurry from algal ponds was settled by using a settling tank. Settled algal biomass was assumed to have a concentration of 20 g L^{-1}. The settled algal biomass was dewatered and concentrated to a 40% solid concentration through centrifugation (Chowdhury et al. 2012). Water recovered during dewatering was recycled to the algal growth system. After dewatering, thermal drying was carried out and the remaining water was evaporated. (iv) Oil extraction from dried algal mass was carried out using hexane. However, the hexane production-related energy and emissions were not included, owing to insignificant contribution from hexane production. (v) After oil extraction, oil was transesterified using methanol and an acid catalyst. Biodiesel and glycerol were recovered as the products after transesterification. (vi) Residual biomass after oil extraction was processed through the anaerobic digester, and biogas was used as an energy source. Nutrients were recovered and used for the algal growth (Figure 21.1).

The results discussed here are after one year of the whole process being commissioned, that is, after completion of eight cycles. Different scenarios were compared in term of net energy ratio (NER) of the biodiesel and methane production. NER is defined here as the chemical energy contained in one unit of fuel divided by total primary energy required to produce one unit of fuel. Three different scenarios that have been considered in the present study are given in Table 21.1.

The mathematical model that takes energy of each process into account can be described in Equations 21.1–21.3. The nomenclature has been described later in the chapter. The parameters that helped construct our model are given in Table 21.2, along with their respective constraints. Different data used in our study, which have been incorporated through different studies, are taken from Chowdhury and Freire (2015). The input, given to the system, at a particular stage can be given as

$$I_j = \sum \left(P_{ij} \cdot E_{ij} \right) + Q_j + R(j-1) \tag{21.1}$$

The output, obtained from the system, at a particular stage can be given as

$$O_j = \sum \left(X_{ij} \cdot E'_{ij} \right) + W_j + R_j \tag{21.2}$$

TABLE 21.1

Different Scenarios Considered in This Study

Scenario 1	Scenario 2	Scenario 3
The biomass density in the algal pond can vary from 0.1 to 6 g/L.	The biomass in the algal pond is constant at 1 g/L.	The biomass in the algal pond is constant at 2 g/L.

TABLE 21.2

Constraints of the Parameters

Parameters	Lower Bounds	Upper Bounds
Solubilization efficiency of nitrogen in the first anaerobic digester	20%	80%
Solubilization efficiency of phosphorous in the first anaerobic digester	20%	80%
Solubilization efficiency of nitrogen in the second anaerobic digester	20%	80%
Solubilization efficiency of phosphorous in the second anaerobic digester	20%	80%
Efficiency of conversion of biomass to biodiesel from the process of transesterification	20%	80%
Concentration of nitrogen in the algal pond (mg/L)	60	250
Concentration of biomass effluent from the algal pond (g/L)	0.1	6
Lipid percentage in the biomass	30%	70%
Phosphorous weight in the incoming dairy waste (tonne)	0.2	0.6

The net energy value of the biodiesel, which is, in turn, optimized, is given by

$$NEV = Min(Ij - Oj) \tag{21.3}$$

The net energy after each cycle of operation was minimized through optimization. The optimization of the net energy function was done through MATLAB. Non linear optimization was carried out with a fixed bound for certain parameters (Table 21.2).

The optimum value of the parameters were used to find the energy associated with each of the processes involved. The energy associated with the nutrients was also considered. Energy associated with the processes was considered "direct," and embodies energy associated with the nutrients was considered "indirect."

21.3 Results and Discussions

Each scenario has been optimized using same set of parameters. The optimum values of each scenario are given in Table 21.3. The "direct" and "indirect" energies associated with each scenario are shown in Figures 21.2 through 21.7. In Figures 21.2 through 21.7, different input energies have been plotted against primary Y-axis, whereas the output energy has been plotted against secondary Y-axis. The energy, associated with each process, has been reported in the form of primary energy. CO_2 sparging in the algal pond, to maintain the required pH, has been considered with other inputs to the system. The energy associated with transportation of materials has not been considered in the present study. The output comprises bio-diesel and methane, generated from each anaerobic digester.

Scenario 1: In this case, the biomass density in the algal pond can vary from 0.1 to 6 g/L. The direct and indirect energies associated with each cycle are given in Figure 21.2 and Figure 21.3, respectively. The optimum value of the net energy function (Equation 21.3) is estimated to be −3184.861 GJ (Giga-Joule) at the given optimum values. The energy utilized in the production process and the useful energy gained from the production, without considering nutrients, were 2477.254 GJ and 5917.663 GJ, respectively. The present scenario has the highest NER value of 2.39, thus having more energy recovery than the other two scenarios.

TABLE 21.3

Optimum Value of the Parameters in Each Scenario

Parameters	Scenario 1	Scenario 2	Scenario 3
Solubilization efficiency of nitrogen in the first anaerobic digester	80%	80%	80%
Solubilization efficiency of phosphorous in the first anaerobic digester	40%	30%	80%
Solubilization efficiency of nitrogen in the second anaerobic digester	80%	80%	80%
Solubilization efficiency of phosphorous in the second anaerobic digester	80%	65%	78%
Efficiency of conversion of biomass to biodiesel from the process of transesterification	80%	80%	80%
Concentration of nitrogen in the algal pond (mg/L)	228	60	76
Concentration of algal biomass in the algal pond (g/L)	6	1	2
Lipid percentage in the biomass	70%	70%	70%
Phosphorous weight in the incoming dairy waste (tonne)	0.6	0.35	0.6
Net energy value (GJ) (Input–output)	−3184.861	−2359.505	−2921.811
Net energy value (NER)	2.39	1.91	2.14

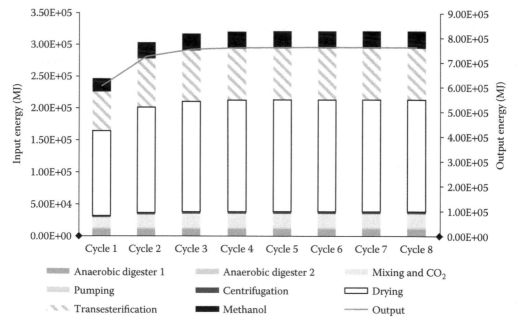

FIGURE 21.2
Direct input and output energy (scenario 1).

In the eighth cycle, the input and output energies increased by 30% and 26%, respectively, of the initial commissioning value, and both the input and output attained steady state by the eighth cycle. The steady-state biodiesel and methane energies that can be extracted in this scenario are 586.259 GJ and 179.063 GJ, respectively. Energy utilized in different processes took similar trend of steadying itself over the course of time. Drying came out

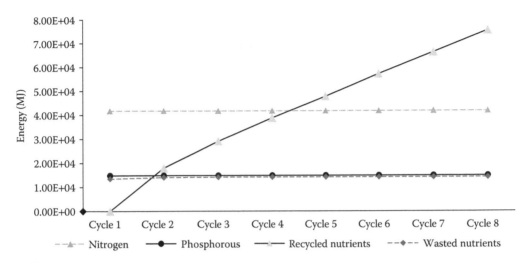

FIGURE 21.3
Indirect input and output energy (scenario 2).

to be most energy-intensive process among all others, accounting for 54% of the direct input energy. It increased by 32% from the initial value and steadied itself at 175.312 GJ. Transesterification is the other prominent energy process; it contributes to 25% of the direct input energy.

In Figure 21.3, indirect energy associated with the system is shown for scenario 1. In this case, the nutrient used for recycling increased with each cycle; however, the increment decreased from 63% (1st cycle) to 14% at the last cycle. Other indirect energies steadied themselves over the course of the eight cycles.

Scenario 2: In this case, the biomass density in the algal pond is constant at 1 g/L.

The direct and indirect energies associated with each cycle are given in Figure 21.4 and Figure 21.5, respectively. Estimated optimum value of the net energy function is –2359.505 GJ (Table 21.3). The energy utilized in the production process and the useful energy gained from the production, without considering nutrients, are 2477.254 GJ and 5917.663 GJ, respectively. The NER for scenario 2 is calculated to be 1.91.

The direct input and output energies show consistent increment during each cycle. At the end of eighth cycle, input and output energies were increased by 101% and 78%, respectively, of the initial commissioning value. Increment was far greater than the previous case and had not attained steady state by the end of the eighth cycle. However, the biodiesel and methane production steadied themselves by the eighth cycle. The steady-state biodiesel and methane energies are 583.301 GJ and 178.735 GJ, respectively. In the output, biodiesel and biogas amount to 3479.127 GJ and 1184.190 GJ, respectively, at the end of one year (aggregated values of first to eighth cycles). Drying comes out to be the most energy-intensive process, accounting for 43% of the direct input energy. Similar observation was also reported in literature, where various authors reported that drying is the most energy-intensive process for algal bioenergy production (Lardon et al. 2009, Batan et al. 2010, Chowdhury et al. 2012, Sander and Murthy, 2010). The other prominent energy consumer, in the present case, was mixing and CO_2 sparging in the algal pond. It consumed 26% of the direct input energy.

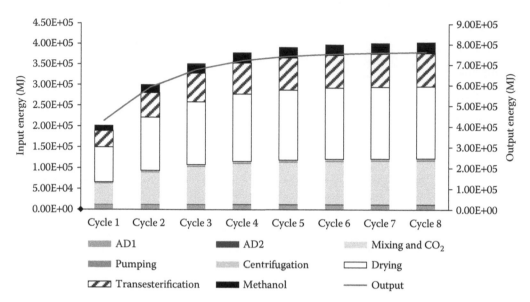

FIGURE 21.4
Direct input and output energy (scenario 2).

In Figure 21.5, each indirect energy (energy values of nutrients), associated with the current scenario, steadies itself. Unlike scenario 1, the recycled nutrients steady themselves over the proposed time period.

Scenario 3: In this case, the biomass density in the algal pond is constant at 2 g/L.

The direct and indirect energies associated with each cycle are given in Figure 21.6 and Figure 21.7, respectively. Estimated optimum value of the net energy function was −2921.811 GJ for the optimum values of various parameters (Table 21.3). The energy utilized in the

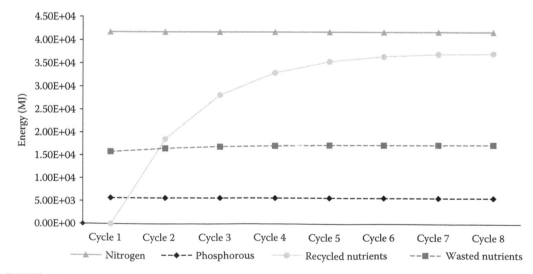

FIGURE 21.5
Indirect input and output energy (scenario 2).

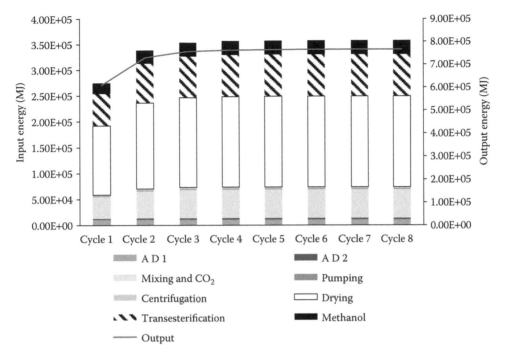

FIGURE 21.6
Direct input and output energy (scenario 3).

production process and the useful energy gained from the production, without considering nutrients, are 2740.749 GJ and 5917.878 GJ, respectively. The NER for the present scenario was calculated to be 2.16. In the output, biodiesel and biogas contributed to 4505.801 GJ and 1412.076 GJ, respectively (aggregated values for one year).

Direct and indirect energies in the third scenario follow a similar trend to scenario 1. All direct and indirect energies associated with the system steadied themselves during the course of eight cycles, except recycled nutrients. At steady state, the input and output energies increased by 30% and 26%, respectively, of the initial commissioning value. The steady-state biodiesel and methane energies that can be extracted in each cycle are 586.259 GJ and 179.063 GJ, respectively. Drying came out to be most energy-intensive process among all others, followed by transesterification, both accounting for 49% and 29%, respectively, of the direct input energy.

In Figure 21.7, indirect energy associated with the system is shown for scenario 3. In this case, the nutrient used for recycling is increasing with each cycle; however, the increment decreased from 62% (first cycle) to 13% in the last cycle.

NER values of our study were assessed against other LCA biofuel studies (Table 21.4). It is difficult to compare between algal biofuel LCA results from different studies, owing to the differences in technical assumptions, types of data used, and the life cycle processes modeled within the system boundaries (Malca and Freire, 2011). In Table 21.4, most of the studies have included anaerobic digestion in their system boundary. However, many of them coupled wet extraction, in their system, over dry extraction.

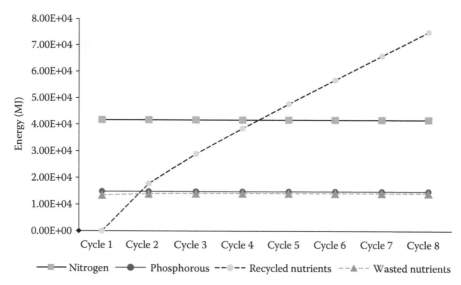

FIGURE 21.7
Indirect input and output energy (scenario 3).

TABLE 21.4

Comparison to Other LCA Results

NER Values	Source
1.25–2.8	Azadi et al. 2014
1.7–2.4	Chowdhury et al. 2012
1.4–2.7	Chowdhury and Freire, 2015
4.3	Jorquera et al. 2010
0.6–2.2	Sawayama et al. 1999
1.5–2.3	Sills et al. 2013
2.38	Xu et al. 2011
1.91–2.39	Our study

Anaerobic digestion and nutrients supplied through dairy waste play an important part in ensuring higher NER value for our study and feasibility of algal biofuel production. NER value of scenario 1, 2, and 3 would decrease by 17%, 43%, and 30%, respectively, if they are not integrated to the system. Unlike other algal biofuel LCA, our study does not supply nutrients to the algal pond from fertilizers, thus neutralizing their energy demand. Biomass drying, followed by transesterification, is the major energy-intensive processes. Adopting wet extraction technique over the dry extraction would considerably reduce the energy burden. However, the lipid conversion efficiency in the wet extraction process is still not on par with its counterpart (Xu et al. 2011).

The results of the present study show potential of algal biofuel as a fuel source; however, more study is required on the production process to further increase the NER value and to make algal biofuel as a prominent energy source. Much work needs to be done to scale up the microalgae cultivation and harvesting systems, so as to sustain these biofuel refineries.

21.4 Conclusion

Overall, scenario 1 has the least net energy (input energy–output energy) than the other scenarios, as well as the highest NER value. Increased biomass concentration and high lipid content are found to be most influential parameters for achieving maximum output. Other parameters have little effects on the overall energy production. NER is varied from 1.91 to 2.39, depending on the biomass concentration and lipid content. Drying energy, followed by the transesterification energy, is found to be the most energy-demanding process.

Acknowledgments

This research has been supported by several funding sources administered by Government of India (GOI) and Indian Institute of Technology, Roorkee (IITR). The Postgraduate Scholarship of Varun Sanwal was supported by the Ministry of Human Resources, Government of India. Other funding sources were Department of Biotechnology, GOI, (DBT 893–CED) and faculty initiation grant by IITR.

Nomenclature

Abbreviations

E_i	Input energy of the particular utility
E'_i	Output energy associated with a process or a product
I	Input
I	Utility
J	Processing stage
LCA	Life cycle analysis
NER	Net energy ratio
NEV	Net energy value
O	Output
P	Process
Q	Energy of nutrients input
R	Energy of nutrients recycle
W	Energy of nutrients wasted
X	Products

References

Azadi, P., Brownbridge, G., Mosbach, S., Smallbone, A., Bhave, A., Inderwildi, O., and Kraft, M. (2014). The carbon footprint and non-renewable energy demand of algae derived biodiesel. *Applied Energy, 113*, 1632–1644.

Batan, L., Quinn, J., Willson, B., and Bradley, T. (2010). Net energy and greenhouse gas emission evaluation of biodiesel derived from microalgae. *Environmental Science and Technology, 44,* 7975–7980.

Boxtel, V. A., Perez-Lopez, P., Breitmayer, E., and Slegers, P. (2015). The potential of optimized process design to advance LCA performance of algae production systems. *Applied Energy, 154,* 1122–1127.

Caporgno, M. P., Purvost, J., Legrand, J., Lepine, O., Tazerout, M., and Bengoa, C. (2016). Hydrothermal liquefaction of Nannochloropsis oceanica in different solvents. *Bioresource Technology, 214,* 404–410.

Castanheira, E., Grisoli, R., Coelho, S., Silva, G. A., and Freire, F. (2015). Life-cycle assessment of soybean-based biodiesel in Europe: comparing grain, oil and biodiesel import from Brazil. *Journal of Cleaner Production, 102,* 188–201.

Chowdhury, R. and Franchetti, M. (2016). Life cycle energy demand from algal biofuel generated from nutrients present in the dairy waste. Sustainable Production and Consumption, *Official journal of the European Federation of Chemical Engineering: Part E.*

Chowdhury, R., and Freire, F. (2015). Bioenergy production from algae using dairy manure as a nutrient source: Life cycle energy and greenhouse gas emission analysis. *Applied Energy, 154,* 1112–1121.

Chowdhury, R., Viamajala, S., and Gerlach, R. (2012). Reduction of environmental and energy footprint of microalgal biodiesel production through material and energy integration. *Bioresource Technology, 108,* 102–111.

Cruz, J. B., Tan, R. R., Culaba, A. B., and Ballacillo, J.-A. (2009). A dynamic input–output model for nascent bioenergy supply chains. *Applied Energy, 86,* S86–S94.

Hende, S. V., Laurent, C., and Begue, M. (2015). Anaerobic digestion of microalgal bacterial flocs from a raceway pond treating aquaculture wastewater: Need for a biorefinery. *Bioresource Technology, 196,* 184–193.

Jorquera, O., Kiperstok, A., Sales, E. A., Embirucu, M., and Ghirardi, M. L. (2010). Comparative energy life-cycle analyses of microalgal biomass production in open ponds and photobioreactors. *Bioresource Technology, 101*(4), 1406–1413.

Lardon, L., Helias, A., Sialve, B., Steyer, J.-P., and Bernard, O. (2009). Life-cycle assessment of biodiesel production from microalgae. *Environ Sci Technol, 43,* 6475–6481.

Malça, J., Coelho, A., and Freire, F. (2014). Environmental life-cycle assessment of rapeseed-based biodiesel: Alternative cultivation systems and locations. *Applied Energy, 114,* 837–844.

Malça, J., and Freire, F. (2011). Life-cycle studies of biodiesel in Europe: A review addressing the variability of results and modelling issues. *Renewable and Sustainable Energy Reviews, 15,* 338–351.

Mehrabadi, A., Craggs, R., and Mohammed, F. M. (2015). Wastewater treatment high rate algal ponds (WWT HRAP) for low-cost biofuel production. *Bioresource Technology, 184,* 202–214.

Park, J., Craggs, R., and Shilton, A. (2011). Wastewater treatment high rate algal ponds for biofuel production. *Bioresource Technology, 102,* 35–42.

Sander, K., and Murthy, G. (2010). Life cycle analysis of algae biodiesel. *The International Journal of Life Cycle Assessment, 15*(7), 704–714.

Sawayama, S., Minowa, T., and Yokoyama, S.-Y. (1999). Possibility of renewable energy production and CO_2 mitigation by thermochemical liquefaction of microalgae. *Biomass and Bioenergy, 17*(1), 33–39.

Schlagermann, P., Gottlicher, G., Dillschneider, R., Sastre, R., and Posten, C. (2012). Composition of algal oil and its potential as biofuel. *Journal of Combustion,* 50–64.

Sills, D. L., Paramita, V., Franke, M. J., Johnson, M. C., Akabas, T. M., Greene, C. H., and Tester, J. W. (2013). Quantitative uncertainty analysis of life cycle assessment for algal biofuel production. *Environmental Science* and *Technology, 47,* 687–694.

Sutherland, D., Williams, C., Turnbull, M., Broady, P., and Craggs, R. (2015). Enhancing microalgal photosynthesis and productivity in wastewater treatment high rate algal ponds for biofuel production. *Bioresource Technology, 184,* 222–229.

Wan, Y. K., Sadhukan, J., and Ng, D. (2015). Techno-economic evaluations for feasibility of sago based biorefinery, Part 2: Integrated bioethanol production and energy systems. *Chemical Engineering Research and Design.*

Xu, L., Derk, B. W., Withag, J. A., Brem, G., and Kerstan, S. (2011). Assessment of a dry and a wet route for the production of biofuels from microalgae: Energy balance analysis. *Bioresource Technology, 102*(8), 5113–5122.

22

Sustainable Production and Utilization Technologies of Biojet Fuels

Philip E. Boahene and Ajay K. Dalai

CONTENTS

ABSTRACT The urgency to critically consider biofuels as potential alternative transportation fuels for engine systems and their sustainability for air transportation purposes is a great step toward global climate control, mitigation of our dependence on fossil fuels, and the overall reduction of greenhouse gas (GHG) emissions by the aviation industry. Production and utilization of biojet fuels by the Fischer–Tropsch synthesis (FTS) approach in a sustainable manner will present a plausible solution to mitigate the current overdependence on fossil fuel energy resources for this sector. In this approach, the bio-waxes generated via catalytic bio-syngas conversion are further hydroconverted in the presence of suitable catalysts to produce biojet fuels. Though commercial sources of waxes are mainly petroleum-based, synthetic waxes commonly derived by the Fischer–Tropsch (FT) approach also contributes significantly to the current global wax supply.

    This chapter discusses sustainable options for biofuels production, their utilization for transportation applications, and cutting-edge technologies utilized for jet fuels production. Moreover, recent advances in the FTS technology toward jet fuels production and catalysts employed for the hydroconversion of FT waxes into aviation (biojet) fuels production are also reviewed. Finally, perspectives are evaluated on both the current and future production and utilization of biojet fuels for the aviation purposes.

22.1 Introduction

Transport sector predominantly relies on a single fossil resource, petroleum, which supplies about 95% of the total energy used by world transport [1]. A crucial challenge facing the transportation sectors (including air, land, and water transport), which directly affects operating cost and environment, is associated with fuel and its utilization [2]. That notwithstanding, the unpredictable price volatility of conventional crude oil derived from fossil fuel makes future planning and budgeting quite a daunting task [3]. A possible way to tackle this problem is to explore viable options for the sustainable production of renewable alternative fuels such as biofuels. In this regard, a more holistic approach can be adopted, owing to the possibility of their potential worldwide production. Biofuels are renewable and readily produced by using a variety of different crops (biomass) and thus present a plausible solution to mitigating fuel price volatility that arises with the overdependence on fossil fuel energy resources [4]. The production of biofuels can also provide economic benefits to parts of the world that not only have large amounts of marginal or unviable land for food crops but are also suitable for growing second-generation biofuel crops. Biofuels are most commonly used for transport, home heating, power generation from stationary engines, and cooking. Developing countries with marginal lands could benefit greatly from a new industry such as sustainable aviation biofuels [5].

The urgency to critically consider biofuels as potential alternative transportation fuels for engine systems and their sustainability for air transportation purposes is a great step toward global climate control, mitigation of our dependence on fossil fuels, and the overall reduction of greenhouse gas (GHG) emissions by the aviation industry. First-generation (food-based; e.g., derived from crops such as corn, rapeseed, and sugarcane) biofuels production has attracted strong criticism owing to its direct utilization of food sources, possibly for humans' and animals' consumption, and its utilization of viable agricultural lands [6]. In addition, not only are some of these first-generation biofuels (e.g., biodiesel and ethanol) unsuitable for powering commercial aircrafts, in most cases, they do not meet the high performance regulations or safety specifications for jet fuel [7]; thus, the urgent need for the aviation industry is to explore options, including second- and/or next-generation biofuels, that are sustainable.

As a result of their less threat imposition on the scarce human resources (agriculture land and crops), the new generation of biofuels derived from non-food crop (second- and next-generation biofuels) sources has gained immense attention [4]. Although second-generation biofuels can be produced in large quantities in a range of locations, including deserts and salt water, they have the potential of producing green and cheaper fuels for aviation applications. The future outlook of the aviation industry will likely resort to a blend of feedstocks that constitutes sizeable fractions of various biofuels, along with jet fuels [7]. Jet fuel or aviation turbine fuel (ATF) is a colorless to straw-coloured liquid, typically designed for use in gas-turbine-powered engines of aircrafts [8]. It is composed of a mixture of large number of different hydrocarbons, with carbon numbers ranging from 8 to 16 (for kerosene-type jet fuel, i.e., Jet A and Jet A-1) or 5 to 15 (for wide-cut or naphtha-type jet fuel, i.e., Jet B). It should be noted that the freezing point of Jet A-1 (−47°C) is lower than that of Jet A (−40°C); however, Jet B has the lowest freezing point among all three of them (i.e., −60°C). In addition, jet fuels have a typical boiling range of 150°C–270°C, which is somewhere between the boiling ranges of conventional gasoline and diesel [1].

Owing to their critical applications for air transport purposes, chemical properties of biojet fuels, including the freezing and smoke point, need to be closely monitored to meet stringent standardized international aviation fuels specifications as well as governmental/ federal regulations. When derived from bioresources, jet fuels production can be a sustainable pathway by the conversion of bio-syngas to oils and waxes and subsequent hydroconversion of the waxes, in the presence of suitable catalysts, into valuable products and chemicals, including biojet fuels. This route holds a lot of promise owing to the readily availability of biomass in Canada and most parts of the world. The challenge will be to produce commercial quantities of the bio-syngas, preconditioning it to remove contaminants, exploiting its application in the FTS reaction, and hence, biojet fuels production.

22.2 Background and Scope

Currently, the inevitable combustion of conventional fossil fuels for transportation purposes and the concomitant stringent environmental regulations in North America and most parts of the world have spurred research into the production of less harmful fuels, with minimal tailpipe emissions, such as SO_x and NO_x. Thus, the generation of "green" gasoline, diesel, and alcohols as alternative liquid transportation fuels and fuel additives via the FTS process has attracted significant attention for the scientists [9]. This technology requires feedstock such as synthesis gas or bio-syngas (a mixture of H_2 and CO) obtained from gasification of biomass/coal and natural gas reforming. In FTS process, the syngas is converted into a range of hydrocarbon products, which are mainly composed of long-chain sulfur- and nitrogen-free normal paraffins and olefins, utilized to generate the liquid fuel [10]. Furthermore, as part of the variety of product streams derived from syngas conversion, wax production and its subsequent upgrading into aviation fuels look attractive, as far as long-term jet fuel production is concerned. It is noteworthy that test runs in North America, Europe, and other parts of the world have succeeded in blending 50% renewable Jet A-1 with fossil and conventional Jet A-1 [7]. Thus, there are no technical barriers concerning the use of these biofuels as a result of extensive test flight programs conducted using 50–50 blends of jet fuel for commercial flights [5,7]. However, renewable Jet A-1 is currently produced in small quantities at high costs, usually as part of research or for test flights. The challenges include the development and applications of novel catalysts for efficient conversion of bio-syngas into long-chain hydrocarbons, alcohols, and waxes, leading to the generation of commercial quantities of "green" gasoline, diesel, jet fuels, and other biofuels for future applications.

22.2.1 Biofuels and Sustainability

Biofuels can be produced sustainably, with positive effects on nature and society. The whole biomass to biofuel value chain has to be sustainable in every step, and feedstock type is usually the most significant parameter [4]. The main goal with biofuels for transportation application is to reduce GHG emissions and provide alternative fuel resources for energy security. The biofuel climate effect is therefore important to identify. For sustainable biofuel production, some factors to monitor and manage include [11]:

- *Nature, environment, and climate*: Reduction of GHG and local pollution, prevention of the loss of biological diversity and ecosystem services, as well as high agricultural standards.
- *Social issues*: Human rights; working rights; and rights to cultivate land, health, and food security.
- *Economics*: Profitability, long-term economic development, new jobs, and an optimal societal resource management.

Large-scale utilization and production of biofuels by the FTS approach are still in an early development phase and require that critical specific regulations and methodologies are taken into consideration.

22.3 Jet Fuels Production Technologies

Renewable Jet A-1 can be produced with various technologies and feedstocks. The pathways are usually divided into three main categories: biochemical, thermochemical, and chemical, but they can also be a combination of these categories [11].

Figure 22.1 presents an overview of possible feedstocks, conversion technologies, and various products from these bioresources. For a given route to be sustainable, the various feedstocks, as well as all individual steps, need to interplay effectively toward the finished products generation. Moreover, the quantity and quality of final products will be dependent on the type and quality of feedstock utilized. Thus, it suggests that a great degree of products flexibility may exist and must therefore be factored into plant design and operation schedules for practical purposes. Also, there exists the possibility to explore several value chains for sustainable biofuels in almost every part of the world. For instance, in Norway,

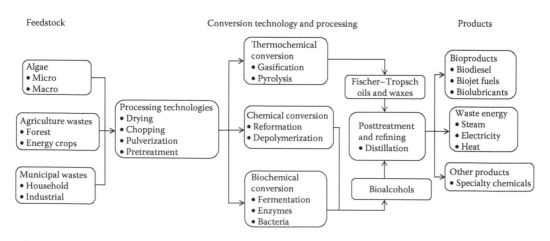

FIGURE 22.1
Biofuels feedstock, process routes, and end products. (Adopted from Rambol, Summary Sustainable Aviation Biofuels, 2013.)

Sintef Energy has accessed possible sustainable Jet A-1 biofuel production technologies suitable for Norwegian value chains toward 2020/25 [11]. It is quite evident that the FT process via the thermochemical conversion of forest biomass (by gasification) not only holds the potential of producing bio-syngas for the FT oils and waxes formation but also produces bioalcohols that can be further processed and refined to jet fuels. This is quite a promising route that has the potential of being sustainable, if all the by-product streams are further refined for sale to offset part of the production costs (Figure 22.1).

22.3.1 Thermochemical Routes for Jet Fuels Production via Fischer–Tropsch Products

As the name implies, in the thermochemical process, heat is utilized to convert the hydrocarbons present in the bio-based feedstock into more valuable products. In some cases, catalysts may be required, which can play the role of rearranging the various molecules into biofuels of desired carbon chain length. In the FT process, the bio-syngas generated (from biomass gasification) gets converted into a range of hydrocarbons of variable carbon chain lengths and waxes. By using a suitable catalyst, the waxes can further be subjected to isomerization and alkylation process to produce hydrocarbons of carbon chain lengths in the range of jet fuels. It is noteworthy to mention that Jet A-1 based on coal has been produced with FT-technology since the 1960s [12]. However, the key issues are the biomass gasification and subsequent preconditioning of the bio-syngas before its utilization for the FTS reaction. Typical by-products from the FT process that can be generated include bio-diesel, bionaphtha, biolubricants, specialty chemicals, and heat and energy.

22.3.1.1 Advances in Fischer–Tropsch Synthesis Technology

The FTS product spectrum consists of hydrocarbon gases (C_1–C_4), gasoline (C_5–C_{12}), diesel (C_{13}–C_{20}) and waxes (C_{21+}) fractions, CO_2, and water [13,14]. The product distribution depends on the catalyst choice and process conditions. The most active transition metals used for FT process are Fe and Co [15–17]. Fe favors water–gas shift (WGS) activity and is the major choice of industries. During FTS, a CO molecule accepts electrons from Fe and K; this facilitates CO chemisorptions, since it can donate electrons to Fe [18]. Co is more resistant to deactivation but is more expensive than iron [19,20]. Better Co dispersion and smaller nanoparticles help to reduce metal loading and result in lower catalyst cost while maximizing activity and durability [21]. Improvements in catalyst activity, selectivity, and stability are needed to improve FT process economy [22,23]. The catalyst supports act to (1) stabilize the active species and promoters; (2) promote hydrogen or oxygen donation or exchange; (3) modify the dispersion, reducibility, electron-donating, or accepting effects of metal particles; (4) provide porosity; and (5) enhance mass transfer [24]. Strong metal support interactions can result in fine dispersion but reduce metal reducibility. An ideal support for FTS requires the following chemical and physical characteristics [19,25]: (1) well-defined chemical composition; (2) high purity; (3) well-defined surface chemistry (inertness to undesired reactions); (4) no sintering at high temperatures; (5) excellent abrasion resistance; (6) high mechanical strength; (7) well-defined porosity, pore size distribution, and pore volume; (8) easy separation from the reactants; and (9) low cost. SiO_2 and γ-Al_2O_3 are mostly used as the support for Fe and Co, as they are inexpensive and structurally stable.

Other metal oxides (zirconia, magnesia, and ceria) have been investigated as supports for FT activity, selectivity, and stability of Co catalysts [26,27]. However, these supports show strong metal support interactions, resulting in the formation of mixed oxides, which are difficult to reduce. In addition, their non-uniformities in the pore size distributions lead to differences in the hydrocarbon products distribution [21]. Compared with conventional oxide supports, inert carbon materials such as carbon nanotubes (CNTs), carbon nanofibers (CNFs), carbon nanohorns (CNHs), carbon nanospheres (CSs), and mesoporous carbons display unique properties such as high mechanical strength, large surface area, uniform pore size, and unique pore structures, thus, eliminating potential mass-transfer resistances associated with microporous supports. They are increasingly attractive as potential supports for FT catalysts [28,29]. Dalai and coworkers have extensively worked on the Fe/CNT for FT synthesis [30,31]. Bahome et al. [32,33] studied the effects of Fe and promoters (K and Cu) loadings on the activity and selectivity of Fe/CNT in a fixed-bed reactor. Addition of K led to decreased hydrogenation and increased chain growth during FTS, producing higher-molecular-weight products. Abbaslou et al. [30,31] studied the effects of different promoters (K, Mo, and Cu) on Fe/CNT and achieved CO conversion >90%. Further details are presented in their patent [31]. Also, Trepanier et al. studied Co/CNT catalyst for FTS. Addition of K promoter increased the wax and alkene yields [34]. Recently, a novel type of nanometric catalysts family has been proposed in the literature by Abatzoglou and coworkers for the conversion of syngas with a composition similar to that obtained from air gasification of urban waste [35]. This nanoiron carbide synthesized by plasma-spraying technology mimicked the core-shell matrix structure, with FeC-rich core inside a graphitic carbon matrix, to enhance the protection of the air-sensitive carbides, thus preventing susceptible catalyst oxidation during their handling and storage. Comparison of their nanoiron carbide catalyst with Nanocat commercial nanoiron-powdered catalyst showed the former to exhibit higher activity and robustness that its counterpart. Moreover, nanoiron carbide catalyst also reported 6% CH_4 selectivity as compared with the Nanocat commercial catalyst, which recorded 10% CH_4 selectivity at similar reaction conditions, using a continuous slurry reactor.

22.3.1.2 Hydroconversion of Fischer–Tropsch Waxes for Aviation (Biojet) Fuels Production

For synthesizing ATF by the FT approach, the produced waxes require hydrocracking to selectively generate shorter-chain hydrocarbons for jet fuels [2]. It is composed of C_5–C_{15} hydrocarbons, with the freezing point below −47°C to −60°C and with typical boiling range of 150°C–270°C [8].

The production of these paraffins in a specified carbon number range is not easily achievable by the conventional FTS reaction but may be achieved by directing the synthesis towards heavy paraffins, which are subsequently cracked selectively [11]. The cracking process should proceed in such a way that fragments already in the desired carbon number range do not undergo further cracking; however, the target should be longer-chain hydrocarbons. This can only be possible with the help of bifunctional catalysts with reaction sites capable of both isomerization and hydrogenation reactions. These bifunctional hydrocracking catalysts are characterized by the presence of acidic sites (for cracking functionalities) and metal sites (for hydrogenation–dehydrogenation functionalities) (Figure 22.2).

As shown in Figure 22.2, preconditioned bio-syngas derived from biomass gasification undergoes catalytic reaction in the FT reactor to yield a variety of products, including light and heavy hydrocarbons and water. Products separation proceeds in the distillation unit to separate the heavier (wax) products, which further undergo hydroconversion, isomerisation, and alkylation reactions in the presence of a suitable catalyst to yield jet fuels. The light hydrocarbons (FT oils) can then be utilized directly as drop-in renewable transportation fuels.

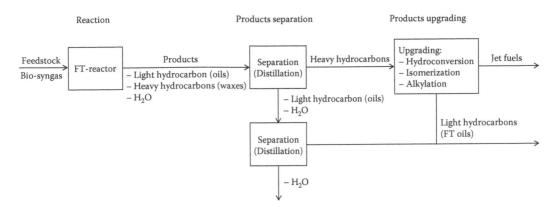

FIGURE 22.2
Overview of the FT process for jet fuels production.

22.3.1.3 Catalysts for Hydroconversion of Fischer–Tropsch Waxes for Aviation (Biojet) Fuels Production

Typical acidic supports are amorphous oxides or mixtures of oxides, zeolites, and silica-aluminophosphates. Conditions favorable for typical hydrocracking reaction are 3–5 MPa total pressure and 550°C–600°C [36,37]. The zeolites, chlorinated alumina, and metal-promoted acidic zeolites (Ni/H-ZSM-5) are acidic and are used for hydrocracking reactions [38]. The dual-functional catalysts such as Pt/chlorinated-Al$_2$O$_3$ and Pt/zeolites are used for isomerization. The acidic components (chlorinated alumina or zeolite) participate in the reaction steps where carbenium ions are involved, while the metallic component offers hydrogenation–dehydrogenation activity. In case of Pt/zeolite catalysts, the chlorination is not required; however, higher reaction temperature has to be applied [39]. This catalyst is resistant to water and sulfur compounds in the feed, but it is less active than Pt/chlorided alumina. Another super acidic bifunctional catalyst is Pt/sulfated zirconia, which is more active than chlorinated alumina [40]. For wax feedstocks, the octane number needs enhancement through isomerization and alkylation reactions and the freezing point adjustment requires chemical depressants. The resulting fuel can be blended with other light hydrocarbons, such as butane and isopentane, to meet the minimum vapor pressure requirement for jet fuels. Finally, the additives such as tetraethyl lead and ethylene dibromide mixture for anti-knocking properties; diethylene glycol monomethyl ether (di-EGME) for icing inhibition; 2,6-ditertiary butyl-4-methyl phenol for preventing gum formation, and dyes for color coding of aviation fuels can be added to meet specifications for jet fuels or ATF applications [2].

22.3.2 Biochemical Routes for Jet Fuels Production Using Second-Generation Wastes

Another route worth exploring for the production of biojet fuels is via the biochemical conversion pathway [11,41,42]. Cellulosic biomass constituting the fibrous, nonedible part of plants serves as renewable feedstock for next-generation biofuels and bioproducts. Biochemical conversion involves the breakdown of bulky biomass molecule into free carbohydrates for further processing into sugars and their subsequent conversion into biofuels and bioproducts by using enzymes (biocatalysts) [41]. Owing to the abundance of biomass worldwide, the biochemical route can be promising for the generation of biofuels. However, the key bottleneck apart from cost factor is the difficulty associated with the breaking down of the complex biomass structure to gain access to the useful cellulose components, which is housed in the rigid cell wall. Furthermore, the efficient conversion of these sugars and starch components into biofuels and their consequent purification before conversion into biojet fuels could also pose another challenge.

22.3.2.1 Processes in Biochemical Conversion

Biochemical conversion uses biocatalysts, such as enzymes, in addition to heat and other chemicals, to convert the carbohydrate portion of the biomass (hemicellulose and cellulose) into an intermediate sugar stream [41,42]. These sugars are intermediate building blocks, which can then be fermented or chemically catalyzed into a range of advanced biofuels, bioalcohols, and other value-added chemicals. The overall process can be simplified into the following three essential steps: feedstocks supply and pretreatment, reactions and products separation, and products and by-products generation, as depicted in the Figure 22.3.

Typically, feedstocks for biochemical processes are selected for optimum composition, quality, and size. Feedstock handling and pretreatment are essential steps toward the production of bioproducts of high quality. In the pretreatment step, the biomass undergoes heat treatment in the presence of acid or base to enhance the breaking down of the rigid and fibrous cell walls to release cellulose and hemicellulose for easier hydrolysis to proceed. Incorporation of enzymes in the hydrolysis step facilitates the separation and release of sugars present in the cellulose and hemicellulose. Depending on the process conditions, hydrolysis of the pretreated products can proceed over a period of hours or days [4,41] to generate smaller molecules. Addition of microorganisms can proceed via fermentation to convert the sugars into other molecules, which can be used as chemical building blocks for biofuels generation. Further products separation ensures the separation of the desired products from water, solvents, and any residual solids. The biocrude can then undergo distillation to separate bioalcohols from the other bioproducts, which can further be converted into biojet fuels via heterogeneous catalytic pathway.

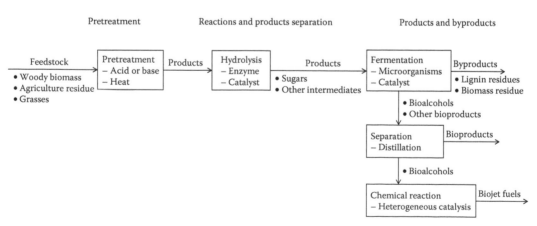

FIGURE 22.3
Biochemical pathway for the production of biojet fuels and other bioproducts.

22.3.2.2 *Biochemical Conversion of Bioalcohols into Jet Fuels*

In the biochemical conversion route, bioalcohols can also be obtained via the conversion of second-generation feedstocks by using enzymes, which could be further processed in the presence of suitable catalyst matrices to produce jet fuels [11]. For instance, the Alcohol-to-Jet (AtJ) process by the Biojet Norway Inc. does not include the bioalcohols production but uses alcohols as feedstock. Moreover, it is anticipated that blending of AtJ fuels with conventional Jet A-1 fuels could be possible after successful certification standards is attained. That notwithstanding, the AtJ process is currently not competitive with bioalcohols as feedstocks compared with conventional fossil-based products; however, the development of novel catalysts could make this process more selective and can change the outlook of the entire process to make it more sustainable.

An advantage of the AtJ process is the global availability of bioalcohols, which can directly be used in this technology. For instance, bioalcohols can be produced from various feedstocks, including lignocellulose from trees or energy crops, agricultural and municipal wastes, and sugars and starch from edible plants. Microalgae or macroalgae (alginate from seaweed) can also be used as feedstock, depending on the volumes and availability. Also, forest wastes, municipal wastes, and other industry residuals can be subjected to thermochemical reaction at high temperature in a gasification process. This conversion technology is feasible and performs well environmentally. Other options for sustainable fuel production are the hydroprocessed esters and fatty acids (HEFA). These could be a good second-generation bioresource if feedstocks are derived from waste cooking oils and other animal fats. Although sustainable HEFA resources can be produced at high costs, 50% blending proportions can be achieved with their petroleum counterpart.

22.4 Sustainability: Current and Future Perspective of Jet Fuels

Today, renewable diesel and jet fuels have proven to be sustainable resources with great potentials for future energy demands and security. That notwithstanding, in order to meet the growing demand for renewable biofuels, which is driven by stringent biofuel mandates and high consumer demands, a wide variety of feedstocks, cutting-edge conversion technologies, as well as the by-products upgrading into more economically valuable products have to be taken into consideration. Unlike other biofuels, including jet fuels, HEFA fuels are chemically similar to their petroleum counterparts and are technically hydrocarbons rather than alcohols or esters. Drop-in fuels from bioalcohols can also be produced during the upgrading process to remove the oxygen present in the bioalcohols to generate hydrocarbons, which could be used directly as fuels or be reformed to generate longer-carbon-chain molecule to produce jet fuels. Currently, HEFA fuels are approved by ASTM International for use in jet engines at up to a 50% blend rate with petroleum jet fuel [43]. HEFA fuels are the most common drop-in biofuels; they can be used in diesel engines, without the need for blending with petroleum diesel fuel.

In the United States, renewable diesel, a common HEFA biofuel, has been used as alternate fuel in place of the conventional diesel. Similarly, in other parts of the world, these alternate HEFA biofuels are marketed as hydrotreated vegetable oil (HVO) and are used for a variety of applications [44]. HEFA fuels are produced by reacting vegetable oil or animal fat with hydrogen in the presence of a catalyst. Currently, biojet fuels can be blended with petroleum jet fuel in proportions up to 50%. As with any alternative jet fuel, biojet fuel derived from the upgrading of FT waxes has to meet stringent specifications that ensure excellent performance under a wide range of conditions [45].

In Canada, aerospace giant Boeing, with support from Canada's aviation industry and other stakeholders, has initiated plans of converting leftover forestry residues and mill wastes into sustainable aviation biofuel to be used alongside their conventional utilization as wood pellets for heating and electricity applications. A consortium that includes Boeing, Air Canada, WestJet, Bombardier, research institutions, and industry partners is anticipated to assess whether forest waste could also be harnessed to produce sustainable aviation biofuel by using thermochemical processing [46] and probably via the FT synthesis approach. It was estimated that by using sustainably produced biofuel, one can reduce life cycle CO_2 emissions by 50%–80% compared with conventional petroleum fuel, according to the U.S. Department of Energy [47].

Finally, potential consumer of this bio-derived jet fuel includes the Canadian Department of Defense, which commonly uses the versatile JP-8 jet fuel in military vehicles, stationary diesel engines, and jet aircraft. Furthermore, agreements between BioFuelNet Canada, Airbus, and Air Canada to evaluate biomass-derived biojet fuels, as well as exploring new technologies and renewable alternative fuels, are such promising attempts toward meeting Air Canada's goal of carbon-neutral growth by 2020 and are giant steps in the right direction [48].

In conclusion, biofuels for jet fuels application can be sustainable if bottlenecks ranging from adequate feedstock supplies, innovative conversion technologies, and making economic gains with the sale of by-products are all taken into consideration to make the process sustainable.

References

1. Suzana, K.R., Kobayashi, S., Michael, B. et al. 2007. Transport and its infrastructure. In *Climate Change*. Cambridge University Press, Cambridge, UK.
2. Greg, H., Tracy, B., John, B. et al. 2007. *Aviation Fuels Technical Reviews*. Chevron Products, San Ramon, CA.
3. Goetz, K., Kearney, A.T. 2012. *Sustainable Transportation Ecosystem: Addressing Sustainability from an Integrated Systems Perspective*. World Economic Forum, Geneva, Switzerland.
4. Naik, S.N., Vaibhav, V.G., Prasant, K.R., Dalai, A.K. 2010. Production of first and second generation biofuels: A comprehensive review. *Renewable and Sustainable Energy Reviews* 14:578–597.
5. Helena, L.C., Francisco, E.B.N., Robert, M. et al. 2015. Conversion technologies for biofuels and their use. In: G.M. Souza, R.L. Victoria, C.A. Joly and L.M. Verdade (eds.), *Bioenergy & Sustainability*. National Renewable Energy Laboratory, Golden, CO. pp. 374–467.
6. Beginner's Guide to Aviation Biofuels. 2009. Air Transport Action Group. Geneva, Switzerland.
7. ICAO Secretariat. 2013. Overview of Biojet Fuels. IATA 2013 Report on Alternative Fuels.
8. https://en.wikipedia.org/wiki/Jet_fuel.
9. Kimberly, L.J., Menard, R.J. English, B.C. 2014. AgResearch. 1–26.
10. de Klerk, A. 2011. *Fischer–Tropsch Synthesis, in Fischer-Tropsch Refining*. Wiley-VCH Verlag GmbH & Co., Weinheim, Germany
11. Rambol. 2013. Sustainable Aviation Biofuels-Summary Report, Norway.
12. Joosung, L., Jeonghoon, M. 2011. Analysis of technological innovation and environmental performance improvement in aviation sector. *Int. J. Environ Res Public Health* 8(9):3777–3795.

13. Van der Laan, G.P., Beenackers, C.M. 1999. Kinetics and selectivity of the Fischer–Tropsch synthesis: A literature review. *Catal. Rev.-Sci. Eng.* 41:255–318.
14. Lee, S., Speight, J., Loyalka, S. 2007. *Handbook of Alternative Fuel Technologies*. CRC Press, Boca Raton, FL.
15. Dry, M.E. 2002. The Fischer–Tropsch process: 1950–2000. *Catal. Today* 71:227–241.
16. Luque, R., de la Osa, A.R., Campelo, J.M., Romero, A.A., Valverde, J.L., Sanchez, P. 2012. *Energy Environ. Sci.* 5:5186–5202.
17. Luque, R., Campelo, J., Clark, J. 2011. *Handbook of Biofuels Production: Processes and Technologies*. Woodhead Publishing Limited, Cambridge, UK.
18. Yang, Y., Xiang, H., Xu, Y., Bai, L., Li, Y. 2004. Effect of potassium promoter on precipitated iron-manganese catalyst for Fischer–Tropsch synthesis. *Appl. Catal. A Gen.* 266:181–194.
19. Steynberg, A., Dry, M. 2004. *Fischer–Tropsch Technology*. Elsevier B.V., Amsterdam, the Netherlands.
20. Perego, C., Bortolo, R., Zennaro, R. 2009. Gas to liquids technologies for natural gas reserves valorization: The Eni experience. *Catal. Today* 142:9–16.
21. Khodakov, A.Y., Chu, W., Fongarland, P. 2007. Advances in the development of novel cobalt Fischer–Tropsch catalysts for synthesis of long-chain hydrocarbons and clean fuels. *Chem. Rev.* 107:1692–1744.
22. Tijmensen, M.J., Faaij, P.C., Hamelinck, C.N., Van Hardeveld, M.R.M. 2002. Exploration of the possibilities for production of Fischer Tropsch liquids and power via biomass gasification. *Biomass Bioenergy* 23:129–152.
23. Bartholomew, C.H. 1990. Recent technological developments in Fischer–Tropsch catalysis. *Catal. Lett.* 7:303–315.
24. Hindermann, J.P., Deluzarche, A., Kieffer, R., Kiennemann, A. 1983. Characterization of chemisorbed species in CO/H2 and CO2/H2 reactions. Evolutive behaviour of the species. *Can. J. Chem. Eng.* 61:21–28.
25. Regalbuto, J. 2007. *Catalyst Preparation: Science and Engineering*. CRC Press, Taylor & Francis Group, Boca Raton, FL.
26. Pan, Z., Bukur, D.B. 2011. Fischer–Tropsch synthesis on Co/ZnO catalyst—Effect of pretreatment procedure. *Appl. Catal. A Gen.* 404:74–80.
27. Kraum, M., Baerns, M. 1999. Fischer–Tropsch synthesis: the influence of various cobalt compounds applied in the preparation of supported cobalt catalysts on their performance. *Appl. Catal. A Gen.* 186:189–200.
28. Tavasoli, A., Trépanier, M., Dalai, A.K., Abatzoglou, N. 2010. Effects of confinement in carbon nanotubes on the activity, selectivity, and lifetime of Fischer–Tropsch Co/carbon nanotube catalysts. *J. Chem. Eng. Data.* 55:2757–2763.
29. De Jong, K.P., Geus, J.W. 2001. Carbon nanofibers: catalytic synthesis and applications. *Catal. Rev.* 42:481–510.
30. Abbaslou, R. 2010. Iron catalyst supported on carbon nanotubes for Fischer–Tropsch synthesis: Experimental and kinetic study. PhD Thesis, University of Saskatchewan.
31. Abbaslou R., Dalai, A.K. 2013. Promoted Iron catalysts supported on carbon nanotubes for Fischer–Tropsch synthesis. US 2013/0116350 A1.
32. Bahome, M.C., Jewell, L.L., Hildebrandt, D., Glasser, D., Coville, N.J. 2005. Fischer–Tropsch synthesis over iron catalysts supported on carbon nanotubes. *Appl. Catal. A Gen.* 287: 60–67.
33. Bahome, M.C., Jewell, L.L., Padayachy, K. 2007. Fe-Ru small particle bimetallic catalysts supported on carbon nanotubes for use in Fischer–Tröpsch synthesis. *Appl. Catal. A Gen.* 328:243–251.
34. Trépanier, M. 2010. Promoted Co-CNT nano-catalyst for green diesel production using Fischer–Tropsch synthesis in a fixed bed reactor. PhD Thesis, University of Saskatchewan.
35. Blanchard, J., Abatzoglou, N., Eslahpazir-Esfandabadi, R., Gitzhofer, F. 2010. Fischer–Tropsch synthesis in a slurry reactor using a nanoiron carbide catalyst produced by a plasma spray technique. *Ind. Eng. Chem. Res.* 49:6948–6955.

36. Sie, S.T., Senden, M.M.G., van Wechem, H.M.H. 1991. Conversion of natural gas to transportation fuels via the shell middle distillate synthesis process (SMDS). *Catal. Today* 8:371–381.
37. Pellegrini, L., Locatelli, S., Raselle, S., Bonomi, S., Calemma, V. 2004. Modeling of Fischer–Tropsch products hydrocracking. *Chem. Eng. Sci.* 59:4741–4785.
38. De Klerk, A. 2010. Fischer-Tropsch Jet fuel Process. US 2010/0108568 A1.
39. Ono, Y. 2003. A survey of mechanism in catalytic isomerisation of alkanes. *Catal. Today* 81:3–16.
40. Loften, T. 2004. *Catalytic Isomerization of Light Alkanes.* Norwegian University of Science and Technology, Trondheim, Norway.
41. US Department of Energy Report—Energy Efficiency and Renewable. Energy DOE/EE-0948. July 2013.
42. Ralph, S., Michael, T., Jack S., Warren, M. International Energy Agency—Bioenergy. November, 2008.
43. Rosalie, S. 2015. HEFA Biofuels Eliminate Need for Blending with Petroleum Fuels. Energy Global Hydrocarbon Engineering. http://www.energyglobal.com/downstream/clean-fuels/09112015/HEFA-biofuels-eliminate-need-for-blending-with-petroleum-fuels-1707/.
44. Sergios, K., James, D.M., Jack N.S. 2014. The potential and challenges of drop-in biofuels. IEA Bioenergy Task 39.
45. Andre, F. 2012. White Paper on Sustainable Jet Fuel. SkyNRG.
46. Green Car Congress. 2015. Boeing, Canadian aviation industry launch sustainable aviation biofuel project using forestry waste. http://www.greencarcongress.com/2015/12/20151203-boeing.html#more.
47. Aaron, C. 2015. Boeing, Japan's Aviation Sector Develop Biofuel for 2020 Olympics. https://www.flightglobal.com/news/articles/boeing-japans-aviation-sector-develop-biofuel-for-414446/.
48. GreenAir. 2013. Airbus and Air Canada Sign Research Agreement to Develop Advanced Aviation Biofuels Solutions in Canada. http://www.greenaironline.com/news.php?viewStory=1888.

Index

Note: Page numbers followed by f and t refer to figures and tables, respectively.